# Reproduction in Sheep

# Reproduction in Sheep

Supervising Editors:
D.R. Lindsay and D.T. Pearce
School of Agriculture (Animal Science)
University of Western Australia

Australian Wool Corporation Technical Publication

CAMBRIDGE UNIVERSITY PRESS
Cambridge
London New York New Rochelle
Melbourne Sydney

CAMBRIDGE UNIVERSITY PRESS
Cambridge, New York, Melbourne, Madrid, Cape Town,
Singapore, São Paulo, Delhi, Tokyo, Mexico City

Cambridge University Press
The Edinburgh Building, Cambridge CB2 8RU, UK

Published in the United States of America by Cambridge University Press, New York

www.cambridge.org
Information on this title: www.cambridge.org/9780521116268



First published 1984
First paperback edition 2011

*A catalogue record for this publication is available from the British Library*

*Library of Congress Cataloguing in Publication Data*
Reproduction in Sheep
1. Sheep - Reproduction
1. Lindsay, D.R. II. Pearce, D.T.
III. Australian Academy of Science
636.3´08926 SF373

ISBN 978-0-521-30659-1 Hardback
ISBN 978-0-521-11626-8 Paperback

# CONTENTS

# FOREWORD

The Wool Research Trust Fund is the oldest of the Australian rural industry research funds and has been in operation since 1936. It obtains its income from a levy on woolgrowers and a matching contribution from the Commonwealth Government. This money is allocated for use by the Commonwealth Scientific and Industrial Research Organisation, universities, relevant government departments, other organisations and individuals to conduct research meeting the needs of the wool industry.

Part VI of the Wool Industry Act 1972 covers the Australian Wool Corporations's responsibility in relation to the administration of the Fund. To assist it, the Corporation appoints ad hoc committees to advise on framing recommendations and advice to the Minister for Primary Industry on the allocation of support from the Fund for the various fields of wool research — wool production and harvesting, measurement, handling and distribution, textile and economic research.

Almost since the Fund's inception, research into sheep reproduction has been supported with the objectives of raising woolgrowers' productivity and lowering costs. There has been a gradual increase in lamb marking percentages, but it seems likely that much of the increase has been due to improved pastures in the 1950s and 1960s and a change in the breed structure to more fecund breeds. Nevertheless, lamb marking percentages in Australia remain low in comparison to our potential and there is a continuing need to search for ways of improving the reproductive performance of our national flock.

Against that background, the Production Research Advisory Committee, one of the Corporation's committees assisting in the allocation of support for research, requested the Australian Society for Reproductive Biology and the Australian Society of Animal Production to be joint sponsors of this symposium on sheep reproduction research as part of a review to determine the future direction of relevant research.

The Corporation acknowledges and is appreciative of the help of the two societies in sponsoring the symposium, of the Australian Academy of Science and of the various co-ordinators and editors who have contributed to the symposium and publication.

This book is the first in a new series of technical publications of the Australian Wool Corporation.

D.J. Asimus
Chairman,
*Australian Wool Corporation*

# NEUROENDOCRINE CONTROL OF THE OVINE OESTROUS CYCLE

I.J. Clarke, *Medical Research Centre, Prince Henry's Hospital, St. Kilda Road, Melbourne 3004, Australia.*

## INTRODUCTION

The oestrous cycle is a dynamic endocrine continuum. It is initiated by ovulation that leads to the formation of a corpus luteum and the secretion of progesterone. During this *luteal* phase of the cycle *tonic* gonadotrophin secretion prevails. Regression of the corpus luteum initiates the *follicular* phase which is dominated by oestrogen and endocrine changes that lead to the preovulatory surge in gonadotrophin secretion (*phasic* secretion) which, in turn, results in another ovulation.

Neural control of the oestrous cycle is exerted primarily by the secretion of gonadotrophin-releasing hormone (GnRH) from the hypothalamus. By action on the pituitary gland, this neuropeptide plays an obligatory role in the synthesis and release of the gonadotrophins, luteinizing hormone (LH) and follicle stimulating hormone (FSH). In turn, the secretion of GnRH, LH and FSH are affected by hormonal feedback signals from the ovaries. Oestradiol, progesterone and inhibin regulate gonadotrophin release by actions on the hypothalamus and pituitary gland. Neural systems influenced by external stimuli, for example light, presumably alter the reproductive status by changes in GnRH secretion.

This paper will review the state of knowledge regarding hypothalamic centres that control GnRH synthesis and release and will then discuss the relationships between GnRH and gonadotrophin secretion. With this background the patterns of gonadotrophin secretion seen throughout the oestrous cycle will be outlined and the roles of the hormonal feedback signals will be addressed.

## LOCALISATION OF GnRH PRODUCING CELLS IN THE HYPOTHALAMUS

GnRH is synthesized in hypothalamic neurones, transported along axons and stored in nerve terminals in the median eminence. In sheep, Wheaton (1979) and Crowder and Nett (1984) measured GnRH levels in the 'preoptico-suprachiasmatic' region, the medio-basal hypothalamus and the pituitary stalk/median eminence by radioimmunoassay. Highest concentrations were found in the stalk/median eminence where GnRH is stored. It is significant that GnRH was measured in the anterior hypothalamus which probably contains GnRH producing neurones. Polkowska *et al.* (1980) have studied the immunohistochemical localisation of GnRH in nerve fibres within the hypothalamus of the ewe, but GnRH immunostaining did not occur in perikarya and this work gave no indication of the hypothalamic nuclei that produce GnRH. Most recently Lehman *et al.* (1984) have reported the presence of GnRH containing perikarya in the medial preoptic area of the ewe hypothalamus. If we are to locate the hypothalamic centres responsible for generating GnRH pulses more extensive mapping of the GnRH neurones is necessary.

## HYPOTHALAMIC CENTRES CONTROLLING TONIC AND PHASIC RELEASE OF GONADOTROPHINS

Work on rodent species led to the concept that there are two major hypothalamic centres controlling the secretion of GnRH; a 'tonic centre' in which GnRH pulses are generated and a 'phasic' or 'surge centre' responsible for initiating the ovulatory LH surge (Halasz and Pupp 1965). Early work in the sheep showed that destruction of anterior hypothalamic nuclei did not affect ovarian cyclicity whereas lesions in the medio-basal hypothalamus caused anovulation (Clegg *et al.* 1958; Radford 1967). Halasz knife[(1)] cuts rostral to the suprachiasmatic nucleus did not prevent the occurrence of regular oestrous cycles (Domanski *et al.* 1980). These data appeared to rule out the existence of a 'phasic' centre in the anterior extremities of the sheep hypothalamus.

More recent work in which cuts have been made at various levels rostral to the arcuate nucleus have further delineated the centres involved in LH secretion. Complete anterior hypothalamic deafferentation between the posterior region of the suprachiasmatic nucleus and the rostral portion of the arcuate nucleus reduced basal gonadotrophin secretion (Jackson *et al.* 1978; Thiery *et al.* 1978, 1979), suggesting a dependence upon GnRH synthesis in hypothalamic regions slightly anterior to the arcuate nucleus. To determine whether or not GnRH synthesizing neurones, important for the maintenance of tonic gonadotrophin secretion, do in fact arise from retrochiasmatic regions, cuts could be made at the levels described by Jackson *et al.* (1978) and Thiery *et al.* (1979) and the concentration of GnRH in the median eminence could be monitored (see above). The 'pulse'

(1) The 'Halasz' knife is a specialised curved knife that can be lowered into the brain and and then rotated to cut in an arc and thus sever fibres in a specific region (Halasz and Pupp 1965).

generator for tonic GnRH/LH secretion appears to reside within the mediobasal hypothalamus, rostral to the arcuate nucleus but posterior to the suprachiasmatic nucleus.

The preovulatory LH surge is evoked by oestrogen (Goding *et al.* 1969), and is due to an increase in GnRH secretion from the hypothalamus (Sarkar *et al.* 1976) resulting from neural mechanisms (Radford and Wallace 1974). Various studies in sheep have indicated that neural connections between anterior hypothalamic areas and the median eminence are essential for the phasic release of LH, but not tonic release. Halasz knife cuts in the hypothalamus between the arcuate and the suprachiasmatic nuclei prevented the phasic discharge of LH in response to oestrogen (Jackson *et al.* 1978; Thiery *et al.* 1978). These authors concluded that inputs from anterior hypothalamic nuclei into the medio-basal hypothalamus were essential for the generation of the LH surge. From these recent studies and earlier work involving electrical stimulation of the anterior hypothalamic nuclei (Przekop and Domanski 1970b), it can be concluded that the areas rostral to the suprachiasmatic nucleus are involved in generating the preovulatory surge.

A greater understanding of endocrine effects at the hypothalamic level could eventually lead to better methods of controlling and manipulating reproduction. In particular, knowledge of the location of the pulse generator for 'tonic' LH secretion and the area responsible for generating the LH surge (phasic release) would allow us to study the central mechanisms that control seasonal patterns of breeding. Such studies would also allow us to understand the endocrinology of the oestrous cycle more fully.

## THE SECRETION OF GnRH

Prior to definitive studies on the relationship between GnRH and LH secretion in 1982, it had been assumed that the pulsatile pattern of gonadotrophin secretion from the pituitary gland was a reflection of episodic discharges of GnRH from the median eminence. The development of models to measure GnRH secretion in sheep (Clarke and Cummins 1982; Levine *et al.* 1982), confirmed that GnRH pulses did generate LH pulses. Evidence that the GnRH pulses are the result of neural activity in the hypothalamus has been obtained by studying plasma LH levels during anaesthesia (Radford and Wallace 1974; Clarke and Doughton 1983) and following treatment with specific neural blocking agents (Jackson 1975, 1977).

Records of the multiunit activity of hypothalamic neurones in sheep (Thiery and Pelletier 1981) have identified changes in neural firing rate concomitant with LH secretion. It is also possible to record from single cells in the supraoptic nucleus of the sheep hypothalamus which have projections into the median eminence (Jennings *et al.* 1978). Unfortunately, the recordings made by Jennings *et al.* (1978) were not accompanied by the collection of blood samples to measure hypothalamic or pituitary hormone secretions. This type of work could yield information regarding the hypothalamic centres which generate GnRH pulses and the preovulatory LH surge.

## NEURAL SYSTEMS INVOLVED IN GnRH SECRETION

The catecholaminergic systems that are implicated in the control of GnRH secretion have been extensively studied in the laboratory rat (Barraclough and Wise 1982), but there is a lack of data on sheep. Gonadotrophin secretion has been measured when catecholamines have been infused into ventricles (Riggs and Malven 1974; Domanski *et al.* 1975; Przekop *et al.* 1975) or into the bloodstream (Deaver and Dailey 1982). These methods do not provide meaningful data regarding the hypothalamic mechanisms that control GnRH secretion. Catecholamines introduced to the ventricles could act upon a variety of neuronal systems which could be either inhibitory or stimulatory with respect to GnRH secretion, and the sites of action could be widespread. Various lines of evidence suggest that GnRH secretion is under the control of inhibitory as well as stimulatory neuronal systems (Przekop and Domanski 1970a; Domanski *et al.* 1980; Pau *et al.* 1982). Introduction of catecholamines into the peripheral bloodstream is not relevant to central mechanisms since these compounds do not cross the blood brain barrier and thus any effects on gonadotrophin secretion would be the result of actions at the level of the median eminence or the pituitary gland (Oldendorf 1971).

Specific neural blocking agents have been used to study the mechanisms controlling GnRH/gonadotrophin secretion, indicating the involvement of noradrenergic, dopaminergic and serotoninergic pathways in the control of GnRH release (Jackson 1975, 1977). This information is of little value also, unless we pinpoint the centres involved in controlling GnRH and determine which types of neuronal systems have inhibitory and stimulatory inputs. Furthermore, the question of neuromodulators needs serious attention. One group of neurally active peptides that appears to be involved in the regulation of GnRH secretion is the opiates and preliminary experiments (Brooks *et al.* 1983; Ebling *et al.* 1983) have suggested opioid participation in the control of GnRH/gonadotrophin secretion in sheep. Manipulation of GnRH/gonadotrophin secretion with neuromodulators is a potentially fruitful area of research.

## PATTERNS OF GONADOTROPHIN SECRETION DURING THE OESTROUS CYCLE

### Luteinizing hormone

During the luteal phase of the oestrous cycle, when the corpus luteum is secreting progesterone, LH pulses occur at a frequency of once every 3-10 h (Baird *et al.* 1976; Rodgers *et al.* 1980; Baird *et al.* 1981; Karsch *et al.* 1983).

As the corpus luteum regresses and plasma progesterone levels fall, the frequency of LH pulses increases to around 1 pulse per h (Baird 1978; Baird *et al.* 1981; Karsch *et al.* 1983). There is a gradual increase in pulse frequency throughout the follicular phase (Baird 1978) until oestrus. There is also a gradual increase in the secretion of oestrogen from the ovarian follicles (Baird 1978; Baird *et al.* 1981; Karsch *et al.* 1979) which triggers two central events, oestrus and the LH surge, which both involve neural mechanisms (Clegg *et al.* 1958; Radford 1967).

Following the LH surge, ovulation and oestrus, there is a period of metoestrus, when oestrogen and progesterone secretion is low (Baird *et al.* 1981). During metoestrus, the tonic LH pulse frequency is higher (once per h) relative to that during the luteal phase of the oestrous cycle (Hauger *et al.* 1977).

There is some disagreement regarding the changes in LH pulse amplitude during the oestrous cycle. Baird (1978) found that the amplitudes of pulses during the luteal phase were greater than during the follicular phase. In 2 other studies (Baird *et al.* 1981; Karsch *et al.* 1983) the pulse amplitudes were found to be similar during the luteal and follicular phases. To understand the relationships between pulse frequency and amplitude we (Clarke *et al.* 1983) perfected a method for the surgical isolation of the pituitary gland from the hypothalamus (hypothalamo-pituitary disconnection, HPD) in the sheep. This removes endogenous GnRH inputs to the pituitary but LH and FSH secretion can be restored by exogenous GnRH pulses. With this model we have partly mimicked the transition from the luteal phase to the follicular phase and found that increasing GnRH pulse frequency reduced LH pulse amplitude. Thus, we could expect the LH pulse amplitude to decrease throughout the follicular phase, as the LH pulse frequency increases.

### Follicle Stimulating Hormone

There is no good evidence that FSH secretion is pulsatile in the sheep although some circhoral fluctuations are apparent (Salamonsen *et al.* 1973; McNeilly *et al.* 1976). It may not be possible to detect discrete pulses of FSH in plasma because of the long half-life of the hormone (102 min; Akbar *et al.* 1974). However, FSH secretion is dependent upon GnRH. After surgical isolation of the pituitary gland from the hypothalamus FSH is no longer detectable in the circulation, but this can be restored by pulsatile GnRH replacement (Clarke *et al.* 1984).

Throughout the luteal phase of the oestrous cycle plasma FSH levels remain relatively constant (L'Hermite *et al.* 1972; McNeilly *et al.* 1976; Pant *et al.* 1977). During the follicular phase FSH levels fall gradually (Salamonsen *et al.* 1973; Baird *et al.* 1981). At the time of the preovulatory LH surge there is a concomitant FSH surge (L'Hermite *et al.* 1972; Salamonsen *et al.* 1973) and then a second FSH rise occurs 18-24 h later (Salamonsen *et al.* 1973).

## FEEDBACK CONTROL OF GnRH AND GONADOTROPHIN SECRETION

### Oestrogen

During the luteal phase of the oestrous cycle a low level of oestrogen secretion is maintained (Baird 1978) which appears to synergise with progesterone to limit LH secretion (Hauger *et al.* 1977; Karsch *et al.* 1977; Karsch *et al.* 1980). FSH secretion during the luteal phase may be partly inhibited by oestrogen but the data of Goodman *et al.* (1981) suggest that an additional factor (inhibin) is required (see below). Oestrogen acts to increase LH pulse frequency during the follicular phase of the cycle (Karsch *et al.* 1983).

The neural mechanisms that initiate the endocrine events leading to the LH surge are not yet understood. It is known that the LH surge in cycling animals is caused by an increase in the secretion of GnRH (Sarkar *et al.* 1976). A surge in LH secretion can also be elicited by giving oestrogen to ovariectomised (OVX) ewes (Scaramuzzi *et al.* 1971) and this model has been used to study neuroendocrine events triggered by oestrogen. At the time of the LH surge, a high GnRH pulse frequency is seen (Clarke and Cummins 1985).

In addition to effects on the hypothalamus, oestrogen can act on the pituitary. The biphasic response of LH to a single injection of oestrogen into an OVX ewe (decrease and then increase) (Scaramuzzi *et al.* 1971), is partly due to changes in the responsiveness of the pituitary gland to oestrogen (Clarke and Cummins 1984; Coppings and Malven 1976). The increased responsiveness does not produce an LH surge of the magnitude seen in normal ewes. From this it can be concluded that a rise in GnRH secretion is also necessary to elicit a full LH surge.

The fall is FSH secretion during the follicular phase is due, in part, to a negative feedback effect of oestrogen (McNeilly *et al.* 1984); this is most likely a pituitary effect (Clarke and Cummins 1984).

Progesterone

Progesterone reduces the LH pulse frequency and increases the pulse amplitude (Goodman and Karsch 1980; Martin *et al.* 1983). This is most likely due to a hypothalamic effect, reducing GnRH pulse frequency. Thus, the relatively low frequency of LH pulses during the luteal phase is probably due to progesterone feedback.

Inhibin

Inhibin is a secretion of the gonads that exerts selective feedback control over the secretion of FSH. Work is in progress to isolate and purify this important hormone but until now we have had to perform experiments with crude material. Using ovine follicular fluid as a source of inhibin, selective inhibition of FSH secretion has been shown in OVX ewes (Cummins *et al.* 1983). FSH secretion in OVX ewes cannot be brought down to levels equal to those of cycling ewes by steroid treatment alone (Goodman *et al.* 1981) and this provides evidence that inhibin acts throughout the cycle to limit FSH secretion. Recent work with GnRH-pulsed OVX-HPD ewes has shown that the effects of inhibin are wholly accounted for by action on the pituitary gland (Clarke and Cummins, unpublished). Since the major source of inhibin is thought to be the ovarian follicles, one might expect increasing secretion during the follicular phase of the cycle accounting, partly, for the fall in FSH secretion at this time.

Steroid receptors

Oestrogen action is effected via specific, high affinity oestrogen receptors that are present in the hypothalamus and pituitary gland of the ewe (Clarke *et al.* 1981). The distribution of oestrogen receptors throughout the ovine hypothalamus has not been examined thoroughly and work in this area would enable us to localise the sites at which oestrogen affects gonadotrophin secretion. This would be particularly important in furthering our understanding of the seasonal changes in responsiveness to oestrogen (Karsch 1984). There has been no comprehensive study of progesterone receptors in the sheep hypothalamus.

## CONCLUSIONS

Localisation of hypothalamic centres that produce GnRH and regulate its secretion could be obtained using immunocytochemistry and by the measurement of oestrogen receptors in defined regions. In these ways it would be possible to locate and then to characterize hypothalamic areas responsible for tonic and phasic GnRH release. More light could then be shed on regulation of GnRH secretion by sampling hypothalamo-hyopohyseal portal blood and measuring neuronal secretions to define the neural systems involved.

Patterns of gonadotrophin secretion have been studied extensively, but recent evidence shows that more definitive information can be obtained if samples are taken at very frequent intervals (≤4-5 min) (Karsch *et al.* 1983; Thomas 1983). It is possible that ovulation rate is determined by a particular *pattern* of gonadotrophin secretion that we have not identified.

Manipulation of feedback signals to the hypothalamus and pituitary by changing the circulating levels of steriods (Scaramuzzi and Martin 1984) can improve ovulation rate but the neuroendocrine mechanisms responsible for this effect are poorly understood. Research into methods of regulating inhibin could be effective. To design appropriate ways of manipulating sheep reproduction with inhibin (or antagonists), basic knowledge is required on the mechanism of action of inhibin at the hypothalamic and/or pituitary level.

It may seem that neuroendocrine research is far removed from the processes governing on farm productivity of sheep. A more careful appraisal would indicate that a thorough knowledge of neuroendocrinology is fundamental to our understanding the reproductive processes and how to manipulate them to improve productivity. Such phenomena as seasonal and post-partum acyclicity (anoestrus), onset of breeding cycles at puberty, the ram effect and clover disease involve neuroendocrine mechanisms that are poorly understood. For example, what is the exact mechanism whereby the pineal gland can control the breeding season of sheep? Answers to such questions as this would enable applied animal scientists to design more efficient methods of dealing with acyclicity.

With regard to the control of the oestrous cycle we know virtually nothing about the way in which progesterone controls LH secretion. Before we can begin to address this issue we need to locate the hypothalamic areas that are responsible for generating GnRH/LH pulses. Then, by studying neural systems at this locus we could perhaps improve on current methods of synchronisation of the oestrous cycle.

Central questions such as the role of prolactin have not been addressed in this chapter, but again, this is an area where neuroendocrine expertise is vital. Does prolactin play a role in the oestrous cycle and the determination of ovulation rate? Before we begin to address these questions we need to identify the physiological prolactin releasing factors and inhibiting factors that are presumed to arise from the hypothalamus.

These are long-term goals requiring basic research. At this stage there is a prerequisite for fundamental knowledge in neuroendocrinology which will, in time, provide new ideas for the manipulation of reproductive processes.

## REFERENCES


Akbar, A.M., Nett, T.M. and Niswender, G.D., 1974. *Endocrinology, 94*, 1318-1324.
Barraclough, C.A. and Wise, P.M., 1982. *Endocrine Reviews, 3*, 91-119.
Baird, D.T., 1978. *Biol. Reprod., 18*, 359-364.
Baird, D.T., Swanston, I.A. and McNeilly, A.S., 1981. *Biol. Reprod., 24*, 1013-1025.
Baird, D.T., Swanston I. and Scaramuzzi, R.J., 1976. *Endocrinology, 98*, 1490-1496.
Brooks, A.N., Lees, P.D., Cheeks, R.E., Lamming, G.E. and Haynes, N.B., 1983. *Proc. Ann. Confr. Soc. for the Study of Fertility. Manchester (Abstract).*
Clarke, I.J., Burman, K., Funder, J.W. and Findlay, J.K., 1981. *Biol. Reprod., 24*, 323-331.
Clarke, I.J. and Cummins, J.T., 1982. *Endocrinology, 111*, 1737-1739.
Clarke, I.J. and Cummins, J.T., 1984. *Neuroendocrinology, 39*, 267-274.
Clarke, I.J. and Cummins, J.T., 1985. *Endocrinology*, (Submitted)
Clarke, I.J. and Cummins, J.T., and De Kretser, D.M., 1983. *Neuroendocrinology, 36*, 376-384.
Clarke, I.J., Cummins, J.T., Findlay, J.K., Burman, K.J. and Doughton, B.W., 1984. *Neuroendocrinology, 39*, 214-221.
Clarke, I.J. and Doughton, B.W., 1983. *J. Endocr, 98*, 79-89.
Clegg, M.T., Santalucito, J.A. Smith, J.D. and Ganong, W.F., 1958. *Endocrinology, 62*, 790-797.
Coppings, R.J. and Malven, P.V., 1976. *Neuroendocrinology, 21*, 146-156.
Crowder, M.E. and Nett, T.M., 1984. *Endocrinology, 114*, 234-239.
Cummins, L.J., O'Shea, T., Bindon, B.M., Lee, V.W.K. and Findlay, J.K., 1983. *J. Reprod. Fert., 67*, 1-7.
Deaver, D.R. and Dailey, R.A., 1982. *Biol. Reprod., 27*, 624-632.
Domanski, E., Przekop, F. and Polkowska, J., 1980. *J. Reprod. Fert., 58*, 493-499.
Domanski, E., Przekop, F., Skubiszewski, B. and Wolinska, E., 1975. *Neuroendocrinology, 17*, 265-273.
Ebling, F.J.P., Lincoln, G.A., Cunningham, R.A., Anderson, N. and Lambert, L., 1983. *Proc. Ann. Confr. Soc. for the Study of Fertility, Manchester (Abstract).*
Goding, J.R., Catt, K.J., Brown, J.M., Kaltenbach, C.C., Cumming, I.A. and Mole, B.J., 1969. *Endocrinology, 85*, 133-142.
Goodman, R.L., and Karsch, F.J., 1980. *Endocrinology, 107*, 1286-1290.
Goodman, R.L., Pickover, S.M. and Karsch, F.J., 1981. *Endocrinology, 108*, 772-777.
Halasz, B. and Pupp, L, 1965. *Endocrinology, 77*, 553-562.
Haugher, R.L., Karsch, F.J. and Foster, D.L., 1977. *Endocrinology, 101*, 807-817.
Jackson, G.L., 1975. *Endocrinology, 97*, 1300-1307.
Jackson, G.L., 1977. *Biol. Reprod., 16*, 543-548.
Jackson, G.L., Kuehl, D., McDowell, K. and Zaleski, A., 1978. *Biol. Reprod., 17*, 808-819.
Jennings, D.P., Hastings, J.T. and Rodgers, J.M., 1977. *Brain Research, 149*, 347-364.
Karsch, F.J., 1984. This volume, 10-15.
Karsch, F.J., Foster, D.L., Bittman, E.L. and Goodman, R.L., 1983. *Endocrinology, 113*, 1333-1339.
Karsch, F.J., Foster, D.L., Legan, S.J., Ryan, K.D. and Peter, G.K., 1979. *Endocrinology, 105*, 421-426.
Karsch, F.J., Legan, S.J., Hauger, R.L. and Foster, D.L., 1977. *Endocrinology, 101*, 800-806.
Karsch, F.J., Legan, S.J., Ryan, K.D. and Foster, D.L., 1980. *Biol. Reprod., 23*, 404-413.
Lehman, M.N., Silverman, A-J., Robinson, J.E. and Karsch, F.J. 1984. *Proc. 7th Int. Congr. Endocrinology, Quebec. Abstract 1245.*
Levine, J.E., Pau, K-Y, Ramirez, V.D. and Jackson, G.L., 1982. *Endocrinology, 111*, 1449-1455.
L'Hermite, M., Niswender, G.D., Reichert Jr., L.E. and Midgley Jr., A.R., 1972. *Biol. Reprod., 6*, 325-332.
Martin, G.B., Scaramuzzi, R.J. and Henstridge, J.D., 1983. *J. Endocr., 96*, 181-193.
McNeilly, A.S., Fraser, H.M. and Baird, D.T., 1984. *J. Endocrin., 101*, 213-219.
McNeilly, J.R., McNeilly, A.S., Walton, J.S. and Cunningham, F.J., 1975. *J. Endocr., 70*, 69-79.
Oldendorf, W.H., 1971. *Am. J. Physiol., 221*, 1629-1639.
Pant, H.C., Hopkinson, C.R.N. and Fitzpatrick, R.J., 1977. *J. Endocr., 73*, 247-255.
Pau, K-Y.F., Kuehl, D.E. and Jackson, G.L., 1982. *Biol. Reprod., 27*, 999-1009.
Polkowska, J., Dubois, M.-P. and Domanski, E., 1980. *Cell Tissue Res., 208*, 327-341.
Przekop, F., and Domanski, E., 1970a. *Acta Physiol. Pol., 21*, 34-49.
Przekop, F. and Domanski, E., 1970b. *J. Endocr., 46*, 305-311.
Przekop, F., Skubiszewski, B., Wolinska, E. and Domanski, E., 1975. *Acta. Physiol. Pol., 26*, 433-437.
Radford, H.M., 1967. *J. Endocr., 39*, 415-422.
Radford, H.M., and Wallace, A.L.C., 1974. *J. Endocr., 60*, 247-252.
Rodgers, R.J., Clarke, I.J., Findlay, J.K., Brown, A., Cumming, I.A., Muller, B.D. and Walker, S.K., 1980. *Aust. J. Biol. Sci., 33*, 213-20.
Salamonsen, L.A., Jonas, H.A., Burger, H.G., Buckmaster, J.M., Chamley, W.A., Cumming, I.A., Findlay, J.K. and Goding, J.R., 1973. *Endocrinology, 93*, 610-618.

Sarkar, D.K., Chiappa, S.A., Fink, G. and Sherwood, N.M., 1976. *Nature (Lond.), 264,* 461-463.
Scaramuzzi, R.J. and Martin, G.B., 1984. This volume 316-325.
Scaramuzzi, R.J., Tillson, S.A., Thorneycroft, I.J. and Caldwell, B.V., 1971. *Endocrinology, 88*, 1184-1189.
Thiery, J.C., Blanc, M., Ravault, J.P. and Pelletier, J., 1979. *Ann. Biol. Anim. Bioch. Biophys., 19*, 1081-1097.
Thiery, J.C., and Pelletier, J., 1981. *Neuroendocrinology, 32*, 217-224.
Thiery, J.C., Pelletier, J. and Signoret, J.P., 1978. *Ann. Biol. Anim. Bioch. Biophys., 18*, 1413-1426.
Thomas, G.B., 1983. *Proc. Aust. Soc. Rep. Biol., 15*, 47.
Wheaton, J.E., 1979. *Endocrinology, 104*, 839-844.

# USE OF ACTIVE IMMUNIZATION TO EVALUATE THE ROLES OF PROGESTERONE DURING THE OESTROUS CYCLE OF THE EWE

G.B. Thomas and C.M. Oldham, *School of Agriculture (Animal Science), University of Western Australia, Nedlands, Western Australia, 6009.*
R.M. Hoskinson and R.J. Scaramuzzi, *CSIRO, Division of Animal Production, Prospect, New South Wales, 2148.*

*Summary* Two experiments used immunization to evaluate the roles of progesterone in controlling the oestrous cycle of the ewe. In experiment 1 the cyclic activity of immunized ewes was determined and in experiment 2 the secretion of LH and progesterone was studied. Immunization against progesterone-HSA caused (i) shortening of the interval between LH surges and between ovulations, (ii) increased ovulation rate, (iii) reduced oestrous behaviour, (iv) persistence of corpora lutea, (v) increased frequency of LH pulses. This study highlights the variety of control mechanisms which directly involve progesterone.

## INTRODUCTION

Progesterone has many important functions in the non-pregnant ewe, which include priming of the central nervous system for the induction of oestrus (Robinson 1954), blockage of the positive feedback effect of oestradiol (Scaramuzzi *et al.* 1971), and a major role in controlling the tonic secretion of luteinizing hormone (LH) during the breeding season (Hauger *et al.* 1977). This paper describes the use of immunization as a technique to evaluate the way in which progesterone controls the timing of ovulation.

## MATERIALS AND METHODS

Twenty seasonally anovular Merino ewes were immunized against progesterone-11$\alpha$-hemisuccinate-human serum albumin (progesterone-HSA). Ten control ewes were immunized against human serum albumin (control-HSA). The primary immunization consisted of 1.5 mg steroid-protein conjugate made up in 3 ml of Freund's complete adjuvant and injected at multiple subcutaneous and intramuscular sites. The initial injection was given in October followed by two booster injections identical to the primary injection 4 and 12 weeks later in November and January.

Experiment 1 began fourteen days after the last injection when all ewes were cyclic. The ewes were synchronized using intravaginal sponges impregnated with 60 mg of Medroxy Progesterone Acetate (Repromap, Upjohn). At sponge withdrawal (day 0) the ewes received an injection of 250 $\mu$g of a prostaglandin analogue (Estrumate, ICI) to ensure complete luteal regression. The ewes were then placed with two harnessed vasectomized rams and oestrous activity was checked twice daily. The ovaries of the ewes were examined by laparoscopy on days 4, 6, 11, 14, 18, 21, 26 after sponge removal and the number of corpora lutea, their location on the ovaries and their estimated age (Oldham and Lindsay, 1980) were recorded.

The same ewes were used for experiment 2, except that one ewe from each group was culled because abdominal adhesions prevented further examination of the ovaries. The ewes received a third booster injection in March and were re-synchronized 14 days later. Beginning at sponge withdrawal (day 0) blood was sampled every four hours for 10 days, and, at 15 minute intervals for four hours on days 5, 7 and 9 after sponge withdrawal. During days 5, 9 and 12 the ovaries of the ewes were observed by laparoscopy. The plasma was stored at -20°C until it was assayed for LH and progesterone as described by Martin *et al.* (1983), with the modification that samples for progesterone were treated with 100 $\mu$l of 1 M nitric acid before extraction in hexane to denature the endogenous antibody.

## RESULTS

The antibody titre was estimated by the dilution at which 50% of 20 pg of labelled progesterone was bound by the plasma. All ewes immunized with progesterone-HSA produced antibodies against progesterone. The mean titre at the beginning of experiments 1 and 2 were 1:19,644 and 1:20,016 respectively. All control-HSA ewes had titres of less than 1:100.

The ovarian response for experiment 1 is shown in Table 1. Only 75% of the ewes immunized against progesterone-HSA displayed oestrus at the first ovulation after synchronization, and the incidence of oestrus and interval between ovulations were then reduced ($P < 0.01$) for the following three ovulations. All control ewes displayed normal oestrous behaviour and cycle length.

All immunized ewes had ovulated by day 4 after sponge withdrawal. In one ewe twin corpora lutea regressed after 10 days and this ewe remained anovular throughout the experiment. In the remaining 19 ewes no corpora lutea regressed even though the ewes had ovulated for a second time by the laparoscopy on day 14. A further 13 ovulated for a third time and seven of these had a fourth ovulation at the laparoscopy on day 26 (Table 1). Eleven out of 20 of the third and fourth ovulations occurred even though the original corpora lutea still persisted on the ovaries. During the first, second and fourth ovulations the progesterone-HSA ewes had significantly ($P < 0.05$) higher ovulation rates than the control-HSA ewes.

**TABLE 1. Ovarian and oestrous activity of ewes immunized against HSA or progesterone-HSA.**

| Treatment | Ovulation number | Ewes Ovulating | Estimated ovulation interval (days) | Ovulation rate | Oestrous ewes (%) |
|---|---|---|---|---|---|
| Control-HSA | first | 10/10 | | 1.30 | 9 (90) |
| | second | 10/10 | 17.7 | 1.20 | 10 (100) |
| Progesterone-HSA | first | 20/20 | | 1.75* | 15 (75) |
| | second | 19/20 | 6.4** | 1.89* | 2** (10) |
| | third | 13/19 | 9.9** | 1.38 | 7* (54) |
| | fourth | 7/13 | 6.0** | 1.86* | 1** (14) |

Significance levels *P < 0.05 **P < 0.01

In experiment 2 18/19 progesterone-HSA ewes and 9/9 control-HSA ewes ovulated but the ovulation rates (OR) were not different (1.44 vs 1.22). By day 12, 10/18 progesterone-HSA ewes had reovulated (OR = 1.0) although the original corpora lutea were still present on the ovaries. The mean time between preovulatory surges of LH was 151.8 ± 11.7 hours (6.3 days). No control ewes reovulated during this time.

The frequency of LH pulses and concentration of total progesterone were significantly (P < 0.001) higher in ewes immunized against progesterone-HSA (Table 2). Treatment of control-HSA plasma with nitric acid did not affect the concentration of progesterone.

**TABLE 2. The secretion of LH and progesterone in ewes immunized against HSA or progesterone-HSA (mean ± SEM).**

| Days from sponge withdrawal | LH[1] (Pulses/h) | | Progesterone[1] (ng/ml) | |
|---|---|---|---|---|
| | Control-HSA | Progesterone-HSA | Control-HSA | Progesterone-HSA |
| 5 | 0.00 | 0.41 ± 0.10 | 1.1 ± 0.3 | 9.4 ± 1.0 |
| 7 | 0.06 ± 0.03 | 0.50 ± 0.11 | 1.8 ± 0.1 | 7.1 ± 0.8 |
| 9 | 0.10 ± 0.06 | 0.38 ± 0.09 | 2.7 ± 0.2 | 10.2 ± 1.1 |

[1] Between columns differences (P < 0.001)

## DISCUSSION

Immunization of Merino ewes against progesterone-HSA decreased but did not eliminate the incidence of behavioural oestrus. Since Robinson (1954) has shown that progesterone priming is essential for oestrous behaviour, our results suggest that in some immunized ewes there was sufficient progesterone unbound to prime them for oestrus but insufficient to block ovulation. The synthetic progestagen, Medroxy Progesterone Acetate, was not bound by the antibody as evidenced by the high incidence of oestrus at the first ovulation.

The reduced ovulatory interval in experiment 1 suggested a reduced interval between preovulatory surges of LH, a result confirmed in experiment 2. These results agree with the hypothesis that a reduction in biologically active progesterone has reduced the inhibition of positive feedback (Scaramuzzi *et al.* 1971). Short ovulatory cycles in ewes immune to progesterone were also seen by French and Spennetta (1981) and Hoskinson *et al.* (1982).

Immunization against progesterone resulted in an increased frequency of LH pulses, probably through a reduced negative feedback action of progesterone on the hypothalamus. This increase in LH may be responsible for the increased ovulation rate seen in experiment 1 since Merino ewes consistently having two ovulations also have a higher frequency of LH pulses during days 2-12 of the oestrous cycle than Merino ewes consistently having one ovulation (Thomas *et al.* 1984). The elevated levels of LH may also have stimulated luteal function since progesterone levels were higher in these animals. Alternatively, immunization *per se* and the increased number of corpora lutea may also have increased the secretion of progesterone.

The high number of new ovulations and the absence of luteal regression led to an accumulation of corpora lutea on the ovaries and indicated that prostaglandin $F_{2\alpha}$ was either not being synthesised or not being released. Since ovariectomized ewes treated with progesterone have high levels of prostaglandin $F_{2\alpha}$ (Scaramuzzi *et al.*

1977), the present findings confirmed that either the synthesis or release of prostaglandin $F_{2\alpha}$ is in response to exposure of the endometrium to elevated levels of progesterone.

The present study highlights the importance of progesterone in control of oestrus, ovulation, ovulation rate and the secretion of LH. It also shows that for research into the control of the oestrous cycle immunization offers a viable alternative to the use of ovariectomized ewes and steroid replacement therapy.

## ACKNOWLEDGEMENTS

This work was supported by the Australian Meat Research Committee. We thank Peter Moore and David Suckling for their technical assistance.

## REFERENCES

French, L.R. and Spennetta, B., 1981. *Theriogenology, 16*, 407-418.

Hauger, R.L., Karsch, F.J. and Foster, D.L., 1977. *Endocr., 101*, 807-817.

Hoskinson, R.M., Scaramuzzi, R.J., Downing, J.A., Hinks, N.T. and Turnbull, K.E., 1982. *Proc. Aust. Soc. Reprod. Biol., 14*, 92.

Martin, G.B., Scaramuzzi, R.J. and Henstridge, J.D., 1983. *J. Endocr., 96*, 181-193.

Oldham, C.M. and Lindsay, D.R., 1980. *Anim. Reprod. Sci., 3*, 119-124.

Robinson, T.J., 1954. *J. Endocr., 19*, 117-124.

Scaramuzzi, R.J., Tillson, S.A., Thorneycroft, I.H. and Caldwell, B.V., 1971. *Endocr., 88*, 1184-1189.

Scaramuzzi, R.J., Baird, D.T., Boyle, H.P., Land, R.B. and Wheeler, A.G., 1977. *J. Reprod. Fert., 49*, 157-160.

Thomas, G.B., Oldham, C.M. and Martin, G.B., 1984. This volume, 102-104.

# ENDOCRINE AND ENVIRONMENTAL CONTROL OF OESTROUS CYCLICITY IN SHEEP

F.J. Karsch, *Reproductive Endocrinology Program and Department of Physiology, The University of Michigan, Ann Arbor, Michigan, USA 48109.*

## INTRODUCTION

The reproductive process of mammals, both domestic and feral, is marked by alternating periods of breeding activity and inactivity. In the female, these alternations are organised into a number of distinct time frames. In the short-term, these changes include the period of sexual arousal and quiescence associated with the oestrous and dioestrous stages of the oestrous cycle. In the longer term, the alternations of fertility and infertility are associated with changes in season, and with pregnancy and lactation. On the even longer term, there is a continuum of sexual change associated with maturation, adulthood and senescence. These types of reproductive shifts do not occur in isolation but are superimposed upon one another such that the underlying mechanisms must interact and co-exist.

Recently, considerable insight has been gained into the endocrine interactions which lead to generation of both the 16-day oestrous cycle and the annual reproductive cycle of female sheep (Legan and Karsch 1979; Karsch *et al.* 1980, 1984a). The present overview will describe some of these interactions in the context of the endocrine requirements for cyclicity, how environmental signals exert their impact, and how this basic information might lead to new approaches to fertility regulation. At the end, similarities and differences will be identified in the fundamental biological processes which underlie transitions in reproductive activity throughout the year, during sexual maturation, and following lactational anoestrous.

## ENDOCRINE MECHANISMS ASSOCIATED WITH SEASONAL ONSET AND CESSATION OF OESTROUS CYCLES

Although the occurrence of the oestrous cycle reflects an intricate balance of a wide variety of endocrine interactions, a focal point for regulation is the LH pulse generator (Figure 1). This ensemble of neural and endocrine tissues produces the pulsatile pattern of GnRH, and hence LH secretion which is evident in nearly all mammals studied to date. The activity of this LH pulse generator is highly susceptible to a wide variety of signals arising in both the external and internal environments. The importance of this regulation to the oestrous and seasonal reproductive cycle of the ewe will now be reviewed; later on the concept will be developed that the LH pulse generator constitutes a final common mechanism for determination of onset or cessation of oestrous cyclicity in a number of other circumstances.

The importance of the pulsatile pattern of LH secretion in the ewe, and its changes throughout the year, are summarised schematically in the lower portion of Figure 2. The main point to be developed is that, for ovulation to occur, the frequency of LH pulses must be sufficiently high that it produces the sustained increase in circulating LH which is needed to stimulate the oestradiol rise which elicits the surge of LH that induces ovulation. During the *breeding season*, pulses of LH occur at a relatively low frequency (every 3-4 hr) in the mid-late luteal phase of the oestrous cycle when progesterone is maximal (Figure 2, BREEDING SEASON). The low frequency of pulses at this time is primarily a function of the inhibitory action of progesterone secreted by the corpus luteum, although oestradiol seems to sensitise the pulse generator to progesterone and thus itensify its frequency-lowering effect (Goodman and Karsch 1980; Goodman *et al.* 1981a; Martin *et al.* 1983a). Between pulses, circulating LH decreases markedly, generally reaching undetectable levels; this is especially true in the mid-late luteal phase when progesterone is maximal and frequency minimal (Hauger *et al.* 1977; Baird 1978). Because LH is important to follicular development and steroidogenesis, the low frequency pulses of LH do not provide sufficient gonadotrophic support for the final stages of follicular development and a sustained increase in oestradiol secretion. When the corpus luteum regresses, the progestational blockade of the LH pulse generator is lifted, and pulse frequency increased markedly (Baird 1978; Karsch *et al.* 1983). This provides important gonadotrophic drive to the developing ovarian follicle and promotes the sustained increase in oestradiol production needed to induce oestrous behaviour and the LH surge which causes ovulation (Goodman *et al.* 1981b; McNatty *et al.* 1981 McNeilly *et al.* 1982). The occurrence of ovulation during the breeding season is thus closely linked to the increase in activity of the LH pulse generator.

During the *anoestrous season*, LH pulses occur infrequently (every 8-12 hr) despite the absence of a corpus luteum and the virtual absence of circulating progesterone (Figure 2, ANOESTRUS). Between pulses, circulating LH plummets to an undetectable level. Thus, there is an insufficient gonadotrophic stimulus for the final stages of follicular maturation and for the preovulatory oestradiol rise. This precludes the LH surge and ovulation.

At the *transitions* into and out of the anoestrous season, there is little information as to changes in the activity of the LH pulse generator. This deficiency exists because extremely frequent blood samples are needed to characterise the LH pulse pattern in the follicular phase, and there is no reliable method for pinpointing the

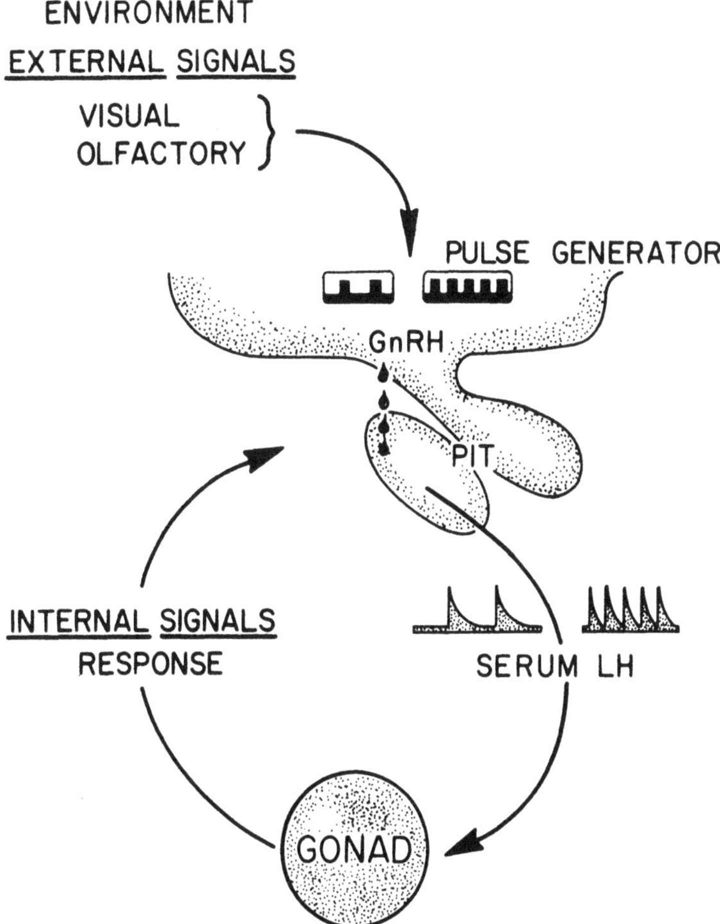


**FIGURE 1. The concept to be developed is that a pulse generator produces episodic secretion of gonadotrophin-releasing hormone (GnRH) from the hypothalamus which causes pulsatile secretion of luteinizing hormone (LH) from the anterior pituitary gland (PIT). The activity of the pulse generator is regulated by factors in both the external and internal environments.**

time of the seasonal transitions as they occur in individual animals. The patterns of LH pulses at the transitions, therefore, must be considered speculative (Figure 2, TRANSITION TO ANOESTRUS, TRANSITION TO BREEDING SEASON). Nonetheless, the notion that the transitions into and out of anoestrus are due to respective decreases and increases in the frequency of LH pulses is supported by a preliminary report of LH pulse patterns (I'Anson 1983) and by indirect evidence from studies utilising relatively infrequent blood collections or pulsed injections of LH or GnRH (Legan and Karsch 1979; McNatty *et al.* 1981; McNeilly *et al.* 1982; Legan 1982; McLeod *et al.* 1983).

## TRANSDUCTION OF ENVIRONMENTAL CUES INTO ENDOCRINE MESSAGES THAT CONTROL SEASONAL BREEDING

### Feedback Change

One important variable which mediates the seasonal pattern of oestrous cyclicity in the ewe is the ability of oestradiol to inhibit tonic gonadotrophin secretion. Seasonal changes in the potency of oestradiol negative feedback have been documented in a number of laboratories utilising breeds ranging from highly seasonal Welsh Mountain and Suffolk ewes to breeds such as Dorset Horn and Merino which have less distinct breeding and anoestrous seasons (Legan *et al.* 1977; Webster and Haresign 1983; Martin *et al.* 1983a). These seasonal swings in the negative feedback potency of oestradiol are profound; they are regulated by environmental photoperiod, and they reflect a dramatic change in the ability of oestradiol to inhibit the frequency of pulsatile LH secretion (Legan and Karsch 1980; Goodman *et al.* 1982; Martin *et al.* 1983a).

The nature of the feedback shift is illustrated in the upper portion of Figure 2. During the breeding season, the response to oestradiol is low; physiological levels of the steroid by itself cannot reduce the frequency of LH

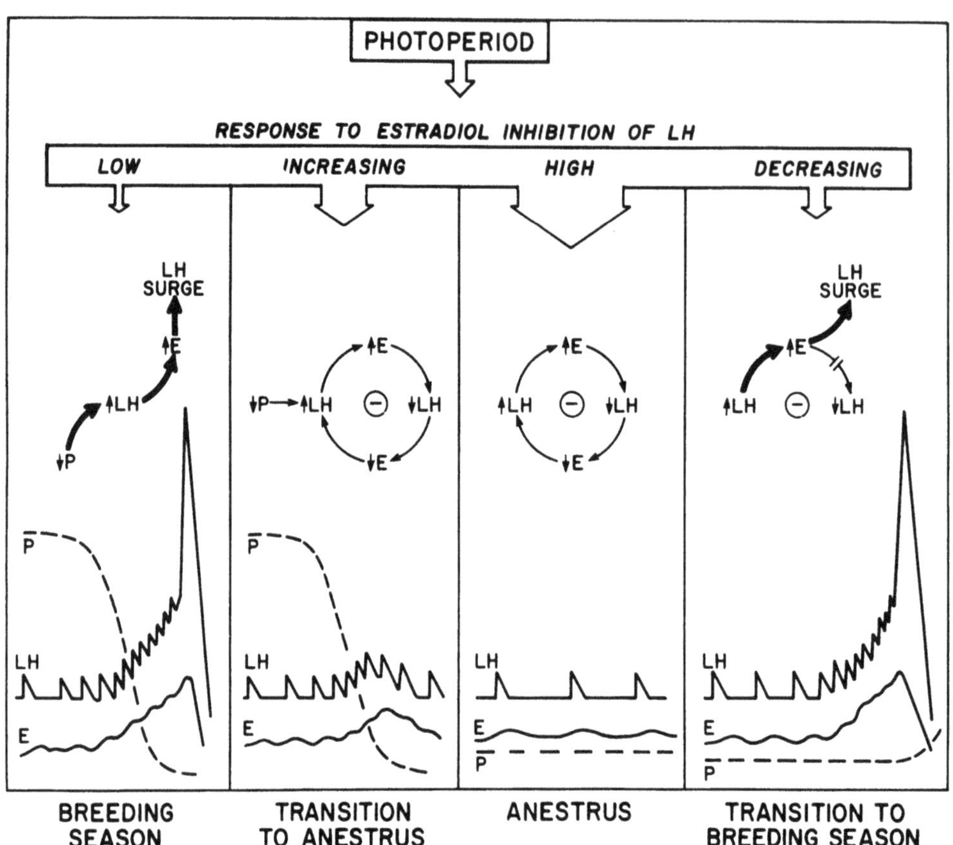


**FIGURE 2. Model for the photoneuroendocrine control of the oestrous and the seasonal reproductive cycle of the ewe. The essence of this model is that the periods of breeding activity and anoestrus, and the transitions between these periods, are determined by changes in the ability of oestradiol to inhibit frequency of the LH pulse generator (from Karsch *et al.* 1980).**

pulses (Goodman and Karsch 1980; Karsch *et al.* 1983; Martin *et al.* 1983a). This explains why LH pulse frequency and oestradiol can rise in parallel between luteal regression and onset of the preovulatory LH surge. At the transition into anoestrus, the ability of oestradiol to inhibit tonic LH secretion increases. Once the anoestrous condition is established, the steroid can evoke a powerful suppression of LH secretion, manifest as a reduction in LH pulse frequency (Goodman *et al.* 1982; Martin *et al.* 1983a). The existence of this highly effective negative feedback loop between LH and oestradiol explains why LH pulse frequency remains minimal during anoestrus and why follicular oestradiol secretion is held in check. At the transition to the breeding season, the negative feedback potency of oestradiol wanes, thus permitting sustained increases in LH pulse frequency, the preovulatory oestradiol rise, and the restoration of oestrous cyclicity.

It is important to note that the LH pulse generator of the ewe is sensitive to environmental variables other than photoperiod. A prime example of another input is the olfactory stimulus provided by the ram. Within minutes after introducing the ram to previously isolated anoestrous ewes, the pulse generator is activated, the oestradiol blockade of LH pulse frequency is lifted, and a follicular phase is initiated which can culminate in the LH surge and ovulation (Poindron *et al.* 1980; Martin *et al.* 1983b). This serves to illustrate the lability of the LH pulse generator, and the likelihood that its activity at any point in time reflects an interplay of multiple stimulatory and inhibitory inputs arising from both the external and internal environments.

### Photoneuroendocrine Mechanisms

The specific neural changes which determine whether or not oestradiol can address the pulse generator, and depress the frequency of LH pulses, are not known. Nevertheless, a great deal is known about the pathway whereby photic cues are picked up and relayed to the LH-pulse generating system of sheep. The proposed photoneuroendocrine pathway, which is illustrated in Figure 3, is the thrust of several recent reviews (Lincoln and Short 1980; Legan and Winans 1981; Bittman 1984; Karsch *et al.* 1984a). According to this model, light cues are picked up by photoreceptors located in the eyes and relayed over a monosynaptic nerve tract to the suprachiasmatic nuclei of the hypothalamus. After receiving input from the circadian system, the photoperiodic message is transmitted by way of the superior cervical ganglia to the pineal gland. The pineal converts this neural input into a hormonal signal which takes the form of a circadian rhythm of melatonin secretion. The duration of elevated melatonin secretion, which is directly proportional to the length of the night, is interpreted as either inductive or suppressive. Inductive melatonin signals stimulate the pulse generator and render it resistant to the frequency slowing action of oestradiol; suppressive melatonin signals inhibit the pulse generator and sensitise it to inhibition by oestradiol. It is important to note that the melatonin pattern itself determines the reproductive response, rather than merely permitting daylength to be measured by some other neural timekeeping device. Once the specific melatonin pattern is set, therefore, the photo-sexual response is independent of daylength (Bittman and Karsch 1984).

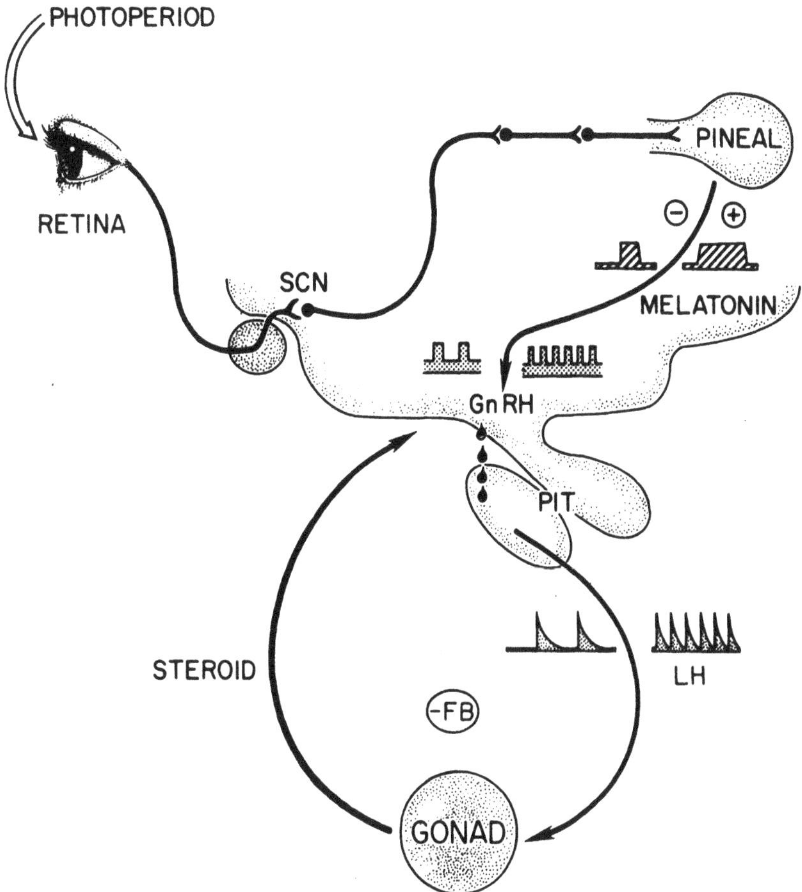


**FIGURE 3. Postulated photoneuroendocrine pathway to the LH pulse generator of the ewe. It is through this pathway that an external signal (light) serves to modulate the capacity of the pulse generating mechanism to respond to an internal signal (steroid negative feedback). This interaction then determines the reproductive state (from Karsch *et al.* 1984a).**

Practical Considerations

It is not surprising that as we gain more and more insight into the mechanisms whereby photoperiod regulates the alternations between breeding and anoestrous conditions, novel approaches aimed at enhancing the efficiency of breeding programs have begun to be developed. One case in point here is the use of melatonin. It is now clear that the onset of the breeding season can be advanced by daily injection or feeding of melatonin to anoestrous ewes (Kennaway *et al.* 1982; Nett and Niswender 1982). The rationale for these treatments is that the exogenous melatonin is provided in the afternoon, thus permitting it to summate with endogenous melatonin and extend the period that melatonin is elevated each day. The photoperiod is thereby effectively "shortened" although the ewes continue to be exposed to the relatively long days typical of the anoestrous season.

Despite the widespread practical benefits which may be realised by utilising and manipulating this response system, several characteristics of the photoneuroendocrine mechanism of the ewe limit the unqualified applicability of this approach. For example, the reproductive response to a given daylength cannot be maintained indefinitely. Over time, the ewe becomes refractory to photoperiods which once had been either stimulatory or inhibitory (Thwaites 1967; Wodzicka-Tomaszewska *et al.* 1967; Ducker *et al.* 1973; Robinson and Karsch 1984). A similar refractoriness can develop with time to a given melatonin pattern, at least for a melatonin pattern which was once inductive (Nett and Niswender 1982; Karsch *et al.* 1984b). Further, it is now known that in the absence of photoperiodic input, such as following blinding or pinealectomy, alternations between breeding activity and quiescence persist although the timing of these shifts no longer conforms to the normal seasonal pattern (Bittman *et al.* 1983; Legan and Karsch 1983). Therefore, either the ewe can cue on another environmental variable which has access to the pulse generator, or she has an endogenous circannual rhythm of reproduction which is normally entrained by the annual photoperiodic cycle but which "free-runs" in its absence.

The point to be stressed here, is that the ewe is likely to be genetically programmed to undergo switches between periods of reproductive activity and inactivity. This awareness must be incorporated in the development of methods to circumvent the natural period of infertility. Of special interest would be an understanding of the genetics of the photoneuroendocrine mechanisms in those breeds of sheep which either have particularly long breeding seasons, or which are claimed to be "nonseasonal".

## A COMMON FINAL MECHANISM FOR DETERMINATION OF CYCLICITY

At the outset, the concept was introduced that the process of reproduction is marked by alternating periods of breeding activity and inactivity which are organised into several distinct time frames. These include the relatively short-term fluctuations associated with the oestrous cycle, the longer term changes associated with season, and the evolution of changes of reproductive potential during the course of a lifetime. The concept was then developed that a common mechanism increased frequency of the LH pulse generator producing the periods of overt reproductive activity in two of these situations — the oestrous and the seasonal reproductive cycles. The last point to be made here is that changes in the activity of the LH pulse generator in general, and its susceptibility to frequency inhibition by steroids in particular, are likely to be a major determinant of fertility in other reproductive transitions as well.

The concept that the seasonal pattern of reproductive condition is a function of the ability of oestradiol to inhibit tonic gonadotrophin secretion is remarkably similar to the 'gonadostat' hypothesis of puberty, proposed for the rat by Ramirez and McCann (1963) and embellished for the female lamb by Foster and his colleagues (Foster and Ryan 1981). In fact, 10 years ago Geschwind (1974) proposed that the yearly onset of the breeding season is simply a matter of undergoing an annual puberty. The extensive mechanistic similarities for onset of reproductive function in sheep in these two situations has already been described (Karsch and Foster 1981). In the developing lamb, for example, pulses of LH occur less frequently than during the follicular phase of the cycle and oestradiol exerts a profound inhibition of tonic gonadotrophin secretion. As the time of first ovulation approaches, the response to oestradiol negative feedback wanes, tonic LH secretion increases, and a sustained increase in oestradiol secretion occurs which induces the first preovulatory gonadotrophin surge (Foster and Ryan 1981). In fact, attempts to demonstrate substantial differences in the neuroendocrine mechanisms leading to first ovulation in the developing lamb and to onset of the breeding season of the ewe have only served to stress the similarities of the two processes. Nevertheless, the factors which lead to activation of the LH pulse generator during sexual maturation in the lamb may be more complex than those which operate at the onset of the breeding season, because the relevant changes in the lamb reflect an interaction of body growth and nutrition in addition to the ambient photoperiod (Foster *et al.* 1984a,b).

Still another physiological situation in which the same general mechanisms seem to apply is lactational anoestrus. The arrest of oestrous cycles in ewes at this time is associated with an increased ability of oestradiol to inhibit tonic LH secretion, a reduced frequency of LH pulses, and the absence of complete follicular maturation (Wright *et al.* 1981; Wright *et al.* 1983). Further, pulsed delivery of GnRH can precipitate endogenous LH surges, ovulation and normal luteal function in ewes during postpartum anoestrus (Wright *et al.* 1984). It has not yet been demonstrated, however, that the recovery from lactational anoestrus is temporally associated with a decreased response to oestradiol negative feedback and an attendant increase in LH pulse frequency. As with sexual maturation, the inhibitory inputs to the pulse generator during lactational anoestrus are likely to be more complex than those during seasonal anoestrus, including signals arising from the suckling stimulus and/or from nutritional deficits in addition to photoperiod.

At the risk of being overly simplistic, I have attempted to develop the concept that a final common mechanism exists for the regulation of certain transitions into and out of reproductive activity which occur over both the short and long term. Many signals impinge upon the LH pulse generating mechanism and determine its susceptibility to sex-steroid negative feedback. These inputs convey visual, olfactory, tactile, developmental, hormonal and nutritional information, and almost certainly other cues which have yet to be identified. It is through the interaction and integration of these inputs that the reproductive potential is realised.

ACKNOWLEDGEMENTS

Supported by grants from the National Institutes of Health (HD-11311 and HD-18337) and from the National Science Foundation (PCM 8316364).

REFERENCES

Baird, D.T., 1978, *Biol. Reprod.*, 18, 359-364.
Bittman, E.L., 1984. *In* Reiter, R.J. (ed) *The Pineal Gland,* Raven Press, New York, 155-192.
Bittman, E.L. and Karsch, F.J. 1984. *Biol. Reprod.*, 30, 585-593.
Bittman, E.L., Karsch, F.J. and Hopkins, J.W., 1983. *Endocrinology*, 113, 329-336
Ducker, M.J., Bowman, J.C. and Temple, A. 1973. *J. Reprod. Fert., Suppl., 19,* 143-150.
Foster, D.L. and Ryan, K.D. 1981. *J. Reprod. Fert., Suppl., 30,* 75-90.
Foster, D.L., Olster, D.H. and Yellon, S.M., 1984a. *In* Venturoli, S., Flamingi, C., Givens, S. (eds) *Adolescence in Females: Endocrinological Development and Implications on Reproductive Function.* Yearbook Medical Publishers, Inc., Chicago, in press.
Foster, D.L., Yellon, S.M. and Olster, D.H., 1984b. *Proceedings 10th International Congress on Reproduction and Artificial Insemination, Vol.IV,* Plenary and Symposium Papers, pp.VII 16-VII 23.
Geschwind, I.I., 1974. *In* Grumbach, M.M., Grave, G.D. and Mayer, F.E. (eds) *Control of Onset of Puberty.* John Wiley, New York, 27-28.
Goodman, R.L., Bittman, E.L., Foster, D.L. and Karsch, F.J., 1981a. *Endocrinology, 109,* 1414-1417.
Goodman, R.L., Bittman, E.L., Foster, D.L. and Karsch, F.J., 1982. *Biol. Reprod., 27,* 580-589.
Goodman, R.L. and Karsch, F.J., 1980. *Endocrinology, 107,* 1286-1290.
Goodman, R.L., Reichert, L.E., Jr., Legan, S.J., Ryan, K.D., Foster, D.L. and Karsch, F.J., 1981b. *Biol. Reprod., 25,* 134-142.
Hauger, R.L., Karsch, F.J. and Foster, D.L., 1977. *Endocrinology, 101,* 807-817.
I'Anson, H., 1983. *Biol. Reprod., 28,* Suppl. 1, Abstract No 64.
Karsch, F.J., Bittman, E.L., Foster, D.L., Goodman, R.L., Legan, S.J. and Robinson, J.E., 1984a. *Rec. Prog. Horm. Res.,* 40, 185-232.
Karsch, F.J. and Foster, D.L., 1981. *In* Gilmore, D. and Cook, B. (eds) *Environmental Factors in Mammalian Reproduction.* MacMillan, London, 30-53.
Karsch, F.J., Foster, D.L., Bittman, E.L. and Goodman, R.L., 1983. *Endocrinology, 113,* 1333-1339.
Karsch, F.J., Goodman, R.L. and Legan, S.J., 1980. *J. Reprod. Fert., 58,* 521-535.
Karsch, F.J., Robinson, J.E., Yellon, S.M., Wayne, N.L., Olster, D.H. and Kaynard, A.H., 1984b. *Biol. Reprod., 30,* Suppl. 1, Abstract No. 155.
Kennaway, D.J., Gilmore, T.A. and Seamark, R.F., 1982. *Endocrinology, 110,* 1766-1772.
Legan S.J., 1982. *Biol. Reprod., 26,* Suppl. 1, Abstract No. 31.
Legan, S.J. and Karsch, F.J., 1979. *Biol. Reprod., 20,* 74-85.
Legan, S.J. and Karsch, F.J., 1980. *Biol. Reprod., 23,* 1061-1068.
Legan, S.J. and Karsch, F.J., 1983. *Biol. Reprod., 29,* 316-325.
Legan, S.J., Karsch, F.J. and Foster, D.L., 1977. *Endocrinology, 101,* 818-824.
Legan, S.J. and Winans, S.S., 1981. *Gen. Comp. Endocrinol., 45,* 317-328.
Lincoln, G.A. and Short, R.V., 1980. *Rec. Prog. Horm, Res., 36,* 1-43.
Martin, G.B., Scaramuzzi, R.J. and Henstridge, J.D., 1983a. *J. Endocr., 96,* 181-193.
Martin, G.B., Scaramuzzi, R.J. and Lindsay, D.R., 1983b. *J. Reprod. Fert., 67,* 47-55.
McLeod, B.J., Haresign, W. and Lamming, G.E., 1982. *J. Reprod. Fert., 65,* 215-221.
McNatty, K.P., Gibb, M., Dobson, C. and Thurley, C., 1981. *J. Endocr., 90,* 375-389.
McNeilly. A.S., O'Connell, M. and Baird, D.T., 1982. *Endocrinology, 110,* 1292-1299.
Nett, T.M. and Niswender, G.D., 1982. *Theriogenol., 17,* 645-653.
Poindron, P., Cognie, Y., Gayerie, F., Orgeur, P., Oldham, C.M. and Ravault, J.P., 1980. *Physiol. Behav., 25,* 227-236.
Ramirez, V.D. and McCann, S.M., 1963, *Endocrinology, 72,* 452-464.
Robinson, J.E. and Karsch, F.J., 1984. *Biol. Reprod., 30,* Suppl. 1, Abstract No. 153.
Thwaites, C.J., 1965. *J. Agric. Sci., 65,* 57-64.
Webster, G.M., and Haresign, W., 1983. *J. Reprod. Fert., 67,* 465-471.
Wodzicka-Tomaszewska, M., Hutchinson, J.C.D. and Bennett, J.W., 1967. *J. Reprod. Fert., 61,* 61-67.
Wright, P.J., Geytenbeek, P.E., Clarke, I.J. and Findlay, J.K., 1981. *J. Reprod. Fert., 61,* 97-102.
Wright, P.J., Geytenbeek, P.E., Clarke, I.J. and Findlay, J.K., 1984. *J. Reprod. Fert., 71,* 1-6.
Wright, P.J., Stelmasiak, T. and Anderson, G.A., 1983. *J. Reprod. Fert., 67,* 197-202.

# A STUDY OF THE REPRODUCTIVE PERFORMANCE OF MATURE ROMNEY AND MERINO RAMS THROUGHOUT THE YEAR

W.J. Bremner, *Department of Medicine, University of Washington, Seattle, Washington, U.S.A.*
I.A. Cumming[1] and C. Winfield, *Animal Research Institute, Department of Agriculture, Werribee, 3030.*
D.M. de Kretser, *Department of Anatomy, Monash University, Melbourne, 3168.*
D. Galloway, *The Veterinary Clinical Centre, University of Melbourne, Werribee, 3030.*

[1] Present address: Health Commission of Victoria, 555 Collins Street, Melbourne, 3000

*Summary* The reproductive performance of Romney and Merino rams was studied over twelve months. Seasonal changes in reproductive function were noted in both species but were more marked in Romney rams. Indices of libido were lowest in November and were associated with low testosterone and high prolactin levels. Testicular size was greatest in May and reached a nadir in November. FSH and LH levels peaked in March and were low from July to November in Romney rams but no definite changes were seen in Merino rams. Testosterone levels in both breeds were greatest in March. Episodes of LH and testosterone secretion were most frequent from January to May and in Romney rams pulsatile LH and testosterone secretion was not seen from July to November. However in Merino rams pulsatile secretion of LH testosterone persisted throughout the year.

## INTRODUCTION

The effect of season on the reproductive physiology of the ram has been recognized and studied by a number of investigators (Ortavant *et al.* 1964, Lincoln and Short 1980). Recently, the effect of photoperiod on endocrine factors controlling the testes of Soay rams have been studied by Lincoln and co-workers (see review, Lincoln and Short 1980) who demonstrated striking changes in episodic luteinizing hormone (LH) and testosterone (T) secretion with season. However no extensive studies of the endocrine factors involved in seasonal control of reproduction in rams have been made under field conditions particularly in Australia. This paper describes the results of a study of the reproductive physiology in Merino and Romney rams during one year under normal environmental conditions in southern Australia.

## MATERIALS AND METHODS

*Animals* Twenty normal adult rams (10 Merinos and 10 Romneys) were kept together on irrigated pasture with free access to salt and water at the Animal Research Institute Farm, Werribee, Victoria.

*Experimental Protocol* At two-monthly intervals commencing in March 1975, testicular size, libido and sperm count were measured and blood samples were collected for subsequent measurement of reproductive hormones. Testicular size was determined after tests of libido, by caliper measurement of length, width and depth and the product of these values was used as an index of testicular volume.

*Libido testing* Eight rams of each breed were exposed to ovariectomized ewes in induced oestrus and the intervals from exposure to first mount and from exposure to first service were recorded. Each ram was exposed to three ewes. At these services, semen was collected in an artifical vagina. The next day, each ram was exposed to five ewes in induced oestrus for 15 minutes in a 5 × 7 m pen and the total number of mounts and services wes recorded (Mattner *et al.* 1971).

*Semen collections and sperm counts* Prior to the beginning of the study the rams were trained to serve an artificial vagina. Several days prior to each study period over the course of the year, the rams were brought into the shed for two retraining sessions with the artificial vagina. The seminal fluid from the artificial vagina was diluted in a 0.1% formalin 0.6 M sodium bicarbonate solution and the sperm was counted in a Fuchs-Rosenthal counting chamber. Due to laboratory technical difficulties sperm counts were only available from March and November, 1975.

*Blood sampling* On the day after libido testing, blood sampling was commenced at 0800 hours through jugular venous cannulae and continued at 20 min intervals until 2000 hours.

*Hormone assays* Follicle stimulating hormone (FSH), LH and prolactin were measured by radioimmunoassays as previously described (Bremner *et al.* 1976). Testosterone was measured following extraction of plasma in hexane by a radioimmunoassay with a cross-reactivity of 50% with dihydrotestosterone. Mean hormonal concentrations were obtained by pooling equal aliquots from the multiple samples over 12 hours from the same ram.

## RESULTS

Testicular size and sperm count

For both breeds testicular size was lowest in July to November (late Winter to Spring) and greatest in March to May (Autumn) ($P < 0.01$ Figure 1). Sperm counts were extremely variable and the mean was lower, though not statistically different, in November ($1.685 \pm 0.469 \times 10^9$/ml [mean ± S.E.], n = 18) than in March 1975 ($2.036 \pm 0.295 \times 10^9$/ml [mean ± S.E.], n = 37). Due to decline in libido in November, semen was not always obtained, two animals showing no interest in the ewes.

Libido

For both breeds, the time taken either to mount or serve was shortest in May and longest in November ($P < 0.01$, Figure 1). A similar pattern was seen in the number of services recorded per 15 min trial, being greatest ($P < 0.01$) in May and lowest in November.

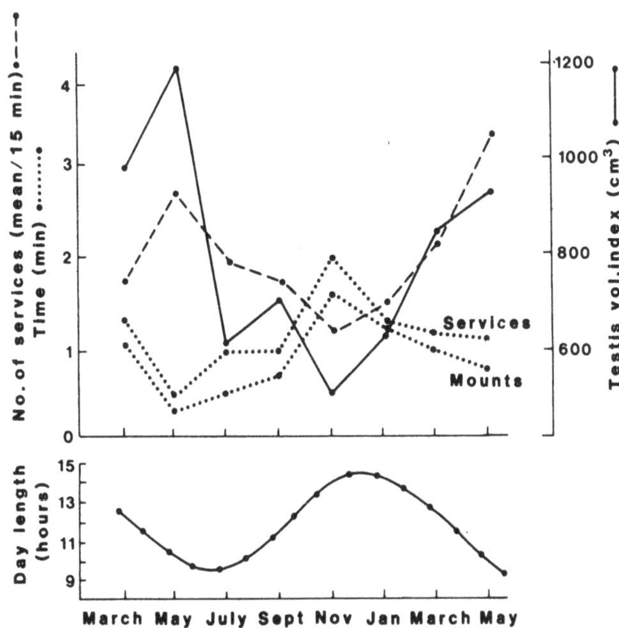


**FIGURE 1. The variation in testicular size and parameters of sexual activity in Romney rams (n = 10). Time indicates the time to first service and first mount.**

Hormonal Changes

*Integrated levels from pooled samples* (i) LH: Plasma levels of LH in Romneys exhibited distinct seasonal changes ($P < 0.01$) with lowest levels in July and September and highest levels in January and March (Figure 2). Levels of LH in the Merinos showed less evidence of a seasonal pattern, several rams showing relatively high levels in July.

(ii) FSH: As for LH, clear seasonal changes in the concentration of FSH were evident in the Romney rams but not in the Merinos (Figure 2). In the Romneys FSH levels were lowest from July to November and highest in March ($P < 0.01$). In Merino rams, FSH levels were low between May and November, reaching their highest levels in March.

(iii) Prolactin: Plasma prolactin levels in both breeds were low from March to September, peaked in November and were still high in January ($P < 0.01$) in comparison to March.

(iv) Testosterone: Both breeds exhibited clear seasonal changes ($P < 0.01$) with low levels from July to November (Figure 2) though Merinos showed low levels in May as well. High levels were seen from January, and in Romneys these persisted to May.

*Multiple Sampling Studies* There was evidence of pulsatile secretion of both LH and testosterone and, in general, each LH pulse induced a testosterone secretion response. In Romney rams these episodes occurred about every 4 hours in January and March but were much less frequent in May and November with no episodes in July and September (Figure 3). In Merinos, episodic secretion of LH and testosterone persisted throughout the year but there were less episodes in July and September. While not shown, FSH levels revealed little evidence of episodic secretion even when large LH episodes of secretion occurred.

## DISCUSSION

The study demonstrates that both Romney and Merino rams showed distinct seasonal changes in reproductive performance which were easily detectable but, at this latitude (37°S), there was no period of infertility. Nearly all rams were willing to serve ewes in induced oestrus throughout the year and the sperm counts suggested that the rams were fertile. However a marked decrease in parameters of sexual behaviour, namely the latency to serve and number of services, was noted from May to November in general correlating with low testosterone levels.

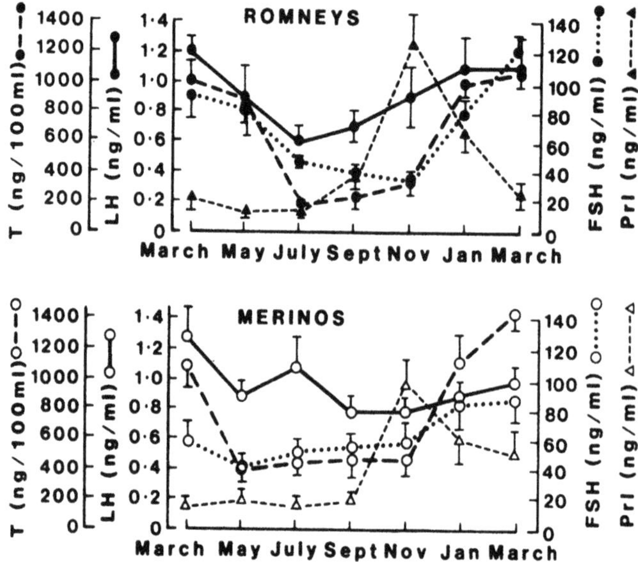


**FIGURE 2. The changes in hormone concentrations in the rams (n=20) with season.**

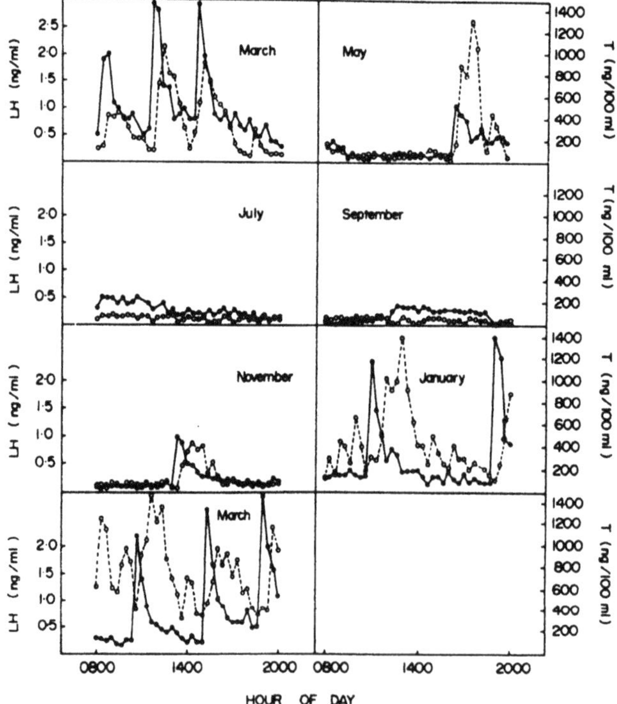


**FIGURE 3. LH (● —— ● and T (○ —— ○) concentrations in a Romney ram over the 12 hour sampling period for each study period.**

The seasonal changes in these breeds, both endocrinological and behavioural, were not as distinct as those noted by Lincoln and Short (1980) in the Soay ram and the differences may be attributed to the latitude (37°S, 50°N) and the relative domestication of the breeds used in the present study. The seasonal changes were most marked in the Romney rams which showed alteration in the levels of the pituitary hormones thought to be most important in transmitting environmental cues to the gonads, namely FSH, LH and prolactin. However the Merino rams showed definite change only in the mean levels of prolactin. The concentrations of LH and FSH were lower during the months of increasing daylength whereas prolactin levels increased with increasing day length, confirming the results of earlier studies (Pelletier 1973; Lincoln and Short 1980). The changes in plasma concentrations of testosterone were larger in both breeds than the changes in LH and FSH, particularly in the Merino where concentrations of testosterone decreased greatly despite no comparable change in LH and FSH. These results suggest that the seasonal change may be amplified at the testicular level and may be related to a changing sensitivity of the testis to gonadotrophic stimulation. It has been reported in rats, that prolactin can increase the sensitivity of the testis to LH stimulation (Hafiez *et al.* 1972), and the observations that the rise in prolactin precedes the rise in testosterone lends support to this concept. However such a change would need to be relatively slow in the ram, since one month separates the peak prolactin concentrations and the initial rise in testosterone levels.

Lincoln and Short (1980) have shown striking changes in episodic secretion of LH and testosterone secretion in the Soay ram under artificial photoperiod such that decreasing day length increased the frequency of LH pulses probably indicating increased pulsatile release of GnRH from the hypothalamus. The present results demonstrate that identical changes in episodic secretion of LH, most likely reflecting episodic GnRH release, are seen under field conditions. The relative insensitivity of the Merino to seasonal changes is reflected by the persistence of episodic secretion of LH, though of decreased frequency during the spring. The reason for this difference remains unknown.

While the present study demonstrates retention of testicular function throughout the year, the results suggest a diminution in spermatogenic capacity during the spring, the extent of which requires further definition. This may be of significance, particularly with English breeds of ram, when mating is undertaken in these months and suggests the need for studies to develop means of overcoming the depression.

## REFERENCES

Bremner, W.J., Findlay, J.K., Cumming, I.A., Hudson, B. and de Kretser, D.M., 1976. *Biol. Reprod., 15*, 141-146.

Hafiez, A.A., Lloyd, C.W. and Bartke, A., 1972. *J. Endocrinol., 52*, 327-332.

Lincoln, G.A. and Short, R.V., 1980. *Recent Prog. Horm. Res., 36*, 1-43.

Mattner, P., Braden, A. and George, J., 1971. *Aust. J. Exp. Agr. Anim. Husb., 11*, 473-477.

Ortavant, R., Maulson, P. and Thibault, C., 1964. *Ann. N.Y. Acad, Sci., 117*, 157-193.

Pelletier, J., 1973. *J. Reprod. Fert., 35*, 143-147.

# RESPONSE OF SEASONALLY ANOESTROUS EWES TO 23 HOURS CIRCADIAN DARKNESS

B.J. McDonald and P.S. Hopkins, *Queensland Department of Primary Industries, Animal Research Institute, Yeerongpilly, Q. 4105.*

*Summary* Exposure of seven Suffolk ewes to 23 hours dark each day at the vernal equinox, and for 90 days thereafter, induced ovulation in one ewe before day 53 and four ewes between days 53 and 90. No ewes in natural light exhibited ovarian activity and none in either group was marked by vasectomised rams. Measurement of pineal $\beta$-adrenoceptor parameters suggested that dark treatment increased $\beta$-receptor density and decreased ligand binding affinity. This work shows that exposure of seasonally anoestrous ewes to excess dark (1) necessitates prolonged treatment to induce ovulation, (2) may not induce oestrus and (3) may affect the onset and magnitude of the response so that it differs from melatonin administration.

## INTRODUCTION

It has been established that day length (photoperiod) is an important stimulus for the initiation of the breeding season of sheep. The role of the pineal gland in mediating the response to photoperiod has been demonstrated by Lincoln (1979) and Karsch *et al.* (1981). In a natural day length environment there are important differences in the short day length (oestrus) and long day length (anoestrus) circadian pattern of plasma melatonin concentration in Suffolk-cross ewes (Rollag *et al.* 1978). This suggests that the pineal indoleamine, melatonin, may be involved in the reproductive response of sheep to photoperiod.

Kennaway *et al.* (1982) showed that feeding melatonin accelerated the onset of oestrus in Border Leicester × Merino ewes in the presence of rams. Melatonin fed to anoestrous Suffolk-cross ewes in the absence of rams induced changes in plasma progesterone indicative of oestrus 2 to 8 weeks before such changes in untreated ewes (Arendt *et al.* 1983). It has been shown that the duration of elevated plasma melatonin concentration in sheep is directly related to the length of darkness (Kennaway *et al.* 1977; Rollag *et al.* 1978; Arendt *et al.* 1981). These studies raise the question of whether increased hours of darkness (and elevated plasma melatonin) affect the time to onset of oestrus in seasonally anoestrous ewes regardless of their stage of anoestrus.

The present experiment was designed to test an extreme example of this hypothesis. We measured the impact of extending the period of dark to 23 hours per day on the reproductive status of anoestrous ewes between the vernal equinox and summer solstice.

## MATERIALS AND METHODS

### Sheep, housing and feed

Fourteen maiden Suffolk ewes, approximately 22 months of age, were allocated to two groups of seven ewes each by stratified randomisation of their live weights. One group was housed indoors in a light-proof room while the other was housed outdoors in a partially enclosed covered pen with a concrete floor. Two vasectomised Wiltshire Horn × Merino rams, one in each group of ewes, were used continually to detect oestrus. The rams were fitted with a Sire-sine harness and alternated between groups every four days to reduce individual ram effects. Each group was offered 6.5 kg pelleted lucerne daily.

### Treatments and measurements

The experiment was conducted at Brisbane (27° 28'S) where the ewes received either natural day length (NDL) for the latitude or excess dark (EDK) treatment. EDK ewes were exposed to 23 hours dark each day with light (340 lux) provided by cool white fluorescent tubes between 1100 and 1200 hours. The treatments commenced at the vernal equinox (day 0) and continued for 90 days.

Ovarian activity was assessed by laparotomy at days 18, 35, and 53. A ewe was classed as ovular if a corpus luteum or corpus albicans was present, or anovular if absent. At day 90 all ewes were weighed and killed by exsanguination and ovaries were examined *in situ.* The ewes were killed between 1100 and 1200 hours. Pineal and adrenal glands and ovaries were removed and weighed immediately. Pineal $\beta$-adrenoceptors were quantified by the method of Foldes *et al.* (1982).

## RESULTS

No ewes were marked by the rams in either treatment group, regardless of their ovarian status and none was ovular at day 18 or 35. One ovular ewe was observed at laparotomy on day 53 in EDK. That ewe and three others in EDK were ovular at post-mortem examination on day 90, whereas all ewes in NDL were still anovular.

Exposure to EDK increased the maximum density of pineal β-receptor sites ($\beta_{max}$) and reduced binding affinity as illustrated by the dissociation constant ($K_D$) for the ligand-receptor complex, although these changes were not statistically significant (Table 1). The variance of these parameters was increased by EDK. This greater variability was a reflection of the large but non-significant differences between the means (± s.e.m.) of ovular ($K_D$ = 19.8 ± 5.90 nM; $\beta_{max}$ = 9.5 ± 2.98 pmol $mg^{-1}$) and anovular ($K_D$ = 8.6 ± 4.31 nM; $\beta_{max}$ = 5.4 ± 2.12 pmol $mg^{-1}$) ewes within the EDK treatment.

Photoperiod had no effect on the wet weight of pineal or adrenal glands. There were no statistically significant differences between ovarian weights; however, the mean weight of ovaries from EDK ewes was 56% greater than from NDL ewes (Table 1), and ovaries from ovular ewes were 43% and 80% heavier than from anovular ewes in EDK and NDL respectively.

**TABLE 1. Changes in pineal β-receptors, selected organ weights, live weight and ovarian activity resulting from exposure to 23 hours dark daily for 90 days.**

| Parameter | | Natural day length | Dark 23 hours $day^{-1}$ |
|---|---|---|---|
| β-receptor ± s.e.m. | $K_D$[1](nM) | 7.3 ± 0.91 | 15.0 ± 4.21 |
| | $\beta_{max}$[2](pmol $mg^{-1}$) | 3.9 ± 0.51 | 7.7 ± 1.97 |
| Organ wt. ± s.e.m. (mg) | Pineal | 88.7 ± 13.29 | 77.1 ± 8.70 |
| | Adrenals | 2410.3 ± 17.42 | 2428.3 ± 12.07 |
| | Ovaries | 1318.6 ± 12.03 | 2063.1 ± 33.83 |
| Live wt. ± s.e.m. (kg) | Day 0 | 34.9 ± 0.73 | 34.6 ± 0.55 |
| | Day 90 | 37.8 ± 0.64 | 37.1 ± 0.74 |
| No. ovular ewes in group of 7 | | 0 | 4 |

(1) $K_D$ = dissociation constant for the ligand-receptor complex.
(2) $\beta_{max}$ = β-receptor density.

## DISCUSSION

The results of this work suggest that EDK does not enhance reproductive performance (ovulation with detectable oestrus) in anoestrous Suffolk ewes between the vernal equinox and summer solstice. This indicates the absence of an inverse relationship between the length of the dark period and the time to onset of oestrus. It follows that increasing the duration of elevated plasma melatonin levels to 23 hours per day by melatonin supplementation may not produce relative improvements in reproductive performance.

Arendt *et al.* (1983) proposed that the use of physiological quantities of melatonin was critical in gaining a response. They also suggested that artificial lengthening of the rhythmic melatonin secretory pattern by increasing the dark period should induce a physiological response appropriate to the duration of melatonin secretion. We suggest that neither the technique of feeding melatonin (Kennaway *et al.* 1982; Arendt *et al.* 1983) nor our work with extended dark simulates a typical physiological state of the sheep. These techniques produce a physiological state which mimics constant day length. If the rate of change of day length is a component of the circannual changes in reproductive status, then the treatments under discussion produce a physiological state which is only experienced at a solstice where the rate of change of day length approaches zero.

The present experiment suggests that the duration of the constant dark period and its effect on pineal function and/or the time of initiation of the treatment may be important. Only four of the seven ewes in EDK ovulated before day 90, whereas Arendt *et al.* (1983) showed that the mean time to onset of oestrus with melatonin feeding was 61 days, and with 16 hours dark was 69 days. However, the latter treatments were applied between the summer and winter solstices where control ewes under natural light did cycle after 110 days. The partial response in the present experiment may have been due to exposure to EDK between the vernal equinox and summer solstice. Time to onset of oestrus may be breed-dependent, since all the Border Leicester × Merino ewes treated by Kennaway *et al.* (1982) responded 65 to 88 days after melatonin feeding began. Although that experiment was conducted in the same season as our experiment, control ewes did show some ovulatory activity after 78 days.

The results suggest that EDK did produce biologically important changes in pineal function although not statistically different. These pineal changes were confined to ewes which showed an ovarian response. When EDK increased β-receptor density and reduced binding affinity, ovulation occurred.

Arendt *et al.* (1983) suggested that the response reported by Kennaway *et al.* (1982) was confounded by the presence of rams. The effect of rams on seasonally anovular ewes may depend on breed and/or stage of anovulation (Oldham 1980) and therefore categorical assumptions regarding the 'ram effect' in this type of work tend to be misleading. In the present work there was no evidence of ram-induced ovulation or oestrus in those ewes that did ovulate; however, the ram may not have responded to oestrus since visibility was poor. This experiment differed from those of Kennaway *et al.* (1982) and Arendt *et al.* (1983) in breed and stage of seasonal anoestrus respectively, but it does raise the possibility that ovulation as measured by those workers may not have been associated with oestrus.

The present experiment illustrates that photo-induced ovulation is associated with changes in pineal $\beta$-receptor parameters, is not necessarily accompanied by oestrus and may be dependent on the stage of anoestrus and breed of sheep. Although melatonin feeding has produced desirable responses, greater progress in the improvement of photo-sensitive reproductive performance may depend on a better understanding of the interaction between intrinsic cyclicity and photoperiodic rhythms.

ACKNOWLEDGEMENT

We wish to thank Dr Andrew Foldes and Mr Colin Maxwell of CSIRO Division of Animal Production for conducting the pineal assays reported in this work.

REFERENCES

Arendt, J., Symons, A.M. and Laud, C.A., 1981. *Experientia, 37*, 548-586.

Arendt, J., Symons, A.M., Laud, C.A. and Pryde, S.J., 1983. *J. Endocr., 97*, 395-400.

Foldes, A., Maxwell, C.A., Hinks, N.T., Hoskinson, R.M. and Scaramuzzi, R.J., 1982. *Biochem. Pharmacol., 31*, 1369-1374.

Karsch, F.J., Bittman, E.L. and Legan, S.J., 1981. *In* Ortavant, R., Pelletier, J. and Ravault, J.P. (eds) *Photoperiodism and Reproduction*. Nouzilly, INRA Publications, 213-218.

Kennaway, D.J., Frith, R.D., Phillipou, G., Matthews, C.D. and Seamark, R.F., 1977. *Endocr., 101*, 119-127.

Kennaway, D.J., Gilmore, T.A. and Seamark, R.F., 1982, *Endocr., 110*, 1766-1772.

Lincoln, G.A., 1979. *J. Endocr., 82*, 135-147.

Oldham, C.M., 1980. *Proc. Aust. Soc. Anim. Prod., 13*, 73-86.

Rollag, M.D., O'Callaghan, P.L. and Niswender, G.D., 1978. *Biol. Reprod., 18*, 279-285.

# CHANGES IN THE CLEARANCE RATE OF IMMUNOREACTIVE LH AFTER OVARIECTOMY IN ILE-DE-FRANCE EWES

G.W. Montgomery and S.F. Crosbie, *Invermay Agricultural Research Centre, Private Bag, Mosgiel, New Zealand.*
G.B. Martin, *M.R.C. Reproductive Biology Unit, Centre for Reproductive Biology, Edinburgh, Scotland.*
J. Pelletier, *I.N.R.A., Station de Physiologie de la Reproduction, 37380 Monnaie, France.*

*Summary* Secretion of LH was monitored for up to 12 months in ewes ovariectomized in either anoestrus or the breeding season. Clearance rate assessed by either concentration dependent (Gompertz decay) or independent (exponential decay) indices decreased significantly with time. There was no evidence of interactions between season and time after ovariectomy. Assuming exponential decay, the apparent half life of LH in plasma varied from 28 minutes on day 1 to 63 minutes on day 365 after ovariectomy.

## INTRODUCTION

Luteinizing hormone (LH) is released from the pituitary gland in pulses and the frequency of pulses is related to different physiological states (for review see Martin 1984). We have shown recently (Montgomery *et al.* 1985) that the pattern of change in LH pulse frequency and amplitude after ovariectomy depends on season. In contrast the progressive increases in nadir concentrations with time are not influenced by season of ovariectomy, suggesting they are a consequence of more than just the composite effects of pulse frequency and amplitude. Furthermore, following gonadectomy in other species, there is a decrease in the clearance rate of LH from plasma (Peckham and Knobil 1976 a,b; Weick 1977; McCreery and Licht 1983).

The ovariectomized ewe is frequently used as model for studies of effects of season and steroid hormone feedback on LH release. Therefore, the clearance rate of immunoreactive LH from plasma after endogenous pulses was examined in profiles obtained from ewes for up to one year after ovariectomy.

## MATERIALS AND METHODS

Data from two experiments with mixed age Ile-de-France ewes, designed to study changes in LH release following ovariectomy and described in detail elsewhere (Montgomery *et al.* 1985) were used to analyse clearance rate of immunoreactive LH.

Groups of 10 ewes (short term study) were ovariectomized under general anaesthesia on 22 October, 1982 (breeding season) or 23 March, 1983 (anoestrus). Blood samples were collected every 10 minutes for 6 h on days 1, 3, 7 and 15 after ovariectomy (day of ovariectomy = day 0). Groups of 5 ewes (long term study) were ovariectomized under general anaesthesia on 17 March, 1982 (anoestrus) and 29 September, 1982 (breeding season). Blood samples were collected every 10 minutes for 6 h on days –1, 7, 15, 30, 60, 90, 120, 150, 180, 270, 365 after ovariectomy (day of ovariectomy = day 0). LH was measured in duplicate in a double anti-body radioimmunoassay (Pelletier *et al.* 1968). The inter and intra-assay coefficients of variation were less than 12%.

### Data Analysis

Pulses of LH in individual profiles were identified according to the method of Goodman and Karsch (1980). Clearance rates of immunoreactive LH were quantified in two ways to determine whether consistent results were obtained with both methods. The models used were exponential decay and an alternative form used by MacMillan *et al.* (1984), sometimes referred to as Gompertz decay. Decay rates (k) were determined in the following manner:— consider a pulse within a profile where $C_o$ denotes the LH concentration (pg/ml) at the peak concentration attained. Given a subsequent sample taken after time t (multiples of 10 min) has a concentration denoted by C, it can be shown that

$$\log C_o - \log C = k_e t$$

and $$\log(\log C_o) - \log(\log C) = k_g t$$

for exponential and Gompertz decay respectively. By calculating the differences in transformed concentrations and performing a regression analysis against their respective t values, pooled across pulses within a profile, estimates of $k_e$ and $k_g$ were thereby attained for the profile.

Split-plot analyses of variance on these estimates of rates of decay across profiles were then performed to ascertain the change in the nature of decay over time.

## RESULTS

Irrespective of whether clearance rate was assessed by concentration dependent (Gompertz decay) or independent (exponential decay) indices, clearance rate of immunoreactive LH decreased significantly

($P < 0.01$) with time after ovariectomy (Figure 1). There was no evidence of interactions between season and time after ovariectomy, and data for the two seasons have been combined. In the short term experiment there was a marked decrease in clearance rate between days 1 and 3. In the long term experiment clearance rate declined progressively from day 120 to day 365.

The apparent half life of LH in plasma calculated from exponential decay varied from 28 minutes on day 1 to 63 minutes on day 365 after ovariectomy. Estimates ($k_e$) of clearance rate for ewes in the long term experiment (day −1) in anoestrus or during the luteal phase of the oestrous cycle did not differ significantly (0.325 and 0.403 respectively, s.e. of difference = 0.058). The pooled estimate for apparent half life of LH in intact ewes in our study was 19 min.

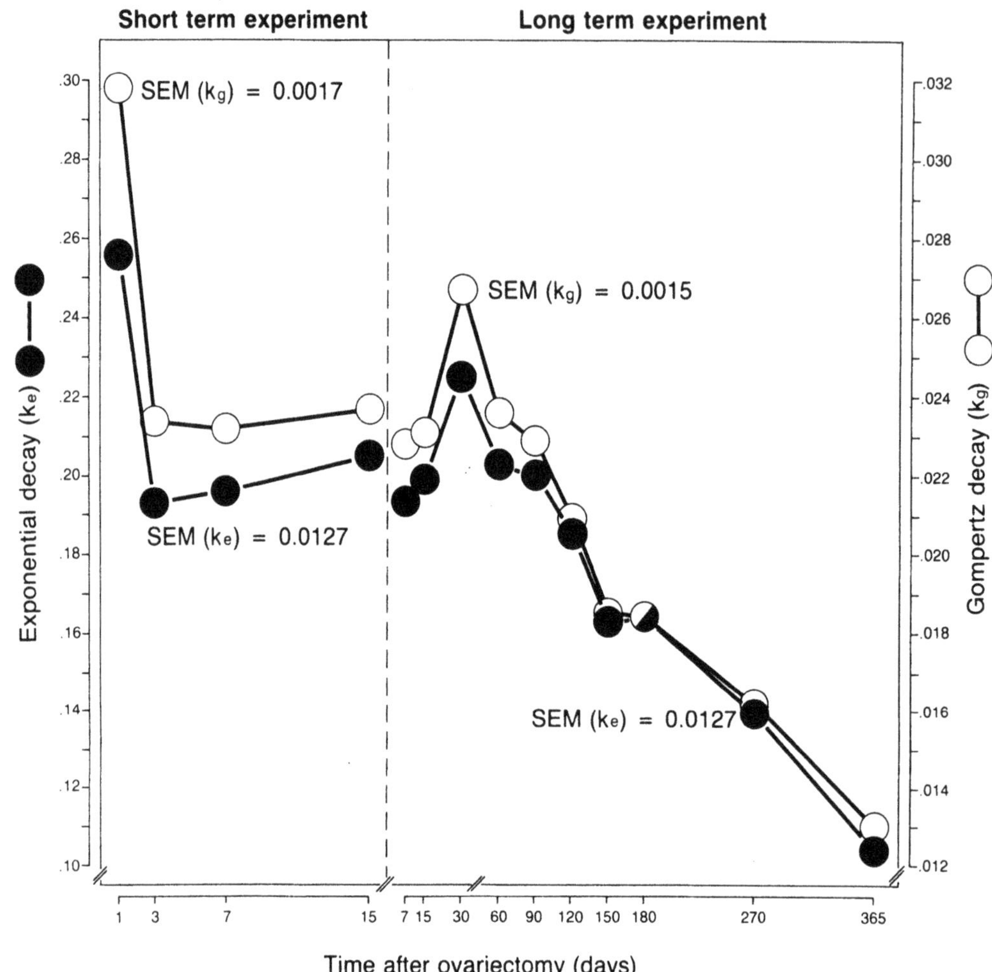


**FIGURE 1. Effect of time after ovariectomy on clearance rate (k) of immunoreactive LH from plasma assessed by concentration dependent (Gompertz decay) or independent (exponential decay) indices.**

## DISCUSSION

These data clearly demonstrate that ovariectomy in ewes results in changes in the clearance rate of immunoreactive LH after endogenous LH pulses. Further, they show that there was a rapid decrease in clearance rate during the first 3 days from values observed in intact ewes followed by a progressive decline from day 120 up to one year after ovariectomy. There was no interaction between clearance rate and season of ovariectomy and the changes were highly correlated (long term experiment, $r = 0.89$, $P < 0.01$) with changes in the nadir values previously described (Montgomery *et al.* 1984). Since we have also demonstrated interactions between season and time after ovariectomy for LH pulse frequency, the data suggest that the decrease in clearance rate is not related to the profile pulse frequency, but rather a progressive decline with time.

Since the results describe clearance after endogenous LH pulses, the decrease in *apparent* clearance rate may result from one or more mechanisms: an increase in the duration of LH release at each pulse; an increase in basal LH release; a change in the mechanisms of clearance of immunoreactive LH from plasma; a change in the structure of circulating LH affecting the rate of clearance.

An increase in the duration of LH release at each pulse appears unlikely. There was no evidence that the proportion of LH pulses, where the rise in LH concentration occurred over two successive sampling periods, increased significantly with time after ovariectomy. Further, when the regression analyses were modified by assigning $t = O$ and $C = C_O$ to the second (rather than first) post-pulse sample a similar decline in clearance rate occurred with time after ovariectomy.

In agreement with the present results changes in the clearance rate of LH occur following gonadectomy in monkeys, rats and bullfrogs (Peckham and Knobil 1976a, b; Weick 1977; McCreery and Licht 1983). When injected into recipient rats the clearance rates of LH from gonadectomized monkeys and rats have been shown to differ from that of intact animals (Peckham and Knobil 1976b; Weick 1977), and therefore, the changes in clearance rate are related to changes in circulating LH. In bullfrogs, the clearance profiles for LH were the same in intact and gonadectomized animals infused with the same pituitary preparation (McCreery and Licht 1983). The reduction in clearance rate of LH is accompanied by an increase in the apparent molecular size of circulating LH in gonadectomized monkeys and bullfrogs (Peckham and Knobil 1976a, b; Licht *et al.* 1983).

An understanding of the mechanism(s) causing changes in the clearance rate of LH after ovariectomy in sheep must await further experiments, but it appears that changes in the clearance rate may result from changes in the structure of circulating LH. The results described suggest limitations to the use of ovariectomized ewes as models to study control of seasonal breeding and the oestrous cycle.

## REFERENCES


Goodman, R.L. and Karsch, F.J., 1980. *Endocrinology, 107*, 1286-1290.

Licht, P., McCreery, B.R. and Papkoff, H., 1983. *Biol. Reprod., 29*, 646-657.

MacMillan, K.L., Fielden, E.D., McNatty, K.P. and Henderson, H.V., 1984. *J. Reprod. Fert.* (in press).

Martin, G.B., 1984. *Biol. Rev., 59*, 1-87.

McCreery, B.R. and Licht, P., 1983. *Biol. Reprod., 29*, 637-645.

Montgomery, G.W., Martin, G.B. and Pelletier, J., 1985. *J. Reprod. Fert.* (in press).

Peckham, W.D. and Knobil, E., 1976a. *Endocrinology, 98*, 1054-1060.

Peckham, W.D. and Knobil. E., 1976b. *Endocrinology, 98*, 1061-1064.

Pelletier, J., Kann, G., Dolais, J. and Rosselin, G., 1968. *C.r. hebd. Seanc. Acad. Sci., Paris D 266*, 2291-2294.

Weick, R.F., 1977. *Endocrinology, 101*, 157-161.

# THE RAM EFFECT, ITS MECHANISM AND APPLICATION TO THE MANAGEMENT OF SHEEP

D.T. Pearce and C.M. Oldham, *School of Agriculture, University of Western Australia, Nedlands, WA.*

## INTRODUCTION

There are many ways in which hormones affect the behaviour of animals but few where behaviour, through changes in the secretion of hormones, can manipulate reproductive processes. The induction of ovulation following the association of anovular ewes with rams, known as the 'ram effect', is probably the most spectacular example so far described and has been the subject of a number of recent reviews (Oldham 1980a; Knight 1983; Martin and Scaramuzzi 1983) or included in other reviews (Lindsay and Signoret 1978, 1980; Martin 1984).

Underwood *et al.* (1944) first drew attention to the 'ram effect' when they reported that the Australian Merino had a period of anoestrus in spring that could be interrupted by the introduction of rams. Ten years later Schinckel (1954a) demonstrated that the introduction of rams to ewes could induce ovulation within six days. In recent years our understanding of its mechanism has expanded and this review describes how the introduction of rams increases the secretion of luteinising hormone which leads to the induction of ovulation. It is now established that the introduction of rams will induce ovulation in seasonally and lactationally anovular ewes and will induce puberty in ewe lambs. It has been investigated in many breeds including the Merino, Romney, Dorset, Southdown, Awasi, Corriedale, Prealpes, Ile de France, and Berechon. This review also describes how the 'ram effect' can be an important tool in the management of sheep.

## PRECONDITIONS FOR USE OF THE 'RAM EFFECT'

### Isolation of ewes from rams

Ewes must be anovular and have been isolated from rams for a period before they will ovulate in response to the re-introduction of rams (Underwood *et al.* 1944). The precise requirements of isolation have not been quantified, but isolation from the sight and smell of rams for one month is a general recommendation (Oldham 1980a).

### Perception of the ram by ewes

It is not essential for ewes to be in visual or physical contact with rams to be induced to mate (Watson and Radford 1960) but most ewes with their sense of smell impaired surgically fail to mate (Morgan *et al.* 1972). These findings have led to the hypothesis that the response is mediated by pheromones. Recent studies by Knight and Lynch (1980a,b) have shown that the putative pheromones are present in the wool and wax of rams' fleeces and that both aqueous and solvent extracts are effective stimuli (Knight *et al.* 1983). Further purification and determination of their structure has yet to be done.

Wethers or ewes which have received short-term treatments with androgens or oestrogens are as effective as rams at inducing ovulation in ewes (Lishman *et al.* 1969; Fulkerson *et al.* 1981; Croker *et al.* 1982; Signoret *et al.* 1982/83). Indeed the use of testosterone-treated wethers to induce ovulation in ewes has received widespread commercial acceptance in Australia. These findings strongly suggest that the production of the putative pheromones is dependent on the sex steroids. However Dorset rams, with lower peripheral concentrations of testosterone than Romney rams, are more effective at stimulating Romney ewes to ovulate (Tervit and Peterson 1978) so factors other than the absolute amount of androgen are involved.

Watson and Radford (1960) have demonstrated that the visual and physical components of perception of the ram are not essential to the induction of ovulation but their importance should not be ignored. For example in a recent experiment (J.L. Reeve, 1983, pers. comm.), a flock of crossbred ewes were kept in a paddock with flocks of rams kept in paddocks adjoining three of the four boundaries. Under these circumstances preconditioning by isolation should not have been possible. Nevertheless 72 per cent of ewes were induced to ovulate and mate after the introduction of other rams of the same breed to the flock of ewes compared with 68 per cent of ewes which had been totally isolated from rams. Thus the visual and physical components of perception and the presence of different pheromones may be more important in the response of ewes than has previously been recognised.

## ENDOCRINE MECHANISM OF THE 'RAM EFFECT'

The following sections describe how the introduction of rams affects the tonic secretion of gonadotrophins and the preovulatory surge of luteinising hormone which lead to ovulation. We also present evidence to support the hypothesis that the presence of rams directly affects the hypothalamo-pituitary axis to override and not reverse the negative feedback effects of the ovarian steroids.

Tonic secretion of gonadotrophins

The frequency of pulses of luteinising hormone (LH) increases dramatically within minutes in seasonally anovular ewes (Martin *et al.* 1980), lactationally anovular ewes (Poindron *et al.* 1980) and prepubertal ewes (J.J. Murtagh, unpublished) after they have been introduced to rams. By contrast the secretion of follicle stimulating hormone has been reported either to remain unchanged or decrease immediately after the introduction of rams (Poindron *et al.* 1980; Atkinson and Williamson 1984).

Mechanism for increasing secretion of LH

*Evidence for a direct mechanism* The increase in the frequency of pulses of LH in anovular ewes within minutes of introduction to rams (Martin *et al.* 1980) strongly suggested a direct mechanism for increased secretion of LH independent of the negative feedback effects of the ovarian steroids. In addition, passive immunisation of seasonally anovular ewes against oestradiol-17$\beta$ did not increase the frequency of pulses of LH until 12 hours after treatment (Thomas *et al.* 1982). Thus the effect of negative feedback of oestradiol-17$\beta$ on the frequency of pulses of LH cannot be rapidly reversed. By contrast, in experiments using ovariectomised ewes implanted with oestrogen in spring or oestrogen and progesterone in autumn (treatments which reduce the frequency of pulses of LH by negative feedback), the introduction of rams rapidly increased the frequency of pulses of LH (Martin *et al.* 1983a,b; Pearce and Oldham 1983; see Figure 1a). Ovariectomised ewes with no implants in spring, or implanted with oestrogen alone in autumn do not have negative feedback on the frequency of pulses of LH and the frequency of pulses of LH was around one pulse per hour. The introduction of rams to these ewes increased the basal concentration of LH but did not significantly increase the frequency of pulses (Martin *et al.* 1983a,b; Pearce and Oldham 1983). Nevertheless individual profiles such as shown in Figure 1b, together with the data cited by Martin and Scaramuzzi (1983) suggest that consistent increases in the frequency of pulses of LH might have been detected if blood samples had been collected every five minutes instead of every fifteen minutes.

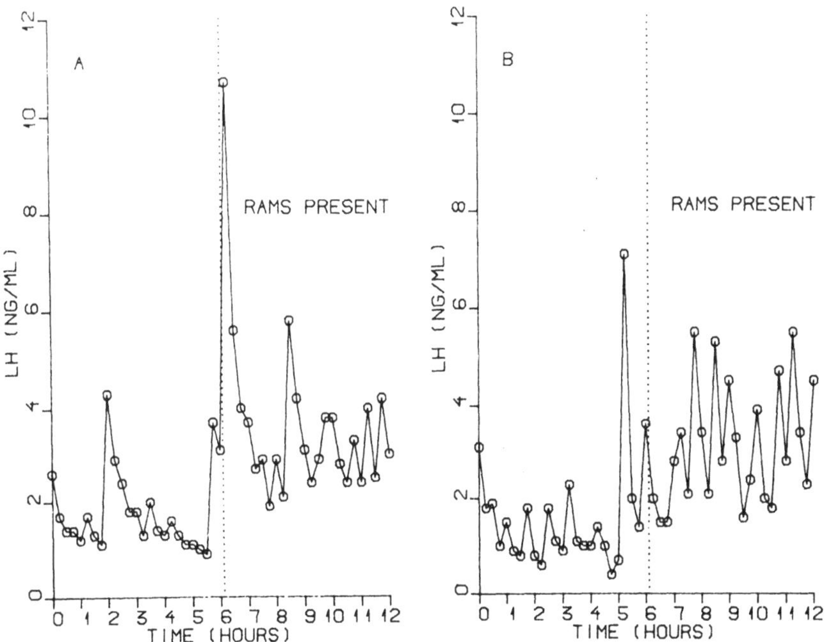


**FIGURE 1. The effect of the introduction of rams during the breeding season on the secretion of LH in two ovariectomised ewes bearing (a) oestrogen and progeterone implants, and (b) an oestrogen implant alone (from data used by Pearce and Oldham 1983).**

Together these results support the hypothesis that the introduction of rams stimulates the rapid increase in the pulsatile secretion of LH by a direct mechanism on the hypothalamo-pituitary axis and thus overrides any influence of the ovarian steroids.

*Effect of withdrawal of rams* If the presence of rams directly affects the secretion of LH then the subsequent withdrawal of rams should reduce the secretion of LH. This hypothesis was tested by introducing rams to ewes for a period of only six hours (Oldham and Pearce 1983). Of the four ewes which increased their secretion of LH, totals of zero, 10 and two pulses were observed in the four hours before, six hours during and four hours after contact with rams. The secretion of LH from one of these ewes is shown in Figure 2 and illustrates that the secretion of LH was high only during the period when rams were present.

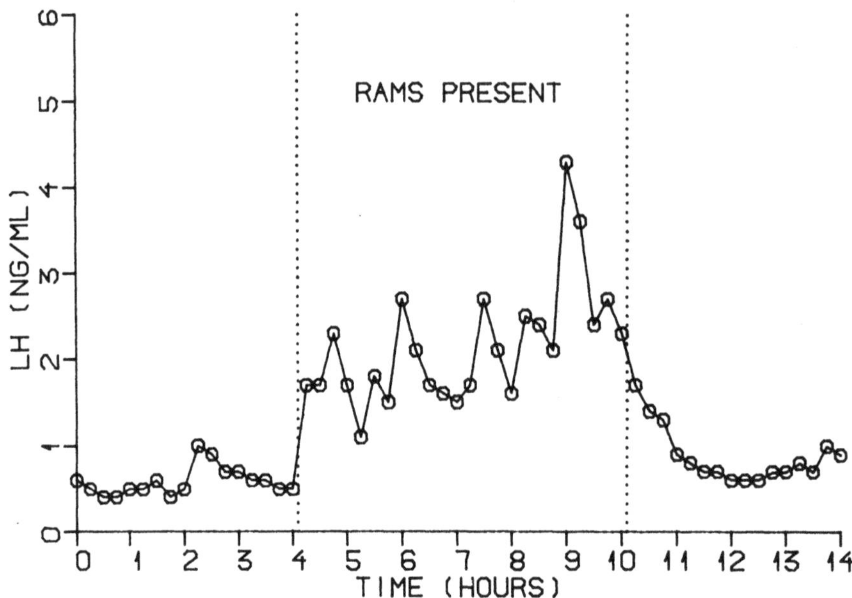


**FIGURE 2. The effect of the presence or absence of rams on the secretion of LH in an anovular Merino ewe in spring (from data used by Oldham and Pearce 1983).**

Variability of responses in secretion of LH

The existing definition of the 'ram effect' uses the induction of ovulation as the end point for response to the introduction of rams. Since it has been shown that the secretion of LH is stimulated in both the breeding season and spring there is the potential to use an increase in frequency of pulses of LH as the end point to identify ewes which respond to the introduction of rams. Martin *et al.* (1980) reported that ewes ovulated only if the rams induced a sustained increase in the secretion of LH but more recent data suggests that the secretion of LH in the first hours after the introduction of rams may not always be a reliable indicator of those ewes destined to ovulate. Oldham and Pearce (1983) found that of a group of ten ewes exposed to rams for 72 hours, eight ovulated. All six with sustained increases in secretion of LH ovulated but two of the remaining four ewes did not increase their secretion of LH in the four hours after the introduction of rams yet ovulated within 72 hours. The differential pattern of secretion of LH for two of the ewes induced to ovulate is shown in Figure 3. Consequently the pattern of secretion of LH in the first hours after the introduction of rams may underestimate those ewes which ovulate after the introduction of rams in spring.

The preovulatory surge of LH

After stimulation Merino ewes have a preovulatory surge of LH, usually within 48 hours of the introduction of rams, but the mean time has varied widely between experiments (27 h, Oldham *et al.* 1978; 37 h, Oldham and Pearce 1983; 11.5 h, Pearce *et al.* 1984b). The incidence of surges of LH in individual ewes within six hours of introduction of rams (Oldham *et al.* 1978; Pearce *et al.* 1984b) and the absence of high peripheral concentrations of oestrogen (Knight *et al.* 1978) suggested that rams could elicit a preovulatory surge by directly affecting the hypothalamo-pituitary axis. More recent evidence, reviewed by Martin (1984), suggests that the normal preovulatory surge of LH is induced by increased secretion of oestrogen and that the very early surges are large increases in the secretion of LH caused directly by the introduction of rams, which are in themselves of sufficient magnitude to induce ovulation.

*Effect of withdrawal of rams* If in most cases the preovulatory surge of LH is dependant upon ram-induced increases in the secretion of oestrogen (by stimulating the pulsatile release of LH) then premature withdrawal of rams would interfere with this mechanism. This hypothesis has been tested by exposing ewes to rams for varying periods. Signoret *et al.* (1982/83) did not measure the time of the preovulatory surge of LH but found that the proportion of ewes ovulating after the introduction of rams was highest after four days of exposure compared with six or 24 hours of exposure. Oldham and Pearce (1983) found that the proportions of ewes induced to ovulate were 5/20 and 4/20 after six and 24 hours contact with rams but increased significantly to 13/20 after 72 hours contact with rams. The mean time of the preovulatory surge of LH was 37 ± 5 hours. These observations support the hypothesis that the continued presence of rams is necessary to stimulate the preovulatory surge of LH since their early withdrawal reduced the incidence of ovulation. Nevertheless the definitive experiment in which the secretion of oestradiol-17$\beta$ is measured reliably after the introduction of rams is yet to be conducted. In such an experiment the concentration of oestradiol-17$\beta$ in ovarian venous blood would have to be measured.

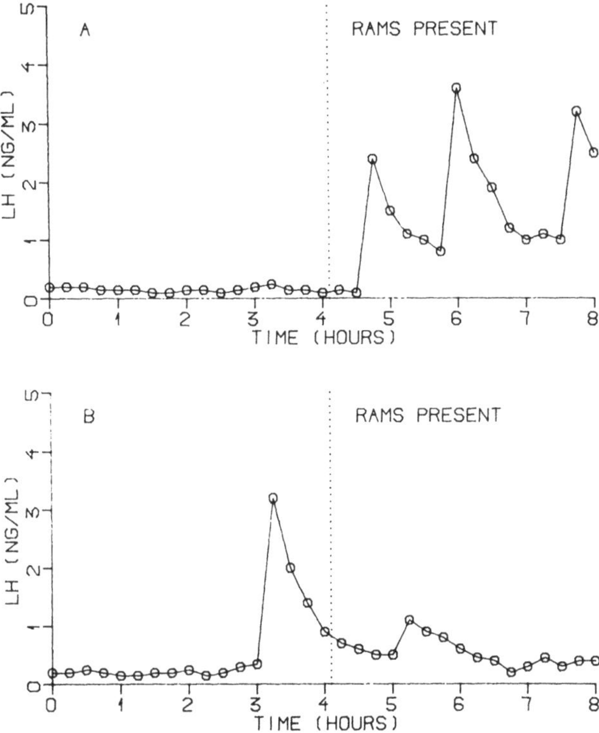


**FIGURE 3. The differential response of the secretion of LH in two seasonally anovular Merino ewes induced to ovulate by the introduction of rams (from data used by Oldham and Pearce 1983).**

*Sensitivity of the brain to oestrogen* In addition to the stimulation of secretion of oestrogen by rams, the finding that the preovulatory surge of LH in ovariectomised ewes injected with oestrogen is advanced by exposure to rams compared with isolated ewes (Signoret 1975) demonstrates that rams heighten the sensitivity to oestrogen.

## OVULATION

The ovulation induced by the introduction of rams is not usually accompanied by oestrus (Schinckel 1954b) but ewes may have an ovulation rate that is higher than normal (Cognie *et al.* 1980; Pearce *et al.* 1984a). The ovulation rate at subsequent ovulations is lower.

The continuation of ovulation in 17 day cycles depends on perception of the ram by ewes. In a recent experiment, published in full in this volume, Murtagh *et al.* (1984b) showed that the withdrawal of rams from a group of ewes induced to ovulate reduced the proportion ovulating to 48% within 30 days compared with 97% among flockmates maintained in contact with rams.

## FUNCTION OF CORPORA LUTEA AND DISTRIBUTION OF OESTRUS

The corpora lutea from ovulations induced by the introduction of rams persist either for the period which produces a normal oestrous cycle or regress prematurely (Oldham and Martin 1978; Knight *et al.* 1981; Pearce *et al.* 1984b). Ewes with normal corpora lutea display oestrus for the first time about 18-19 days after the introduction of rams or one cycle after the induced, but silent, ovulation. Corpora lutea with a short life-span regress after 5-6 days and the ewes have a second silent ovulation soon after. The corpora lutea following this second ovulation are almost always normal and the ewes therefore display oestrus for the first time approximately 24 days after the introduction of rams: that is one cycle after the second silent ovulation. Consequently the distribution of oestrus in flocks in which the ewes have either short or normal cycles is spread over 9-10 days (Fairnie 1976; Martin and Scaramuzzi 1983) and is illustrated in Figure 4.

## POTENTIAL LIMITATIONS OF THE 'RAM EFFECT'

There are three limitations to the efficacy of the 'ram effect'. First, in a flock of ewes a proportion may be cycling spontaneously and are therefore unavailable to be induced to ovulate by the introduction of rams. For example, Oldham (1980b) reported that the proportion of Merino ewes cyclic at the lowest point in spring may still be 20 per cent and subsequently Pearce *et al.* (1984a) reported three experiments where the proportion of

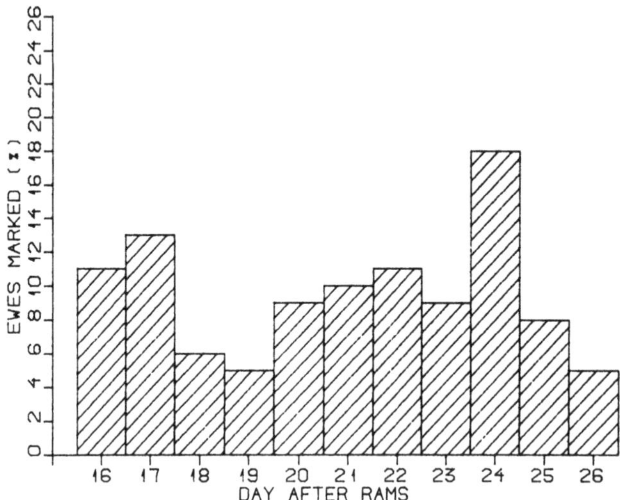


**FIGURE 4. The mean distribution of oestrus of three commercial flocks of ewes after an injection of vegetable oil immediately before introduction to rams.**

cyclic ewes in spring was 36, 24 and 12 per cent. Second, not all anovular ewes may be induced to ovulate, and, third, a proportion of those induced to ovulate in spring may do so only once and become anovular again and never display oestrus (Oldham and Cognie 1980; Oldham *et al.* 1984).

## MANIPULATION OF LUTEAL FUNCTION TO SYNCHRONISE OESTRUS

Oestrus induced by rams can be synchronised if corpora lutea with a short life-span can be eliminated, thus concentrating oestrus after the introduction of rams. Corpora lutea with a short life-span can be eliminated by priming with progestagens (reviewed by Oldham and Pearce 1984) and there are two alternative strategies. If it is sufficient that oestrus is synchronised into a period extending over four to five days then a single injection of progesterone can be used. If more precise synchrony is required then progestagen impregnated intravaginal sponges or subcutaneous implants must be used.

### The single injection of progesterone

A single injection of 20 mg progesterone given before the introduction of rams does not affect the proportion of ewes which ovulate or which display oestrus but ensures that all corpora lutea which result persist for the period of a normal oestrous cycle.

*Efficacy* The efficacy of the injection of progesterone depends on the dose (Cognie *et al.* 1982; Oldham *et al.* 1984) but the time of injection seems to be unimportant. It may be varied from immediately before, to two days or even five days before the introduction of rams (Lindsay *et al.* 1984; D.T. Pearce, C.M. Oldham, W. Haresign and S.J. Gray, unpublished).

*Mechanism* Injection of progesterone immediately before the introduction of rams delays the preovulatory surge of LH and lengthens the period of gonadotrophin priming of follicles (Pearce *et al.* 1984b) but injection five days before rams does not delay the surge (D.T. Pearce *et al.* unpublished). These findings suggest that progesterone may act on the ovary both directly and through the pituitary.

These results have been confirmed in three field experiments in which groups of 50, 50 and 100 ewes were injected with progesterone either five days or immediately before the introduction of rams or injected with vegetable oil vehicle alone. The data for all three experiments have been pooled to calculate the mean distribution of oestrus and show that most ewes displayed oestrus over four days after injection of progesterone (Figure 5) compared with nine days after injection of oil (Figure 4). Among groups of ewes injected with progesterone, advancing the time of injection advanced the mean time of detection of oestrus relative to the introduction of rams (Lindsay *et al.* 1984). Thus after injection of progesterone five days before rams, 65 per cent of ewes were detected in oestrus 16 to 19 days after the introduction of rams. After injection of progesterone immediately before the introduction of rams the same proportion of ewes was detected in oestrus over 18 to 21 days, or about two days later.

*Usefulness* The single injection of progesterone is an inexpensive method to synchronise the oestrous cycles of ewes induced to ovulate by the introduction of rams and has now been extensively tested (Lindsay *et al.* 1984). The method does not synchronise cyclic ewes in the flock nor does it influence the proportion of ewes in the flock that ovulate once and then become anovular again. The numbers of ewes in these categories are not easily predicted but our experience so far suggests that between 50 to 70 per cent of treated ewes can be expected to display oestrus.

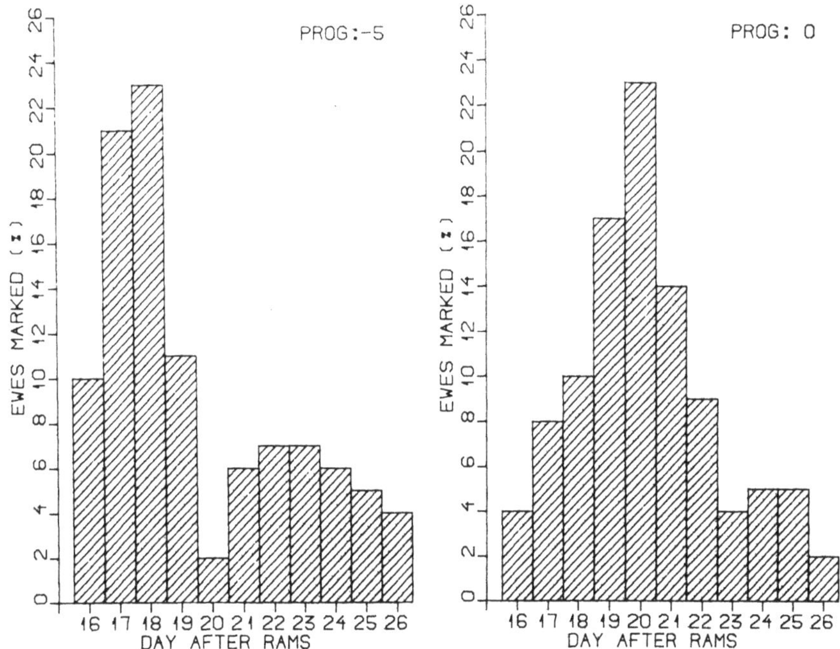


**FIGURE 5. The mean distribution of oestrus of three flocks of ewes after an injection of progesterone 5 days or immediately before the introduction of rams.**

Progestagen impregnated intravaginal sponges and implants

Progestagen impregnated intravaginal sponges or implants can also be used to prime anovular ewes, and insertion may be for a period as little as six days (Reeve and Chamley 1982, 1984) or more conventionally 12-16 days (Hunter *et al.* 1971; Oldham *et al.* 1980; Pearce *et al.* 1984a).

After six days of priming with progestagen impregnated intravaginal sponges a high proportion of ewes display oestrus when they ovulate in response to the introduction of rams and thus allow advantage to be taken both of any associated increase in ovulation rate and fertility associated with a shorter period of insertion (Reeve and Chamley 1982, 1984). This method will not synchronise all of the spontaneously ovulating ewes so it is more appropriate for use in flocks in which no ewes are ovulating such as crossbred ewes rather than in Merinos.

In Merinos insertion of intravaginal sponges for 12-16 days is generally the most efficient system to synchronise oestrus in the flock. The priming with progestagens facilitates the expression of oestrus in ewes at the ram-induced ovulation thus utilising any associated increase in ovulation rate. In an experiment comparing the efficacy of four progestagens, the ovulation rate in ewes induced to ovulate by the introduction of rams was significantly higher than in spontaneously ovulating control ewes (1.17) and was not affected by the type of exogenous progestagen (Table 1). After the introduction of rams eighty per cent of unprimed ewes ovulated and 77 and 87 per cent of ewes were detected in oestrus after priming with Repromap (Upjohn) or Chronogest (Intervet) intravaginal sponges. These data suggest that progestagens administered intravaginally do not affect the proportion of ewes ovulating. The poor results after using progestagen impregnated implants were due to large losses of implants from the ewes.

A further advantage of using intravaginal sponges for 12 to 16 days is that all spontaneously ovulating ewes are synchronised by the treatment. This may be especially important in Merino ewes in spring where the proportion of ewes ovulating spontaneously can be high (Pearce *et al.* 1984a).

## EFFICACY OF RAM-INDUCED BREEDING ACTIVITY

Effect of breed and reproductive state

The variability of response of ewes, both between and within breeds, to the introduction of rams is well documented (Knight 1983; Martin 1984) and has been equated with the 'depth of anoestrus' of ewes. The 'ram effect' appears to be independent of the negative feedback effects of the ovarian steroids but the 'depth of anoestrus' of ewes, although difficult to define, appears to be related not only to the sensitivity to the negative feedback effects of the ovarian steroids but also to the direct effects of photoperiod, both between and within breeds (Karsch 1984; Thomas *et al.* 1984). Thus the proportion of ewes in the flock that are anovular or anoestrus may be a indicator of the 'depth of anoestrus' of the flock and may also provide an indirect measure of the 'depth of anoestrus' amongst anovular ewes. For example, Lindsay and Signoret (1980) in reviewing the

**TABLE 1. The effect of progestagens on the ovarian responses of Merino ewes to the introduction of rams in spring. (Pearce *et al.* 1984, Oldham and Pearce, unpublished data).**

| Treatment | N | Ovulation rate[1] | Change in Ovulation rate[2] | Ewes Marked |
|---|---|---|---|---|
| Control | 133 | 1.44 (106) | 0.27 | – |
| Repromap[3] | 248 | 1.48 (46) | 0.31* | 77% |
| Chronogest[4] | 250 | 1.58 (50) | 0.41* | 87% |
| Norgestomet[5] | 249 | 1.76 (46) | 0.59** | 66% |
| Silestrus[6] | 238 | 1.76 (46) | 0.59** | 55% |

(1) ovulation rate of ewes ovulating. Numbers in parentheses are the numbers of ewes examined.
(2) Change in OR from that of spontaneously ovulating flockmates
(3) Medroxy progesterone-acetate intravaginal sponge, 60 mg, Upjohn
(4) Flugestone-acetate intravaginal sponge, 30 mg, Intervet
(5) Norgestomet implant, 1.3 mg, Intervet
(6) Progesterone implant, 365 mg, Ceva Chemicals
* $P < 0.05$, ** $P < 0.01$

literature found a positive correlation between the proportion of Merino and Romney ewes ovulating spontaneously or 'depth of anoestrus' of the flock and responsiveness of anovular flockmates to the introduction of rams.

It is also possible that the presence of ovulating ewes potentiates the response of anovular ewes to the introduction of rams. The ability of steroid treated ewes to induce ovulation (Croker *et al.* 1982; Signoret *et al.* 1982/83) and of ewes to stimulate secretion of LH in rams (Tilbrook *et al.* 1983) both suggest possible routes of augmentation of the 'ram effect' by cyclic ewes. Field experiments using oestrous ewes to augment the 'ram effect' (Pearce *et al.* 1984a; Reeve and Chamley 1984) have been equivocal but an experiment investigating the effect of deliberately varying the proportion of oestrous to anoestrous ewes has not been conducted. A recent experiment (T.W. Knight 1984, pers. comm.) showed that rams stimulated more ewes to ovulate if they were exposed to oestrous ewes beforehand. This suggests that any effects of the presence of oestrous ewes may operate through stimulating rams and not their anovular flockmates.

Within breeds of ewes the immediate past reproductive history also affects responsiveness to the introduction of rams. Geytenbeek *et al.* (1984) have shown a clear positive relationship between responsiveness and the interval from lambing. The time of ovulation was advanced but the interval from parturition to first oestrus was not advanced. However the use of synthetic progestagens to prime ewes before the introduction of rams would probably facilitate oestrus coincident with any ram-induced ovulations.

There are differences between breeds and between strains of rams in their ability to stimulate ewes (Knight 1983). These differences may be because some rams produce more of the proposed pheromone than others but we cannot be certain of this until there is a technique for measuring the pheromone.

Efficacy in Merino ewes

Surveys of the reproductive performance of unprimed paddock mated Merino ewes in Western Australia have shown that only four per cent of ewes fail to be mated, irrespective of whether the ewes were joined in autumn or spring (Lindsay *et al.* 1975; Knight *et al.* 1975b). Despite this there are differences in the pattern of fertility of ewes during the joining period. During the first two weeks of a ram-induced breeding season fewer ewes are mated and more ewes fail to lamb than in the spontaneous breeding season (Knight *et al.* 1975a). The reasons for this are that in the absence of progesterone priming most ewes induced to ovulate by the introduction of rams have silent ovulations (Schinckel 1954b), but the fertility of the small but significant proportion of ewes which display oestrus coincident with ram-induced ovulations is very low (Knight *et al.* 1975a; Oldham *et al.* 1976). This abnormal pattern of fertility can be rectified by stimulating ewes to ovulate with vasectomised rams or with steroid treated wethers two weeks before joining with entire rams (Oldham *et al.* 1976).

Another potential source of reproductive inefficiency is that ewes induced to ovulate by the introduction of rams may become anovular again before becoming pregnant (Oldham and Cognie 1980; Oldham *et al.* 1984). The likelihood of this apparently increases as the date of introduction of rams approaches the start of the spontaneous breeding season (Oldham and Cognie 1980). There are two consequences of ewes becoming anovular again. First, it increases the porportion of ewes which fail to display oestrus. For example, in a recent experiment Murtagh *et al.* (1984a) found 80 per cent of adult ewes had ovulated within five days of the introduction of rams but in a joining period of 45 days only 62 per cent of ewes were marked by rams. Second, of ewes that are mated by rams, some may not conceive. These anovular ewes cannot be distinguished from pregnant ewes by rams in spring and probably contributed heavily to the 22 per cent of ewes marked by rams that did not return to oestrus but failed to lamb in the Western Australian flocks studied by Knight *et al.* (1975b). Reproductive wastage due to ewes becoming anovular after being induced to ovulate by rams can be minimised by identification of non-pregnant ewes at a later date with harnessed wethers which have been injected with testosterone, harnessed vasectomised rams or by one of the pregnancy testing machines.

### Induction of ovulation in young ewes

The introduction of rams to nine to 11-month-old pre-pubertal Merino ewes in January, May or June caused many to ovulate for the first time (Murtagh *et al.* 1984a; Oldham and Gray 1984). In January the induction of puberty was independant of liveweight but in May or June those ewes induced to ovulate were heavier than those that did not ovulate. The ability of rams to induce ovulation in 14 to 15 month old ewes in spring has not been consistent. Murtagh *et al.* (1984a) found young ewes to be less responsive than adult ewes, but young ewes which had been previously in contact with rams were more responsive than their flockmates previously kept in total isolation from males. By contrast, Oldham *et al.* (1984) found that groups of maiden 18-month-old ewes ovulated with a similar frequency to adult flockmates after the introduction of rams, but more adult ewes subsequently displayed oestrus. Regardless of these conflicting results, it is apparent that the reproductive performance of young ewes after the introduction of rams in spring may be poor and is typically worse than that of adults.

## CONCLUSIONS

The introduction of rams is a powerful and cheap way to control the time of mating of sheep and is an integral part of the reproductive performance of flocks joined in spring. The 'ram effect' appears to operate by directly stimulating the secretion of LH but the factors predisposing responsiveness to the introduction of rams are not well understood. We neither know what precise physical circumstances determine isolation nor the minimum period of isolation required. Furthermore if spontaneously ovulating ewes could be forced to become anovular and high responsiveness of ewes to the introduction of rams could be guaranteed, then using rams to induce breeding activity would be highly efficient. The reproductive performance of ewes joined in a ram-induced breeding season would be further improved if the tendency of ewes to become anovular again could be eliminated or at least controlled. The explanation for this occurrence is not known but because rams appear to stimulate reproductive activity directly we hypothesise that ovulating ewes which subsequently become anovular lose perception of the presence of rams.

## REFERENCES

Atkinson, S. and Williamson, P., 1984. *J. Reprod. Fert.*, In press.

Cognie, Y., Gayerie, F. Oldham, C.M. and Poindron, P., 1980. *Proc. Aust. Soc. Anim. Prod., 13*, 80-81.

Cognie, Y., Gray, S.J., Lindsay, D.R., Oldham, C.M., Pearce, D.T. and Signoret, J.P., 1982. *Proc. Aust. Soc. Anim. Prod., 14*, 519-522.

Croker, K.P., Butler, L.G., Johns, M.A. and McColm, S.C., 1982. *Theriogenology, 17*, 349-354.

Fairnie, I.J., 1976. *In* Tomes, G.J., Robertson, D.E. and Lightfoot, R.J. (ed) *Sheep Breeding*, West. Aust. Inst. Tech., Perth, 500-508.

Fletcher, I.C. and Lindsay, D.R., 1971. *J. Endocr., 50*, 685-696.

Fulkerson, W.J., Adams, N.R. and Gherardi, P.B., 1981. *Appl. Anim. Ethol., 7*, 57-66.

Geytenbeek, P.E., Gray, S.J. and Oldham, C.M., 1984. *Proc. Aust. Soc. Anim. Prod., 15*, 353-356.

Hunter, G.L., Belonje, P.C. and van Niekirk, C.H., 1971. *Agroanimalia, 3*, 133-140.

Karsch, F.J., 1984. This volume, 10-15.

Knight, T.W., 1983. *Proc. NZ Soc. Anim. Prod. 43*, 7-11.

Knight, T.W., and Lynch, P.R., 1980a. *Anim. Reprod. Sci., 3*, 133-136.

Knight, T.W., and Lynch, P.R., 1980b. *Proc. Aust. Soc. Anim. Prod., 13*, 74-76.

Knight, T.W., Peterson, A.J. and Payne, E., 1978. *Theriogenology, 10*, 343-348.

Knight, T.W., Oldham, C.M., Lindsay, D.R. and Smith, J.F., 1975a. *Aust. J. Agric. Res., 26*, 577-583.

Knight, T.W., Oldham, C.M., Smith, J.F. and Lindsay, D.R., 1975b. *Aust. J. Exp. Agric. Anim. Husb., 15*, 183-188.

Knight, T.W., Tervit, H.R. and Fairclough, R.J., 1981. *Theriogenology, 15*, 183-190.

Knight, T.W., Tervit, H.R. and Lynch, P.R., 1983. *Anim. Reprod. Sci., 6*, 129-134.

Lindsay, D.R., Cognie, Y., Pelletier, J. and Signoret, J.P., 1975. *Physiol. and Behav., 15*, 423-426.

Lindsay, D.R., Gray, S.J., Oldham, C.M., and Pearce, D.T., 1984. *Proc. Aust. Soc. Anim. Prod., 15*, 159-161.

Lindsay, D.R., Knight, T.W., Smith, J.F., and Oldham C.M., 1975. *Aust. J. Agric. Res., 26*, 189-198.

Lindsay, D.R., and Signoret, J.P., 1978. *Proc. 1st. Wld. Cong. Ethol. Appl. Zootech.*, 513-521.
Lindsay, D.R., and Signoret, J.P., 1980. *Proc. 9th Int. Cong. Anim. Reprod. and AI, 1*, 83-92.
Lishman, A.W., de Lange, G.M., and Filjoeu, J.T., 1967. *Proc. Sth. Afric. Soc. Anim. Prod., 5*, 137-139.
Martin, G.B., 1984. *Biol. Rev., 59*, 1-87.
Martin, G.B., Oldham, C.M., and Lindsay, D.R., 1980. *Anim. Reprod. Sci., 3*, 125-132.
Martin, G.B. and Scarmuzzi, R.J., 1983. *J. Steroid Biochem., 19*, 867-876.
Martin, G.B., Scaramuzzi, R.J., and Lindsay, D.R., 1983a. *J. Reprod. Fert., 67*, 47-55.
Martin, G.B., Scaramuzzi, R.J., Oldham C.M., and Lindsay, D.R., 1983b. *Aust. J. Biol. Sci., 36*, 369-378.
Morgan, P.D., Arnold, G.W., and Lindsay, D.R., 1972. *J. Reprod. Fert., 30*, 151-152.
Murtagh, J.J., Gray, S.J., Lindsay, D.R. and Oldham, C.M., 1984a. *Proc. Aust. Soc. Anim. Prod., 15*, 490-493.
Murtagh, J.J., Gray, S.J., Lindsay, D.R., Oldham, C.M., and Pearce, D.T., 1984. This volume, 37-38.
Oldham, C.M., 1980a. *Proc. Aust. Soc. Anim. Prod., 13*,73-86.
Oldham, C.M., 1980b. PhD Thesis, University of Western Australia.
Oldham, C.M., and Cognie, Y., 1980. *Proc. Aust. Soc. Anim. Prod., 13*, 82-85.
Oldham, C.M., Cognie, Y., Poindron, P. and Gayerie, F., 1980. *Proc. 9th Int. Congr. Reprod. and AI, 3*, 50.
Oldham, C.M., and Gray, S.J., 1984. *Proc. Aust. Soc. Anim. Prod., 15*, 727.
Oldham, C.M., and Martin, G.B., 1978. *Anim. Reprod. Sci., 1*, 291-295.
Oldham, C.M., Martin, G.B. and Knight, T.W., 1978. *Anim. Reprod. Sci., 1*, 283-290.
Oldham, C.M., and Pearce, D.T., 1983. *Proc. Aust. Soc. Reprod. Biol., 15*, 72.
Oldham, C.M., and Pearce, D.T., 1984. *Proc. Aust. Soc. Anim. Prod., 15*, 158-170.
Oldham, C.M., Pearce, D.T. and Gray, S.J., 1984. *J. Reprod. Fert.* (submitted)
Pearce, D.T., Gray, S.J., Oldham, C.M. and Wilson, H.R., 1984a. *Proc. Aust. Soc. Anim. Prod., 15*, 164-168.
Pearce, D.T., Martin, G.B. and Oldham, C.M., 1984b. *J. Reprod. Fert.*, (submitted)
Pearce, D.T., and Oldham, C.M., 1983. *Proc. Aust. Soc. Reprod. Biol, 15*, 49.
Poindron, P., Cognie, Y., Gayerie, F., Orgeur, P., Oldham, C.M., and Ravault, J.P., 1980. *Physiol. and Behav., 25*, 227-236.
Reeve, J.L., and Chamley, W.A., 1982. *Proc. Aust. Soc. Reprod. Biol., 14*, 89.
Reeve, J.L., and Chamley, W.A., 1984. *Proc. Aust. Soc. Anim. Prod., 15*, 161-164.
Schinckel, P.G., 1954a. *Aust. J. Agric. Res., 5*, 465-469.
Schinckel, P.G., 1954b. *Aust. Vet. Journal, 30*, 189-195.
Signoret, J.P., 1975. *J. Endocr., 64*, 589-590.
Signoret, J.P., Fulkerson, W.J. and Lindsay, D.R., 1982/83. *Appl. Anim. Ethol., 9*, 37-45.
Tervit, H.R., and Peterson, A.J., 1978. *Theriogenology, 9*, 271-277.
Thomas, G.B., Martin, G.B., and Pearce, D.T., 1982. *Proc. Aust. Soc. Reprod. Biol., 14*, 90.
Thomas, G.B., Pearce, D.T., Oldham C.M., and Martin, G.B., 1984. *Proc. Aust. Soc. Reprod. Biol., 16*, 95.
Tilbrook, A.J., Pearce, D.T., Akari, L. and Lindsay, D.R., 1983. *Proc. Aust. Soc. Reprod. Biol., 15*, 50.
Underwood, E.J., Shier, F.L., and Davenport, N., 1944. *J. Agric. Western Australia, 11*, 135-143.
Watson, R.H., and Radford, H.M., 1960. *Aust. J. Agric. Res., 11*, 65-71.

# MANAGEMENT OF MALES FOR RAM INDUCED BREEDING OF CROSSBRED EWES IN SPRING

J.L. Reeve, *Department of Agriculture, Rutherglen Research Institute, Rutherglen, Victoria, 3685.*

*Summary* The effects of substituting testosterone treated Merino wethers for Poll Dorset rams in 'male' teams were observed when progestagen primed Border Leicester × Merino ewes were joined in Spring. Conceptions were delayed at lower proportions of entire males. There were no differences in ewes conceiving or in lambs born between 8% and 4% rams, but at 2% rams 12 fewer ewes conceived per 100 ewes joined.

## INTRODUCTION

Some prime lamb producers in North-East Victoria are priming Border Leicester × Merino (BL × M) ewes with progestagen for six days prior to joining in Spring. This results in oestrous behaviour at the ram induced ovulation, the formation of normal corpora lutea, and reduction in the proportion of ewes which respond to the 'ram effect' but return immediately to the anovular state (Reeve and Chamley 1982).

During development of the progestagen priming management system 8% of Poll Dorset (P.D.) rams were joined. This high proportion was intended to guarantee adequate numbers of sexually active rams, adequate ewe stimulation, and adequate detection of oestrus and insemination at the synchronized oestrus 2-3, and to a lesser extent, 19-20 days after joining.

Fulkerson *et al.* (1981) have shown that testosterone treated Merino wethers (T-wethers) can stimulate the 'ram effect' in anovular ewes and detect ewes in oestrus. As the first step in a program to develop a 'male' management system to optimize the 'ram effect' while keeping 'male' costs to a minimum, an experiment was conducted to examine the effect of replacing P.D. rams with T-wethers while holding the proportion of 'males' joined to progestagen primed BL × M ewes at 8%.

## MATERIALS AND METHODS

On Rutherglen Research Institute a flock of 212 mature BL × M ewes were allocated randomly to groups of 65, 62 and 85 ewes, after stratification for lambing and weaning dates. Mature Merino wethers were given 150 mg testosterone propionate subcutaneously three times at weekly intervals prior to joining. The ewes were primed with REPROMAP® (Upjohn) 60 mg intravaginal sponges for 6 days and joined on 25/10/82 to the following 'male' teams, 8% P.D. rams, 4% P.D. rams + 4% T-wethers, 2% P.D. rams + 6% T-wethers, for 6 weeks. All groups were joined in isolation from each other and from all other adult sheep. Mating was recorded daily. The ewes had a mean condition score of 3.5 (1 to 5) which did not change during joining. Due to subsequent drought conditions it was necessary to slaughter about 30% of the ewes during the third month of gestation. All foetuses were recovered. Daily lambing observations were made on the remaining ewes.

## RESULTS

The total ewes mating, the ewes conceiving at the ram induced ovulation and subsequent cycles, the total ewes conceiving, the total conceptions, and the fecundity are presented in Table 1. Conceptions were delayed when the proportion of entire rams in the 'male' team was reduced ($P < .001$). There was no significant difference in ewes conceiving and number of conceptions between 8% P.D. rams and 4% P.D. rams + 4% T-W. There was a significant reduction in ewes conceiving ($P < .05$) when only 2% P.D. rams were in the 'male' team.

**TABLE 1. Ewes mating, ewes conceiving, lambs or foetuses counted, per 100 ewes joined.**

| Joined 25/10/82 | Ewes Mating | Ewes Conceiving (lamb, foetus/ewe) | | | Ewes Conceiving | Lambs + Foetuses | L + F/ Ewe |
|---|---|---|---|---|---|---|---|
| | | 2-3 d[2] | 19-20 d | 36-37 d | | | |
| 8% Rams | | 69 | 25 | – | | | |
| | 99 | | | | 94 | 111 | 1.18 |
| n = 65 (21)[1] | | (1.2) | (1.13) | – | | | |
| 4% Rams | | 50 | 39 | 3 | | | |
| + 4% T-W | 95 | | | | 92 | 119 | 1.29 |
| n = 62 (22) | | (1.39) | (1.21) | (1.0) | | | |
| 2% Rams | | 29 | 40 | 11 | | | |
| + 6% T-W | 91 | | | | 80 | 94 | 1.18 |
| n = 85 (25) | | (1.32) | (1.09) | (1.11) | | | |

[1] ewes slaughtered [2] Days post joining

## DISCUSSION

Without treatments in which ewes were joined to 4% and 2% P.D. rams alone, conclusions can not be drawn from this experiment on the value to T-wethers in the male team. However general observations, as yet unsupported by detailed behavioural studies, suggest that the T-wethers detect oestrous ewes and could help to disturb the harems and attract the entire rams to the ewes as they come into oestrus.

The 'ram-effect' is important in all sheep production systems but especially when ewes are joined in spring and early summer. T-wethers can be provided at such minimal cost, that research to determine the proportion of T-wethers to maximize the 'ram effect' and the role of T-wethers in ensuring an even distribution of inseminations among oestrous ewes, should be given high priority.

Even where T-wethers are providing the ewe stimulation, the entire rams must be sexually active. Additional work is required on the development of a practical serving capacity test (Blockey 1982) to identify sexually active rams, and the use of oestrous ewes to stimulate inactive rams (Tillbrook *et al.* 1983) to be sexually active before the flock joining date.

The satisfactory performance of the 4% P.D. + 4% T-wether group in this experiment is of particular relevance for prime lamb production. However the concept of mixed 'male' teams may be even more important in extensively run wool producing flocks where poor detection of oestrus and small harem formation may contribute to a wide spread of conception dates, with consequent management inefficiencies.

## ACKNOWLEDGEMENTS

I wish to acknowledge the financial support of the Australian Meat Research Committee, and the technical assistance of A.E. Henderson.

## REFERENCES

Blockey, M.A. deB., 1982. *Proc. Aust. Vet. Assoc., 59.* 60.

Fulkerson, W.J., Adams, N.R. and Gherardi, P.B., 1981. *App. Anim. Ethology, 7*, 57-66.

Reeve, J.L. and Chamley, W.A., 1982. *Proc. Aust. Soc. Reprod. Biol., 14*, 89.

Tilbrook, A.J., Pearce, D.T., Akkari, Laura, and Lindsay, D.R., 1983. *Proc. Aust. Soc. Reprod. Biol., 15*, 50.

# THE EFFECT OF THE PRESENCE OF 'RAMS' ON THE CONTINUITY OF OVARIAN ACTIVITY OF MAIDEN MERINO EWES IN SPRING

J.J. Murtagh, S.J. Gray, D.R. Lindsay, C.M. Oldham and D.T. Pearce, *School of Agriculture (Animal Science), University of Western Australia, Nedlands, Western Australia 6009.*

*Summary* An experiment was designed to test the hypothesis that the continued contact with rams is required to maintain ovarian activity induced by the introduction of rams to Merino ewes. The introduction of testosterone primed wethers to a flock of 187, 19 to 20-month-old Merino ewes on 21 November 1983 (day 0) induced 92% (132/143) of anovular ewes to ovulate within 4 days compared with an incidence of 30% (29/96) of spontaneous ovulations in isolated flockmates. Twenty-nine days after the testosterone primed wethers were introduced, 97% (93/96) of ewes maintained with them had ovaries with functional corpora lutea compared with 48% (43/90) of flockmates reisolated from males on day 4 or 67% (64/95) of flockmates who remained isolated from males throughout the experiment.

## INTRODUCTION

The breeding season of the Merino ewe can be advanced from January to October by the introduction of rams or testosterone primed wethers (TW) (Underwood *et al.* 1944; Fulkerson *et al.* 1981). Approximately half of the Merino ewes in Western Australia are mated in spring in a 'ram-induced breeding season' and about 20% of the mated ewes do not return to oestrus but fail to lamb (Knight *et al.* 1975 a,b). A major reason for this is that a proportion of non-pregnant ewes return to anoestrus (Oldham and Cognie 1980). Recently, Oldham *et al.* (1984) conducted an experiment in which 95% of ewes were ovulating five days after the introduction of rams but only 60% of adult ewes and 33% of 18-month-old ewes were detected in oestrus over the next 21 days. Similarly, Murtagh *et al.* (1984) reported that although 80% of a flock of adult Merino ewes had corpora lutea five days after the introduction of rams, only 62% of the flock were marked in the next 40 days.

We have assumed in the past that the role of the male was simply to stimulate endocrine activity leading to ovulation after which his presence no longer affected the ovulatory activity of the female. This may not be so and an experiment was designed to investigate the role of the continued presence of rams on the continuity of ovarian activity of young Merino ewes in spring.

## MATERIALS AND METHODS

A flock of Merino ewes, 19-20 months of age, were kept on 'Allandale Farm', Wundowie, 70 km east of Perth, Western Australia in isolation from rams from 12 months of age. Between 12 months of age and the beginning of the experiment in November 1983, the ewes were weighed and their ovaries were examined by laparoscopy (Oldham *et al.* 1976) each month. In November 1983 the ewes were randomly allocated to three groups after stratification for live weight and ovarian activity in the preceding three months. Group 1 remained isolated from TW or rams throughout the experiment. Groups 2 and 3 were introduced to 3% of TW on 21st November (day 0). Thereafter the ewes of group 2 remained in continuous contact with TW for 29 days while group 3 was reisolated from TW or rams after four days.

The ovaries of all the ewes were examined by laparoscopy on days −3, 4, 18 and 29. The incidence and position on the ovaries of corpora haemorrhagica (CH), corpora lutea (CL) or corpora albicantia (CA) were recorded.

## RESULTS

Before allocation to their treatment groups the ewes weighed 49 ± 0.3 kg. During the experiment the ewes lost approx 0.5 kg but this loss was equal in the three treatment groups. The ovarian activity of the ewes is presented in Table 1. The ewes in groups 2 and 3 responded to the introduction of the TW with some 92% of anovular ewes and 94% of all ewes ovulating at four days compared with 4% of new ovulations in the control ewes in isolation from the TW (group 1). Thereafter there was a return to anovulation among the ewes of group 3 after they were reisolated from TW. The ovulatory activity in the isolated control ewes of group 1 increased significantly during the experiment.

## DISCUSSION

The continued presence of TW had a profound effect on the continuity of ovarian activity of the young Merino ewes in this experiment. The introduction of TW induced a large ovulatory response in seasonally anovular ewes within four days, but the ewes continued to cycle as long as the TW remained present, whereas the ewes reisolated from TW in this experiment rapidly became anovular. The percentage of ewes returning to an anovular state among the ewes of group 3 was such that by the end of the experiment less were ovulating than in the isolated ewes of group 1. The increase in ovulatory activity of the isolated ewes probably reflected the start of their spontaneous breeding season.

**TABLE 1. Ovarian activity in young Merino ewes isolated from rams or testosterone-primed wethers (TW) (group 1), in continuous contact with TW from day 0 (group 2): or exposed to TW for 4 days and then reisolated (group 3) during 29 days in November and December, 1983.**

| Day of laparoscopy | Proportion of ewes observed with CH, CL or prominent CA (%) Group 1 | Group 2 | Group 3 |
|---|---|---|---|
| − 3 | 25/92 (26)[1a] | 22/96 (23)[1a] | 22/91 (24)[1a] |
| + 4 | 29/96 (30)[1a] | 91/96 (95)[2b] | 85/91 (93)[2b] |
| +18 | 34/96 (35)[1a] | 91/96 (97)[2b] | 63/89 (71)[3c] |
| +29 | 64/95 (67)[2a] | 93/96 (97)[2b] | 43/90 (48)[4c] |

Superscripts 1, 2, 3 and 4 signify within group differences over the four times of laparoscopy ($P < 0.005$) and a, b and c signify between group differences ($P < 0.005$) at each laparoscopy, using $\chi^2$ analysis

Clearly the presence of males does not guarantee continued ovarian cyclicity in ewes in spring (Oldham and Cognie 1980; Murtagh *et al.* 1984; Oldham *et al.* 1984). However, it appears equally clear that in the absence of males the rate at which ewes revert to anovulation is increased. In practice, rams are left with ewes throughout the joining period but it is important to understand the mechanism responsible for the rams stimulating sexual activity in spring and for maintenance of cyclical ovarian activity until pregnancy intervenes and why this seems to vary from experiment to experiment. The first consideration is whether rams override or whether they reverse the effects of photoperiod. Pearce and Oldham (1984) demonstrate that the continued presence of rams is required, by most ewes who ovulate in response to the introduction of rams, until the time of the preovulatory surge of LH. The results of this experiment strongly support the argument that continued perception by young Merino ewes of rams or TW is necessary to maintain ovarian cyclicity, in successfully stimulated ewes, until they reovulate with oestrus 18 to 24 days after their initial introduction to rams (Oldham 1980).

This evidence for the 'ram effect' overriding the influence of an unfavourable photoperiod must be balanced against reports that a significant proportion of ewes become anovular even when the ewes remain with rams or TW (Oldham and Cognie 1980; Oldham *et al.* 1984). We hypothesise that the latter observation is explained by a variable number of ewes becoming refractory to the positive stimulus provided by the presence of rams.

ACKNOWLEDGEMENTS

We are grateful to the AMRC for financial support.

REFERENCES

Fulkerson, W.J., Adams, N.R. and Gherardi, P.B., 1981. *Appl. Amin. Ethol., 7*, 57-66.

Knight, T.W., Oldham, C.M., Lindsay, D.R. and Smith, J.F., 1975a. *Aust. J. Agric. Res., 2*, 577-583.

Knight, T.W., Oldham, C.M., Smith J.F. and Lindsay, D.R., 1975b. *Aust. J. Exp. Agric. Anim. Husb., 15*, 183-188.

Murtagh, J.J., Gray, S.J., Lindsay, D.R. and Oldham, C.M., 1984. *Proc. Aust. Soc. Anim. Prod., 15*, 490-493.

Oldham, C.M., 1980. *Proc. Aust. Soc. Anim. Prod., 13*, 73-86.

Oldham, C.M. and Cognie, Y., 1980. *Proc. Aust. Soc. Anim. Prod., 13*, 82-85.

Oldham, C.M., Knight, T.W. and Lindsay, D.R., 1976. *Aust. J. Exp. Agric. Anim. Husb., 16*, 24-27.

Oldham, C.M., Pearce, D.T. and Gray, S.J., 1984. *J. Reprod. Fert.* (submitted).

Pearce, D.T. and Oldham, C.M., 1984. This volume, 26-34.

Underwood, E.J., Shier, F.L. and Davenport, N. 1944. *J. Agric, Western Australia, 11*, 135-143.

# REPRODUCTIVE BEHAVIOUR OF RAMS

D.G. Fowler, *Department of Agriculture, Agricultural Research and Advisory Station, Glen Innes, New South Wales, 2370.*

## INTRODUCTION

In its simplest form, mating behaviour is seen as the motor patterns displayed during courtship between a ram and a ewe. These patterns of behaviour undergo considerable modification when the individual ram and ewe become part of a group. Patterns of mating behaviour in the group are still further modified by environmental factors such as climate, paddock size and the ratio of rams to ewes. Intrinsic factors such as age and genotype superimpose yet another range of influences. These factors are capable of operating individually or together in any set of combinations to produce patterns of behaviour that can often become quite distorted. The interactions of these factors are responsible for producing marked variations in flock reproductive performances. One of the main aims of the study of mating behaviour is to understand how variations in reproductive performance are produced. Such an understanding enables the development of management strategies for reproducing flocks.

## DEVELOPMENTS IN THE STUDY OF MATING BEHAVIOUR

Early studies were aimed at developing mating strategies for paddock size and ram percentage (Inkster 1956; Lambourne 1956; Haughey 1959). There were clear suggestions that commercial producers were under-utilising their rams and joining their flocks in paddocks that were too large. Attempts were made to examine these strategies under Australian conditions. The improvement in reproductive performance by using smaller paddocks was demonstrated in one study (Lindsay and Robinson 1961b) but not in another (Lindsay and Robinson 1961a). The difficulty of conducting controlled experiments in an uncontrolled environment was becoming apparent.

Highly controlled studies were being conducted in the U.S.A. The motor patterns displayed during courtship and the statistical parameters of mating activity were being described (Hulet *et al.* 1962a, b, c; Banks 1964). Factors affecting sexual satiation in the ram received considerable interest (Pepelko and Clegg 1965; Bermant *et al.* 1969; Beamer *et al.* 1969). Such studies were useful but difficult to interpret in a field context.

In Australia, the emphasis ranged from closely controlled studies in pens to semi-controlled field studies. By the late 1960's there was clear evidence of the importance of mating behaviour in the development of management strategies. It was shown that ewes served thrice or more were more likely to be pregnant than ewes served only once (Mattner and Braden 1967). The need to use young rams at higher percentages than older rams was demonstrated (Lightfoot and Smith 1968).

The decade of the 60's also saw an increasing interest in the flock as a whole, or in factors that might influence mating behaviour of flocks in the field. The three major ram breeds in Australia (Merino, Border Leicester, Dorset Horn) were shown to be sexually active throughout the year (Lindsay and Ellsmore 1968). However, the mating activity of Dorset and Border Leicester rams was shown to be reduced during mid summer (Lindsay 1969). The importance of sight to the sexual activity of the ram and partner-seeking activity of the ewe and of smell to the partner-seeking activity of the ewe was demonstrated (Fletcher and Lindsay 1968).

The dominant theme of the decade to 1980 was the influence of mating behaviour of the ram on the reproductive performance of the flock. As the decade drew to a close there were two distinct schools of thought on the subject. The most recent work in this field is reviewed at the end of this paper.

## MATING BEHAVIOUR IN FLOCKS

Sheep are mostly mated in groups or flocks and the influence of the group on the mating behaviour of the individuals is large. It is considered normal to join 33 to 100 ewes per ram. Variations in this ratio can cause major changes in mating behaviour.

As the number of ewes per ram is increased, the total number of mounts and services will be increased and more ewes will be mounted and served. However, the number of services per ewe tends to fall in both pen and paddock situations. The amount of time taken for the rams to serve all of the oestrous ewes increases and the time taken from first to last mount for each ewe tends to fall. The number of rams mating with each ewe also tends to fall. These patterns of behaviour have been observed from a number of different studies (Hulet *et al.* 1962 a, b; Mattner and Braden 1967; Mattner, *et al.* 1967; Croker and Lindsay 1972; Tomkins and Bryant 1972; Bryant and Tomkins 1975; Raadsma 1981). With mature rams and ewes where there are 33 to 100 ewes per ram, variations in mating behaviour are without any influence on flock reproductive performance (Lightfoot and Smith 1968; Dawe *et al.* 1970, 1974).

As the number of ewes per ram exceeds 100, there is increasing competition between ewes. The number of oestrous ewes in proximity with rams increases and there can be jostling and pushing for much of the day (Fowler 1982a). There are reductions in the services and mounts per ewe, number of ewes served and in flock fertility. Reductions in the numbers of ewes mated as the number of ewes per ram exceeds 100 have also been observed by others (Haughey 1959; Lightfoot and Smith 1968; Croker and Lindsay 1972). Breeds other than the Merino seem less susceptible in terms of reproductive performance as the number of ewes per ram exceeds 100 (Allison 1975).

As the number of ewes per ram falls below 30 and as the number of rams in the syndicate increases, there is increasing competition between rams. This aspect of mating behaviour has been the subject of a number of conflicting reports and is dealt with, more fully, in the following section.

## INTERACTIONS BETWEEN RAMS

Interactions between individual rams are responsible for marked variations in mating activity. Competition between rams is increased as ram syndicate size increases (4 to 6), as the mating arena becomes restricted (large paddocks versus small paddocks versus pens) and as the number of oestrous ewes per ram declines (to about 2/ram/day). Dominance and serving capacity are not related but, in the appropriate circumstances, dominance will exert a modifying influence on mating activity. Dominant rams become more efficient in the presence of subordinate rams than when in isolation (Lindsay *et al.* 1976). Conversely, subordinate rams work less in the presence of dominant rams than when in isolation.

In confined areas, dominant rams will suppress the mating activity of subordinate rams (Hulet *et al.* 1962b; Lindsay 1966; Marincowitz *et al.* 1966; Lindsay *et al.* 1976; Allison and Davis 1976). Where there is ample space, many oestrous ewes per ram and small ram syndicates, social interactions of the dominance subordinance type seem to be unimportant (Lindsay and Robinson 1961a, b; Mattner, *et al.* 1967; Mattner, *et al.* 1971). This trend is particularly noticeable where one or both rams of a two-ram syndicate are sexually inactive (Mattner, *et al.* 1973).

When ram syndicate size is increased (to 3 rams) and the number of oestrous ewes per day starts to become limited (2.5/ram/day) dominant rams begin to suppress the service activity of subordinate rams (Bourke 1967; Tomkins and Bryant 1972). As ram syndicate size is increased even further (4 to 5 rams) and the supply of oestrous ewes remains limited, the effects of dominance become clearly demonstrated. Marked reductions in the service rate of subordinate rams have been observed in paddocks (2 to 5 versus 12 to 15 per day; Lambourne 1956; 0.15 versus 0.82 services per hour; Fowler 1982b).

An important question is whether or not dominance interactions influence productivity. Dominant rams, when infertile, will cause reduced flock fertility (Fowler and Jenkins 1976). Where there is competition between rams, the mating preferences of rams are modified in pens (Synnott and Fulkerson 1984) and paddocks (A.J. Tilbrook, pers. comm.). It is not known if there are genetic correlations between dominance and production parameters of economic significance, but if they should exist, then these could cause uncontrolled variations in selection programmes.

## VARIATIONS IN MATING BEHAVIOUR DURING THE DAY AND DURING JOINING

The patterns of mating behaviour just described are not distributed randomly throughout the day nor throughout the joining period. There are peaks of mating activity in the morning and evening (Hulet *et al.* 1962a; Mattner and Braden 1967; Croker and Lindsay 1972; Tomkins and Bryant 1972; Cahill *et al.* 1974). From sunset to midnight mating activity in the field falls and it remains at very low levels until sunrise (Fowler unpublished). Mating activity in pens is known to remain at very low levels during the hours of darkness (Hulet *et al.* 1962a). The overall pattern is thus one of inactivity before sunrise and activity thereafter, reduced activity around noon and increased activity in the late afternoon. After sunset, activity gradually begins to subside and by midnight, is almost absent.

During the first days of joining, mating activity is at high levels (Hulet *et al.* 1962a) but it falls, sometimes very quickly (Allison 1978). The overall pattern is for a gradual decline in mating activity during joining with slight increases towards the end of joining.

## ENVIRONMENTAL FACTORS AND MATING BEHAVIOUR

When flock mating activity is examined carefully it becomes obvious that certain environmental influences are associated with particular patterns of behaviour.

### Rainfall

Reduced temperatures and heavy rain will cause a drop in conception rates (Davis 1973). This is consistent with observations that heavy rain can cause mating inactivity for prolonged periods (1 to 2 days) (Fowler unpublished). Reduced mating activity during heavy rain and heightened activity the following day has also been observed (Blockey unpublished). Rainfall has been implicated in the loss of one or more of the fertilized ova shed by Blackface ewes (Doney and Gunn 1972).

### Temperature

Mating activity is usually lower during the hours surrounding noon (Hulet *et al.* 1962a; Mattner and Braden 1967; Croker and Lindsay 1972; Tomkins and Bryant 1972; Cahill *et al.* 1974). Perhaps the higher noonday temperatures are partly responsible. During periods of high temperature rams and ewes will seek shade. Such behaviour can be a critical factor in promoting flock dispersion and reduced contact between rams and ewes (Fowler 1972). The availability of shade was a critical factor in promoting flock dispersion on a grassland myall block (Squires 1976). As ambient temperature approached 43°C, Dorset Horn and Border Leicester rams showed reduced mating activity (Lindsay 1969) whereas Merino rams appeared unaffected.

Type and size of paddock

Small, open paddocks should promote improved contact between rams and ewes and, as a result, improved reproductive performance. Large paddocks with timber belts or with low pasture availability, many watering points and perhaps interlaced with gullies, should have the reverse effect. A source of concern has been the conflicting reports suggesting that these concepts do not always hold. Some studies show that small paddocks will improve flock reproductive performance (Inkster 1957; Lindsay and Robinson 1961b; Marincowitz *et al.* 1966; Davis and Allison (1976a,b). Other studies show that small paddocks will not always improve reproductive performance (Lindsay and Robinson 1961a) nor will large paddocks always cause reduced reproductive performance (Fletcher and Nicolson 1976). A possible explanation for the variable responses from one study to another has been suggested (Fowler 1982b) and is presented in the following paragraph.

In large paddocks where feed is plentiful, the sheep tend to remain as one flock. The time devoted to grazing is reduced and ram-ewe contact, mating activity and reproductive performance are high. Where feed availability falls, sheep in large paddocks disperse and devote much time to grazing. Contact between rams and ewes diminishes and mating activity and reproductive performance falls. Where flock dispersion is prevented in such paddocks, ram-ewe contact is maintained and mating activity and reproductive performance remain high. When sheep are in small paddocks with plentiful feed, reproductive performance might be at normal levels. However, if a small paddock contains factors which reduce the contact between rams and ewes, reproductive performance is reduced. In experimental situations, this can be achieved with a maze (Fowler 1982b) but in nature by any number of factors which will reduce contact. The 'paddock-size effect' may therefore be due to an interaction involving paddock size, feed availability, flock dispersion, ram-ewe contact and mating activity.

This explanation may account for the conflicting effects of paddock size in the work of others. In one experiment (Lindsay and Robinson 1961a), size of paddock (6.9 v 0.08 ha) had no effect on mating performance but in that experiment, the sheep lost only 1.9 kg during joining. In a second experiment (Lindsay and Robinson 1961b) there was an effect of paddock size and ewes were observed to lose 4.0 kg (mature ewes) and 5.7 kg (maiden ewes) during joining. These data suggest that pasture availability was much lower in the second study. If such was the case, increased levels of dispersion and reduced ram-ewe contact might have helped to produce the size of paddock effect that was observed in the second experiment.

These studies show that the paddock-size effect can operate in very small areas (Lindsay and Robinson 1961a,b; Fowler 1982b). In commercial enterprises, where paddocks are much larger, the size of the effect may be much greater. Environmental heterogeneity (broken scrub, partially removed fence lines, tall grass, timber belts, gullies and ridges etc.,) can split a flock into many sub-groups (Squires 1976). Poor forage conditions will also cause sub-grouping (Dudzinski and Arnold 1967; Lynch 1967). Where there are large numbers of sub-groups, the percentage of rams in association with each group can vary markedly (Fowler 1982b). Sub-grouping will tend to occur where there is a multitude of watering points (bore drains, creeks) but not where there is only one watering point (Fowler 1982b).

## INTRINSIC FACTORS AND MATING BEHAVIOUR

Intrinsic factors such as age and genotype influence productivity through costs of production and output. However, other intrinsic factors such as psychological effects are not well understood.

Age

The reproductive performance of young ewes is often lower than that of the older ewes in the flock. An important contributing factor may be the grazing and mating behaviour of young ewes. Young ewes tend to disperse throughout a paddock regardless of its level of feed supply (Arnold and Pahl 1967). By increasing the number of rams joined with young ewes, the contact between rams and ewes is also increased. As a result of this there are more services per ewe and more ewes served. Furthermore, it is known that young ewes respond to increased ram numbers with improved reproductive performance (Lightfoot and Smith 1968; Dawe *et al.* 1974).

Ideally, young ewes would be joined at a high ram percentage in a small paddock and this strategy had been tested by Davis and Allison (1976a). The effect of paddock size was, as might be expected, minimal at ram-ewe ratios of 3:100 with ewes being mated by 2.2 versus 2.6 rams.

However, with ram-ewe ratios of 3:300, two tooth ewes in a small paddock were mated by an average of 1.7 rams compared with 1.2 rams in a large paddock. Corresponding values for mature ewes were 2.2 and 1.3 respectively. In a second study, both maiden and mature ewes had higher proportions of ewes mated in smaller (1.1 ha) than in larger (18 ha) paddocks (Davis and Allison 1976b).

Where young rams are used, reproductive performance will be improved by increasing the number of rams joined (Lightfoot and Smith 1968).

### Genotype

There are breed differences in the mating behaviour of rams but they tend to be small by comparison to the differences between individuals within the breeds. Genetic differences express themselves in the age at which mating activity commences (Land and Sales 1977) and at which differences in mating activity ceases (Barwick *et al.* 1985a). There are also reports of breed (Hulet *et al.* 1962a) and strain (Mattner *et al.* 1973) differences persisting after the onset of sexual maturity.

In most breeds, rams remain sexually active throughout the year. However, it has been observed that Border Leicester rams show reduced mating activity during the summer by comparison to Dorset Horn and Merino rams (Lindsay and Ellsmore 1968). It is known that Merino rams will maintain mating activity at higher ambient temperatures than Border Leicester and Dorset Horn rams (Lindsay 1969).

Flock dispersion is known to influence flock reproductive performance (Fowler 1982b) and there are breed differences in this character. At any given age it is known that British breeds of sheep tend to form smaller sub-groups than do Merinos (Arnold and Pahl 1967)

## SEXUALLY INACTIVE RAMS

Sexually inactive rams are those which show no sexual interest during short term exposures to oestrous ewes. Such rams have been identified during semen collection (Hulet *et al.* 1964) and during pen mating tests (Pretorius 1967; Mattner *et al.* 1971; Cahill 1973; Zenchak *et al.* 1973; Le Roux and Barnard 1974; Bryant 1975; Fletcher 1976; Walkley and Barber 1976: Zenchak and Anderson 1980; Kilgour 1983; Barwick *et al.* 1985a). The incidence of inactive rams varies from one study to another but, of the 3060 rams involved in the above studies, 826 (27%) were found to be inactive on their first exposure to ewes.

Some rams remain inactive after prolonged exposures to ewes (Hulet *et al.* 1964; Mattner *et al.* 1973; Kilgour *et al.* 1985). Most inactive rams become active after a further short exposure to oestrous ewes but they tend to be below average performers when they commence mating (Mattner *et al.* 1973; Kilgour *et al.* 1985). Such rams undoubtedly represent a significant source of loss to the sheep and wool industry and a simple means of identifying them could be of immense benefit. Practical methods for the identification of inactive rams have been suggested (Lindsay 1976; Fowler 1982b) but would require further testing prior to possible adoption by industry.

## MATING BEHAVIOUR AND REPRODUCTIVE PERFORMANCE

There have been conflicting reports for over a decade on the relationship between mating behaviour and reproductive performance. A possible reason for this is that the factors important to the relationship may not have been included in all the studies. Furthermore, the various components that might be important have been measured in different ways. Perhaps the important components are serving capacity, testicular size, ram liveweight and a situation where it is unlikely ewes would be served more than thrice.

### Serving capacity

Serving capacity is a measure of the ram's ability to inseminate ewes. It should not be confused with terms such as libido, worker, non worker, dexterity etc. Such terms are valid in their appropriate context but are not, as has sometimes been the case, to be confused with serving capacity.

Tests of serving capacity have been conducted under a variety of conditions. There have been open and closed pens, prolonged and shortened test periods, restrained and unrestrained ewes, oestrous and non-oestrous ewes and one or more rams in the test arena (Mattner *et al.* 1971; Lindsay *et al.* 1976; Kilgour 1980; Blockey 1982). A test about which much is known involves a single ram with a group of oestrous ewes in a private test arena for an hour. Clearly, this is not a commercially viable test but more expedient tests cannot be accepted while their relationship to production traits is unknown.

A pen test of 20 minutes in duration (single ram, oestrous ewes) is highly correlated with tests of longer duration (Kilgour 1980). A test of 20 minutes (Blockey 1982) using single rams with restrained non-oestrous ewes was found to be closely correlated with more prolonged tests (Kilgour and Whale 1980).

### Testicular size

It has been suggested that testicular size is related to conception rate in sheep (Gherardi *et al.* 1980; Burton *et al.* 1982) but clear relationships have not yet been demonstrated. Available data suggests that 400g of testicular tissue per 100 ewes should ensure 'normal' flock fertility. Combined testicular weight of 400g might correspond to a scrotal circumference of 28 to 32 cm or a mean single testis diameter of 5 to 6 cm. However, these are very much approximations, and there is an urgent need to correlate the various measurements of testicular size which have been utilised and to determine operator differences and repeatabilities (Wilkins pers. comm.).

Reproductive performance

Mattner *et al.* (1971) found that the total number of services in three pen tests were highly correlated with service activity during flock mating in the field. Mating activity of both controlled and freely mating flocks was found to be related to reproductive performance (Mattner and Braden 1967; Fowler 1975). A relationship between serving capacity and flock fertility was observed (Kilgour 1980) but some rams of low serving capacity performed better in the field than their pen test would have suggested.

Other workers have failed to show a relationship between pen mating activity tests and flock fertility (Cahill *et al.* 1975; Kelly *et al.* 1975; Walkley and Barber 1976; Allison 1978; Winfield and Cahill 1978). Kilgour (1980) has suggested that the pen tests used by these workers did not discriminate between rams sufficiently well enough. Also, the measures of fertility used by these workers may not have been precise enough to detect differences between individual rams.

In a recent study, the relationships between serving capacity, testicular diameter, ram liveweight and reproductive performance were examined (Barwick *et al.* 1985b). As serving capacity increased more ewes were raddled ($P < 0.05$). Whether or not these ewes became pregnant was influenced by an interaction also involving testicular size ($P < 0.05$). The increase in number of ewes pregnant as serving capacity increased was greatest where testicular diameters were largest (> 6.0 cms) and least where testicular diameter was smallest (< 5.0cms). To explain these relationships we suggest that high serving capacity rams have greater requirements for large testes. Such rams might quickly exhaust the spermatogenic capacity of small testes.

Testicular diameter was related to the number of ewes raddled ($P < 0.06$). Rams with intermediate testicular diameters (5 to 6 cms) being superior to those with either small (4 to 5 cm) or large (6 to 7 cm) testes. In explaining this relationship we earlier observed (Kilgour *et al.* 1985) a relationship exists between inactivity and small testes. Reduced mating activity in rams with large testes was not expected. Perhaps such rams experience mating discomfort when joined with many ewes and so reduced their mating activity.

The number of ewes becoming pregnant of those that were raddled was related to both ram liveweight ($P < 0.05$) and testicular size ($P < 0.05$). Rams of intermediate liveweight were superior to both lighter and heavier rams. We think the explanation for this relationship is that small and large rams can easily mount ewes but may experience difficulty with intromission. The ability to copulate easily and the breeding capacity of a ram are related. The anatomical characteristics of the ram and the ewe are considered to be of major importance for successful intromission (Hulet *et al.* 1962a). The relationship with liveweight held for all values of testicular size but as testicular size increased at constant liveweight, more ewes became pregnant per 100 ewes raddled. This relationship proclaims that rams with bigger testes impregnate more of the ewes they raddle than do rams with small testes and we feel this is quite realistic.

The absolute number of ewes pregnant, as distinct from pregnancies per 100 ewes raddled, was also influenced by liveweight and testicular diameter. This relationship was superficially similar to that described in the previous paragraph but with one important exception. At constant liveweight, rams of intermediate testicular size were superior to rams with both smaller and larger testes.

Testicular diameter was also related to the number of ewes pregnant through its interaction with serving capacity ($P < 0.05$). Where testicular diameter was low, (eg. 4.5 to 5.0 cm) variations in the number pregnant, for rams of differing serving capacity, were small and few ewes became pregnant (eg. 25). At intermediate testicular diameters (5.6 to 5.8 cm) more ewes became pregnant, especially as service capacity increased, but the number of pregnancies increased still further as serving capacity increased eg. 32 pregnancies at 1.5 serving capacity to 39 pregnancies at 4.92 serving capacity. Where testes were large the number of pregnancies tended to fall, particularly at low serving capacities (32 to 23 at serving capacity 1.58).

These relationships highlight the interdependence of all the traits. Clearly, when examining the relationship between mating behaviour and reproductive performance it is necessary that all of the important traits be considered jointly.

Ova wastage

In two flocks where fertilization rate and pregnancy rate were measured (Mattner and Braden 1967) the losses of fertilized ova were greater in ewes served once than in ewes served thrice or more. A similar pattern of losses was observed in two of eight flocks that were studied and reduced numbers of services per ewe were implicated (Blockey *et al.* 1975).

Ova shed by multiple-ovulating ewes appear to be either all fertilized or not, there being very few ewes which shed two eggs but have only one fertilized. This phenomenon has been noted by a number of workers (Cumming and McDonald 1967; Mattner and Braden 1967; Restall *et al.* 1976). If variation in number of services does not influence fertilization rate perhaps its effect is on survival rate of fertilized ova. An increased number of services per ewe was associated with higher numbers of multiple births in super-ovulated ewes (Newton and Betts 1972). An increased twinning rate has resulted from a second insemination in Merino ewes (Dunlop and Tallis 1964).

Sporadic outbreaks of ova wastage in young ewes have been observed in three separate studies involving 19 different flocks (Connors and Giles 1970; Dawe *et al.* 1974; Blockey *et al.* 1975). A possible explanation is the serving activity of rams. Perhaps young ewes have a shorter length of oestrus than older ewes (Lambourne

1956; Blockey and Cumming 1970). Rams with high serving capacity could perhaps provide young ewes with more services than could be provided by rams having low serving capacity.

Increased mating activity might influence the survival rate of fertilized ova by increasing the likelihood of the correct timing of insemination relative to ovulation. Ewes served twice in the period 11 to 15 hours before ovulation had a higher pregnancy rate (81% compared with 31% to 43%) than ewes served twice in any other 4 hour period prior to ovulation (Dzuik 1970).

Where there are many ewes per ram the opportunity for services per ewe to be in the range of 1 to 3 is greater than where there are fewer ewes per ram. At a low ram percentage, the loss rates of fertilized ova might be expected to be high. Where there was one ram with 400 ewes non-random losses of ova were observed (Fowler 1982a).

## PRESENT AND FUTURE NEEDS

On an industry level, reproductive performance can be raised firstly by ensuring that only physically sound rams that are free of brucellosis are used. The use of small paddocks and the adoption of strategies that promote contact between rams and ewes (Fowler 1972, 1982b) would also contribute to improved performance. These aspects constitute sound management and should be adopted by industry on the broadest possible basis.

Research shows that industry is currently under-utilising its rams. Recommendations that would enable more effective use of rams have been developed (Lightfoot and Smith 1968; Dawe *et al.* 1970, 1974; Fowler 1972, 1982b; Blockey 1980). In essence, the industry has opted for a high margin of safety but at a cost. The buying and selling of rams, however, is healthy trade and this aspect of the overall question must also be considered. The reduction of ram percentages which could be achieved may be biologically feasible but the macroeconomics might be less convincing.

Serving capacity tests are in commercial usage and our knowledge about them is far from complete. Where serving capacity, testicular size and ram liveweight take a normal range of values, reproductive performances can vary markedly where mating loads are high (Barwick *et al.* 1984). Reduced reproductive performance might equally well occur at normal mating loads where testicular size, serving capacity or ram liveweight take abnormally high values. Currently, it is fashionable to equate big testes, high serving capacity and higher liveweights with productive excellence. These concepts are clearly misdirected while ever the interdependence between the factors is not fully appreciated.

Ideally, each crop of young rams should be tested for their serving capacity and testicular size. Serving capacity may not be heritable (Purvis *et al.* 1984) so the testing will be an annual task. Consequently there is a need for a reliable low-cost test of ram serving capacity, but no such test exists at present. The aim of such a test would be to eliminate inactive rams and identify those rams whose higher serving capacity enable them to make better use of larger testes.

The development of such a test might be no more difficult than a thorough evaluation of the test developed by Blockey. This test can be performed quickly to test large numbers of rams. However, its relationship to production parameters is unknown. Also, there are some areas of concern surrounding the Blockey test. Rams that were dominated in the pen test might be rejected as having low serving capacity. Those same rams could well be high performers in the field where the additional space and the greater supply of oestrous ewes eliminated competition between rams.

Inactive rams could possibly also be identified and eliminated by other means (Lindsay 1976; Fowler 1982b) and if this were done, benefits would flow to industry. A reliable serving capacity test, however, would allow the use of rams according to their capabilities. Even without a test, industry could drop ram percentages to 1.0 to 1.5 for mature sheep. If ram percentages were dropped and a serving capacity test was conducted the industry would again have the margin of safety that has existed with the current ram percentage strategy.

## REFERENCES

Allison, A.J., 1975. *N.Z. J. Agric. Res., 18*, 1-8.
Allison, A.J., 1978. *N.Z. J. Agric. Res., 21*, 187-195.
Allison, A.J., and Davis, G.H., 1976. *N.Z. J. Agric., 4*, 259-267.
Arnold, G.W. and Pahl, P.J., 1967. *Proc. Ecol. Soc. Aust., 2*, 183-189.
Banks, E.M. 1964. *Behav., 23*, 249-278.
Barwick, S.A., Kilgour, R.J. and Gleeson, A.C., 1985a. *Aust. J. Exp. Agric. Anim. Husb.* (in press).
Barwick, S.A., Kilgour, R.J., Fowler, D.G. and Wilkins, J.F., 1985b. *Aust. J. Exp. Agric. Anim. Husb.* (Submitted).
Beamer, W., Bermant, G. and Clegg, M.T., 1964. *Anim. Behav., 17*, 706-711.
Bermant, G., Clegg, M.T. and Beamer, W., 1969. *Anim. Behav., 17*, 700-705.
Blockey, M.A. de B., 1982. *Proc. Aust. Vet. Assoc. Conf. Adelaide, 60.*
Blockey, M.A. de B., 1983. *In* Veterinary post graduate refresher course proceedings — sheep management *67*, 199-127.
Blockey, M.A. de B. and Cummings, I.A., 1970. *Proc. Aust. Soc. Anim. Prod., 8*, 344-347.
Blockey, M.A. de B., Parr, R.A. and Restall, B.J., 1975. *Aust. Vet. J., 51*, 298-302.

Bourke, M.E., 1967. *Aust. J. Exp. Agric. Anim. Husb., 7*, 203-205.
Bryant, M.J., 1975. *Anim. Prod., 21*, 97.
Bryant, M.J. and Tomkins, T., 1975. *Anim. Prod., 20*, 381-390.
Burton, K.N., Keogh, E.J. and Fairnie, I.J., 1982. *Proc. Aust. Soc. Anim. Prod., 14*, 539-542.
Cahill, L.P., 1973. *J. Agric. Vic., 71*, 438-441.
Cahill, L.P., Kearins, R.D., Blockey, M.A. de B. and Restall, B.J., 1974. *Aust. J. Exp. Agric. Anim. Husb., 15*, 337-341.
Connors, R.W. and Giles, R.J., 1970. *Proc. Aust. Soc. Anim. Prod., 8*, 331-336.
Croker, K.P. and Lindsay, D.R., 1972. *Aust. J. Exp. Agric. Anim. Husb., 12*, 13-18.
Cumming, I.A. and McDonald, M.F., 1976. *N.Z. J. Agric. Res., 10*, 226.
Davis, G.B., 1973. *N.Z. Vet. J., 21*, 54.
Davis, G.H. and Allison, A.J., 1976a. *N.Z. J. Agric., 132*, 77-79.
Davis, G.H. and Allison, A.J., 1976b. *N.Z. J. Agric., 132*, 79-80.
Dawe, S.T., Archer, W.R., Bennett, N.W., Brunskill, A., Cahill, R.J., Donnelly, F.B., Roberts, B.D. and Trimmer, B.I., 1974. *Proc. Aust. Soc. Anim. Prod., 10*, 274-277.
Dawe, S.T., Bennett, N.W., Donnelly, F.B., Rive, J.P., Roberts, B.C. and Trimmer, B.I., 1970. *Proc. Aust. Soc. Anim. Prod., 8*, 317-320.
Doney, J.M. and Gunn, R.G., 1972. *Int. Cong. on Anim. Reprod. and A., Munich*, 421.
Dudziniski, M.L. and Arnold, G.W., 1967. *J. Range Mgt., 20*, 77-83.
Dunlop, A.A. and Tallis, G.M., 1964. *Aust. J. Agric. Res., 15*, 282-288.
Dzuik, P.J., 1970. *J. Reprod. Fert., 22*, 277-282.
Fletcher, I.C., 1976. *In* Tomes, G.J., Robertson, D.E. and Lightfoot, R.J. (eds) *Sheep Breeding*, 345-351.
Fletcher, I.C. and Lindsay, D.R., 1968. *Anim. Behav., 16*, 410-414.
Fletcher, I.C. and Nicolson, A., 1976. *Proc. Aust. Soc. Anim. Prod., 11*, 145-148.
Fowler, D.G., 1972. Proc. Ann. Seminar Livestock Officers Sheep and Wool, Wagga, July 1972 pp 26-51.
Fowler, D.G., 1975. *App. Anim. Ethol., 1*, 357-368.
Fowler, D.G., 1982a. *Aust. J. Exp. Agric. Anim. Husb., 22*, 268-273.
Fowler, D.G., 1982b. Final report to the *Australian Wool Research Committee.*
Fowler, D.G. and Jenkins, L.D., 1976. *App. Anim. Ethol., 2*, 327-337.
Gherardi, P.B., Lindsay, D.R. and Oldham, C.M., 1980. *Proc. Aust. Soc. Anim. Prod., 13*, 48-50.
Haughey, K.G., 1959. Ashburton County. *Sheep Fmg. Ann.*, 17-26.
Hulet, C.V., Ercanbrack, S.K., Price, D.A., Blackwell, R.L. and Wilson, L.O., 1962a. *J. Anim. Sci., 21*, 857-864.
Hulet, C.V., Ercanbrack, S.K., Blackwell, R.L., Price, D.A. and Wilson, L.O., 1962b. *J. Anim. Sci., 21*, 865-869.
Hulet, C.V., Blackwell, R.L., Ercanbrack, S.K., Price, D.A. and Wilson, L.O., 1962c. *J. Anim. Sci., 21*, 870-874.
Hulet, C.V., Blackwell, R.L. and Ercanbrack, S.K., 1964. *J. Anim. Sci., 23*, 1095-1097.
Inkster, I.J., 1957. *Sheep. Fmg. Ann.* 163-169.
Kelly, R.W., Allison, A.J. and Shackell, G.H., 1975. *Proc. N.Z. Soc. Anim. Prod., 35*, 204-211.
Kilgour, R.J., 1980. *In* Reviews in Rural Science, IV Behaviour, 43-46.
Kilgour, R.J. and Whale, R.G., 1980. *Aust. J. Exp. Agric. Anim. Husb., 20*, 5-8.
Kilgour, R.J., and Wilkins, J.F., 1980. *Aust. J. Exp. Agric. Anim. Husb., 20*, 662-666.
Kilgour, R.J., Barwick, S.A. and Fowler, D.G., 1985, *Aust. J. Exp. Agric. Anim, Husb.*, (in press).
Kilgour, R.J., 1983, *Aust. J. Exp. Agric. Anim. Husb., 24*. (In press).
Lambourne, L.J., 1956. *Proc. Rurakura Farm Conf.* 16-20.
Land, R.B. and Sales, D.I., 1977. *Anim. Prod., 24*, 83-90.
Le Roux, P.J. and Barnard, J.P., 1974. *South Afr. J. Anim. Sci., 4*, 171.
Lightfoot, R.J. and Smith, J.A.C., 1968. *Aust. J. Agric. Res., 19*, 1029-1042.
Lindsay, D.R., 1966. *Anim. Behav., 14*, 73-83.
Lindsay, D.R., 1969. *J. Reprod. Fert., 18*, 1-8.
Lindsay, D.R., 1976. *In* Tomes, G.J., Robertson, D.E. and Lightfoot, R.J. (eds), *Sheep Breeding*, 338-345.
Lindsay, D.R. and Ellsmore, J., 1968. *Aust. J. Exp. Agric. Anim. Husb., 8*, 649-652.
Lindsay, D.R. and Robinson, T.J., 1961a. *J. Agric. Sci., 57*, 137-140.
Lindsay D.R. and Robinson, T.J., 1961b. *J. Agric. Sci., 57*, 141-145.
Lindsay, D.R., Dunsmore, D.G., Williams, J.D. and Syme, G.J., 1976. *Anim. Behav., 24*, 818-821.
Lynch, J.J., 1967. *Proc. Ecol. Soc. Aust., 2*, 167-169.
Marincowitz, G., Pretorius, P.S. and Herbst, S.N., 1966. *South Afr. J. Agric. Sci., 9*, 971-980.
Mattner, P.E. and Braden, A.W.H., 1967. *Aust. J. Exp. Agric. Anim. Husb., 7*, 110.
Mattner, P.E., Braden, A.W.H., and George, J.M., 1971. *Aust. J. Exp. Agric. Anim. Husb., 11*, 473-477.
Mattner, P.E., Braden, A.W.H. and George, J.M., 1973. *Aust. J. Exp. Agric. Anim. Husb., 13*, 35-41.

Mattner, P.E., Braden, A.W.H. and Turnbull, K.E., 1967. *Aust. J. Exp. Agric. Anim. Husb., 7*, 103-109.
Newton, J.E. and Betts, J.E., 1972. *Anim. Prod., 14*, 363-365.
Pepelko, W.E. and Clegg, M.T., 1965. *Anim. Behav., 13*, 249-258.
Pretorius, P.S., 1967. *Proc. South Afr. Soc. Anim. Prod., 6*, 208-212.
Purvis, I.W., Kilgour, R.J., Edey, T.N. and Piper, L.R., 1984. *Proc. Aust. Soc. Anim. Prod., 15*, 545-548.
Raadsma, H.W., 1984. *M.Sc. Agr. thesis.* Uni. of New England.
Restall, B.J., Brown, G.H., Blockey, M.A. de B., Cahill, L. and Kearins, R.D., 1976. *Aust. J. Exp. Agric. Anim. Husb., 16*, 329-335.
Squires, V., 1976. *Rural Res., 93*, 4-10.
Synnott, A.L. and Fulkerson, W.J., 1984. *Appl. Anim. Ethol., 11*, 283-299.
Tomkins, T. and Bryant, M.J., 1972. *Anim. Prod., 15*, 203-210.
Walkley, J.R.W., and Barber, A.A., 1976. *Proc. Aust. Soc. Anim. Prod., 11*, 141-144.
Winfield, C.G., Bremner, W.J., Cumming, I.A., Galloway, D.B. and Makin, A.W., 1978. *Proc. Aust. Soc. Anim. Prod., 12*, 248
Winfield, C.G. and Cahill, L.P., 1978. *App. Anim. Ethol., 27*, 361-364.
Zenchak, J.J. and Anderson, G.C., 1980. *J. Anim. Sci., 50*, 167.
Zenchak, J.J., Zenchak, S.H. and Anderson, G.C., 1973. *J. Anim. Sci., 37*, 228.

# RAM MATING PREFERENCES

A.J. Tilbrook, *School of Agriculture (Animal Science), University of Western Australia, Nedlands, 6009.*
A.W.N. Cameron, *Muresk Agricultural College, Northam, 6401.*

## INTRODUCTION

When a number of ewes are in oestrus at one time, rams will preferentially mate with some ewes to the exclusion of others (Hayman 1964; Jennings and Crowley 1972; Jennings 1976; Synnott *et al.* 1981). In this paper we will discuss the basis of these perferences, and the effect that they have on flock fertility.

## REASONS FOR MATING PREFERENCES

Mating preferences for particular ewes by rams would be explained if ewes differ in their sexual 'attractiveness' to the rams. Beach (1976) recognised that females differed in their 'attractivity' or 'stimulus value' to the male. Hafez (1951) maintained that ewes differed in their 'excitatory value' to rams. He presented a number of factors such as age, breed, live-weight, size, colour of the fleece and face, the general appearance, and the coarseness of the wool of the ewes, that effect the 'excitatory value' of the ewe, but he did not quantify their effects.

Recently a specific and highly repeatable test that quantifies the differences in sexual 'attractiveness' between ewes was developed (Tilbrook 1984). In this test a group of six ovariectomised Merino ewes induced into oestrus through treatment with progesterone and oestrogen, was presented to a ram and the time spent by the ram courting and mating each ewe was recorded. Every five minutes the ewe with which the ram had spent most time was removed from the pen. The first ewe removed was classed as the most 'attractive' (ranked, 1) and the last remaining ewe was the least 'attractive' (ranked, 6). Ewes differed in sexual 'attractiveness' and 10 rams ranked the same ewes in a similar order of 'attractiveness'. This order was similar over two oestrous periods (Table 1), indicating that the components that contribute to the 'attractiveness' of an oestrous ewe are relatively stable. It was necessary for ewes to be in oestrus to get maximum discrimination and sexual response from the rams. The oestrous ewes were then tethered and the rams again permitted to rank them for 'attractiveness'. The order of 'attractiveness' of ewes when they were tethered was similar to when they were free (Table 2) indicating that the 'soliciting' behaviour of these ewes was not a major determinant of their 'attractiveness' and that the rams actively discriminated between ewes. Tilbrook went on to investiage the factors that influenced the 'attractiveness' of ewes. Stage of oestrus, the dose of oestradiol benzoate used to induce oestrus, previous exposure of the ewes to rams and the immediate mating history of the ewes were unimportant, but ewes with 12 months wool were more 'attractive' than ewes which had been recently shorn.

**TABLE 1. Mean ranks of 'attractiveness' of six oestrous ewes at consecutive oestruses.**

| Ewe number | Mean ranks of 'attractiveness' | |
|---|---|---|
| | First oestrus | Second oestrus |
| 10 | 1.8 | 2.1 |
| 15 | 2.6 | 2.8 |
| 13 | 3.1 | 2.2 |
| 16 | 4.1 | 3.4 |
| 17 | 4.3 | 5.6 |
| 12 | 5.1 | 4.9 |

(Spearman's rank correlation coefficient = 0.91 $P < 0.05$)

**TABLE 2. Mean ranks of 'attractiveness' of six free or tethered oestrous ewes.**

| Ewe number | Mean ranks of 'attractiveness' | |
|---|---|---|
| | Free | Tethered |
| 10 | 2.1 | 3.2 |
| 13 | 2.2 | 2.3 |
| 15 | 2.8 | 3.8 |
| 16 | 3.4 | 3.4 |
| 12 | 4.9 | 4.5 |
| 17 | 5.6 | 4.3 |

(Spearman's rank correlation coefficient = 0.91 $P < 0.05$)

## THE INFLUENCE OF PREFERENCES ON FERTILITY

Mating preferences lead to an uneven distribution of mating by rams among available oestrous ewes and consequently some ewes are not mated at all (Jennings and Crowley 1972; Jennings 1976; Synnott *et al.* 1981). Rams confined to pens with 8 oestrous ewes achieved an average of 12 ejaculations per day, but these were distributed among only 5 ewes (Synnott *et al.* 1981). We have found that when rams were joined to 6 oestrous ewes for periods of 21 hours, in 0.4 ha paddocks, 39% of ewes were not mated and 25% of ewes were not mounted (Tilbrook 1984). That the rams were actively making a choice to mate with particular ewes, is indicated by the fact that they investigated 87% of those ewes that were not mated. The rams average 9 ejaculations each per 21 hours so in theory were capable of mating all the available ewes. In practise an average of only 3.6 ewes were mated, indicating that the rams' mating preferences restricted the number of ewes that were mated. The Spearman's rank correlation coefficients between the ejaculations received by ewes and the number of observable components of ram behaviour such as nudges and mounts, were all highly significant and all about 0.6. Therefore, ewes that were mated the most were those that received the most attention from the rams. Tilbrook's (1984) test for sexual 'attractiveness' described above ranked ewes on the basis of the amount of time that a ram spent directing total sexual behaviour towards them, and therefore the more 'attractive' a ewe the more frequently she will be mated. If, as suggested by the data of Synnott *et al.* (1981) and Fulkerson *et al.* (1982), the ejaculates of continually mated rams contain fewer spermatozoa than are necessary for maximum rates of conception, then the ewes that are mated most frequently will have the highest chances of conceiving. When Synnott *et al.* (1981) joined rams to either 4 or 8 oestrous ewes for 24 hours, they found that ejaculates, and therefore spermatozoa, were distributed unequally amongst the ewes (Figure 1). They concluded that if such unequal distribution of spermatozoa prevails under field conditions, then only 3-4 ewes would receive a sufficient number of spermatozoa to permit maximum conception rates. Therefore, the sexual 'attractiveness' of a ewe will be important in determining her chances of becoming pregnant in a competitive situation.

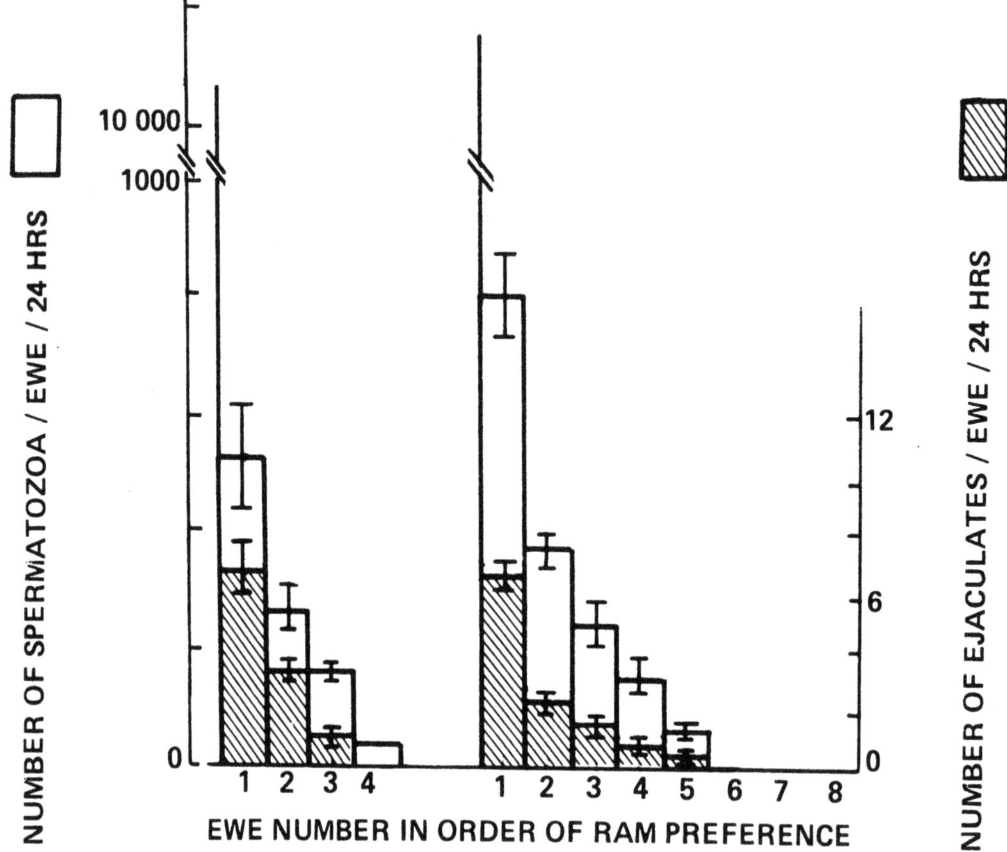


**FIGURE 1. Number of ejaculates and estimated number of spermatozoa received by ewes per day in order of ewe preference from continually mating rams in the presence of 4 or 8 oestrous ewes (from Synnott *et al.* 1981).**

## MODIFICATION OF MATING PREFERENCES BY SOCIAL INTERACTIONS BETWEEN RAMS

When a group of rams is exposed to oestrous ewes a dominance order is quickly established (Lindsay 1966) and this can modify the manner in which rams distribute their ejaculates amongst ewes (Synnott and Fulkerson 1984). When subordinate and dominant rams were mated individually they both preferred the same ewes, but when mated together in 0.23 ha paddocks ewes were chosen randomly by the subordinate ram, while the dominant ram still mated with the most 'preferred' ewes (Synnott and Fulkerson 1984). We have extended these observations to 0.4 ha paddocks with rams mated in groups of 2 or 3 with a constant joining rate of six ewes per ram (Tilbrook 1984). We found that the amount of competition between the rams is important for modifying the expression of their mating preferences. Similarly to Synnott and Fulkerson (1984), when rams were joined in groups of two, the preferences of the subordinate rams were modified and ewes were chosen randomly. Consequently both the dominant and subordinate ram mated about half of the available oestrous ewes. But, when in groups of three, where there was more competition, preferences were not modified. The dominant ram still mated with half of the ewes, but the rams that were subordinate to him mated with only about a quarter of the ewes each. Overall, there was a higher proportion of ewes mated (70% v 57%) and subsequently pregnant (55% v 43%) when rams were mated in groups of two instead of groups of three.

## CONCLUSION

The work to date has shown that ram mating preferences are important in determining whether individual ewes are mated, and how frequently they are mated and consequently can have an important effect on their fertility. This emphasises the need for an understanding of the nature of the effects of competition between rams because it may permit the development of mating stratagies that will reduce the detrimental effects of preferences by rams. An alternate and more effective way of minimising the effects of preferences would be through control of the components that make a ewe more or less 'attractive'. A more complete and detailed study of 'attractiveness' is an integral part of such an approach.

## REFERENCES

Beach, F.A., 1976. *Horm. Behav., 7*, 105-138.
Fulkerson, W.J., Synnott, A.L. and Lindsay, D.R., 1981. *J. Reprod. Fert., 66*, 129-132.
Hafèz, E.S.E., 1951. *Nature, 254*, 777-778.
Hayman, R.H., 1964. *Nature, 203*, 160-162.
Jennings, J.J. and Crowley, J.P., 1972. *Vet. Rec., 90*, 495-498.
Jennings, J.J., 1976. *Ir. J. Agric. Res., 15*, 301-307.
Lindsay, D.R., 1966. *Anim. Behav., 14*, 73-83.
Synnott, A.L. and Fulkerson, W.J., 1984. *Appl. Anim. Ethol., 11*, 283-289.
Synnott, A.L., Fulkerson, W.J. and Lindsay, D.R., 1981. *J. Reprod. Fert., 61*, 355-361.
Tilbrook, A.J., 1984. *PhD Thesis*, University of Western Australia.

# DYNAMICS OF PADDOCK-MATING OF RAMS IN CONVENTIONAL AND INTENSIFIED MATING SYSTEMS

H.W. Raadsma[1] and T.N. Edey, *Department of Animal Science, University of New England, Armidale, NSW, 2351.*

[1] Present Address: *Agricultural Research Centre, Trangie, NSW, 2823.*

*Summary* Three paddock-mating groups were studied in which Poll Dorset rams were mated to Border Leicester × Merino ewes and three in which Merino rams were mated to Merino ewes. Interest was focussed on the effects of mating on the testicular, semen and behavioural characteristics of the rams in conventional 6-weeks mating systems and on the effect on rams of joining them to a second group of ewes after an initial 17-day first mating. It was concluded that in conventional systems, with 2% rams, even though testicular volume and ejaculate volume fall during the first cycle there are no temporal effects on fertility. When a second group of ewes is introduced, ram performance is sufficiently high to achieve mating with the same level of fertility. Rams with larger testes (5.5 g testicular tissue per ewe) or smaller testes (4.1 g tissue per ewe) gave similar performance at this mating pressure. The work emphasises the surplus ram capacity present in most conventional mating systems.

## INTRODUCTION

More efficient use can be made of males in natural mating systems if their fertility can be assessed prior to breeding and their mating load adjusted accordingly. Under field conditions the rate of depletion of reserves of spermatozoa, the subsequent levels of output of spermatozoa and the distribution of spermatozoa amongst the ewes are thought to be the major factors influencing male fertility during mating (Mattner and Braden 1967; Synnott *et al.* 1981). Fertility would thus directly depend on the production rate, reserves and quality of spermatozoa and the libido and mating competency of the male. Testicular size and production of spermatozoa are significantly correlated (Knight 1972), and if production and output of spermatozoa are limiting factors in conventional mating systems, testicular size could therefore be a very useful indicator of ram fertility.

Although production and output of spermatozoa and also ram mating-behaviour have been studied individually in relation to flock fertility, no studies have been reported where these three factors have been taken into consideration simultaneously. In particular, changes in these male fertility indicators and possible changes in flock fertility during the joining period warranted further investigation to determine if any one was a limiting factor and whether or not more efficient use could be made of rams by intensifying the work-load in the second cycle.

## MATERIALS AND METHODS

Field experiments were conducted involving six groups, mated between mid-March and mid-May. Three groups of Poll Dorset rams, aged 3, 1 and 2 years respectively, were joined for 42 days to 110 (2 rams), 112 (2 rams) and 204 (3 rams) mixed age Border Leicester × Merino ewes respectively. This represented 7.6, 10.3 and 6.3 g of testicular tissue per ewe. In a fourth group, mated in 15 ha paddocks, 6 two-tooth Merino rams were joined for 51 days to 336 Merino ewes aged 3½ years, representing 6.5 g of testicular tissue per ewe. Merino rams (2-tooth) of the remaining two groups of 3 rams each, were selected for the highest testicular volume (High TV) and lowest testicular volume (Low TV) out of a group of 14 rams. These groups were joined to 152 and 150 Merino ewes (3½ years) each at a rate of 5.5 and 4.1 g of testicular tissue per ewe for 16 days, following which they were immediately rejoined with a second group of 140 Merino ewes (2½ years) for 34 days.

Measurements of testicular volume using beads, and of liveweight, and two sequential semen collections into an artificial vagina were taken from the working rams and unmated control rams (2-4 per mating group) at regular 6-17 day intervals prior to, during and after the joining period. Semen samples were evaluated for volume, for density using a calibrated colorimeter (spermatozoa × $10^9$/ml), for total number of spermatozoa (volume × density) and for the ejaculate quality parameters of motility score ('swirl' in undiluted drop), per cent live and morphologically abnormal sperm.

Raddle harnesses were used to indicate number of ewes marked. In each experiment mating activity was observed continuously through daylight hours on several days. Reaction time, the interval from arrival of the ram on the service platform, to service, was recorded for each ram at each semen collection.

## RESULTS

In the four conventional mating groups the proportion of ewes raddled in cycle I was 92%. At the end of mating 97-99% of ewes had been raddled. The proportion of ewes returning to oestrus following their first service was from 9.1-18.3% ($P < 0.05$). Because most of the ewes were raddled in cycle I, only data for this period were analysed for variation in daily proportion of ewes served and returning to service. In two of the four mating groups the number of ewes mated per day differed from the expected frequency, but no trend in the mating pattern could be detected (Figure la). The variation in return to service rate in relation to time of mating, corrected for the difference in the number of ewes raddled per day, was significant for one flock only. However, no trend in the return to service rate was observed (Figure la).

The mating performance of the High and Low TV mating groups was similar for the number of ewes mated in cycles I and II (Figure lb). The distribution of ewes raddled per day during each cycle was similar for the two mating groups. The Low TV mating group had a significantly lower return to service rate in cycle II than cycle I. There was no significant difference in the proportion of ewes returning to service between the High and Low TV groups. Also, the pattern of ewes returning to service in relation to day of service was stable and did not differ significantly between the two mating groups. Mating resulted in a significant change of ejaculate volume, ejaculate density and total spermatozoa per ejaculate. For the conventional mating groups three phases were apparent in ejaculate quantity: a high and relatively stable prejoining level, a sharp decrease during the first nine days of mating when mean volume, density and total spermatozoa reached 36-40%, 30-32% and 10-14% of prejoining mean levels respectively and finally a gradual increase during late joining and the postjoining period (Figure 2). Semen motility followed a similar pattern, decreasing to 60-64% of the prejoining score and returning to 86-97% of prejoining means. For ejaculates of the working rams, motility was significantly correlated with ejaculate density ($r_{68} = 0.55$, $P < 0.001$). Although the percentage of live and abnormal spermatozoa changed significantly during the experimental period, no consistent trend could be detected which was common to all rams.

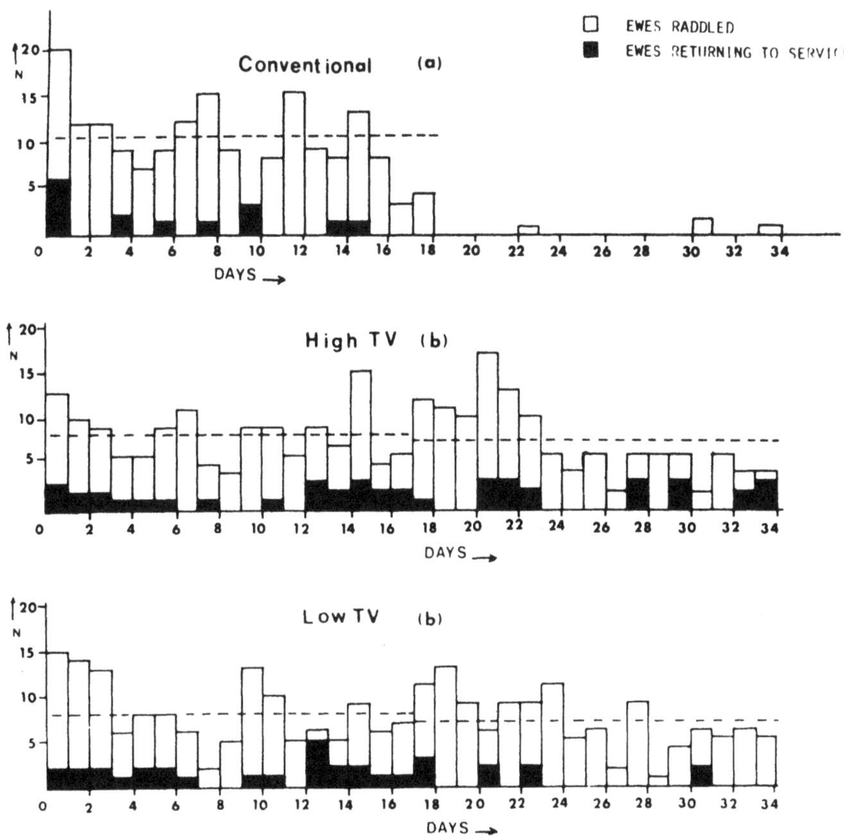


**FIGURE 1. Number of ewes raddled per day (□) and returning to service (■) in a conventional (a) and an intensified mating system (b).**

For the two intensified mating groups, the change in ejaculate characteristics was similar to that the conventional groups except that the period of low sperm output per ejaculate was extended for a further 17 days after the first cycle (Figure 2). The total spermatozoa per ejaculate was not significantly different between the High and Low TV rams during the first 34 days or during the period of recovery. For all rams combined, prejoining testicular size was significantly correlated with ejaculate density and total spermatozoa per ejaculate recorded during cycle I of the mating period. Mean testicular size in the various mating groups decreased by 19-65% during the joining period. The rate of decrease in testicular size was significantly related to prejoining testicular size, and was least in the LTV group.

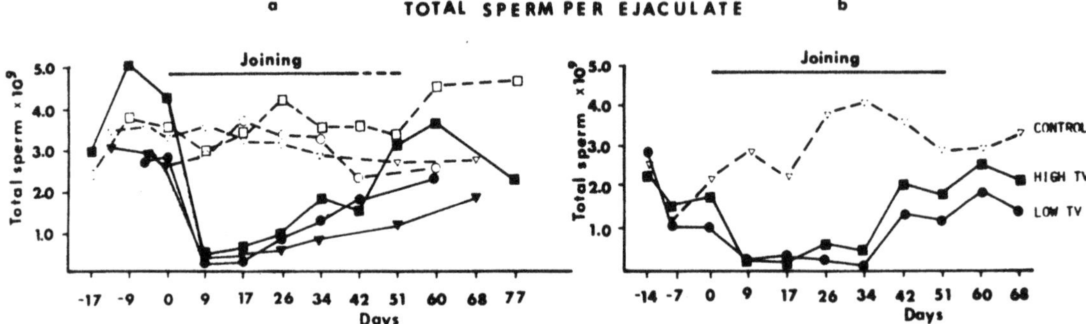


**FIGURE 2. Changes in mean total sperm per ejaculate collected from mated (■—■) and control rams (□—□) in conventional (a) and intensified (b) mating systems.**

The average number of services per ewe ranged from 1.4 to 2.8 in different mating groups during the first cycle. Mount: service ratio varied considerably between mating groups and decreased significantly as joining progressed. The observations suggested that rams became more dexterous in serving oestrous ewes. However, reaction time to collect semen from the working rams increased tenfold ($P<0.001$) during joining.

## DISCUSSION

Fertility and mating activity in the conventional mating groups indicate that the majority of ewes had regular oestrous cycles and the detection of oestrous ewes by the rams was high. No trend in the return rate over time was observed for any of the flocks, indicating that the incidence of fertilisation failure or early embryonic death did not change substantially during cycle I. This suggests that the supply of semen at $0.20\text{-}0.40 \times 10^9$ sperm/ ejaculate with an average of 1.4-2.8 services per ewe was above the level needed to maintain high fertility. Although changes in ejaculate quantity characteristics have been reported (Lightfoot 1968), the levelling of output of spermatozoa to a minimum of 10-14% of the prejoining level, independently of previous mating load, demonstrates the ability of rams to remain fertile during prolonged periods of high sexual activity. Very similar levels of output of spermatozoa have been recorded for rams subjected to far greater mating loads (1 ram:200 ewes) without loss of flock fertility (Allison 1978). Only when mating activity of the rams was low after the first 17 days of mating, did daily production of spermatozoa apparently exceed daily sperm output and a recovery in ejaculate volume took place. From the flock mating results and ejaculate characteristics of the rams in the conventional mating system, it was evident that the rams were under-utilised after the first 17 days of mating. The experiments on intensified mating showed that the level of fertility and activity observed during the first cycle could be maintained for an additional 17 days. If management systems were modified, this would allow more ewes to be mated by a given complement of rams without reducing ram percentage. However, a greater spread of lambing is a potential disadvantage.

Under conventional or intensified mating systems, if 2% of rams are used, selection for prejoining testicular size does not seem warranted as it did not change flock fertility. Variation is daily production of spermatozoa per gram of testis tissue and/or the manner of depletion of reserves of spermatozoa during periods of peak mating activity could account for this result. When rams are joined at 2.0% and generally serve ewes twice, it is unlikley that production of spermatozoa and volume or density of the ejaculate will limit fertility. However, with lower ram percentage, testicular size could become more important.

## REFERENCES

Allison, A.J., 1978. *N.Z. J. Agric. Res., 21*, 187-195.
Knight, T.W., 1972. Ph.D. Thesis, Univ. of Western Australia, Perth, W.A.
Lightfoot, R.J., 1968. *Aust. J. Agric. Res., 19*, 1043-1057.
Mattner, P.E. and Braden, A.W.H., 1967. *Aust. J. Exp. Agric. Anim. Husb., 7,* 110-116.
Synnott, A.L., Fulkerson, W.J. and Lindsay, D.R., 1981. *J. Reprod. Fert., 61*, 355-361.

# FIELD APPLICATION OF THE RAM SERVING CAPACITY TEST

M.A. de B. Blockey, *Veterinary Consultant, Blockey Fertility Services Pty. Ltd., P.O. Box 552, Hamilton, Victoria, 3300.*
J.F. Wilkins,[1] *Department of Agriculture, Agricultural Research Centre, R.M.B. 944, Tamworth, N.S.W., 2340.*

## INTRODUCTION

Numerous studies on the sexual activity of rams have been reported (see reviews by Chenoweth 1981 and Fowler 1984). Many behavioural and physiological components interact during mating but the single most important factor affecting conception is the achievement of insemination. The experiments of Mattner *et al.* (1971) were the first to show a high correlation between observations in a pen test and service activity in paddock mating. Pen test procedures for predicting serving capacity (S.C.) were further investigated and modified by Kilgour and Whale (1980), but were laborious and remained largely restricted to research use. The test developed by Blockey (1983), which is similar to his procedure for testing bulls (Blockey 1984), requires less labour input per ram tested and has therefore been proposed as more commercially acceptable.

The purpose of this paper is to outline the principles and field application of S.C. testing and is presented in three parts:

i) the concept of S.C.
ii) the development of a commercially usable test.
iii) using the S.C. test in the field.

## THE CONCEPT OF S.C.

S.C. is defined as the number of services achieved in a specified mating situation. It can be measured either during paddock mating or in a yard test designed to predict service activity in the mating flock. Some rams do not serve ewes at all and amongst those that do, there are large differences in service activity. Some have been observed to achieve 38 services during 11 hours' observation (Mattner *et al.* 1967), while others may serve only once or twice over 33 hours (Kilgour 1979).

Service activity during short periods of observed paddock mating was shown by Mattner *et al.* (1971) to be well predicted by their pen test procedure ($r$ = 0.71–0.85). Subsequently the test was modified and lengthened by Kilgour and Whale (1980) who found a correlation of $r = 0.88$ between number of services during their test and number of ewes inseminated during flock mating. The test proposed by Blockey (1983) was similar to the above in concept but requiring less input per ram tested. Alternative methods of assessment of sexual drive and capacity have been reported (Rival and Chenoweth 1982) but have not shown any better prediction of paddock performance.

Increasing service activity in the paddock has been shown to be correlated with increased pregnancy rate per oestrus. The relationship has been found by Fowler (1975) at conventional ram to ewe ratios ranging from 1 : 33 to 1 : 83 ($r = 0.81$), and at increased loads of 1 : 200 ($r = 0.63$) by Kilgour (1979). Some results providing comparative performance of rams differing in S.C. rating are shown in Table 1.

**TABLE 1. Relationship between S.C. rating of rams and pregnancy rate (as % of ewes joined) in the first cycle of mating.**

| Ram:ewe ratio | S.C. rating High | Medium | Source |
|---|---|---|---|
| 1 : 50 | 77 | 58 | Kilgour and Wilkins (1980) |
| 1 : 100 | 82 | 68 | M.A. de B. Blockey and J.W. Gough (unpublished) |
| 1 : 200 | 50 | 33 | Kilgour (1979) |

The higher pregnancy rates above appeared to result from more oestrous ewes mounted and more mounted ewes inseminated as supported by the data in Table 2.

While the exact relationship is unknown, it is likely that using rams of increasing S.C. will result in increased number of serves per oestrous ewe at any given mating load. This was suggested by the data of Kilgour and Whale (1980) and was found in cattle by Blockey (1984). In the latter work 80% of oestrous cows were served

(1) The original version of this paper, submitted by Dr. Blockey was modified and redrafted by Mr. Wilkins at the request of the editorial committee. Mr. Wilkins has been included as a co-author by the committee in response to his input.

two or more times by S.C. bulls compared to 57% by medium S.C. bulls. There is evidence that fertility is improved by increasing the number of services per ewe. In the two flocks reported by Mattner and Braden (1967), ewes served two or more times had a higher pregnancy rate than those served once only. However they also reported that this was not the case in a third flock (A.W.H. Braden, unpublished). Examining data from cattle studies, Blockey (1984) reported a 15% higher pregnancy rate over three experiments in cows served two or more times compared to those served once. Results from other experiments with sheep (Kilgour 1979; Kilgour and Wilkins 1980; S. A. Barwick *et al.* pers. comm.) have shown either no effect or have been inconclusive on this point. In the study of Fulkerson *et al.* (1982), it was proposed that more than one insemination per ewe may often be required for adequate fertility as a consequence of depletion of sperm output per ejaculate during paddock mating.

**TABLE 2. Relationship between S.C. rating of rams and proportions of ewes mounted and inseminated during paddock mating.**

| | S.C. rating High | S.C. rating Medium | Ram:ewe ratio | | Source |
|---|---|---|---|---|---|
| % of ewes | 92 | 87 | 1 : 50 | (1) | I.D. Killeen (unpublished) |
| mounted of | 91 | 83 | 1 : 50 | (2) | Kilgour & Wilkins (1980) |
| those joined | 90 | 55 | 1 : 160 | (3) | Kilgour & Whale (1980) |
| | 73 | 52 | 1 : 200 | (4) | Kilgour (1979) |
| % of ewes | 98 | 71 | 1 : 50 | (1) | |
| inseminated | 91 | 76 | 1 : 50 | (2) | |
| of those | 94 | 65 | 1 : 160 | (3) | |
| mounted | 88 | 86 | 1 : 200 | (4) | |

There is some evidence that embryonic survival may be affected by timing and/or number of services (Mattner and Braden 1967; Blockey *et al.* 1979). If this is true, and it is also true that using higher S.C. rams will result in more services per ewe, it is possible that there may be a greater survival of multiple embryos in ewes joined to higher S.C. rams. However, such an effect may only be seen at high mating loads where number of serves per ewe is limiting. This aspect is purely speculative on current evidence, but warrants further examination.

Thus in some studies it has been clearly shown that differences in pregnancy rate have been due to failure of insemination and this has been correlated with S.C. However, in others there were differences in pregnancy rate amongst inseminated ewes and it is unclear whether this was due to fertilization rate, embryonic survival or both. Our understanding of timing and number of services in relation to fertilization and subsequent embryonic survival is still far from complete, as is the effect on those parameters that varying S.C. might bring about.

## DEVELOPMENT OF A COMMERCIALLY USABLE TEST

To have field application, a S.C. test must be simple to conduct, of short duration and, most importantly, be an accurate predictor of paddock performance. The regime suggested by Mattner *et al.* (1971) required 3 × 20 minute individual test periods for each ram, using five unrestrained oestrous ewes in small pens. This procedure was further examined by Kilgour and Whale (1980) who made some modifications and lengthened the testing time required to 2 × 1 hour periods per ram. They found that paddock service activity was better predicted by this procedure than by combinations of the shorter periods suggested above. Pregnancy rate achieved by individuals was also well predicted in most cases, but underestimated in some with low pen test results (Kilgour 1979). However, in the words of Kilgour and Whale (1980), their procedure 'requires considerable input in terms of spayed ewes, yard facilities and labour', which severely limits its application.

To further examine the S.C. test procedure, M.A. de B. Blockey used 47 rams that had been tested using the 2 × 1 hour regime as described by Kilgour and Whale (1980). They were then put through the following modified S.C. test:

Oestrous ewes were restrained in service crates with one ewe per 2 m × 1 m pen. Rams were sexually stimulated by allowing them to watch other rams mounting these restrained ewes. Each stimulated ram was then placed in a pen with a restrained ewe and after 10 minutes changed to a second pen with a different ewe. The number of services achieved in each of the two 10 minute periods was counted. The correlation between the total number of services achieved in 20 minutes in the modified S.C. test and mean number of services completed in the two 1 hour pen tests was high ($r = 0.91, n = 47$). Ewes stood quietly in the service crates and a later experiment showed that rams achieved similar numbers of services using either oestrous or non-oestrous ewes (M.A. de B. Blockey unpublished). The reduction in duration allowed a reasonable throughput of tests, however, further observations indicated that rams unfamiliar with the regime may not have reliably expressed their potential. The test was then changed to a multiple testing procedure, similar to that developed by Blockey (1984) for testing beef bulls.

The multiple test used non-oestrous ewes given 2 cc. of obstetrical lubricant in the vagina restrained in service crates around the perimeter of a large yard. (Service crates should be at least 5 m apart and the yard at least 10 m × 10 m). Rams were sexually stimulated by 5 minutes observation of others mounting. The same number of rams as restrained ewes were allowed into the yard for each 20 minute test period and the number of services achieved by each ram counted. Using this procedure with 500 rams on 15 properties, the variation in S.C. was from 0 to 8 (M.A. de B. Blockey unpublished). Rams fought very little and usually only at the end of the test. No vaginal damage in ewes was observed.

The major difference between this test and others is that rams are tested in a group rather than individually. This may introduce the 'audience effect' (Lindsay *et al.* 1976), in which socially subordinate rams may show depressed S.C. in the presence of rams higher in the social order, or some rams may physically interfere with others attempting service. To minimise these effects, it is recommended that young rams are tested separately from old rams, service crates are placed at least 5 m apart, rams are thoroughly sexually stimulated before testing, rams are spread about the yard evenly and the occasional bullying ram is removed from the test very quickly. Social dominance relationships will exist in every group of rams tested as they will in paddock mating. Since it is not often practical to test rams in the identical groups in which they are to be used for mating, the dominance effect may involve some compromise in the predictive value of the test. However, the extent of prevention of service in the paddock due to dominance is unlikely to depress flock fertility (Mattner *et al.* 1967). Thus the effect, if any, during pen tests is likely to cause underestimates of paddock potential.

Regrettably there is no published information on the repeatability of performance in the multiple S.C. test for working rams and only a little field data for non-working rams (M.A.de B. Blockey unpublished). Rams tested in 1983, all having prior sexual experience but failing to serve were retested in 1984. Of 26 such rams, 23 achieved 0 or 1 service in the 1984 tests.

Further data are required to determine accurate estimates of within and between year repeatability of performance in the multiple S.C. test for both working and non-working rams. The crucial evaluation of the procedure involves its correlation with paddock fertility parameters for which there is yet no published data.

## Procedural recommendations

The following points are considered important to the efficient operation of the multiple ram S.C. test (Blockey 1983). Those not supported by published experimental data have a sound basis in the practical execution of tests in commercial flocks.

*Selection of ewes* Some ewes even restrained in service crates are very difficult for rams to serve. These include ewes with long tails as they forcibly cover the vulva. Uncrutched ewes and those with more than 10 months' wool may also present difficulties to the ram. Ewes should be replaced if they jump about violently in the crate and in any case after three 20 minute tests. Ewes need not be in oestrus as immobility is sufficient to attract mounting. Ewes should be a little smaller than the rams tested. (A plan of the service crate for restraining ewes is available from M.A. de B. Blockey).

*Environmental effects* Rams may show depressed service activity when tested outside the breeding season recognised for ewes of their breed. (Lindsay and Ellsmore 1968; M.A. de B. Blockey unpublished). Merino and Dorset rams can be tested from October to June, Corriedales and other Comeback types from December to June and British breed rams from February until June.

British breed rams have shown poor S.C. on days hotter than 30°C whilst Merinos appear less affected by heat. Tests should not be conducted in heavy rain.

*Sexual inhibition in virgin rams* Up to 50% of young virgin rams may fail to serve in their initial S.C. test but most have been found to start serving after 1 – 2 weeks of paddock mating (Hulet *et al.* 1964; Mattner *et al.* 1973). Field evidence (M.A. de B. Blockey and J.W. Gough unpublished) suggests that such initial inhibition can be eliminated by providing prior exposure. A group of 26 young (1½ year old) Corriedale rams with no prior experience produced eight (31%) which failed to serve in 20 minutes, the remainder achieving a mean of 2.4 services/ram.

In a contemporary group of 26 given 2½ weeks exposure to cycling ewes, there were none which failed to serve and the mean was 3.5 services/ram. Results from two other flocks support these findings. In flock 1, only 6 of 21 rams tested before sexual experience achieved service, whereas when tested after 2 weeks exposure only 1 failed to serve. In flock 2 only 5 of 54 rams served before 3 weeks exposure compared to 47 after. The provision of virgin rams with 2 to 3 weeks exposure to cycling ewes at a ratio of 1 ram per 2 to 4 ewes prior to test is now recommended.

*Test conduct* Thorough sexual stimulation can be enhanced by the provision of a clear view of the testing yard to the waiting rams. The older rams should be tested first whilst the younger ones observe. A minimum of 5 to 10 minutes stimulation before test is essential. Movement of the operator amongst the rams during the test promotes an even distribution of rams over the ewes available.

## USING THE S.C. TEST IN THE FIELD

The S.C. test is an important contributor to the overall assessment of breeding soundness. Its usefulness and limitations will vary considerably for different situations. The following points are discussed on the specific uses of S.C. testing:

1. Relative importance in 'fertility testing'.
2. Identification and use of highly fertile rams.
3. Assessment of 'mating potential'.
4. S.C. testing for sale rams.
5. S.C. in breeding programmes.

### Fertility testing

A complete 'fertility test' comprises a physical examination of testicles, epididymes, penis, feet, legs, eyes and jaws as well as the S.C. test (Blockey 1983). It should also include freedom from infection with blood testing where necessary. Table 3 illustrates the importance of S.C. testing in identifying 'unsound' rams in commercial flocks. The significant contribution of locomotor problems detected during S.C. testing that would otherwise be unnoticed would not be generally appreciated.

**TABLE 3. Incidence of unsound rams in commercial flocks (M.A. de B. Blockey and P.R. Holmes unpublished).**

| Survey | No. of flocks | No. of rams examined | % unsound rams | % rams detected unsound in | | % rams detected with low S.C. |
|---|---|---|---|---|---|---|
| | | | | physical exam. | S.C. test[1] | |
| Blockey (Victoria) | 38 | 1775 | 21.5 (381) | 8.2 (145) | 6.3 (111) | 7.0 (125) |
| Holmes (N.S.W.) | 30 | 1756 | 18.2 (321) | 6.4 (113) | 5.9 (104) | 5.9 (104) |

[1] locomotor problems.

Of the rams found 'low S.C.' in Table 3, about half (3 to 3.5%) did not serve. This is a similar proportion to that found by others (Hulet *et al.* 1964; Mattner *et al.* 1971) in rams having previous opportunity for experience. However, it has also been shown (Kilgour 1979) that some rams of low pen test perform considerably better in the paddock which signals a caution before culling such animals. Rams that are unable or unwilling to work in the paddock (for any reason) should be culled. Their disastrous effect on single sire matings is obvious. The effect of unsound rams in a syndicate is also likely to result in depressing the overall performance as demonstrated by the results of Fowler and Jenkins (1970). However, that effect was produced by including completely infertile rams and may be larger than we might expect with a 'lesser degree of unsoundness'.

### Identification and use of highly fertile rams

Highly fertile (H.F.) rams are defined as those achieving 4 or more services in the 20 minute multiple test and also having a scrotal circumference of 30cm or greater (Blockey 1983). It is proposed that the use of highly fertile rams may be of specific benefit to maiden ewe matings, to flocks with high ovulation rates and for genetic improvement of fertility.

Performance of maiden ewes in particular is likely to benefit from increased service activity since they have shorter duration of oestrus than adults and are less aggressive in seeking rams (Blockey and Cumming 1970). It is suggested that lower pregnancy rates often found may be due to fewer being served at the optimum time relative to ovulation (Blockey *et al.* 1979). Maiden ewe performance has been improved by increasing ram percentage (Lightfoot and Smith 1968) and a similar effect may be achieved by using high S.C. rams. In comparing maiden ewes joined to rams of high or medium S.C., I.D. Killeen (unpublished) found an increased proportion of oestrous ewes served by the high S.C. rams and Galloway (1984) reported an increase in conception rate. These results have been supported by field data from commercial flocks (M.A. de B. Blockey and P.R. Holmes unpublished). I.D. Killeen's data (unpublished) also indicated that the effects of increasing service activity may be accentuated in maidens of low liveweight. It is generally accepted that maiden ewes require specific attention to optimise performance and the use of H.F. rams at no less than 2% is proposed as a worthwhile strategy.

H.F. rams are recommended for use in flocks with high ovulation rate on the basis that increased service activity may improve multiple conception rate as previously discussed. There is further support for this hypothesis in Fowler's (1982) data but the effect was significant only at very low ram percentages. Further experiments are required to resolve this issue. In the meantime, various techniques of stimulating ovulation rate have become increasingly popular and all approaches to maximise the increased potential must be considered. The specific use of H.F. rams may well be one worthwhile.

Assessment of mating potential

The 'mating potential' of a ram is the number of ewes to which he can be successfully joined. It is a function of the ram's actual service activity in the paddock and his sperm production. The exact nature of this function is unknown but it is proposed that we can predict the components with reasonable reliability by the S.C. test and some measurement of testicular size. The simplest means of measuring these traits are the 20 minute (multiple) pen test (Blockey 1983) and scrotal circumference (Knight 1977). In considering the available published and field data, the following (Table 4) is proposed as a guide to mating potential.

**TABLE 4. Proposed mating potentials for rams of varying S.C. and scrotal circumference (Blockey 1983).**

| Mating potential (no. of ewes) | S.C. test performance | Scrotal circumference (cm) |
|---|---|---|
| 50 | 2 | >28 |
| 75 | 3 | >30 |
| 100 | 4 or 5 | >30 |
| 125 | 6 or more | >32 |

On the above criteria (Table 4), rams assigned 100 ewes or more are classified as H.F. and those with 75 or 50 as medium fertile (M.F.). In principle, increasing S.C. and increasing testicular size should allow successful mating with a greater number of ewes. Recent data from a large experiment conducted by S.A. Barwick *et al.* (pers. comm.) provided general support for the above principle and also demonstrated the interdependence of S.C., testicular size and liveweight in affecting paddock mating. These results and the implications of their interactions are more fully discussed by Fowler (1984). The data clearly showed a positive relationship between pregnancy rate and S.C. with the effect increasing with increasing testicular size. Rams of high S.C. needed large testicles to express their potential. However, the independant effect of increasing testicular size was not always beneficial which posted a warning in this regard.

The beneficial effect of using high S.C. rams compared to medium on first cycle conception rate (Kilgour and Wilkins 1980) was previously discussed. Further data is available from field observations (M.A. de B. Blockey and J.W. Gough unpublished) which compares the performance of H.F. and M.F. rams in Corriedale flocks. When H.F. rams were joined at 1% there was a mean of 82% for ewes lambing to the first cycle (range 75 — 87% over 8 flocks) and a mean dry ewe rate of 6% (range 0 — 12% in 7 of those recorded). In comparison, when M.F. rams were also joined at 1%, the corresponding figures were 68% for the first cycle (range 53 — 83% over 5 flocks) and 12% for dry ewes (range 8 — 19% in 4 of those recorded). These data and the experiments of others previously discussed suggest that M.F. rams should not be joined at 1% and that the potential of H.F. rams would be wasted if joined at 2% (Table 4).

The concept of mating potential can be used in conjunction with objective and subjective ram classing procedures so that maximum use can be made of rams ranked highest on production characters. Details of implementing this approach have been outlined elsewhere (Blockey 1983).

S.C. testing for sale rams

The provision of objective data for sale rams is now widespread and likely to increase. However, to date there has been little done to include any measures of fertility in such information. The guarantee of a ram's breeding soundness has obvious benefits to both buyer and vendor. The assessment of mating potential would allow the buyer further options in his purchasing policy as the price can be considered against the likely usage in his flock (Blockey 1983). As one example he may choose to buy rams of superior production grade if confident of needing fewer. Unfortunately, there are logistic problems in providing reliable S.C. test data for young sale rams. These included the provision of prior sexual experience, attainment of sufficient age, liveweight and condition before test, and testing within the breeding season. Such constraints must be evaluated for individual situations to determine feasibility. The problems above however, are usually soluble given commitment to the principle.

S.C. in breeding programmes

The estimate of heritability of S.C. in beef bulls reported by Blockey *et al.* (1981) was 0.67 (± 0.19). Kilgour (1979) proposed that S.C. in rams was at least moderately heritable but the only published study to date (Purvis *et al.* 1984) produced a very low heritability estimate. However, in that study rams were tested without sufficient prior sexual experience, which was reflected in the high proportion (36.5%) failing to serve.

Further studies are required to improve the data used for estimating heritability before we can assess the true genetic component.

Indications of correlated responses to selection for male and female reproductive traits have been reported (Land 1973). This area is very much one of speculation but the report of Wilkins and Kilgour (1978) indicated superior early reproductive performance in the daughters of high S.C. rams. This was shown in both proportions of ewes cycling and conception rate and may have been associated with duration and/or intensity of oestrus (J.F. Wilkins and R.J. Kilgour unpublished).

The importance of selection for traits affecting fertility in both males and females and correlated responses if they exist is an area of continuing interest and undoubted value. However, we do not yet have sufficient data in many cases (including S.C.) to predict the outcome of selection.

## CONCLUSION

The concept of S.C. and its field application is in its infancy. The potential benefits of S.C. testing to the improvement of flock mating efficiency have been well demonstrated but some areas require further clarification. While some aspects of the test itself also require further experimental examination, it provides the most efficient means of identifying rams of high mating potential. The options offered to both the ram breeder and buyer have been widened.

## ACKNOWLEDGEMENTS

Support for several of the concepts discussed here have been drawn from the unpublished work of others. Permission to quote their data is gratefully acknowledged.

## REFERENCES

Blockey, M.A. de B., 1983. *Univ. Syd. Post-grad. Comm. Vet. Sci. Proc. No 67 (Werribee)*, 119-127.
Blockey, M.A. de B., 1984. *Univ. Syd. Post-grad. Comm. Vet. Sci. Proc. No 68 (Sydney)*, 509-528.
Blockey, M.A. de B. and Cumming, I.A., 1970. *Proc. Aust. Soc. Anim. Prod., 8*, 344-347.
Blockey, M.A. de B., Holst, P.J., Makin, A.W. and Cahill, L.P., 1979. *Aust. J. Exp. Agric. Anim. Husb. 19*, 150-155.
Blockey, M.A. de B., Straw, W.M. and Jones, L.P., 1981. *Proc. Aust. Assoc. Anim. Breed. Genet., 2*, 244.
Chenoweth, P.J., 1981. *Theriogenology, 16*, 155-177.
Fowler, D.G., 1975. *Appl. Anim. Ethol., 1*, 357-368.
Fowler, D.G., 1982. *Aust. J. Exp. Agric. Anim. Husb., 22*, 268-273.
Fowler, D.G., 1984. This volume, 39-46.
Fowler, D.G. and Jenkins, L.D., 1970. *Proc. Aust. Soc. Anim. Prod., 8*, 321-325.
Fulkerson, W.J., Synnott, A.L. and Lindsay, D.R., 1982. *J. Reprod. Fert., 66*, 129-132.
Galloway, D.B., 1983. *Univ. Syd. Post-grad. Comm. Vet. Sci. Proc. No 67 (Werribee)*, 163-195.
Hulet, C.V., Blackwell, R.L. and Ercanbrack, S.K., 1964. *J. Anim. Sci., 23*, 1095-1097.
Kilgour, R.J., 1979. *In, Reviews in Rural Science IV Behaviour*, pp 43-46.
Kilgour, R.J. and Whale, R.G., 1980. *Aust. J. Exp. Agric. Anim. Husb., 20*, 5-8.
Kilgour, R.J. and Wilkins, J.F., 1980. *Aust. J. Exp. Agric. Anim. Husb., 20*, 662-666.
Knight, T.W., 1977. *N.Z. J. Agric. Res., 20*, 291-296.
Land, R.B., 1973. *Nature, 241*, 208-209.
Lightfoot, R.J. and Smith J.A.C., 1968. *Aust. J. Agric. Res., 19*, 1029-1042.
Lindsay, D.R., Dunsmore, D.G., Williams, J.D. and Syme, G.J., 1976. *Anim. Behav., 24*, 818-821.
Lindsay, D.R. and Ellsmore, J., 1968. *Aust. J. Exp. Agric. Anim. Husb., 8*, 649-652.
Mattner P.E. and Braden, A.W.H., 1967. *Aust. J. Exp. Agric. Anim. Husb.*, 7, 110-116.
Mattner P.E., Braden, A.W.H. and George, J.M., 1971. *Aust. J. Exp. Agric. Anim. Husb., 11*, 473-481.
Mattner, P.E., Braden, A.W.H. and George, J.M., 1973. *Aust. J. Exp. Agric. Anim. Husb., 13*, 35-41.
Mattner, P.E., Braden, A.W.H. and Turnbull, K.E., 1967. *Aust. J. Exp. Agric. Anim. Husb., 7*, 103-109.
Purvis, I.W., Kilgour, R.J., Edey, T.N. and Piper, L.R., 1984. *Proc. Aust. Soc. Anim. Prod., 15*, 545-548.
Rival, M.D. and Chenoweth, P.J., 1982. *Proc. Aust. Soc Anim. Prod., 14*, 174.
Wilkins, J.F. and Kilgour, R.J., 1978. *Proc. Aust. Soc. Reprod. Biol., 10*, 22.

# THE VALUE OF TESTING YOUNG RAMS FOR SERVING CAPACITY

I.W. Purvis and T.N. Edey, *Department of Animal Science, University of New England, Armidale, New South Wales, 2351.*
R.J. Kilgour, *N.S.W. Department of Agriculture, Trangie Agricultural Research Centre, Trangie, New South Wales, 2823.*
L.R. Piper, *CSIRO, Division of Animal Production, Armidale, New South Wales, 2350.*

*Summary* The repeatability of pen-test serving capacity (SC) of young Merino rams has been evaluated. Repeatability is low (0.18) for rams designated as high SC. Rams initially sexually inactive but active in a second test, display a wide range of SC scores. It is concluded that the benefits arising from pen SC testing of young rams may not offset the costs.

## INTRODUCTION

With increasing attention being focussed on improving the reproductive efficiency of sheep breeding enterprises, more consideration is being given to the rams' contribution to overall flock reproductive efficiency. An important component of male reproductive performance is serving capacity (SC) which is a measure of both libido and mating dexterity.

Several studies have examined the relationship between pen-test SC and paddock mating performance (reviewed by Kilgour 1980), but the majority of these studies involved adult rams of varying and unquantified experience. However a large proportion of flock rams purchased from studs by commercial breeders have their first joining at approximately 18 months of age and have had no prior heterosexual experience.

This paper reports the results of a study of the efficiency with which pen-tests identify the SC of 18-month-old Merino rams and discusses the potential benefits of such tests to breeders.

## MATERIALS AND METHODS

The genetic background of the animals, experimental techniques and statistical analysis relevant to this study have been described in detail by Purvis *et al.* (1984a, 1984b). Rams from a wide spectrum of Merino genotypes born in 1978 - 1981 (n = 837) were weaned at 5 months of age and run as monosexual year of birth groups until pen-tested for SC at 18 months of age using the method described by Kilgour and Whale (1980). In this procedure, rams were individually penned with 4 oestrous ewes and given two 20-minute introductory tests for training purposes and two 1-hour tests over a period of two to three weeks. The number of mounts and serves was recorded for each ram during each test. The results of the introductory tests are not included in these analyses.

## RESULTS

### Distribution of serving capacity scores

Table 1 shows the frequency distribution of the number of serves for each ram in the two one-hour tests. The predominant feature of these distributions is the large number of rams which did not serve in either or both tests.

**TABLE 1. Distribution of the number of serves in the two tests for all 837 rams.**

| | | Number of serves | | | | | | | | | |
|---|---|---|---|---|---|---|---|---|---|---|---|
| Test | n | 0 | 1 | 2 | 3 | 4 | 5 | 6 | 7 | 8 | Mean |
| First | 837 | 447 | 107 | 103 | 89 | 51 | 25 | 11 | 4 | - | 1.20 |
| Second | 837 | 389 | 98 | 139 | 86 | 61 | 30 | 23 | 7 | 4 | 1.49 |

### Identification of rams of high serving capacity

The efficiency with which the pen-test identifies high SC (HSC) rams should ideally be verified by comparison with their performance in a paddock mating. However, given the large numbers of rams involved in this study, such a procedure was not possible. We were therefore restricted to an examination of the repeatability of the tests for those rams designated as HSC on the basis of the results of the first test.

The choice of the criterion for designation of rams as HSC is complicated by the shape of the distribution of number of serves in the first test. This distribution is approximately Poisson and adopting the criterion that HSC rams should have a score greater than two standard deviations above the mean, results in the cut-off value for HSC rams being 4 or more serves in the first test. The SC data for the 91 (10.9%) rams thus classified are presented in Table 2.

**TABLE 2. Distribution (%) of the number of serves in the two tests for the 91 rams classified as having high serving capacity.**

| Test | n | No. of serves 0 | 1 | 2 | 3 | 4 | 5 | 6 | 7 | 8 | Mean |
|---|---|---|---|---|---|---|---|---|---|---|---|
| 1st test | 91 | - | - | - | - | 56 | 27 | 12 | 4 | - | 4.65 |
| 2nd test | 91 | 8 | 5 | 25 | 14 | 18 | 11 | 12 | 3 | 3 | 3.46 |

Of the 91 rams classified as HSC from the first test, only 47% would have been similarly classified on the basis of their second test. In more formal terms, the repeatability of the number of serves in the first and second tests was 0.18 which is not significantly different from zero.

Identification of sexually inactive rams and rams of low serving capacity

Of the 447 rams (53% of the total) with zero serves in the first test, 317 (71%) did not serve in the second test. However, as Table 3 indicates, the remaining 29% of rams had SC scores covering the complete range shown by all rams in the first test and 22 of these rams would have been classified as HSC. Of the 317 rams which failed to serve in the second test, 98 (31%) exhibited mounting behaviour and thus cannot be regarded as being sexually inactive.

**TABLE 3. Distribution (%) of the number of serves in the second test for rams with zero (447) or one (107) or two (103) serves in the first test.**

| Score in first test | n | Number of serves 0 | 1 | 2 | 3 | 4 | 5 | 6 | 7 |
|---|---|---|---|---|---|---|---|---|---|
| 0 | 447 | 71 | 10 | 10 | 4 | 3 | 1 | .5 | .5 |
| 1 or 2 | 210 | 24 | 18 | 26 | 15 | 8 | 5 | 3 | 1 |

Examination of the second test performance of those rams achieving only one or two serves in their first test, reveals a similar variation in performance (Table 3) with 17% of rams improving to the extent of being classified as HSC. In contrast to the first test HSC group (Table 2) where the second test mean was lower than the first, the mean score for these rams was higher in the second test ( 1.49 vs 1.98, $P < 0.01$).

## DISCUSSION

In order to achieve 'current flock' benefits from selection of sires on the basis of pen measures of SC, the SC test must be a good predictor of paddock mating capacity and this latter trait must have a direct influence on flock fertility.

Earlier studies which have examined the repeatability of pen-test SC have invariably used mature rams and have usually involved small numbers of rams. For example, Kilgour and Wilkins (1980), in a study of seven adult rams found that the repeatability of one-hour pen-tests was 0.77 and that the correlation between pen-test SC and paddock mating performance (effective ram:ewe ratio of 1:170 ) was 0.88. By contrast the findings of this study suggest that when the same test procedure is used to determine the SC of 18 month old heterosexually inexperienced Merino rams the results are not highly repeatable. This is particularly evident for those rams categorized, on the basis of the first pen-test, as being of high serving capacity (HSC). Under these circumstances, the 'current flock' benefits of SC testing of young Merino rams are unlikely to offset the costs.

Mattner *et al.* (1973) and Kilgour *et al.* (1984) have shown that a large proportion (> 85%) of mature rams initially sexually inactive in pen-tests began mounting and serving when continuously exposed to oestrous ewes, and at a level which would have allowed mating at percentages normal in commercial flocks. On the basis of these reports, and because some of the initially inactive rams began to serve during the second test, it could be expected that a large proportion of these rams in our study would subsequently develop into rams satisfactory for flock mating. Further studies are required to establish methods of identifying those animals which may permanently remain sexually inhibited.

In respect of 'future flock' performance, estimates of the heritability of SC derived from these same flocks are not different from zero (Purvis *et al.* 1984b), and there seems little prospect of successful breeding for improved SC.

On balance therefore, SC testing of young rams, using the procedures outlined in this study, would not appear to be worthwhile in either 'current flock' or 'future flock' terms. There may be some benefit from SC testing in the case of sexually mature rams particularly in respect of maiden ewe performance (Killeen I.D. unpublished, quoted by Blockey 1983) or where short joining periods (< one cycle) are desired.

ACKNOWLEDGEMENTS

The authors wish to thank the staff of the Trangie Agricultural Research Centre. I.W. Purvis held an Australian Wool Corporation Postgraduate Scholarship whilst these studies were conducted.

REFERENCES

Blockey, M.A. de B., 1983. Proc. Sheep and Wool Conf. Department of Agriculture, Victoria, June.

Kilgour, R.J., 1980. *In* Wodzicka-Tomaszewska,M., Edey, T.N., and Lynch, J.J. (eds) *Behaviour*, Reviews in Rural Science IV, University of New England, Armidale, 43-46.

Kilgour, R.J., Barwick, S.A. and Fowler, D.G., 1984. *Proc. 10th Inter. Congr. Anim. Reprod. and A.I.*, Urbana-Champaign, USA.

Kilgour, R.J. and Whale, R., 1980. *Aust. J. Expl. Agric. Anim. Husb., 20*, 5-8.

Kilgour, R.J. and Wilkins, J., 1980. *Aust. J. Expl. Agric. Anim. Husb 20*, 662-666.

Mattner, P.E., Braden, A.W.H. and George, J.M., 1973. *Aust. J. Expl. Agric. Anim. Husb.,13*, 35-41.

Purvis, I.W., Kilgour, R.J., Edey, T.N. and Piper, L.R., 1984a. *Proc. Aust. Soc. Anim. Prod., 15*, 545-549.

Purvis, I.W., Kilgour, R.J., Edey, T.N. and Piper, L.R., 1984b. *Proc. 4th Conf. Aust. Assoc. Anim. Breed. and Genet., 4*, 190-191.

# THE FUNCTIONS OF THE TESTIS AND EPIDIDYMIS IN RAMS

B.P. Setchell, *Department of Animal Sciences, Waite Agricultural Research Institute, University of Adelaide, Glen Osmond, South Australia, 5064.*

## INTRODUCTION

The testes of rams are comparatively large organs (each testis is about 0.5 per cent of body weight, compared with 0.02 per cent in humans) concerned with the formation of spermatozoa and secretion of the male sex hormones. These hormones are responsible for male sexual behaviour and the function of the accessory sexual organs such as the prostate, seminal vesicles and Cowper's glands. The testis consists mostly of the long, convoluted seminiferous tubules (each testis contains about 7000 m of tubules) in which the spermatozoa are formed; these tubules contain no blood vessels or nerves but between them is the interstitial tissue, with the blood and lymphatic vessels, the Leydig cells, which secrete the male sex hormones, and some other cells, including a substantial number of macrophages. The spermatozoa formed in the testis are matured in the epididymis and then held there awaiting ejaculation. In this review, both hormone secretion and sperm production will be described, as well as the various factors which affect these processes.

## ENDOCRINE FUNCTION

### Steroid hormones

*Hormones secreted* Testosterone is believed to be quantitatively the most important steroid secreted by the mature ram testis, although appreciable amounts of androstenedione are secreted before puberty (Lindner 1961). In other species, particularly the boar, considerable amounts of other steroids, including oestrogens, and conjugated androgens and oestrogens (Setchell *et al* 1983) are secreted, and the situation in the ram needs to be re-examined. The testes of the boar also secrete some important steroid pheromones which are concentrated by the salivary glands into the saliva (see Gower 1972). In view of the recent interest in pheromones in sheep, it will be important to see whether the compounds involved are secreted as such by the testis, or formed elsewhere in the body either from precursors secreted by the testes or as a result of stimulation from the testis.

*Time course of secretion* It has been known now for some time that in conscious rams testosterone levels in the blood rise to a peak between 1 and 8 times in 24 h and then fall back towards a baseline value. A pulse of LH precedes each testosterone pulse, usually by about 40-45 min (Katongole *et al.* 1974; Purvis *et al.* 1974; Sanford *et al.* 1974; Lincoln 1978; Setchell 1978; D'Occhio *et al.* 1982) but it is not yet clear whether the change in testosterone secretion is due to release of preformed hormone or a sudden change in the rate of synthesis by the Leydig cells. There are no changes in testicular blood flow (Laurie and Setchell 1978) or blood velocity (Amann *et al.* 1978) during the pulses and their significance is obscure.

*Route and rate of secretion* Hormones can leave the testis in three possible ways, in the venous blood, in the lymph or into the peritoneal fluid in the scrotal sac. The last-mentioned is not often considered to be serious possibility although there is some evidence in man that the concentration of testosterone in fluid contained in the tunica vaginalis may be about 20 times as high as that in arterial blood plasma, and about one-third of that in spermatic venous blood (Karpe *et al.* 1982). No equivalent data are available for the ram but in this species, it has been reported that the concentration of testosterone is very similar in testicular lymph collected from a lymphatic in the spermatic cord, and in spermatic vein blood, both in normal animals and in those stimulated with hCG (Lindner 1963, 1967). Because testicular lymph flow in rams in only about 6 to 10 ml/h (Cowie *et al.* 195[illegible]; Lindner 1963; Galil *et al.* 1981), compared with a blood flow of about 1600 ml/h/testis, (see Setchell 1978) the proportion of testosterone returning to the general circulation via the lymph is only a small fraction of the total. The situation may be quite different for conjugated steroids, as it has recently been shown that in the pig, where testicular lymph flow is quite a lot faster in relation to blood flow than in the ram, as much as 80% of the DHA sulphate and oestrone sulphate produced by the testis leaves in the lymph as does about 20% of the testosterone. In the ram by multiplying testicular blood flow by the venous-arterial concentration difference for testosterone, it can be calculated that the secretion rate for the two testes is about 3.5 mg/day (Setchell 1978). However, it should be remembered that most of these measurements have been made in animals under general anaesthesia or within 48 h of anaesthesia, when pulses of LH and testosterone are much reduced, and therefore they should be regarded as base-line values.

### Factors affecting steroid secretion

*Hormones* The secretion of testosterone by the ram testes, as in other mammals, is increased following treatment with luteinizing hormone (LH) (Amann *et al.* 1978; Schanbacher 1980) or when endogenous LH levels are raised by injection of LHRH (Galloway *et al.* 1974; Galloway and Pelletier 1975; Wilson and Lapwood 1978; Lincoln 1979; Chandrasekhar *et al.* 1984). While the LH response to LHRH is dose-responsive over the range of doses used, the testosterone response was similar at all doses, suggesting that the testis is responding maximally to a normal LH pulse.

Testosterone secretion is also increased after injection of human chroionic gonadotrophin (hCG) (Lindner 1967; Garnier and Saez 1980; Chandrasekhar *et al.* 1984). However, in contrast to the responses to LH, testosterone levels show a biphasic response to hCG, with an initial peak between 1 and 2 h, and a second peak between 2 and 4 days later (Garnier and Saez 1980, Figure 1). This is similar to the response seen in some other species (Saez and Forest 1979; Padron *et al.* 1980; Martikainen *et al.* 1980, 1982; Risbridger *et al.* 1981). The testosterone response to hCG is maximal at a dose of about 20 i.u. hCG/kg body weight. Repeated injections of hCG can be used to produce sustained increased in testosterone levels, although the response to the individual injections becomes blunted. After about 2 weeks, testosterone concentrations fall back towards baseline (Figure 2), presumably because of the formation of antibodies, as already described in rats (Risbridger *et al.* 1982).

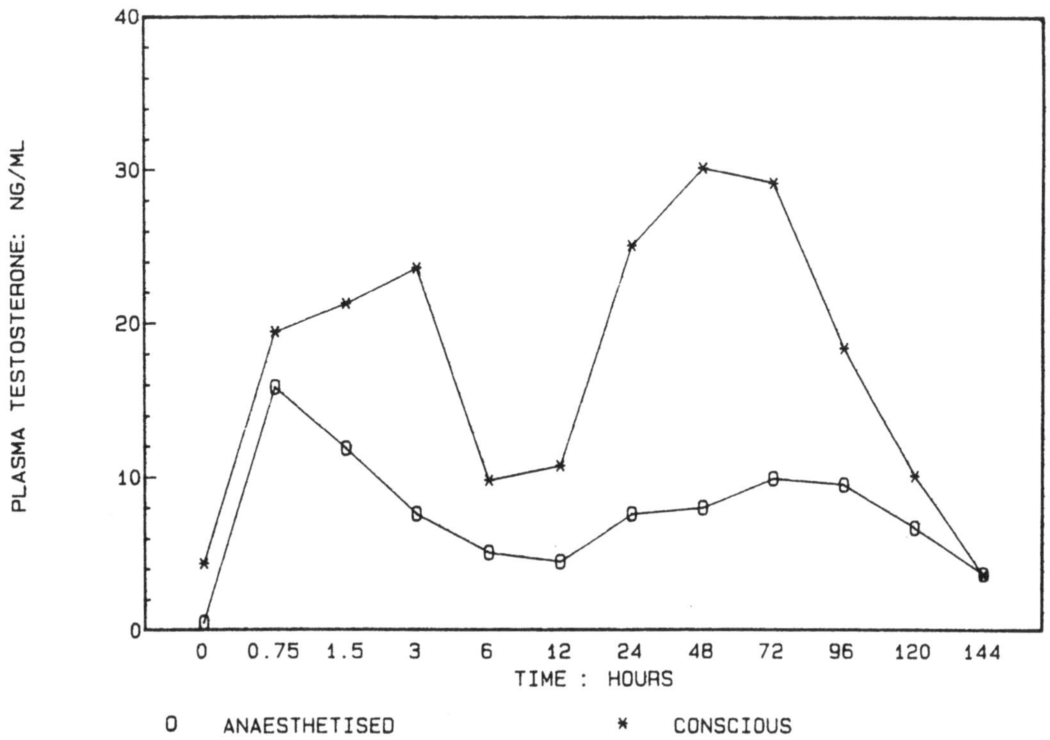


**FIGURE 1. The response of anaesthetized and conscious Merino rams to a single intravenous injection of hCG (15 I.U./kg body weight). (Unpublished data of Y. Chandrasekhar, M.R. Jones, and B.P. Setchell.)**

*Season and time of day* There are wide variations in testosterone levels in sheep depending on the season of the year. Much of this variation is a consequence of differences in the activity of the hypothalamus and pituitary (Lincoln and Short 1980; Lincoln 1981; Haynes and Schanbacher 1983). There are considerable differences between breeds in this seasonal variation, but all breeds studied show increased testosterone levels during the time of the year when the days are shortening. Similar changes in hormone secretion can be induced by changing the light: dark periods artifically (Lincoln 1981) and again there are significant breed differences (D'Occhio *et al.* 1984). While changes in the extent of hormonal stimulation are of paramount importance, the responsiveness of the testis (in terms of testosterone secretion following challenge with a fixed amount of LH) does vary at different times of the year. The testosterone response of the testis is greatest when the testis has reached its maximal size, but it is greater when the testis is developing than when it is fully regressed (Lincoln 1976). Presumably this is a reflection of the amount of stimulation to which the testis has been subjected in the immediately preceding period, but it is not known whether this involves changes in Leydig cell numbers or in enzymatic activity of individual cells, or a combination of both. There is very little evidence for any diurnal variation in plasma testosterone in sheep (Purvis *et al.* 1974; Setchell 1978) in contrast to the situation in many other species (see Setchell 1978).

*Interactions with other animals* There were no differences in plasma testosterone profiles in rams from birth to 21 months of age when reared in isolation, in an all-male group or in a mixed sex group (Illius *et al.* 1976b). However, mature rams kept near ewes were found to have higher plasma testosterone levels, larger testes and greater sexual and aggressive activity (Illius *et al.* 1976a).

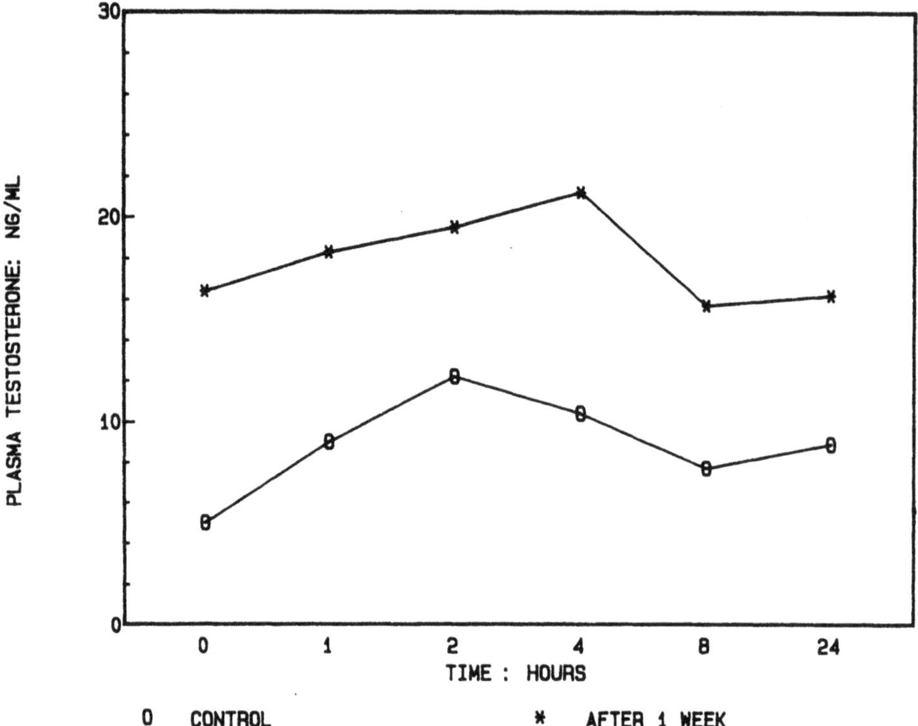


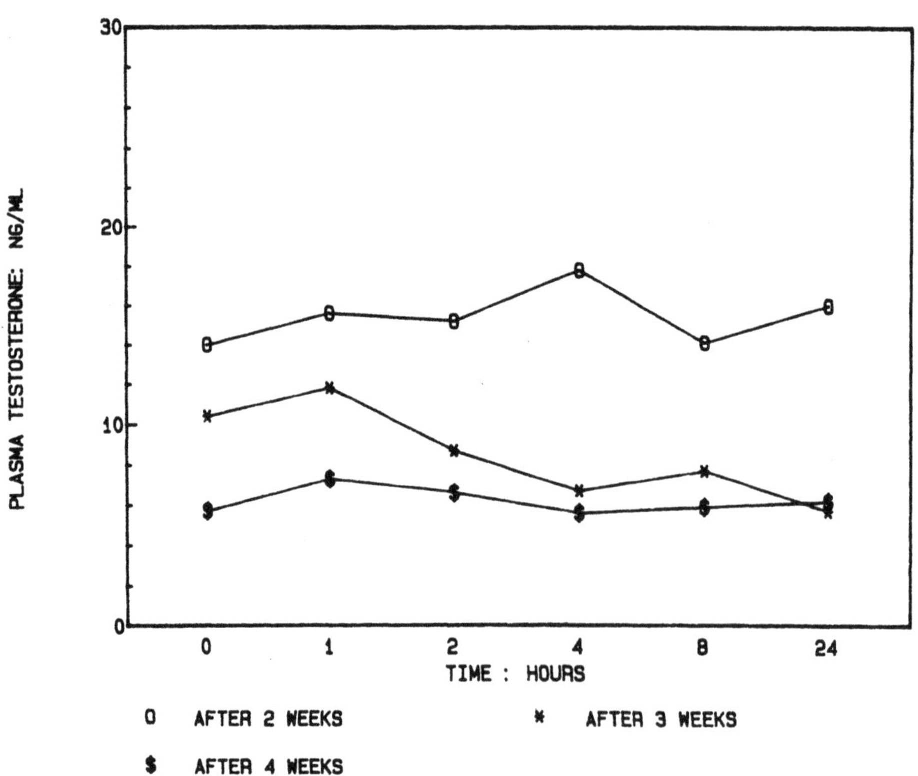


**FIGURE 2.** **The response of Clun Forest rams to a single intravenous injection of hCG (375 I.U. approx. 5 I.U./kg body weight) before, and after 1, 2, 3 and 4 weeks of twice daily injections of the same dose of hormone. (Unpublished data of M.S. Laurie and B.P. Setchell.)**

*Nutrition* Undernutrition of adult rams leading to a substantial loss in body weight was associated with an appreciable fall in testosterone secretion by the testis (Setchell *et al.* 1965). This probably resulted from a decrease in gonadotrophin secretion by the pituitary and the changes in hormone production were much greater than the changes in testicular size.

*Stress* A number of studies in other species have suggested that stress can lead to reductions in testosterone secretion. In some instances, this is associated with falls in LH secretion (Thibier and Rolland 1976; Gray *et al.* 1978; Collu *et al.* 1979) but in other cases, testosterone levels in plasma and testis fall without any change in LH (Charpenet *et al.* 1981). In humans, severe exercise leads to a decrease in testosterone levels in blood, without any appreciable change in LH (Aakvaag *et al.* 1978; Aakvaag *et al.* 1978; Nieschlag 1984). Because of difficulty in quantitating stress, administration of ACTH has been used to simulate stress. Injections of ACTH in men lowers testosterone concentrations in blood with no effect on LH (Schaison *et al.* 1978).

Comparatively little information is available for the ram; some recent studies have shown that repeated injections of ACTH suppressed both LH and testosterone, an effect probably mediated by cortisol released by the ACTH (Mohamed *et al.* 1984).

Exposure of humans to anaesthesia and surgery causes a rise in LH and T during anaesthesia, and then LH returns to normal while T falls to very low levels (Carstensen *et al.* 1973, 1974; Wang *et al.* 1978). In rats, there are conflicting results (Bardin and Petersen 1967; Farriss *et al.* 1969; Fee *et al.* 1980; Frankel and Ryan 1981). In rams, baseline levels of testosterone secretion do not appear to be affected (Setchell *et al.* 1965) although pulses of LH and T are suppressed for several days (Laurie and Setchell 1978).

*Aspermatogenesis* It was usually believed that although tubular function depends on testosterone secretion by the Leydig cells, the function of the latter is largely independent of tubular activity. This may not be true. The spermatogenic function of the rat testis can be damaged by a number of procedures which apparently do not affect the Leydig cells, as judged by normal circulating levels of testosterone. Nevertheless, under these circumstances the Leydig cells often appear hypertrophic and produce normal or greater than normal amounts of testosterone *in vitro* (see Setchell and Galil 1983). In these animals LH is often increased (Main *et al.* 1978) suggesting that the Leydig cells are being "driven" harder to maintain normal steroid levels. In contrast, the *in vivo* testosterone production by the aspermatogenic testes is substantially reduced, largely as a result of decreased testis blood flow (Main and Setchell 1980; Setchell and Galil 1983; Wang *et al.* 1983). Equivalent studies have not yet been done in rams.

### Inhibin

Inhibin is a water-soluble hormone originating in the testis which regulates FSH secretion by the pituitary (see Setchell and Main, 1974; Baker *et al.* 1976; Setchell *et al.* 1977; Franchimont *et al.* 1979). Inhibin-like activity, defined as something which preferentially decreases FSH secretion, was demonstrated in ram rete testis fluid (Setchell and Jacks 1974) and subsequently in testicular lymph (Hudson *et al.* 1979); recently the isolation and structure of a 31 amino-acid peptide from human seminal plasma with inhibin-like activity has been reported (Ramasharma *et al.* 1984).

However the physiological significance of testicular inhibin is still far from clear. Diversion of rete testis fluid in rams had little or no effect on FSH secretion (Walton *et al.* 1978; Blanc *et al.* 1979; Davies *et al.* 1979) and retention of rat rete testis fluid for 24 h inside the testis by ligation of the efferent duct similarly had no effect on plasma FSH (Davies *et al.* 1979) whereas castration or the administration of antiserum against testosterone caused obvious rises within that time (Main *et al.* 1980). However, these experiments may be irrelevant because inhibin, being a peptide, probably should leave the testis in the lymph. Furthermore in sheep made aspermatogenic by heat or efferent duct ligation, there was no evidence for a selective rise in FSH (S.J. Main and B.P. Setchell, unpublished observations), in contrast to the situation in the rat (Main *et al.* 1978, 1979).

Both LH and FSH are increased in cryptorchid rams, although testosterone is normal (Schanbacher and Ford 1977; Blanc *et al.* 1978). This has been interpreted as circumstantial evidence for inhibin in rams, but no attempt has been made to show that there are preferential effects on FSH, and clearly the LH-testosterone system is also perturbed. In any case, cryptorchidism is not a good model for assessing the importance of inhibin because the treatment used to induce the tubular lesion persists throughout the endocrinological studies and it is therefore difficult to exclude direct effects on hormone secretion.

## SPERMATOGENESIS

The production of spermatozoa by the testis is an exceedingly complicated process involving interactions between the somatic Sertoli cells on the one hand and the germinal cells at various stages of development on the other. These stages are: 1. stem cells; 2. diploid, mitotically-dividing spermatogonia usually subdivided into A, Intermediate and B types (sometimes with subscripts 1,2,3 or 4); 3. primary spermatocytes during the long meiotic prophase; 4. short-lived secondary spermatocytes between the two meiotic divisions; 5. haploid spermatids, during their long development from simple-looking round cells to the highly elaborate spermatozoa.

Rates of production of spermatozoa

These can be determined either directly by counting the spermatozoa in frequently-collected ejaculates or in 24 h urine samples or in the fluid collected from catheters in the ductus deferens or in the rete testis. They can also be calculated from a knowledge of the kinetics of the spermatogenic cycle, and a determination of the numbers of spermatid nuclei which resist homogenization. All these techniques give values of about $10 \times 10^9$/day/ram, or $20 \times 10^6$/day/g testis (Amann 1970; Braden *et al.* 1974; Oldham *et al.* 1978); the latter is similar for other mammals studied with the notable exception of the human, in which sperm production per g testis is substantially less (Johnson *et al.* 1980, 1981). Therefore testis size is probably the best index of sperm production in individuals of one species.

Spermatogenic cycles and waves

At any one point in a seminiferous tubule, there are germ cells belonging to four or five different generations, all in close association with the Sertoli cells. From the first mitosis of the spermatogonia to the liberation of the spermatozoa, the development of all cells of one generation is synchronized and linked with development of the other generations. Consequently the cells in one part of the tubule go through a series of stages of a cycle before a similar cell association recurs, but by then each generation is one generation further ahead. From the first synchronized spermatogonial mitoses to free spermatozoa almost four cycles are needed; from the beginning of meiotic prophase to free spermatozoa takes three cycles. The length of each stage of the cycle can be estimated by estimating their frequency in a large number of tubular cross-sections. The length of one cycle can be estimated by autoradiography of testicular sections after injection of tritiated thymidine. In the ram, the length of one cycle is 10.1 days and is constant under all circumstances so far studied (see Courot *et al.* 1970). There is considerable confusion about the classification of different stages of the cycle, with several schemes in current use (see Courot *et al.* 1970; Setchell 1978, 1982).

The length of the spermatogenic cycle and hence the whole spermatogenic process is potentially an important determinant of sperm production but as it does not appear to vary, there seems to be little scope for improvement or selection in this character.

The spermatogenic cycle should not be confused with the wave of spermatogenesis. The cycle involves changes with time in cell associations at one point in one seminiferous tubule. The wave was originally described by Regaud (1901) and subsequently elaborated by Perey *et al.* (1961) and Hochereau (1963), and involves a spatial arrangement along each tubule of successive stages of the cycle. The significance of the wave is still obscure, and it may simply reflect the way spermatogenesis is initiated at puberty.

Spermatogonial renewal

The $A_1$ spermatogonia divide synchronously just after the release of an earlier generation of spermatozoa. The origin of these $A_1$ spermatozonia is a topic which has, and still does arouse considerable dispute. Clermont and Bustos-Obregon (1968)) on the one hand and Oakberg (1971), Huckins (1971) and de Rooij (1973) on the other have proposed two different schemes, one involving an origin from stem cells (sometimes called $A_O$ or $A_S$ spermatogonia) and the other involving maintenance of the population of $A_1$ spermatogonia with daughter cells from one of the later spermatogonial divisions. Hochereau-de Reviers (1981) holds an intermediate view, believing that both schemes may operate. Lok *et al.* (1982) have recently extended to rams earlier studies in that laboratory on mice and hamster (de Rooij 1968, 1973; Oud and de Rooij, 1977) and believe that there is good evidence in all three species that all the $A_1$ spermatogonia originate from stem cells.

The question is an important one to resolve because the regulation of the input of spermatogonia into the system is important in determining quantitatively the level of spermatogenesis established at puberty. Furthermore, once spermatogenesis is established, elimination of degeneration (see next section) could enable the full potential to be realized, but the only way that the potential could be increased would be to increase the number of cells feeding in to the system. This is because the cell associations seem to be immutable and the length of each cycle appears to be constant for each species. There is also some recent evidence (Suzuki and Withers 1978) that in mice, the number of spermatogonial stem cells decreases with age; so the potential for sperm production even under the most favourable conditions may decrease as the animal ages. Further evidence is needed on this point.

Degeneration

During spermatogenesis, there is appreciable degeneration of cells at many stages. In the rat, it has been estimated that only 25% of the maximum possible number of preleptotene spermatocytes are formed from each $A_1$ spermatogonium (Huckins 1978). In the ram, Ortavant (1959a) showed that the number of $A_1$ spermatogonia was reduced by 10% if the animals were exposed to long days, whereas there was a 23% decrease in intermediate (In) spermatogonia, 36% in pachytene spermatocytes and 43% in cells in meiosis, indicating increased degeneration at various steps in spermatogenesis. Degeneration of spermatogonia, pachytene spermatocytes and cells during meiosis is appreciable even in normal rams, although in contrast to other species, degeneration of spermatids is minimal. Pituitary gonadotrophins are probably important in regulating degeneration at several stages (Courot 1971).

The Sertoli cells, the blood testis barrier and the exocrine secretion of the testis

As already stated, the germinal cells develop in close association with the somatic Sertoli cells, which are diploid cells that do not appear to divide after puberty (see Orth 1982). Seasonal changes in testicular weight in Soay rams are not accompanied by any change in the numbers of Sertoli cells (Hochereau-de Reviers *et al.* 1984), although there does appear to be some change in the stallion (Johnsson and Thompson 1983) and the red deer (Hochereau-de Reviers and Lincoln 1978).

The Sertoli cells also form very specialized junctions with each other and these junctions effectively exclude many large water-soluble molecules from the lumen of the seminiferous tubule (Dym and Fawcett 1970). They thus constitute an important part of the blood-testis barrier which allows the composition of the fluid inside the seminiferous tubules and rete testis to be kept different from blood plasma and testicular lymph (see Setchell and Waites 1975; Setchell 1980). In the ram, there is good evidence for restriction of the entry of many substances from blood into rete testis fluid. Lipid soluble molecules enter the tubules more rapidly than hydrophilic compounds, but there appear to be specific carrier systems somewhere in the rat testis for glucose, some amino acids and testosterone (Setchell *et al.* 1978; Setchell *et al.* 1984). Ram rete testis fluid also contains a number of specific peptides which are found in much lower concentrations elsewhere in the body. These include androgen-binding protein, a stimulator of mitosis, inhibin, clusterin and many others (see Fritz 1984). The functions of most of these substances are still being studied.

Factors affecting sperm production

*Hormones* Hypophysectomy stops the growth of testes of impuberal lambs (45 to 60 days of age), accompanied by a 30% reduction in the number of Sertoli cell precursors and a cessation in the increase in the number of gonocytes. LH injections induce an increase in testicular weight and the number of Sertoli cells. Large doses of testosterone had a similar action but FSH alone was without effect (Courot 1967; Hochereau-de Reviers and Courot 1978). However, FSH and LH together had a much greater effect than either alone (Courot 1970). A role for FSH was also indicated from passive immunization of young lambs from 4 to 20 weeks of age with antisera to $\beta$-FSH which retarded testicular growth and the increase in numbers of Sertoli cells and germ cells (Courot *et al.* 1984). Furthermore, in young lambs, hemicastrated at 1 week of age, there was accelerated growth of the remaining testis associated with increased plasma FSH levels and no change in LH or testosterone (Walton *et al.* 1978; Walton *et al.* 1980; Waites *et al.*1983).

Hypophysectomy of adult rams is followed by a sharp fall in testicular size. Treatment with hCG, PMSG or testosterone had little effect on testicular weight, although all treatments maintained normal concentrations of testosterone in rete testis fluid. Differentiation of $A_1$ spermatogonia from $A_0$ was maintained by PMSG or hCG, the transition from intermediate spermatogonia to primary spermatocytes was maintained by PMSG, and meiotic prophase and spermiogenesis were maintained by all 3 treatments although there were some abnormalties in spermatids (Monet-Kuntz *et al.* 1976; Courot *et al.* 1979).

The situation in the ram therefore appears to be different from that in the rat in which spermatogenesis in the adult can be maintained with just testosterone (see Setchell 1978). The adult ram, like the adult monkey (Wickings *et al.* 1980) does appear to have a specific requirement for FSH, probably in the later stage of development of spermatogonia, and possibly also of LH as such (not mediated via testosterone) on the recruitment of stem spermatogonia, with a requirement for testosterone as in all species for meiotic prophase.

There is still no direct evidence for a quantitative control of sperm production in the adult, and certainly no treatment has been shown to increase the rate of sperm production once spermatogenesis is established. Nevertheless, unilateral orchidectomy in adult rams has been reported to cause compensatory hypertrophy of the remaining testis (Hochereau-de Reviers *et al.* 1976) so perhaps the correct treatment has not yet been tested; continued investigation appears warranted.

*Season* The testes of rams of most breeds increase in size as day length begins to shorten and decreases when days are long (Ortavant 1959a,b; Setchell 1978; Lincoln and Short 1980; Lincoln 1981; Haynes and Schanbacher 1983). These effects can be induced experimentally by altering the light: dark ratio and are mediated via the hypothalamus and pituitary, with an important input from the pineal.

*Nutrition* Undernutrition of previously well-fed adult rams which caused losses of about 30 per cent of body weight over 3 months led to some tubular degeneration, a decrease of about 30% in tubular diameter and a reduction in testicular size of about 50% in absolute terms or about 30% fall as a percentage of body weight; epididymal sperm reserves fell by about 60%. However these falls are much less than the 80% decrease in testosterone secretion in the same rams (Setchell *et al.* 1965).

Braden *et al.* (1974) also found that epididymal sperm reserves were reduced to the same extent as testicular weight in 2 year old rams whose testis were reduced to 50% of controls by feeding a low energy low protein diet for 11 weeks at a level which arrested body growth. High energy low protein and low energy high protein diets which produced smaller effects on body weight also had smaller effects on testicular weight and sperm reserves. In a second experiment, these authors also showed that daily sperm production in adult rams was increased by raising the energy intake or both energy and protein intake, whereas increasing just protein intake had no consistent effect.

On the other hand, supplementation of a roughage diet with protein rich lupin grain for 9 weeks led to a 32 per cent increase in body weight and a 67 per cent increase in testicular size; the effects were reversed when the groups were interchanged for a further 9 weeks. At the end of the experiment there were also significant differences in sperm production per gram of testis which amplified the effects on testicular weight (Oldham *et al.* 1978). It will be important to determine whether this effect is due to the high protein content of the lupin grain or the increased energy intake, or some specific factor in the lupins.

Many of the nutritional effects are probably mediated via the hypothalamus and pituitary, and many of the older studies are difficult to interpret because of the lack of endocrinological data.

*Heat* It has been known for many years that the scotal testis is sensitive to high temperature and even to core temperature (see Setchell 1978). The ram appears to be particularly vulnerable in this regard, and a short period of local heating of the testes not only causes death of most of the pachytene spermatocytes, and many young spermatids but also induces abortive divisions of the B-spermatogonia (Waites and Ortavant 1968). The damage to the pachytene spermatocytes is reflected in a fall in the concentration of spermatozoa in rete testis fluid about 20 days after one short period of local heating and a further fall at about 30 days probably due to the effect on B-spermatogonia (Setchell *et al.* 1971). However, the sperm leaving the testis before 20 days after heating are already abnormal, perhaps indicating that spermatids are affected even if they are not killed (Voglmayr *et al.* 1971).

There are important differences in the susceptibility of strains of Merino rams selected for different amounts of skin fold. This appears to be largely due to variations in the ability of the scrotal skin to keep the testes cool under given conditions (Fowler 1968).

Cryptorchidism occurs spontaneously in rams, but it is not as common as in pigs, horses or men. When it does occur, or if the testis is returned surgically to the abdomen, then the testis is much reduced in size and spermatogenesis is arrested (Schanbacher and Ford 1977; Cahoureau *et al.* 1979; Hochereau-de Reviers *et al.* 1979). This appears to be exclusively due to temperature effects because if the cryptorchid pig testis is cooled artificially, it develops normally (Frankenhuis and Wensing 1979).

*Chemicals* A wide range of drugs and other chemicals have been found to have deleterious effects on the testis. These include drugs used for treatment of a variety of bacterial diseases, amoebocides, schistosomicides, nematocides, and grain fumigants as well as cytotoxic drugs and inhibitors of DNA synthesis such as hydroxyurea which could be expected to interfere with spermatogenesis (see Setchell 1978; Davies 1980).

*Immunological aspermatogenesis* The germ cells in the testis from mid-pachytene onwards as well as the spermatozoa possess antigens which are foreign to the body's immunological system. Normally, the blood-testis barrier prevents these cells from stimulating an antibody reaction, and also keeps circulating antibodies from reacting with the germ cells. Antibodies against spermatozoa can be produced by injecting these cells with adjuvant, and under certain conditions, immunological aspermatogenesis can follow (see Mancini 1976; Setchell 1978).

The ram is relatively more resistant to experimentally induced immunological spermatogenesis than other species (M.H. Johnson and B.P. Setchell, unpublished results). The interesting recent finding of spermatozoa in lymph from vessels in the spermatic cords of vasectomized rams (Ball and Setchell 1983) may help to explain the high incidence of sperm antibodies in vasectomised males, and the occurrence of testicular atrophy in some vasectomized rams (see Ball and Setchell 1983).

## EPIDIDYMIS

When the spermatozoa leave the testes, they are immotile and infertile and acquire both these capacities during their passage through the long, convoluted duct of the epididymis.

In the ram, this duct is approximately 50 m long, and it takes between 10 and 14 days for the spermatozoa to pass along it (Ortavant 1959b). This time is not appreciably affected by sexual activity although contractions of the duct can be increased by various pharmacological agents (Pholpramool and Triphron 1984). The capacity for motility is acquired by the spermatozoa in the distal part of the corpus (Fournier-Delpech *et al.* 1977; Dacheux and Paquignon 1980) and the ability to fertilize ova is achieved at about the same time. Sperm from the proximal caput exhibit some motility, but when ewes were inseminated with these sperm some eggs were fertilized but none developed to term (Fournier-Delpech *et al.* 1979).

The histology of the cells lining the epididymal duct of the ram has been described by Nicander (1958). He found that, as in other species there was considerable variation in appearance of the cells along the duct. (Note that Figures 11 and 21 in Nicander's paper were inadvertently transposed, L. Nicander, *pers comm*).

There are variations in blood flow per unit weight of tissue in different areas along the epididymis (Setchell *et al.* 1964) with the areas of highest blood flow corresponding approximately to those parts of the duct with the highest cells.

It is known that in other species, particularly the rat, that there are substantial changes in the ionic and organic composition of the epididymal luminal fluid (see Setchell and Hinton 1981). Less information is available for the ram, except for the composition of rete testis fluid entering the epididymis (Setchell *et al.*

1969; Setchell 1970) and the luminal fluid from the ductus deferens (Scott *et al.* 1963). As in other species, the sodium and chloride content of ductal fluid is much lower, and the potassium, phosphate and protein content much higher than in rete testis fluid. There are also high concentrations of glycerophosphocholine (GPC) in ram ductal fluid (Scott *et al.* 1963) and also of carnitine (Hinton *et al.* 1979). Both these compounds are secreted by the epididymis, GPC being synthesized there and carnitine accumulated from the blood. However, the ram differs from the rat in that the inositol concentration in the ductal fluid is lower than in rete testis fluid, whereas in the rat, it is about 10 times higher (Hinton *et al.* 1980). The significance of this difference is obscure, as indeed is the function of most of the compounds secreted by the epididymis. The protein composition of epididymal luminal fluid is currently of considerable interest, particularly as it seems likely that some of the changes in protein constitution of the spermatozoa may be achieved by interactions with the fluid surrounding them. A number of specific proteins are synthesised by the ram epididymis, under the influence of androgens (Jones *et al.* 1982) and some are secreted *in vitro* by some regions (Carabott and Brooks 1984). In contrast to rodents (see Brooks 1983), the epididymis-specific proteins are only a small ($< 5\%$) part of the total secreted. The immature ram epididymis secretes a much larger number and amount of protein, but lacks the regional pattern of secretion. Administration of testosterone or DHT to immature ram lambs results in precocious growth of the epididymis, especially of the cauda, but without the appearance of the regional pattern of protein secretion. Likewise, as the lambs age, testosterone levels in peripheral blood rise to adult values at between 20 and 24 weeks, sperm appear in the epididymis at about 28 weeks, but the regional pattern of protein secretion does not appear until about 35 weeks of age. The factor or factors responsible have not yet been identified (Holland and Brooks 1984).

A number of proteins of uncertain specificity have been identified in rete testis fluid and ductal fluid (Voglmayr *et al.* 1980) and in epididymal luminal fluid from 10 sites along the epididymis (Dacheux and Voglmayr 1983) and some of these also appear on the spermatozoa. Iodinated rete testis fluid proteins bind to testicular spermatozoa *in vitro* in a relatively non-specific fashion and most of the adherent protein can be removed by treatment with tryspin. In contrast, proteins from caudal epididymal plasma bind selectively and a substantial proportion rapidly becomes inaccessible to subsequent treatment with tryspin (Voglmayr *et al.* 1980; Voglmayr *et al.* 1982). Epididymal proteins and glycoproteins labelled with colloidal gold also bind specifically to the plasma membrane of epididymal spermatozoa, and the part of the spermatozoa to which binding occurs changes as the spermotozoa move through the epididymis (Courtens *et al.* 1982).

However, no-one has yet demonstrated a function for any of these proteins, and it will be important to try in the near future to determine whether or not there are specific effects of any of the proteins under investigation.

## REFERENCES


Aakvaag, A., Sand, T., Opstad, P.K. and Fonnum, F., 1978. *Eur. J. Appl. Physiol., 39*, 283-291.

Aakvaag, A., Bentdal, O., Quigstad, K., Walstad, P., Ronningen, H. and Fonnum, F., 1978. *Int. J. Androl.*, 23-31.

Amann, R., 1970, *In* Johnson, A.D., Gomes, W.R. and Vandemark, N.L. (eds) *The Testis Vol 1*, Academic Press, New York, 433-472.

Amann, R.P., Nett, T.M. and Niswender, G.D., 1978. *J. Anim Sci., 47*, 1307-1313.

Baker, H.W.G., Bremner, W.J., Burger, H.G., de Kretser, D.M., Dulmanis, A., Eddie, L.W., Hudson, B., Keogh, E.J., Lee, V.W.K. and Rennie, G.C., 1976. *Recent Prog. Horm. Res., 32*, 429-476.

Ball, R.Y. and Setchell, B.P., 1983. *J. Reprod. Fert., 68*, 145-153.

Bardin, C.W. and Petersen, R.E., 1967. *Endocrinology, 80*, 38-44.

Barenton, B., 1981. *J. Endocr, 88*, 451-454.

Barenton, B., Hochereau-de-Reviers, M.T., Perreau, C. and Poirier, J.C., 1982. *Reprod. Nutr. Develop., 22*, 621-630.

Barenton, B. and Pelletier, J., 1980. *Biol. Reprod. 22.*, 781-790.

Blanc, M.R., Dacheux, J.L., Cahoreau, C., Courot, M., Hochereau-de-Reviers, M.T., Lacroix, A. and Pisselet, C., 1979. *Ann. Biol. Biophys., 19*, 1027-1032.

Braden, A.W.H., Turnbull, K.E., Mattner, P.E. and Moule, G.R., 1974. *Aust. J. Biol. Sci., 27*, 67-73.

Brooks, D.E., 1983. *Aust. J. Biol. Sci., 36*, 205-221.

Cahoreau, C., Blanc, M.R., Dacheux, J.L., Pisselet, C. and Courot, M., 1979. *J. Reprod. Fert., Suppl. 26*, 97-116.

Carabott, M.J.J. and Brooks, D.E., 1984. *Proc. Aust. Soc. Reprod. Biol. 16*, 87.

Carstensen, H., Amer, B., Amer, I. and Wide L., 1973. *J. Ster. Biochem., 4*, 45-55.

Carstensen, H., Amer, I., Wide, L. and Amer, B., 1973. *J. Ster. Biochem., 4*, 605-611.

Chandrasekhar, Y., Holland, M.K., D'Occhio, M.J. and Setchell, B.P., 1984. *J. Endocrinol.* (in press).

Charpenet, G., Tache, Y., Forest, M.G., Haour, F., Saez, J.M., Bernier, M., Ducharme, J.R. and Collu, R., 1981. *Endocrinology, 109*, 1254-1258.

Clermont, Y. and Bustos-Obregon, E., 1968. *Am. J. Anat., 122*, 237-248.

Collu, R., Tache, Y. and Ducharme, J.R., 1979. *J. Ster. Biochem., 11*, 989-1000.
Courot, M., 1967. *J. Reprod. Fert., Suppl. 2*, 89-101.
Courot, M., 1970. *In* Rosemberg, E. and Paulsen, C.A. (eds) *The Human Testis*, Plemum Press, New York. 355-364.
Courot, M., 1971. These de doctorat d'état es-sciences naturelles. Université Paris VI.
Courot, M., Hochereau-de-Reviers, M.T., Kuntz, C.M., Locatelli, A., Pisselet, C., Blanc, M.R. and Dacheux, J.L., 1979. *J. Reprod. Fert., Suppl. 26*, 165-173.
Courot, M., Hochereau-de-Reviers, M.T. and Ortavant, R., 1970, *In* Johnson, A.D. Gomes, W.R. and Vandemark, N.L. (eds) *The Testis Vol 1*, Academic Press, New York, 399-411.
Courot, M., Pisselet, C., Kilgour, R.J., Sairam, M.R. and Dubois, M.P., 1984. *Proc. 3rd European Workshop on the Testis*. p. 19.
Courtens, J.L., Rozinek, J. and Fournier-Delpech, S., 1982. *Andrologia, 14*, 509-514.
Cowie, A.T., Lascelles, A.K. and Wallace, J.C., 1964. *J., Physiol., 171*, 176-187.
Dacheux, J.L. and Paquignon, M., 1980. *Reprod. Nutr. Develop., 20*, 1085-1099.
Davies, A.G., 1980. *Effects of Hormones, Drugs and Chemicals on Testicular Function Vol. 1.* Churchill Livingston 276 pp.
Davies, R.V., Main, S.J. and Setchell, B.P., 1979. *J. Reprod. Fert., Suppl., 26*, 87-95.
D'Occhio, M.J., Schanbacher, B.D. and Kinder, J.E., 1982. *Endocrinology, 110*, 1547-1554.
D'Occhio, M.J., Schanbacher, B.D. and Kinder, J.E., 1984. *Biol. Reprod.* (in press).
Dym, M. and Fawcett, D.W., 1970. *Biol. Reprod., 3*, 308-326.
Fariss, B.L., Hurley, T.J., Hane, S. and Forhsam, P.H., 1969. *Endocrinology, 84*, 940-942.
Fournier-Delpech, S., Colas, G., Courot, M. and Orvant, R., 1977. *Ann. Biol. Anim. Bioch. Biophys., 17*, 897-990.
Fournier-Delpech, S., Colas, G., Courot, M., Ortavant, R. and Brice, G., 1979. *Ann. Biol. Anim. Bioch. Biophys., 19*, 597-605.
Fowler, D.G., 1968. *Aust. J. Exp. Agric. Anim. Husb., 8*, 142-150.
Franchimont, P., Verstraelen-Proyard, J., Hazee-Hagelstein, M.T., Renard, Ch., Demoulin, A., Bourguignon, J.P. and Hustin, J., 1979. *Vitamins and Hormones, 37*, 243-302.
Frankel, A.I. and Ryan, E.L., 1981. *Biol. Reprod., 24*, 491-495.
Frankenhuis, M.T. and Wensing, C.J.G., 1979. *Fertil. Steril., 31*, 428-433.
Free, M.J., Jaffe, R.A. and Cheng, H.C., 1980. *J. Androl., 1*, 182-196.
Fritz, I.B., 1984. *Proc. 3rd European Workshop on the Testis* (in press).
Galil, K.A.A., Laurie, M.S., Main, S.J. and Setchell, B.P., 1981. *J. Physiol., 319*, 17P.
Galloway, D.B., Cotta, Y., Pelletier, J. and Terqui, M. 1974. *Acta Endocrinologica, 77*, 1-9.
Galloway, D.B. and Pelletier, J. 1975., *J. Endocr., 64*, 7-16.
Garnier, F. and Saez, J.M., 1980. *Biol. Reprod., 22*, 832-836.
Gower, D.B., 1972. *J. Steroid Biochem., 3*, 45-103.
Gray, G.D., Smith, E.R., Damassa, D.A., Ehrenkranz, J.R.L. and Davidson, J.M., 1978. *Neuroendocrinology, 25*, 247-256.
Haynes, N.B. and Schanbacher, B.D., 1983. In W. Haresign (ed) *Sheep Production* Butterworths, London. 431-451.
Hinton, B.T., Snoswell, A.M. and Setchell, B.P., 1979. *J. Reprod. Fert., 56*, 105-111.
Hinton, B.T., White, R.W. and Setchell, B.P., 1980. *J. Reprod. Fert., 58*, 395-399.
Hochereau, M.T., 1963. *Ann. Biol. Anim., Bioch. Biophys., 3*, 5-20.
Hochereau-de-Reviers, M.T., 1981. *In* McKerns, K.W. (ed) *Reproductive Processes and Contraception* Plenum. 307-331.
Hochereau-de-Reviers, M.T., Blance, M.R., Cahoreau, C., Courot, M., Dacheux, J.L. and Pisselet, C., 1979. *Ann. Biol. Anim. Bioch. Biophys., 19*, 1141-1146.
Hochereau-de-Reviers, M.T. and Courot, M., 1978. *Ann. Biol. Anim. Bioch. Biophys., 18*, 573-583.
Hochereau-de-Reviers, M.T. and Lincoln, G.A., 1978. *J. Reprod. Fert., 54*, 209-213.
Hochereau-de-Reviers, M.T., Loir, M. and Pelletier, J., 1976. *J. Reprod. Fert., 46*, 203-209.
Hochereau-de-Reviers, M.T. Ortavant, R. and Courot, M., 1976. *Prog. Reprod. Biol., 1*, 13-19.
Hochereau-de-Reviers, M.T., Perreau, C. and Lincoln, G., 1984. *Proc. 3rd. European Workshop on the Testis* p. 74.
Holland, M.K. and Brooks, D.E., 1984. *Proc. Aust. Soc. Reprod. Biol. 16*, 86.
Huckins, C., 1971. *Anat. Rec., 169*, 533-558.
Huckins, C., 1978. *Anat. Rec., 190*, 905-926.
Hudson, B., Baker, H.W.G., Eddie, L.W., Higginson, R.E., Burger, H.G., de Kretser, D.M., Dobos, M. and Lee, V.W.K., 1979. *J. Reprod. Fert., Suppl. 26*, 17-29.
Illius, A.W., Haynes, N.B. and Lamming, G.E., 1976a. *J. Reprod. Fert., 48*, 25-32.
Illius, A.W., Haynes, N.B., Purvis, K. and Lamming, G.E., 1976b. *J. Reprod. Fert., 48*, 17-24.

Johnson, L., Petty, C.S. and Neaves, W.B., 1980. *Biol. Reprod, 22*, 1233-1243.
Johnson, L., Petty, C.S. and Neaves, W.B., 1981. *Biol. Reprod., 25*, 217-226.
Johnson, L. and Thomson, D.L., 1983. *Biol. Reprod., 29*, 777-789.
Jones, R., Fournier-Delpech, S. and Willadsen, S.A., 1982. *Reprod. Nutr. Develop., 22*, 495-504.
Karpe, B., Fredricsson, B., Svennson, J. and Ritzen, E.M., 1982. *Int. J. Androl., 5*, 549-556.
Katongole, C.B., Naftolin, F. and Short, R.V., 1974. *J. Endocr., 60*, 101-106.
Laurie, M.S. and Setchell, B.P., 1979. *J. Physiol., 298*, 10P.
Lincoln, G.A., 1976. *J. Endocr., 69*, 213-226.
Lincoln, G.A., 1978. *J. Reprod. Fert., 53*, 31-37.
Lincoln, G.A., 1979. *J. Endocr., 80*, 133-140.
Lincoln, G.A., 1981. *In* Burger H. and de Kretser D.M. (eds) *The Testis*, Raven Press, New York, 255-302.
Lincoln, G.A. and Short, R.V., 1980. *Rec. Progr. Horm. Res., 36*, 1-52.
Lindner, H.R., 1961. *J. Endocr., 23*, 171-178.
Lindner, H.R., 1963. *J. Endocr., 25*, 483-494.
Lindner, H.R., 1967. *Proc. Int. 2nd Cong. Horm. Ster. Exc. Med. Int. Con. series No. 132*, 821-827.
Lok, D., Weenk, D. and de Rooij, D.G., 1982. *Anat. Rec., 203*, 83-99.
Main, S.J., Davies, R.V. and Setchell, B.P., 1978. *J. Endocr., 79*, 255-270.
Main, S.J., Davies, R.V. and Setchell, B.P., 1979. *J. Reprod. Fert., Suppl., 26*, 3-14.
Main, S.J., Davies, R.V. and Setchell, B.P., 1980. *J. Endocr., 86*, 135-146.
Main, S.J., and Setchell, B.P., 1980. *J. Endocr., 87*, 445-454.
Mancini, R.E., 1976. *Immunologic Aspects of Testicular Function*, Springer-Verlag, Berlin Heidelberg, New York. 114pp.
Martikainen, H., Huhtaniemi, I., Lukkarinen, O. and Vihko, R., 1982. *J. Ster. Bioch., 16*, 187-191.
Martikainen, H., Huhtaniemi, I. and Vihko, R., 1980. *Clin. Endocr., 13*, 157-166.
Mohamed, F.H.A., Cox, J.E. and Moonan, V., 1984. *J. Endocr.* (in press).
Monet-Kuntz, C., Terqui, M., Locatelli, A., Hochereau-de-Reviers, M.T. and Courot, M., 1976. *C.r. hebd. Seanc. Acad. Sci. Paris D 283*, 1763-1766.
Nicander, L., 1957. *Acta. Morph. Neerlando-Scand., 1*, 337-362.
Nieschlag, E. 1984. *Proc. 3rd European Workshop on the Testis* (in press).
Oakberg, E.F., 1971. *Anat. Rec., 169*, 515-532.
Oldham, C.M., Adams, N.R., Gheradi, P.B., Lindsay, D.R. and MacIntosh, J.B., 1978. *Aust. J. Agric. Res., 29*, 173-179.
Ortavant, R., 1959a. *Ann. Zootech, 8*, 183-244 and 271-321.
Ortavant, R., 1959b. *In* Cole H.H. and Cupps, P.T. (eds) *Reproduction in Domestic Animals* 1st. ed. Academic Press, New York. 1-50
Orth, J.M., 1982. *Anat. Rec., 203*, 485-492.
Oud. J.L. and de Rooij, D.G., 1977. *Anat. Rec., 187*, 113-124.
Padron, B.S., Wischusen, J., Hudson, B., Burger, H.G. and de Kretser, D.M., 1980. *J. Clin. Endocr. Metab., 50*, 1100-1103.
Perey, B., Clermont, Y. and Leblond, C.P., 1961. *Am. J. Anat., 108*, 47-77.
Pholpramool, C. and Triphrom, N., 1984. *J. Reprod. Fert., 71*, 181-188.
Purvis, K., Illius, A.W. and Haynes, N.B., 1974. *J. Endocr., 61*, 241-253.
Ramasharma, K., Sairam, M.R., Seidah, N.G., Chretien, M., Manjunath, P., Schiller, P.W., Yamashiro, D. and Li, C.H., 1984. *Science, 223*, 1199-1202.
Regaud, C., 1901. *Arch. Anat. Microscop. Morphol. Emptl. 4*, 101-155 and 231-380.
Risbridger, G.P., Hodgson, Y.M. and de Kretser, D.M., 1981. *In* H. Burger and D. de Kretser (eds) *The Testis* Raven Press, New York. 195-211.
Risbridger, G.P., Robertson, D.M. and de Kretser, D.M., 1982. *Endocrinology, 110*, 138-145.
de Rooij, D.G., 1968. *Z. Zellforsch, 89*, 133-136.
de Rooij, D.G., 1973. *Cell Tissue Kinet., 6*, 281-287.
Saez, J.M. and Forest, M.G., 1979. *J. Clin. Endocr. Metab.*, 278-283.
Sanford, L.M., Winter, J.S.D., Plamer, W.M. and Howland, B.E., 1974. *Endocrinology, 95*, 627-631.
Schaison, G., Durand, F. and Mowszowicz, I., 1978. *Acta Endocr., 89*, 126-131.
Schanbacher, B.D., 1980. *J. Reprod. Fert., 59*, 155-154.
Schanbacher, B.D. and Ford, J.J., 1977. *Endocrinology, 100*, 387-393.
Scott, T.W., Wales, R.G., Wallace, J.C. and White, I.G., 1963. *J. Reprod. Fert., 6*, 49-59.
Setchell, B.P., 1970. *In* Johnson, A.D., Gomes, W.R. and Vandemark, N.L. (eds). *The Testis Vol. I*, Academic Press, New York, 101-239.
Setchell, B.P., 1978. *The Mammalian Testis* Elek Books, London and Cornell, University Press, Ithaca, 452pp.
Setchell, B.P., 1980. *J. Androl, 1*, 3-10.

Setchell, B.P., 1982. *In* Austin, C.R. and Short, R.V. (eds) *Reproduction in Mammals : Germ cells and Fertilization*, Cambridge University Press, Cambridge, 63-101.
Setchell, B.P., Bustamante, J.C. and Niemi, M., 1984. *In* Courtice, F.C., Garlick, D.G. and Perry, M.A. (eds), *Progress in Microcirculation Research II*, University of New South Wales, Sydney. (in press)
Setchell, B.P., Davies, R.V. and Main, S.J., 1977. *In* Johnson, A.D. and Gomes, W.R. (eds) *The Testis Vol. 4*, Academic Press, New York. 189-238.
Setchell, B.P. and Galil, K.A.A., 1983. *Aust. J. Biol. Sci., 36*, 285-293.
Setchell, B.P. and Hinton, B.T., 1981. *Prog. Reprod. Biol., 8*, 58-66.
Setchell, B.P. and Jacks, F., 1974. *J. Endocr., 62*, 675-676.
Setchell, B.P., Laurie, M.S., Flint, A.P.F. and Heap, R.B., 1983. *J. Endocr., 96*, 127-136.
Setchell, B.P., Laurie, M.S., Main, S.J. and Goats, G.C., 1978. *Int. J. Androl., Suppl. 2*, 506-512.
Setchell, B.P. and Main, S.J., 1974. *Bibliog. Reprod., 24*, 245-252 and 361-367.
Setchell, B.P., Scott, T.W., Voglmayr, J.K. and Waites, G.M.H., 1969. *Biol. Reprod., Suppl. 1*, 40-66.
Setchell, B.P., Voglmayr, J.K. and Hinks, N.T., 1971. *J. Reprod. Fert., 24*, 81-89.
Setchell, B.P., Voglmayr, J.K. and Waites, G.M.H., 1969. *J. Physiol., 200*, 73-85.
Setchell, B.P. and Waites, G.M.H., 1975. *In* Hamilton, D.W. and Greep, R.O. (eds) *Handbook of Physiology, Section 7 Endocrinology, Vol. V. Male Reproductive System*. American Physiological Society, Washington D.C. 143-172.
Setchell, B.P., Waites, G.M.H. and Lindner, H.R., 1965. *J. Reprod. Fert., 9*, 149-162.
Setchell, B.P., Waites, G.M.H. and Till, A.R., 1964. *Nature, 23*, 317-318.
Suzuki, N. and Withers, H.R., 1978. *Science, 202*, 1214-1215.
Thibier, M. and Rolland, O., 1976. *Theriogenology,* 5, 53-60.
Voglmayr, J.K., Fairbanks, G., Jackowitz, M.A. and Colella, J.R., 1980. *Biol. Reprod., 22*, 655-667.
Voglmayr, J.K., Fairbanks, G., Vespa., D.B. and Colella, J.R., 1982. *Biol. Reprod., 26*, 483-500.
Voglmayr, J.K., Setchell, B.P. and White, I.G., 1971. *J. Reprod. Fert., 24*, 71-80.
Waites, G.M.H. and Ortavant, R., 1968. *Ann. Biol. Anim. Bioch. Biophys, 8*, 323-331.
Waites, G.M.H., Wenstrom, J.C., Crabo, B.G. and Hamilton, D.W., 1983. *Endocrinology, 112*, 2159-2167.
Walton, J.S., Evins, J.D., Hillard, M.A. and Waites, G.M.H., 1980. *J. Endocr., 84*, 141-152.
Walton, J.S., Evins, J.D. and Waites, G.M.H., 1978. *J. Endocr., 77*, 75-84.
Wang, C., Chan, V. and Yeung, R.T.T., 1978. *J. Clin. Endocr., 9*, 255-266.
Wang, J., Galil, K.A.A. and Setchell, B.P., 1983. *J. Endocr., 98*, 35-46.
Wickings, E.J., Usadel, K.H., Dathe, G. and Nieschlag, E., 1980. *Acta Endocr., 95*, 117-128.
Wilson, P.R. and Lapwood, K.R., 1978. *Theriogenology, 9*, 417-428.

# THE EFFECTS OF SPECIFIC NEURECTOMIES AND CREMASTER MUSCLE SECTIONING ON SEMEN CHARACTERISTICS AND SCROTAL THERMOREGULATORY RESPONSES OF RAMS

I.C.A. Martin, *Department of Veterinary Physiology, University of Sydney N.S.W., 2006, Australia.*
K.R. Lapwood, *Department of Physiology and Anatomy, Massey University, Palmerston North, New Zealand.*
R.L. Kitchell, *Department of Anatomy, School of Veterinary Medicine, University of California, Davis, 95616, U.S.A.*

*Summary* Neurectomy of supply to scrotal skin, tunica vaginalis or testicular parenchyma or transection of cremaster muscles was performed bilaterally on adult NZ Romney rams kept at a controlled ambient temperature of 15°C and in lighting conditions simulating autumn. The rams had been trained for semen collection using an artificial vagina and were randomised into four pairs of rams for each of three classes of neurectomy or transection of cremaster muscles, and three rams were allocated as controls.

None of the treatments had detectable effects in the post-operative period of six weeks on capacity to ejaculate, semen characteristics or testicular size. Acute tests of responses to raised (25°C) or lowered (5°C) ambient temperatures showed that all treated rams retained reflexes associated with thermoregulation of the testes. After section of the cremaster muscles, activity in the dartos muscle alone altered the size of the scrotum moving the testes toward, or away from, the abdomen.

## INTRODUCTION

In the course of studies of the responses of rams to local heating of the scrotum and its contents, Waites (1962) and Waites and Voglmayr (1963) denervated scrotal skin and tunica vaginalis by sectioning the genitofemoral nerve and the scrotal branches of the pudendal nerve. In one ram they removed the lumbar sympathetic chains L2 — L5 to interrupt the efferent innervation of the sweat glands of the scrotal skin as well sectioning the genitofemoral nerve. However, semen was not collected from these rams to ascertain whether these neurectomies also affected spermatogenesis or maturation of spermatozoa in the epididymis.

Both the cremaster and dartos muscles function to position the testes in the scrotum in response to a range of stimuli and of these the dartos appears to be the principal modifier of the extensibility of the scrotal skin and, therefore, is important in the thermoregulation of the testis (Setchell 1978).

In this paper we report on the responses of rams to neurectomy of the scrotal skin, tunica vaginalis or testicular parenchyma or transection of cremaster muscles. Capacity to ejaculate, semen characteristics and acute responses to hot and cold environments related to thermoregulation of the testes were all observed in a post-operative period of six weeks.

## MATERIALS AND METHODS

Eleven Romney rams, aged about 20 months, were housed in a light and temperature controlled room in the third week of April 1982. Temperature was set at 15°C for both the preparation and experimental periods but lighting was controlled with an initial daily photoperiod of 16 hr for four weeks, reduction to 14 hr for two weeks, and thereafter the daily photoperiod was reduced by 30 min at weekly intervals. This pattern of lighting simulated summer and autumn lighting conditions (Barrell and Lapwood 1979), and when surgery was performed the period of illumination was 8½ hr per day. From then until the end of the experiment the daily photoperiod was eight hours.

Rams were trained for semen collection by artificial vagina and when assessment of sperm output rates were made, three semen collections per day were made on four successive days after an initial day when semen was collected as often as rams would serve during a period of two hours.

Each nerve or muscle transection described below was performed on two rams.

I. To produce scrotal skin denervation, three branches of the pudendal nerve (the superficial perineal, scrotal and cranial preputial, and scrotal branches from the dorsal nerve of the penis) were located and sectioned. Great care was taken to avoid damage to the main trunk of the dorsal nerve of the penis.

II. The cranial branch of the genitofemoral nerve as located on the lateral surface of the cremaster muscle through an incision in the inguinal region. This branch was traced to its origin from the main nerve and the genitofemoral nerve was transected proximal to both its cranial and caudal branches.

III. The cremaster muscle was located through an inguinal incision similar to that used for II and transected after its origin was dissected away from the inguinal ligament. The cranial and caudal branches of the genitofemoral nerve were left intact.

IV. To section the testicular nerves, the testicular artery and vein were located through a lateral abdominal incision as they ran across the lateral edge of the pelvis in a fold of peritoneum. An incision was made in the peritoneum of the fold with iridectomy scissors and all nerves on, or around the vessels were sectioned but lymphatic vessels in the region were avoided.

One of the control rams was subjected to bilateral laparotomy as above (IV) but vessels and nerves were not touched.

Photographs were taken of the scrota of rams standing unrestrained for 10 min in air temperatures of 5°, 15° and 25°C, both before and after surgery so that changes in scrotal dimensions could be measured. Rams in treatments I, II and III were anaesthetised and responses in nerves supplying the cutaneous fields of the perineal, inguinal and scrotal regions were recorded electrophysiologically and the denervated areas mapped (Kitchell *et al.* 1980). Electromyographs were recorded from the distal portion of the cremaster muscle during stimulation of its nerve supply for the two rams in treatment III. Testes, epididymides, spermatic vessels and nerves and cremaster muscles were obtained within minutes of death of the rams induced by an overdose of barbiturate.

## RESULTS

Post-operative recovery was uneventful and semen collection was resumed 10 days after surgery. Table 1 contains a summary of the characteristics of semen measured 3 weeks before, and 6 weeks after, surgery. The means presented are for semen collected in 3 ejaculates from each ram on the fourth day of semen collection after an initial day of frequent collection (i.e. after depletion of sperm reserves). The weight of the testes was obtained *post mortem* in the seventh week after surgery and mean weights for each treatment group are also shown in Table 1.

**TABLE 1. Total volume of, and sperm contained in, three ejaculates of semen collected in one day from each ram three weeks before and six weeks after surgery. The testicular weight was obtained at autopsy seven weeks after surgery.**

| Treatment | Semen Characteristics | | | | |
|---|---|---|---|---|---|
| | Pre-Surgery | | Post-Surgery | | Weight of testes (g) |
| | Volume (ml) | Spermatozoa (millions) | Volume (ml) | Spermatozoa (millions) | |
| I. Denervation of scrotal skin (n = 2) | 0.80 | 2277 | 1.20 | 4397 | 264.2 |
| II. Section of genitofemoral nerves (n = 2) | 1.30 | 3025 | 1.11 | 3909 | 275.3 |
| III. Section of cremaster muscles (n = 2) | 1.05 | 2664 | 0.98 | 3953 | 245.0 |
| IV. Section of testicular nerves (n = 2) | 1.50 | 2867 | 1.45 | 5278 | 220.6 |
| Controls (n = 3) | 1.42 | 2754 | 1.50 | 5519 | 237.8 |
| Overall means | 1.25 | 2721 | 1.27 | 4693 | 247.6 |

The volume of semen obtained did not differ significantly between treatment groups nor were there any significant changes in semen volume from pre- to post-surgery. However, the number of spermatozoa increased significantly ($P < 0.001$) from pre- to post-surgery; there was no significant interaction between treatment and time of sampling. Testicular weights were not significantly affected by treatment. These measurements corresponded with repeated observations made *in vivo* throughout the experiment using palpation and measurement of scrotal circumference when no changes in testicular size were detected.

Testing for sensitivity of the scrotal skin by light touch or pinching the skin with forceps indicated that denervation of the body of the scrotum had been achieved with method I. Electrophysiological records confirmed this except for one ram where a narrow field from extension of the genitofemoral innervation was detected on one side of the scrotum.

The two rams subjected to transection of the cremaster muscles (III) reacted to stimulation of scrotal skin by movement of kicking but were incapable of elevating their testes which was the response of control animals tested in the same way. Electromyograph responses were recorded in part of one of the cremaster muscles below its point of section but the contractile force generated was insufficient to elevate the testis.

Electrophysiological mapping of the skin of the inguinal region and around the neck of the scrotum indicated that the genitofemoral nerves had been sectioned in II, so that it was assumed that the tunica vaginalis was also denervated.

Histological examination of the spermatic vessels and nerves taken from IV showed degenerative changes in all major nerve bundles in this region. Some small groups of nerves closely applied to, or embedded in, the vessels were intact. There was no evidence that treatment IV had disrupted blood flow through the testes.

The rams with denervated scrotal skin showed relevant responses to elevated or reduced ambient temperatures by raising or lowering their testes. Although the scrotal skin is richly innervated with themoreceptors normally giving specific input to the reflex, it appeared that other skin surfaces supply sensory input to the testicular thermoregulatory reflex. In the same conditions, the rams without functioning cremaster muscles were still able to position their testes in response to hot or cold. At 5°C strong ridges formed in the scrotal skin, indicating activity of the dartos muscles.

## DISCUSSION

Barrell and Lapwood (1979) showed that the NZ Romney responded to decreasing daily illumination with an increase in semen production. As preparation of rams for this experiment began in mid-autumn, it was decided to subject them initially to a long (summer) photoperiod followed by tapering of the daily illumination to simulate light conditions of autumn as part of the initial conditioning in the controlled environment. It was expected that this would bring the rams to optimal condition in mating activity and sperm production for the post-operative period. The manipulation of the photoperiod is the most likely explanation for the increase in daily sperm output observed in the course of the experiment.

Reflexes involved in ejaculation of semen were not impaired by any of the surgical treatments. Additionally, the neurectomies did not adversely affect daily sperm outputs. As these rams tolerated these procedures so well, multiple neurectomies on individual rams could be considered and future studies could include observation of rams at pasture. Although there was no evidence of gross alteration in spermatogenesis, there may be neuroendocrine responses to environmental changes or to associating and mating with ewes, which would only be displayed by experiments of longer duration and, possibly, in field conditions.

Histological examination of the testicular vessels and their surroundings taken from the site of the neurectomies indicated that some nerve fibres were still intact. However, they were embedded in the vessel walls and are thought to have been supplying this region of the vessels whereas the nerve bundles transected were most likely to have included the innervation of the testicular parenchyma. Future attempts at this denervation should include transection of the vessels, dissection of an area around them and then reanastomosis using microsurgical techniques. The gross changes following neurectomy of the testis described by Setchell (1978) did not occur but the work he reviewed consisted of reports of more extensive removals of parts of the lower sympathetic chain than were attempted by us and many of these would also impair the capacity to ejaculate as well as causing reduction in control of blood flow in the region.

The post-operative period of study was not long enough to detect from daily sperm output whether spermatogonial division for replenishment of the basal cells and maturation toward 'committed' cells for spermatogenesis was affected by these surgical procedures. Preliminary studies of the histology of testes obtained at *post mortem* examination indicates that spermatogenesis was continuing normally but quantitative studies will be made on these tissues to check on this point.

## ACKNOWLEDGMENTS

ICAM and RLK thank the members of the Department of Physiology and Anatomy, Massey University, for providing help and facilities to undertake this work.

## REFERENCES

Barrell, G.K. and Lapwood, K.R., 1979. *Anim. Reprod. Sci.*, *1*, 213-228.
Kitchell, R.L., Whalen, L.R., Bailey, C.S. and Lohse, C.L., 1980. *Am. J. Vet. Res.*, *41*, 61-76.
Setchell, B.P., 1978. *The Mammalian Testis*, Paul Elek, London, pages 58, 86 and 98.
Waites, G.M.H., 1962. *Quart. J. Exp. Physiol.*, *47*, 314-323.
Waites, G.M.H. and Voglmayr, J.K., 1963. *Aust. J. Agric. Res.*, *14*, 839-851.

# EFFECT OF LUPIN FEEDING ON LH, TESTOSTERONE AND TESTES

A.J. Ritar, N.R. Adams and M.R. Sanders, *Division of Animal Production, CSIRO, Private Bag, P.O., W.A. 6014.*

*Summary* The secretion of luteinizing hormone (LH) and testosterone (T) was examined in Merino rams fed a diet supplemented with lupins in the anoestrous season. Testicular size increased in lupin-fed animals, but not in controls. The mean concentration of T increased ($P < 0.005$) in both control and lupin-fed rams whereas mean LH concentration did not change during the 6 week experiment, and LH and T responsiveness to Gn-RH was also similar. The intervals between pulses of LH declined 5 days after the start of lupin feeding and was thereafter maintained. In control rams the pulse interval was similar to pre-bleed values after 5 days but declined after 6 weeks, although no change in testicular size was evident. This change in the secretory pattern in control rams indicates that these hormones alone are not responsible for testicular growth during lupin feeding.

## INTRODUCTION

Several weeks of lupin supplementation of the diet increases the testicular size of Merino rams with a consequent increase in sperm production (Oldham *et al.* 1978) and this management technique is used routinely prior to mating. Further, in the ewe the gonadal response, in the form of increased ovulation rate, is rapid and occurs within 9 days of the start of lupin feeding (Knight 1979). It is assumed that these responses occur via a hormonal pathway. It has been suggested that in the ram it is due to an increase in the frequency of LH peaks (Sutherland and Martin 1979) although the evidence is inconclusive.

In this study we investigated whether LH and/or T were primarily responsible for the mediation of testicular growth during lupin supplementation of Merino rams.

## MATERIALS AND METHODS

Ten mature Merino rams were divided into two groups of five by stratified random sampling based on the circumference of their scrotum, and liveweight. The mean initial liveweight ranged from 55 to 60 kg. Animals were weighed at weekly intervals before feeding. At the same time the maximum scrotal circumference was measured. Animals were housed individually from three weeks before the commencement of and then throughout the experiment (June-August). The clean wool weight grown on a 10 cm × 10 cm mid-side patch was determined at the end of the experiment.

A 1 kg pelleted ration consisting of oats, barley, wheat, oaten hay, minerals and vitamins, was fed to all animals before and during the experiment. The lupin supplemented group received an additional 750 g lupin/day.

Serial blood samples (5 ml) were collected by jugular venepuncture at hourly intervals for 30 h 2 days before (—2 d), 5 days after (+5 d) and 6 weeks after (+6 w) the start of lupin feeding. Hourly blood sampling is considered suitable for detecting 93.5% and 98.6% of all possible LH and T peaks respectively (Pelletier *et al.* 1982). Gonadotrophin-releasing hormone (Gn-RH; Relefact, Hoechst Australia Ltd.) was injected (1 $\mu$g in 1 ml saline) 2 weeks after the start of lupin feeding. Five ml blood samples were collected before (at −30 and −10 min) and at 20 min intervals for 3 hr after Gn-RH injection. Blood was centrifuged and the plasma stored at −20°C until assay. Concentrations of LH and T in plasma were determined by radioimmunoassay as described by Martin *et al.* (1980) and Corker and Davidson (1978).

Concurrent elevations in both LH and T as defined by Pelletier *et al.* (1982) were identified as pulses. Pulsatility and the mean concentrations during the 30 hr blood samplings were examined by split-plot analysis of variance, with time before and after the start of lupin feeding being the sub-plot. Other differences between means were compared by Student's *t* test.

## RESULTS

The analysis of the diets and the growth responses of the animals are presented in Table 1. During the 6 week experiment lupin feeding increased liveweight ($P < 0.01$), wool growth ($P < 0.05$) and scrotal circumference ($P < 0.05$).

**TABLE 1. Effect of dietary regime on liveweight change, woolgrowth and scrotal circumference of Merino rams.**

| Dietary regime | Intake of Energy (MJ ME/d) | Intake of Crude Protein (g/d) | Liveweight change (g/d) | Wool growth (mg/cm²/d) | Scrotal circumference (mm/d) |
|---|---|---|---|---|---|
| Control | 11.0 | 85 | −7 | 1.08 | −0.15 |
| + Lupins | 18.6 | 300 | 222** | 1.73* | 0.71* |

Values significantly different from controls are indicated; *P<0.05, **P<0.01

The analysis of variance showed a significant interaction ($P < 0.025$) between dietary regime and day of blood sampling on pulse interval (Table 2).

**TABLE 2. Effect of dietary regime on LH and testosterone concentration and pulse interval (mean ± S.E.).**

| Dietary regime | Time of serial blood sampling | Pulse interval (hr) | Mean concentration (ng/ml) LH | T |
|---|---|---|---|---|
| Control | −2d | 15.8±3.9 | .41±.06 | .70±.10 |
| | +5d | 17.9±3.9 | .49±.05 | .71±.04 |
| | +6w | 7.6±1.9 | .80±.30 | 2.47±.88 |
| + Lupins | −2d | 16.1±4.8 | .48±.09 | 1.53±.50 |
| | +5d | 7.1±1.5 | .49±.06 | 1.83±.56 |
| | +6w | 6.7±1.7 | .66±.16 | 4.06±1.22 |

By +5 d pulse interval had declined and was thereafter maintained at a steady level in lupin-fed rams whereas in control rams no decline was apparent until +6 w. The dietary regime did not affect the mean LH concentrations of blood samples collected serially during the experiment. T concentrations increased ($P < 0.005$) over the 6 week period in both control and lupin-fed rams. Within times of serial blood sampling, T levels were not significantly different between groups.

**TABLE 3. Mean (± S.E.) LH and testosterone response to Gn-RH.**

| Dietary regime | Peak height (ng/ml) LH | T | Area units under peak[1] LH | T |
|---|---|---|---|---|
| Control | 5.2 ± 0.9 | 6.8 ± 3.5 | 18.2 ± 2.6 | 24.4 ± 7.2 |
| + Lupins | 3.5 ± 0.9 | 8.4 ± 2.4 | 12.9 ± 3.2 | 33.3 ± 5.3 |

[1] Area under Gn-RH-stimulated peak was assigned arbitrary units.

The response data for a single injection of Gn-RH 2 weeks after the start of lupin feeding are presented in Table 3. There were no significant differences between dietary regimes.

## DISCUSSION

Testicular size, liveweight and wool weight for rams fed lupins increased steadily during the experimental period. Thus, the rams showed a classic response to lupin supplementation, as described by Oldham *et al.* (1978).

Seasonal growth and regression of testes is associated with variations in the LH secretion pattern (Lincoln 1979) and increased pulsatility of LH secretion has also been implicated in testicular growth during lupin feeding (Sutherland and Martin 1980). However our study failed to obtain conclusive support for such involvement. There was a rapid decline in pulse interval which was maintained throughout the experimental period in lupin-fed rams. In control rams there was no change in testicular size yet they also showed a reduced pulse interval and an increased testosterone concentration by +6 w.

No reason for this confounding effect in control rams can be given but it serves to illustrate that the increase in LH pulsatility alone was not responsible for testicular growth. It is also unlikely that seasonal influences were involved in the altered pulsatility in control rams since the experiment was conducted at the onset of the anoestrous season. The apparent trends in LH and T peak height and peak area following Gn-RH injection (Table 3) were examined in a further experiment (data not shown) but there were no differences between control and lupin-fed rams. The absence of any difference in LH and T responsiveness to Gn-RH challenge suggests that neither LH production nor secretion is altered by lupin supplementation. Hence other hormones, possibly FSH or prolactin, may need to be examined as the primary mediators of gonadal growth during lupin feeding.

## ACKNOWLEDGEMENTS

We are indebted to J.B. Rowe and P.J. Murray for the nutritional aspects of the study, and assistance with blood sampling, and to G.B. Thomas for the LH assay. Testosterone antiserum was generously provided by Dr. R.I. Cox.

## REFERENCES

Corker, C.D. and Davidson, D.W., 1978. *J. Steroid. Biochem.*, *9*, 373-374.

Knight, T.W., 1979. *Proc. Aust. Soc. Reprod. Biol.*, *11*, 42.

Martin, G.B., Oldham, C.M. and Lindsay, D.R., 1980. *Anim. Reprod. Sci.* *3*, 125-132.

Oldham, C.M., Adams, N.R., Gherardi, P.B., Lindsay, D.R. and Mackintosh, J.B., 1978. *Aust. J. Agric. Res.*, *29*, 173-179.

Pelletier, J., Garnier, D.H., De Riviers, M.M., Terqui, M. and Oravant, R., 1982. *J. Reprod. Fert.*, *64*, 341-346.

Sutherland, S.R.D. and Martin, G.B., 1980. *Proc. Aust. Soc. Anim. Prod.*, *13*, 459.

# SEMEN QUALITY, QUANTITY AND FLOCK FERTILITY

A.W.N. Cameron, I.J. Fairnie *Muresk Agricultural College, Northam, 6401.*
E.J. Keogh *Department of Clinincal Biolchemistry, University of Western Australia, Nedlands, 6009.*

## INTRODUCTION

Given 100% conception rates, the maximum number of lambs produced by a flock of ewes is determined by the mean ovulation rate of the flock. Various forms of reproductive wastage intervene to ensure that this potential lamb production is never achieved. Optimum usage of rams involves keeping those sources of wastage that are attributable to rams to a minimum, using the least possible number of rams. To achieve optimum joining rates, the ability of rams to mate and induce satisfactory lambing rates must first be reliably predicted. The ram can contribute to reproductive wastage by failing to inseminate all the available oestrous ewes, or by failing to achieve maximum lambing rates in those ewes that are inseminated. Poor lambing rates in ewes that are mated can be due to a low proportion of ewes that lamb (termed — fertility), or a low number of lambs born per ewe lambing (termed — prolificacy).

In this paper, the relationships between the quality and quantity of spermatozoa that are inseminated and lambing performance will be considered. Some of the discussion will involve data derived from artifical insemination but the central theme will be one of how the semen that is produced by rams will affect flock fertility under conditions of natural mating. Factors that may influence the ability or rams to provide satisfactory semen will then be discussed.

## THE RELATIONSHIP BETWEEN THE NUMBER OF SPERMATOZOA INSEMINATED AND FERTILITY

The relationship between the number of spermatozoa inseminated and fertility is most clearly demonstrated though the use of artificial insemination, where the number of spermatozoa inseminated can be precisely controlled. Using artificial insemination Salamon (1962) found a linear relationship between the number of spermatozoa inseminated and lambing rates, over the range of 28 million to 128 million sperm. He predicted that within this range, each additional 25 million spermatozoa would increase the lambing rate by approximately 13%. In an earlier study, no improvement in lambing rate was seen as the number of spermatozoa inseminated increased beyond 150 million sperm (Salamon and Robinson 1962). These observations suggest that the relationship between the number of spermatozoa inseminated and subsequent fertility is described by an asymptotic model. With such a model fertility should first increase linearly as the number of spermatozoa inseminated increases from zero. Then, as the level of fertility approaches the asymptotic or maximum value, the curve relating fertility to numbers of spermatozoa inseminated should flatten, until no further increase in fertility can be gained through increasing the number of spermatozoa inseminated. Such a model is also consistant with results obtained with cattle and the domestic fowl (Schwartz *et al.* 1981). While the asymptotic model holds true, the estimated number of spermatozoa that are necessary to achieve maximum fertility varies between reports. For example Jones *et al.* (1969) found no improvement in fertility as the number of spermatozoa that were inseminated rose from 25 to 100 million and Entwistle and Martin (1972) found no differences in fertility when 50 or 100 million spermatozoa were inseminated. These results are contrary to those of Salamon (1962), cited above, and Martin and Watson (1976), which show clearly the inadequacy of a dose of 25 million spermatozoa. The reason for these differences is not clear. There may have been qualitative differences in the semen that was used, although only semen of good motility was inseminated. Alternatively differences in the physiological characteristics of the ewes have lead to altered requirements for semen. A variety of factors can influence the reproductive function of ewes including intrinsic factors such as age and breed, and extrinsic factors such as nutrition and photoperiod. Whether these can influence the relationship between the number of spermatozoa inseminated and fertility is not known.

When determining how the number of spermatozoa inseminated affects the fertility of naturally mated ewes, the numbers of spermatozoa that are inseminated cannot be determined directly but instead must be estimated from the number of spermatozoa present in ejaculates collected during the mating period. The accuracy of these estimates may vary according to the methods used to collect the ejaculates. Intuitively one would expect ejaculates collected through electroejaculation or with an artificial vagina, to be less likely to reflect those of naturally mated rams, than would ejaculates collected using the methods of Synnott *et al.* (1981) or Cameron *et al.* (1984a). These workers allowed rams to mate naturally. The ejaculates were recovered from the ewes, either from condoms inserted into their vaginas prior to mating, or by flushing the ejaculates directly from the vagina. A further error in estimating the number of spermatozoa received by naturally mated ewes arises because consecutive ejaculates of continuously mated rams fluctuate in their content of spermatozoa (Cameron *et al.* 1984a), making it impossible to calculate exactly how many spermatozoa individual ewes receive. Accepting this limitation, it is still evident that the number of spermatozoa that are required to give maximum fertility under conditions of natural mating are similar to those required when using artificial insemination. For example Fulkerson *et al.* (1982) found that 95% of ewes receiving more than 60 million spermatozoa were pregnant, whereas only 35% of ewes receiving less than this amount were pregnant. There was no further improvement in ewe fertility when more than 60 million spermatozoa were inseminated. Data from A.W.N. Cameron and A.J. Tilbrook (1984, unpublished) supports the proposition that beyond a threshold level, fertility will not be improved by increasing the number of spermatozoa inseminated. Their experiment involved recording the sexual activity of mating groups in which 10 rams were allotted six oestrous ewes each per day. To allow determination of the numbers of spermatozoa being delivered to ewes, individual ejaculates were collected on a daily basis from each ram. Ejaculates were collected by flushing the vaginas of ewes that the rams had mated. The mean number of spermatozoa per ejaculate was calculated for each ram. Using these data, along with information on how many times each ewe was mated, the total number of spermatozoa received by each ewe was estimated. Lambing rates were subsequently determined, and related to the numbers of spermatozoa received by the ewes. The lowest mean number of spermatozoa in the ejaculate of any ram was 150 million, but individual ejaculates contained as few as 30 million spermatozoa. The lambing results (Figure 1) show that the proportion of ewes lambing did not vary according to the number of spermatozoa that were inseminated.

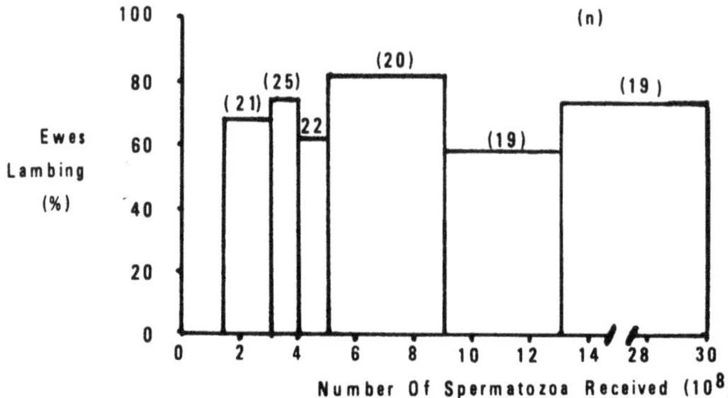


**FIGURE 1. The effect of number of spermatozoa inseminated on the fertility of naturally mated ewes (A.W.N. Cameron and A.J. Tilbrook, unpublished data).**

The ability of rams to supply ewes with adequate numbers of spermatozoa under conditions of natural mating depends on the frequency with which ewes are served and on the number of spermatozoa contained in each ejaculate. More frequent service can improve the fertility of naturally mated ewes (Mattner and Braden 1967; Cameron *et al.* 1984b). The ejaculates collected in these experiments contained sufficient spermatozoa to make it seem unlikely that the fertility of ewes that were served once was limited by the number of spermatozoa they received. It is possible that the effects of multiple service are not mediated solely through the provision of additional spermatozoa. The number of spermatozoa delivered by different rams varies, being dependent on their testicular weights and on the frequency with which they ejaculate (Synnott *et al.* 1981; Cameron *et al.* 1984a,c). One might hypothesize that where testicular weights are low or the levels of mating activity high, the numbers of spermatozoa per ejaculate would fall to levels that would limit ewe fertility. Support for this hypothesis comes from Gherardi *et al.* (1980) who demonstrated that joining rams of differing testicular weights can affect lambing rates. The absolute amount of testicular tissue required for maximum fertility varied between farms. For example, on two farms on which the ram:ewe ratio was 1:100, reducing the amount of testicular tissue from 625g to 275g per ram, lead to a 17% reduction in the proportion of ewes lambing on one farm, but to only a 3% reduction on the other. On a third farm, 700g of testicular tissue were required by each ram to allow maximum fertility. It is not surprising that the absolute amount of testicular tissue required for maximum fertility varies between farms, as the number of spermatozoa that are necessary for maximum fertility also varies. Also, a large proportion of the variation in the numbers of spermatozoa received by ewes is not attributable to variation in testicular weight. For example, the number of times ewes are served is obviously important in determining how many spermatozoa they receive, and as stated earlier, there is considerable fluctuation in the numbers of spermatozoa within the ejaculates of individual rams, even when they are continuously mated.

## THE INFLUENCE OF SEMEN QUALITY ON FLOCK FERTILITY

With currently available techniques, the most meaningful assay of semen quality is the proportion of ewes lambing as the result of insemination of a standard number of spermatozoa. This assay is cumbersome because of the number of females required and is of limited practical value because the quality of the semen is only known some time after it has been used. Thus the practice has developed of examining various characteristics of semen with a view to predicting the fertility that would result from its use. The relationships between the characteristics measured and fertility is best assessed after artificial insemination where the number of spermatozoa inseminated, and the number and timing of inseminations can be standardized. Thorough investigations of this type have only been carried out using cattle in which none of the semen characterisics have consistantly explained more than about 20% of the variation in conception rates (Johnson *et al.* 1952; Bishop *et al.* 1955; Pace *et al.* 1981). Bulls included in these studies had already proved to be reasonably fertile. Where bulls representing all stages of fertility were deliberately included in a similar study, simple tests such as motility detected ejaculates that yielded low conception rates (Swanson and Herman 1944).

Seminal characteristics of rams have been measured prior to natural mating and related to subsequent lambing performance. Their characteristics include motility, morphological abnormalities, the proportion of dead spermatozoa, and pH. A sample of the correlations found between these seminal characteristics and fertility is shown in Table 1. The correlations vary according to whether rams of poor fertility are included or excluded from each study. The lowest correlations between semen characteristics and fertility have been found by Wiggins *et al.* (1953), who generally only studied rams expected to be of good fertility. In contrast Hulet *et al.* (1964), used an unselected ram population and obtained higher correlations. The highest correlations were found by Hulet and Ercanbrack (1962), who deliberately included rams expected to have a wide range of fertility. The data from these three reports along with those of Edgar (1959), and Hulet *et al.* (1965), indicate that the relationship between these seminal characteristics and fertility is described by an asymptotic model and is thus analagous to the relationship between the number of spermatozoa inseminated and fertility. Up to certain threshold value, these seminal characteristics are important in determining fertility. Beyond this value other factors become more important. This model is exemplified by the work of Colas and Guerin (1981) who examined seasonal changes in the semen quality of Ile-de-France rams. Lambing rates following artificial insemination were negatively related to the proportion of abnormal sperm in the ejaculates during spring. During this period the proportion of abnormalities ranged as high as 59.3%. During autumn, the proportion of abnormalities was never greater than 20%, and were then unrelated to lambing rate. As was the case with the numbers of spermatozoa, the threshold levels of the characteristics of semen that are consistent with maxmum fertility are difficult to define. For example Hulet *et al.* (1965) found that fertility dropped markedly as the proportion of abnormal spermatozoa rose above 40%. Following a similar study, Edgar (1959) concluded that for semen to be considered satisfactory, it should contain fewer than 20% abnormal spermatozoa.

**TABLE 1. Correlations between semen characteristics and fertility.**

| Experiment | Hulet & Ercanbrack (1962) | Hulet *et al.* (1964) | Wiggins *et al.* (1953) |
|---|---|---|---|
| Semen Trait: | | | |
| pH | −0.66 | −0.12 | −0.06 |
| Motility % | 0.67 | 0.35 | 0.06 |
| Live normal % | 0.70 | 0.61 | 0.29 |
| Abnormal % | −0.66 | −0.75 | — |
| Normal % | — | — | 0.43 |

There are numerous tests of semen quality that have not yet been evaluated in the ram. These include metabolic tests such as the rate of fructolysis or methylene blue reduction time, and techniques for examining the structure and function of acrosomes. When evaluated in the bull these tests have provided no more information about the fertility of semen than have more simple tests such as those relating to motility (Bishop *et al.* 1955; Pace *et al.* 1981).

Differences in the quality of semen may exist that cannot be detected by the currently available techniques for examining semen. For example Dun and Hamelton (1965) and Robinson *et al.* (1967) used only semen of good motility for artificial insemination in sheep, and found conception rates to vary widely according to the rams used. The semen of rams derived from strains of sheep selected for prolificacy gives higher fertilization rates in ewes into which it is inseminated, than does semen from lines of lower prolificacy, despite no apparent differences in the conventional semen analysis (Moore and Whyman 1980). Bishop (1964) comments that conception rates achieved through using semen from different bulls, may vary significantly, without differences in their seminal characteristics being apparent. Bishop suggested that some of these unexplained differences in fertility are due to differences in the level of embryonic death. The currently available techniques for assessing the quality of semen are more likely to be assessing the ability of spermatozoa to fertilize ova than other aspects of its ability to participate in the formation of viable embryos.

Because differences in the quality of semen are not always demonstrated by conventional semen analysis, semen should be assessed directly through the insemination of ewes whenever this is practical. Such a direct assay is the method of choice for those who wish to include an assessment of seminal quality in their experiments. Where fertility is to be assessed in this way, artificial insemination should be used so as to eliminate variation in fertility arising from differences in the numbers of spermatozoa inseminated or differences in mating behaviour, which could arise if natural mating were used.

## THE ROLE OF BEHAVIOUR AND SEMINAL CHARACTERISTICS IN DETERMINING JOINING RATES

Under conditions of natural mating the limitations imposed on the fertility of ewes by the semen produced by rams will be influenced by mating behaviour. This is because behavioural factors will affect the number of times ewes are served within each oestrus and the number of oestrous periods during which the ewes are exposed to the rams. Thus the relationships between seminal characteristics and fertility may be less clearly defined under conditions of natural mating than they are when artificial insemination is practised.

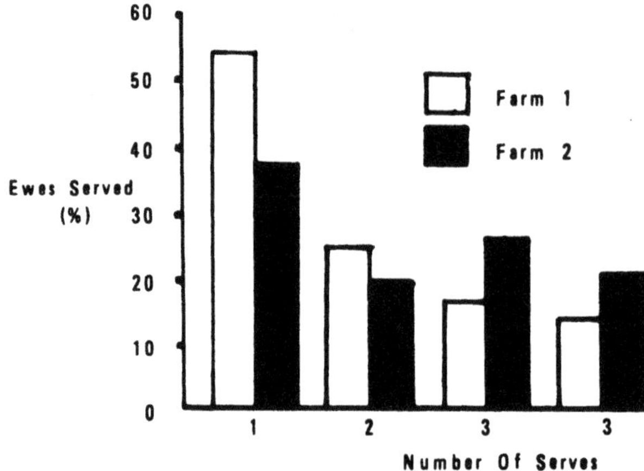


**FIGURE 2. Number of times ewes are served under conditions of natural mating (A.J. Tilbrook and A.W.N. Cameron, unpublished data).**

In the field the majority of ewes that are served, are served more than once during a single oestrus. A.J. Tilbrook and A.W.N. Cameron (unpublished data) have determined this through direct observation of flocks in which 6 oestrous ewes were joined to each ram, each day. Approximately half of those ewes that were served once went on to receive multiple serves (Figure 2). In these experiments approximately 40% of the available oestrous ewes were not served at all, so a high incidence of multiple mating occurred even though the joining rates were sub-optimal. If the rams are able to cover all the available ewes, the proportion of ewes that are served more than once would presumably be greater. Where rams are fitted with sire-sine harnesses in a multi-sire joining system, a large proportion of ewes are marked by more than one ram, even where the joining rates used are low enough to prevent all the available oestrous ewes from being marked (Lightfoot and Smith 1968). These observations could be attributable to the phenomenon of 'ram mating preferences' that has been described by Tilbrook (1984). Tilbrook observed that when confronted with a group of oestrous ewes, rams will consistently direct most of their sexual behaviour towards certain 'attractive' ewes. Furthermore different rams tended to favour the same ewes. These behavioural phenomena will restrict the number of ewes that rams serve during the period of joining. By way of compensation, they will also tend to improve the fertility amongst ewes that are served. Firstly most ewes will be served at least twice, thus improving the chances of ewes receiving adequate numbers of spermatozoa. Secondly, in a multi-sire joining most ewes are served by more than one ram, so the impact of a low incidence of infertile rams on ewe fertility should be slight. Finally there may be additional mechanisms through which multiple service affects fertility. For example, the timing of insemination can affect fertility (Amir and Schindler 1973) and this would be more likely to be optimal with multiple insemination.

The importance of the seminal characteristics in affecting fertility will also depend on what opportunities are available for ewes to return to service. Poor conception rates at first service will be less important if ewes can be impregnated at a subsequent service. The opportunities for ewes to return to service will be limited by a short joining period or if the ewes are prone to enter anoestrus, for example if joining is being carried out in spring. The chances of ewes returning to service are also affected by the stage at which pregnancy fails, with delayed return to service being expected where embryos are lost more than 12 days after conception (Edey 1967).

A reduction in joining rates may lead to reduced fertility by causing either a decrease in the proportion of ewes that are mated, or a lower conception rate among ewes that are mated. Australian work suggests that the potential to reduce joining rates may at times be limited by the ability of rams to achieve optimum lambing rates in ewes that are mated. For example when Dawe *et al.* (1970) compared the effect of using ram:ewe ratios of 1:33 or 1:100 on thirteen properties the incidence of unmarked ewes was never greater than 7.5%, but up to 49.8% of ewes failed to lamb when the lower joining rate was used. Lightfoot and Smith (1968), noted reductions in flock fertility that were due to a decrease in the proportion of ewes marked, as well as lower in conception rates, when ram:ewe ratios were reduced from 1:25 to 1:100. One cannot say whether these reductions in conception rates were due to deficiencies in the quality or number of spermatozoa delivered, or due to an increased incidence of ewes that were mounted but not inseminated. In a survey into reproductive wastage in south Western Australia, Knight *et al.* (1975) concluded that ewes that were raddle-marked, but failed to lamb, made the largest contribution to reproductive wastage even though joining rates of two to four rams per 100 ewes were being used. Further work is required to define whether inadequacies in the quality and quantity of semen produced by rams contributed to this high level of wastage. New Zealand work suggests that the ability of rams to produce satisfactory semen will not necessarily limit the rate at which they can be joined. For example Allison and Davis (1976) found that the decrease in fertility as joining rates declined was due to ewes failing to mate. In other work, conception rates in the first cycle did not fall as ram:ewe ratios were reduced from 1:70 to 1:210 (Allison 1975).

## THE INFLUENCE OF RAMS ON FLOCK PROLIFICACY

The discussion so far has centred on how rams influence the number of lambs born through their effects on the fertility of flocks. There is evidence that rams can also influence the prolificacy of the flock to which they are mated. For example, seminal characteristics prior to joining such as the percentage of live, motile and morphologically normal spermatozoa were correlated with the number of lambs born per ewe that subsequently lambed (Hulet *et al.* 1965). Differences in the proportion of ewes lambing, and in the proportion of twin births were obtained following the insemination of ewes by rams selected either for or against skinfold (Dun and Hamelton 1965). A reduction in joining rates from 1% to 0.25% of rams, led to a lower proportion of twin-bearing ewes among ewes that were impregnated (Fowler 1982). Finally, Burfening *et al.* (1977), found that prolificacy in a random population of ewes was higher when they were mated to rams whose dams were selected for high, rather than low prolificacy. Such observations suggest that the ram can influence the prolificacy of ewes to which they are mated. This conclusion needs confirming in an experiment in which the results are corrected for differences in ovulation rates between the different treatments.

## FACTORS AFFECTING THE FERTILITY OF RAMS

Factors which are likely to influence all the rams within a flock should have the greatest impact on flock fertility. These include environmental factors such as temperature, photoperiod, and nutrition. Research in this area is made difficult because these factors may also be affecting the ewes. Thus it can be difficult to establish whether the effects of these factors on the fertility of ewes is due to direct effects on ewes or whether they are being mediated through effects on rams.

### Temperature

Severe deteriorations in seminal characteristics with concommittant reductions in fertility can be induced by placing rams in hotrooms (Fowler and Dun 1966; Braden and Mattner 1970). There are a number of reports suggesting that high temperatures may have a deleterious effect on the quality of semen of rams maintained at pasture. For example poor fertilization rates have been reported from north-west Queensland where the environmental temperature is often high (Entwistle 1970). High ambient temperatures have clear effects on the fertility of rams derived from flocks selected for a high degree of skinfold (Fowler and Dun 1966). Gunn *et al.* (1942) reported that exposure to normal summer temperatures can reduce the quality of semen although the significance of their findings are questionable, since ejaculates seldom contained more than 10% of abnormal forms. Despite the above observations, the general importance of environmental temperature effects on seminal quality under natural mating conditions is equivocal, and will remain so until experiments are carried out that exclude the confounding effects that temperature may have on ewes.

### Season

In breeds that are strongly influenced by photoperiod such as the Ile-de-France or Suffolk, seminal characteristics tend to be better in autumn than in spring and there is evidence that the fertility of rams is also affected by photoperiod in these breeds (Colas and Brice 1976; Schanbacher 1979; Colas and Geurin 1981). The conception rates that were achieved by naturally mated Suffolk rams were influenced by their photic environment during the 10 weeks prior to mating (Schanbacher 1979). These effects of photoperiod are mediated through the release of melatonin (Bittman *et al.* 1983). Thus pharmacological means of manipulating reproductive function in rams may soon be practicable. It therefore becomes a matter of practical importance to establish whether photoperiod significantly affects Merino rams. No clear relationship between seminal characteristics and changes in daylength have been noted for the Merino (Fowler 1965; Skinner and VanHeerdon 1971). However lambing rates in Merino flocks are reported to be higher following artificial insemination in autumn than in spring (Dun *et al.* 1960; Salamon and Robinson 1962). The relative contributions of ewes and rams to this phenomenon needs to be defined.

The testicular volume of Merino rams is also influenced by photoperiod. The testicular weights of rams maintained on a constant nutritional regimen were approximately 80g higher in autumn than in spring (Skinner and VanHeerdon 1971). However nutrition is a more powerful seasonal influence on testicular weight than is photoperiod. Manipulation of the protein and energy intake of rams can have profound effects on their rate of sperm production. Oldham *et al.* (1978) achieved rates of production of spermatozoa of 13.6 billion versus 7.5 billion per day, through providing differing levels of nutrition. Thus with pasture-fed Merino rams in winter rainfall zones, testicular size may be higher in late spring than in autumn (Masters and Fels 1984).

Age

Testicular development appears to be more closely related to the liveweight rather than age of ram lambs (Watson *et al.* 1956; Dyrmundsson and Lees 1972). Watson *et al.* (1956) first detected spermatozoa in the lumen of the seminiferous tubules of Merino rams when they reached a liveweight of 27kg. By increasing the plane of nutrition of ram lambs, the age at which puberty is reached can be reduced but the liveweight at puberty is increased, suggesting that both age and liveweight influence testicular development prior to puberty (Pretorius and Marincowitz 1968). Ejaculates that are collected shortly after puberty may contain a high proportion of abnormal spermatozoa which exhibit poor motility, but several weeks later these characteristics will be similar to those of adult rams (Louw and Joubert 1964; Skinner and Rowson 1968).

Growth of the testes continues after puberty, but the age at which maximum testicular weight is reached is not known. With pasturefed Merino rams testicular growth continued until rams were at least 18 months of age with the extent of this growth being governed by nutrition (Masters and Fels 1984). Thus testicular weight is less likely to be adequate when rams are joined at less than 18 months of age compared to when joined as adults, especially if their nutrition is poor.

There are a variety of additional causes of poor quality semen in rams. Some will have a sporadic incidence such as epididymitis, flystrike, footrot and inherited abnormalities, while others such as vitamin A deficiency may exert a more general influence. These have been discussed fully in a review by Galloway (1973).

## CONCLUSIONS

The fertility, and probably the prolificacy of ewes are influenced by characteristics of the semen that they receive, but the absolute requirements for maximum fertility vary. Thus high levels of reproductive wastage can result from the inability of rams to provide satisfactory semen, but it is difficult to predict when this will occur. In this review, two reasons have been proposed as to why this is so, and are suggested as being worthy of further research. The first is that the relationships between seminal characteristics and fertility are affected by mating behaviour and the physiological characteristics of the ewes that are inseminated. These in turn may be influenced by many factors including the age and breed of both ewes and rams, and environmental factors such as temperature, nutrition and photoperiod. Secondly, qualitative differences between the semen of rams that induce variation in fertility, may not be detected by conventional semen analysis. This not only makes it difficult to evaluate individual ejaculates, but more importantly has retarded research into how fertility may be affected by such factors as photoperiod, temperature and nutrition. Future workers in this area should remember that there is a more meaningful assay of semen that is already available; the fertility resulting from the insemination of ewes.

## ACKNOWLEDGEMENT

The advice of Messers John Beilby and Colin Carati and Professors Lindsay and Curnow on the preparation of this review is gratefully acknowledged. AWNC was supported on a grant from the Wool Research Trust Fund during the writing of this review.

## REFERENCES

Allison, A.J., 1975. *N.Z. J. Agric. Res., 18*, 1-8.

Allison, A.J., and Davis, G.H., 1976. *N.Z. J. Exp. Agric.,* 4, 259-267.

Amir, D. and Schindler H., 1972. *J. Reprod. Fert., 28*, 261-264.

Bishop, M.W.H., 1964. *J. Reprod. Fert.,* 7, 383-396.

Bishop, M.W.H., Campbell, R.C., Hancock, J.L. and Walton, A., 1954. *J. Agric. Sci., Camb., 44*, 227-248.

Bittman, E.L., Dempsey, R.J. and Karsch F.J., 1983. *Endocr., 113*, 2276-2283.

Braden, A.W.H. and Mattner, P.E., 1970. *Aust. J. Agric. Res., 21*, 509-518.

Burfening, P.J., Friedrick, R.I. and Vanhorn, J.L., 1977. *Theriogenology, 7*, 285-291.

Cameron, A.W.N., Fairnie, I.J., Curnow, D.H., Keogh, E.J. and Lindsay, D.R., 1984a. *Proc. Tenth Int. Congr. Anim. Reprod. A.I. Chicago, 2*, 266.

Cameron, A.W.N., Fairnie, I.J., Curnow, D.H., Keogh, E.J. and Lindsay, D.R., 1984c. *Proc. Tenth Int. Anim. Reprod. A.I. Chicago., 2*, 267.

Cameron, A.W.N., Hall, R., Fairnie, I.J., Curnow, D.H., Keogh, E.J. and Lindsay, D.R., 1984b. *Proc. Sixteenth Conf. Aust. Soc. Reprod. Biol.*, 79.

Colas, G. and Brice, G., 1976. *Proc. 8th Int. Cong. Anim. Reprod. and A.I., Krakow, IV*, 977-980.
Colas, G. and Guerin, Y., 1981. *Reprod. Nutr. Develop., 21*, 399-407.
Dawe, S.T., Bennet, N.W., Donnelly, F.B., Ferguson, B.D., Rive, J.P., Roberts, B.C. and Trimmer, B.I., 1970. *Proc. Aust. Soc. Anim. Prod., 8*, 317-320.
Dun, R.B., Ahmed, W. and Morrant, A.J., 1960. *Aust. J. Agric. Res., 11*, 805-826.
Dun, R.B. and Hamelton, B.A., 1965. *Aust. J. Exp. Agric. Anim. Husb.*, 5, 805-826.
Dunlop, A.A., Tallis, G.M., Brown, G.H. and Gream, B.D., 1972. *Aust. J. Agric. Res., 23*, 295-307.
Dyrmundsson, O.R. and Lees, J.L., 1972. *J. Agric. Sci., 79*, 83-89.
Edey, T.N., 1967. *J. Reprod. Fert., 13*, 437-443.
Edgar, D.G., 1959. *N.Z. Vet. J., 7*, 61-63.
Entwistle, K.W., 1970. *Proc. Aust. Soc. Anim. Prod., 8*, 353-357.
Entwistle, K.W. and Martin, I.C.A., 1972. *Aust. J. Agric. Res., 23*, 467-472.
Fowler, D.G., 1965. *Aust. J. Exp. Agric. Anim. Husb.*, 5, 247-251.
Fowler, D.G., 1982. *Aust. J. Exp. Agric. Anim. Husb., 22*, 268-273.
Fowler, D.G. and Dun, R.B., 1966. *Aust. J. Exp. Agric. Anim. Husb., 6*, 121-127.
Fulkerson, W.J., Synnott, A.L. and Lindsay, D.R., 1982. *J. Reprod. Fert., 66*, 129-132.
Galloway, D.B., 1973. *A.M.R.C. Review*, No. 10, 1-20.
Gherardi, P.B., Lindsay, D.R. and Oldham, C.M., 1980. *Proc. Aust. Soc. Anim. Prod., 13*, 48-50.
Gunn, R.M.C., Sanders, R.N. and Granger, W., 1942. *C.S.I.R., Aust, Bulletin No. 148*.
Hulet, C.V. and Ercanbrack, S.K., 1962. *J. Anim. Sci., 21*, 489-493.
Hulet, C.V., Foote, W.C. and Blackwell, R.L., 1964. *J. Anim. Sci., 23*, 418-424.
Hulet, C.V., Foote, W.C. and Blackwell, R.L., 1965. *J. Reprod. Fert., 9*, 311-315.
Johnson, J.E., Branton, C. and Hathorn, F., 1952. *J. Anim. Sci., 11*, 740-741.
Jones, R.C., Martin, I.C.A. and Lapwood, K.R., 1969. *Aust. J. Agric. Res., 20*, 141-150.
Knight, T.W., Oldham, C.M., Smith, J.F. and Lindsay, D.R. 1975. *Aust. J. Exp. Agric. Anim. Husb., 15*, 183-188.
Lightfoot, R.J. and Smith, J.A.C., 1968. *Aust. J. Agric. Res., 19*, 1029-1042.
Louw, D.F. and Joubert, D.M., 1964. *S. Afr. J. Agric. Sci., 7*, 509-520.
Martin, I.C.A. and Watson, P.F., 1976. *Theriogenology, 5*, 29-35.
Masters, D.G., and Fels, H.E., 1984. *Proc. Aust. Soc. Anim. Prod., 15*, 444-447.
Mattner, P.E. and Braden, A.W.H., 1967. *Aust. J. Exp. Agric. Anim. Husb., 7*, 110-116.
Moore, R.W. and Whyman, D., 1980. *J. Reprod. Fert., 59*, 311-316.
Oldham, C.M., Adams, N.R., Gherardi, P.B., Lindsay, D.R. and Macintosh, J.B., 1978. *Aust. J. Agric. Res., 29*, 173-179.
Pace, M.M., Sullivan J.J., Elliot F.I., Graham, E.F. and Coulter, G.H., 1981. *J. Anim. Sci., 53*, 693-701.
Pretorius, P.S. and Marincowitz, G., 1968. *S. Afr. J. Agric. Sci., 11*, 319-334.
Robinson, T.J., Salamon, S., Moore, N.W. and Smith, J.F., 1967. *In* Robinson, T.J. (ed.) *The Control of the Ovarian Cycle in the Sheep*. Sydney University Press, Sydney, 208-236.
Salamon, S., 1982. *Aust. J. Agric. Res., 13*, 1137-1150.
Salamon, S. and Robinson; T.J., 1962. *Aust. J. Agric. Res., 13*, 52-68.
Schanbacher, B.D., 1979. *J. Anim. Sci., 49*, 927-932.
Schwartz, D., Macdonald, P.D.M. and Heuchel, V., 1981. *Reprod. Nutr. Develop., 21*, 978-988.
Skinner, J.D., and Rowson, L.E.A., 1968. *J. Reprod. Fert., 16*, 479-488.
Skinner, J.D. and VanHeerdon, J.A.H., 1971. *Sth. Afr. J. Anim. Sci., 1*, 77-80.
Swanson, E.W. and Herman, H.A., 1944. *J. Dairy. Sci., 27*, 297-301.
Synnott, A.L., Fulkerson, W.J. and Lindsay, D.R., 1981. *J. Reprod. Fert., 61*, 355-361.
Tilbrook, A.J., 1984. *PhD Thesis*, University of Western Australia.
Watson, R.H., Sapsford, C.S. and McCance, I., 1956. *Aust. J. Agric. Res., 7*, 574-590.
Wiggins, E.L., Terrill, C.E. and Emik, L.O., 1953. *J. Anim. Sci., 12*, 684-696.

# CHANGES IN PLASMA CONCENTRATIONS OF CORTISOL AND PROLACTIN IN RAMS ASSOCIATED WITH EJACULATION OF SEMEN

I.C.A. Martin, *Department of Veterinary Physiology, University of Sydney, N.S.W. 2006.*
K.R. Lapwood and H.J. Elgar, *Department of Physiology and Anatomy, Massey University, Palmerston North, New Zealand.*

*Summary* Five rams with chronic, jugular cannulae and conditioned to frequent handling for blood sampling were used to study release of cortisol, prolactin and luteinising hormone in response to preparation for, and then semen collection by artificial vagina or electroejaculation. Blood samples were collected at 15 min intervals from 1 hr before, to 2 hr after ejaculation.

Plasma concentrations of cortisol rose significantly after all treatments and were highest wehn electroejaculation was the method of collection. Cortisol returned to pre-treatment levels within 90 min of any of the stimuli. Prolactin was elevated by semen collection using either method and remained higher than pre-treatment levels for 2 hr. Again electroejaculation caused a greater release of prolactin than was evoked by semen collection by artificial vagina.

None of the treatments applied had effects on the secretion of luteinising hormone.

## INTRODUCTION

Although the technique of electroejaculation (EE) has been used extensively to obtain samples of ram semen, very little is known about the way in which EE stimuli induce ejaculation. Furthermore, there do not appear to be any reports of studies of more general reactions of rams following the application of the electrical pulses to cause ejaculation.

Modern circuits employing low amplitude stimuli of controlled, short duration delivered through bipolar rectal electrodes optimally placed in the pelvic cavity are extremely unlikely to cause injury to rams. However, the technique is potentially stressful. Welsh and Johnson (1981) reported that the concentration of corticosteroids rose sharply in the peripheral blood of bulls 15 min after EE and remained elevated for 2 to 4 hr.

This paper reports the levels of cortisol (C), prolactin (PRL) and luteinising hormone (LH) measured in samples of blood obtained from rams before and after EE and, using the same sampling frequency, from control rams, rams restrained as if for EE and rams allowed to mount a ewe and ejaculate in an artificial vagina.

## MATERIALS AND METHODS

### Rams

Merino × Romney rams, aged 27 months, were housed in individual pens in a light and temperature controlled room providing conditions simulating autumn and were trained for semen collection by artificial vagina (AV). Jugular cannulae were inserted bilaterally in 5 rams and taken subcutaneously to the dorsal surface of the neck and fitted with taps. Daily blood samples were taken for 6 days before the experiment and, in this period, the rams were trained to permit blood sampling without physical restraint.

### Experimental design

The design was a 4 × 4 Latin square to permit the testing of 4 situations (treatments) of each of 4 rams over a period of time, and the same test situations were applied to the fifth ram in a sequence different to any expressed in the rows or columns of the Latin square. The treatments were applied at the same time of day (0930 to 1230 hr) on days 6, 10, 13 and 17 after insertion of the cannulae.

The test situations were:

I: Control. Ram left in pen but not isolated from others.

II: A single ejaculate obtained when the ram mounted the decoy ewe (AV).

III: The ram held in lateral recumbency and the rectal electrode inserted for 1 minute but no electrical stimulation.

IV: Restrained as in III and EE of 10 volleys of impulses each of 3 sec duration with 3 sec rest between volleys applied. This pattern of stimulation induced ejaculation in all 5 rams.

EE and AV collections were made in the same area of the room so that no pattern of activity could indicate in advance to rams what treatment was to be applied. Blood samples ere taken before and after treatment at 15 min intervals as shown in Table 1.

### Assays

C, PRL and LH in plasma samples harvested from heparinised blood were measured by radioimmunoassay (RIA) in validated assays for ovine hormones. The method for LH was published (Barrell and Lapwood 1979) and details of the other assays are available from KRL. The sensitivities were 0.1, 1 and 0.06 ng/ml for C, PRL and LH respectively.

Analysis of data

The data were transformed to logarithms to achieve homogeneity of variance over the range of levels of C and PRL observed. Accordingly, the values presented in Table 1 are geometric means. The analyses of variance were performed following a split plot model accounting for between ram variation and sampling within rams.

## RESULTS

Table 1 summarises the mean plasma concentrations of C and PRL assayed in plasma samples from the 5 rams in 4 treatment situations.

Cortisol

When the rams served as controls (I) they could see and hear all of the pattern of activity in treating the other rams and these are thought to have been sufficient stimuli to cause the mild, but just significant ($P<0.05$), elevation of plasma C observed in the first hour after other rams had been treated. Three of the rams (one each used for treatments II, III and IV) on the second day of the experiment had elevated C levels before treatment. Management procedures were kept constant in terms of lighting regime, feeding and commencement of sampling throughout the experiment so that the stimulus to secrete C cannot be identified. A single natural ejaculation (II) caused a significant ($P<0.05$) increase in mean C concentration. Handling rams for EE but not stimulating them electrically (III) caused a highly significant ($P<0.01$) increase in concentration of circulating C and, when EE was applied (IV), C levels rose further and remained significantly elevated for 60 min.

**TABLE 1. Plasma concentrations (ng/ml) of cortisol and prolactin before and after collection of semen from rams either naturally (AV) or by electrical stimulation (EE). (Values are geometric means of observations from 5 rams).**

| Time relative to treatment (min) | Cortisol | | | | Prolactin | | | |
|---|---|---|---|---|---|---|---|---|
| | I[1] | II | III | IV | I | II | III | IV |
| –60 | 5.1 | 7.8 | 6.2 | 8.5 | 5.3 | 9.0 | 6.5 | 13.0 |
| –45 | 3.6 | 7.5 | 5.1 | 12.6 | 6.1 | 8.6 | 5.2 | 11.5 |
| –30 | 5.1 | 7.9 | 8.2 | 14.5 | 4.8 | 8.5 | 4.4 | 10.3 |
| –15 | 3.7 | 9.4 | 8.6 | 9.3 | 4.4 | 21.0 | 4.4 | 9.9 |
| 0 | 5.0 | 11.9 | 8.1 | 14.1 | 4.3 | 20.0 | 23.4 | 36.8 |
| 15 | 7.9 | 12.3 | 26.9 | 45.7 | 4.3 | 17.6 | 34.3 | 42.0 |
| 30 | 8.9 | 9.2 | 26.0 | 52.5 | 5.5 | 22.0 | 20.9 | 49.4 |
| 45 | 7.9 | 7.1 | 18.4 | 30.2 | 4.1 | 21.2 | 29.7 | 52.5 |
| 60 | 9.0 | 7.3 | 10.0 | 21.6 | 5.0 | 22.2 | 27.5 | 56.5 |
| 75 | 3.3 | 4.8 | 8.6 | 13.2 | 4.1 | 17.6 | 18.1 | 44.1 |
| 90 | 3.1 | 4.9 | 9.9 | 9.1 | 6.1 | 23.4 | 15.1 | 32.3 |
| 105 | 3.9 | 4.6 | 11.5 | 10.7 | 7.0 | 21.8 | 13.0 | 24.1 |
| 120 | 3.7 | 4.5 | 6.7 | 9.5 | 6.7 | 20.7 | 10.3 | 20.9 |

[1] Treatments:
- I Control, ram remained in pen
- II Semen collection by AV
- III Control for EE. Ram restrained but not stimulated.
- IV EE

Prolactin

The patterns of PRL release were similar to those observed for C but there was no significant change in the PRL concentrations when the animals served as controls. A single natural service caused PRL levels to rise significantly ($P<0.01$) and persist at elevated levels for 2 hr. On the third day the ram allocated to AV collection had a pre-treatment (–15 min) concentration of 84.1 ng/ml which accounts for the anomalous pre-treatment mean level of 21.0 ng/ml, but again no extraneous stimulus could be identified. However, the concentration of PRL in this ram increased further to 121.4 ng/ml 15 min after mating so that the pattern of secretion of this hormone in response to mating and ejaculation is consistent. Restraint in preparation for EE (III) stimulated a significant ($P<0.01$) increase in circulating PRL concentrations but this was not significantly different to that following mating. The stimuli of EE were followed by high levels of PRL for 90 min afterwards and this response was significantly greater than those observed after II or III ($P<0.001$).

Luteinising hormone

Changes in plasma concentrations of LH were unrelated to the application of any treatment. Concentrations of over 1 ng/ml (compared with a mean $\pm$ s.e. of $0.31\pm0.01$ ng/ml pre-treatment) were classified as indicative of secretion of LH and occurred at a mean frequency of 0.15 per hr.

## DISCUSSION

This experiment has shown that both C and PRL are secreted in response to stimuli that are part of normal sexual behaviour or to those that are associated with restraint and stress-related stimuli. Comparable PRL responses were observed when rams associated with ewes and mated (Yarney and Sanford 1983), but the Finnish Landrace rams used in that research had much higher PRL concentrations overall. The stimulation of EE raised circulating concentrations of both hormones to relatively high levels but these declined toward pre-treatment values within 105 min of EE. These changes in hormone levels were not accompanied by any alteration in behaviour nor were any longer term responses observed in behaviour. The rams were used in studies beyond this experiment, involving cross-over of semen collection methods of AV and EE, sometimes on the same day, and always at the same collection site with the same operator. All had been trained to walk unrestrained from their pens to the semen collection area and no conditioned aversive responses were observed in a period of 4 weeks of frequent semen collections by either method.

However, the concentrations of C and PRL measured after EE were higher than after AV collection, suggesting that EE is stressful. The use of analgesics or anaesthetics might reduce or eliminate these indications of stress, but further research is necessary to ascertain whether premedication with such drugs alters the efficiency of EE.

## ACKNOWLEDGMENTS

ICAM is grateful for the facilities and assistance given by the Department of Physiology and Anatomy, Massey University, New Zealand, for these experiments.

## REFERENCES

Barrell, G.K. and Lapwood, K.R., 1979. *Anim. Reprod. Sci., 1*, 213-228.
Welsh, T.H. and Johnson, B.H., 1981. *Arch. Androl., 7*, 245-250.
Yarney, T.A. and Sanford, L.M., 1983. *Horm. Behav., 17*, 169-182.

# TOWARD QUANTIFICATION OF PROCEDURES FOR COLLECTION OF SEMEN FROM RAMS BY ELECTROEJACULATION

I.C.A. Martin, *Department of Veterinary Physiology, University of Sydney, Sydney, 2006.*
I.W. Purvis, *Department of Animal Science, University of New England, Armidale, 2350.*

*Summary* Semen was collected by electroejaculation from a total of 430 Merino rams (including rams from CSIRO 'Booroola' and Trangie 'D' flocks) ranging in age from 7 to 27 months using stimulator and timing circuits which automatically delivered 9 volleys of stimuli each of 3 sec duration in a total of 51 sec to each ram examined. The same stimulus was repeated about 2 hr later in the day and the total output of spermatozoa was measured. This procedure has shown that 29% of 7-8 month old rams produced more than 1500 million spermatozoa. This number of spermatozoa or more was ejaculated by two thirds of rams tested at 14-15 months, and by over 80% of 27 month old rams. Sperm output from 61% of this oldest group of rams was greater than 4500 million. These data will be correlated with information on the strains and pedigrees represented to try to identify rams with high reproductive potential as early as possible.

## INTRODUCTION

The major factor prompting the original investigations on methods of electrical stimulation of rams to induce ejaculation was that a technique was needed for collection of semen from large numbers of flock rams in circumstances where the use of the artificial vagina was not practical (Gunn 1936). Edgar (1959 a,b) used electroejaculation extensively in New Zealand for testing rams for fertility and Hulet *et al.* (1964) in the U.S.A. studied the relative merits of natural and electrically-induced ejaculation for obtaining semen for evaluation and prediction of fertility. These workers concluded that natural ejaculates were preferable as indicators of semen output and potential fertility and indicated that the variability of response to electroejaculation was a problem. This was not surprising considering the simplicity of the stimulator circuits which were then available for field use.

The radical changes in circuit design made possible in the last 20 years by the development of transistors and integrated circuits have permitted the construction of stimulators which are much more reliable as well as being highly portable and capable of delivering a pre-programmed set of impulses on every use. Martin (1978) defined the characteristics of the stimuli required to cause ejaculation by rams using such circuits. This paper outlines a method which has been developed for delivery of stimuli, minimal in amplitude and duration, for the electroejaculation of large numbers of flock rams, where every ram examined receives a constant, defined stimulus.

## MATERIALS AND METHODS

### Rams

Details of the flocks from which the rams were drawn and of ages of rams are given in Table 1. The D flock rams were examined at Trangie Agricultural Research Station in October, 1981. The 14-15 month old rams at 'Longford' via Armidale, NSW, were examined in December, 1983, and the 7-8 month age group were tested at the same location in May, 1984.

### Electroejaculation

The stimulator was a development from that described by Martin (1978). It incorporated timing circuits to control automatically the duration of the volleys of stimuli. Voltage was 6.5V, frequency 32 Hz of square-wave AC and 9 volleys of 3 sec of stimulation were used with a 3 sec rest between each volley. A bipolar, rectal electrode carrying 4 longitudinal electrodes with opposite pairs having the same polarity was used. Rams were held manually in lateral recumbency on a table during stimulation.

### Semen evaluation

The volume obtained, the motility and concentration of spermatozoa (determined photometrically using a diluted sample) were observed within minutes of collection for each semen collection made. Total sperm output in response to the two periods (about 2 hr apart) of stimulation in one day was used as the variable for ranking the responsiveness of rams.

## RESULTS

Table 1 shows the mean number of spermatozoa obtained from the ram flocks together with the highest individual sperm output recorded for each flock of rams. As the distribution of sperm outputs from rams 14 months and older was skewed, the data were transformed to logarithms to normalise distributions. Accordingly, the geometric means are presented for each flock.

**TABLE 1. Summary of the ages and numbers of rams examined together with mean numbers of spermatozoa obtained in one day from two standardised sets of stimulation for electroejaculation.**

| Location | Flock | Age (months) | Number of rams | Number of spermatozoa ($10^6$) *Geometric mean* | *Greatest individual sperm output* |
|---|---|---|---|---|---|
| Longford | Control | 7-8 | 38 | 232 | 6165 |
| | Booroola | 7-8 | 99 | 499 | 12103 |
| Longford | Control | 14-15 | 77 | 1862 | 13921 |
| | Booroola | 14-15 | 113 | 1650 | 19038 |
| Trangie | D | 15 | 54 | 1027 | 23410 |
| Trangie | D | 27 | 49 | 3087 | 23130 |

Although flocks have been examined at different times, the results show a consistent trend of increased sperm output with greater age. The distribution of sperm yields in the different age groups (Table 2) is also indicative of development toward a sperm production rate typical of an adult ram.

The differences in sperm output between the control and Booroola rams, within each age group, were not significant and the greatest individual sperm output did not increase after the rams had rached 15 months of age. However, as Table 2 shows, the proportion of rams falling in the category of ejaculating 4500 million spermatozoa or more rose with increasing age. Nevertheless, even in the 27 month old group of rams, no spermatozoa were collected from 3 rams and another 6 produced less than 1500 million spermatozoa.

**TABLE 2. Distribution of rams within flocks according to the total number of spermatozoa collected by electroejaculation from each ram.**

| Location | Flock | Age | Number of rams with output of spermatozoa ($10^8$) in the ranges of: 0 | $n < 15$ | $15 < n < 45$ | $n > 45$ |
|---|---|---|---|---|---|---|
| Longford | Control | 7-8 | 6 | 24 | 5 | 3 |
| | | | (15.8)[1] | (63.1) | (13.2) | (7.9) |
| | Booroola | 7-8 | 7 | 60 | 27 | 5 |
| | | | (7.1) | (60.6) | (27.3) | (5.1) |
| Longford | Control | 14-15 | 5 | 13 | 28 | 31 |
| | | | (6.5) | (16.9) | (36.4) | (40.3) |
| | Booroola | 14-15 | 4 | 32 | 43 | 34 |
| | | | (3.5) | (28.3) | (38.1) | (30.1) |
| Trangie | D | 15 | 7 | 12 | 13 | 22 |
| | | | (13.0) | (22.2) | (24.1) | (40.7) |
| Trangie | D | 27 | 3 | 6 | 10 | 30 |
| | | | (6.1) | (12.2) | (20.4) | (61.2) |

(1) percentage of flock

## DISCUSSION

Although the choice of criteria for classification of output of spermatozoa by electroejaculation may seem arbitrary, the two levels of 1500 and 4500 million spermatozoa have practical bases. In the youngest age group of rams studied, when sperm numbers in this range were produced, these were most frequently contained in volumes of between 0.5 and 1.0ml, i.e. the specimens were approaching the characteristics of concentration and motility observed in semen from adult rams. Older animals produced semen with somewhat higher concentration of spermatozoa but the volume produced increased substantially. The daily sperm production of an adult ram is of the order of 4500 million spermatozoa. If a 7- to 8-month ram can produce this number of spermatozoa even once a week, then it is technically feasible to use the earliest maturing ram lambs in artificial insemination programmes. From the testicular size of young rams, it is likely that such ejaculates represent more than their daily sperm production. The repeatability of electroejaculation of young rams under field conditions has not been tested, but it is an obvious factor for future examination.

The identity of every animal in these field trials has been recorded, but this information has not yet been used to extract strain and pedigree information to ascertain whether sperm production rates from electroejaculation are related to factors associated with reproductive performance. To date no characteristics of

Booroola rams have been identified which are correlated with their fecundity genotype. In this study, the sperm output from Booroola and control rams was similar. Further examination of the rams born in 1983 will be undertaken which, together with pedigree analysis and progeny testing, should give an indication of the prognostic value of testing by electro-ejaculation.

In all flocks, a proportion of rams produced no spermatozoa. This was not surprising with 7-8 month Merinos but it is a matter of far greater concern to find that 6% of 27 month old rams failed to ejaculate and another 12% produced fewer than 1500 million sperm. Before defining such rams as having inadequate sperm production, and therefore culling them, further research on management (e.g. period of isolation from ewes before testing) and technical factors (e.g. the effect of lush pasture on the physical form of the rectal contents so that these do not obstruct the delivery of stimuli) is needed. Limited observations have already shown that natural service earlier in the day sharply reduces yields to later electroejaculation.

In all of these studies, rams have been kept under observation close to the collection area for at least 30 min after stimulation. None showed symptoms of distress nor were there any subsequent reports of malaise or injury which could be attributed to electroejaculation.

## ACKNOWLEDGEMENTS

This research is supported by a grant (L/9/919) from the Australian Wool Corporation.
We thank officers of the N.S.W. Department of Agriculture and of the Division of Animal Production, CSIRO, for access to these flocks and for assistance in the conduct of the field work.

## REFERENCES

Edgar, D.G., 1959a. *N.Z. Vet. J.*, *7*, 61-63.
Edgar, D.G., 1959b. *N.Z. Vet. J.*, *7*, 113-115.
Gunn, R.M.C. 1936. *Bull. Comm. Sci. Industr. Res. Aust.*, No. 94, 116pp.
Hulet, C.V., Foote, W.C. and Blackwell, R.L., 1964. *J. Anim. Sci.*, 418-424.
Martin, I.C.A., 1978. *In* Watson, P.F. (ed). *The Artificial Breeding of Non-domestic Animals*. Symp. Zool. Soc. Lon., *43*, 127-152.

# FOLLICULOGENESIS AND OVULATION RATE IN SHEEP

L.P. Cahill, *Animal Research Institute, Department of Agriculture, Werribee, Victoria. 3030.*

## INTRODUCTION

Folliculogenesis can perhaps be best described as the growth and development of follicles from the pool of primordial follicles, through the pre-antral and antral phases and culminating usually in atresia or rarely in ovulation. In the sheep follicular growth is first observed in the foetal ovary at about Day 70 and thereafter folliculogenesis is a continual process with new follicles commencing growth daily.

This review will endeavour to describe the current mechanisms governing folliculogenesis particularly in antral follicles and then relate this to ovulation rate in sheep.

## PRE-ANTRAL FOLLICLES

### Initiation of Follicular Growth

Considerable controversy exists in defining when a follicle has entered the growth phase with many criteria being used. There are two schools of thought on whether the initial stimulus for a follicle to commence growth and development comes from the oocyte or granulosa cells. Mariana (1978) suggests that the first signs of growth occur in the oocyte whilst Lintern-Moore and Moore (1979) suggest that the stimulus comes from granulosa cells. Gonadotrophins may not be essential agents in initiating follicular growth since following hypophysectomy, follicles with 4-5 or more layers of granulosa cells have been observed in the rat (Paesi 1949) hamster (Moore and Greenwald 1974) and sheep (Dufour, Cahill and Mauleon 1979). Prolactin could play a role in the initiation of follicular growth as prolactin receptors have been observed in the oocyte of primordial and small follicles (Nolin 1978; Dunaif *et al.* 1982), and in sheep, treated with bromocriptine to reduce prolactin levels the number of growing follicles was greatly reduced (Cahill *et al.* 1984a).

The study of De Reviers and Mauleon (1979) investigated in detail the first 'wave' of follicles that commence growing in the infant rat; the number of follicles entering the growth phase increased greatly until Day 20 when there was a dramatic decrease in the number of follicles commencing growth. This decrease in follicles commencing growth is coincident with both the first follicles secreting fluid to form an antrum and a decrease in FSH secretion. This suggests there may be some factor from antral follicles which, through local feedback, influences the number of follicles initiating growth as suggested by Peters *et al.* (1973) or alternatively FSH controls the onset of follicular growth.

### Duration of follicle growth

The first follicles to begin growing in the ovine foetus commence development at Day 70 of gestation and by birth, at Day 148, antral follicles are well developed. This suggests that the growth time in the foetal ovary is slightly more than 78 days which contrasts with the estimate of 6 months in mature ewes (Cahill and Mauleon 1980). This marked difference in development time may be explained by a lack of local growth inhibiting factors, (see below) and thus an accelerated growth in the first 'wave' of follicular growth in the foetus.

## ANTRAL FOLLICLES

Antrum formation occurs in ovine follicles at 0.2mm diameter and further increases in follicular size are due to both an increase in granulosa cell number and follicular fluid accumulation. Soon after antrum formation follicles enter a phase of rapid growth with division of granulosa cells reaching a maximum growth rate at approximately 1.0mm diam. Thereafter there is a progressive decrease in granulosa cell multiplication as the follicle continues to increase in size until the mitotic division in granulosa cells is almost nil in preovulatory follicles (Turnbull *et al.* 1977; Cahill and Mauleon 1980). This general pattern of follicular growth in sheep, occurs independently of the hormonal changes associated with the oestrous cycle or after superovulation with PMSG (Turnbull *et al.* 1977).

### Follicular Fluid Components

Recent research in folliculogenesis has focused on the components of follicular fluid. Analyses of follicular fluid from both human and sheep follicles led McNatty *et al.* (1981) to conclude that most hormones and blood metabolites are present in follicles; moreover no two follicles contain the same hormonal mileau and the local endocrine environment of the ovary is markedly different from that in plasma.

*Steroids* In sheep, oestrogens are present in both pre-antral and antral follicles although most of the oestrogen is secreted from the ovary containing the largest one or two follicles (Baird and Scaramuzzi 1976). The mitogenic action of oestrogens on granulosa cells increases *in vitro* with additions of oestradiol and FSH (McNatty *et al.* 1979). However *in vivo* oestrogen levels are at a maximum when mitotic division of pre-ovulatory follicles is at a minimum. This is further evidence of a local mitotic inhibiting agent in the follicular fluid of large pre-ovulatory follicles.

The two-cell hypothesis which postulates that androgens of thecal origin are aromatized in granulosa cells and the roles of LH and FSH in this system has been well documented (Armstrong *et al.* 1980). This mechanism may not be applicable to smaller antral follicles (1-3mm diam.) as androgens in these follicles are present in a higher concentration than oestrogens (Carson *et al.* 1979). It is also interesting to note that the highest incidence of atresia occurs in follicles 1-3mm diam. (Turnbull *et al.* 1977).

*Follicular Fluid Proteins* Numerous compounds contained in follicular fluid of many species are thought, by many workers, to be of ovarian origin and have a regulatory effect on follicular growth (Table 1).

**TABLE 1. Some non-steroidal follicular fluid compounds thought to regulate follicular growth.**

| | Substance | Action | Reference |
|---|---|---|---|
| 1. | FSHBI | Inhibits binding of FSH to granulosa cells | Daume *et al.* (1979) |
| 2. | P.A. | Plasminogen activator in follicular rupture | Beers (1975) |
| 3. | O.M.I. | Oocyte maturation inhibitor prevents final meiotic division | Tsafriri and Channing (1975) |
| 4. | L.I. | Luteinization inhibitor that prevents LH action on granulosa cells | Channing (1979) |
| 5. | L.S. | Luteinization stimulator that promotes LH receptors | Channing *et al.* (1982) |
| 6. | Inhibin | Suppress FSH secretion from pituitary | deJong and Sharpe (1976) |
| 7. | Anti aromat. protein | Blocks aromatization of androgens | diZerega *et al.* (1982) see below |
| 8. | Follicle growth inhibitor | Blocks granulosa cell division and increases atresia | Cahill *et al.* (1984b, c) see below |

The ability of steriod-free follicular fluid to prevent follicular growth and to decrease both oestrogen production and ovarian weight has been shown by many authors. The mechanism of action can act at the level of the pituitary, eg. inhibin, to suppress FSH secretion. Such a mechanism has been used to enhance the superovulatory response to PMSG administration by combining PMSG with a 'rebound' effect in FSH secretion after a period of bovine follicular fluid (bFF) pre-treatment (Cummins 1983).

More recent work in hypophysectomized animals suggests there is a local ovarian effect. DiZerega *et al.* (1982) found in immature hypophysectomized rats that the ovarian weight and oestrogen response following PMSG was blocked by steriod free bFF. In acutely hypophysectomised ewes where follicles were maintained by PMSG, the administration of steroid-free ovine follicular fluid (oFF) caused a cessation of follicular growth, a decrease in mitotic division of granulosa cells and an increase in atresia (Cahill *et al.* 1984b). More recently we have treated chronically-hypopyhysectomised ewes with 2250 i.u. PMSG to stimulate follicular growth and again 2ml oFF b.i.d. blocked the follicular growth (Cahill *et al.* 1984c; Table 2) as seen by both the visual appraisal and histological examination of ovaries 48h post PMSG treatment.

**TABLE 2. Effects of PMSG (2250 iu) and steroid-free oFF (4ml/day) on the number of large antral follicles.**

| Treatment of hypox ewes | Number of Follicles per size | | | | | |
|---|---|---|---|---|---|---|
| | Visible on ovaries (per ewe)[1] | | | Histol. exam (per ovary)[1] | | |
| | <2mm | 2-4 | >4 | <0.05mm | 0.5-2 | >2.0 |
| Control | 29.3 | 0 | 0 | 53.8 | 8.8 | 0 |
| PMSG | 10.8 | 4.8 | 15.8 | 53.8 | 16.3 | 7.8 |
| PMSG+oFF | 4.8 | 5.3 | 2.8 | 51.2 | 9.3 | 1.3 |
| P (Anovar) | <0.01 | <0.05 | <0.01 | NS | <.1 | <0.01 |

[1] Histological preparation reduces fresh ovarian dimension by × 0.5-0.6. (Taken from Cahill *et al.* 1984c).

Since 4ml/day of steroid-free oFF administered systemically can override such a large dose of PMSG (2250i.u.) it would appear that the follicle growth inhibiting factor in oFF is a very potent agent. In addition these data suggest a mechanism whereby the large or dominant follicle in the hierachy secretes a factor which can locally suppress the growth of smaller follicles (Figure 1) in addition to any feed-back effects via the pituitary. Such a model has important ramifications in our understanding of ovulation rate as the presence of a local agent modulating the action of gonadotrophins would explain why attempts to correlate gonadotrophin levels and ovulation rate have been generally unsuccessful.

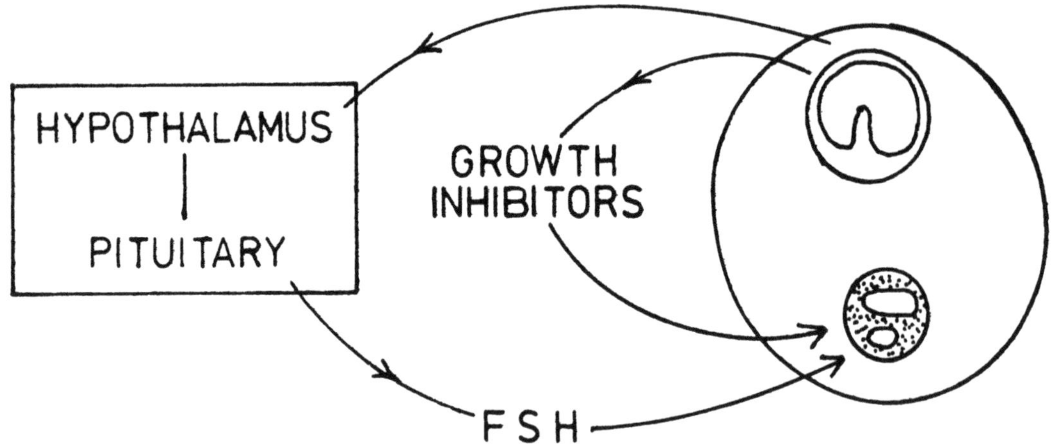


**FIGURE 1. Diagramatic representation of the dual action of follicular fluid components.**

Recruitment and Selection of Follicles to Ovulate

Recruitment of follicles is defined as the mechanism whereby the final grouping of follicles occurs and then from among these follicles some are selected to ovulate. In sheep, follicles are thought to be recruited for ovulation soon after luteolysis with selection or fine tuning occurring later. This hypothesis is based on three pieces of evidence: (i) Findlay and Cumming (1977) found that unilateral ovariectomy of ewes before luteolysis (Day 14) caused the remaining ovary to ovulate at the normal rate 3 days later. However when unilateral ovariectomy was carried out after luteolysis (Day 16) the time of ovulation was delayed. (ii) Tsonis (1984) extended this finding by ablating follicles of various sizes and he was able to show that the follicles that ovulated came from amongst those >2mm (thus readily visible on the ovary) at the time of luteolysis. (iii) Driancourt and Cahill (1984) found that at the start of luteolysis, atresia among follicles >2mm diam. brought about the formation of the recruitment pool of follicles, and at 40h *post* PG there was additional atresia which resulted in the selection of follicles that went on to ovulate (Figure 2). A notable exception was the Booroola where after luteolysis follicles less than 2mm continued to grow and enter the recruitment pool and hence their high ovulation rate is due to this late follicular growth and not only to an increased follicular number (see next section).

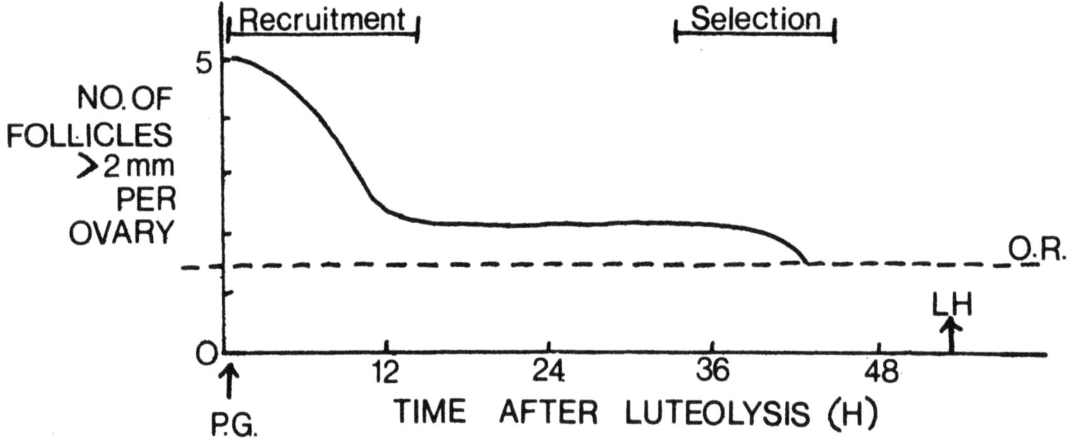


**FIGURE 2. Diagramatic representation of the decrease in the number of healthy follicles in the recruitment pool (> 2mm diam.) between luteolysis and the LH release in ewes with a low ovulation rate (adapted from Driancourt and Cahill 1984).**

Gonadotrophin receptors in ovulatory follicles

Identification of the pre-ovulatory follicle has been carried out using various criteria with the LH/hCG receptor population in thecal and granulosa cells proving highly successfuly. Webb and England (1982) found in Finn and Suffolk ewes (ovulation rates of 2.7 and 1.3 respectively) that during the follicular phase follicles which are 'activated' i.e. acquire LH/hCG receptors in thecal and granulosa cells, respond to the LH surge and ovulate. The number of 'activated' follicles during the luteal phase (England *et al.* 1981) and following PMSG stimulation during anoestrus (England and Webb 1979) has also been found to equal the ovulation rate (O.R.). The induction of LH/hCG receptors has been shown to be influenced by oestradiol production (Richards 1980) which in turn can be influenced by LH pulse rate, aromatase activity, androgen precursor and FSH. Unlike LH/hCG receptors, FSH binding to granulosa cells does not increase during the follicular phase in sheep but remains constant per granulosa cell (Carson *et al.* 1979).

## OVULATION RATE IN SHEEP

In general terms O.R. in sheep is influenced by five factors:- Genetic effects, age of ewe, liveweight-nutrition complex, season and hormonal therapies. Each factor shall be discussed in terms of our knowledge of folliculogenesis.

### Genetic effects

The genetic effects on O.R. and their endocrine basis are reviewed by Bindon *et al.* (1983) and Scaramuzzi and Radford (1983). The five genotypes recognized as having high O.R. (3.0 or greater) are the Booroola Merino, Finn, Dahman, Romanov and Hu Yang. The Romanov and Booroola ewe have been studied in detail but they differ fundamentally in how their high O.R. is achieved. The high O.R. in Romanov ewes, thought to be due to many genes, is additive, hence O.R. of the offspring is half that of the sire's plus the dam's line (Ricordeau *et al.* 1978). In contrast the high O.R. in the Booroola is thought to be due to a single gene (Piper and Bindon 1983) thus in crossbreeding the O.R. is not additive but dependent on the segregation of this gene. Given these basic differences between the breeds it is perhaps not too surprising that the mechanisms to achieve their high ovulation rates also differ.

*Romanov: case-study 1* Quantitative histological examinations of ovaries from mature Romanov (O.R., 3.1) and Ile-de-France (O.R., 1.5) ewes have shown no difference between the breeds in atresia or follicular growth rates but the number of growing follicles at all stages appears to be directly related to O.R. (Cahill *et al.* 1979). Hence most of the difference in O.R. can be explained simply by a difference in number of follicles.

*Booroola: case-study 2* Quantitative histological studies of Booroola and Merino ovaries have failed to show that the observed difference in O.R. can be explained in terms of the number of growing follicles. Driancourt (1984) found no difference between Booroola and Merino ewes in the number of growing follicles and Cahill *et al.* (1982) found a 2 fold difference but the difference in O.R. was 3 to 4 fold. Thus Booroola ewes do not have a higher O.R. simply because there are more follicles but rather by growth of additional follicles into the recruited pool of follicles (i.e. > 2mm) after luteolysis (Driancourt 1984). This phenomena does not appear to occur in ewes with low O.R. or Romanov ewes (M.A. Driancourt, pers. comm.). It is possible therefore that this additional growth in the follicular phase in Booroolas could be due to a different action of local follicle growth inhibitors. Thus two different mechanisms have been demonstrated in groups of Booroola and Romanov ewes and this raises the possibility that for a given individual both mechanisms may operate simultaneously depending on supply of follicles at recruitment.

### Age of Ewe

The low O.R. of maiden 1.5 y.o. ewes and its subsequent increase in mature ewes is well documented (McKenzie and Terrill 1937 and many others). At about 3-5 y.o. ewes reach their peak in O.R. which is then maintained until at least 10 y.o. (Bindon *et al.* 1980).

The physiological mechanism associated with the effect of age on O.R. is poorly understood. The study of Al-Obaidi *et al.* (1983) where immunization against a crude inhibin fraction of bFF resulted in a dramatic advancement in the age of puberty, would suggest that young ewes may be highly sensitive to oFF either at the pituitary or ovarian level. A similar mechanism could explain the decreased O.R. in young ewes.

The quantitative histological examination of 2 and 8 year old ovaries of Merino ewes (Cahill *et al.* 1982, Table 3) shows that the older ewes had an 8% and 21% increase in the number of growing (pre antral and antral) follicles in Booroola and AB 20 ewes respectively which compares with an 11% and 26% increase in O.R. (Bindon *et al.* 1980).

Although these results suggest that changes in O.R. with age can be explained primarily on the basis of follicle number, further work is required to be carried out as the follicle data is based on only 12 ewes.

### Liveweight-Nutrition Complex

Several studies have reviewed at length the relationships of liveweight and nutrition on O.R. in sheep; furthermore the interaction of the other factors (season, genotype, age and hormone therapies) with liveweight in terms of ovulation rate have also been studied at length. The review of Morley *et al*, (1978) concluded that a positive correlation exists with a 2-2.5% increase in O.R. per 1kg increase in mean liveweight. The relative contributions to this relationship from the so called 'static' effects (liveweight *per se*) and 'dynamic' effects ('flushing') have been variously attacked by many studies and remains controversial. Specific supplements

**TABLE 3. Comparison of number of growing follicles and ovulation rates and the percentage increase in each between 2 and 8 year old Booroola and AB 20 Merino ewes.**

| *Strain* | *Age* | *Growing Follicles*[1] | | *Ovulation Rate*[2] | |
|---|---|---|---|---|---|
| | | *No.* | *% increase* | Rate | *% increase* |
| Booroola | 2 | 184 | — | 2.49 | — |
| | 8 | 200 | 8% | 2.88 | 11% |
| AB20 | 2 | 87 | — | 1.03 | — |
| | 8 | 111 | 21% | 1.40 | 26% |

(1) data taken from Cahill *et al.* (1982).
(2) data taken from Bindon, Piper and Evans (1980).

such as lupins given over time intervals as short as 8 days prior to joining have resulted in substantial increases in O.R. (Knight *et al.* 1975). Since this effect of lupins occurs over such a short period of time it would seem that lupins may influence the recruitment or selection of follicles leading up to ovulation. Further studies proposed by Dr. J.B. Rowe and his team (W.A. Dept. of Agric. — see this review) to investigate the energy or protein requirement in this phenomenon should be most useful.

Researchers have investigated the relationship between gonadotrophin levels and liveweight with little success. At the ovarian level, Haresign (1981) found 'flushing' increased the number of follicles > 2mm diam. whereas others have found nutritional effects to have no influence on the number of follicles but rather on the size of follicles in the hierachy (Dufour and Matton 1977). A study of the effect of liveweight on the sensitivity to oFF has shown that the ovarian response to oFF may be closely related to liveweight of ewes but this requires further verification (R.F. Fry and L.P. Cahill, unpublished data). These effects of liveweight and nutrition, supposedly not mediated via gonadotrophins require a great deal of research particularly at the ovarian level since complex nutritional studies, whose end-point is the binomial nature of O.R. are costly and difficult.

Season

The time of joining and the consequential effects on O.R. are well documented. In most breeds ewes commence cyclic oestrous activity in summer and O.R. rises to a maximum in mid-autumn followed by a decline as the ewes approach anoestrus.

The mechanisms responsible for the seasonal change in ovulation rate are not well understood although the endocrine changes with season are well documented (see Karsch 1984). At the ovarian level Gherardi and Lindsay (1980) showed that the O.R. response in Merinos to PMSG was lowest in spring and highest in autumn which mimics the seasonal variation observed in the number of large follicles (Kammerlade *et al.* 1952). These seasonal variations could be explained by the influence that season (and photoperiod) has been reported to have on the incidence of atresia (cow, Rajakoski 1960; sheep, Cahill *et al.* 1984a) such that the incidence of atresia in large follicles falls in spring before the onset of the breeding season and then rises again at the end of the season (Figure 3). It remains unexplained whether the incidence of atresia changes slowly and so accounts for changes in O.R. within the breeding season. Furthermore, endocrine factors which vary with season, such as melatonin and LH pulse rate may be worthy of investigation in respect to follicular atresia. The 'ram effect' can advance the onset of the breeding season possibly through an increase in the LH pulse rate (see Pearce and Oldham 1984). This example may provide further evidence that follicles which are normally destined for atresia grow through to ovulation due to a change in LH pulse rate brought about by introduction of the ram.

Hormone Therapy

Exogenous hormones which can increase O.R. include PMSG (seasonal effects: Gherardi and Lindsay 1980), FSH (Bondioli and Wright 1981) and hCG (see Radford *et al.* 1984). Oestrogens, particularly coumestrol can depress O.R. (Smith, *et al.* 1979). In addition immunization against androstenedione and other ovarian steriods has led to an increase in O.R. (see Scaramuzzi and Martin 1984). In general exogenous gonadotrophins are highly variable in response and hence are more useful in super-ovulating ewes whilst immunization against androstenedione is most useful in inducing twin ovulations.

The mode of action of exogenous gonadotrophins is poorly understood. It has been suggested that PMSG prevents atresia (Hay and Moor 1978) and increases the growth rate of large antral follicles (Turnbull *et al.* 1977). Both these actions could contribute to a higher O.R. Immunization against androstenedione leads to a change in the distribution of atretic follicles and more healthy antral follicles but it is difficult to correlate these changes to the hormonal profile (Scaramuzzi 1984).

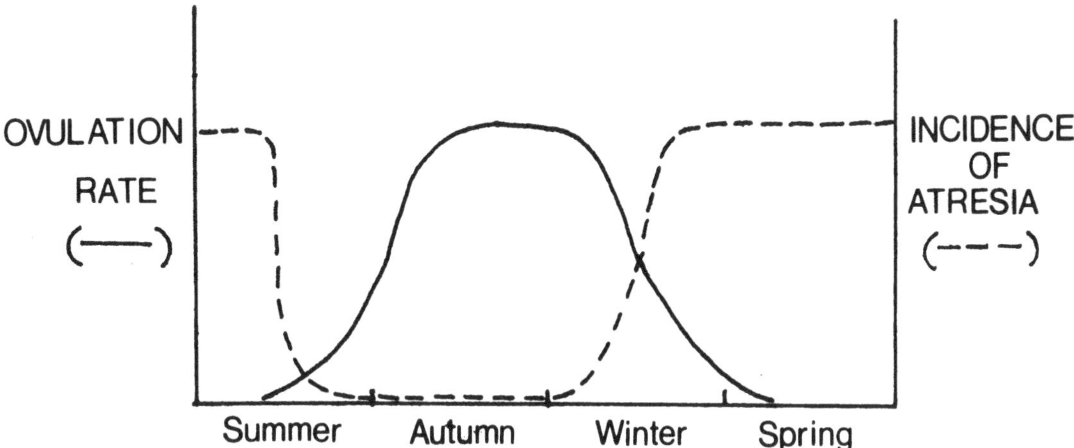


**FIGURE 3. A model of seasonal interaction of follicular atresia and ovulation rate.**

## CONCLUSIONS

The effects on O.R. of such factors as genetics, age, liveweight-nutrition, season and hormonal therapy and their inter-relationships have been studied in detail and are well documented. In contrast the endocrine and physiological mechanisms that determine ovulation rate are poorly understood, although new areas of work look promising.

*Gonadotrophin studies* Numerous attempts to correlate hormonal parameters with O.R. have been generally unsuccessful. The role of gonadotrophins, particularly FSH, in the late luteal and early follicular phase, presumably to present more follicles at a stage whereupon recruitment can occur, could be a profitable area of research. In this respect the Booroola is an ideal experimental model. Furthermore, a deeper understanding of factors that influence a follicle's acquisition of LH receptors, which denotes it as an ovulatory follicle, could also be most useful.

*Nutrition — O.R. studies* Detailed studies of the lupin effect are greatly adding to our knowledge of the important feed components. Unfortunately most such studies to date have been dominated by nutritionists (who have investigated many and varied nutrients in respect to a single end point, O.R.) or by physiologists (who have studied crude liveweight effects on indepth follicular dynamics). An investigation incorporating detailed nutrition partitioning and follicular studies, simultaneously, could prove to be far more productive.

*Follicle growth inhibitors* The discovery of follicle growth inhibitor(s) which can directly suppress follicular growth and can override the effects of gonadotrophins opens up new horizons in our understanding of O.R. It would be of interest to know how production of, or sensitivity to, this growth inhibitor varies according to breed, age, liveweight, time of joining and hormonal treatment of ewes. It is not known where and what stage of follicle growth that the follicle growth inhibitor is produced. Its mode of action in relation to steroids and other ovarian metabolites is also unknown. Ultimately it would be hoped to purify this follicle growth inhibitor with a view to measuring it in blood, lymph and tissue; and finally immunization, either active or passive against this inhibitor could improve the control of follicular growth and ovulation rate in sheep to a far greater extent than presently exists.

## REFERENCES

Al-Obaidi, S.A.R., Bindon, B.M., O'Shea, T., Hillard, M.A. and Cheers, M.A., 1983. *Proc. Aust. Soc. Rep. Biol. 15*, 80.

Armstrong, D.T., Weiss, T.J., Selstam, G. and Seamark, R.F., 1980. *J. Reprod. Fert.* Suppl. 30, 143-154.

Baird, D.T. and Scaramuzzi, R.J., 1976. *Acta. endocr. 83*, 402-409.

Beers, W.H., 1975. Cell *6*, 379-386.

Bindon, B.M., Piper, L.R., Cummins, L.J., O'Shea, T., Millard, M.A., Findlay, J.K. and Robertson, D.M. (1984). *In* R.B. Land (ed), *The Genetics of Sheep Fecundity*, Butterworth (Lond.), pp217-235.

Bindon, B.M., Piper, L.R. and Evans, R., 1980. *Proc. Booroola Workshop,* Armidale C.S.I.R.O. (Melb) pp21-33.

Bondioli, K.R. and Wright, R.W., 1981. *Theriogenology. 15*, 118.

Cahill, L.P., Clarke, I.J., Cummins, J.T., Driancourt, M.A., Carson, R.S. and Findlay, J.K., 1984b. Proc. *Soc. Study. Repro*. Laramie, Wy. U.S.A.

Cahill, L.P., Clarke, I.J., Cummins, J.T. and Findlay, J.K., 1984c. *Proc. Aust. Soc. Reprod. Biol. 16*, 21.

Cahill, L.P., Loel, T.A., Turnbull, K.E., Piper, L.R., Bindon, B.M. and Scaramuzzi, R.J., 1982. *Proc. Aust. Soc Rep. Biol. 13*, 76.

Cahill, L.P., Mariana, J.C. and Mauleon, P., 1979. *J. Reprod. Fert.* 55, 27-36.
Cahill, L.P. and Mauleon, P., 1980. *J. Reprod. Fert. 58*, 321-328.
Cahill, L.P., Oldham, C.M., Cognie, Y., Ravault, J.P. and Mauleon, P., 1984a. *Aust. J. Biol. Sci. 37*, 71-77.
Carson, R.S., Findlay, J.K., Burger, H.G. and Trounson, A.O., 1979. *Biol. Reprod. 21*, 75-87.
Channing, C.P., 1979. *In* Channing, C.P., Marsh, J.M., and Sadler (eds), *Ovarian Follicular and Corpus Luteum Function*, pp 327-343. W.A., Plenum Press, N.Y.
Channing, C.P., Anderson, L.D., Hoover, D.J., Kolena, J., Osteen, K.G., Pomerantz, S.H. and Tanabe, K., 1982. *Rec. Prog. Horm. Res. 38*, 331-408.
Cummins, L.J., 1983. Ph.D. thesis, Univ. of New England.
Daume, E., Chari, S., Hopkinson, C.R.N. and Sturm, G., 1979. *Arch. Gynecol. 227*, 289-290.
De Jong, F.H. and Sharpe, R.M., 1976. *Nature, 263*, 71-72.
De Reviers, M.M. and Mauleon, P., 1979. *Annls. Biol. Anim. Biochim. Biophys. 19*, 1745-1756.
Dizerega, G.S., Goebelsmann, U. and Nakamura, K., 1982. *J. Clin. Endocr. Metab. 54*, 1091-1096.
Dufour, J.J., Cahill, L.P. and Mauleon, P., 1979. *J. Reprod. Fert.* 57, 301-309.
Dufour, J.J. and Matton, P., 1977. *Can. J. Anim. Sci.* 57, 647-652.
Dunaif, A.E., Zimmerman, E.A. Friesen, H.G. and Frantz, A.G., 1982. *Endocrinology 110*, 1465-1471.
Driancourt, M.A., 1984. M.Sc. Thesis, Monash University.
Driancourt, M.A. and Cahill, L.P., 1984. *J. Reprod. Fert. 71*, 205-211.
England, B.G. and Webb, R., 1979. *Proc. Amer. Soc. Anim. Soc. 71*, Abstr.
England, B.G., Webb, R. and Dahmer, M.K., 1981. *Endocrinology* 109, 881-887.
Findlay, J.K. and Cumming, I.A., 1977. *Biol. Reprod. 17*, 178-183.
Gherardi, P.B. and Lindsay, D.R., 1980. *J. Reprod. Fert. 60*, 425-429.
Haresign, W., 1981. *Anim. Prod. 32*, 197-202.
Hay, M.F. and Moor, R.M., 1978. *In* D.B. Crighton, G.R. Foxcroft, N.B. Haynes and G.E. Lamming (eds), *Control of Ovulation* Butterworths (Lond.) pp176-196.
Kammerlade, W.G., Welch, J.A., Nalbandov, A.V. and Norton, H.W., 1952. *J. Anim. Sci. 11*, 646-655.
Karsch, F.J., 1984. This volume, 10-15.
Knight, T.W., Oldham, C.M. and Lindsay, D.R., 1975. *Aust. J. Agric. Res. 26,* 567-575.
Lintern-Moore, S. and Moore, G.P.M., 1979. *Biol. Reprod. 20*, 773-778.
Mariana, J.C., 1978. *Annls. Biol. Anim. Biochim. Biophys. 16*, 545-559.
McKenzie, F.F. and Terrill, C.E., 1937. Univ. Mo. Agric. Res. Bull. 264.
McNatty, K.P., Gibbs, M., Dobson C., Thurley, D.C. and Findlay, J.K., 1981. *Aust. J. Biol. Sci. 34*, 67-80.
McNatty, K.P., Markis, A. Degrazia, C., Osathanondh, R. and Ryan, K.J., 1979. *J. Clin. Endocrinol. Metab. 49*, 687-699.
Moore, P.J., and Greenwald, G.S., 1974. *Am. J. Anat. 139*, 37-48.
Morley, F.H.W., White, D.H., Kenney, P.A. and Davis, I.F., 1978. *Agric. Syst. 3*, 27-45.
Nolin, J.M., 1978. *J. Cell. Biol. 79*, 177 (Abstract).
Paesi, F.J., 1949. *Acta. Endocr. 3*, 173-180.
Pearce, D.T. and Oldham, C.M., 1984. This volume, 26-34.
Peters, M., Byskov, A.G. & Faber, M., 1973. *Excerpta Medica, Int. Congress Series,* No. 267.
Piper, L.R. and Bindon, B.M., 1983. *Proc. Workshop Ovine Prolific Breeds.* Edinburgh (in press).
Radford, H.M., Avenell, J.A. and Szell, A., 1984. This volume, 342-344.
Rajakoski, E., 1960. *Acta. Endocr.* Suppl. 52, 7-68.
Richards, J.S., 1980. *Physiol. Rev.* 60, 51-72.
Richordeau, G., Tchamitchian, L., Thimonier, J., Flamant, J.C. and Theriez, M., 1978. *Livestock Prod. Sc.* 5, 181-201.
Scaramuzzi, R.J., 1984. *Proc. Aust. Soc. Anim. Prod. 15*, 191-193.
Scaramuzzi, R.J. and Martin, G.B., 1984. This volume, 316-325.
Scaramuzzi, R.J. and Radford, H.M., 1983. *J. Reprod. Fert. 69,* 353-367.
Smith, J.F., Jagusch, K.T., Brunswick, L.F.C. and Kelly, R.W., 1979. *N.Z.J. Agric. Res. 22*, 411-416.
Tsafriri, A. and Channing, C.T., 1975. *Endocrinology 96*, 922-927.
Tsonis, C.G., 1984. Ph.D. Thesis, Monash University.
Turnbull, K.E., Braden, A.W.H. and Mattner, P.E., 1977. *Aust. J. Biol. Sci. 30*, 229-241.
Webb, R. and England, B.G., 1982. *Endocrinology* 110, 873-881.

# SEASONAL OESTROUS AND OVULATION PATTERNS OF BORDER LEICESTER CROSSBRED EWES FROM THREE STRAINS OF MERINO EWES AND THEIR BOOROOLA CROSSES

E.A. Dunstan and D. Phillips, *Department of Agriculture, Kybybolite Research Centre, Box 2, Kybybolite, South Australia, 5262.*

*Summary* Six groups of Border Leicester cross ewes were produced from South Australian Strongwool (Bungaree), Medium Wool Peppin and Victorian Finewool ewes, and from Booroola Fl crosses of all three. Lambs were born in May 1982. Oestrous and ovulation patterns were determined in the continuous presence of rams for the first oestrous season which extended from March until July, 1983.

No differences were detected between the three non Booroola lines in either oestrus or ovulation. All three lines containing the Booroola genotype had higher ovulation rates than their respective non Booroola counterparts.

These data do not favour any of the three main crossbred types in terms of potential reproductive ability as prime lamb mothers. In all three lines there is a potential to improve lamb production by incorporating the Booroola genotype, although this necessitates a more complex breeding system for the production of prime lamb mothers and increases the chance of higher order (3 or more) multiple births which many producers would see as undesirable.

## INTRODUCTION

By far the most popular ewe genotype for prime lamb production in Southern Australia is the Border Leicester × Merino cross, a strong wool type renowned for high fertility, fecundity, mothering ability, hardiness and easy care.

Many producers have reported marked differences in the reproductive performance of Border Leicester × Merino cross ewes purchased for their fat lamb enterprises. This is not surprising when one considers the marked variation between strains of Merinos in seasonal oestrous and ovulation patterns, with widely differing litter sizes (Bindon and Turner 1974) and early spring joining performance (Dun *et al.* 1966). In one of the few studies published comparing the performance of crossbred ewes from different types of Merinos, Beard and Hodge (1978) reported that Border Leicester × Bungaree ewes were 9-10 kg heavier and produced fleeces 0.4-0.5 kg heavier than Border Leicester × Peppin ewes as two and three year olds. While there were no consistent differences in ovulation rate or lambing percentages in the first two years with a March mating, significantly more of the Border Leicester × Bungaree ewes showed oestrus at 12 months of age and in early summer before their first mating. These results suggest that the known differences between the Merino Strains, namely that South Australian strains are larger and less seasonal, may carry through into the first cross ewe.

The main components of profitability in a prime lamb enterprise lie in the fertility, fecundity and mothering ability of the ewe and in the ability of the producer to finish lambs to meet market demands. One way of improving the fecundity of crossbred ewes is to incorporate the Booroola Merino genotype.

This paper reports oestrous and ovulation rates in the continuous presence of rams for their first breeding season of Border Leicester cross ewes derived from South Australian Strongwool (Bungaree), Medium Wool Peppin and Victorian Finewool and crosses incorporating the Booroola Merino.

## MATERIALS AND METHODS

500 ewes of each of the three genotypes of Merino ewes, South Australian stongwool (SAS), Medium Wool Peppin (MWP) and Victorian Finewool (VF) were mated in January 1980 at Struan Research Centre to rams of their own breed or to Booroola rams to produce three lines of base Merinos (SAS, MWP and VF), and three lines of Fl Booroola × Merino (Bo/SAS, Bo/MWP and Bo/VF) ewe lambs. These were all born and reared at the same time and under the same conditions.

These ewes were mated in December 1981 and Border Leicester rams to produce six groups of crossbred ewes (ranging in number from 34 to 61) for this study.

Ewes were transferred to Kybybolite Research Centre in September 1982, and run as a single flock until early December when they were separated into their respective groups and kept separate until March, when all groups were cycling. Vasectomised rams fitted with Sire-Sine harnesses were introduced on 16/12/82 and run continuously with the ewes from that time. After most ewes in all groups were cycling, the groups were mixed and run as a single mob until July when they were again run separately until all oestrous activity ceased.

Ovulation rates (OR) of all ewes were recorded at monthly intervals from December 1982 until September 1983 by laparoscopy. Oestrous activity was determined by weekly recording of Sire-Sine marks.

Oestrous data were analysed by analysis of variance and ovulation rate data by chi-square analysis of contingency tables as described by Bhapkar (1980).

## RESULTS

No ewes exhibited oestrous or ovulatory activity until March, 1983. A similar proportion of each group exhibited oestrus during the period March 18 — April 13 (Table 1). A slightly lower proportion of BL × Bo/MWP and BL × Bo/VF ewes exhibited oestrus in the next period, April 14 — May 11. However, these two groups tended to stay in season longer and a high proportion showed oestrous activity in May.

**TABLE 1. Percentage of Border Leicester × Merino and Border Leicester × (Booroola × Merino) ewes born in 1982, exhibiting oestrus at various times of the year in 1983.**

| TIME OF YEAR | STRAIN | | | | | |
|---|---|---|---|---|---|---|
| | BL × SAS | BL × MWP | BL × VF | BL × Bo/SAS | BL × Bo/MWP | BL × Bo/VF |
| Jan 1 – Mar 17 | 0 | 0 | 0 | 0 | 0 | 0 |
| Mar 18 – Apr 13 | 41a[(1)] | 38a | 53a | 33a | 56a | 31a |
| Apr 14 – May 11 | 98a | 98a | 100a | 92ab | 82b | 85b |
| May 12 – Jun 16 | 87a | 91a | 82a | 79a | 69a | 85a |
| Jun 17 – Jul 13 | 7abc | 9ab | 3bc | 0c | 16ab | 20a |
| Jul 14 – Dec 31 | 0 | 0 | 0 | 0 | 0 | 0 |

(1) Figures in a row with differing notation differ significantly, $p < 0.05$

There were significant differences between groups in OR per ewe ovulating at the time of each recording while ewes were in season (Table 2). The OR for the three non Booroola lines started at around 1.0 in March and increased to a peak of 1.25 — 1.4 before falling away towards anoestrus.

The OR of the Booroola crosses were significantly higher than their 'normal' counterparts in all cases except BL × MWP and BL × Bo/MWP in March. This was largely due to an increase in the number of twin ovulations in these groups, although a considerable proportion of higher order ovulations (mainly three or four) were recorded (17% for BL xBo/SAS, 14% for BL × Bo/MWP and 28% for BL × Bo/VF in the month of peak ovulation — May).

**TABLE 2. Ovulation rate per ewe ovulating during 1983 of Border Leicester × Merino and Border Leicester × (Booroola × Merino) ewes born in 1982.**

| MONTH | STRAIN | | | | | |
|---|---|---|---|---|---|---|
| | BL × SAS | BL × MWP | BL × VF | BL × Bo/SAS | BL × Bo/MWP | BL × Bo/VF |
| Number of ewes per mob | 46 | 45 | 34 | 52 | 45 | 61 |
| Jan | 0 | 0 | 0 | 0 | 0 | 0 |
| Feb | 0 | 0 | 0 | 0 | 0 | 0 |
| Mar | 1.0a[(1)] | 1.14ab | 1.00a | 1.31b | 1.18ab | 1.38b |
| Apr | 1.03a | 1.24ab | 1.26b | 1.30bc | 1.54cd | 1.71d |
| May | 1.25a | 1.38a | 1.41ab | 1.87c | 1.67bc | 1.96c |
| Jun | 1.14a | 1.13a | 1.17a | 1.59b | 1.64b | 1.79b |
| Jul | 0a | 0a | 0a | 2.0b | 1.25b | 1.87b |
| Aug–Dec | 0 | 0 | 0 | 0 | 0 | 0 |

(1Figures in a row with different notation differ significantly, $p < 0.05$.

## DISCUSSION

From these results, reproductive rates would not be expected to differ markedly between the three non Booroola lines. This confirms the results of Beard and Hodge (1978). Considerations of wool quantity and quality may make one cross more profitable than the others.

Incorporating the Booroola genotype into the crossbred ewe by the method described in this paper can only lead to a maxmum of 50% of the ewes carrying the Booroola's major gene for high fecundity, provided that all Booroola rams used were homozygous. It is clear, however, that the use of this genotype can improve breeding potential of all three crossbreds. Disadvantages to its use would be the extra step involved in producing the Fl Booroola × Merino ewe, and the number of multiple births, triplets or greater, which might result. Under extensive grazing conditions, survival of these lambs is not likely to be high and more intensive care would be necessary.

This work has recorded the reproductive performance of ewes in the continuous presence of rams. Most producers in Southern Australia mate their ewes in mid-summer and the first oestrus is induced by the introduction of rams. Ovulation rates under these conditions could be different from those recorded here.

This study will continue for a second season in the continuous presence of rams. After the second anoestrus has been established, rams will be removed, and reintroduced in mid-summer to determine oestrous and ovulation rates relating to the more commonly found mating treatment.

## REFERENCES

Beard, K.T., and Hodge, R.W., 1976. *Proc. Aust. Soc. Anim. Prod., 12*, 262.

Bhapkar, V.P., 1980. *In, Handbook of Statistics Volume 1*, North-Holland Publishing Co., Amsterdam, Chapter 11.

Bindon, B.M. and Turner, H.N., 1974. *J. Reprod. Fert., 39*, 85-88.

Dun, R.B., Alexander, R. and Smith, M.D., 1966. *Proc. Aust. Soc. Anim. Prod., 6*, 66-68.

# FREQUENCY OF LUTEINIZING HORMONE RELEASE IN MERINO EWES WITH ONE AND TWO OVULATIONS

G.B. Thomas and C.M. Oldham, *School of Agriculture (Animal Science), University of Western Australia, Nedlands, Western Australia, 6009.*
G.B. Martin, *MRC Reproductive Biology Unit, Centre for Reproductive Biology, 37 Chalmers Street, Edinburgh, EH3 9EW, Scotland.*

*Summary* This study compared the secretion of LH during the oestrous cycle of Merino ewes consistently having one or two ovulations. Ewes with two corpora lutea released more pulses of LH but produced less progesterone during days 2-12 of the oestrous cycle than ewes with one corpus luteum. There were no differences at other times of the cycle. Neither the mean concentration of LH nor the mean amplitude of pulses of LH varied with the number of ovulations.

## INTRODUCTION

The role of luteinizing hormone (LH) in the regulation of ovulation rate is not known (Cahill *et al.* 1981; Scaramuzzi and Radford 1983). McLeod *et al.* (1983) suggested the final stages of follicular development was dependent on an increase in the concentration of LH, although other studies have failed to find any correlation with the concentration of LH and ovulation rate (Cahill *et al.* 1981). However, ovulatory responses are often correlated with increases in the frequency of the pulses of LH. For example, immunization of ewes against ovarian steroids and the introduction of rams to seasonally anoestrous ewes both result in increases in the frequency of pulses of LH and ovulation rate (Martin 1984). The present study tested the hypothesis that ewes consistently having two ovulations have a higher frequency of pulses of LH than ewes consistently having one ovulation.

## MATERIALS AND METHODS

Two groups of six Merino ewes with similar mean liveweights (47.5 ± 1.08 vs 48.3 ± 1.96 kg; $\bar{x}$ ± SEM) which had either one or two ovulations respectively for seven consecutive oestrous cycles were selected between January and April 1981 (C.M. Oldham, unpublished). Oestrus was synchronised using a synthetic prostaglandin (Estrumate, ICI) given in two 125 $\mu$g injections 10 days apart in May. On the day of the second injection one jugular vein of each ewe was cannulated and the ewes were placed with two vasectomised rams wearing Sire-Sine harnesses and crayons. The ewes were then checked for crayon marks every 8 hours. From the onset of oestrus (day 0) blood was collected every 4 hours for 48 hours. During days 2, 4, 6, 8, 10, 12, 14 and 16 blood was also collected every 20 minutes for 8 hours. From day 17 blood was sampled again every 4 hours for a further 5 days. On days 5 and 24 the ovaries of the ewes were observed by laparoscopy and the number of corpora lutea recorded.

The plasma was stored at −20°C until assayed for LH and progesterone as described by Martin *et al.* (1983). Pulses of LH were defined using the definition of Martin *et al.* (1983). Statistical analysis of mean concentration and pulse amplitude was performed using analysis of variance. Pulse frequency was analysed using logistic regression analysis.

## RESULTS

At both laparoscopies all ewes had either one or two corpora lutea in accord with their previous history, and there was no difference between the groups in the mean interval between the preovulatory surges of LH (17.4 ± 0.4 days).

Overall, the patterns of secretion of LH and progesterone through the oestrous cycle were similar in both groups. During the period of increasing concentration of progesterone (day 2–12), the luteal phase, the frequency of pulses of LH was low, and independent of the concentration of progesterone ($r^2$=0.05). However, ewes with two corpora lutea released more pulses of LH ($P<0.05$) but secreted less progesterone ($P<0.05$) than ewes with one corpus luteum (Figure 1). There were no differences at other times of the cycle.

The mean concentration of LH and the mean amplitude of pulses of LH were similar during both the luteal and follicular phases of the oestrous cycles of the two groups of ewes (Table 1).

## DISCUSSION

Merino ewes consistently having two ovulations released approximately twice as many pulses of LH during the luteal phase of their oestrous cycle than flockmates consistently having a single ovulation. The ewes in our experiment were a highly selected sample and therefore the frequency of pulses of LH may have been associated with selection rather than with the difference in ovulation rate *per se*. However, further evidence that the frequency of pulses of LH, during the luteal phase, controls ovulation rate has been reported in brief by Rhind and McNeilly (1983). In their study the frequency of pulses of LH in the luteal phase of Scottish

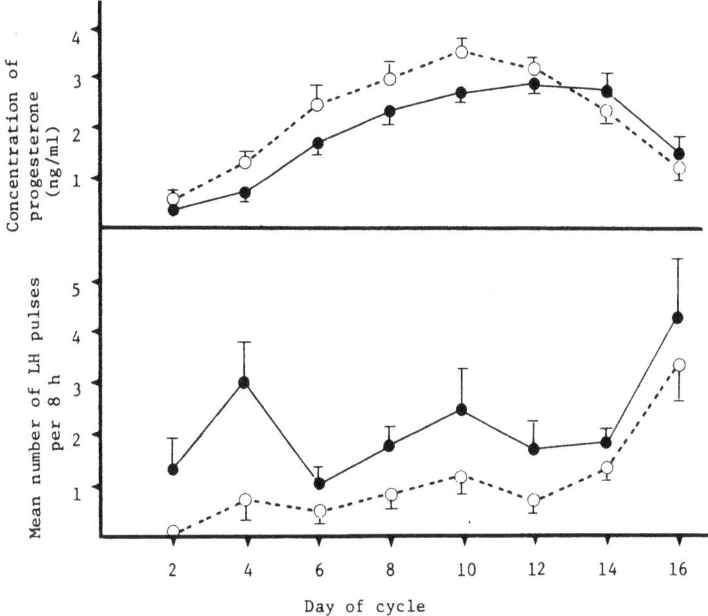


**FIGURE 1. The secretion of LH and progesterone in peripheral plasma of Merino ewes with one (○ -- ○) or two (● — ●) ovulations. Each point represents the mean ±SEM of 6 animals. (Day 0 = day of oestrus).**

**TABLE 1. The mean amplitude of pulses of LH and mean concentration of LH during the oestrous cycle in ewes with one or two ovulations (mean ±SEM).**

| | Phase of the oestrous cycle | | | |
|---|---|---|---|---|
| | Luteal (Day 2–12) | | Follicular (Day 14–16) | |
| Number of ovulations | one | two | one | two |
| Pulse amplitude (ng/ml) | 0.71±0.23 | 0.83±0.17 | 0.84±0.12 | 1.09±0.11 |
| Concentration of LH (ng/ml) | 0.67±0.10 | 0.69±0.14 | 0.94±0.08 | 1.24±0.10 |

Blackface ewes was higher in ewes in high body condition with a mean ovulation rate of 1.8 ± 0.18 then flockmates in low body condition with a mean ovulation rate of 1.0 ± 0.13 ($P<0.05$). Similar associations between the frequency of pulses of LH and ovulation rate have also been observed in ewes immunised against ovarian steroids (Martin 1984). An attempt to stimulate ovulation rate by artificially elevating the frequency of LH pulses with exogenous pulses of gonadotrophin releasing hormone for 72 hours beginning on day 11 of the oestrous cycle was unsuccessful (Martin and Clapin 1982). However, this experiment used flockmates of those used in our study, that is ewes with a history of either one or two ovulations at a steady and similar liveweight. Clearly the definitive experiment linking increased secretion of LH with increased ovulation rate in ewes has yet to be carried out.

The absence of any differences in the secretion of LH between ewes with one and two ovulations during days 14 and 16 of the cycle agree with previous studies showing ovulation rate is independent of the secretion of LH in the follicular phase (Scaramuzzi and Radford 1983; Bindon 1984).

Our observation that ewes with two corpora lutea produce less progesterone than flockmates with one corpus luteum is in contrast with Thorburn *et al.* (1969) who reported the inverse relationship, but other studies have shown the concentration of progesterone can be independent of the number of corpora lutea in some breeds (Quirke *et al.* 1979; Bindon 1984). The secretion of LH is affected by the ovarian steroids, oestradiol and progesterone, both of which act principally to reduce the frequency of the pulses. Martin (1984) argues strongly that during the luteal phase these two steroids act synergistically to exert this effect. While our results showed an increase in the frequency of pulses of LH after the concentration of progesterone decreased around day 14, the frequency of pulses before day 14 was clearly independent of the concentration of progesterone alone, in both ewes with one or two corpora lutea. Nevertheless, the increased secretion of LH in ewes with two corpora lutea may be the direct result of the lower concentration of progesterone reducing negative feedback on the hypothalamus (Karsch *et al.* 1979). However, this hypothesis is not supported by Denamur *et al.* (1973) who reported that LH and prolactin act additively on the corpus luteum to increase its secretion of progesterone.

Low ovulation rates and hence a low incidence of multiple births continue to be a major problem of the sheep industry in Australia (Fogarty 1984). While there are now two pharmacological methods (Pregnant Mare Serum Gonadotrophin and Fecundin) available to increase the ovulation rate of ewes neither the physiological pathways whereby these methods affect ovulation rate, nor the pathways whereby the ewe controls her spontaneous ovulation rate are understood. A deeper understanding of the basis of control of ovulation rate will eventually lead to more effective methods of manipulating ovulation rate and hence improvement in the overall reproductive efficiency of Australia's sheep flocks. In this paper we have demonstrated a relationship between the spontaneous release of LH of ewes and their ovulation rate, but its precise role in the control of ovulation rate remains to be determined.

## ACKNOWLEDGEMENTS

This work was supported by the Australian Meat Research Committee. We thank Peter Moore and David Suckling for their technical assistance.

## REFERENCES

Bindon, B.M., 1984. *Aust. J. Biol. Sci., 37*, 163-189.

Cahill, L.P., Saumande, J., Ravault, J.P., Blanc, M., Thimonier, J., Mariana, J.C. and Mauleon, P., 1981. *J. Reprod. Fert., 62*, 141-150.

Denamur, R., Martinet, J. and Short, R.V., 1973. *J. Reprod. Fert., 32*, 207-220.

Fogarty, N.M., 1984. *Proc. Aust. Soc. Anim. Prod., 15*, 66-79.

Karsch, F.J., Foster, D.L., Legan, S.L., Ryan, K.D. and Peter, G.K., 1979. *Endocr., 105*, 421-426.

McLeod, B.J., Haresign, W. and Lamming, G.E., 1983. *J. Reprod. Fert., 68*, 489-495.

Martin, G.B., 1984. *Biol. Rev., 59*, 1-87.

Martin, G.B. and Clapin, D.T., 1982. *Ann. Con. Soc. Study Fert.*, (Abstract 22).

Martin, G.B., Scaramuzzi, R.J. and Henstridge, J.D., 1983. *J. Endocr., 96*, 181-193.

Quirke, J.F., Hanrahan, J.P. and Gosling, J.P., 1979. *J. Reprod. Fert., 55*, 37-44.

Rhind, S.M. and McNeilly, A.S., 1983. *Ann. Con. Soc. Study Fert.*, (Abstract 82).

Scaramuzzi, R.J. and Radford, H.M., 1983. *J. Reprod. Fert., 69*, 353-367.

Thorburn, G.B., Bassett, J.M. and Smith, I.D., 1969. *J. Endocr., 45*, 459-469.

# MATERNAL RECOGNITION OF PREGNANCY

J.K. Findlay, *Medical Research Centre, Prince Henry's Hospital, St. Kilda Road, Melbourne, Victoria, 3004.*

## INTRODUCTION

During the preimplantation period, the embryo and the uterus develop in a coordinated synchronous manner. The growing embryo becomes increasingly dependent on the uterine environment for survival and growth, and in turn, the uterine environment must undergo continual modification to cope with the needs of the embryo before implantation. Coordination of this synchrony of development rests with the maternal hormones, particularly the ovarian steroids and with an influence of the embryo on the maternal environment. It is the modification of the maternal environment induced by the presence of an embryo which can be called maternal recognition of pregnancy.

The objective of this review is to summarize our current knowledge of maternal recognition of pregnancy in the ewe, to highlight recent advances made in this area and to indicate areas of further research.

## IMMUNOLOGICAL RECOGNITION OF PREGNANCY

The conceptus is considered to be a natural allograft since it normally possesses some paternal genetic characteristics that differ from the maternal host. Yet it escapes immunological destruction for the duration of pregnancy, despite its antigenicity and the immunological competence of the mother. The theories proposed to explain the lack of rejection of the conceptus (Beer and Sio 1982) fall essentially into two main categories; (a) lack of transplantation antigens (or lack of their expression) on the trophoblast, and (b) alterations in the maternal immune response, either by local release of immunosuppressive agents in the uterus or by peripheral circulation of specific antipaternal immunosuppressive agents.

To my knowledge, no one has published data on expression of alloantigens by ovine trophoblast prior to implantation or in early pregnancy. There is conflicting evidence that cell mediated immunity is suppressed *in vivo* in sheep pregnancy either locally or peripherally (Burrells *et al.* 1978; Miyasaka and McCullagh 1981). An alternate view might be that there are no differences in cell-mediated immune response due to pregnancy *per se*, particularly during the early stages. There may be local suppression of cell-mediated immunity within the uterus, a property which might be unique to lymphocytes in that organ, even in non-pregnant ewes (Segerson and Libby 1984). In mice, there is recent evidence for maternal non-T suppressor cells recruited from bone marrow to the decidua which may act to protect the trophoblast from effector cells of the mother's immune system (Clark *et al.* 1984).

The conceptus is responsible, directly or indirectly, for producing substances with *in vitro* immunosuppressive properties that could act peripherally or locally. Morton *et al.* (1979) reported that the ability of antilymphocyte serum to inhibit spontaneous erythrocyte rosetting by sheep lymphocytes was enhanced by a factor present in sheep serum shortly after conception, and claimed that this early pregnancy factor (EPF) is a peripheral immunosuppressive agent (Noonan *et al.* 1979). Crude extracts of 25-day sheep conceptus (Staples *et al.* 1983a) (but not oLH, oPRL, oGH or human placental lactogen) and uterine flushings from pregnant ewes particularly on Day 14 (Segerson 1981) suppressed the response of peripheral sheep lymphocytes to various mitogens. One factor present in uterine flushings at that time is a pregnancy-associated antigen (oPAA; Staples *et al.* 1978; Staples 1980) which has a dose-dependent inhibition of lymphocyte activity *in vitro* (Staples *et al.* 1983a).

The putative immunosuppressive factor could be progesterone, which can suppress lymphocyte blastogenesis *in vitro* provided it is added at $> \mu M$ concentrations (Staples *et al.* 1983b). While it is unlikely that such concentrations were present in the study of Segerson and Libby (1984), ovine utero-ovarian lymph does contain progesterone concentrations in the $\mu M$ range in early pregnancy (Staples *et al.* 1982). Any hypothesis implicating progesterone as an immunosuppressive agent at these concentrations in early pregnancy must account for the successful establishment of pregnancy in ovariectomized ewes treated with exogenous progesterone at doses which result in peripheral concentrations in the nM range (see Miller and Moore 1983). Studies on uterine infiltration of polymorphonuclear leucocytes implicated progesterone, directly or indirectly, as a local anti-inflammatory agent in the gravid uterus (Staples *et al.* 1983c). Total expulsion of lymphocytes from uterine epithelia at implantation does not occur in pregnancy (Staples *et al.* 1983c) and is therefore not a mechanism for immunoprotection of the conceptus.

The extent to which the conceptus is analagous to an allograft has been questioned (Cooper 1980) on the basis that there is no unequivocal evidence that the trophoblastic elements possess transplantation antigens which are expressed and recognized by the mother *in vivo*. Furthermore, if it is not clear that the trophoblast presents alloantigens, the search for substances in the mother which would prevent attack of these alloantigens can be considered premature. Cooper (1980) also considers that in Darwinian terms, immunosuppression which effects the immune system of a pregnant mammal in any general way is an unlikely idea. If it is found that the trophoblast does express alloantigens on its surface, then antigen-specific suppression, e.g. by specific antibody production or masking by glycoproteins (Guillomot *et al.* 1982), is a much more plausible hypothesis which should be tested. Future work should involve the isolation and characterization of the putative immunosuppressive factors and their presence and activity should be established *in vivo*. Both oPAA (Staples 1980) and EPF (Clarke *et al.* 1980) have been only partially characterized. The suggestion that oPAA was a progesterone-dependent protein in the endometrium (Staples *et al.* 1978) should also be examined using the isolated cell culture systems recently established (O *et al.* 1982).

## UTERINE BLOOD FLOW

The sheep blastocyst is reported to exert a local influence on the uterine vascular bed to cause transient increases in uterine blood flow and increased vascular permeability in the caruncular endometrium prior to attachment (Greiss and Anderson 1970). Fleet and Heap (1982) were unable to show a difference in total uterine blood flow between anaesthetized pregnant and non-pregnant ewes on Day 15, although the vasodilatory response of the pregnant ewes to adenosine was reduced compared with non-pregnant ewes. The authors concluded that their results were consistent with the failure to demonstrate oestrogen production by the conceptus at this stage (see Findlay 1983). The question of whether or not the ovine blastocyst can influence endometrial blood flow still requires confirmation.

Neither the nature of the substance(s) produced by the conceptus (or gravid uterus) nor the mechanism causing vasodilation are understood. It had been postulated that oestrogens are involved and that they either increased the local concentrations of vasoactive substances such as adenosine (Fleet and Heap 1982) or prostaglandins in the uterine vasculature, or decreased $\beta$-adrenergic receptor activity in the arterial vasculature to increase uterine blood flow (Ford 1982).

$PGE_2$ and $PGI_2$ increase total uterine blood flow in sheep, in contrast to $PGF_{2\alpha}$ which is a vasoconstrictor (Ford 1982). However, blood flow *within* the endometrium measured by the Kr technique, was increased only by $PGD_2$ and $PGI_2$ and there were no significant effects of either $PGF_{2\alpha}$ or $PGE_2$ (Schramm *et al.* 1984). The reduction in neurotransmitter effects on arterial smooth muscle from gravid uteri of sheep were not altered by *in vitro* treatment with $PGE_2$ (Ford 1982); the effect of $PGI_2$ was not examined. A role for PGs in controlling local endometrial blood flow is supported by the capacity of the ovine blastocyst to synthesize PGs (Findlay 1983; Silvia *et al.* 1984). Also, $PGI_2$ is a major product of the sheep uterus, and there is evidence for a role of PG in increased vascular permeability at implantation sites in rodents (see Findlay 1983).

## ENDOMETRIAL PROTEIN SYNTHESIS

It is well established that protein synthesis by the sheep endometrium is under the control of ovarian steroids (see Findlay 1981; 1983). Studies of incorporation of labelled leucine into explant tissues and secreted protein *in vitro* have shown that oestradiol stimulates protein synthesis, that progesterone alone or in combination with oestradiol is either without effect or mildly inhibitory, and that protein synthesis is generally higher in the intercaruncular than caruncular endometrium.

In view of the need for synchronous development between the blastocyst and the maternal endometrium as the blastocyst becomes increasingly dependent on the uterine environment for survival and development (Findlay 1981, 1983);

1. Does the sequence of exposure of the endometrium to oestradiol and progesterone observed during the oestrous cycle determine the pattern of protein synthesis and secretion and, if so, is that pattern qualitatively or quantitatively important for embryonic growth and survival?
2. What is the cellular origin and steroid-dependence of the proteins synthesized and secreted by endometrial cells?
3. Does the blastocyst modify the pattern of protein synthesis and secretion by the adjacent endometrial cells to facilitate its survival?
4. What are the relative contributions of the blastocyst, the endometrium and serum to the proteins which make up the environment in the uterine lumen?

### Pattern of protein synthesis, steroids and survival of embryos

A sequence of steroid hormone treatments which simulate ovarian secretion of progesterone during the luteal phase of the cycle before oestrus, of oestradiol around oestrus and of progesterone during the subsequent luteal phase, is necessary for normal development of embryos transferred to the uterus of ovariectomized ewes a few days after the induced oestrus (see Miller and Moore 1983). When one or more of the steroids was omitted from the treatment regime, embryos failed to show normal development. Oestradiol may influence embryonic development by regulating endometrial protein synthesis including progesterone receptor levels which would determine the sensitivity of the endometrium to the progesterone of pregnancy. A mechanism by which exposure of the endometrium to progesterone prior to oestradiol might effect embryonic development is not apparent.

Recently, Miller and Moore (1983) showed that the sequence of treatment *in vivo* with steroids determines the rate and pattern of protein synthesis and secretion by explants of endometrium harvested from ovariectomized ewes up to 10 days after the induced oestrus. Oestradiol increased protein secretion by both caruncular and intercaruncular endometrium, whereas progesterone given before oestradiol inhibited secretion of protein by intercaruncular but not caruncular endometrium at about 4-7 days after oestrus. The rate of protein synthesis and the tissue RNA:DNA and protein:DNA ratios were generally higher in entire than in ovariectomized ewes receiving the total steroid regime, probably reflecting the continued oestradiol stimulation of the uterus in entire ewes during the early part of the cycle. These data show that the deleterious effects of omitting progesterone preceding oestradiol on subsequent embryo development in the uterus are more likely to be related to the pattern of protein secretion around days 4-7 after oestrus than the overall rate of protein synthesis. These important observations require further studies to isolate and characterize the proteins secreted in response to different steroid treatments, particularly at about the time the blastocyst enters the uterus and to test the effects of these proteins on embryonic development.

Cellular origin and steroid-dependence of endometrial protein secretion

The sheep endometrium can be conveniently divided into the caruncular (attachment) and intercaruncular (glandular) regions. Caruncles consist of a luminal epithelium separated by a basement membrane from a layer of dense connective tissue composed of stromal fibroblasts, endothelial cells of the vasculature, lymphocytes, polymorphonuclear leucocytes etc. The coiled and branched tubular glands consisting of secretory epithelium are confined to the intercaruncular endometrium.

The higher levels of protein sythesis observed *in vitro* in intercaruncular than caruncular tissue (Findlay *et al.* 1982; Miller and Moore 1983) are more likely to be due to differences in the relative abundance of luminal and glandular epithelial cells in the two regions than differences in function of the two epithelial cell types. Luminal epithelial cells isolated from the caruncular region had similar rates of protein synthesis *in vitro*, to a mixture of luminal and glandular epithelium cells from the intercaruncular area (Salamonsen *et al.* 1984a). Ultrastructurally, the luminal epithelial cells from both regions are highly defferentiated cells with structures consistent with protein secretion (Salamonsen and Findlay, unpublished observations).

It is now possible to characterise and contrast the proteins synthesised and secreted by the luminal and glandular epithelial cells using a procedure to isolate and enrich epithelial and stromal cells from the sheep endometrium with analysis of the products by 2 dimensional gel electrophoresis (O *et al.* 1982; Salamonsen *et al.* 1984a). The steroid-dependence of the individual proteins synthesised and secreted by endometrial cells can also be determined. Of interest in this regard is the observation (Findlay *et al.* 1982) that the concentration of oestrogen receptors per mg protein or per g tissue is approximately 2-fold higher in caruncular then in intercaruncular endometrium. The cellular localisation of these oestrogen receptors is not known, but if the receptor concentration reflects the predominant cell type, then they are likely to be in the stromal cells. This raises the possibility of an interaction between stromal and epithelial cells to control oestradiol and progesterone-dependent synthesis and secretion of endometrial proteins.

Effect of the blastocyst on endometrial protein synthesis

The blastocyst can influence endometrial protein synthesis and secretion. Findlay *et al.* (1981, 1982) have demonstrated a quantitative increase in the rate of leucine incorporation into cellular protein by endometrium of pregnant ewes prior to implantation. The time course and cellular origin of this increase is still unclear because in the first study on Day 15 the increase was only significant in caruncular tissue whereas in the second study, the increase was significant on Day 11 in intercaruncular tissue. Secreted proteins were not examined in either study, mainly because $< 1\%$ of incorporated leucine was present in the medium after a 4 h incubation. Analysis of the cellular protein by SDS-PAGE confirmed the overall increase in protein synthesis and revealed the synthesis of a protein(s) with a molecular weight $> 50{,}000$ by caruncular tissue of pregnant ewes on Day 15 (Findlay *et al.* 1981). Recently, Godkin *et al.* (1983) demonstrated that a protein (trophoblast protein 1; TP1) secreted by the preimplantation blastocyst binds specifically to ovine endometrium and increases the rate of protein release by endometrial explants from non-pregnant ewes on Day 12.

The relationships between ovarian steroids, factors of conceptus origin and the secretions of protein by the endometrial cells is an area requiring further study.

Origin of proteins in the uterine lumen

It had previously been thought that the increase in protein content of the uterine lumen, particularly after Day 13 of pregnancy, was due mainly to proteins of endometrial origin, together with serum proteins and a small contribution from the blastocyst (Roberts *et al.* 1976; Ellinwood *et al.* 1979). It is of interest therefore that a 2 dimensional PAGE analysis of proteins in flushings of pregnant and non-pregnant ewes on Day 13-16 (Salamonsen *et al.* 1984b) has revealed that quantitatively, the major proteins present in pregnancy originate from the maternal serum and the blastocyst, while no proteins secreted by the endometrium were detected. This should not be taken to mean that the endometrial proteins, if present, are not qualitatively important, but it does emphasize the contribution made by the blastocyst to the uterine environment. The function of most of these uterine proteins, with a few exceptions (see below) is not known.

## CONTROL OF LUTEAL FUNCTION IN EARLY PREGNANCY

Sheep require luteal progesterone of the establishment and maintenance of pregnancy. The period of dependence on luteal progesterone exceeds the length of the luteal phase of the infertile cycle, necessitating an extension of the lifespan of the corpus luteum (CL), until the placenta takes over that function at around Day 50 of pregnancy. Moore and Rowson demonstrated that the signal to extend the lifespan of the CL is given directly or indirectly by the blastocyst by Day 12, 4-5 days before implantation and was called maternal recognition of pregnancy by R.V. Short (see Findlay 1981).

$PGF_{2\alpha}$ of uterine origin is the luteolytic agent responsible for regression of the CL of non-pregnant sheep (Findlay 1981). The sheep conceptus may counteract the luteolytic action of uterine $PGF_{2\alpha}$ in several ways. It may alter the synthesis, metabolism and release of $PGF_{2\alpha}$ or other PGs from the uterus, it may modify the local transport of $PGF_{2\alpha}$ from the uterus to the ovary by the countercurrent system, or it may counteract the luteolytic action of $PGF_{2\alpha}$ at the ovarian level. Before discussing these mechanisms and the blastocyst agents likely to be responsible, it is necessary to review recent data on control of luteolysis in non-pregnant ewes.

### Luteolysis

$PGF_{2\alpha}$ is released from the ovine uterus into the uterine vein in episodic surges lasting about 1 h, beginning on Days 12-13 and increasing in frequency to about 5-6 pulses/day during luteolysis on Days 14-16 (McCracken *et al.* 1984). $PGF_{2\alpha}$ passes from the uterine vein to the ovarian artery via a countercurrent transfer mechanism in the ovary pedicle, and acts in the ovary resulting in regression of the CL. The recent demonstration of episodic pulses of $PGF_{2\alpha}$ in utero-ovarian lymph (Staples and Wylie 1984) suggests an alternative pathway for the luteolysin to pass locally from the uterus to the ipsilateral ovary containing the CL.

The pulses of $PGF_{2\alpha}$ in the uterine vein are accompanied by pulses of its primary metabolite 13,14-dihydro-15-keto $PGF_{2\alpha}$ (PGFM) in peripheral blood which is thought to be formed as a result of metabolism of $PGF_{2\alpha}$ by the lung. Recent studies raise the possibility that some peripheral PGFM might arise from metabolism of $PGF_{2\alpha}$ by epithelial and stromal cells of the endometrium (Findlay *et al.* 1981, 1983; O *et al.* 1982b).

The pulsatile secretion of $PGF_{2\alpha}$ by the uterus is now believed to involve regulation of endometrial cells by ovarian oestradiol and progesterone and oxytocin, probably of luteal origin (McCracken *et al.* 1984; Sheldrick and Flint, 1984). Briefly, oestradiol-17$\beta$ may influence the endometrial secretion of $PGF_{2\alpha}$ in response to oxytocin by inducing receptors for oxytocin on the stromal or epithelial cells of endometrium. The endometrium requires a minimum of 5-7 days priming with progesterone to activate the PG synthetase system thought to reside mainly in the stromal cells of the caruncles. The oestrogen-primed endometrium will respond to oxytocin in the presence of progesterone (Fairclough *et al.* 1983) but maximal PG responses are only seen after progesterone withdrawal. The demonstrated capacity of the endometrium to elicit a PG response to oxytocin in the presence of progesterone may be important for initiating luteolysis around Days 13-15 and has to be considered in any antiluteolytic action of the blastocyst (see below). Progesterone withdrawal cannot itself be the initiating event for PG release by uterus as suggested by McCracken *et al.* (1984), because regression of the CL occurs in ewes treated with progestagens over the time of expected luteolysis to synchronize oestrus (see Fairclough *et al.* 1983). Since a minimum of 5-7 days priming by progesterone is needed to activate the PG synthetase system (by what mechanism is not known), significant endometrial PG production does not begin before Day 10-11 Findlay *et al.* 1983). Circulating concentrations of oxytocin rise in the luteal phase by about Day 10-11 (Sheldrick and Flint 1984), however, little is known about the pattern of oxytocin release by the CL before Day 13. In the endometrium receptors for oxytocin do not increase significantly until Days 13-14 (Fairclough *et al.* 1984; McCracken *et al.* 1984). Thus, it would appear that the presence of endometrial receptors for oxytocin is the factor limiting PG *release* by the uterus, and by inference, luteolysis. Why such relatively long periods of hormonal priming are required to stimulate PG production and release by the endometrium and which cell types are involved is not fully understood and is an area requiring further study.

A recent important observation has been the capacity of the ovine CL to synthesise (Rodgers *et al.* 1983a) and secrete oxytocin in response to both $PGF_{2\alpha}$ and a $PGE_2$ analogue, suggesting a positive feedback loop between ovarian oxytocin and uterine PG to bring about luteolysis (Sheldrick and Flint 1984). There is considerable evidence that circulating oxytocin is involved in the induction of luteolysis in the ewe, but the proportion of oxytocin accounted for by luteal secretion, as apposed to neural secretion, is uncertain (Sheldrick and Flint 1984). The role of uterine $PGF_{2\alpha}$ in releasing luteal oxytocin *in vivo* has been questioned (Fairclough *et al.* 1984) on the basis that non-pregnant ewes given oxytocin on Day 14 of the cycle did not show a rise in oxytocin-associated neurophysin despite an increase in PGFM in peripheral blood. The action of PGs in controlling release of oxytocin by the CL is likely to be on the large luteal cells which secrete oxytocin (Rodgers *et al.* 1983a). These large luteal cells contain significant numbers of PG receptors, whereas LH receptors reside mainly on the small luteal cells (Silvia *et al.* 1984) which are primarily responsible for progesterone production by the CL (Rodgers *et al.* 1983b; Silvia *et al.* 1984). This raises the unresolved problem of the mechanism of the luteolytic action of $PGF_{2\alpha}$ on the CL and implies an interaction between the

2 luteal cell types, such that $PGF_{2\alpha}$ induces the large luteal cell to release the luteolytic factor which acts on the small luteal cell (Silvia *et al.* 1984).

Endocrine recognition of pregnancy

The primary mechanism by which the embryo prevents luteolysis in the ewe is not known. Two hypotheses, not mutually exclusive, are currently fashionable. One involves an effect of the embryo on the synthesis and release of PG by the endometrium, whereas the other hypothesis predicts an action at the ovarian level to prevent luteolysis. A third hypothesis would be suppression of release of luteal oxytocin responsible for PG release by the endometrium.

The pattern of release of $PGF_{2\alpha}$ by the uterus is changed in early pregnancy (Findlay 1981), although the question of actual uterine secretion rates of $PGF_{2\alpha}$ in pregnant and non-pregnant ewes remains controversial (Silvia *et al.* 1984). The capacity of the endometrium to synthesise PG *in vitro* is not decreased (Findlay *et al.* 1981) and the content and concentration of PG in the endometrium increases in the preimplantation period (Findlay 1981; Findlay *et al.* 1983; Silvia *et al.* 1984). However, pulses of $PGF_{2\alpha}$ in the uterine vein and PGFM in the peripheral circulation are reduced or absent (McCracken *et al.* 1984), although there is a significant increase in the basal level of PGFM in peripheral blood or pregnant ewes (Fairclough *et al.* 1984) consistent with increased production by the uterine contents. Some of this extra PG production probably originates from the blastocyst (Silvia *et al.* 1984), which is more likely to be responsible for the increased content of $PGF_{2\alpha}$ in the uterine lumen than the endometrium (Findlay *et al.* 1983).

The absence of a pulsatile pattern of PG release by the endometrium could result from the inability of the endometrium to respond to oxytocin (Fairclough *et al.* 1984; McCracken *et al.* 1984) due to the very low concentrations of oxytocin receptors in the endometrium of pregnant ewes in the preimplantation period (McCracken *et al.* 1984). It has been postulated that because oxytocin receptors are under oestrogenic control, the continued presence of progesterone in pregnancy would reduce the nuclear accumulation of oestrogen-receptor complex in endometrial cells and inhibit synthesis of the oxytocin receptors (McCracken *et al.* 1984). This postulate is untenable because the uterus can still respond to oxytocin in the presence of progesterone (Fairclough *et al.* 1983). An alternative explanation is that the blastocyst may interfere locally with the action of oestradiol, evidenced by the fact that oestrogen receptor concentrations in the caruncular endometrium are reduced by a local action of the blastocyst between Days 11-14 (Findlay *et al.* 1982). Finally, the presence of an embryo may interfere with the release of oxytocin by the CL. Moore *et al.* (1982) showed that transfer of an embryo, on either Day 12 or Day 13, to the uterus of a non-pregnant ewe inhibited pulsatile release of oxytocin-associated neurophysin.

The process by which the blastocyst modulated PG production and release by the endometrium is a critical one in the establishment of pregnancy and requires more investigation. In particular, we need to know which endometrial cells are involved, how oxytocin controls PG synthesis and release and what the blastocyst factor(s) are that inhibit this response and by what mechanism.

The luteal concentrations of oxytocin of pregnant ewes begin to decline at the time at which regresion would normally take place, and the circulating concentrations of oxytocin decrease on Days 14 and 15 (Sheldrick and Flint, 1984). Before that time, peripheral concentrations of oxytocin do not differ between pregnant and non-pregnant ewes. The factor responsible for the decrease in luteal oxytocin content in the CL of pregnancy is not known, although it has been suggested that $PGE_2$ of embryonic origin is responsible (Sheldrick and Flint 1984). Two mechanisms operating to maintain luteal function in early gestation have been proposed, *viz.* the embryonic antiluteolysin acting on $PGF_{2\alpha}$ between Days 12-18 and the absence of luteal oxytocin after Day 18 (Sheldrick and Flint 1984).

$PGE_2$ has been promoted as the agent acting at the ovarian level to prevent luteal regression (Silvia *et al.* 1984). It is proposed that the increased uterine secretion of $PGE_2$ in early pregnancy is the result of synthesis of $PGE_2$ by the endometrium under the influence of the blastocyst or by the blastocyst itself. $PGE_2$ may prevent luteal regression by inhibiting release of the luteolytic factor by large luteal cells, by stimulating small luteal cells to secrete progesterone, and/or increase blood flow to the ovary, or by increasing luteal resistance to the luteolytic influence of $PGF_{2\alpha}$. Further research is required to elucidate the role of $PGE_2$ in early pregnancy in the ewe.

The antiluteolysin

The factor secreted by the blastocyst which ultimately results in maintenance of the CL, is a protein which must act through the uterus to exert its effect (Findlay 1981, 1983; Silvia *et al.* 1984). An antiluteolysin, called trophoblastin, has been extracted from the trophoblast from Days 12-25 and is active when infused into the uterine lumen of non-pregnant ewes (Martal 1981). More recently, Godkin *et al.* (1982, 1983) have isolated and purified a major protein secreted *in vitro* by the ovine blastocysts harvested on Days 13-21, but not on Day 23. This protein called TP1 has a molecular weight of 17K and a pI around 5.5, and is localised in the trophectoderm cells. TP1 binds to the surface epithelium of endometrial cells and increases protein secretion by endometrial explants, but it did not bind to luteal membranes or stimulate progesterone production by dispersed luteal cells. TP1 prolongs luteal function when infused into the uterine lumen of non-pregnant ewes

(Godkin *et al.* 1984) and suppresses uterine release of PGFM in response to oestradiol and oxytocin (Fincher *et al.* 1984). It appears, therefore, that TP1 fulfils a number of the criteria of the antiluteolysin, but further studies are required to fully substantiate its role in early pregnancy. There are a number of proteins secreted by the blastocyst in culture (Godkin *et al.* 1982; Salamonsen *et al.* 1984b; Wallace *et al.* 1984) which are present in the uterine flushings of pregnant ewes (Salamonsen *et al.* 1984b), and whose function is not yet known.

## SIGNIFICANCE TO THE SHEEP INDUSTRY

The application of this knowledge to sheep production falls into two areas. It will provide a basis on which to design future research on the causes and cures of embryonic growth retardation and mortality. Twenty to 30% of fertilized ova are normally lost during pregnancy with most losses occurring in the first month (Edey 1969) when maternal recognition of pregnancy is occurring. It is difficult to relate the extent of embryonic mortality to the various facets of maternal recognition and establishment of pregnancy because further definition of the timing of embryonic mortality during the first few weeks is not yet possible (Kelly 1984). Also it is necessary to distinguish basal mortality due to genetic defects and immune rejection from that which occurs in response to identifiable environmental influences such as nutrition, age, ovulation rate, etc. It is likely that most basal mortality due to genetic defects occurs at the very early stages of blastogenesis (Long and Williams 1978). The timing and extent to which immune rejection of the embryo contributes to embryonic loss is unknown. If the embryo fails to signal its presence prior to Day 12 because of mortality or growth retardation and lack of synchrony between the uterine environment and the blastocyst (Lawson *et al.* 1983), there is no extension of the lifespan of the CL and the animal returns to service at the expected times thus giving no indication that fertilization may have taken place. Embryonic death after Day 12 is accompanied by extension of CL function for periods varying up to Days 40-50 and a significant decrease in fertility at the next oestrus (Edey 1970). The ways in which environmental factors influence embryonic growth and mortality are not properly understood and can be investigated using the approaches outlined in this review.

The second application involves development of tests for pregnancy based on measurement of factors of embryonic origin in the ewe. This would provide not only further definition of the timing of embryonic death, particularly if the test detected the presence of an embryo from the time of fertilization, but also a quantitative, non-invasive method of measuring embryonic viability and growth.

## CONCLUSIONS

Maternal recognition of pregnancy extends to the immune, vascular and endocrine systems. It is initiated at various times in the preimplantation period by embryological agents which are not well characterized and it is brought about by processes which are not fully understood. Although this review has concentrated on the influence of the conceptus on the maternal systems, the reverse situation should not be ignored. Embryo transfer studies in sheep have clearly shown that the signalling capacity and growth of the blastocyst is determined by its uterine environment (Parr *et al.* 1982; Lawson *et al.* 1983). Until we fully understand the complex interactions between the conceptus and its maternal environment, it will not be possible to achieve major reductions in embryonic growth retardation and mortality.

## ACKNOWLEDGMENTS

I wish to thank Fuller Bazer, Michael Roberts, Robert Fairclough, Linton Staples and Lois Salamonsen for sharing data in press and for their helpful discussions. Jill Volfsbergs and Diane Hollingsworth provided invaluable secretarial assistance and the financial support of the Wool Research Trust Fund of Australia is gratefully acknowledged.

## REFERENCES

Beer, A.E. and Sio, J.O., 1982. *Biol. Reprod., 26*, 15-27.

Burrells, C., Wells, P.W. and Sutherland, A.D., 1978. *Clin. Exp. Immunol., 33*, 410-415.

Clark, D.A., Slapsys, R., Croy, B.A., Krcek, J. and Rossant, J., 1984. *Am. J. Reprod. Immunol.,* 5, 78-83.

Clarke, F.M., Morton, H., Rolfe, B.E. and Clunie, G.J.A., 1980. *J. Reprod. Immunol., 2*, 151-162.

Cooper, D.W., 1980. *In* Hearn, J.P. (ed) *Immunological Aspects of Reproduction and Fertility Control*, M.T.P., Lancaster, 33-63.

Edey, T.N., 1969. *Anim. Breed. Abstr., 37*, 173-190.

Edey, T.N., 1970. *J. Agric. Sci., Camb., 74*, 199-204.

Ellinwood, W.E., Nett, T.M. and Niswender, G.D., 1979. *Biol. Reprod., 21*, 834-856.

Fairclough, R.J., Moore, L.G., McGowan, L.T., Smith, J.F. and Watkins, W.B., 1983. *Biol. Reprod., 29*, 271-277.

Fairclough, R.J., Moore, L.G., Peterson, A.J. and Watkins, W.B., 1984. *Biol. Reprod., 31*, 36-44.

Fincher, K.B., Hansen, P.J., Thatcher, W.W., Roberts, R.M. and Bazer, F.W., 1984. *Proc. Ann. Meeting, Amer. Soc. Anim. Sci.*, Missouri.

Findlay, J.K., 1981. *J. Reprod. Fert., Suppl., 30*, 171-182.
Findlay, J.K., 1983. *In* Martini, L. and James, V.H.T. (eds) *The Endocrinology of Pregnancy and Parturition*, Current Topics in Exptl. Endocr., Vol 4. Academic Press, New York, 35-67.
Findlay, J.K., Ackland, N., Burton, R.D., Davis, A.J., Maule Walker, F.M., Walters, D.E. and Heap, R.B., 1981. *J. Reprod. Fert., 62*, 361-377.
Findlay, J.K., Clarke, I.J., Swaney, J., Colvin, N. and Doughton, B., 1982. *J. Reprod. Fert., 64*, 329-339.
Findlay, J.K., Colvin, N., Swaney, J. and Doughton, B., 1983. *J. Reprod. Fert., 68*, 343-349.
Fleet, I.R. and Heap, R.B., 1982. *J. Reprod. Fert., 65*, 195-205.
Ford. S.P., 1982. *J. Anim. Sci., 55 (Suppl. II)*, 32.
Greiss, F.C. and Anderson, S.G., 1970. *Am. J. Obstet. Gynec., 106*, 30-38.
Godkin, J.D., Bazer, F.W., Moffat, J., Sessions, F. and Roberts, R.M., 1982. *J. Reprod. Fert., 65*, 141-150.
Godkin, J.D., Bazer, F.W. and Roberts, R.M., 1983. *Endocrinology, 114*, 120-130.
Godkin, J.D., Bazer, F.W., Thatcher, W.W. and Roberts, R.M., 1984. *J. Reprod. Fert., 71*, 57-64.
Guillomot, M., Flechon, J.E. and Wintenberger-Torres, S., 1982. *J. Reprod. Fert., 65*, 1-8.
Kelly, R.W., 1984. This volume, 127-133.
Lawson, R.A.S., Parr, R.A. and Cahill, L.P., 1983. *J. Reprod. Fert., 67*, 477-483.
Long, S.E. and Williams, C., 1978. *Vet. Rec., 102*, 153.
Martal, J., 1981. *J. Reprod. Fert. Suppl., 30*, 201-210.
McCracken, J.A., Schramm, W. and Okulicz, W.C., 1984. *Anim. Reprod. Sci., 7*, 31-56.
Miller, B.G. and Moore, N.W., 1983. *J. Reprod. Fert., 68*, 137-144.
Miyasaka, M. and McCullagh, P., 1981. *J. Reprod. Immunol., 3*, 15-27.
Moore, L.G., Watkins, W.B., Peterson, A.J., Tervit, H.R., Fairclough, R.J., Havik, P.G. and Smith, J.F., 1982. *Prostaglandins, 24*, 79-88.
Morton, H., Nancarrow, C.D., Scaramuzzi, R.J., Evison, B.M. and Clunie, G.J.A., 1979. *J. Reprod. Fert., 56*, 75-80.
Noonan, F.P., Halliday, W.J., Morton, H. and Clunie, G.J.A., 1979. *Nature, 278*, 649-650.
O, W.S., Findlay, J.K., Colvin, N., Swaney, J. and Doughton, B., 1982b. *Proc. Aust. Soc. Med. Res.* Abstr. 43.
O, W.S., Findlay, J.K., Swaney, J., Colvin, N. and Doughton, B., 1982. *Proc. Aust. Soc. Reprod. Biol., 14*, 16.
Parr, R.A., Cumming, I.A. and Clarke, I.J., 1982. *J. Agric. Sci., Camb., 98*, 39-46.
Roberts, G.D., Parker, J.M. and Symons, H.W., 1976. *J. Reprod. Fert., 48*, 99-107.
Rodgers, R.J., O'Shea, J.D., Findlay, J.K., Flint, A.P.F. and Sheldrick, E.L., 1983a. *Endocrinology, 113*, 2302-2304.
Rodgers, R.J., O'Shea, J.D. and Findlay, J.K., 1983b. *J. Reprod. Fert., 69*, 113-124.
Salamonsen, L.A., Doughton, B. and Findlay, J.K., 1984a. *Proc. Aust. Soc. Reprod. Biol., 16*, 27.
Salamonsen, L.A., Doughton, B. and Findlay, J.K., 1984b. This volume, 115-117.
Schramm, W., Einer-Jensen, N., Brown, M.B. and McCracken, J.A., 1984. *Biol. Reprod., 30*, 523-531.
Segerson, E.C., Jr., 1981. *Biol. Reprod., 25*, 77-84.
Segerson, E.C., Jr., and Libby, D.W., 1984. *Biol. Reprod., 30*, 126-133.
Sheldrick, E.L. and Flint, A.P.F., 1984. *Anim. Reprod. Sci., 7*, 101-114.
Silvia, W.J., Fitz, T.A., Mayan, M.H. and Niswender, G.D., 1984. *Anim. Reprod. Sci., 7*, 57-74.
Staples, L.D., 1980. *Biol. Reprod., 22*, 675-685.
Staples, L.D., Binns, R.M. and Heap, R.B., 1983b. *J. Endocr., 98*, 55-69.
Staples, L.D., Brown, D., Binns, R.M. and Heap, R.B., 1983a. *Placenta, 4*, 125-132.
Staples, L.D., Fleet, I.R. and Heap, R.B., 1982. *J. Reprod. Fert., 64*, 409-420.
Staples, L.D., Heap, R.B. Wooding, F.B.P. and King, G.J., 1983b. *Placenta, 4*, 339-350.
Staples, L.D., Lawson, R.A.S. and Findlay, J.K., 1978. *Biol. Reprod., 19*, 1076-1082.
Staples, L.D. and Wylie, P., 1984. This volume, 122-124.
Wallace, A.L., Nancarrow, C.D. and Sutton, R., 1984. This volume, 118-121.

# THE USE OF THE OVARIECTOMIZED EWE IN STUDIES ON SURVIVAL OF EMBRYOS DURING EARLY PREGNANCY : A ROLE FOR PROGESTERONE PRECEDING CONCEPTION

N.W. Moore and B.G. Miller, *Department of Animal Husbandry, University of Sydney, Camden, N.S.W., 2570.*

*Summary* The results of three experiments using entire and ovariectomized Merino ewes: I. Confirmed previous findings of the need for a period of progesterone priming prior to induced oestrus for the subsequent survival of embryos in ovariectomized ewes. II. Suggested in the entire ewe that a luteal phase prior to oestrus is required for survival of embryos and that 4 days treatment with progesterone might provide a 'luteal phase' of adequate duration.

## INTRODUCTION

In the ewe the extent and timing of prenatal loss has been documented (Henning 1939; Robinson 1951; Moore and Rowson 1960; Quinlivan *et al.* 1966), but little is known of its cause apart from the rather broad consideration that loss could be due to faults within the embryo, or to failure of the maternal environment to support embryos. No doubt faults, such as genetic aberrations, occur within embryos (Bishop 1964), but there seems to be general acceptance that failure of the maternal environment plays the major role (Edey 1969) and it is reasonable to assume an endocrine involvement in these losses. Most loss occurs within the first 3 weeks after mating (Robinson 1951) and it takes one of two forms; that in which all embryos die and that in which one or more of those present die, leaving the remaining embryos to progress to term (Moore and Rowson 1960). When the single embryo present, or all of several present, die, the ewe simply returns to oestrus in most instances at the expected time for non-pregnant animals.

The classic approach of removal of ovaries followed by replacement hormone therapy has been used to establish the steroid hormone requirements of pregnancy. Early studies in which ewes were ovariectomized within a few days after mating to entire rams (Foote *et al.* 1957) or had embryos transferred to their uteri at the time of ovariectomy carried out a few days after mating to vasectomized rams (Moore and Rowson 1959) showed that progesterone alone at a dose rate of around 10 mg/day would maintain pregnancy. The transfer of embryos removed the uncertainty of whether or not fertilization had occurred, but these studies could not provide details on any effects that ovarian steroid hormones secreted before ovariectomy might have on survival and development of embryos.

In a series of experiments, ewes ovariectomized several months prior to experimentation were used to investigate the roles played by ovarian hormones secreted before, around the time of, and after oestrus and ovulation (Miller and Moore 1976, 1983; Miller *et al.* 1977). The ewes were treated with all, or part of, a progesterone/oestradiol (P/E) regime (see Table 1), designed to simulate ovarian secretion during the luteal phase prior to oestrus (Priming P, days 1-12 of treatment), during the subsequent follicular phase (Oestrous E, Days 13 and 14) and during the luteal phase following oestrus and ovulation (Maintenance P, Days 16-40). In most experiments, embryos collected from donor ewes 3 and 4 days after oestrus (Day 3 and Day 4 embryos) were transferred synchronously to the uteri of ovariectomized recipients 3 and 4 days after oestrus induced by Oestrous E (or equivalent stage of Maintenance P treatment in ewes which did not receive Oestrous E). Survival and development of embryos were assessed at slaughter at around 3 weeks after transfer.

All components of the regime are required for survival and development of embryos suggesting that a follicular phase and a prior luteal phase are required for survival of embryos in the entire ewe.

Three further experiments have been carried out to examine in more detail the role of Priming P.

## MATERIALS AND METHODS

Merino ewes of 6-7 years of age were used in all experiments, each ewe had 1 or 2 embryos transferred to the uterus and the survival and development of embryos was assessed at slaughter carried out at an embryo age of 20-21 days.

### Experiment 1

Day 4 embryos were transferred synchronously to ovariectomized ewes which were treated with all components of the regime shown, but Priming P was given for 1, 4 or 8 days before Oestrous E instead of the 12 days shown in Table 1.

### Experiment 2

Entire anoestrous ewes were used and the work was done during November-December. The ewes were treated with pregnant mare serum gonadotrophin (PMSG) and Day 4 embryos were transferred to 13 ewes 3 days after the induced ovulation. At the time of transfer 8 ewes were ovariectomized and then given Maintenance P, the remainder were left entire and received no further treatment.

**TABLE 1. Steroid hormone treatment regime.**

| Day of treatment | Time | Dose | | Day of treatment | Time | Dose | |
|---|---|---|---|---|---|---|---|
| | | P (mg) | E (μg) | | | P (mg) | E (μg) |
| 1-12 | a.m. | 12 | – | 19 | a.m. | 2.0 | – |
| 13 | 1600 | – | 3.5 | | p.m. | 2.5 | – |
| | 2400 | – | 7.0 | 20 | a.m. | 3.0 | – |
| 14 | 0800 | – | 14.0 | | p.m. | 4.0 | – |
| | 1600 | – | 7.0 | 21 | a.m. | 5.0 | – |
| | 2400 | – | 3.5 | | p.m. | 6.0 | – |
| 15 | Oestrus | – | – | 22 | a.m. | 8.0 | – |
| 16 | a.m. | 0.50 | – | | p.m. | 8.0 | – |
| | p.m. | 0.625 | – | 23 | a.m. | 8.0 | – |
| 17 | a.m. | 0.75 | – | | p.m. | 8.0 | – |
| | p.m. | 1.0 | – | 24-40 | a.m. | 12.0 | – |
| 18 | a.m. | 1.25 | – | | | | |
| | p.m. | 1.5 | – | | | | |

Experiment 3

Day 4 or Day 8 embryos were transferred to ovariectomized ewes which received Oestrous E and Maintenance P, but *not* Priming P. Day 4 embryos were transferred to six Day-8 and five Day-12 recipients (where Day for recipient ewes = Day after oestrus induced by Oestrous E), while Day 8 embryos were transferred to eight Day-8 and eight Day-12 recipients.

## RESULTS

Experiment 1

Thirty-five of the 63 ovariectomized ewes had one or two normal embryos at slaughter (Table 2).

**TABLE 2. Proportion of ewes with normal embryos at slaughter.**

| | Duration of Priming P (days) 1 | 4 | 8 |
|---|---|---|---|
| Proportion of ewes | 6/20 (30%) | 14/21 (67%) | 15/22 (68%) |

There was no difference between the proportion of ewes with normal embryos following 4 and 8 days of Priming P, but 1 day's treatment gave inferior results to 4 ($\chi_2 = 5.51$; $P < 0.02$) and 8 ($\chi_2 = 6.11$; $P < 0.02$) days treatment.

Experiment 2

At slaughter no ewe, irrespective of whether entire or ovariectomized, had normal embryos. Two of the 5 entire ewes and 5 of the 8 ovariectomized ewes had embryonic membranes (probably trophoblast), but all were in an advanced stage of resorption.

Experiment 3

At slaughter no ewe had normal embryos and there was no evidence in any ewe of retarded and/or resorbing embryos or membranes.

## DISCUSSION

Although numbers of animals were small, Experiment 2 provides further evidence for the need in the entire ewe of a luteal phase prior to oestrus and ovulation for the survival and normal development of embryos. It is well established that ovulation can be induced in the anoestrous ewe by the unaccustomed presence of rams or by a single injection of PMSG, but oestrus rarely occurs. A period of progesterone treatment prior to the introduction of rams (Martin *et al.* 1981) or injection of PMSG (Robinson 1952) will induce oestrus with ovulation and the minimum period of progesterone treatment required prior to PMSG or introduction of rams is about 6 days (Robinson 1954; Reeve and Chamley 1982). Results of Experiment 1 suggest that 4 days treatment might be adequate for subsequent survival and normal development of embryos.

Experiment 3 showed that an extended period of maintenance P treatment could not compensate for the lack of a Priming P component in the treatment of ovariectomized ewes. Hence, it seems that Priming P (or a prior luteal phase in entire ewes) conditions the ewe to respond to Maintenance P and establish a uterine environment favourable for survival of embryos. Omission of Priming P from the treatment regime results in a small decrease in the concentration of progesterone receptor in the endometrium at 4-6 days after oestrus induced by Oestrous E (Miller *et al.* 1977). Perhaps more importantly the inclusion of Priming P depresses the secretion of proteins of molecular weight of about 42,000 by explants of intercaruncular endometrium collected 6 days after oestrus (Miller and Moore 1983). In entire ewes it may be that progesterone secreted during the luteal phase of the previous oestrous cycle inhibits the secretion after oestrus of particular endometrial proteins.

These studies do not provide definitive evidence for the causes of embryonic death, but they do indicate progesterone priming prior to oestrus has a marked influence on the subsequent capacity of the reproductive tract to support embryos. It remains to be determined if the effect on survival and development of embryos is directly related to protein secretory activity of the uterus.

## REFERENCES

Bishop, M.W.H., 1964. *J. Reprod. Fert.*, *7*, 383-396.
Edey, T.N., 1969. *Anim. Breed. Abs.*, *37*, 173-190.
Foote, W.D., Gooch, L.D., Pope, A.L. and Casida, L.E., 1957. *J. Anim. Sci.*, *16*, 986-989.
Henning, W.L., 1939. *J. Agric. Sci.*, *58*, 565-580.
Martin, G.B., Scaramuzzi, R.J. and Lindsay, D.R., 1981. *Aust. J. Biol. Sci.*, *34*, 569-575.
Miller, B.G. and Moore, N.W., 1976. *Aust. J. Biol. Sci.*, *29*, 565-573.
Miller, B.G. and Moore, N.W., 1983. *J. Reprod. Fert.*, *68*, 137-144.
Miller, B.G., Moore, N.W., Murphy, L. and Stone, G.M., 1977. *Aust. J. Biol. Sci.*, *30*, 379-388.
Moore, N.W. and Rowson, L.E.A., 1959. *Nature*, *184*, 1410.
Moore, N.W. and Rowson, L.E.A., 1960. *J. Reprod. Fert.*, *1*, 332-349.
Quinlivan, T.D., Martin, C.A., Taylor, W.B. and Caviney, I.M., 1966. *J. Reprod. Fert.*, *11*, 379-390.
Reeve, J.L. and Chamley, W.A., 1982. *Proc. Aust. Soc. Reprod. Biol.*, *14*, 89.
Robinson, T.J., 1951. *J. Agric. Sci.*, *41*, 6-63.
Robinson, T.J., *Nature*, *170*, 373.
Robinson, T.J., 1954. *J. Endocr.*, *10*, 117-123.

# PROTEIN SECRETION BY PREIMPLANTATION SHEEP BLASTOCYSTS

L.A. Salamonsen, B. Doughton and J.K. Findlay, *Medical Research Centre, Prince Henry's Hospital, Melbourne, 3004, Australia.*

*Summary* Protein profiles of blastocyst culture medium and of flushings from pregnant and non pregnant sheep show that while preimplantation blastocysts make a significant contribution to the protein content of the uterine lumen, proteins originating from endometrial epithelial cells are not detectable in uterine flushings. Plasma proteins contribute a large proportion of the total protein present in the uterine lumen.

## INTRODUCTION

Although plasma proteins predominate in the uterine lumen, the relative contributions of secretions of the maternal endometrium and of the trophectoderm of the conceptus to the uterine environment during the preimplantation period are not well understood. There is evidence that the blastocyst can modify protein synthesis by endometrial cells before implantation, but the blastocyst factors responsible are not well characterized (Findlay 1984). We have cultured sheep blastocysts and compared the pattern of proteins released into the medium with that found in uterine flushings of pregnant and non pregnant ewes.

## MATERIALS AND METHODS

### Chemicals and reagents

Tissue culture reagents were obtained from Flow Laboratories, Australia. To Dulbecco's modified minimum essential medium was added 200mM glutamine, 1% non essential amino acids, 0.2 U/ml insulin (Actrapid MC, Novo), penicillin, 200 IU/ml and streptomycin, 200 μg/ml, bacitracin ($5 \times 10^{-5}$M, Sigma) and polyvinyl-pyrollidine (6 μg/ml, BDH) (DMEM+).

### Blastocysts and uterine flushings

Blastocysts (days 13-16; day 0 = day of oestrus) were obtained from pregnant Corriedale ewes as previously described (Findlay *et al.* 1982), except that Hanks Buffered Salt Solution (HBSS) was used as flushing medium. The uteri of three non pregnant ewes at day 15 of the oestrous cycle were flushed similarly. Flushings were centrifuged (2000 g) and the supernatants stored at −20°C.

### Culture of blastocysts *in vitro*

Blastocysts were washed, DMEM+ was added, the flask flushed with 5% $CO_2$ in oxygen and incubated in a shaking water bath at 37°C and 36 oscillations per minute. The volume of medium (5-10 ml depending on the size of the embryo), was such that the embryos were exposed partly to medium and partly to the gaseous atmosphere. Medium was changed after 24 hr (D1) and the incubation repeated (D2). After 48 hr incubation, the embryo was blotted dry and weighed. Pieces of control tissue (skeletal muscle, connective tissue and myometrium) were incubated similarly. Medium was centrifuged to remove cellular debris, and stored at −20°C.

### Two dimensional SDS polyacrylamide gel electrophoresis (2D PAGE)

Uterine flushings and samples of culture medium were dialysed (Mr cut-off approx. 3K) against 10mM Tris-HC1 buffer (pH 8.2). Aliquots were put aside for protein estimation (Bradford 1976) and the remainder lyophilised and stored at −20°C. Approximately 100 μg protein were subjected to 2D PAGE (O'Farrell 1975). The pI range in the first dimension was 4.5-7.2 and polyacrylamide concentration in the second dimension 12.5%. enabling separation of proteins of Mr <13K to 113K. The gels were stained with silver stain (Morrissey 1981).

## RESULTS

### Protein secretion

On both D1 and D2 the mean total protein secretion (μg/24 h) by 9 day 15 blastocysts (D1, 545 ± 104; D2, 426 ± 91) was greater ($P < 0.05$) than secretion by 5 day 13-14 blastocysts (D1, 275 ± 18; D2, 218± 12). Differences between D1 and D2 of culture were significant only for the younger blastocysts ($P < 0.01$). Mean wet weight (mg) of embryos were: D15, 117 ± 28; D13, 11 ± 3. Electrophoretic analysis revealed significant contamination with serum proteins of day 13 (data not shown) but not day 15 blastocysts (Figure 1A).

### Analysis of proteins by 2D PAGE

The patterns of protein secreted by day 15 and day 16 blastocysts and on D1 and D2 of culture were similar (data not shown). Of particular note in Figure 1 are numbers (1) a major group of proteins of Mr approximately 36K and pI approximately 6.8., (2) a protein of Mr approximately 20K and pI approximately 4.3 which is probably trophoblast protein 1 (OTP1) described by Godkin *et al.* (1982, 1984a) and (3-5), a group of low molecular weight proteins. A number of other blastocyst proteins are also clearly visible (e.g. nos. 6-9). With the exception of protein 2, these are also present in uterine flushings from pregnant, but not from non pregnant animals. All flushings also contain clearly defined plasma proteins, but no proteins of endometrial epithelial origin (Salamonsen *et al.* 1984) are detectable.

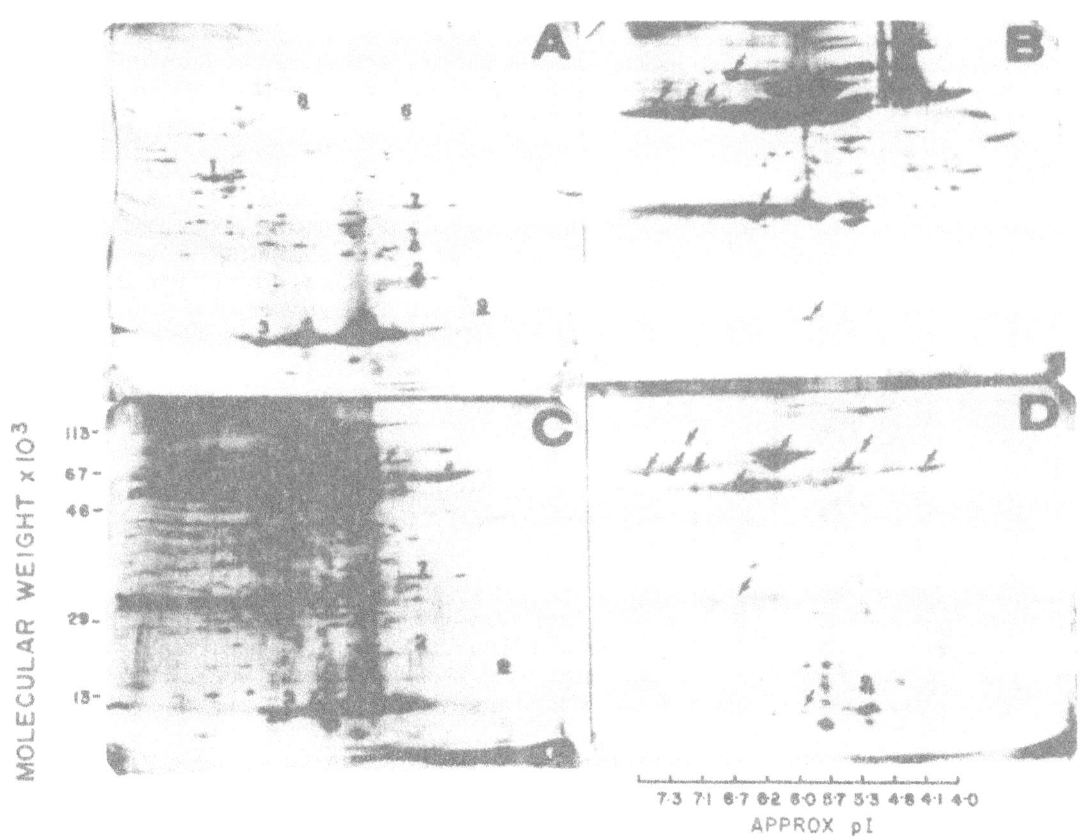


**FIGURE 1. 2D PAGE profiles of proteins present in (A) medium from culture of day 15 blastocysts (B) non pregnant sheep plasma, (C) uterine flushings from day 15 pregnant sheep, and (D) uterine flushings from non pregnant sheep on day 15 of the oestrous cycle. Arrows represent major plasma proteins found in uterine flushings (Mr 67K = albumin), and numbers 1 to 9 represent proteins of blastocyst origin (refer to text).**

## DISCUSSION

This study has shown that preimplantation blastocysts release a range of proteins *in vitro*, many of which are also found in uterine flushings of pregnant ewes. We have identified a protein with characteristics similar to OTPI in secretions from day 15 blastocysts, but have shown many additional proteins to those described by Godkin *et al.* (1982). We cannot rule out the possibility that media from blastocyst culture and pregnant uterine flushings contained some protein from dead or broken cells. Secretions from cultured control tissues showed quite different patterns after 2D PAGE (data not shown). The similarity of patterns of secretions from both days of culture enable us to harvest secretions from day 15-16 blastocysts over 48 h to use in other studies. Analyses of secretions from day 13 blastocysts revealed many plasma proteins, suggesting proteins from the flushing remaining in the flask or taken up by the embryo. Until this is resolved, we do not believe it prudent to calculate protein secreted per day by day 13 blastocysts.

Individual proteins of blastocyst origin were also present in uterine flushings of pregnant ewes but not in flushings of non pregnant ewes, supporting the idea that transmission of information from blastocyst to mother may occur via these proteins (Findlay 1984). Protein 2 is barely visible in pregnant flushings, a finding also observed by Godkin (personal communication). Of particular note is that specific endometrial epithelial secretory proteins, previously demonstrated using primary cultures of isolated cells (Salamonsen *et al.* 1984) were not visible in the gels of uterine flushings from either pregnant or non pregnant animals. Serum proteins constituted a major portion of the protein in uterine flushings of both pregnant and non pregnant ewes as described previously (Roberts *et al.* 1976). The endometrial contribution to the uterine protein milieu is therefore quantitatively small, though not necessarily unimportant.

The functions of most blastocyst proteins are yet to be determined. Godkin *et al.* (1984b) have presented evidence that blastocyst proteins (specifically OTP1) are involved in the maintenance of luteal function during early pregnancy and that this action is probably mediated through interaction with the uterine endometrium. The approach described in this paper together with cultures of isolated endometrial cells (Salamonsen *et al.* 1984) offer a powerful tool for the study of fundamental reasons for the high early embryonic mortality in the ewe.

ACKNOWLEDGEMENTS

This work was supported by a grant from the Australian Wool Research Trust Fund. One of us (LAS) is currently in receipt of an AWC Postgraduate Scholarship. Jill Volfsbergs provided invaluable secretarial assistance.

REFERENCES

Bradford, M.M., 1976. *Anal Biochem., 72,* 248-254.

Findlay, J.K., 1984. This volume, 105-111.

Findlay, J.K., Ackland, N., Burton, R.D., Davis, A.J., Maule-Walker, F., Walters, D.E. and Heap, R.B., 1981. *J. Reprod. Fert, 62*, 361-377.

Findlay, J.K., Clarke, I.J., Swaney, J., Colvin, N. and Doughton, B., 1982. *J. Reprod. Fert., 64*, 329-339.

Godkin, J.D., Bazer, F.W., Moffatt, J., Sessions, F. and Roberts, R.M., 1982. *J. Reprod. Fert., 65*, 141-150.

Godkin, J.D., Bazer, F.W. and Roberts, R.M., 1984a. *Endocrinology, 114*, 120-130.

Godkin, J.D., Bazer, F.W., Thatcher, W.W. and Roberts, R.M. 1984b. *J. Reprod. Fert., 71*, 57-64.

Morrisey, J.H., 1981. *Anal. Biochem., 117*, 307-310.

O'Farrell, P.H., 1975. *J. Biol. Chem., 250*, 4007-4021.

Roberts, G.P., Parker, J.M. and Symons, H.W., 1976. *J. Reprod. Fert, 48*, 99-107.

Salamonsen, L.A., Doughton, B. and Findlay, J.K., 1984. *Proc. Aust. Soc. Reprod. Biol., 16*, 27.

# PROTEINS SECRETED BY THE 16-DAY-OLD SHEEP BLASTOCYST HAVE IMMUNOSUPPRESSIVE PROPERTIES

A.L.C. Wallace, C.D. Nancarrow and R. Sutton, *CSIRO, Division of Animal Production, PO Box 239, Blacktown, N.S.W. Australia, 2148.*

*Summary* Material secreted by the 16-day-old sheep blastocyst has been shown to be highly immunosuppressive *in vitro* using either the stimulation of lymphocyte mitogenesis by phytohaemagglutinin or by mixed lymphocyte culture.

## INTRODUCTION

The mammalian foetus is protected against maternal immunological rejection from conception until parturition. Many different mechanisms have been proposed to explain how this protection operates (Beer and Billingham 1976). Godkin *et al.* (1982) have shown that from at least day 13 sheep blastocysts secrete proteins when incubated *in vitro*. One of these proteins binds to the uterine endometrium and may prolong the function of the corpus luteum (Godkin *et al.* 1984).

However, proteins secreted by the pre-implantation blastocyst might have other properties, such as immunosuppression. They could thus serve as a local protective mechanism against immune rejection at least until the trophoblast implants. In this paper we describe the suppressive effect of proteins secreted by the 16-day-old sheep blastocyst on the *in vitro* response of sheep peripheral lymphocytes, as measured by [$^3$H]-thymidine uptake to phytohaemagglutinin (PHA) stimulation and to mixed lymphocyte culture (MLC).

## MATERIALS AND METHODS

### Animals

Adult Merino ewes were induced to superovulate by the use of progesterone implants (Silestrus, Ceva, NSW) followed 12 days later by 800 i.u. of pregnant mare serum gonadotrophin (Folligon, Intervet, NSW). After a further 2 days the implants were removed and the ewes mated over the next 4 days. The ewes were slaughtered on day 10 or 16 after mating and the uterus removed.

### Incubation

Conceptuses were flushed from the uterus with sterile physiological saline and transferred to medium. The medium and incubation conditions were essentially as described by Godkin *et al.* (1982), except that phenol red and [$^3$H] leucine were omitted and the embryos were not shaken during incubation. After centrifugation the medium was dialysed (M.W. cut off 6000) against distilled water and freeze dried. A total of 39 16-day-old blastocysts were incubated. In most experiments the combined freeze dried material from 24 of the 39 16-day-old blastocysts was used.

### PHA induced lymphocyte transformation

Peripheral lymphocytes (PBL) were obtained as described by Grewal *et al.* (1984). Culture conditions were essentially as described by Seegerson (1981) using 3 x $10^5$ PBL, 0.4 $\mu$g of PHA, 0.5 $\mu$Ci [$^3$H]-thymidine and 4% pooled sheep serum per well. The degree of transformation was measured by the amount of [$^3$H]-thymidine taken up by lymphocytes.

### Mixed lymphocyte culture

Culture conditions and medium were similar to those used for PHA stimulation. A two way MLC (Hudson and Hay 1980) was performed with 3.75 $\times$ $10^5$ PBL from each of two sheep in each well.

### Gel filtration chromatrography

Forty mg of material from 16-day-old conceptuses was applied to a Sephadex G-100 superfine column (Pharmacia, Sweden), with a phospho-saline buffer, pH 7.2. Tube contents were combined into 7 fractions, dialysed and freeze dried.

### Electrophoresis

Isoelectric focussing was carried out in an 0.9 mm, 1% agarose gel containing 2.5% Pharmalite, pH 3-10 (Pharmacia, Sweden). The dried gels were stained with Coomassie Blue, R-250. SDS electrophoresis was carried out according to the method of Laemmli (1970).

### Total carbohydrate

Total hexoses were determined by the method of Dubois *et al.* (1956) using galactose as a standard.

## RESULTS

The weight of secreted material per blastocyst was 3.5 mg per 24 h. Some blastocysts were incubated at 10 days of pregnancy and each blastocyst secreted on average 1 mg of material into the medium per 24 h. The total hexose content of the material was estimated to be 12% by weight. Isoelectric focussing of material from 16-day-old blastocysts revealed at least 10 proteins, 4 of which were present in relatively high concentration (Figure 1A). The most intensely staining protein had an iso-electric point of about 4.5. In SDS gels stained with Coomassie blue there was one major component of 17,000 MW and several minor components. Silver staining of the gel revealed many more minor components (Figure 1B).

The material from 16-day-old blastocysts was highly immunosuppressive in both PHA and MLC tests (Figure 2a and b). PHA stimulation was suppressed to 50% by 68 µg/ml and MLC stimulation 70% by 25 µg/ml.

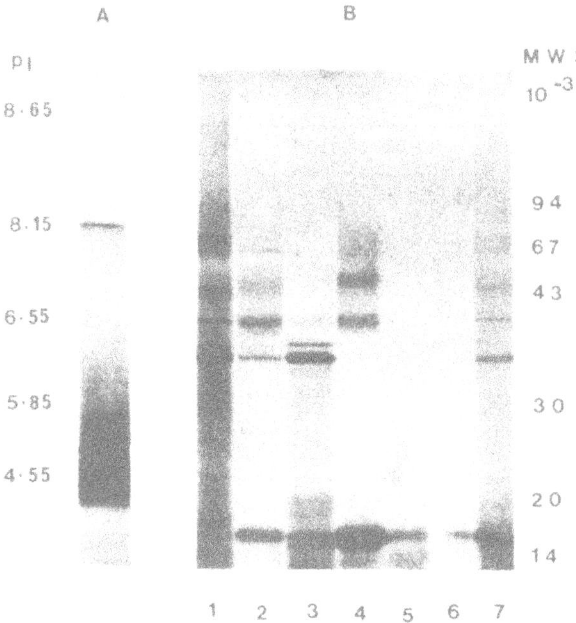


**FIGURE 1. Electrophoresis gels of proteins secreted by 16-day-old blastocysts. (A) Isoelectric focussing gel of unfractionated material stained with Coomassie blue. (B) SDS electrophoresis of Sephadex G-100 fractions 1-6 (Lanes 1-6) and unfractionated material (Lane 7) and stained with Coomassie blue.**

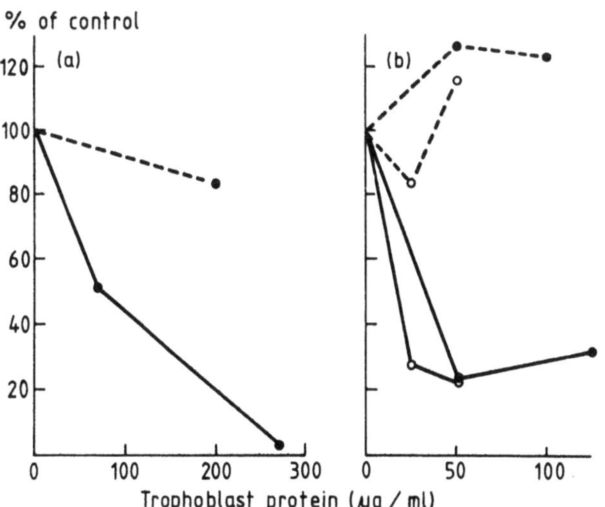


**FIGURE 2. Effect of blastocyst protein on [$^3$H]-thymidine uptake by lymphocytes.**
**(a) PHA stimulation, (b) MLC.**
**---------------- pooled sheep serum ———————— trophoblast proteins**

The 7 fractions obtained by separation on Sephadex G100 superfine (figures 1b and 3) were also tested in the MLC. Two fractions, 4 and 5, gave 40-50% suppression of MLC stimulation at a concentration of 25 μg/ml. Fraction 4 contained a major protein with a molecular weight of 17,000. Two other fractions, F1 and F6 at the same concentration showed a stimulation of MLC response. No fraction had more suppressive activity than the unfractionated material.

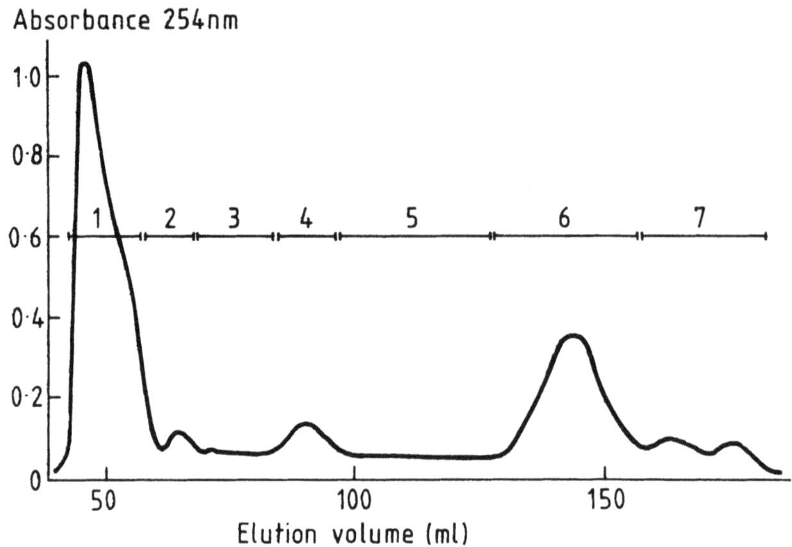


**FIGURE 3. Fractionation of blastocyst proteins on Sephadex G100 superfine column. 1.6 × 90 cm. 0.01 M sodium phosphate buffer, 0.15 M NaCL, pH 7.2.**

## DISCUSSION

It has been suggested (Nancarrow *et al.* 1981) that the sheep embryo secretes a substance 'zygotin' which stimulates the production of an early pregnancy factor (EPF). By using an *in vitro* immunosuppressive test, EPF has been demonstrated in sheep plasma as early as 24 h after mating (Morton *et al.* 1979; Nancarrow *et al.* 1981). It has also been claimed that EPF has immunosuppressive properties *in vivo* (Noonan *et al.* 1979). Although we cannot assume the immunosuppressive activity of our material is either EPF or zyogotin it is interesting that 1) the 10-day-old blastocyst is secreting detectable amounts of material and 2) using a rosette inhibition test similar to that used by Morton *et al.* (1979), Smart *et al.* (1981) have shown immunological activity in culture medium from a 48 h old human pre-implantation embryo. However, Chen *et al.* (1983) could not repeat this observation. This leaves uncertain the question whether immunosuppressive material is released from the early human embryo.

Fractions 4 and 5 from gel filtration appear to share 2 low M.W. proteins. One of these, M.W. 17,000, may be similar to that described by Godkin *et al.* (1984). Why no fraction was more active than the starting material remains unexplained.

Seegerson (1981) showed that uterine flushings obtained from sheep on day 4, 9, and 14 of the oestrous cycle or pregnancy were immunosuppressive in the PHA test. Flushings from day 14 pregnant ewes were much more active than those flushings obtained at other times. Seegerson (1981) suggested that the embryo may be contributing to this increased immunosuppressive activity and our findings would support this interpretation.

Immunosuppressive activity of the material secreted by the blastocyst has so far been demonstrated only *in vitro*. Whether it is active *in vivo* remains unproved. However, with species such as sheep in which implantation is considerably delayed it is a plausible hypothesis that during this time the blastocyst protects itself by secreting immunosuppressive material. This protection is probably reinforced by oviduct and uterine secretions (Wallace *et al.* 1984; Seegerson 1981) until the blastocyst implants, at which time other mechanisms may take over.

This work may also be of practical significance since immunoassay for the proteins secreted by the blastocyst may serve as a basis for a early pregnancy test, thus enabling the problem of early embryo mortality to be thoroughly investigated.

## REFERENCES

Beer, A.E. and Billingham, R.E., 1976. The Immunobiology of Mammalian Reproduction, Prentice-Hall, NJ.

Chen, C., Jones, W.R., Bastin, F. and Forde, C., 1983. *Proc. Fert. Soc. Aust., Abs.,* 45.

Dubois, M., Gilles, K.A., Hamilton, J.K., Rebers, P.A. and Smith, F., 1956. *Anal. Chem., 28*, 350-56.
Godkin, J.D., Bazer, F.W., Moffatt, J., Sessions, F. and Roberts, R.M., 1982. *J. Reprod. Fert., 65*, 141-150.
Godkin, J.D., Bazer, F.W. and Roberts, R.M., 1984. *Endocrinology, 114*, 120-130.
Grewal, A.S., Wallace, A.L.C., Pan, Y.S., Rigby, N.W., Donnelly, J.B., Eagleson, G.K. and Nancarrow, C.D., 1984. *J. Reprod. Immunol.* (in press).
Hudson, L. and Hay, F.C., 1980. Practical Immunology, Blackwell, Oxford.
Laemmli, U.K., 1970. *Nature, 227*, 680-685.
Morton, H., Nancarrow, C.D., Scaramuzzi, R.J., Evison, B.M. and Clunie, G.J.A., 1979. *J. Reprod. Fert., 56*, 75-80.
Nancarrow, C.D., Wallace, A.L.C. and Grewal, A.S., 1981. *J. Reprod. Fert. Suppl. 30*, 191-199.
Noonan, F.P., Halliday, W.J., Morton, H. and Clunie, G.J.A., 1979. *Nature, 278*, 649-651.
Seegerson, E.C., 1981. *Biol. Reprod., 25*, 77-84.
Smart, C.Y., Cripps, A.W., Clancy, R.L., Roberts, T.K., Lopata, A. and Shutt, D.A., 1981. *Med. J. Aust., 1*, 78-79.
Wallace, A.L.C., Nancarrow, C.D. and Sutton, R., 1984. *Proc. Aust. Soc. Reprod. Biol., Abs.*, 101.

# CONTRIBUTION OF THE UTERO-OVARIAN LYMPHATIC NETWORK TO THE CONTROL OF CORPUS LUTEUM FUNCTION IN THE EWE

L.D. Staples and P.A. Whylie, *Department of Agriculture, Animal Research Institute, Werribee, 3030, Victoria.*

*Summary* The anatomical disposition of lymphatic ducts in the region of the utero-ovarian vascular pedicle suggests that they may contribute to the local transfer of uterine prostaglandin (PG) F2$\alpha$ to the ovary. This hypothesis is strengthened by the present data which show that the utero-ovarian lymphatic ducts are more closely associated with the ovarian artery than is the utero-ovarian vein, that utero-ovarian lymph contains PGF2$\alpha$ at the expected time of luteolysis, that obstruction of the uterine lymphatic drainage along the pedicle is associated with prolongation of luteal function, and finally that infusion of PGF2$\alpha$ into a major uterine lymphatic duct results in abrupt and premature regression of the ipsilateral corpus luteum (CL). It is concluded that the utero-ovarian lymphatic network can participate in control of ovarian function in the ewe.

## INTRODUCTION

The present hypothesis concerning the mechanism by which the uterus exerts a local luteolytic effect on the CL focuses attention on the transfer PGF2$\alpha$ from the uterus to the ovary. This transfer is thought to occur by countercurrent diffusion from the utero-ovarian vein to the ovarian artery in the region of the vascular pedicle (Goding 1974; Horton and Poyser 1976; McCracken and Schramm 1983). While this concept is supported by a large number of studies, the literature also contains several anomalous reports which could be explained if alternative routes for PGF2$\alpha$ transfer existed. Of particular importance in this regard is the frequent failure to prevent or delay luteal regression when the uterine vein was obstructed or surgically separated from the ovarian artery (Inskeep and Butcher 1966; Kiracofe *et al.* 1966; Dobrowolski and Hafez 1970; Thorburn and Mattner 1971; Baird and Land 1973; Lammond and Drost 1973). One alternative route for PGF2$\alpha$ transfer is via the oviducal vein (Alwachi *et al.* 1981).

Recent anatomical studies have shown that the lymphatic ducts which drain the uterus and ovaries of the ewe form a common utero-ovarian lymphatic network along the vascular pedicle (Staples *et al.* 1982). This disposition of lymphatic ducts raises the possibility that the utero-ovarian lymphatic network (see Figure 1) could also participate in local control of luteal function. This study further evaluates this hypothesis.

## MATERIALS AND METHODS

### Anatomical studies

At surgery india ink was injected just beneath the uterine serosal layer in 6 ewes to permanently identify lymphatic pathways for uterine lymph drainage. The ewes were killed and the vascular pedicles were excised and fixed in Bouin's solution. Full cross sections at 1,4,7 and 10 cm distal to the ovary were studied histologically to determine the relationships between the lymphatics and the blood vessels within the pedicle.

### Prostaglandin concentrations in lymph

The lymphatic duct closest to the ovarian artery was identified, ligated and cannulated at surgery after injection of Evans Blue dye into the uterine serosa. After recovery, lymph was collected continuously from 7 non-pregnant ewes between Days 9-21 of the oestrous cycle. Lymph samples were pooled over 30 or 60 min intervals and analysed for PGF2$\alpha$, PGE2 and progesterone by specific radioimmunoassays. Jugular venous progesterone levels were also measured at 4h intervals in three of these ewes.

### Ligation of lymphatic ducts

Chronic cannulations of utero-ovarian lymphatic ducts (as above) were frequently associated with an extension of the luteal phase. Therefore the effect of lymphatic ligation on cycle length was studied in 25 ewes in which uterine lymphatic ducts were obstructed by ligation within the mesometrium between uterus and ovary of one uterine horn on Days 8 (n = 4), 9(6), 10(2), 11(12) or 12(1) of the cycle. In a further 7 ewes, all visible utero-ovarian lymphatic ducts along one ovarian pedicle were ligated on Days 10(3) or 11(4) of the cycle.

### Infusion of PGF2$\alpha$ into uterine lymphatic ducts

PGF2$\alpha$ was infused into a major uterine lymphatic at day 7 or 8 of the cycle to test whether elevated levels of PGF2$\alpha$ in a lymphatic duct could induce luteolysis. Functional CL were marked with india ink at surgery. In ewes bearing unilateral CL PGF2$\alpha$ was infused in saline at 2.2 ml/h for 3h at doses of 0 (n = 3, control), 0.1(4), 2(3), 20(6) and 100 $\mu$g/h (5). Luteal function was assessed by determination of peripheral progesterone concentrations. Gross morphology of marked CL was observed after subsequent oestrus.

To test for peripheral effects, PGF2$\alpha$ was infused unilaterally at the same doses (n = 1 per dose) in ewes bearing bilateral CL. Both ovaries were collected and fixed 72 h after infusion then examined histologically for evidence of regression.

## RESULTS

### Anatomical studies

At 7 and 10 cm from the ovary the utero-ovarian vascular pedicle contained several large (> 100 μm dia.) lymphatic ducts accompanied by single profiles of the ovarian artery and utero-ovarian vein. In all 6 animals

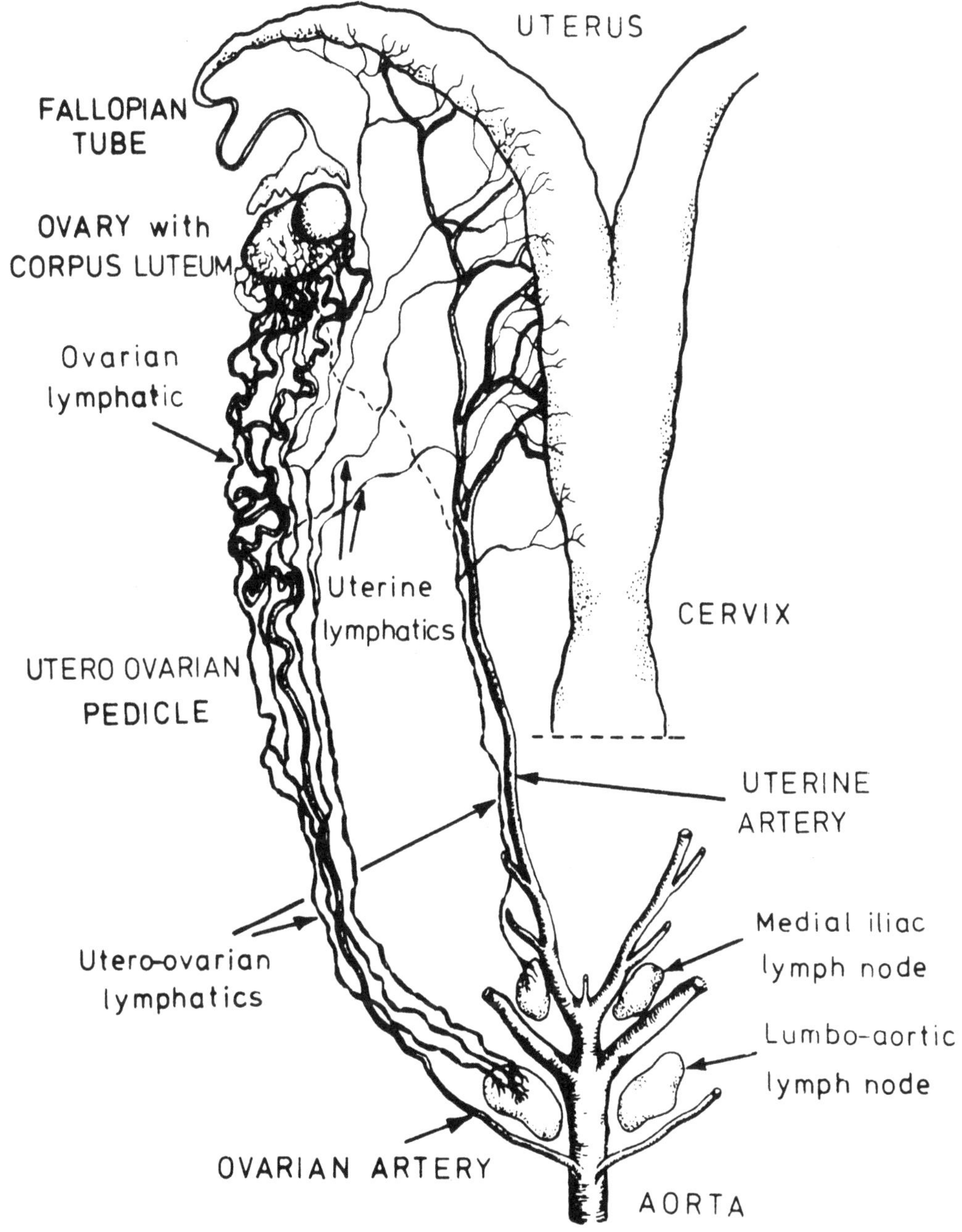


**FIGURE 1. Composite drawing showing the major pathways in sheep of lymphatic drainage from the ovary and uterus along the utero-ovarian pedicle to the lumbo-aortic node(s). The interconnections between uterine and ovarian ducts to form the utero-ovarian lymphatic network are shown and also the connection (broken line) between an ovarian and uterine lymphatic parallel to the uterine artery. Valves in the lymphatic vesels are indicated as irregular narrowing of the black lines while the ovarian and uterine arterial supplies are shown as shaded ducts. The uterus is shown reflected about the broken line near the cervix to allow for clearer portrayal of the lymphatic and vascular pathways. (From Staples, Fleet & Heap (1982), reproduced by courtesy of Journals of Reproduction and Fertility Ltd.)**

the nearest lymphatic was closer, lumen to lumen, to the ovarian artery than was the utero-ovarian vein (221 ± 104 vs 320 ± 152 μm respectively, mean ± sd, $P < 0.05$ by paired t-test). This close association of the lymphatic ducts with the ovarian artery was primarily due to the lymphatic ducts having thin walls.

Prostaglandin concentrations in lymph

Pulses of PGF2α (> twice day 11-12 baseline of 0.23 - 0.79 ng/ml) commenced at Days 12. 13, 13, 13, 14, 14 and 16 in the 7 ewes. Pulse duration was 1-4h and, where recurrent peaks were observed, the interval between peaks was 16.64 ± 0.9 h (mean ± se, n = 22). Peak PGF2 α concentrations ranged widely from 0.53 to 13.0 ng/ml with a mean of 2.55 ± 0.57 ng/ml (n = 27 observed peaks). No peaks were observed in PGE2 concentration.

Ligation of lymphatic ducts

Ligation of uterine lymphatic ducts within the mesometrium was associated with prolongation of the cycle to > 20 days in 7 of 25 ewes. Four of these ewes had bilateral ovulations but in only one case was the CL maintained unilaterally on the side of ligation. The oestrous cycle was extended in 3 of 7 ewes after utero-ovarian lymphatic ducts were ligated along the vascular pedicle.

Infusion of PGF2α into uterine lymphatic ducts

Luteal regression and new ovulations occurred within 72 h in all 5 ewes which received PGF2α into the uterine lymphatic ducts at 100 μg/h. In 4 of 6 ewes infused at 20 μg/h progesterone levels declined and 2 of these ewes ovulated by 72 h post infusion. In contrast, progesterone levels showed a transient decline in one of three ewes infused at 2 μg/h and this ewe, along with all others treated at 2, 0.1 and 0 μg/h showed normal oestrus and ovulation patterns between days 16-18 of the cycle. The immediate luteolytic effect of 100 and 20 μg/h doses of PGF2α was confirmed in ewes bearing bilateral CL and the effect was confined to the CL ipsilateral to the infused lymphatic duct. Thus the luteolytic effect of PGF2α in lymph was exerted locally.

## DISCUSSION

These data provide novel evidence for a functional endocrine role for the utero-ovarian lymphatic network and give new insight into the mechanism of corpus luteum regression in sheep. The anatomical data, PGF2α measurements and luteolytic effects of PGF2α infusions suggest that utero-ovarian lymphatic ducts can provide an additional route for PGF2α transfer to the ovary. That obstruction of lymphatic ducts in the mesometrium or vascular pedicle did not always prolong luteal function and may have been due to rapid regeneration of lymphatic tissue around the obstruction (eg. see Odén 1961) or utilization of alternative lymph pathways which were not visible at surgery. The venous pathway may also have contributed to the transfer of PGF2α and subsequent regression of the CL since blood vessels were not affected by ligation of lymphatic ducts. These studies should be expanded to investigate the role of the lymphatic network in prevention of corpus luteum regression and maintenance of progesterone secretion for successful pregnancy.

## ACKNOWLEDGEMENTS

We wish to thank Dr J.D. O'Shea, Professor G.D. Thorburn, Mrs C. Harper and Miss A. Howse who have contributed to these studies.

## REFERENCES

Alwachi, S.N., Bland, K.P. and Poyser, N.L., 1981. *J. Reprod. Fert., 61*, 197-200.
Baird, D.T. and Land, R.B., 1973. *J. Reprod. Fert., 33*, 393-397.
Dobrowolski, W. and Hafez, E., 1970. *J. Reprod. Fert., 23*, 165-167.
Goding, J.R., 1974. *J. Reprod. Fert., 38*, 261-271.
Horton, E.W. and Poyser, N.L., 1976. *Physiol. Reviews, 56*, 595-651.
Inskeep, E.K. and Butcher, R.L., 1966. *J. Anim. Sci., 25*, 1164-1168.
Kiracofe, Guy, H., Menzies, C.S., Gier, H.T. and Spies, H.G., 1966. *J. Anim. Sci., 25*, 1159-1163.
Lamond, D.R. and Drost, M., 1973. *Prostaglandins, 3*, 691-695.
McCracken, J.A. and Schramm, W. 1983. *In* P.B. Curtis-Prior (ed.), *A Handbook of Prostaglandins and Related Compounds*, Churchill-Livingstons, Edinburgh.
Odén, B., 1961. *Acta Chir. Scand., 121*, 129-232.
Staples, L.D., Fleet, I.R. and Heap, R.B., 1982. *J. Reprod. Fert., 64*, 409-420.
Thorburn, G.D. and Mattner, P.E., 1971. *J. Endocr., 50*, 307-320.

# THE ENDOCRINE CONTROL OF LUTEAL OXYTOCIN AND UTERINE PGF2$\alpha$ RELEASE IN THE EWE

R.J. Fairclough[1] and A.J. Peterson, *Ministry of Agriculture and Fisheries, Ruakura Animal Research Station, Hamilton, New Zealand.*
L.G. Moore and W.B. Watkins, *Postgraduate School of Obstetrics and Gynaecology, University of Auckland, National Womens Hospital, Epsom, Auckland, New Zealand.*
[1] Present address: *Animal Research Institute, Princes Highway, Werribee, 3030. Victoria. Australia.*

## INTRODUCTION

The relatively high embryonic mortality rates (up to 30%) in some flocks is considered to be one of the major factors limiting reproductive performance in sheep. Moreover, extensive field surveys have indicated that most of this embryonic wastage occurs during early pregnancy or over the time at which luteolysis occurs in non-pregnant ewes (Moore *et al.* 1960; Quinlivan *et al.* 1966). These data suggest that the problem may be due, at least in part, to the embryo failing to prevent luteal regression. The present study was undertaken to provide further information on the hormonal control of luteolysis and the mechanism by which the embryo inhibits luteal regression during early pregnancy.

## OXYTOCIN AND UTERINE PROSTAGLANDIN $F_2\alpha$

Although it is now firmly established that $PGF_2\alpha$ is the natural uterine luteolytic factor in the ewe the mechanism of control of uterine $PGF_2\alpha$ secretion during the oestrous cycle and early pregnancy is still uncertain. The ovarian steroids oestrogen and progesterone have been implicated in the hormonal control of uterine $PGF_2\alpha$ release since the administration of large amounts of oestrogens to progesterone primed ewes at midcycle can provoke a marked response in uterine $PGF_2\alpha$ secretion. However, as pointed out by Fairclough *et al.* (1984c) the levels of $PGF_2\alpha$ found in uterine venous blood of ovariectomised ewes supplemented with physiological infusions of progesterone and oestrogen are 5-10 fold less than the maximum levels of utero-ovarian $PGF_2\alpha$ reported in intact ewes over the time of luteal regression. It appears therefore that factors other than oestrogen and progesterone may be involved in regulating the release of $PGF_2\alpha$ from the uterus of ewes at luteolysis. One possible factor is oxytocin, since this hormone has been shown to elicit uterine $PGF_2\alpha$ release *in vivo* and coincident surges of the main metabolite of $PGF_2\alpha$ 13, 14-dihydro-15-keto-PGF (PGFM) and oxytocin or its associated neurophysin (OT-N) have been found in ewes over the time of luteal regression (Fairclough *et al.* 1980, 1983; Flint and Sheldrick 1983). Furthermore, immunization against oxytocin prolongs the luteal phase of the oestrous cycle (Sheldrick *et al.* 1980). The uterine-$PGF_2\alpha$ response to oxytocin varies throughout the oestrous cycle. Fairclough *et al.* (1984b) has shown that oxytocin (250 mIU), given to ewes at mid cycle did not cause any detectable increase in plasma PGFM concentrations whereas over the late luteal phase of the cycle there was a dramatic PGFM response to oxytocin. The data illustrated in Figure 1 indicate that the uterine $PGF_2\alpha$ secretory system increases in sensitivity to oxytocin over the late luteal phase of the oestrous cycle (Fairclough *et al.* 1984b).

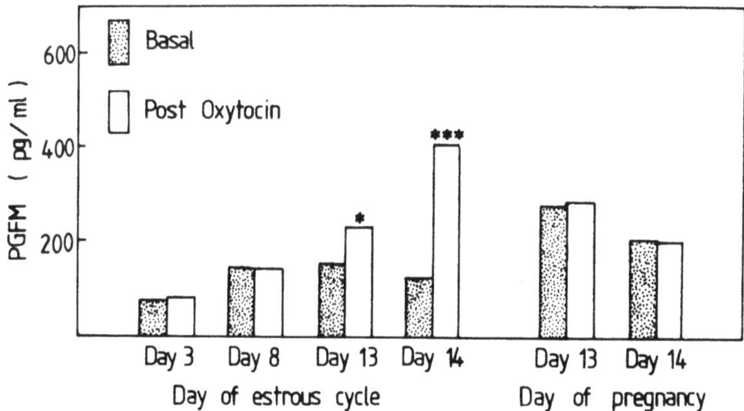


**FIGURE 1. Mean basal and mean peak concentrations of PGFM in ewes injected with oxytocin (250mIU) on Days 3(n = 4), 8(5), 13(4), 14(5) of the oestrous cycle and Days 13(2), 14(4) of pregnancy. *Different from mean basal level, $P < 0.05$; ***Different from mean basal level, $P < 0.01$. (From Fairclough *et al.* 1984b with permission).**

This affect of oxytocin may in turn be influenced by ovarian steroids. In the oestrogen primed ewe it has been shown that the $PGF_2\alpha$ response to oxytocin was initially inhibited by exogenous progesterone but after exposing the ewe to progesterone for a period of around 10 days there was a dramatic amplification of the uterine PGF secretory system to oxytocin (McCracken 1980).

## RELEASE OF OXYTOCIN FROM THE OVARY

The ovary appears to be the source of oxytocin appearing as a series of surges in the peripheral circulation of ewes during luteolysis. Relatively high amounts of oxytocin have been found in the corpora lutea of ewes and active secretion of oxytocin and OT-N by the ovary has been observed on Day 11 of the cycle (Flint and Sheldrick 1982; Watkins *et al.* 1984). The mechanism responsible for pulsatile secretion of ovarian oxytocin is still unclear although there is some evidence implicating PG's in the control of oxytocin release from the ovary. Adminstration of an analogue of $PGF_2\alpha$, cloprostenol, and $PGF_2\alpha$ itself stimulates both OT-N or oxytocin release from the ovary of ewes at mid cycle (Fairclough *et al.* 1984d; Watkins *et al.* 1984) while the administration of the PGF synthetase inhibitor, indomethacin, reduces OT-N surges in the peripheral circulation of ewes over the time of expected luteal regression (Fairclough *et al.* 1984a).

## OXYTOCIN AND PROSTAGLANDIN $F_2\alpha$ RELEASE IN EARLY PREGNANCY

In early pregnancy it has been reported that there is a marked reduction in surges of $PGF_2\alpha$ in utero-ovarian venous plasma (Thorburn *et al.* 1973) and of PGFM in jugular plasma (Peterson *et al.* 1976). This reduction in uterine PGF secretion in early pregnancy also appears to be associated with a suppression of the oxytocin-induced uterine $PGF_2\alpha$ response (Figure 1, Fairclough *et al.* 1984b). In addition, the number and magnitude of OT-N surges in plasma are very much lower over Days 13-15 of pregnancy when compared with non-pregnant ewes (Moore *et al.* 1982). Therefore the embryo seems to affect both oxytocin secretion from the corpus luteum and the uterine-$PGF_2\alpha$ response to oxytocin. The mechanism by which the embryo influences ovarian oxytocin secretion is not clear, although recent data has indicated that intra-ovarian $PGF_2\alpha$ may be involved in regulating oxytocin release from the ovary in non-pregnant ewes (Fairclough *et al.* 1984a). If this is correct then the embryo could suppress ovarian oxytocin secretion by inhibiting ovarian $PGF_2\alpha$ release or the effect of $PGF_2\alpha$ on oxytocin secretion from the ovary. However this latter hypothesis is not supported by the finding that the $PGF_2\alpha$ induced luteal oxytocin response is not blocked during early pregnancy in the ewe (Fairclough *et al.* 1984d). On the basis of this finding the possibility was raised by Fairclough *et al.* (1984d) that the conceptus inhibits not only uterine $PGF_2\alpha$ but also ovarian $PGF_2\alpha$ release. In summary the experiments outlined in this paper have extended knowledge on the underlying mechanisms involved in the hormonal control of luteolysis in sheep and provided further information on the mechanism by which pregnancy is established in the ewe. The very important discovery that the embryo suppresses the uterine $PGF_2\alpha$ response to oxytocin is now being used to identify, isolate and establish an assay for the 'putative' antiluteolytic factor in sheep.

## REFERENCES


Fairclough, R.J., Moore, L.G., McGowan, L.T., Peterson, A.J., Smith, J.F., Tervit, H.R., Watkins, W.B., 1980. *Prostaglandins, 10*, 199-208.

Fairclough, R.J., Moore, L.G., McGowan, L.T., Smith, J.F., Watkins, W.B., 1983. *Biol. Reprod., 29*, 271-277.

Fairclough, R.J., Moore, L.G., Peterson, A.J., Tervit, H.R., Watkins, W.B., 1984a. *Biol. Reprod.* (submitted).

Fairclough, R.J., Moore, L.G., Peterson, A.J., Watkins, W.B., 1984b. *Biol. Reprod.*, 31, 36-43.

Fairclough, R.J., Moore, L.G., Peterson, A.J., Watkins, W.B., 1984c. *Biol. Reprod.* (submitted).

Fairclough, R.J., Peterson, A.J., Moore, L.G., Watkins, W.B., 1984d. *Biol. Reprod.* (in press).

Flint. A.P.F., Sheldrick, E.L., 1982. *Nature, 297*, 587-588.

Flint. A.P.F., Sheldrick, E.L., 1983. *J. Reprod. Fert., 67*, 215-225.

McCracken, J.A., 1980. *In* B. Samuelsson, P.R. Ramwell and R. Poaletti, (eds.). *Advances in Prostaglandin and Thomboxane Research, Raven Press, New York, 8*, 1329-1344.

Moore, N.W., Rowson, L.E.A., Short, R.V., 1960. *J. Reprod. Fert., 1*, 332-349.

Moore, L.G., Watkins, W.B., Peterson, A.J., Tervit, H.R., Fairclough, R.J., Havik, P.G., Smith, J.F., 1982. *Prostaglandins, 24*, 79-88.

Peterson, A.J., Tervit, H.R., Fairclough, R.J., Havik, P.G., Smith, J.F., 1976. *Prostaglandins, 12*, 551-558.

Quinlivan, T.D., Martin, C.A., Taylor, W.B., Cairney, I.M., 1966. *J. Reprod. Fert.*, *11*, 379-390.

Sheldrick, E.L., Mitchell, M.D., Flint, A.P.F., 1980. *J. Reprod. Fert., 59*, 37-42.

Thorburn, G.D., Cox, R.I., Currie, W.B., Restall, B.J., Schneider, W., 1973. *J. Reprod. Fert., Suppl. 18*, 151-158.

Watkins, W.B., Moore, L.G. Flint, A.P.F., Sheldrick, E.L., 1984. *Peptides,* 5, 61-64.

# FERTILISATION FAILURE AND EMBRYONIC WASTAGE

R.W. Kelly, *Sheep and Wool Branch, Department of Agriculture, South Perth, W.A., 6151.*

## INTRODUCTION

Although the proportion of ewes which fail to lamb underestimates the total losses due to failure of fertilisation and wastage of embryos in breeding flocks, it is an easily obtainable and practical measurement of reproductive loss often used by farmers and scientists. Studies involving 186 commercial Merino flocks in the south-west of Western Australia demonstrate a wide variation between flocks in the proportion of ewes failing to lamb per ewe present at lambing (0.03-0.49, mean value = 0.17). Detailed analyses of the sources of wastage (Knight *et al.* 1975b) have shown that the wastage due to marked ewes which fail to lamb contributes at least half of the total reproductive losses between joining and marking. In other regions of Australia, although the average levels of ewe infertility are less than those recorded in Western Australia (e.g. Drinan and Dun 1965; Mullaney and Hyland 1967; Dawe *et al.* 1970; Connors 1971; King 1981), considerable differences still exist between flocks. This paper reviews research into factors contributing to failure of fertilisation and embryonic wastage in commercial sheep flocks, and discusses potential areas for future work.

## FAILURE OF FERTILISATION

The process of fertilisation involves the successful fusion of male and female gametes, and therefore on a flock basis includes the detection of oestrous ewes, mating, deposition of spermatozoa in the vagina and sperm transport. Mating performances of rams and ewes are usually recorded using the sire-sine harness and crayon (Radford *et al.* 1960), but raddle marks may not be accurate indicators of the success of mating since they do not identify all mated ewes (e.g. Radford *et al.* 1960; Wroth and Lightfoot 1976) and neither are all marked ewes inseminated (e.g. Braden 1971; Allison *et al.* 1975; Kilgour and Wilkins 1980). It is therefore incorrect to term marked ewes as mated ewes, yet this source of error is often overlooked in evaluating the performances of rams and ewes during joining. In situations where problems related to infertility are being researched, vaginal swabbing (Allison *et al.* 1975) and the choice of suitable crayons and recording intervals must be used together with recovery of ova to determine fertilisation rates, if a definitive analysis is to be obtained. Oestrus unaccompanied by ovulation, which has been recorded in ewe lambs, will not be considered as a component of failure of fertilisation in this review.

### Failure to be marked

Failure of ewes to be marked can be due to either lack of behavioural oestrus of failure of rams to mount oestrous ewes. Lack of behavioural oestrus is associated with either anovular ewes as occurs frequently when ewe lambs of low liveweight are joined, or when ovulation is not preceded by a progesterone phase as can occur in ewe lambs having only one ovulation during the breeding season and when older ewes are joined in the non breeding season. In Western Australian Merino ewes joined out of season, observations of cessation of ovarian activity in ewes following teasing together with the low lambing performances of inseminated oestrous ewes (Lindsay *et al.* 1984; Pearce *et al.* 1984) suggest that to achieve high lambing performances pretreatment of ewes with progestagens may be necessary. Reeve and Chamley (1984) in Victoria successfully used a programme involving 6 days treatment with progestagen sponges combined with the 'ram effect'.

During the period of joining, other factors known to contribute to failure of a ewe to be marked include ewe age and liveweight (through effects on duration of oestrus and migratory behaviour), ram:ewe ratio, paddock size and topography, ram age, grazing pastures with a high phyto-oestrogen content and high ambient temperatures. Compounding with failure to be marked are effects of many of the forementioned factors on failure of insemination and sperm transport.

Recently some attention has been focused on using tests of libido to predict ram performance during the joining period, with the aim of using animals of high libido to reduce the failure to mate problem. Nevertheless, many workers have failed to demonstrate significant relationships between pen tests and subsequent performance during joining and flock fertility (Cahill *et al.* 1975; Kelly *et al.* 1975; Fletcher 1976; Walkley and Barber 1976; Allison 1978). Kilgour and Wilkins (1980) reported that in flocks of ewes joined with groups of high (H) or low (L) serving capacity rams, the proportions of ewes lambing after 6 weeks of joining were similar for both groups (0.95 in group H and 0.94 in group L). It is important to note that the selection of extremes in serving capacity would maximise differences between joining practices. Under normal flock management, it is likely that groups of rams will contain animals with varying levels of activity, so that effects of individual ram performance on fertility of the flock will be minimal (e.g. Mattner *et al.* 1973). Therefore only under conditions where small numbers of rams are used at joining, or where there is a high incidence of rams with low levels of libido, is the testing of libido likely to be worthwhile.

### Marked but not inseminated

Variations in failure of insemination have been recorded. A.J. Allison and R.W. Kelly (unpublished data) found fewer marked and ovular Romney ewe lambs had spermatozoa present in their anterior vagina than older ewes (0.84 *v*. 0.98, $p < 0.001$). Cahill *et al.* (1974) and Brien *et al.* (1977) also reported 0.06-0.08 fewer maiden Merino ewes with spermatozoa in the vagina than older ewes. However in the Border Leicester × Merino ewe, age (maiden versus mature) had no significant effect on insemination rate (proportion with spermatozoa in anterior vagina) in marked ewes, but liveweight and breed of ram were important (Killeen 1974). In many cases, it is often impossible to separate between the effects of age and liveweight on the performance of ewes. Serving capacity of rams may also effect the insemination rate (Kilgour and Wilkins 1980; Kilgour and Whale 1980).

Recently, the work of A. Tilbrook and A. Cameron (pers. com.) has identified the relative importance of failure to be marked and failure of marked ewes to be inseminated under several joining systems. Rams (n = 10, sexually experienced) were joined either singly, or in groups of two or three, with six oestrous and 30 non-oestrous ewes per ram for 24 hours (equivalent to the number of oestrous ewes expected at a ram:ewe ratio of about 1:100) in 0.5 ha paddocks, and observed continuously. The major sources of wastage were 0.12-0.33 of ewes in oestrus not being mounted in the 24 hours of observation, and of those mounted 0.14-0.21 were not served (Table 1). In total therefore, 0.32-0.52 of oestrous ewes in this experiment within each type of joining would have failed to lamb to that oestrus as a direct result of these factors. Although the ram:ewe ratio was lower and stocking rate higher than that normally used in Western Australia, and ewes joined during the breeding season would have a second and perhaps third opportunity to conceive when they returned to service, it is clear that the losses reported in this work are important. In situations where the number of oestruses may be limited (joining out-of-season) or ovulation rate declines during the period of joining, these sources of wastage achieve added significance. Undoubtedly the work warrants further examination under normal joining conditions, particularly in situations where returns to service or failure to be marked are high.

**TABLE 1. Joining and lambing performance of ewes (A. Tilbrook and A. Cameron, pers. com.).**

| Type of joining | Number of oestrous: non-oestrous ewes present | Proportions of | | |
|---|---|---|---|---|
| | | Oestrous ewes mounted | Mounted ewes served | Served ewes that lambed |
| Single sire | 6:30 | 0.69 | 0.79 | 0.73 |
| Groups of 2 | 12:60 | 0.88 | 0.80 | 0.78 |
| Groups of 3 | 18:90 | 0.67 | 0.86 | 0.74 |

Failure of sperm transport and fertilisation

In studies where fertilisation rates or non returns to service are related to ewes inseminated, the percentage of ova fertilised is high (Table 2). The major exceptions are in flocks with a history of exposure to isoflavone-rich pastures, and possibly following high environmental temperatures immediately prior to joining. Concomitant field studies recording reproductive wastage in commercial sheep flocks grazing oestrogenic and non oestrogenic pastures in South Western Australia (Wroth and Lightfoot 1976; Marshall *et al.* 1976) report average fertilisation rates per ewe of 73% and 93% respectively. Failure of sperm transport (Lightfoot *et al.* 1967) due to changes in cervical mucus (Adams 1977) has been diagnosed as the major factor responsible for the lower fertilisation rates. Similarly the grazing of pastures with a high isoflavone content at joining markedly depresses fertilisation rates (Lightfoot and Wroth 1974). In an attempt to overcome problems related to oestrogenic pastures, some farmers use more rams for joining, but the efficacy of this practice has not been resolved.

With respect to high temperatures, major effects have been recorded on spermatozoa developing in the testes but not on those present in the epididymis, with consequent effects on fertilisation rates to matings 14-47 days after exposure (Braden and Mattner 1970). However Lindsay *et al.* (1975) found no significant correlation between rates of return to service and high temperatures about the time of joining.

General remarks

Clearly failure of insemination and sperm transport are important factors influencing fertilisation rates in commercial sheep flocks, and in many cases are identifiable with specific management or pasture situations. In higher rainfall zones in Australia where feed supply and time of joining related to the onset of the breeding season are not major factors influencing reproductive performance, failure of fertilisation is unlikely to be a major component of reproductive wastage. During joining, despite the fact that the fertility of ewes that return to service is lower than the fertility of ewes mated once (Smith and Lindsay 1972), ewes have a second and sometimes third opportunity to conceive so that failure of fertilisation usually has a small effect on reproductive efficiency. The major exception to this rule is where ewes may only have one or two oestruses during joining and the first oestrus is of low fertility, as occurs when ewes are joined in the non-breeding season without prior teasing (Knight *et al.* 1975a; Oldham *et al.* 1976). Additionally, effects on reproductive performance such as a decline in ovulation rate due to seasonal or liveweight changes during joining should not be overlooked.

**TABLE 2. Fertilisation rates of ova in inseminated ewes.**

| Source | Age of ewe | Fertilisation rate (%) |
|---|---|---|
| A.J. Allison and R.W. Kelly (unpublished data) | ewe lamb | 86 |
| | older ewes | 88 |
| Cahill *et al.* (1974) | 1.5 y.o. maiden | 88[(1)] |
| | mature ewes | 94[(1)] |
| Kilgour and Wilkins (1980) | mature ewes | 90-93 |
| Killeen (1974) | mature ewes | 96[(2)] |
| R.J. Lightfoot (unpublished data) | mixed age | 80-95 |

[(1)] based on non returns within 21 days of marking
[(2)] percentages of ewes yielding fertilised ova

## EMBRYONIC MORTALITY

Losses from the period of mid embryogenesis (days 30-40) through to parturition are small (Robinson 1951; Quinlivan *et al.* 1966) and can be regarded as negligible except in cases of mineral deficiency (e.g. selenium; Hartley 1963), abortion due to disease (vibriosis, toxoplasmosis; Munday *et al.* 1966) or ovulation rates of three and higher (Wilkins *et al.* 1984; J. Owens and G. Hinch pers. com.). These and other experiments where animals have been slaughtered in early pregnancy (reviewed by Edey 1969), and observed by sequential ultrasonic scanning (Wilkins *et al.* 1984) support the conclusion that losses after day 30 of pregnancy are likely to be insignificant in the majority of flocks in Australia where ovulation rate per ewe rarely exceeds two. There are many reports on the extent of wastage that occurs between fertilisation and day 30 of pregnancy. During this period there is a small increase in size of the embryo up to day 10 — 11 of pregnancy, followed by rapid increases and the commencement of implantation. Edey (1969) summarised the estimates of prenatal mortality in the ewe, and concluded that 20 — 30% of embryos are normally lost during pregnancy with most of the loss occurring in the first month.

Further definition of the time of embryonic mortality is likely to be unrewarding until techniques to determine viability of embryos other than size and stage of development are utilised. Blockey *et al.* (1975) suggested that there was little loss of embryos prior to day 12 *post coitum* by comparing return rates up to day 20 with fertilisation rates, but other work (e.g. Moore *et al.* 1960; Mattner and Braden 1967) has indicated that most of the losses occur prior to day 12. There may be a considerable interval between times of experimental treatment and embryonic mortality (e.g. Lawson and Findlay 1977), as it seems that embryonic development can proceed despite defects until about the time of commencement of endometrial attachment. Robinson (1951) suggested that losses after commencement of implantation are merely a continuation of a process initiated long before. Morphological appearance of embryos prior to day 12 of pregnancy is not a reliable indicator of viability (e.g. Killeen and Moore 1971). It would be worthwhile to investigate the fluorescein diacetate technique (Mohr and Trounson 1980) as a tool for examination of early embryo viability.

### Factors influencing embryonic mortality

*Ovulation rate* One of the most important factors affecting embryonic mortality is ovulation rate. The mean reproductive performance of marked ewes at the end of the first cycle of joining related to the number of corpora lutea (CL's) is summarised in Table 3. When all losses from ovulation to parturition are considered, 0.19, 0.24 and 0.34 of ova shed (as recorded by number of CL's) cannot be accounted for as live lambs in ewes with one, two or three CL's respectively. However, these values include losses due to failure of fertilisation. A more accurate estimate of some of the losses due to early embryonic mortality can be obtained by examining the extent of partial failure of multiple ovulations, that is the proportion of ewes lambing that have fewer lambs than number of CL's recorded at mating. This is possible since fertilisation is generally regarded as an all or none process in multiple ovulating ewes (Restall *et al.* 1976; I.D. Killeen, unpublished data, referenced by Kelly *et al.* 1978), although other studies indicate that at high ovulation rates such as may be recorded in Booroola flocks partial failure of fertilisation may occur (e.g. Trounson and Moore 1974). Assuming negligible foetal mortality, the estimated proportion of embryos that are lost through embryonic mortality (as indicated by partial failure of multiple ovulation) in ewes that have two or three CL's and lamb are 0.15 and 0.27 respectively. To further test the dependence of embryonic survival on ovulation rate, the statistical approach described by Restall *et al.* (1976) was used on the data reported in Table 3, assuming a fertilisation rate of 0.9 with and without a partial fertilisation rate in the ewes with two CL of 0.06. In both cases the $\chi^2$ test for independence was highly significant ($p < 0.001$). Therefore, it is concluded that embryonic survival is dependent on ovulation rate, which supports the conclusions reached by Edey (1969) and Geisler *et al.* (1977).

**TABLE 3. Mean reproductive performance of ewes with 1, 2 or 3 corpora lutea (from Kelly, 1980; Kelly and Johnstone, 1983).**

| | | Proportion of ewes | | | | |
|---|---|---|---|---|---|---|
| Number of corpora lutea | Number of ewes observed | Returning to service | Barren | 1 lamb | 2 lambs | 3 lambs |
| 1 | 2,430 | 0.14 | 0.05 | 0.81 | | |
| 2 | 4,087 | 0.08 | 0.03 | 0.26 | 0.63 | |
| 3 | 188 | 0.06 | 0.03 | 0.22 | 0.30 | 0.39 |

Given that differences in ovulation rate are associated with differences in embryonic survival, it is disappointing that many studies pool data across ovulation rates and express losses as the percentage of CL's not represented by embryos at slaughter or lambs at birth, or the percentage of ewes that fail to lamb. Also, much of the work has concentrated on factors affecting embryonic mortality in ewes shedding only one egg. There may well be different degrees of response to the treatments for the loss of one embryo from ewes with one CL (complete failure of pregnancy) compared with the loss of one embryo from two CL's (partial failure of multiple ovulation). Age, nutrition, hormonal conditions, site of ovulation, environment and genotype have all been suggested as influencing embryonic mortality.

*Age* Embryonic mortality is greater in young than older ewes. A.J. Allison and R.W. Kelly (unpublished data) found that although fertilisation rates in mated lambs (7-9 months of age) and older ewes were similar (0.86 *v* 0.88), the proportion lambing following transfer of cleaved ova was markedly lower in lambs (0.39 *v* 0.75, $p < 0.01$) indicating substantial embryonic wastage. Quirke and Hanrahan (1977) recorded a 33% survival rate of cleaved ova from lambs compared with 73% for cleaved ova from adult ewes following their transfer to adult ewes. These results suggest that the cleaved ova from ewe lambs are either of a lower survival capacity inherent in the ovum itself, or there are important differences in conditions in the donor animals prior to collection and transfer. Significant though less pronounced differences in embryonic loss were estimated between maiden 1.5 year old and older parous ewes (Edgar 1962; Blockey *et al.* 1975).

*Nutrition* There is considerable variation in the published work on the influence of nutrition about mating on embryonic wastage. Extensive research by Cumming (1972a,b) and others found increased wastage with submaintenance diets, but Parr *et al.* (1982) failed to record any significant effect in ewes with one CL. Similar variation in the effect of above maintenance feeding has also been recorded. Smith *et al.* (1983) reported no effect of premating nutrition on partial failure of multiple ovulations. Edey (1976a) reviewed nutrition and embryonic survival in the ewe, and pointed out that when detected, mortalities have been less than 15% and with rare exceptions significant results have been associated only with severe undernutrition for periods of 7-21 days in the first month of pregnancy. Under normal farming conditions, it is unlikely that such severe changes in nutrition will occur. Also the short periods of experimental treatment about joining ignore any possible long term carry-over effects on the ewe. It is therefore not totally surprising that Kelly and Johnstone (1982) found that liveweight and liveweight change explained only a very small amount of the variation between New Zealand farms in partial failure of multiple ovulation. Mineral deficiencies, in particular selenium and iodine, have been found to result in higher embryonic mortality.

*Hormonal conditions* Studies have suggested that the ewe reacts to nutritional extremes by an inverse response in concentrations of progesterone thereby influencing embryonic survival (Cumming and Findlay 1977; Brien *et al.* 1981). However other work (Cumming *et al.* 1971) showed little change in embryonic survival despite significant differences in plasma concentrations of progesterone in ewes on differential feeding from days 2-16 post mating. Also, in ewes ovariectomised within several days post mating, followed by various treatments with progesterone, it has been found that pregnancy is maintained over wide ranges of concentrations of progesterone (Bindon 1971; Parr *et al.* 1978, 1982). Therefore the implication that increased embryonic mortality is due to increased concentrations of progesterone in ewes subjected to nutritional restriction has not been borne out experimentally.

In fact exogenous progesterone administered for 14 days from days 10-13 after artificial insemination in ewes bred out-of-season increased the proportion that lambed (0.34 *v.* 0.45; Pearce *et al.* 1984). Given the aberrations observed in luteal function in ewes joined in the non-breeding season, this latter observation may be related to maintaining poorly formed CL's. However other work (Peterson *et al.* 1984) has also recorded increased fertility in ewes treated with progesterone post mating.

Low concentrations of progesterone have also been associated with lower embryo survival (Parr *et al.* 1982). In contrast rates of embryonic mortality in Booroola ewes heterozygous for the F gene, which have lower concentrations of progesterone than ewes without the gene at ovulation rates greater than 2 (Kelly *et al.* 1983), are similar to those recorded in non Booroolas (Kelly and Johnstone 1983; G.H. Davis *et al. pers. com.*). In conclusion then, on the basis of these results the role of variations in concentrations of progesterone during early pregnancy as a factor directly influencing embryonic survival is tenuous. Nevertheless, recent results are sufficiently promising to warrant further studies. Oestrogen post mating does not appear to be essential for embryonic survival (Cumming *et al.* 1974).

More recently attention has been focused on the role of hormones in the preconditioning of the uterus prior to conception. For example in ewes that had two CL Kelly and Smith (1980) reported that the embryonic survival was higher in those ewes that had two than one CL in the previous cycle. Studies with ovariectomised ewes (Miller and Moore 1976) also show that hormonal conditions before transfer of cleaved ova have a marked influence on embryonic survival. Lack of hormonal preconditioning through an effect on embryonic mortality may account for the reduced fertility of Merino ewes at their first oestrus of the breeding season (Oldham *et al.* 1976; Williams *et al.* 1978) and thc highcr wastage rates in maiden than parous ewes. Hormonal conditioning has been suggested as a possible explanation for the negative relationship between ovulation rate and embryonic mortality (Kelly and Johnstone 1982). Obviously in studies on factors influencing embryonic mortality the events preceding the cycle of conception should not be ignored, and could explain a substantial part of the variability in mortality rates that exists in much of the published literature.

*Site of ovulation* Evidence of an effect of site of ovulation on embryonic survival varies, particularly with reference to partial failure of multiple ovulation. White *et al.* (1981) reported more embryonic losses by ewes with two CL's on one ovary as compared with one CL on each, but Kelly and Allison (1976), King (1981), Kelly and Johnstone (1983) and Meyer *et al.* (1983) found no significant influence of the distribution of two CL on the survival of embryos. The extensive data published by Kelly and Johnstone (1983) shows that for ewes with one or two CL, site of ovulation had small, non significant effects on the proportion of ewes returning to service, barren, or suffering partial failure of multiple ovulation. At present the weight of evidence supports the contention that site of ovulation is not an important factor in embryonic wastage, except perhaps when there is a low frequency of distributive migration of embryos between uterine horns in ewes with more than one CL (Doney *et al.* 1973).

*Other environmental effects* Season, wet weather and handling of ewes after mating have all been shown to influence embryonic mortality. However with the exclusion of 'season' many of the experimental treatments used were extreme relative of farming practice and conditions. For instance, the animal handling after mating used by Doney *et al.* (1976) involved the trucking of animals, shepherding and/or handling in yards daily over various periods up to 20 days post mating. On commercial sheep farms in Australia the amount of handling of ewes during joining is minimal, and therefore is unlikely to increase embryonic mortality.

The most significant climatic factor thought to influence embryonic survival under Australian conditions is high temperature at joining. Lindsay *et al.* (1975) reported a significant negative correlation between high summer temperatures and lambing performance, and suggested that the main association was with survival of embryos. To identify the paternal or maternal contribution to reproductive wastage through joining at different times of the year, R.J. Lightfoot (pers. com.) used artificial insemination (A.I.) with fresh and frozen semen. Fertilisation rates (0.84 — 0.93) were not significantly different to A.I. with fresh semen between January (high ambient temperatures), March and May, but the estimate of embryonic mortality (difference between fertilisation and lambing rates) varied ($p < 0.05$) from 0.50 (January) to 0.25 (March) and 0.38 (May). The pattern of lambing after A.I. with frozen semen was similar to that after use of fresh semen, which implicates the ewe as the major contributor to variation in the level of wastage between the three times of joining. Experiments aimed at studying the components of temperature stress on reproductive performance (stage of pregnancy, duration of heating) have identified that the susceptibility of the ewe is greatest preceding or about the time of fertilisation (Thwaites 1971; Sawyer 1979). Embryonic mortality in ewes subjected to high temperatures from day 3 *post coitum* onwards is invariably low, and probably has only a minor contribution to reproductive losses under such conditions. In contrast other studies (see review by Edey 1976b) have failed to implicate heat-induced embryonic mortality as an important source of reproductive wastage.

*Genotype* Significant differences in embryonic mortality have been recorded between breeds of sheep (Cumming *et al.* 1975; Meyer *et al.* 1983) and ewes mated to rams from lines of sheep selected for high versus low prolificacy (Burfening *et al.* 1973).

Despite these observations, recent evidence suggests that there may be little potential for the selection of Merinos for higher embryonic survival. J.P. Hanrahan *et al.* (unpublished data) using data from both Australia and New Zealand have obtained estimates of the repeatability of embryonic survival of 0.11 ($\pm$ 0.04), while the daughter-dam estimate of heritability was 0.06 ($\pm$ 0.07).

## CONCLUSION

Fertilisation failure and embryonic wastage can contribute significantly to the total losses between joining and marking. Where the proportion of barren ewes is high, or lambing is protracted, failure of insemination has largely escaped scientific consideration. It warrants future attention as to its importance and causes..

Undoubtedly genetic factors can contribute to embryonic death (Bishop 1964; Long and Williams 1980), but the magnitude of the loss is debatable and perhaps indefinable. Certainly, with the adoption of techniques to increase ovulation rate, embryonic wastage becomes an increasingly important contributor to the total losses over the period between joining and tailing (Kelly 1980). Just how much of this wastage can be regarded as induced and therefore responsive to manipulation by management or exogenous hormonal therapy is unknown at this time. Certainly wide variations in partial failure of multiple ovulation have been recorded between farms within years, but the repeatability of losses between years is low and largely inexplicable (Kelly and Johnstone 1982). There are three areas of research warranting immediate study. Firstly, there is an urgent need to develop a suitable technique to determine embryonic viability prior to days 12 — 15 of pregnancy to facilitate diagnosis of treatment effects on the embryo. Secondly, an investigation of the reason(s) for the differences in survival rates of embryos from ewe lambs and older animals would appear to be a reasonable area in which to initiate studies, given that the substantial differences may give the greatest opportunity to detect factors inducing embryonic wastage. Thirdly, hormonal treatment pre and post mating could play an important role in embryonic survival as it relates to maiden ewes and out-of-season breeding, and requires further study.

REFERENCES

Adams, N.R., 1977. *Aust. J. Agric. Res.*, *28*, 481-489.
Allison, A.J., 1978. *N.Z. J. Agric. Res.*, *21*, 187-195.
Allison, A.J., Kelly, R.W., Lewis, J.S. and Binnie, D.B., 1975. *Proc. N.Z. Soc. Anim. Prod.*, *35*, 83-90.
Bindon, B.M., 1971. *Aust. J. Biol. Sci.*, *24*, 149-158.
Bishop, M.W.H., 1964. *J. Reprod. Fert.*, *7*, 383-396.
Blockey, M.A. deB., Parr, R.A. and Restall, B.J., 1975. *Aust. Vet. J.*, *51*, 298-302.
Braden, A.W.H., 1971. *Aust. J. Exp. Agric. Anim. Husb.*, *11*, 375-378.
Braden, A.W.H. and Mattner, P.E., 1970. *Aust. J. Agric. Res.*, *21*, 509-518.
Brien, F.D., Cumming, I.A. and Baxter, R.W., 1977. *J. Agric. Sci., Camb.*, *89*, 437-443.
Brien, F.D., Cumming, I.A. Clarke, I.J. and Cocks, C.S., 1981. *Aust. J. Exp. Agric. Anim. Husb.*, *21*, 562-565.
Burfening, P.J., Friedrick, R.L. and Van Horn, J.L., 1977. *Theriogenology*, *7:* 285-291.
Cahill, L.P., Blockey, M.A. deB. and Parr, R.A., 1975. *Aust. J. Exp. Agric. Anim. Husb.*, *15*, 337-341.
Cahill, L.P., Kearins, R.D., Blockey, M.A. deB. and Restall, B.J., 1974. *Aust. J. Exp. Agric. Anim. Husb.*, *14*, 723-725.
Connors, R.W., 1971. *Wool Tech. Sheep Breed.*, *18*, 55-56.
Cumming, I.A., 1972a. *Proc. Aust. Soc. Anim. Prod.*, *9*, 192-198.
Cumming, I.A., 1972b. *Proc. Aust. Soc. Anim. Prod.*, *9*, 199-203.
Cumming, I.A., Baxter, R. and Lawson, R.A.S., 1974. *J. Reprod. Fert.*, *40*, 443-446.
Cumming, I.A., Blockey, M.A. deB., Winfield, C.G., Parr, R.A. and Williams, A.H., 1975. *J. Agric. Sci., Camb.*, *84*, 559-565.
Cumming, I.A. and Findlay, J.K., 1977. *In* J.H. Calaby and C.H. Tyndale-Biscoe, (ed.) *'Reproduction and Evolution'*, Aust. Academy of Science, p. 225-233.
Cumming, I.A., Mole, B.J., Obst, J., Blockey, M.A. deB., Winfield, C.G. and Goding, J.R., 1971. *J. Reprod. Fert.*, *24*, 146-147.
Dawe, S.T., Bennett, N.W., Donnelly, F.B., Ferguson, B.D., Rive, J.P., Roberts, B.C. and Trimmer, B.I., 1970. *Proc. Aust. Soc. Anim. Prod.*, *8*, 317-320.
Doney, J.M., Gunn, R.G. and Smith, W.F., 1973. *J. Reprod. Fert.*, *34*, 363-366.
Doney, J.M., Smith, W.F. and Gunn, R.G., 1976. *J. Agric. Sci., Camb.*, *87*, 133-136.
Drinan, J.P. and Dun, R.B., 1965. *Aust. J. Exp. Agric. Anim. Husb.*, *5*, 345-352.
Edey, T.N., 1969. *Anim. Breed. Abstr.*, *37*, 173-190.
Edey, T.N., 1976a. *Proc. N.Z. Soc. Anim. Prod.*, *36*, 231-239.
Edey, T.N., 1976b. *In* G.J. Tomes, D.E. Robertson and R.J. Lightfoot (ed), *'Sheep Breeding'*, p 400-410.
Edgar, D.G., 1962. *J. Reprod. Fert.*, *3*, 50-54.
Fletcher, I.C., 1976. *In* G.J. Tomes, D.E. Robertson and R.J. Lightfoot (ed.), *'Sheep Breeding'*, p 345-351.
Geisler, P.A., Newton, J.E. and Mohan, A.E., 1977. *J. Agric. Sci. Camb.*, *89*, 309-317.
Hartley, W.J., 1963. *Proc. N.Z. Soc. Anim. Prod.*, *23*, 20-27.
Kelly, R.W., 1980. *Proc. N.Z. Vet. Assoc. Sheep and Beef Cattle Soc.*, *10*, 78-93.
Kelly, R.W., and Allison, A.J., 1976. *In* G.J. Tomes, D.E. Robertson, and R.J. Lightfoot (ed.), *'Sheep Breeding'*, p 418-423.
Kelly, R.W., Allison, A.J. and Johnstone, P.D., 1978. *Proc. N.Z. Soc. Anim. Prod.*, *38*, 80-89.
Kelly, R.W., Allison, A.J. and Shackell, G.H., 1975. *Proc. N.Z. Soc. Anim. Prod.*, *35*, 204-211.
Kelly, R.W., and Johnstone, P.D., 1982. *N.Z. J. Agric. Res.*, *25*, 519-523.
Kelly, R.W., and Johnstone, P.D., 1983. *N.Z. J. Agric. Res.*, *26*, 433-435.
Kelly, R.W., Owens, J.L., Crosbie, S.F., McNatty, K.P. and Hudson, N., 1983. *Anim. Reprod. Sci.*, *6*, 199-207.

Kelly, R.W., and Smith, J.F., 1980. *Proc. Aust. Soc. Reprod. Biol.*, *12*, 68.
Killeen, I.D., 1974. *J. Reprod. Fert.*, *36*, 464.
Killeen, I.D. and Moore, N.W., 1971. *J. Reprod. Fert.*, *24*, 63-70.
Kilgour, R.J. and Whale, R.G., 1980. *Aust. J. Exp. Agric. Anim. Husb.*, *20*, 5-8.
Kilgour, R.J. and Wilkins, J.F., 1980. *Aust. J. Exp. Agric. Anim. Husb.*, *20*, 662-666.
King, C.F., 1981. *Theriogenology*, *15*, 135-147.
Knight, T.W., Oldham, C.M., Lindsay, D.R. and Smith, J.F., 1975a. *Aust. J. Agric. Res.*, *26*, 577-583.
Knight, T.W., Oldham, C.M., Smith, J.F. and Lindsay, D.R., 1975b. *Aust. J. Exp. Agric. Anim. Husb.*, *15*, 183-188.
Lawson, R.A.S. and Findlay, J.K., 1977. *In* J.H. Calaby and C.H. Tyndale-Biscoe (ed.), *'Reproduction and Evolution'*, Aust. Academy of Science, p 349-357.
Lightfoot, R.J., Croker, K.P. and Niel, H.G., 1967. *Aust. J. Agric. Res.*, *18*, 755-765.
Lightfoot, R.J. and Wroth, R.H., 1974. *Proc. Aust. Soc. Anim. Prod.*, *10*, 130-134.
Lindsay, D.R., Gray, S.J., Oldham, C.M. and Pearce, D.T., 1984. *Proc Aust. Soc. Anim. Prod.*, *15*, 159-161.
Lindsay, D.R., Knight, T.W., Smith, J.F. and Oldham, C.M., 1975. *Aust. J. Agric. Res.*, *26*, 189-198.
Long, S.E. and Williams, C.V., 1980. *J. Reprod. Fert.*, *58*, 197-201.
Marshall, T., Beetson, B.R. and Lightfoot, R.J., 1976. *Proc. Aust. Soc. Anim. Prod.*, *11*, 229-232.
Mattner, P.E., and Braden, A.W.H., 1967. *Aust. J. Exp. Agric. Anim. Husb.*, *7*, 110-116.
Mattner, P.E., Braden, A.W.H. and George, J.M., 1973. *Aust. J. Exp. Agric. Anim. Husb.*, *13*, 35-41.
Meyer, H.H., Clarke, J.N., Harvey, T.G. and Malthus, I.C., 1983. *Proc. N.Z. Soc. Anim. Prod.*, *43*, 201-204.
Miller, B.G. and Moore, N.W., 1976. *Aust. J. Biol. Sci.*, *29*, 565-573.
Mohr, L.R. and Trounson, A.O., 1980. *J. Reprod. Fert.*, *58*, 189-196.
Moore, N.W., Rowson, L.E.A. and Short, R.V., 1960. *J. Reprod. Fert.*, *1*, 332-349.
Mullaney, P.D. and Hyland, P.G., 1967. *Aust J. Exp. Agric. Anim. Husb.*, *7*, 304-307.
Munday, B.L., Ryan F.B.., King, S.J. and Corbould, A., 1966. *Aust. Vet. J.*, 42, 189-193.
Oldham, C.M., Knight, T.W. and Lindsay, D.R., 1976. *Proc. Aust. Soc. Anim. Prod.*, *11*, 129-132.
Parr, R.A., Cumming, I.A. and Clarke, I.J., 1982. *J. Agric. Sci., Camb.*, *98*, 39-46.
Parr, R.A., Cumming, I.A., Lawson, R.A.S., Kerton, D.J. and Harris, A.M., 1978. *Proc. Aust. Soc. Anim. Prod.*, *12*, 257.
Pearce, D.T., Gray, S.J., Oldham, C.M. and Wilson, H.R., 1984. *Proc. Aust. Soc. Anim. Prod.*, *15*, 164-168.
Peterson, A.J., Barnes, D., Shanley, R. and Welch, R.A.S., 1984. *Proc. N.Z. Endocr. Soc.*, *21*, 13.
Quinlivan, T.D., Martin, C.A., Taylor, W.B. and Cairney, I.M., 1966. *J. Reprod. Fert.*, *11*, 379-390.
Quirke, J.F. and Hanrahan, J.P., 1977. *J. Reprod. Fert.*, *51*, 487-489.
Radford, H.M., Watson, R.H. and Wood, G.F., 1960. *Aust. Vet. J.*, *36*, 57-66.
Reeve, J.L. and Chamley, W.A., 1984. *Proc. Aust. Soc. Anim. Prod.*, *15*, 161-164.
Restall, B.J., Brown, G.H., Blockey, M.A. deB., Cahill, L.P. and Kearins, R.D., 1976. *Aust. J. Exp. Agric. Anim. Husb.*, *16*, 329-335.
Robinson, T.J., 1951. *J. Agric. Sci., Camb.*, *41*, 6-63.
Sawyer, G.J., 1979. *Aust. J. Agric. Res.*, *30*, 1133-1141.
Smith, J.F., Jagusch, K.T. and Farquhar, P.A., 1983. *Proc. N.Z. Soc. Anim. Prod.*, *43*, 13-16.
Smith, J.F. and Lindsay, D.R., 1972. *Proc. Aust. Soc. Anim. Prod.*, *9*, 65-70.
Thwaites, C.J., 1971. *Aust. J. Exp. Agric. Anim. Husb.*, *11*, 265-267.
Trounson, A.O. and Moore, N.W., 1974. *Aust. J. Biol. Sci.*, *27*, 301-304.
Walkley, J.R.W. and Barber, A.A., 1976. *Proc. Aust. Soc. Anim. Prod.*, *11*, 141-144.
White, D.H., Rizzoli, D.J. and Cumming, I.A., 1981. *Aust. J. Exp. Agric. Anim. Husb.*, *21*, 32-38.
Wilkins, J.F., Fowler, D.G., Bindon, B.M., Piper, L.R., Hall, D.G. and Fogarty, N.M., 1984. *Proc. Aust. Soc. Anim. Prod.*, *15*, 768.
Williams, A.H., Lawson, R.A.S., Cumming, I.A. and Howard, T.J., 1978. *Proc. Aust. Soc. Anim. Prod.*, *12*, 252.
Wroth, R.H. and Lightfoot, R.J., 1976. *Proc. Aust. Soc. Anim. Prod.*, *11*, 225-228.

# PROTEIN SECRETED BY THE ENDOMETRIUM OF THE EWE DURING PREGNANCY

B.G. Miller and X. Zhang, *Department of Animal Husbandry, University of Sydney, Camden, NSW, 2570.*
G.M. Stone, *Department of Veterinary Physiology, University of Sydney, Sydney NSW, 2006.*

After implantation in the ewe the uterine glands increase in length and complexity. Those portions of the chorioallantois facing the mouths of glands become specialised to form absorptive 'areolae' (Davies and Wimsatt 1966; Perry 1981). Between days 20 and 100 of pregnancy there is a striking increase in the *in vitro* rate of protein secretion by explants of intercotyledonary endometrium. This is accompanied by relatively small increases in the rate of protein synthesis and tissue RNA:DNA and protein:DNA ratios, and appears to be related to a marked increase in the nuclear concentration of progesterone receptor but not of oestradiol receptor (Miller *et al.* 1983a). The period of most rapid increase in protein secretion (Days 56-84) coincides with that of increasing placental progesterone synthesis (Ricketts and Flint 1980) and rising peripheral plasma progesterone levels (Bassett *et al.* 1969). In the pregnant sow the substantial endometrial secretion of uteroferrin is involved in the transport of iron to the conceptus (Bazer 1975; Ducsay *et al.* 1982), but little is known of the subsequent distribution or function of secreted endometrial proteins in the pregnant ewe.

When endometrial tissue from ewes at days 84-112 of pregnancy is incubated in Eagle's (Basal) medium under an atmosphere of 95% $O_2$:5% $CO_2$ and in the presence of L-[4,5 − $^3$H] leucine, and the protein secreted or diffusing into the incubation medium is fractionated on Sephadex G-200 in phosphate saline buffer (pH 7.3), a single dominant radioactive peak of 'pregnancy protein' (PP) with a molecular weight (MW) of about 60,000 is observed. Most of the protein in the incubation medium is not radio-labelled and is apparently derived from plasma and tissue extracellular fluid. PP appears not to be secreted by the intercaruncular endometrium of non-pregnant or very early pregnant ewes (Miller *et al.* 1983b). Proteins in the fractions off Sephadex G-200 containing PP were fractionated on a column of DEAE-cellulose with a continuous NaCl gradient in 10 mM Tris-HCl, pH 8.3, in order to separate PP from the principal contaminating protein, albumin (approximate MW = 66,000). This partially purified PP has been subjected to further characterisation.

Samples electrophoresed in SDS-acrylamide (7.5%) gels in parallel with appropriate MW markers yielded a single peak of radioactivity corresponding to a MW of about 60,000, indicating a lack of sub-unit structure in PP. Further samples were subjected to density gradient centrifugration over a linear gradient of 5-20% sucrose for 18 h at 150,000 g (av). A major peak of radioactivity was present at S = 4.30 (albumin, S = 4.6), but minor peaks representing material of larger MW were also present. The apparent heterogeneity in this system reflects an *in vitro* aggregation problem repeatedly encountered with PP. Nevertheless, these data confirm that PP is a substance or group of substances without sub-unit structure and with a MW of about 60,000.

Aliquots of partially purified PP were incubated for 2 h at 37° in the presence and absence of pronase (from *Streptomyces griseus*), then fractionated by two procedures. After density gradient centrifugration the radioactivity in samples treated with pronase remained at the top of the tube. After electrophoresis in 1% agar gels at 4° in acetate-barbitone buffer, pH 8.2, the radioactivity in pronase-treated samples essentially all moved towards the cathode (presumable free $^3$H-leucine) whereas that in untreated samples showed multiple peaks, all toward the anode. These results confirm the proteinaceous nature of PP. The glycoprotein nature of PP was shown by its behaviour on heparin-Sepharose affinity chromatography. When partially purified PP samples were run on a heparin-Sepharose column in 0.15 M NaCl in phosphate buffer, pH 7.4, essentially all of the material absorbing at 280 nm was not retained, whereas the majority of radioactivity was retained and could be eluted with 0.4 M NaCl. This result also shows clearly that only very small quantities of PP are produced by the *in vitro* incubation procedure. Anti-sera against partially purified PP samples have been raised in rabbits and portions absorbed against plasma fron non-pregnant, ovariectomized ewes. Studies employing techniques of immuno-double diffusion and immuno-electrophoresis in agar in 0.1 M barbiturate buffer, pH 8.6, have revealed the presence of at least 3 distinct uterus-specific antigens in partially purified PP (as well as of several proteins of plasma origin). In the immuno-electrophoretic system one uterus-specific antigen remains at the origin, whereas the other two migrate towards the anode. Moffatt *et al.* (1983) have partially characterised two uterine secretory glycoproteins produced *in vitro* by endometrial explants from ewes at days 140-144 of pregnancy, one or both of which are probably the same as the uterus-specific antigens we have observed earlier in pregnancy. Preliminary attempts have been made to detect uterus-specific antigens in allantoic fluid and in foetal and maternal sera. At least one uterine antigen is present in allantoic fluid, but so far none has been detected in either serum. However, conclusions concerning the distribution of these antigens within the conceptus and dam must await the preparation of more substantial quantities of PP and more potent antisera to PP.

Two potential methods of obtaining larger quantities of PP *in vivo* have been examined. In the first, 4 non-pregnant, entire ewes received continuous progestagen treatment for 113 days. A fresh intravaginal pessary containing 60 mg medroxyprogesterone acetate ('Repromap', Upjohn, Rydalmere, NSW) was inserted in each ewe every 10 days, and ewes were killed 3 days after insertion of the final pessary. This protocol attempted to approximate the continuous stimulation of the endometrium by progesterone that occurs during pregnancy. In 2 of these ewes one uterine horn was ligated one month after insertion of the first pessary. At the time of killing there was no significant intra-luminal fluid accumulation in either ligated or non-ligated uterine horns, and 0.9% NaCl flushings of each uterus yielded only 0.6-4.8 mg protein. The appearance of the endometrium was unaltered from that seen during the luteal phase of the oestrous cycle. In the second method for harvesting PP *in vivo* a ligature of umbilical tape was placed around the uterine horn contralateral to the functional corpus luteum of several ewes 5 days after joining; and subsequently the accumulated luminal fluid (uterine milk) in the non-pregnant portion of the ligated uterine horn (NPH, remainder of uterus = PH) was collected (Bazer *et al.* 1979). Recoveries of 135 ± 19 (mean ± SE) ml of uterine milk containing 2.15 ± 0.10 g of protein were obtained from 4 ewes killed on days 91-110 pregnancy (foetal weight = 880 ± 257 g, crown-rump length = 27.4 ± 3.5 cm). The volume of intercaruncular endometrium in NPH increased considerably, whereas the caruncles were unusually small in relation to the endometrium during the luteal phase of the oestrous cycle. Rates of endometrial protein synthesis, total protein secretion and PP secretion *in vitro* (Miller *et al.* 1983a,b) as well as tissue protein:DNA ratios were determined in NPH and PH in these ewes and in the non-pregnant progestagen-treated ewes. Rates of total protein and PP secretion were also measured in the endometrium of some further control ewes killed on days 82-123 of pregnancy (Table 1).

**TABLE 1. Comparisons of endometrial function in different classes of ewes.**

| Class of ewe | n | Protein: DNA | Rate of protein synthesis[1] | Rate of protein secretion: total protein[1] | Rate of protein secretion: 'pregnancy protein'[1] |
|---|---|---|---|---|---|
| (a) Non-pregnant, progestagen treated | 4 | 6.8 ± 0.2[2] | 1936 ± 129 | 16 ± 2 | 3.0 ± 0.4 |
| (b) Pregnant, portion of one horn ligated | | | | | |
| 1. NPH | 4 | 11.7 ± 1.9 | 1813 ± 399 | 264 ± 57 | 95 ± 27 |
| 2. PH | | 15.8 ± 2.9 | 1984 ± 390 | 529 ± 129 | 112 ± 5 |
| (c) Pregnant, control | 7 | n.d.[3] | n.d. | 704 ± 365 | 189 ± 41 |

[1] $^3$H dpm/μg DNA; [2] mean ± SE; [3] not determined.

Rates of *in vitro* PP secretion in NPH and PH were high and similar, indicating that PP secretion is regulated primarily by systemic factors, presumably by continuous progesterone stimulation. Secretion of PP can be induced in non-pregnant ewes by giving progesterone continuously for 120 days (Moffatt *et al.* 1980). In contrast, continuous medroxyprogesterone acetate treatment of non-pregnant ewes caused no increase in the level of total protein secretion as compared to that seen during the oestrous cycle (Miller *et al.* 1983a), suggesting that this particular treatment failed to provide a level of progestagen stimulation sufficient to induce and maintain significant PP secretion.

Fractionations of uterine milk on Sephadex G-200 and heparin-Sepharose columns suggest that it contains significant quantities of PP. Protein from this source should permit further characterisation of PP and the raising of more potent antisera against PP. Since PP is clearly of maternal origin it is possible that some of the protein 'spills' into the uterine vasculature and at some stage of pregnancy between days 30 and 80 becomes identifiable in peripheral maternal plasma. The identification of one or more PP antigens in plasma by an 'on farm' ELISA procedure could provide the basis of a simple pregnancy test for use after the joining period. It seems likely that PP has some important nutritive, immunological or other function in relation to the foetus. Studies are planned to examine the possible influence on several aspects of foetal growth and development of immunising ewes against PP at different stages of pregnancy.

## REFERENCES

Bassett, J.M., Oxborrow, T.J., Smith I.D. and Thorburn, G.D., 1969. *J. Endocr., 45*, 449-457.

Bazer, F.W., 1975. *J. Anim. Sci., 41*, 1376-1382.

Bazer, F.W., Roberts, R.M., Basha, S.M.M., Savy, M.T., Caton, D. and Barron, D.H., 1979. *J. Anim. Sci., 49*, 1522-1527.

Davies, J. and Wimsatt, W.A., 1966. *Acta anat., 65*, 182-223.

Ducsay, C.W., Buhi. W.C., Bazer, F.W. and Roberts, R.M., 1982. *Biol. Reprod., 26*, 729-743.

Miller, B.G., Tassell, R. and Stone, G.M., 1983a. *J. Endocr., 96*, 137-146.

Miller, B.G., Tassell, R. and Stone, G.M., 1983b. *Proc. 15th Ann. Conf. Aust. Soc. Reprod. Biol.*, pp. E11-12.
Moffatt, R.J., Bazer, F.W., Caton, D. and Roberts, R.M., 1980. *J. Anim. Sci., 51*, Suppl.I, 307.
Moffatt, R.J., Roberts, R.M. and Bazer, F.W., 1983. *J. Anim. Sci., 57*, Suppl.I, 361.
Perry, J.S., 1981. *J. Reprod. Fert., 62*, 321-335.
Ricketts, A.P. and Flint, A.P.F., 1980. *J. Endocr., 86*, 337-347.

# REPRODUCTIVE WASTAGE AND CHROMOSOMAL ABNORMALITIES IN EARLY EMBRYOS FROM ANDROSTENEDIONE–IMMUNE AND CONTROL MERINO EWES

M.P. Boland, J.D. Murray, R.J. Scaramuzzi, R. Sutton, R.M. Hoskinson, I.G. Hazelton and C.D. Nancarrow, *CSIRO, Division of Animal Production, P.O. Box 239, Blacktown, NSW, Australia, 2148.*
C. Moran, *Department of Animal Husbandry, University of Sydney, Sydney, NSW, Australia, 2006.*

*Summary* This work is the first of 3 replicate experiments designed to define the extent and nature of embryonic loss in both control and androstenedione-immunized Merino ewes. Fertilization and embryonic recovery rates at day 2 tended to be lower in the immunized group, even though the mean ovulation rate of this group was significantly higher (1.88 v 1.47). The incidence of chromosomally abnormal embryos at day 2 (≅12%) corresponds with the observed drop in pregnancy rate (≅10%) between day 2 and 13 in both groups. There was a further drop in the pregnancy rate of the control ewes by day 30. The mean fecundity of both groups at day 30 was 1.06.

## INTRODUCTION

Lambing rates in sheep are generally lower than fertilization rates especially when multiple ovulations have occurred. This loss is of economic importance to the sheep farmer and has been ascribed to embryonic mortality. Several studies attempting to define the incidence, timing and causes of reproductive failure (Edey 1969; Edey 1976; Lunstra and Christenson 1982) have not assessed the unavoidable reproductive wastage due to fertilized eggs with chromosomal abnormalities. Long and Williams (1980) found 6% of early stage sheep embryos to be chromosomally abnormal. Recently a vaccine (Fecundin® registered trade mark of Glaxo Animal Health) has been marketed to increase the lambing percentage of sheep flocks. Fecundin® confers transient androstenedione-immunity on breeding ewes which, in turn, induces an increase in their ovulation rate. In its practical application Fecundin® has been less successful in Merino ewes than in other breeds (Scaramuzzi *et al.* 1983). This paper compares the egg recovery rates, fertilization rates, embryonic survival and extent of chromosomal abnormalities in embryos and foetuses from androstenedione-immune and control ewes.

## MATERIAL AND METHODS

### Treatment of Ewes

Two groups of 60 Merino ewes were maintained at pasture in a single flock throughout the experiment. One group was immunized against androstenedione (Scaramuzzi *et al.* 1983) with the booster injection given 28 days after the primary and integrated with the synchronization procedures such that mating occurred about 25 days later. The remaining group served as uninjected controls. All ewes were synchronized by a 14-day treatment with an intravaginal sponge (Repromap, Upjohn Pty Ltd) then run with vasectomized rams and later joined with harnessed fertile rams to cover the second oestrus. Two relays of 8 rams per 120 ewes were exchanged twice daily at the morning and evening oestrus checks.

### Embryo Recovery

Twenty marked ewes were selected from each group for recovery about 48 h after the first detection of oestrus and the oviducts were flushed with 5 ml phosphate-buffered saline (PBS) containing 5% heat-treated sheep serum (SS). A further twenty ewes were killed 13 days after being marked and each uterine horn was flushed individually with 15 ml PBS. Remaining ewes not returning to service were slaughtered between days 26 and 30. Data collected were ovulation rates and the number and development of eggs and embryos. Ovulation rates of ewes allocated to day 28 recovery group were assessed by endoscopic methods on day 10 after marking. Egg recovery and fertilization rates were subjected to Chi-square analysis while ovulation rates were compared using an analysis of variance.

### Chromosome Methods

Embryos were cultured in PBS containing 15% SS, 0.8 μg/ml colchicine, 100 i.u./ml penicillin G and 100 μg/ml streptomycin sulphate. Cultures were incubated at 37°C in a 5% $CO_2$ atmosphere. Two-day-old embryos were incubated for 10 h and harvested by the method of Long and Williams (1978). Thirteen and 14 day old embryos were cultured for 1 to 1.75 h. Embryos greater than 15 mm long were processed for chromosomes by the method of Hare *et al.* (1976), while embryos less than 15 mm in length were harvested by a modification of the method of Shaw *et al.* (1976) following 10 min in hypotonic medium. Cells were fixed in 3:1 methanol: acetic acid fixative at the time of dispersal over the slide.

Entire 26 to 30 day old foetuses, dissected free of most membrane, were placed in colchicine medium, finely minced with scissors and aspirated through an 18 g and then a 21 g needle. The cell suspension was incubated for 1 to 1.75 h and harvested by standard cytological methods. Cells were dropped on to cold, wet slides and air dried. Chromosome preparations were stained in 5% giemsa for 3.5 min, air dried and mounted in Depex.

## RESULTS

The mean ovulation rates for 57 control and 56 immunized ewes were 1.49 and 1.88 respectively ($P = 0.0001$). At 30 days after mating the mean number of normal embryos per marked ewe in both groups was 1.06. Examination of eggs recovered on day 2 (Table 1) showed that the immunized ewes had fewer eggs fertilized than the controls (68% v. 89%; $P < 0.05$) and they also tended to have a lower recovery rate (69% v. 85%; $P < 0.1$). By day 13 fertilization and recovery rates were not significantly different between the two groups. The results indicate that in both groups of animals there was a reduction in pregnancy rate of about 10% between days 2 and 13. At day 28 there was a further decrease in pregnancy rate in controls but the differences in pregnancy rate between the two groups was not significant at any time.

**TABLE 1. Assessment of fecundity, fertility and embryonic survival in control (C) and androstenedione-immunized (I) Merino ewes.**

| | Day 2 | | Day 13 | | Day 28 | |
|---|---|---|---|---|---|---|
| | C | I | C | I | C | I |
| No. of marked ewes | 19 | 20 | 20 | 20 | 18 | 16 |
| No. of ovulations | 33 | 36 | 25 | 39 | 27 | 30 |
| No. of eggs/embryos recovered | 28 | 25 | 19 | 28 | 19 | 17 |
| No. of eggs fertilized | 25 | 17 | 18 | 27 | 19 | 17 |
| No. of single or degenerate eggs/embryos | — | — | 1 | 4 | 2 | 0 |
| % ewes pregnant | 89 | 75 | 80 | 65 | 72 | 63 |

Data on the chromosome complement of all embryos analysed are presented in Table 2. The sex ratios for the control (17 ♀:23 ♂) or immunized (21 ♀:19 ♂) groups were not significantly different from 1:1. Fifteen chromosomally normal ($2n=54$) embryos and two aberrant embryos (that is 12%) were observed in the day 2 embryo sample. These were a 1-cell zygote with 4 haploid ($n=27$) metaphase chromosome groups, three of which contained a Y-chromosome and a 4-cell embryo which had spermatozoa on the surface, and contained two haploid metaphases, without Y-chromosomes. Sixteen embryos from the 13-14 day old samples were diploid with a further 16 embryos being 2n/4n mosaics. This contrasts with Long (1977) who reported no 4n cells in a sample of 75 sheep blastocyts. Eleven cells were counted in one chromosomally aberrant embryo; 5 had 53 chromosomes, 4 were normal and 2 were 4n. Karyotypes prepared from both 53 and 54 chromosome cells, revealed that the $2n = 53$ cells were monosomic for an unidentified acrocentric autosome.

Seven 26 to 30 day old foetuses were diploid, while 20 were 2n/4n mosaics and 5 were complex mixaploids. The presence of 6n cells indicates cell fusion is occuring in the cell lines leading to these cells.

**TABLE 2. Number of embyros with the following chromosome complements ($2n = 54$).**

| | | No. embryos | | Chromosome complement | | | | | | | |
|---|---|---|---|---|---|---|---|---|---|---|---|
| Age (days) | Group | Processed | Yielding chromosomes | n | 2n | 2n–1 2n/4n | 4n | 2n/ 4n | 2n/ 4n/6n | 2n/ 4n/8n | 2n/8n |
| 2 | C | 16 | 10 | | 9 | | 1(1) | | | | |
| 2 | I | 14 | 7 | 1 | 6 | | | | | | |
| 13–14 | C | 18 | 16 | | 6 | 1 | | 9 | | | |
| 13–14 | I | 24 | 17 | | 10 | | | 7 | | | |
| 26–30 | C | 18 | 15 | | 2 | | | 10 | 2 | 1 | |
| 26–30 | I | 17 | 17 | | 5 | | | 10 | | 1 | 1 |

(1)This was a 1-cell zygote with 4 haploid metaphase groups
C = Control group; I = Immunized group

## DISCUSSION

Immunization of Merino ewes against androstenedione increased the ovulation rate as previously reported for crossbred ewes (Scaramuzzi *et al.* 1983). The present study indicates that this treatment of Merino ewes, leads to a higher incidence of reproductive wastage, possibly because of the higher ovulation rate (White *et al.* 1981). Part of the losses in the immunized group at day 2 appear to result from fertilization failure but our inability to recover 31% of the eggs (compared to 15% in controls) indicates a possible failure of fimbrial collection, or a loss from the oviduct subsequent to fertilization.

At day 2, two lethal chromosome complements were observed — a zygote with 4 haploid divisions resulted from polyspermy and a haploid embryo which may have been derived by parthenogenesis, gynogenesis or androgenesis (Beatty 1957). The 13 day 2n-1/2n mosaic probably arose by the loss of a chromosome during a mitotic division in an earlier stage cell. The consequences of this type of mosaicism is not known, although it may not necessarily be incompatible with development to term (Gustavsson 1980). The polyploid mosaicism observed in the 26 to 30 day old foetuses can be assumed to be normal as some differentiated tissues (e.g. liver) are known to be polyploid (Brodsky and Uryvaeva 1977).

The level of chromosomally abnormal cleavage stage embryos observed here (12%) is higher than the 6% reported for British breeds by Long and Williams (1980). However, the number of embryos sampled by us is small (17) compared to that used in the above study (89). The low incidence of abnormalities observed in the older embryos agrees with Long's (1977) findings.

## REFERENCES

Beatty, R.A., 1957. *Parthenogenesis and Polyploidy in Mammalian Development.* Cambridge University Press.

Brodsky, W.Y. and Uryvaeva, I.V., 1977. *Int. Rev. Cytol., 50*, 275-332.

Edey, T.N., 1969. *Animal Breeding Abstracts, 37*, 173-190.

Edey, T.N., 1976. *In* Tomes, G.J., Robertson, D.E. and Lightfoot, R.J . (eds), *Sheep Breeding: Proceedings of 1976 International Congress, Muresk and Perth*, Western Australian Institute of Technology, Perth, 400-410.

Gustavsson, I., 1980. *Z. Tierzüchtz.* Züchtgsbiol., 97, 176-195.

Hare, W.C.D., Singh, E.L., Betteridge, K.J., Eaglesome, M.D., Randall, G.C.B., Mitchell, D., Bilton, R.J. and Trounson, A.O., 1980. *Canad. J. Genet. Cytol., 22*, 615-626.

Long, S.E., 1977. *Cytogenet. Cell. Genet., 18*, 82-89.

Long, S.E. and Williams, C., 1978. *Vet. Rec., 102*, 153.

Long, S.E. and Williams, C.V., 1980. *J. Reprod. Fert, 58*, 197-201.

Lunstra, D.D. and Christenson, R.K., 1982. *J. Anim. Sci., 53*, 458-466.

Scaramuzzi, R.J., Geldard, H., Beels, C.M., Hoskinson, R.M. and Cox, R.I. 1983. *Wool Technol. Sheep Breed, 31*, 87-97.

Shaw, D.D., Webb, G.C. and Wilkinson, P., 1976. *Chromosoma (Berl.), 56*, 169-190.

White, D.H., Rizzoli, D.J. and Cumming, I.A., 1981. *Aust. J. Exp. Agric. Anim. Husb., 21*, 32-38.

# BINDING OF SHEEP OVIDUCAL FLUID PROTEINS TO SPERMATOZOA

R. Sutton, A.L.C. Wallace, H. Engel and C.D. Nancarrow, *CSIRO Division of Animal Production, P.O. Box 239, Blacktown, N.S.W., 2148.*

*Summary* Proteins from met-oestrous oviducal fluid were bound to ram spermatozoa to a greater extent than those from di-oestrous fluid. The greatest binding activity was retained in a high molecular weight protein fraction which contains an oestrus-associated glycoprotein.

## INTRODUCTION

The underlying biochemical mechanisms required for early events in the establishment of pregnancy, such as capacitation of spermatozoa, fertilization and embryo cleavage, are poorly understood in domestic animals. This is shown by the relative lack of success in achieving *in vitro* fertilization and subsequent cell division in sheep and cattle compared with humans and mice (Wright and Bondioli 1981). Since these events occur *in vivo* in oviducal fluid, we have studied fluid proteins with the aim of identifying any potentially important components.

Using SDS electrophoresis, a novel glycoprotein (Mr 80-90,000) was detected in ovine oviducal fluid collected on the first three to six days but not for the remaining days of the oestrous cycle (Sutton *et al.* 1984). This oestrus-associated glycoprotein (EGP) is probably synthesized by the oviduct itself and released into the lumen in response to oestrogen since the injection of oestradiol benzoate (ODB) caused the appearance of EGP in oviducal fluid of ovariectomized ewes (Nancarrow *et al.* 1983).

The role of EGP is unknown although the large amount produced and its proposed native molecular weight of several million daltons suggest that EGP may be a structural protein rather than an enzyme or hormone. The aim of the present study was to determine if EGP or any other oviducal fluid proteins bind to spermatozoa, as the first step in determining their importance to fertilization.

## MATERIALS AND METHODS

### Oviducal fluid samples

Oviducal fluid was collected daily from indwelling catheters and stored at -20°C after centrifugation (Sutton *et al.* 1984). The fluid used in this study came from two cyclic Merino ewes and an ovariectomized Merino ewe which had been injected weekly with 25 $\mu$g of ODB. The presence of EGP was established by SDS electrophoresis and silver staining (Laemmli 1970; Merril *et al.* 1981; Sutton *et al.* 1984). Fluid with EGP, taken from the ODB treated ovariectomized ewe, was concentrated five-fold by ultrafiltration with a UM 2 membrane (Amicon, Mass., USA). A 0.75 ml volume was then fractionated on a 1.5 $\times$ 90 cm Sephacryl 300 column (Pharmacia, Uppsala, Sweden) using 50 mM sodium phosphate buffer pH 6.8 with 0.9% NaCl. Pooled fractions were dialysed and freeze-dried. Oviducal fluid samples and fractions were labelled with iodine-125 (Greenwood *et al.* 1963) and separated from free $^{125}$I on a Pharmacia PD10 column presaturated with 200 $\mu$l of oviducal fluid. The protein (50 or 100 $\mu$g) was labelled to give a low specific activity — approximately 1 $\mu$C/$\mu$g for ligands used in the experiment in Table 1 and 3 $\mu$C/$\mu$g for Figure 1 and Table 2. The protein was eluted in Dulbecco's phosphate buffered saline with calcium and magnesium (D-PBS), and this eluate was used directly in binding studies carried out on the same day.

### Binding experiments

Ram semen was diluted 1 in 500 in D-PBS, and spermatozoa were collected by centrifugation (1000 g, 10 min). After rewashing and resuspension, 1 ml aliquots, with approx. 6 $\times$ 10$^6$ spermatozoa, were centrifuged in 1.5 ml Eppendorf tubes (12,000g, 2 min.). The supernatants were removed and the pellets resuspended in 100 $\mu$l (Table 1 data) or 50 $\mu$l (Table 2, Figure 1 data) of labelled protein solution. After the incubation the cells were washed 3 times with 1 ml of D-PBS with 0.05% Tween. After gamma counting the pellets were frozen at −20°C prior to SDS electrophoresis. Sheep peripheral blood lymphocytes prepared according to Grewal *et al.* (1984), were washed in D-PBS, and aliquots of 3 $\times$ 10$^6$ cells were treated in the same way as spermatozoa.

## RESULTS

The hypothesis that EGP binds to spermatozoa was tested in three preliminary experiments, the results of one of which are shown in Table 1. Radioactively labelled met-oestrous oviducal fluid (taken on day 1 of the cycle) bound to ram spermatozoa to a greater extent at both 37°C and 4°C than di-oestrous fluid (day -4) from the same ewe at both 37°C and 4°C. The only detectable difference between these 2 fluids as assessed by SDS electrophoresis and fluorography was the presence of EGP in the met-oestrous sample (not shown). Data from the same experiment also showed a much greater binding to spermatozoa of a high molecular weight fraction of oviducal fluid (Fraction A) which contains only 13% of the fluid protein but almost all of the EGP (Table 1, Figures 2 and 3). These results were confirmed by 2 other smaller experiments using slightly different incubation conditions and pooled oviducal fluid (day 0-3 of the cycle with EGP; days -5 to -1 and 5 to 8 without EGP) from a second cyclic ewe, and freshly labelled Fraction A.

In each of these experiments, we attempted to identify the proteins involved in the binding by SDS electrophoresis and fluorography of the labelled spermatozoa. However, in all cases, particularly with the samples incubated at 37°C for 1 hour and the Fraction A incubations, the $^{125}I$ was associated with a continuum of protein molecular weight sizes, with much more low molecular weight material than in the original labelled ligands, suggesting that extensive degradation of the proteins had occurred during the incubation.

The time course of binding to spermatozoa of EGP-containing oviducal fluid was investigated at both 37°C and 4°C using fluid from an ovariectomized ewe (8124) treated with ODB (Figure 1). There was an initial rapid binding of $^{125}I$ ligand at both temperatures followed at 37°C by a further increase in counts retained in the pellet.

**TABLE 1. Binding of $^{125}I$-labelled oviducal fluid proteins to ram spermatozoa. Comparison of whole oviducal fluid from 2 different stages of the oestrous cycle — either containing EGP (metoestrous) or without it (di-oestrous) and a partially purified EGP preparation.**

| $^{125}I$ ligand | | Percentage of counts retained in the pellet | | |
|---|---|---|---|---|
| | μg protein/ incubation | Incubation conditions 37°C, 5 min | 37°C, 60 min | 4°C, 60 min |
| Fluid with EGP (Day 1) | 3.0 | 3.2 | 3.3 | 2.8 |
| Fluid without EGP (Day −4) | 3.0 | 1.6 | 1.9 | 1.4 |
| Fraction A (EGP) | 1.7 | 22.1 | 20.3 | 20.9 |

Data represent individual determinations.

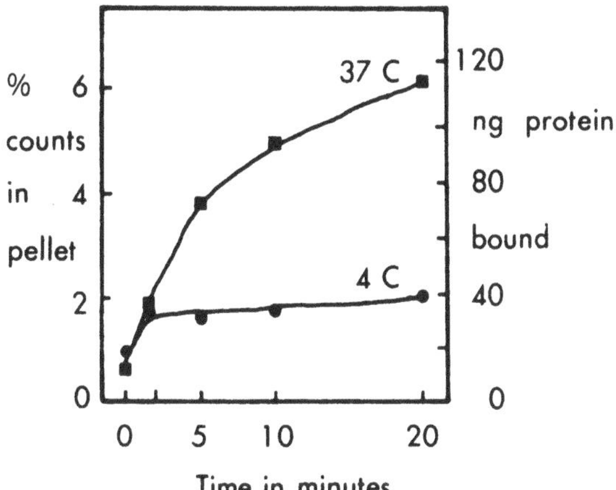


**FIGURE 1. Binding of $^{125}I$ protein to spermatozoa.**

In the final experiment the initial binding properties (at 4°C for 5 min) of semipurified preparations of the 3 major proteins in oviducal fluid were compared with whole oviducal fluid (Table 2). Oviducal fluid containing EGP was fractionated by gel filtration on a Sephacryl 300 column into 7 different fractions (A-G) (Figure 3). Fractions A, D and F were selected for the study on the basis of SDS electrophoresis (Figure 2). Fraction A, which elutes in the void volume of the column, contains EGP (subunit Mr 80-90,000) confirming a previous prediction that non-dissociated EGP has a molecular weight of several million (Sutton *et al.* 1984). The second major glycoprotein in oviducal fluid, which has a subunit size of Mr 50-60,000, eluted in Fractions D and E. Both these fractions also contained a second strong band (Mr 30,000) and Fraction E contained a lot of albumin. Albumin, which is the major protein in oviducal fluid (Mr 67,000) eluted in E, F and G. More Fraction A and D were bound to both spermatozoa and lymphocytes compared with the original complete mixture of fluid proteins while Fraction F showed very little binding. The binding of the different ligands to spermatozoa compared to lymphocytes was very similar except for Fraction A which bound to a greater extent to spermatozoa. The results of SDS electrophoresis and fluorography of these samples (not shown) confirmed that EGP does bind to spermatozoa, but indicated that either proteolysis is still occurring in the samples or that other binding proteins have predominantly low molecular weights.

**TABLE 2. Comparison of the initial binding of $^{125}$I-labelled whole oviducal fluid and three protein fractions.**

| $^{125}$I ligand | | Percentage of counts retained in the pellet | |
|---|---|---|---|
| | μg protein/ incubation | Spermatozoa | Lymphocytes |
| Whole fluid | 1.6 | 2.1 ± 0.1 | 2.5, 2.7 |
| Fraction A | 1.3 | 10.4 ± 0.9 | 5.9, 7.2 |
| Fraction D | 1.6 | 5.3 ± 0.4 | 6.4, 6.1 |
| Fraction F | 1.8 | 0.6 ± 0.0 | 0.5, 1.1 |

Data represent means and standard deviations of 3 determinations for spermatozoa, individual values for lymphocytes.

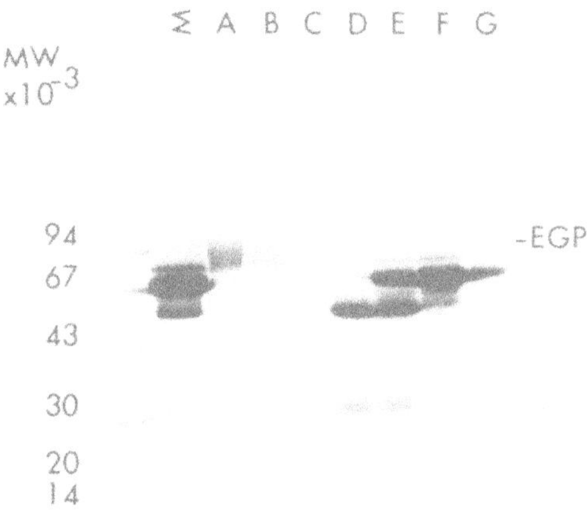


**FIGURE 2. SDS gel of Sephacryl 300 fractions.**

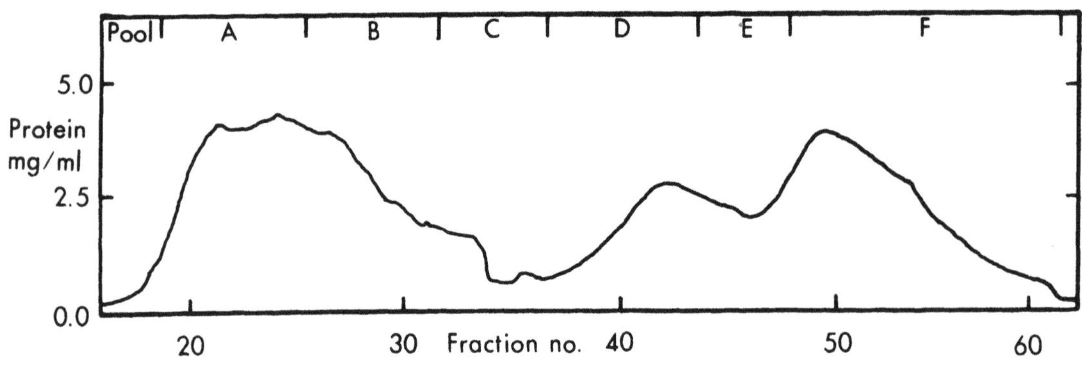


**FIGURE 3. Sephacryl 300 fractions of oviduct fluid with EGP from ewe 8124 (ovx given ODB).**

## DISCUSSION

Oviducal fluid from a variety of mammals has been shown to contain both plasma proteins at a low concentration and proteins produced by the oviducal epithelium, some of which are only released into the fluid at particular stages of the oestrous cycle (Mastroianni and Go 1979; Edwards 1980). Those oviduct-specific proteins which are present only at the oestrous and met-oestrous stages, such as EGP (Sutton *et al.* 1984), are more likely to be involved in the events occurring in the oviduct for the first 3 days after mating. A comparison was therefore made of the ability of $^{125}$I-labelled oviducal fluid proteins from this stage and from later in the cycle to bind to spermatozoa. For fluid from two different cyclic ewes there was about twice as much binding with met-oestrous fluid containing EGP compared with di-oestrous fluid which had none. Further, a semipurified EGP preparation showed a greatly enhanced binding capacity for spermatozoa. These differences were observed at both 37°C and 4°C and following both 5 and 60 minute incubations. However, a time course study indicated that there is an initial rapid binding of oviducal fluid proteins, followed later at 37°C by a continued increase in counts which might well reflect internalization of the label and probably the gradual degradation of the proteins. Consequently, a 5 minute incubation at 4°C was used for the last experiment in order to compare initial binding only.

To further test the hypothesis that EGP binds to ram spermatozoa, the binding of Fraction A was compared with Fractions D and F and whole oviducal fluid using both spermatozoa and lymphocytes which were used as a control. Both Fractions A and D were bound to both cell types much more than F, perhaps not an unexpected result since glycoproteins tend to be more surface active than other proteins. Twice as much Fraction A (EGP) was bound to spermatozoa compared to other glycoprotein (Fraction D), while this effect was not seen with lymphocytes.

These preliminary experiments have established that EGP binds to spermatozoa but do not indicate the extent, nature or physiological significance of this binding. Further experiments are needed to investigate EGP's binding properties such as its saturability, reversibility, specificity and affinity. Other experiments could test whether EGP is responsible for the stimulation of spermatozoal respiration by sheep genital tract fluids (Iritani *et al.* 1969) or the increased numbers of spermatozoa undergoing the acrosome reaction in oestrous compared with luteal cow oviducts (Herz *et al.* 1983). Since the successful *in vitro* fertilization of ruminant oocytes has usually involved prior contact of the eggs or spermatozoa with oviducal fluid (Wright and Bondioli 1981) it is possible that the inclusion of EGP or other oviducal proteins in the medium may facilitate the *in vitro* fertilization of ruminant eggs.

## REFERENCES

Edwards, R.G., 1980. *In*, *Conception in the Human Female.* Academic Press, London, 416-524.

Greenwood, F.C., Hunter, W.M. and Glover, J.S., 1963. *Biochem. J., 89*, 114-123.

Grewal, A.S., Wallace, A.L.C., Pan, Y.S., Rigby, N.W., Donnelly, J.B., Eagleson, G.K. and Nancarrow, C.D., 1984. *J. Reprod. Immunol.* (in press).

Herz, Z., Northey, D., Lawyer, M. and First, N.L., 1983. *J. Anim. Sci., 57*, Suppl. 1, 344.

Iritani, A., Gomes, W.R. and Vandemark, N.L., 1969. *Biol. Reprod., 1*, 77-82.

Laemmli, U.K., 1970. *Nature, 277*, 680-685.

Mastroianni, L. Jr. and Go, K.J., 1979. *In* Beller, F.K. and Schumacher, G.F.B. (eds), *The Biology of the Fluids of the Female Reproductive Tract*, Elsevier, North Holland, 335-344.

Merril, C.R., Goldman, D., Sedman, S.A. and Ebert, M.H., 1981. *Science, 211*, 1437-1438.

Nancarrow, C.D., Sutton, R. and Wallace, A.L., 1983. *Proc. Aust. Soc. Reprod. Biol.*, 103.

Sutton, R., Nancarrow, C.D., Wallace, A.L.C. and Rigby, N.W., 1984. *J. Reprod. Fert., 72*, 415-422.

Wright, R.W. Jr and Bondioli, K.R., 1981. *J. Anim. Sci., 53*, 702-729.

# FACTORS CONTROLLING PLACENTAL AND FOETAL GROWTH AND THEIR EFFECTS ON FUTURE PRODUCTION

A.W. Bell, *School of Agriculture, La Trobe University, Bundoora, Victoria 3083.*

## INTRODUCTION

The effects of various factors on prenatal growth and development in the sheep, and some of their consequences for perinatal lamb mortality and postnatal production, have been recently reviewed (Alexander 1974; Mellor 1983). These reviews highlighted the possible importance of placental size as a determinant of foetal growth and Alexander (1978) has discussed some of the factors that might regulate growth of the ovine placenta. However, the patterns of foetal and placental growth in sheep are quite different, particularly in late pregnancy, when foetal growth is greatest but placental size is declining (Barcroft and Kennedy 1939; Stegeman 1974; Figure 1). This suggests that placental size can be only a crude index of the functional (transport) capacity of the placenta, especially when animals of different gestational ages are compared. Techniques for studying placental nutrient transport and metabolism in vivo in the conscious ewe have been available for almost 20 years (Meschia *et al.* 1967a, 1967b). However, until very recently, there has been no serious attempt to relate placental function quantitatively to foetal growth in the sheep, under conditions in which placental size is likely to vary considerably.

In this review, some of the sources of the large variation in placental size observed in sheep under natural and experimental conditions will be described, if not explained, and evidence for the notion that placental size is an important determinant of foetal growth will be briefly discussed. The following questions will then be considered:

1. How are placental size and function related?
2. What specific placental and non-placental factors limit foetal growth and development (including wool follicle development)?
3. What is the impact of these factors on future production?

## PLACENTAL GROWTH

### Gross morphology

As described by Alexander (1964a) and Stegeman (1974), the non-pregnant uterus contains about 60 to 150 endometrial thickenings, termed 'caruncles'. These are potential sites for chorionic attachment, which takes place at about 30 days after conception and usually occupies 70-80% of available caruncles, depending on litter size and other factors. The points of attachment develop into button-shaped 'cotyledons' or 'placentomes', the foetal and maternal tissues of which together constitute the placenta. Placental growth proceeds rapidly and exceeds foetal growth, until about 90-100 days of gestation; cotyledonary mass then tends to decline slowly until term at 145-150 days (Figure 1). Foetal tissue accounts for about 60% of the placentome from about day 90 until term (Stegeman 1974).

### Control of placental size

Placental weight varies widely in uniformly-treated ewes (Mellor 1983) but the basis for this variation is poorly understood. The following is thus more a list of factors which may affect placental size than a synthesis of available evidence into a general hypothesis for the control of placental growth.

*Endocrine factors* Although luteal progesterone is necessary for maintenance of early pregnancy and implantation in sheep, there is little evidence that it has other than a permissive role in subsequent placental growth and development (Alexander and Williams 1966). The role, if any, of maternal oestrogen is similarly unclear. In some species, oestrogens appear to inhibit placental growth (see Alexander 1978). However, Alexander and Williams (1968) were unable to show such an effect in ovariectomized ewes treated with progesterone and/or oestrogen between days 20 and 60 of gestation. More recently, Miller *et al.* (1983) found very low levels of receptors for progesterone and oestradiol in maternal and foetal placental tissues between days 56 and 112 of gestation, relative to those in caruncular endometrium on day 0. Steroid receptor levels were not related to the time-course of placental cellular growth or protein synthesis, suggesting that neither progesterone nor oestradiol has an important effect on these processes. The influence of other maternal hormones, such as cortisol and the thyroid hormones, on placental growth have not been studied in sheep.

The possible effects of foetal hormones on placental growth and development also need to be investigated. Disruption of the foetal hypothalamic-pituitary-adrenal axis causes ultra-structural changes in the sheep placenta (Barnes *et al.* 1976; Steven *et al.* 1978) but gross morphological effects have not been reported.

Insulin is thought to be an important trophic hormone in foetal sheep (Bassett and Fletcher 1983). It is therefore interesting that Fletcher *et al.* (1982) recently found a significant correlation not only between foetal plasma insulin concentration and foetal weight in rabbits, but also between foetal insulin and placental weight.

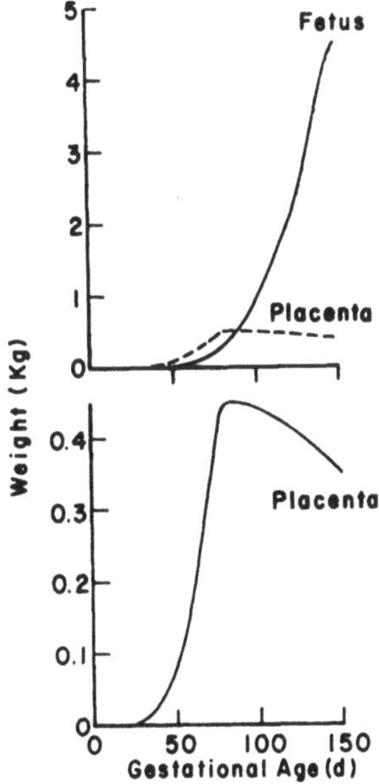


**FIGURE 1. Typical patterns of foetal and placental growth.**

*Nutritional and environmental factors* Moderate to severe undernutrition throughout pregnancy significantly reduced placental weight at or near term (Wallace 1948; Everitt 1968; Alexander and Williams 1971), and this effect was apparent at 90 days of gestation after severe underfeeding during early and mid pregnancy (Everitt 1964). The effect of undernutrition in late pregnancy, in ewes which have previously been adequately fed, is less clear. There is limited evidence that placental weight may be reduced in twin-bearing, but not in single-bearing ewes (Thomson and Thomson 1948; Mellor and Murray 1981). In contrast, the preliminary data of Faichney (1981) suggest that following underfeeding from 50 to 100 days of gestation, a return to an adequate diet may prevent or even reverse the normal decline in placental weight during late pregnancy. However, Mellor (1983) found no evidence of placental compensatory growth in ewes well fed after being moderately underfed between days 35 and 119 of gestation.

Neither the specific nutrients nor the possible mechanisms(s) involved in these nutritional effects have been examined in sheep. In pregnant guinea pigs, protein deficiency appears to be more potent than energy deficiency in reducing placental weight (Young and Widdowson 1975). Uterine blood flow was reduced by 25-30% in fasted (Morriss *et al.* 1980) or chronically undernourished, late-pregnant ewes (Bell *et al.* 1984; Table 1). It is not clear whether this was a cause or an effect of the associated depression in placental metabolism (Bell *et al.* 1982; 1984; Table 1). Although maternal progesterone and oestradiol may not play a primary role in ovine placental growth, Anderson (1975) almost completely reversed the substantial decline in placental size in underfed sows by treating them with these hormones.

**TABLE 1. Effect of undernutrition during late pregnancy on uterine blood flow and uteroplacental metabolism (Bell *et al.*, 1984).**

| | No. ewes | Fed | Underfed |
|---|---|---|---|
| Uterine blood flow (ml $min^{-1}$) | 7 | 1907 | 1297 |
| Uteroplacental oxygen uptake (ml $min^{-1}$) | 5 | 15.8 | 5.9 |
| Uteroplacental glucose uptake ($\mu$mol $min^{-1}$) | 5 | 178 | 60 |

Prolonged environmental heating of pregnant ewes, sufficient to elevate deep body temperature for several hours daily, can cause spectacular reductions in placental size (Alexander and Williams 1971). These were most marked when heat exposure was continued through mid and late pregnancy, but significant reductions also occurred when heating was confined to mid (50-100 days) or late gestation (100-150 days) only. Thus, heating during late pregnancy may accelerate the normal rate of decline in placental weight, but in this study the placenta was also shown to be capable of compensatory growth during this period.

The mechanism by which heat stress reduces placental growth is not known. Chronic heat exposure depresses thyroid activity in sheep (Sanchez and Evans 1972) but administration of thyroxine to heated ewes had no effect on placental size; treatment with growth hormone was similarly ineffective (Alexander and Williams 1971). Blood flow to the maternal cotyledons was reduced in ewes exposed to heat for several weeks (G. Alexander and J.R.S. Hales, personal communication, 1984). The rapid decrease in flow during acute heating suggested that the effect was not due to changes in placental metabolism. Preliminary observations made after ewes were heated from 50 to 120 days gestation indicated a reduction in foetal perfusion of the placenta (umbilical blood flow), which was in proportion with the reduction in placental size, but greater than the reduction in foetal size (G. Meschia and A.W. Bell, unpublished observations).

*Maternal and foetal influences* Placental weight tends to increase with age of the ewe (Alexander 1964a). Stegeman (1974) showed that placental weight is correlated with uterine weight and speculated that inadequate uterine development before mating may explain lower placental weights in young ewes. The combined weight of twin placentae is usually greater than that of single foetuses, due to a greater number of available caruncles being occupied by cotyledons (80% in twins vs 70% in singles) and a greater average weight of individual cotyledons. However, the weight for individual twins is almost always less than that for singles, indicating an important but incomplete compensation for the smaller number of cotyledons associated with a twin (Alexander 1974). These general relationships extend to triplets and quadruplets in highly fecund Finnish Landrace × Dorset Horn ewes (Robinson *et al.* 1977), in which the increase in total placental weight with increasing litter size is apparent as early as 50-60 days post conception. The stimulus for such early manifestation of placental compensatory growth is unknown. Development of known foetal endocrine systems at 50-60 days is probably insufficient for the foetus to be totally responsible, but there is no clear evidence that luteal progesterone is an important mediator of placental growth. Perhaps some other luteal or ovarian growth factor is involved, a possibility that could be checked by luteal or ovarian transplantation in single-pregnant ewes.

Alexander (1964a) reported that male foetuses tend to have heavier placentae than females, but Stegeman (1974) found no significant effect of foetal sex.

Genetic influences on placental size have yet to be systematically studied.

## PLACENTAL SIZE AND FOETAL GROWTH

Three major lines of evidence suggest that placental size limits foetal growth. First is the commonly observed statistical association between placental and foetal weights during late pregnancy or at term (Alexander 1964a; Mellor and Murray 1981, 1982a, 1982b). Mellor (1983) found that in six groups of pregnant ewes subjected to widely different levels and patterns of nutrition, the regression of foetal weight on $\log_e$ placental weight at 142 days of gestation accounted for 69 to 91% of the variance in foetal weight, and 63% of variance when all data were pooled. Significant correlations of this nature were not found at stages of gestation earlier than about 130 days (Stegeman 1974). Both Alexander (1974) and Mellor (1983) interpret the curvilinear relation between foetal and placental weight to indicate decreasing placental 'efficiency' as placental weight increases. However, it is equally conceivable that the placenta has a functional reserve, and that foetal growth is placentally limited only when this reserve is exceeded, either because of small placental size or large foetal growth potential.

Of course, the relation between placental size and foetal growth in late pregnancy may not necessarily be one of cause and effect, because rather than limiting foetal growth, placental size may be regulated by, or together with the foetus. This objection was at least partly answered by experiments in which graded reductions in placental size, achieved by surgical carunclectomy before mating, caused reduced birth weights (Alexander 1964b). It is interesting that in these and subsequent similar studies (Robinson *et al.* 1979; Worthington *et al.* 1981; Maloney *et al.* 1982), a minority of ewes showed considerable hypertrophic compensation in the reduced number of placentomes and little or no reduction in foetal growth. The basis for this compensatory ability warrants investigation.

The decline in foetal growth during the last few weeks of pregnancy, which occurs even in extremely well fed ewes, has also been cited as evidence for placental restriction of foetal growth (Alexander 1978; Mellor 1983). This was difficult to demonstrate in serial slaughter experiments, but has been clearly demonstrated by daily measurement of foetal crown-rump length and thoracic girth in individual foetuses *in utero* (Mellor 1983; Figure 2). While placental insufficiency could explain these observations, foetal growth potential during the last few weeks of gestation may be limited by the profound endocrine changes which occur during the lead-up to parturition (Challis *et al.* 1984; Gluckman 1984).

## RELATIONS BETWEEN PLACENTAL SIZE AND FUNCTION

### Basic principles of placental function

The principal functions of the placenta are exchange of maternal nutrients and foetal metabolites between maternal and foetal circulations, and hormone synthesis and secretion. The latter is considered elsewhere in this volume (Rice *et al.* 1984) and will not be discussed here.

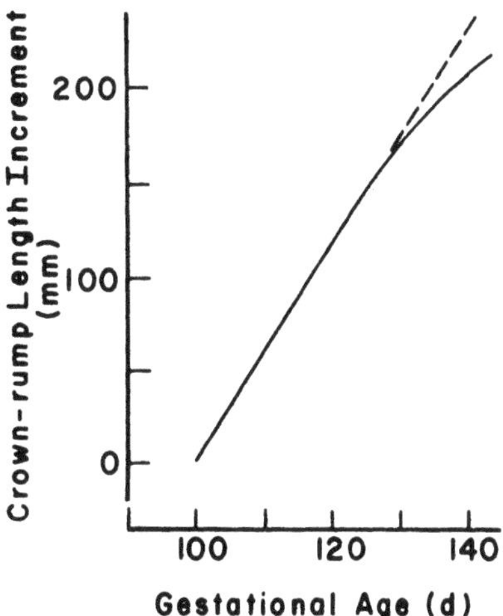


**FIGURE 2. Daily increment in foetal crown-rump length during late gestation in well-fed ewes (modified from Mellor 1983, with the author's permission).**

Placental transport of heat, respiratory gases, nutrients and other substances occurs by processes of varying complexity, including active transport and facilitated and passive diffusion. Transport of certain substances, such as oxygen and glucose, is complicated by their substantial net consumption by the placenta (Meschia *et al.* 1980). An appropriate index of placental transport function is *clearance*, which is defined as the ratio of transfer rate over maternal-foetal arterial concentration difference. Clearance of some substances, such as sodium and chloride ions, may be primarily limited by membrane permeability, while clearance of other substances may be limited by rates of placental perfusion (e.g. antipyrine and tritiated water); diffusion of urea is limited by both placental permeability and blood flow (Meschia *et al.* 1967a). The morphological limiting factors for permeability are exchange surface area and/or thickness, and for blood flow, the total cross-sectional area of maternal and foetal placental capillaries.

Functional versus morphological growth of the placenta

The rapid increase in placental weight until about 90 days of gestation, followed by a slow decline until term, correlates well with similar patterns of change in placental DNA content (Kulhanek *et al.* 1974), total number of placental nuclei (Teasdale 1979), volume of connective tissue in the placental villi (Barcroft and Barron 1946) and total villous surface area (Stegeman 1974). In contrast, there is a continuing increase throughout late pregnancy in the degree of vascularization (measured histometrically) (Stegeman 1974) and numbers of endothelial nuclei (Teasdale 1976) in the foetal but not maternal components of the placenta. This suggests that growth of the placental vascular bed is at least partly responsible for the marked growth of umbilical blood flow during late gestation, whereas the associated increase in maternal placental blood flow is due largely to vascular dilation. It appears that once placental size is determined, progressive changes in maternal and foetal perfusion are crucial in controlling the functional capacity of the placenta.

The continued growth in placental function during the last two months of gestation is exemplified by a logarithmic increase in placental urea permeability per gram of DNA (Kulhanek *et al.* 1974; Figure 3). Similar changes have been observed in placental clearance of antipyrine (Meschia *et al.* 1967b) and tritiated water, and in umbilical uptake (= placental transport) of oxygen and glucose (A.W. Bell, J.M. Kennaugh, F.C. Battaglia and G. Meschia, unpublished observations). In the latter study, uteroplacental uptakes of oxygen and glucose in ewes at 70-80 days of gestation were about 50% and 30%, respectively, of those in animals near term. This indicates a substantial gestational change not only in placental transport of these substances to the foetus but also in placental metabolic activity (Table 2).

**TABLE 2. Uteroplacental uptake of oxygen and glucose in mid versus late gestation.**

| | Mid gestation(1) | Late gestation(2) |
|---|---|---|
| n | 8 | 35 |
| Gestational age (d) | 73-81 | 130-145 |
| Oxygen uptake (ml $min^{-1}$) | 12.7 | 23.5 |
| Glucose uptake ($\mu$mol $min^{-1}$) | 55 | 184 |

(1) Bell *et al.*, unpublished; (2) Sparks *et al.* (1983).

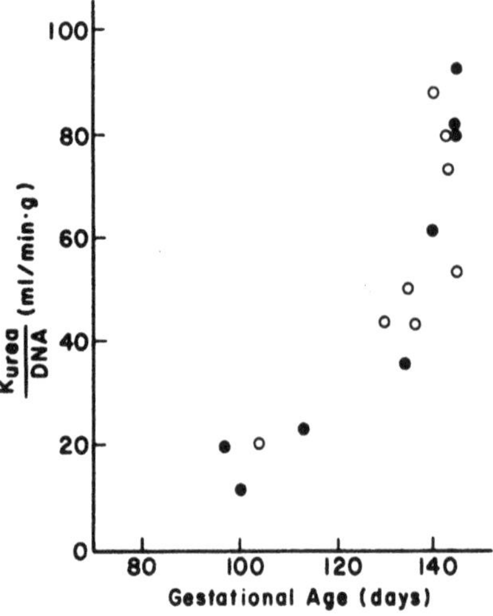


**FIGURE 3. Gestational change in placental urea permeability per gram of placental DNA. ● singles; ○ twins (from Kulhanek *et al.* (1974); reproduced by permission of the American Journal of Physiology).**

Size versus function in late gestation

The relation between placental size and function in late gestation, when the placenta is thought most likely to limit foetal growth is poorly understood. However, there are two pieces of evidence which suggest that placental size is a reasonably good index function at this time.

Firstly, Kulhanek *et al.* (1974) observed that placental urea permeability is directly and linearly related to placental DNA content in ewes at 140-150 days of gestation. The range of placental weights and urea permeabilities were 94 to 380 g and 28 to 161 mlmin$^{-1}$, respectively in these ewes. Urea permeability per kg foetal body weight was also directly related to placental size. Secondly, recent experiments on chronically heat-stressed ewes with abnormally small placentae suggest that this abnormal relation between foetal size and placental function may extend to important substrates, such as oxygen and glucose (G. Meschia and A.W. Bell, unpublished observations). In heat stressed-ewes, placental weights at 130-136 days of gestation ranged from 93 to 253 g, compared with 250 to 400 g in unheated ewes from the same flock. Placental clearances of antipyrine and ethanol and consumption of oxygen and glucose were directly related to placental size, and with the exception of glucose uptake, were similar to values for unheated ewes when expressed in terms of placental weight. Most notably, umbilical blood flow per kg foetal weight was reduced by about 30% in heat-treated ewes, further supporting the idea that placental perfusion is an important determinant of placental function.

## FOETAL GROWTH AND DEVELOPMENT

Placental influences

If foetal growth can be limited by placental size, particularly during late gestation, and placental transport functions are related to placental size, which aspects of placental function are limiting for foetal growth?

Foetal hypoxaemia and hypoglycaemia invariable accompany foetal growth retardation which is achieved by carunclectomy (Robinson *et al.* 1979; Worthington *et al.* 1981), placental embolization (Creasy *et al.* 1972) or chronic maternal heat exposure (G. Meschia and A.W. Bell unpublished observations). This might suggest an impaired umbilical supply of oxygen and glucose, but recent experiments have found little or no effect on foetal uptake of these nutrients, *expressed in terms of foetal weight*, in small foetuses from heat-stressed ewes (Meschia and Bell, unpublished). In fact, arterial hypoxaemia and hypoglycaemia are inevitable consequences for the foetus if it is to maintain an adequate umbilical veno-arterial extraction of oxygen and glucose in the face of a diminished umbilical venous supply of these substances. The fact that growth retardation does occur in foetuses with small placentae, but not to the degree predicted by diminution of placental function, suggests that a compromise is reached between achievement of foetal growth potential and the constraints imposed by reduced placental oxygen and glucose supply.

Under practical conditions, heat stress appears to be the most important potential cause of placental insufficiency leading to retarded foetal growth and development in Australian sheep. It has long been recognised that in parts of Australia where pregnant ewes experience high environmental temperatures and solar radiation, lamb birth weights tend to be low (Moule 1954). Controlled experiments have confirmed a specific effect of heat stress, distinct from the depression of maternal feed intake which usually accompanies prolonged heat exposure (Alexander and Williams 1971; Cartwright and Thwaites 1976a; Hopkins *et al.* 1980).

It seems likely that a smaller placenta contributes to the lower birth weights of twin compared with single lambs because in normal twins as in experimentally growth-retarded foetuses, the ratio of foetal to placental weight is greater than that in normal single lambs.

Maternal nutrition

The sensitivity of lamb birth weight, particularly in twins, to the ewe's plane of nutrition in late pregnancy is well known (Thomson and Aitken 1959; Everitt 1968; Robinson 1977; Mellor 1983). Although foetal growth retardation in undernourished ewes may have a placental component, it is assumed here that maternal supply of nutrients to the pregnant uterus has an important, direct influence on umbilical nutrient uptake and foetal growth.

Nutritional supplementation of pregnant ewes is generally not recommended until the last two months of gestation. While this appears to be practicable under most situations, severe undernutrition during early and mid pregnancy can lead to significant reduction of foetal weight at 90 days of gestation (Everitt 1964; Curet 1973). The degree to which this effect can be reversed by adequate feeding in late pregnancy may depend on when refeeding commences. Everitt (1967) found almost complete compensation in birth weights of lambs from ewes which were severely underfed until 90 days gestation then refed until term, whereas Mellor (1983) reported no such effect in ewes moderately underfed from 35 to 119 days of gestation, and then refed until day 142.

Energy and protein undernutrition in late pregnancy separately caused decreases in birth weight which were linearly related to the degree of underfeeding (Robinson 1977). The absolute effects on maternal body weight were more dramatic than those on foetal weight, which highlights the ability of the ewe to sustain foetal growth at the expense of her own body tissues (Robinson 1977). The detrimental effect of low maternal protein intake on birth weight was accentuated by low levels of energy intake (Sykes and Field 1972).

The specific nutrient(s) that are limiting for foetal growth during maternal undernutrition have not been clearly defined. Although glucose is not the quantitatively most important nutrient for foetal growth and metabolism (Battaglia and Meschia 1978), the evidence indicates that it has a special role. Firstly, maternal hypoglycaemia is an inevitable consequence of undernutrition, and according to Mellor (1983), a reasonable index of foetal growth retardation. Secondly, it is now evident that, contrary to earlier beliefs, both uterine and umbilical net uptakes of glucose are reduced proportionately with maternal glucose production in starved, hypoglycaemic ewes (Hay *et al.* 1983). This decline in exogenous supply of foetal glucose is buffered to some extent by an apparent increase in foetal gluconeogenesis, but foetal glucose utilisation is, nevertheless, significantly reduced (Hay *et al.* 1984). Also, under starvation conditions, umbilical uptake of amino acids is unaltered (Lemons and Schreiner 1983), whereas foetal amino acid catabolism is doubled (Simmons *et al.* 1974), presumably to fill the shortfall in glucose supply. Thirdly, we have preliminary evidence that the significant reduction in foetal weight at 143 days of gestation, caused by severe undernutrition during the last five weeks of pregnancy, can be abolished by foetal gastric supplementation with glucose at levels which maintained approximately normal foetal glycaemia (A.W. Bell, B.J. Leury and A.R. Bird, unpublished observations). In an almost identical study, Charlton (1984) obtained similar results, except that she supplemented foetuses not only with glucose, but also with amino acids.

It could be argued that glucose is merely sparing the utilisation of another, more important nutrient, such as one of the essential amino acids. However, insulin is considered to be an important growth promoter in the foetal lamb (Bassett and Fletcher 1983) and of the various metabolites which are known to stimulate foetal insulin secretion, the concentration of only glucose is seriously reduced by maternal undernutrition.

Other factors

Cold stress during the last 5-6 weeks of gestation can cause a significant increase in lamb birth weight, independent of any effect on feed intake (Thompson *et al.* 1982). This was ascribed to elevated foetal plasma insulin levels associated with moderate maternal and foetal hyperglycaemia. These observations could be relevant to winter lambing systems in southern Australia, particularly where shearing in late pregnancy is practised.

Hallford *et al.* (1982) recently reported a significant increase in lamb birth weights of exercised compared with sedentary ewes, which was attributed to stimulation of feed intake. We have found a similar trend in a small group of ewes exercised daily for 30 min throughout late pregnancy and fed at the same level as non-exercised controls; this trend could be related to foetal hyperinsulinaemia during recovery from maternal exercise (Bell *et al.* 1983).

Foetal development

Restriction of foetal nutrient supply, whether by the placenta or by level of maternal nutrition, affects not only foetal body growth, but the rates of development of wool follicles and tissue energy reserves, and the relative growth of organs and tissues (Wallace 1948; Everitt 1968; Alexander 1974). Each of these effects has consequences for neonatal survival and possibly, longer term growth and wool production.

Density of mature secondary, but not primary wool follicles at birth can be markedly reduced by undernutrition (Schinckel and Short 1961; Alexander and Williams 1971; Cartwright and Thwaites 1976b), heat stress (Alexander and Williams 1971; Cartwright and Thwaites 1976b; Hopkins *et al.* 1980) and previous caruncleotomy (Alexander 1964b) in pregnant ewes. The direct, linear relation between secondary wool follicle density or the ratio of secondary to primary wool follicles, and birth weight in lambs whose prenatal growth was restricted by a variety of means (Alexander 1974; Cartwright and Thwaites 1976b) suggests a common explanation for the inhibition of secondary wool follicle development. However, the limiting factors for foetal skin development in late pregnancy are unknown.

The initiation of most secondary wool follicles occurs between about 95 and 135 days of gestation. In Scottish Blackface foetuses, initiation appears to be less affected by maternal undernutrition during the first half than during the second half of this period (Hutchison and Mellor 1983; Table 3). This pattern of response requires confirmation in the Merino and other wool-producing breeds.

**TABLE 3. Effect of undernutrition at different stages in late pregnancy on foetal weight and S/P ratio at 142 days (Hutchison and Mellor, 1983).**

| Group | n | Weight (kg) | S/P(1) |
|---|---|---|---|
| Well fed | 22 | 5.12 | 2.42 |
| Underfed 112-142 d | 16 | 3.84 | 1.71 |
| 112-131 d | 16 | 3.97 | 1.42 |
| 95-116 d | 16 | 4.10 | 2.10 |

(1) The ratio of secondary/primary wool follicles

Alexander (1974) concluded that newborn lambs, small for whatever reason, tend to have lower total body lipid and liver and muscle glycogen per unit body weight than large lambs. This once again focuses attention on foetal glucose supply because glucose is not only a major glycogenic precursor but also an important lipogenic substrate in the foetal lamb (Vernon *et al.* 1981).

There is considerable disproportion in the degree to which different organs and tissues are affected in growth-retarded foetuses from underfed, heat-stressed or caruncleotomised ewes (Wallace 1948; Everitt 1968; Alexander 1974). In general, the early-maturing vital organs, such as the brain and heart, are less affected than the later maturing skeletal tissues by diminished nutrient supply in late pregnancy, so that small lambs appear malproportioned at birth simply because they are at an earlier stage of the differential growth process. Maloney *et al.* (1982) found impaired function, as well as structure, in the respiratory system of growth-retarded foetuses.

## CONSEQUENCES FOR FUTURE PRODUCTION

The consequences of low birth weight, and associated reductions in energy reserves and thermoregulatory capability, in terms of lamb mortality are reviewed elsewhere in this volume (Alexander 1984). This section is concerned with the long-term effects of foetal growth restriction on postnatal body growth and wool production in surviving lambs.

Early postnatal growth is clearly related to lamb birth weight. Thus in single lambs with a birth weight reduction of 30% due to severe maternal undernutrition throughout pregnancy, weaning weights were reduced by 25% (Schinckel and Short 1961; Everitt 1967). This effect can be difficult to differentiate from that of the depressed milk production which usually occurs in ewes subjected to prenatal nutritional stress. Markedly increasing maternal nutrition at birth can largely overcome inhibitory effects of underfeeding during late pregnancy on lactation and lamb growth rate (Peart 1967). Nevertheless, even when lambs are fed milk *ad libitum* the positive correlation between birth weight and preweaning growth rate persists (Penning *et al.* 1980).

Longer term growth and ultimate mature body size may also be related to birth weight. Lambs that were about 30% lighter at birth than those from well-fed ewes remained 5-10% lighter at 3 years of age (Schinckel and Short 1961; Everitt 1967). This degree of birth handicap is severe, but even in lambs from ewes which were underfed until 90 days of gestation and then well-fed until term, a 10% reduction in birth weight was associated with a 9% lower weaning weight and significant differences from control lambs which persisted until 18 weeks after weaning (Everitt 1967). Such a handicap could be significant in lambs slaughtered for meat production at 4-6 months of age.

Depressed initiation and retarded maturation of secondary wool follicles in growth-restricted newborn lambs can result in reductions of secondary follicle density and total population, and of the secondary: primary ratio which persists for at least 18 months in Merino sheep (Schinckel and Short 1961; Everitt 1967). The resultant effect on wool production is partly compensated by the production of larger, coarser wool fibres, but even so, the clean wool production over three years of sheep whose mothers had been underfed throughout pregnancy was 20% lower than that of sheep born to well-nourished ewes (Everitt 1967). Effects on secondary follicle number and density, but not on total fleece production, persisted until at least 18 months of age in lambs from ewes that were severely underfed in late pregnancy only, while undernutrition in early pregnancy had little subsequent effect on secondary follicle characteristics or on total fleece production (Everitt 1967).

## PRIORITIES FOR FUTURE RESEARCH

This review has identified a number of areas which are in need of new, renewed or continued investigation:

(i) Regulation of placental size. In particular, an in-depth study of the endocrine influences on placental initiation and growth during early and mid pregnancy, under optimal and suboptimal conditions is required.

(ii) The relation between placental size and function. Preliminary evidence suggests that under certain conditions, placental size may be a fair index of the placenta's capacity to transport oxygen and nutrients. However, it is by no means clear whether size and function are related over the wide 'normal' range of placental size.

(iii) Limiting nutrient(s) for foetal growth and wool follicle development. In placentally-induced growth retardation, oxygen and glucose are prime candidates as limiting factors, while in nutritionally restricted foetuses, glucose deficiency alone may be responsible. However, other vital nutrients may, under certain conditions, be limiting for body growth and wool follicle development.

(iv) Maternal capacity for buffering deleterious influences on foetal growth. More information is needed on factors such as breed and strain, previous level of nutrition and body condition, which may influence the ewe's ability to mobilise her body reserves during late pregnancy without detriment to foetal weight.

(v) Consequences for future production. In all but the most extreme cases, much of any handicap to production associated with suboptimal foetal growth and development can probably to overcome by postnatal feeding and management. Applied research, associated with production systems modelling, is required to evaluate the relative merits of pre- and postnatal feed supplementation of pregnant and lactating ewes, and of weaner lambs, in relation to life-time body growth, wool production and reproduction.

## REFERENCES


Alexander, G., 1964a. *J. Reprod. Fertil.*, *7*, 289-305.

Alexander, G., 1964b. *J. Reprod. Fertil.*, *7*, 307-322.

Alexander, G., 1974. *In* Elliott, K. and Knight, J. (eds.) *Size at Birth*. Elsevier, Amsterdam, 215-245.

Alexander, G., 1978. *In* Naftolin, F. (ed) *Abnormal Foetal Growth: Biological Bases and Consequences*. Dahlem Konferenzen, Berlin, 149-164.

Alexander, G. This volume, 199-209.

Alexander, G. and Williams, D., 1966. *J. Endocr.*, *34*, 241-245.

Alexander, G. and Williams, D., 1968. *J. Endocr.*, *41*, 477-485.

Alexander, G. and Williams, D., 1971. *J. Agric. Sci., Camb.*, *76*, 53-72.

Anderson, L.L., 1975. *Am J. Physiol.*, *229*, 1687-1694.

Barcroft, J. and Barron, D.H., 1946. *Anat. Rec.*, *94*, 569-592.

Barcroft, J. and Kennedy, J.A., 1939. *J. Physiol. (Lond.)*, *95*, 173-186.

Barnes, R.J., Comline, R.S., Silver, M. and Steven, D.H., 1976. *J. Physiol. (Lond.)*, *263*, 173-174P.

Bassett, J.M., and Fletcher, J.M., 1983. *In* Jones, C.T. (ed.), *Biochemical Development of the Foetus and Neonate*, Elsevier, Amsterdam, pp. 393-423.

Battaglia, F.C. and Meschia, G., 1978. *Physiol. Rev.*, *58*, 499-527.

Bell, A.W., Bassett, J.M., Chandler, K.D. and Boston, R.C., 1983. *J. Dev. Physiol.*, 5, 129-141.

Bell, A.W., Chandler, K.D. and Bird, A.R., 1984. *Proc. Soc. Gynecol. Invest.*, 31st Annual Meeting, 122 (abstr. 205).

Bell, A.W., Chandler, K.D. and Leury, B.J., 1982. *In* Ekern, A. and Sundstol, F. (eds) *Energy Metabolism of Farm Animals*, Agricultural University of Norway, Aas, 58-61.

Cartwright, G.A. and Thwaites, C.J., 1976a. *J. Agric. Sci., Camb.*, *86*, 573-580.

Cartwright, G.A. and Thwaites, C.J., 1976b. *J. Agric. Sci., Camb.*, *86*, 581-585.

Challis, J.R.G., Mitchell, B.F. and Lye, S.J., 1984. *J. Dev. Physiol.*, *6*, 93-105.

Charlton, V., 1984. *Semin. Perinat.*,*8*, 25-30.

Creasy, R.K ., Barrett, C.T., De Swiet, M., Kahanpaa, K.V. and Rudolph, A.M., 1972. *Am. J. Obstet. Gynecol.*, *112*, 566-573.

Curet, L.B., 1973. *In* Comline, R.S., Cross, K.W., Dawes, G.S. and Nathanielsz, P.W. (eds) *Foetal and Neonatal Physiology*. Cambridge University Press, Cambridge, 342-345.

Everitt, G.C., 1964. *Nature Lond.*, *201*, 1341-1342.

Everitt, G.C., 1967. *Proc. N.Z. Soc. Anim. Prod.*, *27*, 52-68.

Everitt, G.C., 1968. *In* Lodge, G.A. and Lamming, G.E. (eds) *Growth and Development of Mammals*, Butterworth, London, 131-157.

Faichney, G.J., 1981. *Proc. Nutr. Soc. Aust.*,*6*, 48-53.

Fletcher, J.M., Falconer, J. and Bassett, J.M., 1982. *Diabetologia*,*23*, 124-130.

Gluckman, P.D., 1984. *J. Dev. Physiol.*, *6*, 301-312.

Hallford, D.M., Hudgens, R.E., and Morrical, D.G., 1982. *Theriogenology, 18*, 663-670.
Hay, W.W., Sparks, J.W., Wilkening, R.B., Battaglia, F.C. and Meschia, G., 1983. *Am. J. Physiol.,245*, E347-E350.
Hay, W.W., Sparks, J.W., Wilkening, R.B., Battaglia, F.C. and Meschia, G., 1984. *Am. J. Physiol.,246*, E237-E242.
Hopkins, P.S., Nolan, C.J. and Pepper, P.M., 1980. *Aust. J. Agric. Res.,31*, 763-771.
Hutchison, G. and Mellor, D.J., 1983. *J. Comp. Path.,93*, 577-583.
Kulhanek, J.F., Meschia, G., Makowski, E.L. and Battaglia, F.C., 1974. *Am. J. Physiol.,226*, 1257-1263.
Lemons, J.A. and Schreiner, R.L., 1983. *Am. J. Physiol.,244*, E459-E466.
Maloney, J.E., Bowes, G., Brodecky, V., Dennett, X., Wilkinson, M., and Walker, A., 1982. *J. Dev. Physiol.,4*, 279-297.
Mellor, D.J., 1983. *Br. Vet. J., 139*, 307-324.
Mellor, D.J., and Murray, L., 1981. *Res. Vet. Sci.,30*, 198-204.
Mellor, D.J. and Murray, L., 1982a. *Res. Vet. Sci.,32*, 177-180
Mellor, D.J. and Murray, L., 1982b. *Res. Vet. Sci.,32*, 377-382.
Meschia, G., Battaglia, F.C. and Bruns, P.D., 1967a. *J. Appl. Physiol.,22*, 1171-1178.
Meschia, G., Battaglia, F.C., Hay, W.W. and Sparks, J.W., 1980. *Fed. Proc.,39*, 245-249.
Meschia, G., Cotter, J.R., Makowski, E.L. and Barron, D.H., 1967b. *Q. J. Exp. Physiol.,52*, 1-18.
Miller, B.G., Tassell, R. and Stone, G.M., 1983. *J. Endocr.,96*, 137-146.
Morriss, F.H., Rosenfeld, C.R., Crandell, S.S. and Adcock, E.W. III, 1980. *J. Nutr.,110*, 2433-2443.
Moule, G.R., 1954. *Aust. Vet. J.,30*, 153-171.
Peart, J.N., 1967. *J. Agric. Sci.,68*, 365-371.
Penning, P.D., Corcuera, P. and Treacher, T.T., 1980. *Anim. Feed Sci. Technol.,5*, 321-336.
Rice, G.E., Leach Harper, C.M., Hooper, S. and Thorburn, G.D. This volume, 165-17:
Robinson, J.J., 1977. *Proc. Nutr. Soc., 36*, 9-16.
Robinson, J.J., McDonald, I., Fraser, C. and Crofts, R.M.J., 1977. *J. Agric. Sci., Camb., 88*, 539-552.
Robinson, J.S., Kingston, E.J., Jones, C.T., and Thorburn, G.D., 1979. *J. Dev. Physiol., 1*, 379-398.
Sanchez, O. and Evans, J.W., 1972. *Environ. Physiol. Biochem.,2*, 201-208.
Schinkel, P.G. and Short, B.F., 1961. *Aust. J. Agric. Res.,12*, 176-202.
Simmons, M.A., Meschia, G., Makowski, E.L. and Battaglia, F.C., 1974. *Pediatr. Res.,8*, 830-836.
Sparks, J.W., Hay, W.W., Jr., Meschia, G. and Battaglia, F.C., 1983. *Europ. J. Obstet. Gynec. Reprod. Biol.,14*, 331-340.
Stegeman, J.H.J., 1974. *Bijd. Dierk.,44*, 3-72.
Steven, D.H., Bass, F., Jansen, C.J.M., Krane, E.J., Mallon, K., Samuel, C.A., Thomas, A.L. and Nathanielsz, P.W., 1978. *Q.J. Exp. Physiol.,63*, 221-229.
Sykes, A.R. and Field, A.C., 1972. *J. Agric. Sci., Camb.,78*, 127-133.
Teasdale, F., 1976. *Anat. Rec.,185*, 187-196.
Thompson, G.E., Bassett, J.M., Samson, D.E., and Slee, J., 1982. *Br. J. Nutr.,48*, 59-64.
Thomson, A.M. and Thomson, W., 1948. *Br. J. Nutr.,2*, 290-305.
Thomson, W. and Aitken, F.C., 1959. *Tech. Commun. Commonw. Bur. Anim. Nutr.* no. 20.
Vernon, R.G., Robertson, J.P., Clegg, R.A. and Flint, D.J., 1981. *Biochem. J.,196*, 819-824.
Wallace, L.R., 1948. *J. Agric. Sci., Camb., 38*, 93-153, 243-302, 367-401.
Worthington, D., Piercy, W.N. and Smith, B.T., 1981. *Obstet. Gynecol.,58*, 215-221.
Young, M. and Widdowson, E.M., 1975. *Biol. Neonate,27*, 184-191.

# THE ACUTE EFFECTS OF MATERNAL HYPERTHERMIA ON THE FOETAL LAMB.

A.N. Davies, D.W. Walker, I.C. McMillen and G.D. Thorburn, *Department of Physiology, Monash University, Clayton, Vic. 3168, Australia.*

*Summary* Maintenance of pregnant ewes in a hot room at 43°C for 8 hours did not effect foetal arterial oxygen saturation or concentrations of glucose or alter uterine activity. Foetal lactate concentrations increased gradually during the hyperthermia, cortisol concentrations increased approximately 2- fold, and foetal breathing movements were increased for 4-6 hours following the hyperthermia. The results suggest that heat stress has some effect on foetal metabolism and the central nervous system, but it is uncertain whether this is due to changes within foetal tissues or alteration of placental function.

## INTRODUCTION

High environmental temperatures are known to suppress lamb birth weight both in field and laboratory conditions, but the causative mechanisms remain obscure (Alexander 1974, 1978). Acute hyperthermia has been shown to decrease uterine and umbilical blood flows in pregnant sheep (Oakes *et al.* 1976) and to increase uterine activity in baboons (Morishima *et al.* 1975). Thus, it is possible that prolonged hyperthermia causes a decreased transfer of nutrients to the foetus. Morphometric data have suggested that such growth retardation is consistent with foetal undernutrition (Cartwright and Thwaites 1976). Also high temperatures may cause abnormal hormone secretion and metabolic dyscrasias within the foetus, as suggested by Hopkins *et al.* (1980).

To date there are no studies which describe the physiological changes which occur within the foetus during prolonged exposure to heat loads which approximate those experienced in the field. To identify some of the changes which occur during hyperthermia we have studied chronically catheterized pregnant sheep which were housed in a chamber in which the ambient temperature was raised to 43°C for a period of 8 hours.

## MATERIALS AND METHODS

Ten pregnant cross-bred ewes of 110-120 days gestation were used. Using sterile procedures and general anaesthesia (Halothane, $O_2$, $N_2O$), catheters were placed in a carotid artery of the ewe and foetus, and temperature probes were inserted into the neck of the foetus and attached to the outside of the uterus of the ewes. A catheter was placed in the trachea of 7 foetuses for measurement of breathing movements, and a catheter placed in the amniotic sac to measure intrauterine pressure. In 3 sheep a pair of electrodes was attached to the outside of the pregnant horn of the uterus to measure the electromyographic activity associated with non-labour contractions which are normally present throughout gestation (Harding *et al.* 1983).

After the operation the ewes were held in mobile metabolic crates. Two such crates could be housed in a 2.5 m × 2.5 m × 2.5 m insulated room. The temperature of this room was controlled by a radiant heater, fan, and a feedback circuit to within ± 1°C of the preset temperature. Five to seven days after the operation, each ewe was exposed to a temperature of 43°C for 8 hours; humidity was estimated from wet and dry bulb temperatures and was not controlled. Paired arterial blood samples (5 ml) were drawn at hourly intervals into heparinized syringes beginning 1 hour before the onset of the heating period, and continuing throughout and until 2 hours after the end of heating. The samples were stored on ice until centrifugation and recovery of plasma, which was then frozen until subsequent analysis of glucose, lactate and cortisol concentrations. Arterial samples (0.6 ml) were also taken for the immediate measurement of the partial pressure of oxygen ($PaO_2$), carbon dioxide ($PaCO_2$), pH, oxygen saturation and haemoglobin concentration using Radiometer ABL3 and OSM2 instruments. Blood gases and pH were measured at 37°C and corrected for the temperature which prevailed at the time of sample collection using the formulae given by Severinghaus (1966).

Maternal and foetal temperature, blood pressure, and heart rate were recorded continuously on a polygraph, together with foetal tracheal pressure and amniotic pressure. The foetal arterial and tracheal pressures were corrected by electronic subtraction of amniotic pressure. Plasma glucose concentration was determined by the glucose oxidase method, lactate concentration was measured by the method of Gutmann and Wahlefeld (1974), and cortisol was determined by a specific radio-immunoassay which was a modification of the method of Singh-Asa *et al.* (1982).

## RESULTS

Twenty-three heat trials were performed with the 10 ewes at gestational ages ranging from 113 to 142 days gestation (Table 1). The changes in core temperatures, arterial blood gases, pH, and metabolites is summarized in Table 2.

Maternal core temperature rose from 38.7 ± 0.1 to 40.7 ± 0.18°C after 8 hours of heating. Foetal core temperature rose from 39.6 ± 0.14 to 41.3 ± 0.15°C over the same time. A foetal to maternal temperature difference of between 0.6 and 1.0°C was observed at all times. Room humidity fell from between 50 and 60%

to 25-30%. Foetal and maternal temperatures were not different from control values 2 hours after the end of the heating period.

**TABLE 1. Foetal ages at which 23 heating trials were conducted in 10 ewes.**

| Ewe | No. trials | Gestational age (days) |
|---|---|---|
| 1 | 1 | 138 |
| 2 | 4 | 120, 127, 131, 134 |
| 3 | 1 | 134 |
| 4 | 3 | 126, 140, 142 |
| 5 | 3 | 129, 131, 140 |
| 6 | 2 | 113, 119 |
| 7 | 1 | 132 |
| 8 | 4 | 134, 137, 139, 141 |
| 9 | 2 | 121, 123 |
| 10 | 2 | 127, 129 |

We did not measure maternal respiratory rate quantitatively, but observed that the pattern of panting during hyperthermia fell into the two phases originally described by Hales and Webster (1967). Initially, there was an increase of the breathing rate to approximately 200 breaths/min, with the mouth closed and air passing through the nostrils. During the last 2-3 hours of hyperthermia breathing was usually very rapid (around 300/min), shallow, and through an open mouth with the tongue extended.

The ewe developed a respiratory alkalosis within 1 hour of the beginning of the heating period (Figure 1). Arterial pH rose and $PaCO_2$ fell progressively for 5-6 hours in the different sheep. There was no consistent change in the $PaO_2$ or $O_2$ saturation of the ewe. Foetal pH increased and $PaCO_2$ decreased in association with the maternal changes. In 7 of 19 trials foetal $PaO_2$ fell below 15 mmHg, but in the remainder there was no consistent change in $PaO_2$ from the control value of $23 \pm 0.7$ mmHg. Over all of the trials in which foetal $O_2$ saturation was measured there was no consistent or significant change. There were no consistent changes in foetal blood pressure and heart rate.

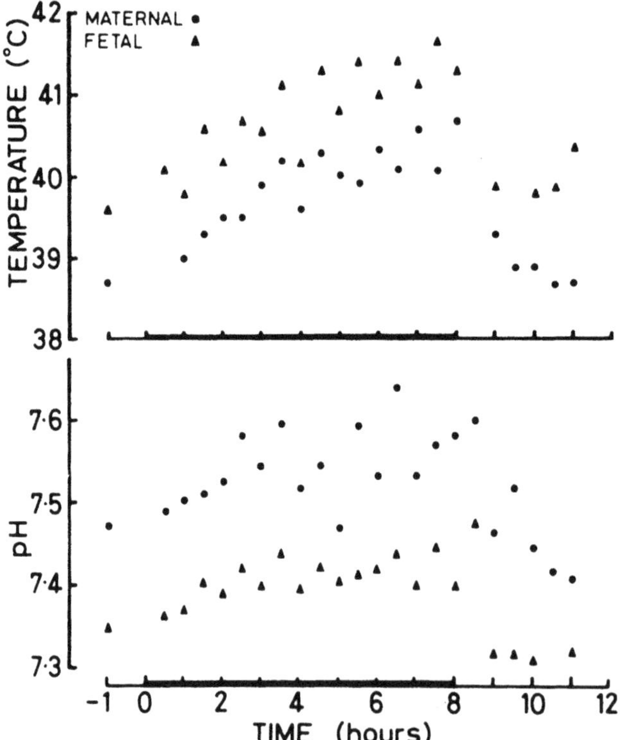


**FIGURE 1. The changes in body temperature and arterial pH before, during and after the period of heating (denoted by the black bar from 0-8 hours). These data demonstrate the maintenance of the foeto-maternal temperature difference, and the respiratory alkalosis observed during meternal hyperthermia. ●, maternal; ▲, foetal temperatures or pH.**

**TABLE 2. Mean (± s.e.m.) temperatures, partial pressures of $O_2$ and $CO_2$, pH, and arterial concentrations of glucose, lactate and cortisol in the mother and foetus before, during and after exposure to an ambient temperature of 43°C for 8 hours.**

| | | Control | +4 Hrs | +8 Hrs | Recovery |
|---|---|---|---|---|---|
| Temperature (°C) | Ewe | 38.7 ± 0.1 (19)[1] | 39.6 ± 0.12 (8) | 40.7 ± 0.18 (16) | 38.9 ± 0.13 (12) |
| | Foetus | 39.6 ± 0.14 (19) | 40.2 ± 0.13 (8) | 41.3 ± 0.15 (16) | 39.8 ± 0.21 (11) |
| pH | Ewe | 7.471 ± 0.010(13) | 7.518 ± 0.025(6) | 7.582 ± 0.036(8) | 7.444 ± 0.017(6) |
| | Foetus | 7.351 ± 0.008(19) | 7.397 ± 0.011(7) | 7.401 ± 0.017(12) | 7.310 ± 0.019(12) |
| $PaCO_2$(mm Hg) | Ewe | 38.5 ± 1.6 (13) | 32.8 ± 1.7 (6) | 26.4 ± 2.8 (8) | 38.2 ± 0.8 (7) |
| | Foetus | 54.6 ± 1.1 (19) | 45.4 ± 1.5 (7) | 38.0 ± 2.1 (12) | 51.0 ± 1.2 (11) |
| $PaO_2$(mm Hg) | Ewe | 115.1 ± 3.7 (13) | 119.8 ± 7.7 (6) | 125.6 ± 6.9 (8) | 123.4 ± 6.1 (6) |
| | Foetus | 23.1 ± 0.67 (19) | 21.6 ± 0.6 (7) | 19.7 ± 1.1 (12) | 23.6 ± 0.8 (10) |
| $O_2$ Saturation (%) | Ewe | 95.3 ± 0.7 (13) | 94.5 ± 1.5 (6) | 95.8 ± 0.5 (8) | 95.2 ± 0.8 (7) |
| | Foetus | 60.6 ± 1.9 (19) | 58.0 ± 2.9 (7) | 50.7 ± 4.0 (12) | 59.4 ± 3.0 (12) |
| Glucose (μmol/ml) | Ewe | 2.45 ± 0.31 (4) | 2.48 ± 0.22 (4) | 2.53 ± 0.30 (3) | 2.50 ± 0.46 (3) |
| | Foetus | 0.89 ± 0.07 (4) | 0.83 ± 0.06 (4) | 0.76 ± 0.09 (3) | 1.15 ± 0.13 (3) |
| Lactate (μmol/ml) | Ewe | 0.12 ± 0.05 (2) | 0.12 ± 0.01 (2) | 0.15 ± 0.00 (2) | 0.18 ± 0.00 (2) |
| | Foetus | 0.43 ± 0.05 (4) | 0.95 ± 0.013(4) | 1.88 ± 1.2 (4) | 0.88 ± 0.47 (4) |
| Cortisol (ng/ml) | Ewe | 17.8 ± 2.7 (4) | 25.8 ± 5.0 (4) | 27.3 ± 7.4 (3) | 10.0 ± 1.1 (3) |

[1] Number of trials.

Foetal arterial glucose concentrations were 0.89 ± 0.07 μmol/ml and did not change consistently during the period of heating. Control lactate concentrations were 0.43 ± 0.05 μmol/ml and increased to 0.95 ± 0.13 μmol/ml after 4 hours, and to 1.88 ± 1.2 μmol/ml after 8 hours of heating. Maternal arterial lactate concentrations (0.12 μmol/ml) did not change.

Foetal breathing movements, recorded as the characteristic negative deflections of tracheal pressure, were present for 35-40% of the time in the control period, and were present in episodes lasting 10-30 mins as is normal for foetal sheep after 115 days gestation. During heat exposure foetal breathing activity was decreased in incidence with occasional brief episodes (5-10 mins) of breathing at high rates (> 3 breaths/sec). However, in the recovery period, breathing activity became continuous, was augmented in amplitude by 30-100%, and persisted in this altered state for between 4 and 8 hours in different trials. In 4 of 5 trials this post-hyperthermic 'hyperventilation' was augmented by and persisted during a 30 min episode of hypoxia produced by allowing the ewe to breath 9% $O_2$; this contrasts with the situation in normal foetal lambs where hypoxia suppresses breathing movements.

Plasma concentrations of cortisol were measured in four ewes before, during and after the eight hour period of heating. In all four ewes there was an increase in the plasma cortisol concentrations during the heating period although the time at which peak cortisol concentrations were reached varied between animals. The mean plasma cortisol concentrations increased from control values of 17.8 ± 2.7 ng/ml (n = 4) to 25.8 ± 5.0 ng/ml (n = 4) at four hours into the heating period and were still increased (27.3 ± 7.4 ng/ml, n = 4) at the end of eight hours of heating. One hour after the end of the heating episode the plasma cortisol concentrations had fallen to 13.8 ± 3.2 ng/ml (n = 4). Plasma concentrations of cortisol were also measured in two foetuses at 139 and 142 days of gestation before, during and after the eight hour heating experiment. There was a 50-80% increase in the plasma cortisol concentrations in the foetuses during the period of heating.

Amniotic pressure (7 sheep) or uterine electromyographic activity (3 sheep) was not changed during or after the period of heating.

## DISCUSSION

In this study exposure of pregnant ewes to a temperature of 43°C without attempting to regulate humidity, was sufficient to raise maternal core temperature above 40°C. This is similar to the hyperthermia which Hopkins *et al.* (1980) noted in sheep in field conditions during the summer months in Northern Australia. There was a similar increase of foetal core temperature so that foetal temperatures in excess of 41°C were often observed.

That this duration and level of hyperthermia may be stressful to the pregnant ewe is indicated by the small but consistent increase in plasma cortisol concentrations during the period of heating. This degree of hyperthermia produced relatively small and inconsistent changes in foetal oxygenation and arterial glucose concentrations. Although foetal $PaO_2$ fell in some trials, this did not occur consistently in each foetus. There was a gradual increase of foetal lactate concentrations, this being greatest in those trials where the foetus became hypoxic. It is possible that release of catecholamines contributed to this metabolic change, although the lack of a significant increase of foetal blood pressure or heart rate indicates that circulating catecholamines concentrations did not change greatly. Acute hyperthermia has been shown to markedly decrease both uterine and umbilical blood flows and to cause foetal $PaO_2$ to fall to 10 mmHg (Oakes *et al.* 1976). Our results indicate that this does not occur in chronic hyperthermia. Based upon observations in 3 foetuses and 1 ewe Hales *et al.* (1972) suggested that the greater fall of foetal $PaCO_2$ during prolonged hyperthermia could be accounted for by an increase of placental blood flow. It will be important to measure uterine and umbilical blood flows under the conditions of our experiments, but the small and inconsistent effects on foetal oxygenation that we observed suggests that uterine and umbilical blood flows were not changed greatly.

At present, it is not possible to determine whether the increase of foetal lactate was the result of increased production by the foetus, or alternatively, by the placenta. Normally, lactate is produced by the placenta and it accounts for approximately 30% of the foetal $O_2$ consumption (Burd *et al.* 1975). Further experiments are required to determine the effects of prolonged hyperthermia on the arterio-venous concentration differences of the principal metabolic substrates in the umbilical circulation, including also amino acids, most of which are acquired by the foetus by selective placental transfer.

Increased uterine activity has been shown to cause transient foetal hypoxia (Harding *et al.* 1983). Morishima *et al.* (1975) found that hyperthermia caused increased uterine activity and clinical signs of foetal distress in anaesthetized pregnant baboons. However we observed no changes in amniotic pressure or uterine muscle activity which suggests that this did not occur under the conditions of our study.

The increase of foetal breathing which followed the heating episode suggest that the central mechanisms which regulate breathing movements are altered for some hours after the return of core temperature to normal. The decrease in the amount of foetal breathing which was observed during hyperthermia was not unexpected in view of the maternal and foetal hypocapnia which existed at this time. Episodes of foetal breathing at very high rates did occur during hyperthermia, and this probably represents the primary panting response of the foetus to high temperatures. It is notable, however, that the post-hyperthermia breathing activity, which although very prolonged and augmented in amplitude, was normal in frequency and like that seen in the control period. In one foetus in which the brainstem had been lesioned at the level of the rostral pons post-hyperthermia hyperventilation did not occur. This suggests that heat stress causes an alteration of suprapontine, possibly hypothalamic mechanisms which persist for some hours after the environmental temperature has been lowered. We propose to investigate the effect of hyperthermia on the foetal hypothalamic-pituitary axis by measuring the plasma concentrations of both ACTH and $\beta$ endorphin in foetuses subjected to heat stress.

Our results suggest that the foetal plasma concentrations of cortisol may increase during the period of heating, thus, we hope to extend this preliminary study and investigate the effects of both acute and chronic hyperthermia on the foetal adrenal.

As hyperthermia is detrimental to sheep production it is important to relate the acute effects on the foetus to growth *in utero*. It is the continuing aim of this project to study the effects of hyperthermia on the development of the foetus. Research in this area presents the possibility of breeding flocks better adapted to hot environments.

## REFERENCES

Alexander, G., 1974. *Ciba Found. Symp., No. 27*, 'Size at Birth', pp 215-239.

Alexander, G., 1978. Abnormal Fetal Growth: Biological Bases and Consequences. *Life Sciences Research Report, No. 10*, Dahlem Konferenzen, . 149-164.

Burd, L.I., Jones, M.D.Jr., Simmons, M.A., Makowski, E.L., Meschia, G. and Battaglia, F.C., 1975. *Nature, 254*, 710-711.

Cartwright, G.A. and Thwaites, C.J., 1976. *J. Agric. Sci., 86*, 573-580.

Hales, J.R.S., Hopkins, P.S. and Thorburn, G.D., 1972. *Experientia, 28*, 801-802.

Hales, J.R.S. and Webster, M.E.D., 1967. *J. Physiol., 190*, 241-260.

Harding, R., Sigger, J.N. and Wickham, P.J.D., 1983. *J. Devel. Physiol.* 5, 267-277.

Hopkins, P.S., Nolan, C.J. and Pepper, P.M., 1980. *J. Agric. Res., 31*, 763-771.

Oakes, G.K., Walker, A.M., Ehrenkranz, R.A., Cefalo, R.C. and Chez, R.A., 1976. *J. Appl. Physiol., 41*, 197-201.

Morishima, H.O., Glaser, B., Niemann, W.H. and James, L.S., 1975. *Amer. J. Obstet. Gynec., 121*, 531-538.

Severinghaus, J.W., 1966. *J. Appl. Physiol., 21*, 3-26.

Singh-Asa, P., Jenkin, G. and Thorburn, G.D., 1982. *J. Endocrinol., 92*, 205-212.

# IMMUNOSYMPATHECTOMY OF THE SHEEP FOETUS

J.A. Schuijers, D.W. Walker, C.A. Browne and G.D. Thorburn, *Department of Physiology, Monash University, Wellington Road, Clayton. Victoria. 3168. Australia.*

*Summary* Nerve growth factor (NGF), a protein regulating the development and normal function of sensory and sympathetic neurones, has been isolated and purified from mouse submaxillary glands using high performance liquid chromatography (HPLC). Antibodies raised in sheep to this NGF was used to immunosympathectomize foetal sheep. Foetuses treated with anti-NGF at 80-90 days had reduced cardiovascular responses to hypoxia and tyramine infusion, and reduced tissue content of catecholamines as shown by fluorescence histochemistry and a specific catecholamine assay. We conclude that treatment of the ovine foetus with anti-NGF resulted in a reduction in the amount of sympathetic innervation.

## INTRODUCTION

Nerve growth factor is a protein found in a large number of tissues in widely varying amounts. Its actions are still relatively uncertain, although it is well known that it plays an important role during growth and differentiation of sensory and sympathetic neurones. Whereas NGF elicits a growth response from sensory nerve cells only during a restricted period of development, sympathetic nerve cells respond to NGF during all developmental stages and also in their fully differentiated state.

There are two main types of sympathetic neurones, both derived from the primitive sympathetic cells. The 'long axon' neurones which innervate the cardiovascular system, intestinal and glandular tissues are sensitive to depletion of their noradrenaline stores by both 6-hydroxydopamine and anti-NGF. The 'short axon' neurones found in the urinary tract, vas deferens, uterus as well as the pheochromoblast, chromaffin and S.I.F. cells found in the adrenal medulla, paraganglia, smooth muscle and heart are generally resistant to noradrenaline depletion by 6-hydroxydopamine and anti-NGF. However, with anti-NGF, the time at which it is administered may determine which neurones are affected. It is possible that treatment of the foetus with anti-NGF will result in a more widespread destruction of adrenergic neurones.

Three tissues have unusually large amounts of NGF. These are the submaxillary gland of the male mouse (Levi-Montalcini and Angeletti 1968), the seminal plasma of bulls (Harper *et al.* 1982), and the guinea-pig prostate (Harper *et al.* 1979) and it is from the mouse that we have isolated and purified NGF. Antibodies were raised against the NGF, and were then used to immunosympathectomize the sheep foetus.

## MATERIALS AND METHODS

### Isolation and Purification of Mouse NGF

Using the method of Bocchini and Angeletti (1969) to isolate mouse NGF from male mouse submaxillary glands, we found that this NGF was heterogeneous when subjected to HPLC using reversed phase chromatography. We used the HPLC system previously described (Elson *et al.* 1984). The impure mouse NGF was applied to a $\mu$-Bondapak C18 reversed phase column (Waters Associates), by trace enrichment, and the column was eluted with a linear gradient of 20-60% acetonitrile containing 0.1% trifluoroacetic acid (TFA) over 2hrs, at a flow rate of 2 ml/min. The active NGF peak was determined by radioimmunoassay (Young *et al.* 1978). This was further purified by repeated reversed phase chromatography (Browne *et al.* 1981), and final purification was achieved by high performance gel filtration on an I-125 protein analytical column (Waters Associates), using an isocratic system of 40% acetonitrile with 0.1% TFA at 1 ml/min. The mouse NGF was finally chromatographed by reversed phase HPLC under the original conditions.

### Antibodies to Mouse NGF

2.0 mg of lyophilized pure mouse NGF was dissolved in 1.5ml 0.1M sodium phosphate buffer, pH 7.0 and 1.5ml complete Freund's adjuvant (C.S.L.). This was repeatedly passed through an 18 gauge needle until it thickened. One ml of this was injected intramuscularly into each hind leg of the ewe. The remaining one ml was injected subcutaneously in the lumbar region over ten sites. After eight weeks 1.0mg of NGF, made up the same way, was injected intramuscularly as a booster. Blood was taken from the jugular vein of the ewe each week and the titre was determined by the amount of binding attainable with dilutions of the serum when used in the mouse NGF radioimmunoassay. The highest titre was usually obtained two weeks after boosting.

### Immunosympathectomy of the Foetal Sheep

Antibodies to mouse NGF raised in sheep were used to immunosympathectomize the foetal sheep. Using sterile procedures, a single injection of the antiserum (5ml) was given to the foetus intramuscularly at several sites at 80-90 days gestational age. This was sufficient antiserum to neutralize 6.5mg of NGF. Control foetuses were given normal sheep serum. At a second operation at 110 days gestational age, the foetal femoral artery and vein were catheterized, as well as the amniotic sac. Maternal carotid artery and jugular vein were also catheterized at this time.

Experiments were performed both *in vivo* and *in vitro* to establish the effect of the anti-NGF on the sympathetic nervous system of the treated foetus. Firstly, at 120 and 130 days gestational age the ewe, and hence the foetus, was subjected to a 60 minute hypoxic episode, induced by allowing the ewe to breathe 9% $O_2$, 3% $CO_2$ in $N_2$. Foetal blood pressure and heart rate were recorded and blood samples were taken before, during and after the hypoxia and assayed for catecholamines. Secondly, tyramine (10 mg over 10 mins) was infused into the foetus at 125 days gestational age and again foetal heart rate and blood pressure were monitored, and blood samples were taken from the foetal femoral artery before, during and after the infusion and assayed for catecholamines. Tyramine is known to cause the release of catecolamines from sympathetic nerves, which results in an increase in both heart rate and blood pressure. Any decrease in this response could be attributed to a decrease in sympathetic innervation. Finally, at 135 days post conception, the foetus was killed and various tissues were taken for fluorescence histochemistry (de la Torre and Surgeon 1967), and direct determination of tissue catecholamine content by the HPLC method (see below).

Assay of Catecholamines Using HPLC-Electrochemical Detection (ECD)

Catecholamines were extracted from plasma by adding 25mg of acid washed alumina (washed with 6M HC1 and dried at 300°C), 10mM sodium metabisulphite (30μl), plasma (1ml), distilled water (1ml) and 0.1M perchloric acid (100μl) containing 2ng of 3,4-dihydroxybenzylamine (DHBA) as an internal standard to glass tubes. The pH was then adjusted to 8.5-8.6 by the addition of 400μl of 1M Tris-Base containing 2% EDTA. After shaking the tubes for 15 minutes, the supernatant was removed with a pasteur pipette and the alumina was washed four times with distilled water. The catecholamines were desorbed from the alumina with 100μl of 0.1M perchloric acid containing 0.1mM sodium metabisulphite. Tissue samples were homogenized in 0.1M perchloric acid with an internal standard, spun at 1500g at 4°C for 10 minutes and the supernatant added to the above extraction procedure in place of the plasma.

The HPLC-ECD mobile phase was prepared by making a solution of 0.07M $Na_2HPO_4$, 0.1mM EDTA and 3.0mM heptanosulphonate in 940ml of distilled, degassed and deionized water. The pH was adjusted to 4.8 with 2M NaOH, and 60ml of degassed filtered methanol (Waters Associates) was then added and the solution filtered through a 0.4μm Fluoropore filter (Millipore). The column used was a 15cm Novapak column (Waters Associates) which was protected with an in-line high pressure filter. The electrochemical detector consisted of a cell and electrode unit from Bio Analytical Systems and a model 410 control unit (Waters Associates). The HPLC-ECD flow rate was 0.7ml/min and the electrochemical detector set at 2mA/V 0.72V applied potential. Absolute sensitivity of the assay was 50pg for adrenaline and noradrenaline.

## RESULTS

Isolation and Purification of Mouse NGF

The purity of the mouse NGF was greatly improved by the use of the HPLC steps. It was necessary to use both reversed phase and gel filtration HPLC in order to obtain homogenenous mouse NGF. The purified NGF eluted as a single peak when chromatographed with HPLC on a variety of columns (Waters C18 μ- Bondapak, Waters Protein Analysis Column I-125 and Ultrapore RPSC column) and mobile phases (acetonitrile with 0.1% trifluoroacetic acid, acetonitrile with 0.2% heptafluorobutyric acid and isopropanol with 0.1% trifluoroacetic acid).

Immunosympathectomy of the Sheep Foetus

All foetuses treated with anti-NGF survived and at autopsy were seen to have grown normally. Resting blood pressure and heart range were within the normal range.

*Responses to hypoxia* When foetuses were made hypoxic, foetal $PO_2$ dropped from 20.0 ± 0.68 to 11.8 ± 0.70 mmHg (mean ± S.E.M.) in both treated and control foetuses. Six untreated (control) foetuses of 120-130 days gestation responded to hypoxia with an initial fall in heart rate followed by a slow recovery during the hypoxia, with a prolonged tachycardia occurring on return of the ewe to breathing room air. Arterial blood pressure rose 5-10 mmHg during hypoxia in all control foetuses. In four foetuses treated with anti-NGF, there was no decrease in heart rate during hypoxia or prolonged post hypoxic tachycardia, and blood pressure did not change significantly. Although basal plasma catecholamine concentrations were similar in the control and anti-NGF treated foetuses, there was a smaller increase in the concentration of adrenaline and noradrenaline in the plasma of the anti-NGF treated foetuses during hypoxia.

*Responses to tyramine infusion* Upon infusion to 10 mg of tyramine over 10 minutes, the six control foetuses showed a marked increase in blood pressure (54 ± 9%) and heart rate (29 ± 0.8%), followed by a gradual recovery (see Figure 1b). The four treated foetuses, however, showed little response and the increase in blood pressure (7 ± 0.6%) and heart rate (5 ± 1.3%) was significantly less, (see Figure 1a) suggesting that the treatment had impaired the response.

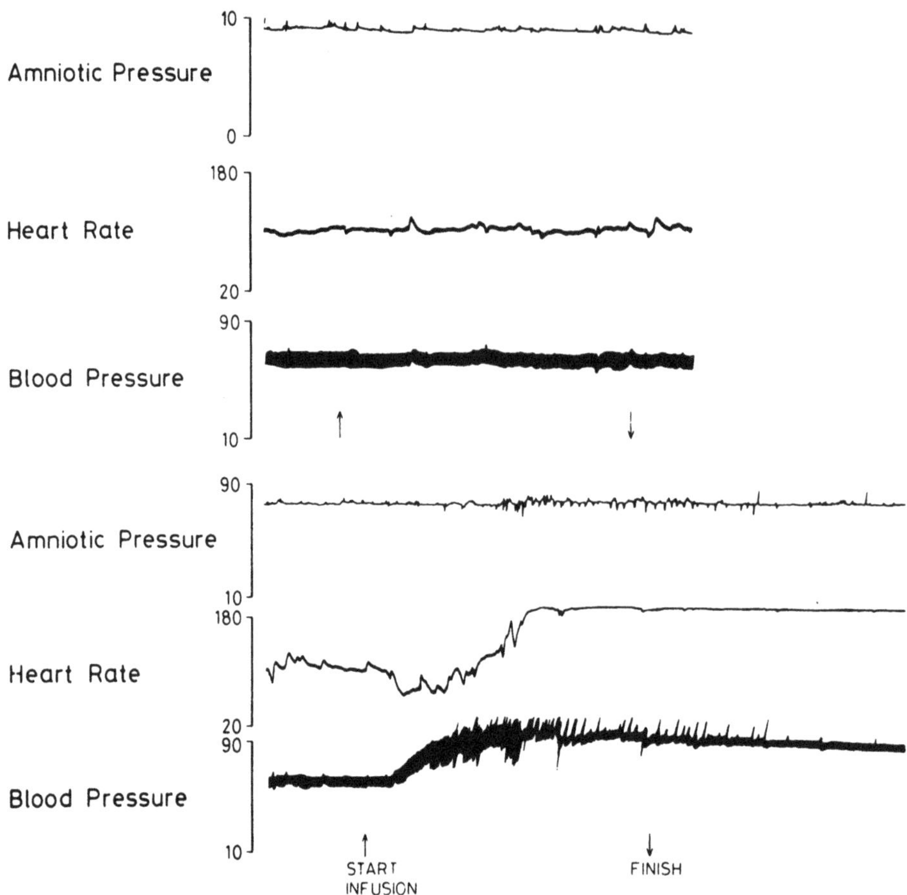


**FIGURE 1a: (Top) Heart rate and blood pressure trace in a typical foetus treated with anti-NGF when infused with tyramine.**

**FIGURE 1b: (Bottom) The response in a typical control foetus under the same conditions.**

*Tissue catecholamine content* At autopsy (135 days gestational age), various tissues were taken from both the control and anti-NGF treated foetuses for direct assay of catecholamine content (Table 1), and for fluorescence histochemistry. The catecholamine content of the heart, spleen and thyroid were markedly reduced in the anti-NGF treated foetuses, when compared to the controls. The adrenal and bladder content were both essentially unchanged. Fluorescence histochemistry of the same tissue showed that adrenergic nerve fibres were reduced in number and showed less intense fluorescence in the heart, spleen and thyroid but not in the adrenal and bladder.

**TABLE 1. Catecholamine content of tissue obtained at autopsy from anti-NGF treated foetuses.**

| | Noradrenaline | Adrenaline | Dopamine |
|---|---|---|---|
| Heart | 5.4 ± 1.3 | 9.2 ± 4.6 | 16.8 ± 4.1 |
| Spleen | 8.3 ± 0.9 | 14.6 ± 6.2 | 24.1 ± 2.8 |
| Thyroid | 18.2 ± 6.1 | 29.4 ± 8.3 | 31.0 ± 4.9 |
| Adrenal | 94.1 ± 2.8 | 96.7 ± 3.9 | 91.4 ± 8.4 |
| Bladder | 102.9 ± 6.4 | 98.3 ± 8.0 | 114.0 ± 4.3 |

The values are given as the percentage of the values seen in control foetuses and are expressed as means ± S.E.M.

## DISCUSSION

These studies indicate that administration of antibodies raised in sheep to mouse NGF is successful in causing immunosympathectomy of foetal sheep *in utero*. The NGF isolated from mouse submaxillary glands had to be purified by supplementing conventional methods with reversed phase and gel filtration HPLC, and once achieved, the pure NGF was used to raise antibodies which were used in the animal experiments.

The foetus, when treated with anti-NGF at 80-90 days, subsequently showed reduced cardiovascular responses to hypoxia and tyramine which indicated that the sympathetic innervation was impaired and the effect persisted throughout gestation. These included a qualitatively different response of blood pressure and heart rate during hypoxia, a reduced response to tyramine infusion and a reduction in both catecholamine content and adrenergic innervation as determined by electrochemical detection and fluorescence histochemistry in those tissues which normally are innervated by long axons. These results suggest that the destruction of the sympathetic nerves occurring at 80-90 days upon infusion of anti-NGF is successful and persists throughout gestation. The observation of only modest increases in plasma concentrations of the catecholamines during hypoxia in the anti-NGF treated foetuses suggests that the sympathetic nerves only contribute partly to the total plasma catecholamines. The catecholamines in the plasma of the anti-NGF treated foetuses presumably are derived to a significant extent from the adrenal medulla and extra adrenal chromaffin tissue. The catecholamine containing cells in these tissues are probably regulated by an unknown trophic factor which is distinct from nerve growth factor. Foetuses survived treatment with anti-NGF and appeared to develop normally until 135 days gestation. We shall use this technique to investigate the role of the sympathetic nervous system in, firstly, the adaptation of the foetus to birth and its early post natal development and secondly, in the development of the fleece and wool growth.

In 1949, Ferguson successfully performed a unilateral thoracic sympathectomy on sheep. He found that there was an increase of 36% in wool growth rate on the sympathectomized side over the control side to 10 weeks. The effect disappeared after this. Ferguson suggested that the initial effect of sympathectomy on wool growth was due to vasodilation of the denervated vessels and that the subsequent disappearance of this was due to the onset of warmer weather causing vasodilation on the control side. Recently, Foldes, *et al.* (1984) have shown that there was an increase in wool production on a body weight basis in a sheep that had a bilateral superior cervical ganglionectomy. Immunosympathectomy has been shown to destroy long axon innervation, the primary source of innervation to the skin and also has been shown to lead to a marked decrease in the weight of the superior cervical ganglia in mice. We are therefore investigating the effect of immunosympathectomy on the development of the fleece and wool growth.

## REFERENCES

Browne, C.A., Bennett, H.P.J. and Solomon, S., 1981. *Biochemistry, 20*, 4538-4546.

de la Torre, J.C. and Surgeon, J.W., 1976. *Histochemistry, 49*, 81-93.

Elson, S.D., Browne, C.A. and Thorburn, G.D., 1984. *Biochem. Internat. 8*, 427-435.

Ferguson, K.A., 1949. *Aust. Series B, Biol. Sci. 2 (4)*, 438-443.

Foldes, A., Scaramuzzi, R.J., Downing, J.A., Maxwell, C.A. and Rintoul, A.J., 1984. *Proc. Aust. Physiol. Pharmacol. Soc. 15 (2)*, 137P.

Harper, G.P., Barde, Y.a., Burnstock, G., Carstairs, J.R., Dennison, M.E., Suda, K. and Vernon, C.A., 1979. *Nature, 279*, 160-162.

Harper, G.P., Glanville, R.W. and Thoenen, H., 1982. *J. Biol. Chem., 257*, 8541-8548.

Levi-Montalcini, R. and Angeletti, P.U., 1968. *Physiol. Rev., 48*, 534-569.

Young, M., Blanchard, M.H. and Saide, J.D., 1978. *In*: Jaffe, B.M. and Behrman, H.R. (eds), *Methods of Hormone Radioimmunoassay*, New York, Academic Press, 941-958.

# THE EFFECT OF RESTRICTION OF PLACENTAL GROWTH ON THE PRODUCTION OF PLACENTAL LACTOGEN IN THE SHEEP

J. Falconer, J.A. Owens, E. Allotta and J.S. Robinson, *Faculty of Medicine, University of Newcastle, Shortland, N.S.W., 2308, Australia.*

*Summary* The effects of restricting placental growth on the production of ovine placental lactogen (oPL) by the placenta in cross-bred Merino ewes with a single foetus were studied. Endometrial caruncles were removed from the uterus of 6 ewes prior to pregnancy. Simultaneous arterial and venous blood samples were obtained via indwelling vascular catheters from both the maternal and foetal side of the placenta of these and 8 control ewes. There were no significant differences across the foetal side of the placenta nor between the concentrations of oPL in foetal plasma from the two groups of foetuses. The concentration of oPL was higher in the uterovarian vein than the carotid artery ($P < 0.01$). Concentrations of oPL in the uterovarian vein were higher in the control ewes than in the carunclectomized ewes (784 ± 116 ng/ml (mean ± SEM, $n = 15$) and 418 ± 108 ng/ml ($n = 10$, $P < 0.05$) respectively). Carunclectomy similarly reduced the concentrations of oPL in maternal arterial plasma (695 ± 104 ng/ml ($n = 15$) and 354 ± 96 ng/ml ($n = 11$, $P < 0.02$)). Production of oPL by the placenta was also reduced by limiting placental growth (132 ± 43 μg/min ($n = 12$) and 30 ± 10 μg/min ($n = 8$, $P < 0.05$)).

## INTRODUCTION

Although there has been much speculation as to the role of the placental lactogens their precise function remains unclear. The roles suggested include trophic support for the corpus luteum of pregnancy, regulating foetal growth and altering maternal intermediary metabolism to make nutrients available to the foetus (Thorburn 1979). In an effort to clarify the role of ovine placental lactogen (oPL) we have studied this hormone in sheep where the growth of the placenta has been restricted by the removal of endometrial caruncles prior to pregnancy (Alexander 1964 ; Robinson *et al.* 1979). The restriction of placental growth causes foetal growth retardation and changes in foetal metabolism including hypoglycaemia, hypoxaemia and hypoinsulinaemia. Growth retardation is associated with increased neonatal mortality and decreased productivity (by reduced numbers of wool follicles and reduced bodyweight) (Alexander 1974).

## MATERIALS AND METHODS

Fourteen Border Leicester-Merino crossbred sheep carrying singleton foetuses were used. Six ewes were operated on before pregnancy and endometrial caruncles removed as previously described (Robinson *et al.* 1979). In five control and six caruncle ewes catheters were implanted into the foetal femoral artery, tarsal vein and common umbilical vein while in the remainder the foetal jugular vein and carotid artery were catheterised (Owens *et al.* 1984). Catheters were inserted into the uterovarian and jugular veins and a carotid artery of all the ewes.

The ewes were housed individually in pens and fed 1 kg alfalfa pellets, water and lucerne hay *ad libitum*. A minimum of ten days recovery from surgery was allowed before experimentation. If possible two experiments were carried out at 120 and 130 days gestation. One day after the last experiment ewes were sacrificed and foetal, placental and uterine weights recorded.

Uterine blood flows were measured by the steady-state transplacental diffusion technique using antipyrine infused into the foetal tarsal vein (Owens *et al.* 1984). Blood samples were obtained simultaneously from the foetal femoral or carotid artery, common umbilical vein if catheterised, maternal carotid artery and utero-ovarian vein using a Technicon Autoanalyser II pump. Plasma was immediately separated and stored at −20°C until assayed.

Placental lactogen was measured using an antiserum and oPL supplied by Dr. M.J. Waters (University of Queensland). Significance was tested using Student's paired and unpaired t-test.

## RESULTS

The mean concentrations of oPL in foetal plasma were not significantly different (Table 1). There was no consistent arterio-venous difference on the foetal side of the placenta.

Concentrations of oPL were higher in the uterovarian vein than the carotid artery in all the paired samples collected ($n = 24$, $P < 0.001$). Restriction of placental growth reduced the concentration of oPL in the uterovarian vein and carotid artery (Table 1, $P < 0.002$ for both). The A-V difference was also reduced in the carunclectomized ewes (Table 1, $P < 0.05$). A significant correlation was found between the concentration of oPL in arterial plasma and placental weight ($r = 0.66$, $n = 22$, $P < 0.001$).

Since both the uterine blood flow (Table 1) and A-V difference in the concentration of oPL in plasma were reduced, the net production of oPL by the placenta was lower in the carunclectomized ewes ($P < 0.05$) with very large variations in the production rate of the control animals (range 9 to 607 μg/min). Production per gram of placenta however was not significantly different although it was only 34% of the control values (Table 1).

**TABLE 1. Summary of data on the concentration of ovine placental lactogen in control and caruncleclectomized ewes and their foetuses. Numbers in parentheses are the number of observations for each point.**

| | CONTROL | CARUNCLE | P |
|---|---|---|---|
| Number of animals | 8 | 6 | |
| Fetal oPL (ng/ml) | 110 ± 11 (26) | 88 ± 6.4 (15) | ns |
| Maternal oPL (ng/ml) | | | |
| Uterovarian vein | 705 ± 106(18) | 283 ± 65 ( 8) | ** |
| Carotid artery | 621 ± 96 (18) | 231 ± 54 ( 9) | ** |
| Arterio-venous difference | 102 ± 24 (17) | 40 ± 10 ( 8) | * |
| Uterine plasma flow (ml/min) | 1430 ± 254(15) | 851 ± 266( 8) | ns |
| Placental oPL production | | | |
| Total (μg/min) | 132 ± 43 (12) | 30 ± 10 ( 8) | * |
| Per gram placenta (μg/g.min) | 451 ± 149(10) | 154 ± 36 ( 8) | ns |

* $P<0.05$
** $P<0.002$

## DISCUSSION

This study extends a previous report where in a smaller number of animals no differences were found in the concentrations of oPL in plasma from carunclectomized and control ewes or their foetuses (Taylor, 1980). In this study we have again found no effects of restricting placental growth on concentrations of oPL in foetal plasma and have also seen large variations in the concentration of oPL between animals. Increasing the number of animals in the group and confining the time of experiments to 120-130 days of gestation has allowed us to demonstrate a significant difference in the concentration of oPL in maternal plasma. We have also confirmed our previous report of the association between oPL and placental mass (Taylor *et al.* 1980). In this study we have shown that this relationship, previously described for multiple pregnancies, also applies to singleton pregnancies.

Restriction of the growth of the placenta significantly reduced the concentration of oPL in maternal blood. Acute administration of oPL has been shown to depress plasma glucose and insulin (Handwerger *et al.* 1976). This data supports the hypothesis that oPL plays a role in regulating maternal intermediary metabolism during pregnancy to mobilise fat stores and make more glucose available to the foetus. If this is true then restriction of the growth of the placenta may not only impair the supply of nutrients to the foetus by limiting placental transport but also decrease the mobilisation of maternal stores and so reduce the supply of nutrients available from the maternal circulation. We have previously shown that restriction of placental growth is associated with higher maternal concentrations of glucose (Falconer *et al.* 1984) which is consistent with this hypothesis.

Perinatal survival of lambs is greatly reduced for the small neonate (Alexander 1974). The postnatal growth and productivity of these small lambs is also greatly impaired. The results presented here suggest that the failure of the small lambs to thrive may be due not only to a failure of growth *in utero* due to placental deficiencies but also to maternal factors. Another proposed role of oPL is to assist in the preparation of the mammary gland for lactation by stimulating its growth during pregnancy (Buttle *et al.* 1979). Inadequate placental growth may not only be accompained by impaired foetal growth but also by a reduced capacity of the ewe for milk production following delivery.

In conclusion we have shown that restriction of placental growth is accompanied not only by reduced foetal growth but also by lower production of oPL. The reduction in maternal oPL may have important effects on maternal metabolism which compromise the already limited supply of nutrients to the foetus and may decrease the capacity of the ewe for lactation after delivery.

## ACKNOWLEDGEMENTS

This work was supported by the National Health and Medical Research Council of Australia.

## REFERENCES

Alexander, G., 1964. *J. Reprod. Fert.*, *7*, 289-305.

Alexander, G., 1974. *In*, K. Elliott and J. Knight (eds) *Size at Birth*, Ciba Foundation Symposium No. 27, pp 215-239.

Buttle, H.L., Cowie, A.T., Jones, E.A. and Turvey, A., 1979. *J. Endocr., 80*, 343-351.

Falconer, J., Owens, J.A., Allotta, E. and Robinson, J.S. 1984. *J. Endocr.*, (submitted).

Handwerger, S., Fellows, R.E., Crenshaw, M.C., Hurley, T., Barrett, J. and Maurer, W.F., 1976. *J. Endocr., 69*, 133-137.

Owens, J.A. Allotta, E., Falconer, J. and Robinson, J.S., 1984. *Am. J. Physiol.* (submitted).

Robinson, J. S., Kingston, E. J., Jones, C.T. and Thorburn, G. D., 1979. *J. Devel. Physiol., 1*, 379-398.
Taylor, M. J. T., 1980. DPhil Thesis, Oxford.
Taylor, M., Jenkin, G., Robinson, J.S., Thorburn, G.D., Friesen, H. and Chan, J.S.D., 1980. *J. Endocr., 85*, 27-34.
Thorburn, G.D., 1977. *Ann. Vet. Res., 8*, 428-437.

# ENDOCRINOLOGY OF PREGNANCY AND PARTURITION

G.E. Rice, C.M. Leach Harper, S. Hooper and G.D. Thorburn, *Department of Physiology, Monash University, Clayton, Victoria, 3168.*

## INTRODUCTION

Elucidation of the fundamental endocrine involvement in pregnancy and parturition is essential for the development of reproductive strategies to facilitate optimal commercial management of domestic animals. This review concerns recent advances in the area of the endocrine regulation of pregnancy and parturition in sheep and highlights their relevance to the industry. The review is divided into four sections: (1) the endocrine changes that occur during early pregnancy; (2) the hormonal regulation of uterine activity; (3) the structural changes that occur in the cervix in preparation for parturition; and (4) the regulation of prostaglandin biosynthesis and their involvement in the initiation of parturition. As the endocrine involvement in the growth and maturation of the foetus has been the subject of recent reviews (Thorburn *et al.* 1984 a, b) it will not be considered in this presentation.

## HORMONAL CHANGES DURING EARLY PREGNANCY

Prostaglandins produced by the endometrium play a major role in dictating the course of pregnancy (Thorburn and Challis 1979; Thorburn 1983). Their release into the uterine vein can cause luteolysis whereas their diffusion into the myometrium stimulates uterine activity. Therefore, suppression of endometrial PG synthesis by the conceptus ensures the maintenance of a functional corpus luteum and inhibition of myometrial activity. In sheep, there is considerable evidence that PGF release is inhibited throughout pregnancy. The pulsatile release of PGF which characterizes luteal regression is not observed between days 12-16 of pregnancy (Thorburn *et al.* 1972, 1973); and low concentrations of prostaglandin $F_{2\alpha}$ ($PGF_{2\alpha}$, in uterine vein plasma) and PGFM (in peripheral plasma) are found throughout pregnancy.

It is now generally accepted that the primary luteolysin in sheep is $PGF_{2\alpha}$. The $PGF_{2\alpha}$ is released by the endometrium into the uterine vein where it is thought to diffuse into the ovarian artery by way of a counter-current mechanism (for review see Horton and Poyser 1976). The mechanisms which regulate endometrial $PGF_{2\alpha}$ synthesis, however, remain enigmatic.

Progesterone has been implicated in the maintenance of the corpus luteum during early pregnancy and acts by suppressing uterine prostaglandin synthesis. Thorburn (1984) suggested that progesterone may induce the synthesis of a phospholipase inhibitor (a lipomodulin-like protein), thus preventing the release of arachidonic acid for PG synthesis. Progesterone priming of the uterus, however, is a prerequisite for the release of significant quantities of $PGF_{2\alpha}$ (Louis *et al.* 1977). Its action may also stimulate the accumulation of lipid droplets in the endometrium (Brinsfield and Hawk 1973; Louis *et al.* 1977; Thorburn and Challis 1979), which may act as a store of precursor for PG synthesis (Thorburn 1977, 1978). Oestradiol, however, is thought to induce the acyl hydrolases necessary for the hydrolysis of this endometrial lipid and, thus, large amounts of arachidonic acid are available for PG synthesis. Oestrogen, particularly after progesterone priming, stimulates $PGF_{2\alpha}$ synthesis and release and the depletion of endometrial lipid droplets in ovariectomised ewes (McCracken *et al.* 1970; Louis *et al.* 1977).

McCracken (1980) proposed that oxytocin stimulates the release of endometrial $PGF_{2\alpha}$. He also proposed that progesterone and oestrogen regulate this system by directly controlling the appearance of oxytocin receptors. This proposal is supported by the finding that progesterone inhibits, while oestradiol stimulates, the appearance of oxytocin receptors on the endometrium (Soloff 1979).

There is now considerable evidence consistent with an ovarian site of oxytocin synthesis (Wathes and Swann 1982; Sheldrick and Flint 1983a; Rice and Thorburn 1984). The functional significance of ovarian oxytocin has been the subject of intensive study in recent years. Estrumate, a $PGF_{2\alpha}$ analogue has been shown to stimulate the release of oxytocin from the corpus luteum (Sheldrick and Flint 1983b). It has been suggested that a positive short-loop feedback may exist whereby oxytocin stimulates the release of $PGF_{2\alpha}$ which in turn induces further release of oxytocin from the corpus luteum. The factors regulating the release of oxytocin from the corpus luteum remains to be established. It is not clear whether $PGF_{2\alpha}$ released from the endometrium causes oxytocin release from the corpus luteum or whether ovarian oxytocin stimulates the release of endometrial $PGF_{2\alpha}$.

To characterize the endocrine changes which occur during luteal regression and early pregnancy, we developed a technique for implanting and maintaining catheters in both utero-ovarian veins (UOV's) for long periods of time (at least 30 days). Blood samples were collected from both UOVs, the carotid artery and the jugular vein. Daily samples were collected from day 6-12 after behavioural oestrus and hourly samples were collected for 12 hours from days 13-16. Large pulses of $PGF_{2\alpha}$ (4— > 20 ng/ml) occurred in UOV plasma of cycling animals during the period of luteolysis. The concentration of oxytocin in UOV plasma was extremely high until day 8 (approximately 1-3 ng/ml), but declined to a lower level on days 12-13. These results are suggestive that the ovary provides the major contribution to the circulating concentration of oxytocin during the oestrous cycle and that oxytocin concentrations in the UOV follow a similar pattern to the concentrations in peripheral blood. Our results further indicate both oxytocin and oxytocin-neurophysin (measured by W.B. Watkins, Post Graduate School of Obstetrics and Gynaecology, University of Auckland), are released into the UOV as short-duration, large-amplitude pulses which coincide with pulses of $PGF_{2\alpha}$. It was not, however, possible to determine the temporal relationship between these pulses.

Both Thorburn *et al.* (1973) and Barcikowski *et al.* (1974) found evidence suggestive that the release of $PGF_{2\alpha}$ into the UOV is greatly reduced between days 12 to 16 in the early pregnant ewe, compared with the cycling ewe. Low concentrations of $PGF_{2\alpha}$ (in uterine vein plasma) and 13,14-dihydro,15keto $PGF_{2\alpha}$ (PGFM; in peripheral plasma) were found throughout pregnancy. We have confimed the findings of Thorburn *et al.* (1972) that in early pregnant ewes, uterine $PGF_{2\alpha}$ release is significantly depressed. We found, furthermore, that oxytocin concentrations in UOV plasma were elevated during days 12-16 after behavioural oestrus. This is suggestive that oxytocin can be released from the corpus luteum in the absence of a $PGF_{2\alpha}$ stimulus and that the suppression of uterine PG synthesis is not a result of a lack of an oxytocin stimulus. Suppression of the endometrial prostaglandin synthesis by the presence of the conceptus, thus ensures the maintenance of a functional corpus luteum and inhibition of myometrial activity.

We suggest that the endometrial lipid droplets described previously, and which are accumulated under the action of progesterone, are utilised by the trophoblast as an energy source. The trophoblast undergoes rapid growth during days 12-20 and would therefore have a large nutritional requirement. If the lipid droplets are utilised by the conceptus this would decrease the amount of precursor available for endometrial PG synthesis. Thus, a low release of $PGF_{2\alpha}$ during early pregnancy may be, in part, due to the redirection of fatty acid utilisation to foetal nutrition. Consequently the use of fatty acids for PG synthesis may only occur when pregnancy does not eventuate.

In addition to facilitating the characterization of fundamental endocrine changes which occur during early pregnancy and luteal regression, chronic cannulation of UOVs will permit investigation of other important aspects of reproductive management. Using this technique, we have recently observed a high frequency (7 out of 9 sheep examined) of prolonged oestrous cycles in maiden ewes. Incomplete luteal regression, which result in persistant corpora lutea, may represent the major cause of the low fertility observed in maiden ewes. This technique will prove useful in elucidating the mechanism(s) involved and the development of treatment to improve maiden ewe fertility.

In conjunction with the chronic cannulation of the ovarian artery (a technique developed by Thorburn and Nicol 1971) the preparation will also facilitate the investigation of factors which affect ovulation rate. For example, follicular fluid inhibitors and androstenedione immunization.

## UTERINE ACTIVITY DURING PREGNANCY AND PARTURITION

The changes that occur in uterine activity during pregnancy and parturition (see Thorburn *et al.* 1984) represent the development of inherent myoelectrical activity into synchronous uterine contractions, which occur at parturition. The suppression of synchronous uterine contractions throughout gestation is a consequence of the prevailing endocrine milieu.

There is considerable evidence which is indicative that progesterone is involved in the suppression of myometrial electrical activity during pregnancy and that progesterone withdrawal is associated with parturition (in sheep). Progesterone increases to high concentrations, in the maternal circulation, during early pregnancy and remains elevated until parturition. Non-labour contractures of the uterus, which are of low intensity and occur at a frequency of 1-4/hour throughout pregnancy, and the responsiveness of the uterus to signals which stimulate myometrial activity (e.g. oxytocin) remains low until near delivery. Marked uterine activity commences only 12-24 hours before delivery, at a time when the plasma levels of progesterone are decreasing. The administration of progesterone or medroxyprogesterone acetate blocks the cortisol induction of delivery. In progesterone-treated animals, neither the characteristic increases in uterine activity nor the increase in maternal and foetal PGFM associated with parturition, were observed (Jenkin *et al.* 1980).

The effects of progesterone withdrawal on ovine uterine activity have been examined using the $3\beta$-HSD inhibitor, trilostane (Jenkin and Thorburn 1984). Trilostane, in doses of 100 mg administered intravenously, reduced progesterone levels and enhanced PG synthesis, as indicated by the plasma concentrations of PGFM. Concomitantly, myometrial EMG activity and the frequency of intrauterine pressure (IUP) elevations increased over this period. Ewes which aborted prematurely did so at a time when PGFM levels were maximal. No consistant change in plasma oestradiol levels was observed after trilostane treatment. It appears that this extent of progesterone withdrawal may be adequate to cause $PGF_{2\alpha}$ release and parturition.

In a further series of experiments, $PGF_{2\alpha}$ was infused extra-amniotically, into pregnant ewes at approximately 125 days gestation. Both foetal and maternal plasma, and amniotic fluid PGFM concentrations were increased to values in excess of those observed at term. There was, however, no sustained increase in EMG activity or IUP during the 84 hours of the $PGF_{2\alpha}$ infusion. Maternal plasma progesterone concentrations were maintained at preinfusion levels during this treatment. When trilostane (100 mg) was injected intravenously, during the extra-amniotic administration of $PGF_{2\alpha}$, there was a precipitous decline in maternal plasma progesterone concentrations and parturition was initiated within 48 hours in three of the four animals treated. These experiments clearly indicate that the myometrium of pregnant ewes, presumably as a result of progesterone, is insensitive to $PGF_{2\alpha}$ and the removal of the progesterone block increases the sensitivity of the myometrium to $PGF_{2\alpha}$ stimulation.

The mechanism by which progesterone suppresses uterine activity has yet to be elucidated. Thorburn (1984) proposed a model for the inhibitory action of progesterone, in which progesterone acts at a level of phosphatidyl inositol turnover by inducing the synthesis of a phospholipase inhibitor (lipomodulin). According to this model, the inhibition of myometrial phospholipase activity results in: (1) a decrease in the capacity of myometrial cells to mobilize calcium, which is essential for muscle contraction; and (2) suppression of prostaglandin-induced enhancement of cell to cell coupling and, thus, the propagation of EMG activity throughout the uterus, which is necessary for synchronous uterine contractions (see Figure 1).

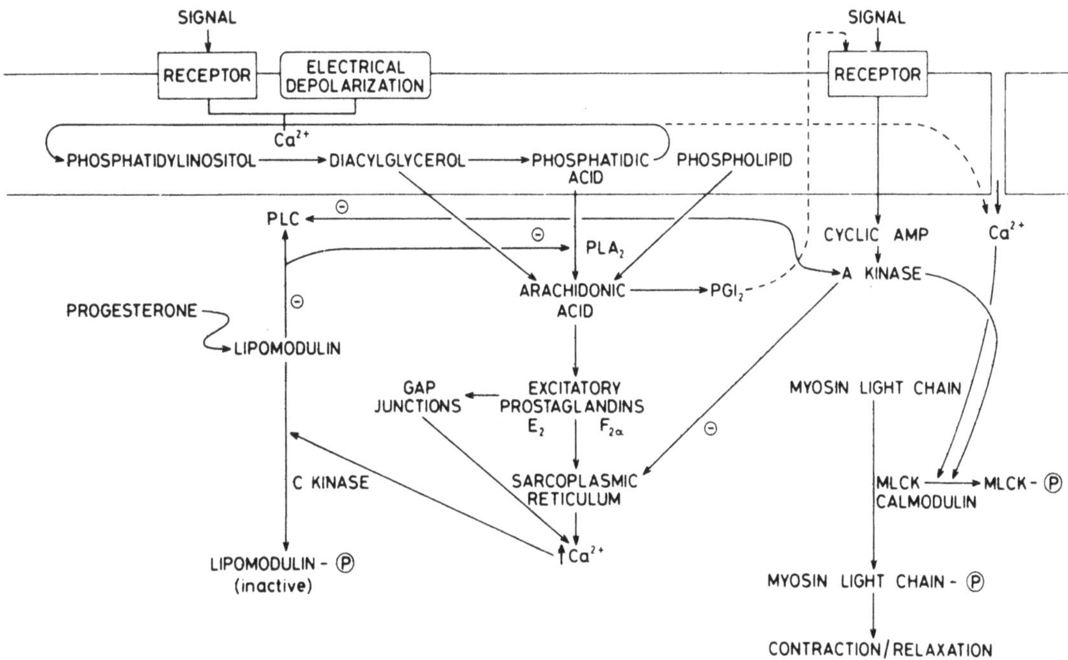


**FIGURE 1. A model of the hormonal regulation of myometrial EMG activity and the involvement of prostaglandins in the contractile-response pathway (Thorburn 1984). The model is comprised of four inter-related processes: (i) the interaction of a signal (e.g. oxytocin, $PGF_{2\alpha}$) with a myometrial cell membrane receptor or electrical depolarization of the membrane; (ii) calcium-stimulated phosphatidylinositol turnover within the membrane and the release of arachidonic acid; (iii) the synthesis of $PGE_2$ and $PGF_{2\alpha}$ which results in an increase in intra-cellular calcium; (iv) calcium-dependent phosphorylation of myosin light chain and muscle contraction.**

Preceding delivery, in sheep, plasma progesterone concentration decreases and oestrogen concentrations increase. As a result of these changes, the synthesis of lipomodulin is suppressed and myometrial phospholipase activity increases. The density of oxytocin receptors increases in response to higher circulating concentrations of oestrogens and, at least in the latter phase of delivery, oxytocin concentrations also increase. The interaction of oxytocin with its receptor and in the electrical depolarization of the cell membrane, stimulate phospholipase C activity. Via the action of phospholipase C (PLC) and phospholipase $A_2$ ($PLA_2$), arachidonic acid (the substrate for PG synthesis) is released and the capacity to mobilize calcium is enhanced. Increased PG synthesis stimulates the formation of gap junctions and enhances cell to cell coupling. Consequent to these changes, labour-associated contractions (where EMG activity is characterized by increased frequency and amplitude and decreased duration of burst) occur.

Recently, some properties of myometrial activity have been compared *in vivo* and *in vitro* (Parkington 1984; Thorburn *et al.* 1984). These data demonstrated that there is considerable spontaneous electrical and mechanical activity (of the non-labour contracture type) during much of pregnancy. Furthermore, *in vitro* study has shown, paradoxically, that spontaneous activity is absent in myometrium removed from animals in labour, a time when circulating progesterone is decreasing. One interpretation of these observations is that the myometrium at parturition is under the control of two opposing systems, with one system tending to inhibit activity whilst the other promotes it. There is evidence from a number of studies to support this view. The dominant inhibitory factor at parturition may be $PGI_2$, which is known to inhibit oestrogen-induced uterine activity in ovariectomized ewes whilst still leaving the myometrium responsive to oxytocin and $PGF_{2\alpha}$ (Lye and Challis 1982).

Furthermore, we have shown that the contractile force generated per action potential during labour was approximately 15 times that recorded from tissues obtained from ewes prior to day 60 of pregnancy (Parkington 1984; Thorburn *et al.* 1984). The increase in the contractile force may have resulted from an increase in the total number of cells active (an expression of cell-to-cell coupling) or possibly from an increase in force generated by each cell. Parkington (1984) has shown that the space constant ($\lambda$) (an index of current spread between cells) increased during the latter part of pregnancy but showed a further increase during labour, the increase could result from an increase in the membrane resistance (Rm) of the muscle cells or an increase in cell-to-cell coupling or both. An increase in cell-to-cell coupling may be the physiological correlate of the increase in the number of gap junctions reported in sheep myometrium at term (Garfield *et al.* 1979). The membrane time constant ($\tau_m$) was unchanged during pregnancy but increased markedly during labour (Thorburn *et al.* 1978; Parkington 1984). The increase in $\tau_m$ during labour was considered to be due to an increase in membrane resistance (Rm).

In addition to the hormonal regulation of uterine activity, an important area of future research will be the involvement of hormones (such as growth factors and relaxin) in regulating the rate of uterine expansion. It is now apparent that insufficient uterine growth can limit the growth and development of the foetus. Furthermore, elucidation of the role of insulin-like growth factors (IGF I and II) and choriosomatotrophin in regulating foetal growth will provide fundamental information which may be applicable to the prevention of foetal under development (e.g. heat-stress induced foetal under development).

## CERVIX

The functional requirements of the uterine cervix changes considerably in pregnancy. As the uterus grows to accommodate the developing conceptus, the cervix forms an essential mechanical barrier. In contrast, at delivery the cervix must become soft and distensible to permit dilatation and passage of the foetus. It is now generally accepted that dilatation and effacement of the cervix are not simply a result of uterine contraction, but also depend upon an active ripening process within the cervix. The sheep cervix is composed of dense connective tissue. The extracellular matrix of this tissue is made up of collagen fibers and elastin, separated by ground substance. The collagen fibers are interspersed with fibroblasts. The ground substance in the extracellular matrix is composed of proteoglycans and various glycoproteins of specific function. Proteoglycans consist of glycosaminoglycans (GAGs) covalently bound to a protein core. A GAG is a long chain of repeating disaccharides containing one hexosamine and one hexuronic acid moiety.

In collaboration with Professor D.A. Lowther's group of the Department of Biochemistry, we have found that the concentration of collagen, proteoglycan and hyaluronate in the cervix decreases with increasing gestational age, accompanied by a small increase in tissue hydration (Fosang *et al.* 1983, 1984). The rate of synthesis of proteoglycan increases during pregnancy, and doubles between 140 days and parturition. The increased rate of proteoglycan synthesis in the presence of a diminishing hexuronate concentration is indicative of enhanced proteoglycan turnover. We have shown that the cervix from ewes at term can synthesize a large proteoglycan which is absent in the synthetic profile from non-pregnant tissue before 140 days gestation (Figure 2). The identification of this labour-specific proteoglycan may represent a new proteoglycan species, or a proteoglycan monomer with large glycosaminoglycan constituents. This is an extremely important finding as no other studies, in any species, have found a change in connective tissue components of the cervix that specifically relates to labour. This proteoglycan, therefore, may be used as a biochemical marker for cervical dilatation.

We suggest that cervical softening and ripening is not only a result of matrix degradation, but may be due to the production of a new matrix with altered physical properties. We postulate that the rapid turnover of tissue proteoglycan is important in the reorganization of the matrix components which occurs during the softening and subsequent ripening processes. Proteoglycan metabolism may be associated with the remodelling of the collagen matrix.

Whether a relaxin-like hormone is involved in the process of ovine cervical ripening has yet to be established. Such a study, however, should provide insight into the cause of dystocia in domestic animals and its associated neonatal mortality.

## REGULATION OF PROSTAGLANDIN BIOSYNTHESIS

Prostaglandins have been implicated in the final chain of events resulting in parturition in many species. In sheep, the concentrations of PGF and 6-keto $PGF_{1\alpha}$ increase in maternal plasma during spontaneous delivery (Thorburn *et al.* 1972; Mitchell *et al.* 1979). Similarly, the concentrations of PGE in foetal plasma (Challis *et al.* 1976) and PGE, PGF and 6-keto $PGF_{1\alpha}$ in amniotic fluid (Mitchell *et al.* 1977, 1978; Challis *et al.* 1978) increase before delivery.

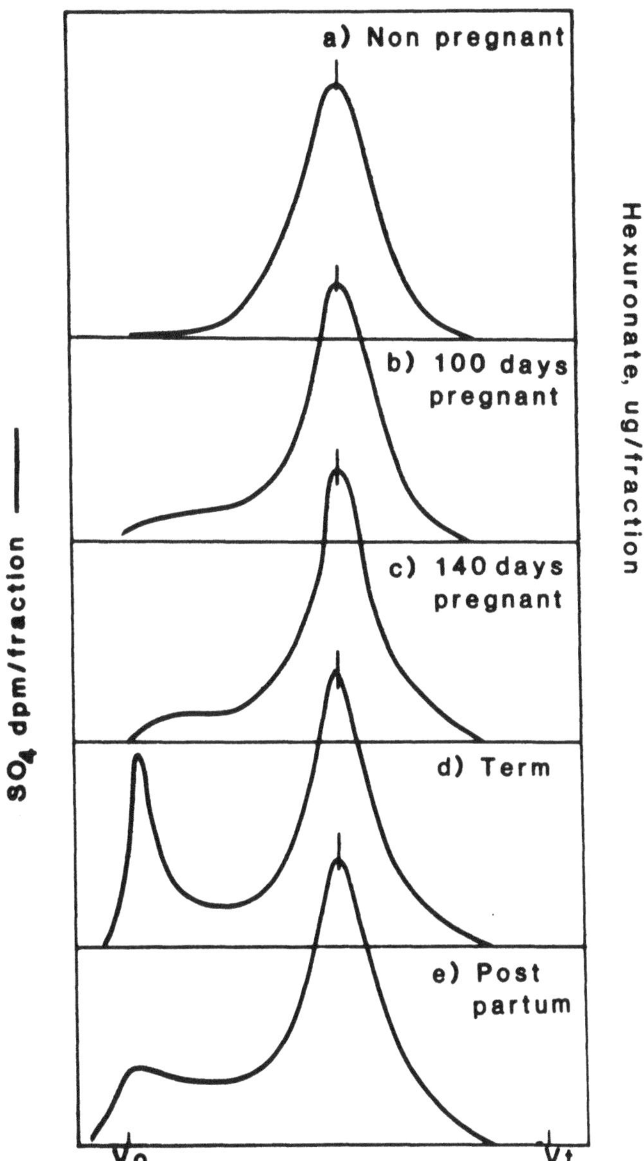


**FIGURE 2. Elution profiles of cervix proteoglycans on Sepharose CL-4B. $^{35}SO_4$-labelled proteoglycans from non-pregnant (a), pregnant (b-d) and postpartum (e) cervix were extracted with 4M GuHC1 in the presence of inhibitors, and purified by ion-exchange chromatography on DEAE-Sephacel. Pooled proteoglycans were chromatographed on Sepharose CL-4B in an elution buffer containing 4M GuHC1. Fractions were assayed for radioactivity.**

In sheep, Evans *et al.* (1981) suggested that the source of elevated PGs observed at parturition, may be the cotyledons and/or chorioallantois for PGF and 6-keto $PGF_{1\alpha}$, and the endometrium for PGE. Mitchell *et al.* (1982) found that ovine cotyledons collected at delivery, produced the greatest concentration of PGE and PGF while the membranes produced the greatest concentration of 6-keto $PGF_{1\alpha}$. The cotyledons and chorioallantois are, thus, important sources of PGs during ovine parturition, the cotyledons being the principle site of placental PG synthesis.

We have shown that foetal trophoblast cells have the capacity to synthesize prostaglandins *in vitro* (Risbridger *et al.* 1982, 1984). This capacity increases markedly with gestational age after 100 days gestation (Figure 3). These cells have the ability to produce $PGF_{2\alpha}$ and $PGE_2$ in approximately equal amounts and a smaller synthetic capacity for $PGI_2$, as shown by the levels of 6-keto $PGF_{1\alpha}$ measured in the incubation medium. These cells also have the ability to metabolize $PGF_{2\alpha}$ to PGFM. PG synthesis by the cells was dependent upon cell number in a linear manner and was significantly inhibited by indomethacin, but not affected by the addition of exogenous arachidonic acid. Our findings are consistent with those of Evans *et al.*

(1982) who showed that the net synthesis of $PGE_2$, $PGF_{2\alpha}$ and 6-keto $PGF_{1\alpha}$ by isolated ovine cotyledon cells was significantly lower on days 50 and 100 than in later stages of pregnancy. The foetal trophoblast cells, thus, appear to be the major site of PG production by the placenta at parturition.

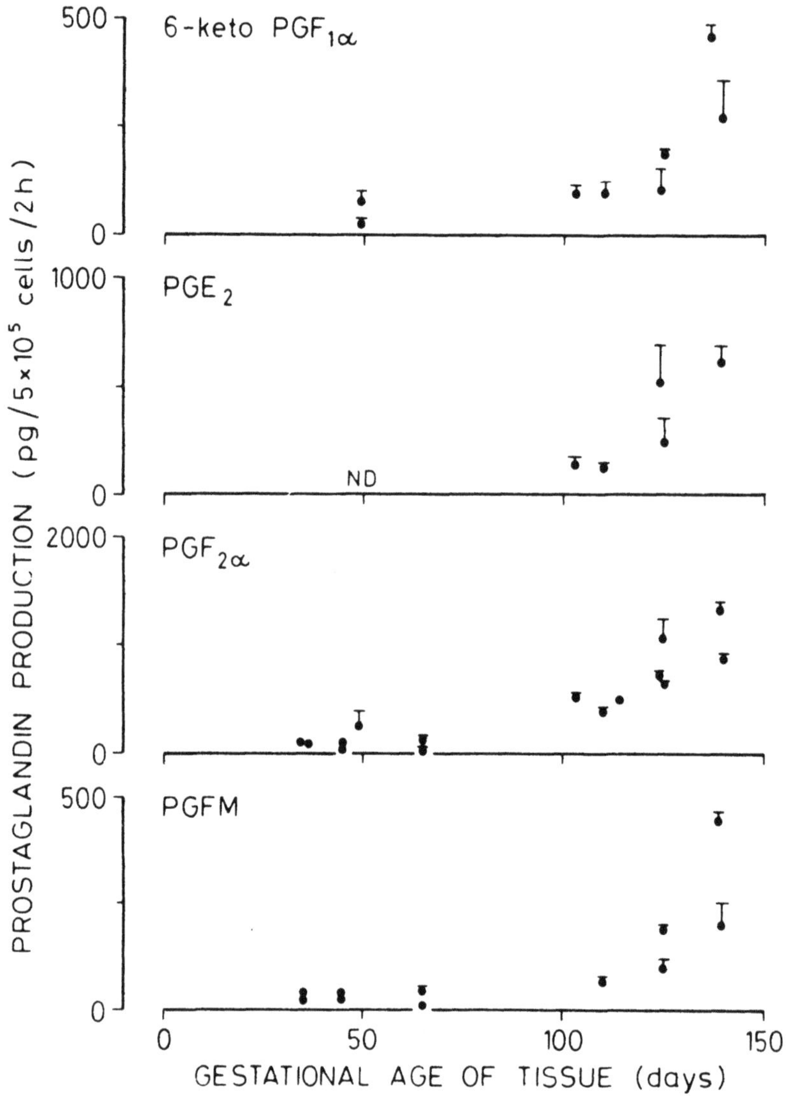


**FIGURE 3. Synthesis of 6-keto $PGF_{1\alpha}$, $PGE_2$, $PGF_{2\alpha}$ and PGFM by isolated foetal trophoblast cells prepared from ewes of 35-145 days gestation. Shown are the mean ± S.E.M. for triplicate determinations for each cell preparation. Each point was obtained from a single ewe.**

In support of the suggestion that intrauterine PG synthesis may be tonically suppressed (Robinson *et al.* 1978; Thorburn 1978), an endogenous inhibitor of prostaglandin synthesis (EIPS) has been found in plasma and serum from several species (Saeed *et al.* 1977; Collier *et al.* 1982). EIPS activity in plasma, however, does not appear to change at parturition in the human (Brennecke *et al.* 1981) or sheep (Mitchell *et al.* 1982). Human amniotic fluid contains an inhibitor of PG synthesis during early gestation which exhibits an acute reduction in potency during labour (Saeed *et al.* 1982). These results are suggestive that, in humans, the onset of labour is associated with a local withdrawal of inhibition of PG biosynthesis. An inhibitor of PG synthesis in allanotic fluid may provide a local mechanism for regulating prostaglandin synthesis by the membranes. We therefore measured the capacity of ovine allantoic fluid to inhibit prostaglandin synthesis during mid and late gestation (Leach Harper *et al.* 1983; Leach Harper and Thorburn 1984).

Allantoic fluid was withdrawn from the allantoic sac of 11 ewes, at post mortem. Allantoic fluid from all ewes of 100-130 days gestation showed a dose-dependent inhibition of $PGF_{2\alpha}$ synthesis by endometrial-cotyledon microsomes (Figure 4). The inhibition ranged from 3.8 ± 2.2% (mean ± S.E.M.) to 36.4 ± 2.6%. The allantoic fluid exhibited a similar capacity to inhibit $PGE_2$, $PGF_{2\alpha}$ and 6-keto $PGF_{1\alpha}$ synthesis.

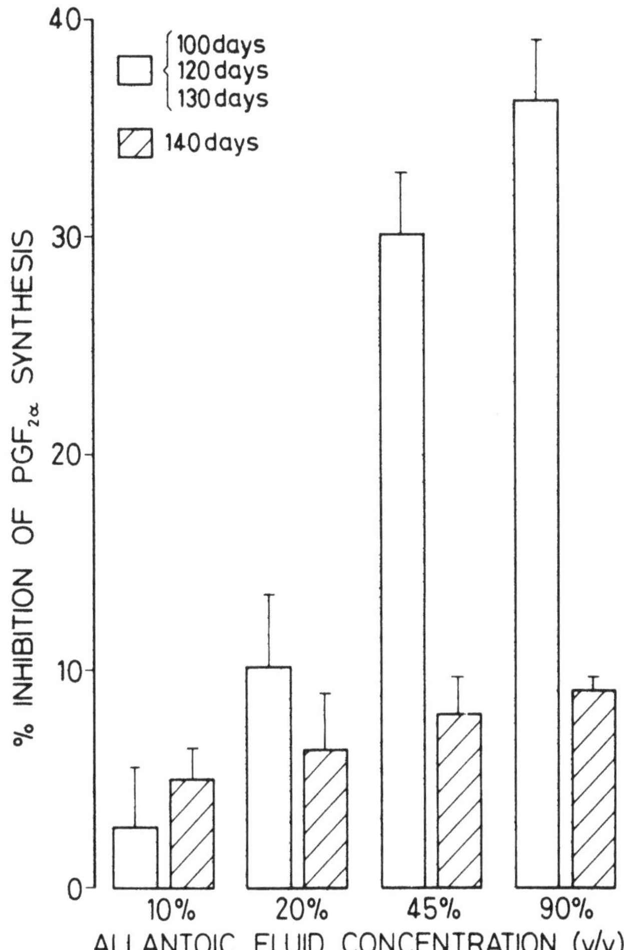


**FIGURE 4. Allantoic fluid samples obtained between 100-130 days gestation showed a significant linear correlation ($P < 0.001$, $r = 0.570$) between the inhibitory activity and albumin concentration. At 140 days gestation there was no significant correlation ($P < 0.1$, $r = 0.105$).**

The ability of allantoic fluid at 140 days gestation to inhibit PG synthesis (5.0 ± 1.4% to 9.1 ± 0.5%) was significantly less than that present at 100-130 days gestation ($p < 0.05$) (Figure 4).

Dialysis or boiling of allantoic fluid did not significantly decrease its inhibitory activity on PG synthesis. The inhibitor, thus is not an electrolyte or small protein, steroid or carbohydrate unbound to a larger protein. Similarly, pre-incubation of allantoic fluid with proteases did not destroy the inhibitory activity, indicating that the active portion of the inhibitor is probably not a protein. There was some indication that the inhibitory activity may have increased following treatment with proteases. A chloroform:methanol extract of allantoic fluid inhibited $PGF_{2\alpha}$ synthesis. No inhibitory or stimulatory activity was present in the residual aqueous phase. We therefore suggest, that the PG synthesis inhibitor present in allantoic fluid is a lipid, possibly complexed to protein.

We have shown that the significant decrease in the inhibitory potency of allantoic fluid from ewes of late gestational age was not related to any change in the albumin concentration as gestation progressed. Thus, we suggest that ovine allantoic fluid contains an inhibitor of prostaglandin synthesis that is distinct from albumin, a known inhibitor of PG synthesis. A significant correlation, however, was observed between concentrations of albumin and the extent of inhibition of prostaglandin synthesis in allantoic fluid obtained at 100-130 days gestation. Thus, the inhibitor may be bound to albumin while in the allantoic fluid, which is an agreement with our studies on the characterization of the inhibitor.

The source of the PG synthesis inhibitor observed in these experiments is unknown, but it may originate from the foetal trophoblast cells and/or another site. The early trophoblast cells exhibit little capacity to produce prostaglandins. Thus, at the final stage of pregnancy the foetal trophoblast cells may gain the capacity to synthesize prostaglandins as well as lose the ability to synthesize the prostaglandin synthesis inhibitor.

## CONCLUDING COMMENTS

In conclusion, the continuation of research efforts to elucidate the endocrine regulation of pregnancy and parturition in domestic animals will facilitate the development of optimal reproductive strategies. Advances in a number of areas may be anticipated. These include:

i) identifying the cause of maiden ewe infertility and the development of appropriate treatment;
ii) elucidation of factors which control ovulation rate;
iii) identification of factors which effect preimplantation mortality;
iv) characterization of the endocrine requirements for foetal growth and maturation; and
v) elucidation of the causes of dystocia and its associated neonatal mortality.

These advances will be ultimately manifested as an increase in flock fecundity. Elucidation of the endocrine parameters involved in these processes, however, is essential to the development of appropriate management strategies.

## REFERENCES

Barcikowski, B., Carlson, J.C., Wilson, L. and McCracken, J.A., 1974. *Endocrinology, 95*, 1340-1349.

Brennecke, S.P., Bryce, R.L., Turnbull, A.C. and Mitchell, M.D., 1981. *The Lancet, 8231*, 1215.

Brinsfield, T.H. and Hawk, H.W., 1973. *J. Anim. Sci., 36*, 919-922.

Challis, J.R.G., Dilley, S.R., Robinson, J.S. and Thorburn, G.D., 1976. *Prostaglandins, 11*, 1041-1052.

Challis, J.R.G., Hart, I., Louis, T.M., Mitchell, M.D., Jenkin, G., Robinson, J.S. and Thorburn, G.D., 1978. *In* Coceani, F. and Olley, P.M. (ed) *Advances in Prostaglandin and Thromboxane Research, Vol. 4*, Raven Press, New York, 115-132.

Collier, H.O.J., Denning-Kendall, P.A., McDonald-Gibson, N.S., Saeed, S.A., Brennecke, S.P. and Mitchell, M.D., 1982. *In* McCann, R. (ed) *Role of Chemical Mediators in the Pathophysiology of Acute Illness and Injury*. Raven Press, New York, 65-80.

Evans, C.A., Kennedy, T.G., Patrick, J.E. and Challis, J.R.G., 1981. *Endocrinology, 109*, 1533-1538.

Evans, C.A., Kennedy, T.G. and Challis, J.R.G., 1982. *Biol. Reprod., 27*, 1-11.

Fosang, A.H., Handley, C.J., Santer, V., Lowther, D.A., Leach Harper, C.M. and Thorburn, G.D., 1983. *Proc. 26th Ann. Con. Endo. Soc. Aust., Canberra, Aust.*, Abs 36.

Fosang, A.J., Handley, C.J., Santer, V., Lowther, D.A. and Thorburn, G.D., 1984. *Biol. Reprod., 30*, 1223-1235.

Garfield, R.E., Rabideau, S., Challis, J.R.G. and Daniel, E.E., 1979. *Biol. Reprod., 21*, 999-1007.

Jenkin, G. and Thorburn, G.D., 1984. *Can. J. Physiol. Pharmacol.*, (in press).

Jenkin, G., Jorgensen, G., Nathanielsz, P.W., Buster, J.E. and Thorburn, G.D., 1980. *Proc. Aust. Soc. Reprod. Biol., 12*, 7.

Leach Harper, C.M. and Thorburn, G.D., 1984. *Can. J. Physiol. Pharmacol.*, (in press).

Leach Harper, C.M., Wong, M.H. and Thorburn, G.D., 1983. *Proc. 26th Ann. Con. Endo. Soc. Aust. Canberra, Aust.*, 28.

Louis, T.M., Parry, D.M., Robinson, J.S., Thorburn, G.D. and Challis, J.R.G., 1977. *J. Endoc., 73*, 427-439.

Lye, S.J. and Challis, J.R.G., 1982. *J. Reprod. Fert., 66*, 311-315.

McCracken, J.A., 1980. *In* Samuelsson, B. and Paoletti, R. (ed) *Advances in Prostaglandin and Thromboxane Research, 8*, Raven Press, New York, 1329-1344.

McCracken, J.A., Glew, M.E. and Scaramuzzi, R.J., 1970. *J. Clin. Endocr. Metab., 30*, 544-546.

Mitchell, M.D., Anderson, A.B.M., Brunt, J.D., Clover, L., Ellwood, D.A., Robinson, J.S. and Turnbull, A.C., 1979. *J. Endoc., 83*, 141-148.

Mitchell, M.D., Ellwood, D.A., Anderson, A.B.M. and Turnbull, A.C., 1978. *Prostaglandins Med., 1*, 265-272.

Mitchell, M.D., Ellwood, D.A. and Brennecke, S.P., 1982. *Proc. 4th Ross Conference on Obstetrics Research, Dallas, TX, U.S.A.*, 34-41.

Mitchell, M.D., Robinson, J.S. and Thorburn, G.D., 1977. *Prostaglandins, 14*, 1005-1011.

Parkington, H.C., 1984. *J. Reprod. Fert.*, (in press).

Rice, G.E. and Thorburn, G.D., 1984. *Can J. Physiol. Pharmacol.*, (submitted).

Risbridger, G.P., Leach Harper, C.M. and Thorburn G.D., 1982. *Proc. Aust. Soc. Reprod. Biol., 14*, Abs. 30.

Risbridger, G.P., Leach Harper, C.M., Wong, M. and Thorburn, G.D., 1984. *Placenta.*, (in press).

Robinson, J.S., Chapman, R.L.K., Challis, J.R.G., Mitchell, M.D. and Thorburn, G.D., 1978. *J. Reprod. Fert., 34*, 307-373.

Saeed, S.A., Macdonald-Gibson, W.J., Cuthburt, J., Copas, J.L., Schneider, C., Gardiner, P.J., But, N.M. and Collier, H.O.J., 1977. *Nature (London), 270*, 32-36.

Saeed, S.A., Strickland, D.M., Young, D.A., Dang, A. and Mitchell, M.D., 1982. *J. Clin. Endocr. Metab., 55*, 801-803.

Sheldrick, E.L. and Flint, A.P.F., 1983a. *J. Reprod. Fert., 68*, 477-480.

Sheldrick, E.L. and Flint, A.P.F., 1983b. *J. Reprod. Fert., 68*, 155-160.
Soloff, M.S., 1979. *Life Sciences, 25*, 1453-1460.
Thorburn, G.D., 1977. *Proc. Int. Cong. Physiol. Sc. (Lyons, 1977), Ann. Rech. Vet., 8*, 428-437.
Thorburn, G.D., 1978. *Sem. Perinatol., 2*, 235-245.
Thorburn, .GD., 1983. *In, Initiation of Parturition: Prevention of Prematurity*, Fourth Ross Conference on Obstetric Research, Ross Conference, Ross Laboratories, Dallas, 2-11.
Thorburn, G.D., 1984. *In, The Physiological Development of the Foetus and Newborn*, Oxford, (in press).
Thorburn, G.D. and Challis, J.R.G., 1979. *Physiol. Rev., 59*, 863-918.
Thorburn, G.D., Nicol, D.H., Bassett, J.M., Shutt, D.A. and Cox, R.I., 1972. *J. Reprod. Fertil., Suppl. 16*, 61-84.
Thorburn, G.D., Cox, R.I., Currie, W.B., Restall, B.J. and Schneider, W., 1973. *J. Reprod. Fert., Suppl. 18*, 151-158.
Thorburn, G.D., Young, I.R., Dolling, M., Walker, D.W., Browne, C.A. and Carmichael, G., 1984a. *In, Growth and Maturation Factors, Vol. 2*, J. Wiley & Sons, (in press).
Thorburn, G.D., Harding, R., Jenkin, G., Parkington, H. and Sigger, J.N., 1984b. *J. Develop. Physiol., 6*, 31-43.
Wathes, D.C. and Swann, R., 1982. *Nature, Lond., 297*, 225-227.

# PHYSIOLOGICAL COST OF PREGNANCY AND LACTATION IN THE EWE

E.F. Annison, J.M. Gooden, G.M. Hough and G.H. McDowell, *Department of Animal Husbandry, University of Sydney, Camden, N.S.W. 2570.*
A.J. Williams, *Agricultural Research Centre, Orange, N.S.W. 2800.*

## INTRODUCTION

In pregnancy and lactation, the development of the foetus and the provision of milk to the lamb requires energy-yielding nutrients, essential amino acids, minerals and micronutrients which are derived from the maternal circulation. The ewe may respond to these increased nutrient demands by eating more, by the mobilization of maternal tissue, by improving the effectiveness of overall digestion or by increasing the efficiency of utilization of nutrients by tissues. These alterations in metabolism occur in a continously changing hormonal milieu dominated in pregnancy by the steroid hormones, and in lactation by insulin, glucagon and growth hormone.

Quantitative data on the utilization of nutrients in pregnancy and lactation, and the regulation of the partitioning of nutrients between maternal tissues and foetal or mammary tissues are reviewed in relation to the effects of pregnancy and lactation on wool growth.

## SUBSTRATE UTILIZATION IN PREGNANCY AND LACTATION

*General considerations* The energy cost of late pregnancy in the ewe is largely met by increased feed intake, except in very late pregnancy when intake may decline. This raised nutrient intake probably accounts for the increase in the weight of the liver and gastro-intestinal tract in late pregnancy, which may be as high as 30% in the ewe (Robinson *et al.* 1978).

In lactation, the requirements of the mammary gland for milk synthesis impose quantitative changes on almost all aspects of metabolism. These include a rise in cardiac output, which is reflected in increased blood flows in the gut, mammary gland and liver (see Mepham 1983). The onset of lactation leads to hypertrophy of the alimentary tract, but in high-yielding ruminants in early lactation food intake fails to keep pace with requirement and there may be substantial mobilization of body tissue. In mid-lactation, when milk yield falls, voluntary feed intake fails to respond immediately to the reduced demands of lactation, and if feed intake is not restricted, animals may become fat.

Pregnancy and lactation both significantly affect digesta flow in the alimentary tract. The increased rate of digesta clearance and of flow of liquor from the rumen observed in late pregnancy continues into early lactation (Weston 1979). These changes in digesta flow act to reduce organic matter digestibility in the alimentary tract, but in some circumstances the resulting loss of energy may be offset by the improved supply of dietary protein to the small intestine (Faichney and White 1980).

*Procedures for the in vivo study of foetal and mammary metabolism* Arterio-venous (AV) difference procedures require the catheterization of the appropriate blood vessels in chronic preparations, a relatively simple procedure for the mammary gland (Linzell 1974), but a much more exacting task when ewes are prepared for the study of foetal metabolism (Meschia *et al.* 1969-70; Young *et al.* 1974). Accurate estimates of substrate uptake require that blood flow, rate of uptake and arterial concentration of substrate all remain constant, and that venous blood samples are representative of total venous drainage. The utility of the AV difference procedure is greatly enhanced when combined with isotope dilution in the study of mammary gland (Linzell and Annison 1965) and foetal (Hodgson *et al.* 1981; Hay *et al.* 1983) metabolism.

## GLUCOSE

Pregnancy and lactation both increase glucose requirements, which are met mainly by raised feed intake but also by increased use of propionate for glucose production. Wilson *et al.* (1983) showed that the percentage of propionate production used for gluconeogenesis rose from 37% in non-pregnant, non-lactating ewes to 55% in late pregnancy and 60% in lactation. Other possible metabolic adaptations to increase glucose supply or conserve glucose in pregnancy and lactation include increased operation of the Cori and alanine cycles, which involve the recycling of glucose-C as lactate, pyruvate and alanine. Wilson *et al.* (1981) demonstrated considerable recycling of glucose in ewes in late pregnancy, but Baird *et al.* (1983) reported much lower levels of glucose recycling, and concluded that, in general, glucose-lactate interconversions are lower in ruminants than in non-ruminants.

*Utilization of glucose in pregnancy* Glucose taken up from the maternal circulation is used by the foetus both as an energy source, and for the synthesis of new tissues (Bassett and Jones 1975). The uterus and placenta also metabolize glucose, and the partition of maternal glucose production between these tissues and the foetus and maternal tissues has been examined by Hay *et al.* (1983) using AV difference and isotope dilution procedures in fed and starved late pregnant, conscious ewes. Uptakes of glucose by the uterus, foetus, uteroplacenta and non-uterine tissues of fed ewes with a mean glucose production rate of 178.7 mg/min were 56.5, 15.7, 40.8 and 122.2 mg/min respectively. Over a wide range of blood concentration of glucose, the fractional distribution of maternally produced glucose was relatively constant. In recent studies with Merino ewes, uterine glucose uptake was 10.5 mg/min/kg, and was independent of both diet and stage of pregnancy. Total uptake of glucose by the uterus increased from 17 to 35 mg/min between days 94 and 125 of pregnancy (Oddy *et al.* 1984).

The use of isotope dilution procedures has confirmed that considerable gluconeogenesis occurs in the ruminant foetus (Hodgson *et al.* 1981; Prior 1982). Glucose accounts for about 46% of oxygen uptake by the sheep foetus, with amino acids (25%) and lactate (20%) supplying the balance of energy yielding substrates (see Faulkner 1983).

*Utilization of glucose in lactation* Glucose is taken up by the lactating mammary gland and utilized for the synthesis of lactose, citrate and glycerol, and for the generation of part of the reducing equivalents used in *de novo* fatty acid synthesis (see Annison 1983). The mammary AV difference for glucose is linearly related to arterial level over a relatively wide range of concentrations (Linzell 1967), suggesting that at low arterial concentrations, glucose uptake will be determined by mammary blood flow.

The lactose content of milk is largely unaffected by plane of nutrition or stage of lactation, although small variations are matched by inverse changes in both sodium and potassium content (Peaker 1983). There is evidence that lactose, after it is secreted, draws water from the secretory cell osmotically and that milk secretion cannot proceed in the absence of lactose secretion (Peaker 1983). In studies with the isolated, perfused mammary gland of the lactating goat, Hardwick *et al.* (1961) showed that milk secretion required the continuous supply of glucose. In the absence of glucose, milk secretion ceased, but was restored when glucose was added to the perfusion medium. The dependence of mammary secretion on glucose supply is almost certainly due to the role of glucose as a precursor of lactose.

As mentioned above, glucose supply to ruminant tissues is correlated with dietary energy, and in lactating dairy cows glucose entry rate is well correlated with ME intake, milk yield and lactose output (Horsfield *et al.* 1974).

Quantitative data on the uptake and oxidation of glucose by the mammary gland and whole animal metabolism are shown in Table 1.

**TABLE 1. Glucose and acetate metabolism in lactating ewes (number shown in brackets).**

| Whole animal | | | | Mammary Gland | | |
|---|---|---|---|---|---|---|
| Milk Yield (kg/day) | Irreversible loss (mg/min.kg$^{0.75}$) | Glucose Oxidized (%) | $CO_2$ from Glucose (%) | Glucose Uptake (mg/min) | Glucose Oxidized (%) | $CO_2$ from Glucose (%) |
| | | | Glucose | | | |
| 2.00[1][3](5) | 9.17 | 23 | 8 | — | — | — |
| 0.67[2] (5) | 4.83 | 66 | 8 | 19 | 16 | 21 |
| | | | Acetate | | | |
| 1.62[4] (6) | 12.4 | — | 22 | 33 | 22 | 20 |

[1] Bergman and Hogue (1967)
[2] Oddy *et al.* (1984)
[3] Calculated
[4] King *et al.* (1983)

## AMINO ACIDS

The biosynthesis of tissues, wool and milk requires different profiles of essential amino acids, and when energy supply is adequate, specific amino acids may be first limiting for production. The high cysteine content of wool, for example, imposes a considerable drain on available cysteine and methionine (converted to cysteine in tissues) and in many dietary situations, wool growth is stimulated by post-ruminal supplementation with sulphur containing amino acids (Reis 1979).

*Utilization of amino acids in pregnancy* Amino acids are transferred from maternal blood across the placenta into foetal blood by active transport against a concentration gradient. Glutamate is the only amino acid transferred from the foetus to the maternal circulation, and there is little net transport of other amino acids (Battaglia and Meschia 1978). This movement of glutamate may supplement the transfer of excretory nitrogen across the placenta as urea and ammonia. In foetal sheep the rate of excretion of urea per unit weight of tissue is four-fold greater than in adult sheep, and Battaglia and Meschia (1978) calculated that in late gestation, 20% of foetal oxygen consumption could be required to sustain observed rates of urea production.

The total quantity of amino acids transferred is in excess of that required for growth, and amino acid oxidation accounts for about 20% of oxygen consumption by the foetal lamb (Girard *et al.* 1979).

*Utilization of amino acids in lactation* The large requirement for essential amino acids to sustain milk protein synthesis in the lactating ruminant poses the question of the possible existence of metabolic adaptations to lactation which spare the utilization of amino acids in non-mammary tissues. At this time, however, only changes which result in the mobilization of tissue protein reserves to support milk protein synthesis have been identified (Bryant and Smith 1982).

Quantitative data on amino acid uptake by the lactating ruminant mammary gland were first obtained by Mepham and Linzell (1966) in AV difference studies with lactating goats. Their data showed that the essential amino acids in milk protein are derived from plasma, but that non-essential amino acid uptake is highly variable, and in many cases is too low to supply all of the corresponding amino acids in milk protein. Subsequent studies showed that there is considerable synthesis of non-essential amino acids in ruminant mammary tissue. The equivalence of carbon and nitrogen taken up by the mammary gland to output in milk is consistent with the view that essential amino acids taken up in excess of their requirements for milk protein synthesis provide most of the carbon and $\alpha$-amino nitrogen used for amino acid synthesis in mammary tissue (Mepham *et al.* 1982).

## ACETATE

Blood acetate is derived mainly from microbial fermentation in the gut, but a substantial fraction is of endogenous origin (see Annison 1983).

*Acetate utilization in pregnancy* Acetate crosses the placenta in ruminants, but makes a relatively small contribution to the metabolic requirements of the pregnant uterus. Char and Creasy (1976) showed that acetate accounted for about 8% of the carbon transferred across the placenta in late pregnancy in ewes, and Girard *et al.* (1979) reported that uptake of acetate accounted for 5% of the oxygen uptake of the foetal lamb.

Acetate concentrations in the blood of sheep are relatively constant during pregnancy (Vernon *et al.* 1981), although Remésy and Demigné (1976) showed a substantial increase in the last 15 days of pregnancy, particularly in twin-bearing animals and those on low planes of nutrition. Recent work in this laboratory supports these findings as shown in Table 2.

**TABLE 2. Mean blood levels of acetate (mM) in pregnant ewes (n = 4), bearing single or twin lambs and fed balanced rations at 0.8 (low) and 1.3 (high) times calculated energy requirement.**

| | | Days from mating | | | |
|---|---|---|---|---|---|
| | Diet | 76 | 97 | 118 | 139 |
| Single | Low | 1.8 | 1.5 | 2.2 | 4.2 |
| | High | 1.8 | 2.2 | 2.2 | 3.2 |
| Twins | Low | 1.2 | 1.9 | 2.2 | 3.6 |
| | High | 1.5 | 1.7 | 1.9 | 2.1 |

In agreement with Remésy and Demigné (1976) the ewes on a low plane of nutrition had the highest acetate levels at day 139 of pregnancy. Raised acetate levels presumably reflect either increased endogenous production, or reduced tissue utilization. The latter suggestion is consistent with reduced lipogenesis from acetate in subcutaneous adipose tissue in late pregnancy (Vernon *et al.* 1981).

*Acetate utilization during lactation* Acetate is extracted by the mammary gland and utilized for milk fat synthesis (Annison 1983). Data on acetate metabolism in lactating ewes are shown in Table 1. Endogenous acetate is produced by the mammary gland, and in ewes the rate of acetate release is related to milk yield (King *et al.* 1983). An important metabolic adaptation to lactation was suggested by Pethick and Lindsay (1982), who reported that the uptake of acetate by hind-limb skeletal muscle in lactating ewes was substantially less than in nonlactating ewes, when measured at similar arterial concentrations.

## FREE FATTY ACIDS

Mobilization or deposition of lipid in most animals largely reflects energy balance (Robinson *et al.* 1971).

*Utilization of FFA during pregnancy* In early pregnancy sheep are usually in positive energy balance, and adipose tissue accumulates. During the last third of pregnancy foetal energy demands increase dramatically and mobilization of lipid reserves gives rise to raised levels of plasma free fatty acids (FFA) (see Baldwin *et al.* 1976; Bell 1981). The extent of fat mobilization depends on plane of nutrition and number of foetuses (Remésy and Demigné 1976; Robinson *et al.* 1978). For example, ewes bearing triplets lost 14% of bodyweight during pregnancy even though energy intake was increased from 6 to 15 MJ ME/day (Robinson *et al.* 1978).

The ketone bodies 3-hydroxybutyrate and acetoacetate are formed in the liver from FFA, and sheep are particularly vulnerable to ketosis in late pregnancy. An inadequate supply of glucose, rapid mobilization of FFA, changes in liver function resulting in a greater proportion of FFA being converted to ketones, and loss of appetite are contributory factors (Katz and Bergman 1969; Faulkner 1983).

The transfer of plasma FFA across the placenta of sheep is small (Girard *et al.* 1979), and fatty acids are synthesized in the foetus from both glucose and acetate (see Faulkner 1983).

*Utilization of FFA during lactation* Plasma FFA remain elevated in early lactation largely because food intake fails to keep pace with nutrient requirements for milk synthesis, resulting in adipose tissue mobilization.

Isotope dilution studies have revealed that plasma FFA equilibrates with FFA liberated during the mammary uptake of blood triacylglycerol, and that there is simultaneous uptake and release of FFA (see Annison 1983). In fed lactating ruminants, when plasma FFA levels are usually low, net uptakes of FFA by the mammary gland are negligible. Recent studies (K.R. King, J.M. Gooden and E.F. Annison, unpublished) based on the continuous infusion of [U-$^{14}$C] palmitate in fed, lactating ewes have shown that FFA may contribute significantly to mammary metabolism (Table 3).

**TABLE 3. Plasma FFA metabolism in lactating sheep.**

| | | | Whole body | | | Udder | |
|---|---|---|---|---|---|---|---|
| Sheep No. | Feed intake (MJ ME/d) | Milk yield (kg/d) | Plasma conc. (μM) | | Entry rate (μmol/min) | Net uptake (μmol/min) | FFA oxidized (% of uptake) |
| | | | A | A-V | | | |
| 1 | 24.8 | 2.4 | 129 | 32 | 377 | 18.8 | 0.1 |
| 2 | 15.0 | 1.8 | 326 | 140 | 630 | 57.6 | 2.8 |
| 3 | 23.2 | 2.1 | 63 | –4 | 115 | –4.0 | 1.2 |

The proportion of lipoprotein triacylglycerol dervied from FFA in sheep 1, 2 and 3 was 7.6, 34.1 and 5.1% respectively: corresponding values for the total recovery of radioactivity in milk fat 13 h after infusion were 14.2, 41.0 and 19.0% respectively.

The data suggest that the extent of plasma FFA utilization for milk fat synthesis, either directly, or indirectly via lipoprotein, is related to concentration.

## THE EFFECTS OF PREGNANCY AND LACTATION ON WOOL GROWTH

Pregnancy and lactation depress wool growth by 3-27% (see Corbett 1979). Ewes bearing single lambs produce about 10-20% less wool than non-pregnant (NP) ewes, and wool production is reduced by a further 5-10% in ewes with twin lambs. Reductions in fibre diameter and length have also been reported (Corbett 1979). The extent and timing of these effects on wool production and fibre characteristics vary considerably with nutritional status, suggesting that the nutrient demands of the pregnant uterus and lactating udder significantly depress wool growth by reducing the supply of energy-yielding substrates and essential amino acids to the skin. A second possible explanation is that the hormonal environment associated with pregnancy and lactation depresses wool growth by reducing the responsiveness of the wool follicles (Oddy and Annison 1979).

In a study conducted by Mr. G. File (personal communication) the effects of level of feeding during the last 9 weeks of pregnancy on midside patch wool production from single and twin-bearing Merino ewes was examined. A pelleted lucerne chaff ration — 7.6 MJ/kg dry matter (DM), 173 g crude protein (CP)/kg DM — was fed at one of three levels during pregnancy; *ad libitum* (H), at a level adjusted to meet requirements based on MAFF (1975) recommendations (M) and at a level to maintain NP ewe liveweight (L). Single-bearing ewes produced considerably more clean wool than twin-bearing ewes in late pregnancy (Table 4). Feed intakes, however, were similar despite twin-bearing ewes being offered more feed in the H and M groups than single-bearing ewes. There was also a positive relationship between feed intake and wool production.

**TABLE 4. Mean feed intake and wool production over last 9 weeks of pregnancy in single and twin-bearing Merino ewes (G. File, unpublished).**

| | Single | | | Twin | | |
|---|---|---|---|---|---|---|
| | H | M | L | H | M | L |
| Intake (g DM/d) | 1840 | 1198 | 848 | 1760 | 1263 | 814 |
| Wool production (μg/cm²/d) | 1126 | 594 | 568 | 695 | 516 | 370 |

These data show that wool growth is depressed when the nutrient requirements for pregnancy and lactation exceed the ability of the ewe to increase feed intake sufficiently to meet these requirements.

Evidence in support of these conclusions was obtained by Williams (unpublished) who fed a ration of sorghum and lucerne (75% digestibility, 2.5% N) to NP, non-lactating (NL) Merino ewes, and to ewes bearing single or twin lambs. The NP, NL ewes were fed to maintenance, whereas the pregnant ewes were fed a ration which was adjusted weekly to maintain constant maternal liveweight throughout pregnancy. After parturition all lactating ewes were offered the diet *ad libitum* for 10 weeks and after weaning received a maintenance ration for 3 months. Overall there were no significant differences between NP, NL animals and those which had reared lambs, but significant interactions between physiological state and wool growth occurred at about day 90 of pregnancy, and at weaning (90 days post-partum). Average fibre diameter tended to be greater in ewes which had carried and reared lambs.

This evidence suggests that wool growth is depressed when the energy requirements for pregnancy and lactation cannot be met by increased feed consumption, and that nutrients are utilized by the pregnant uterus and lactating mammary gland at the expense of the wool follicle. Ceilings to the voluntary intake of twin-bearing ewes, especially during late pregnancy, pose a particular problem in trying to meet these nutrient requirements. If energy requirements are met the availability of rumen undegradable protein may limit wool growth during pregnancy and lactation (see Kempton 1979) but there is evidence that the sulphur containing amino acids (SSA) are not first limiting amino acids in these situations.

Williams *et al.* (1978) observed that supplements of methionine and cysteine administered twic daily into the abomasum during the last month of pregnancy and the first six weeks of lactation were ineffective in increasing wool growth in Merino ewes. Throughout this period feed intakes were adjusted to maintain constancy of maternal liveweight. The sulphur content of the wool was greater, however, in the supplemented ewes. These results, compared with those obtained with non-mated ewes, indicated that availabilities of SAA were not limiting wool growth during late pregnancy and early lactation.

In subsequent experiements (A.J. Williams, R.B. Murison and J.G. Padgett, unpublished), the availability and utilization of methionine and cysteine were measured in pregnant ewes on three occasions during the latter half of pregnancy. Ewes received either 0, 0.5 or 1.0 g/day of organic sulphur (two thirds as methionine) infused into the abomasum. As in the previous experiment, maternal liveweight was maintained constant. Supplementary SAA increased rates of irreversible losses of both methionine and cysteine and their concentrations in plasma. The results showed that neither the availability, or the overall utilization of SAA altered as pregnancy advanced.

The second experiment confirmed the insensitivity of the rate of wool production of pregnant ewes to an increased supply of SAA at the intestines. The biokinetic data did not provide any evidence to account for this lack of response. Furthermore, under the conditions of the two experiments the rate of wool growth did not decrease as pregnancy advanced.

In the studies above wool production was limited by either energy or essential amino acid supply. In an attempt to prevent the depression of wool growth normally observed in pregnancy and lactation we have fed a ration rich in energy and protein (9.59 MJ/kg DM; 210 g CP/kg DM). The main protein source in the ration was fishmeal, which is relatively poorly degraded in the rumen. The ration was fed at 0.8, 1.0 and 1.3 times calculated requirement to 2 genotypes of single and twin-lamb bearing Merino ewes during pregnancy and lactation. The genotypes differed by about 40% in their capacity to produce wool. Preliminary results suggest that with both genotypes the depression of wool growth normally associated with reproduction was absent.

## HORMONAL CONTROL OF NUTRIENT UTILIZATION DURING PREGNANCY AND LACTATION

The hormonal control of nutrient utilization in pregnant and lactating ruminants has been comprehensively reviewed (Bauman and Currie 1980; Hart 1983; McDowell 1983). Bauman and Currie (1980) postulated that hormonal control of metabolism in pregnant and lactating ruminants involves two distinct mechanisms. They considered that homeostatic hormones maintain a constant internal environment whereas homeorhetic hormones co-ordinate metabolism to support specific physiological states. Although the concept of homeorhetic control is an interesting one there are no clear demonstrations that hormones exert such control. Possible homerohetic hormones include placental lactogen, progesterone, prolactin and possible growth hormone.

*Insulin* Insulin inhibits gluconeogenesis and glucose release from the liver, increases incorporation of amino acids into muscle protein and promotes lipogenesis in adipose tissue. Plasma concentrations of insulin decrease as pregnancy advances in the ewe and remain low during early lactation. These changes correlate with increased requirements for energy-yielding nutrients to support foetal growth and subsequently milk synthesis.

*Pancreatic glucagon* Glucagon acts to increase glucose production, principally by promoting hepatic gluconeogenesis and glycogenolysis and it may facilitate lipolysis in adipose tissue. Although information on changes in plasma concentrations throughout pregnancy and lactation is scant it appears that concentrations increase as pregnancy advances. Indeed, plasma concentrations in pregnant ewes increase in response to exercise — a situation where requirements for energy-yielding nutrients increase (Bell *et al.* 1983).

*Insulin:glucagon ratio* The molar ratio of insulin:glucagon appears to be more important for control of nutrient supply than absolute concentrations of either hormone. Concentrations of glucagon remain relatively stable but marked fluctuations occur in insulin concentrations — resulting in shifts in the molar ratio of the hormones. Studies by Gow *et al.* (1981) showed that the ratio of insulin:glucagon for lactating ewes decreased in response to food deprivation, and studies by Bell *et al.* (1983) showed similar shifts in the ratio following exercise of pregnant ewes.

*Growth hormone* Key biological activities of growth hormone are promotion of muscle protein synthesis, and lipolysis and its diabetogenic effects. Characteristically, plasma concentrations are very variable due to episodic release, but, in general, concentrations increase during periods when energy-yielding nutrients are deficient. Plasma concentrations increase throughout pregnancy and are high in early lactation.

There has been renewed interest in the use of exogenous growth hormone for improvement of animal production. Daily injection of growth hormone increases milk yield in sheep (McDowell and Hart 1983) and cows (Bines *et al.* 1980; Peel *et al.* 1981; McDowell *et al.* 1983). Wool production commonly decreases during chronic administration of growth hormone but a prolonged period of accelerated wool growth follows withdrawal of hormone (see Wynn 1982).

These effects are due to marked effects of growth hormone on partition of nutrients. The specific actions of growth hormone appear to differ depending on physiological state. In early lactation in cows the hormone appears to exert diabetogenic effects resulting in increased glucose production whereas in later lactation lipolytic effects predominate (McDowell *et al.* 1983). Effects of the hormone on wool growth appear to be due to promotion of muscle protein synthesis during administration followed by release of amino acids (and possibly other nutrients) after withdrawal (Wynn 1982).

*Insulin:growth hormone ratio* The insulin:growth hormone ratio appears to be important for controlling nutrient availability. Where there is a high demand for energy-yielding nutrients the insulin:growth hormone ratio is low, as, for example, during early lactation in cows (Bines *et al.* 1980) and during feed restriction in lactating ewes (Gow *et al.* 1981). On the other hand, Bassett (1983) reported data showing a marked increase in the ratio in ewes immediately after abrupt weaning of lambs when nutrient availability might be expected to exceed requirements.

*Placental lactogen* Concentrations of placental lactogen increase dramatically during pregnancy in the ewe and there is circumstantial evidence (see McDowell 1983) that placental lactogen regulates maternal metabolism to provide nutrients for the developing foetus. Thordarson *et al.* (1983) infused partially-purified placental lactogen intravenously into non-pregnant ewes. Infusion of placental lactogen led to increased plasma concentrations and whole body irreversible loss of FFA and plasma concentrations of glucose. These preliminary results are consistent with placental lactogen exerting lipolytic and diabetogenic activities.

## FEEDING STANDARDS FOR BREEDING EWES

MAFF (1975) feeding standards are of limited applicability to Australian conditions in view of the extensive reliance on pasture. The standards have been used successfully for experimental animals and could form the basis for recommendations to flock managers faced with the prospect of total hand feeding of a breeding ewe flock (e.g. in a drought). Major problems for a flock manager are to assess the extent to which the requirements of his flock are being met presently by available pasture, and uncertainty as to whether seasonal conditions will allow requirements to be met later in the season.

## PRACTICAL APPLICATION OF RESULTS

The results from physiological studies have complemented those from feeding studies to demonstrate the increased demands for various nutrients (particularly those supplying energy) by pregnant and lactating ewes. In practice, the problem becomes one of attempting to meet these increased demands. Initially, this would involve the choice of a mating period so that the seasonal period of greatest availability of highly digestible pasture would coincide with the period of greatest nutrient demand by the ewe flock. Given an eight week period of maximum demand by a breeding ewe (final weeks of pregnancy and early lactation) and a five week joining period (two oestrous cycles) for a flock, the seasonality of pasture growth in much of Australia is such that flock managers do not have 13 weeks of plentiful, high quality pasture available for the flock of breeding ewes. Their problem is then to allocate a scarce resource so that it is most efficiently utilized. Techniques are now available to assist flock managers to achieve this objective. The use of ram harnesses with crayons enables the non-pregnant ewes to be separated from the flock and the flock to be divided into early and late lambing groups. A more recent development is the use of real time ultrasound imaging (Fowler and Wilkins 1982) for the early diagnosis of ewes with single or multiple foetuses. This makes it possible to separate and treat preferentially those ewes with the greatest requirements.

These procedures should be adopted with the aim of reducing ewe deaths, and increasing both the number and the weight of lambs weaned. The successful accomplishment of these objectives should also ensure that the depression in wool production, associated with the bearing and rearing of lambs, is minimized. With the imminent introduction of measured staple strength into the system for selling wool on the basis of objective measurement, the need to ensure the adequacy of nutritional supply to the breeding ewe flock becomes even greater, although the manager may be able to manipulate time of shearing so that the point of weakness in the staple is at the end of the staple, rather than in its middle. However, the decision to shear pre-lambing to increase staple strength should not be taken lightly. The extra requirements for energy of shorn sheep would be additional to those due to physiological state, and hence shearing could tip the balance of energy from a mild deficit into a major nutritional deficit, with subsequent severe loss of production.

## SUGGESTED GUIDELINES FOR FUTURE RESEARCH

*Agronomic research to increase the supply of pasture* The emphasis in agronomic research should be altered from total dry matter per unit area to evaluation of the effects of species, fertilizers etc. on an increased supply of highly digestible organic matter for a more extended period of time, e.g. the 13 weeks quoted earlier. This changed emphasis for evaluation of pastures could also include an assessment of the degradability of the proteins in the herbage.

*Management* (i) Although we recognize that non-pregnant and ewes pregnant with one or two foetuses have different requirements, the managerial aspects of handling the ewe flock as three separate entities have not been evaluated in the grazing environment. If a manager identifies the twin-bearing ewes and stocks these at a proportionately lower intensity, is production greater? Do behavioural problems arise so that 'mismothering' is greater in a flock of twin-bearing ewes?

(ii) Given the variability in effective rainfall in Australia and consequent variability in expectation of growth of pasture, is there any role in Australia for the pasture-allowance system which is apparently working so well in New Zealand. As this system involves rotational grazing, a practice disavowed by most agricultural scientists in Australia, is prejudice preventing an adequate evaluation in Australia?

(iii) Since ovulation rate is very largely determined by the liveweight of the ewes at joining, the total nutritional management of the breeding flock requires critical re-evaluation. This re-evaluation would include considerations of early weaning and the appropriate allocation of pasture between the weaned lambs and the ewes. This total concept would be difficult and expensive experimentally, but is within the scope of simulation modelling.

(iv) Although the provision of adequate quantities of high quality diet will largely prevent the depressed wool production frequently observed in pregnant/lactating animals, the gross efficiency of conversion of feed into wool is very low under these conditions. There still remains the biological question of the proportion of ME available for wool production in pregnant or lactating ewes and whether the response in wool production to this available energy varies with physiological state. Future studies should consider the conclusion of Graham and Searle (1982) that the partial efficiency of wool growth is 15-20%.

(v) There is a lack of data on responses to supplements by grazing sheep, i.e. substitution effects.

## REFERENCES

Annison, E.F., 1983. *In* Mepham, T.B. (ed), *Biochemistry of Lactation*, Elsevier, Amsterdam-New York, 399-436.
Baird, G.D., Van Der Walt, J.G. and Bergman, E.N., 1983. *Br. J. Nutr., 50*, 249-265.
Baldwin, R.L., Yang, Y.T., Crist, K. and Gritchting, G., 1976. *Fedn Proc., 35*, 2314-2318.
Bassett, J.M., 1983. *Proc. Nutr. Soc., 42*, 34A.
Bassett, J.M. and Jones, C.T., 1975. *In* Beard, R.W., Nathanielsz, P.W. (eds), *Foetal Physiology and Medicine*, Saunders, London, 158-172.
Battaglia, F.C. and Meschia, G., 1978. *Physiol. Rev., 58*, 499-527.
Bauman, D.E. and Currie, W.B., 1980. *J. Dairy Sci., 63*, 1514-1529.
Bell, A.W., 1981. *In* Christie, W.W. (ed), *Lipid Metabolism in Ruminant Animals*, Pergamon Press, London, 363-410.
Bell, A.W., Bassett, J.M., Chandler, K.D. and Boston, R.C., 1983. *J. Dev. Physiol.,* 5, 129-141.
Bergman, E.N. and Hogue, D.E., 1967. *Am. J. Physiol., 213*, 1378-1384.
Bines, J.A., Hart, I.C. and Morant, S.V., 1980. *Br. J. Nutr., 43*, 179-188.
Bryant, D.T.W. and Smith, R.W., 1982. *J. Agric. Sci., Camb., 99*, 319-323.
Char, V.C. and Creasy, R.K., 1976. *Am. J. Physiol., 230*, 357-361.
Corbett, J.L., 1979. *In* Black, J.L., Reis, P.J. (eds), *Physiological and Environmental Limitations to Wool Growth*, University of New England Publishing Unit, Armidale, 79-98.
Faichney, G.J. and White, G.A., 1980. *Proc. Aust. Soc. Amin. Prod., 13*, 455.
Faulkner, A., 1983. *In* Rook, J.A.F., Thomas, P.C. (eds), *Nutritional Physiology of Farm Animals*, Longmans, London, 203-242.
Fowler, D.G. and Wilkins, J.F., 1982. *Proc. Aust. Soc. Anim. Prod, 14*, 636.
Girard, J., Pintado, E. and Ferre, P., 1979. *Annls Biol. Anim. Biochim. Biophys., 19*, 181-197.
Gow, C.B., McDowell, G.H. and Annison, E.F., 1981. *Aust. J. Biol. Sci., 34*, 469-478.

Graham, N. McC. and Searle, T.W., 1982. *Aust. J. Agric. Res., 33*, 607-615.
Hardwick, D.C., Linzell, J.L. and Price, S.M., 1961. *Biochem, J., 80*, 37-45.
Hart, I.C., 1983. *Proc. Nutr. Soc., 42*, 181-194.
Hay, W.W., Sparks, J.W., Wilkening, R.B., Battaglia, F.C. and Meschia, G., 1983. *Am. J. Physiol., 245* (Endocr. Metab. 8), E347-E350.
Hodgson, J.C., Mellor, D.J. and Field, A.C., 1981. *Biochem. J., 196*, 179-186.
Horsfield, S., Infield, J.M. and Annison, E.F., 1974. *Proc. Nutr. Soc., 33*, 9-15.
Katz, M.L. and Bergman, E.N., 1969. *Am. J. Physiol., 216*, 953-960.
Kempton, T.J., 1979. *In* Black, J.L., Reis, P.J. (eds), *Physiological and Environmental Limitations to Wool Growth*, University of New England Publishing Unit, Armidale, 209-222.
King K.R., Gooden, J.M. and Annison, E.F., 1983. *Proc. Nutr. Soc. Aust.*, 8, 166.
Linzell, J.L., 1967. *J. Physiol., Lond., 190*, 347-357.
Linzell, J.L., 1974. *In* Larson, B.L., Smith, V.R. (eds), *Lactation*, Vol.1, Academic Press, New York, 143-225.
Linzell, J.L. and Annison, E.F., 1965. *In* McDonald, I.W., Warner, A.C.I. (eds), *Digestion and Metabolism in the Ruminant*, University of New England Publishing Unit, Armidale, 306-319.
McDowell, G.H., 1983. *Proc. Nutr. Soc., 42*, 149-167.
McDowell, G.H. and Hart, I.C., 1983. *Proc. Aust. Soc. Reprod. Biol., 15*, 96.
McDowell, G.H., Hart, I.C., Bines, J.A. and Lindsay, D.B., 1983. *Proc. Nutr. Soc. Aust., 8*, 165.
MAFF, 1975. *Energy Allowances and Feeding Systems for Ruminants*, H.M.S.O., London, Technical Bulletin 33.
Mepham, T.B., 1983. *In* Mepham, T.B. (ed),, *Biochemistry of Lactation*, Elsevier, Amsterdam-New York, 3-28.
Mepham, T.B., Gaye, P. and Mercier, J.C., 1982. *In* Fox, P.F. (ed), *Developments in Dairy Chemistry, Vol. 1, Proteins*, Applied Science Publishers, London, 115-156.
Mepham, T.B. and Linzell, J.L., 1966. *Biochem. J., 101*, 76-83.
Meschia, G., Makowski, E.L. and Battaglia, F.C., 1967-70. *Yale J. Biol. Med., 42*, 154-165.
Oddy, V.H. and Annison, E. F., 1979. *In* Black, J.L., Reis, P.J. (eds), *Physiological and Environmental Limitations to Wool Growth*, University of New England Publishing Unit, Armidale, 295-309.
Oddy, V.H., Gooden, J.M., Hough, G.M., Teleni, E. and Annison, E.F., 1984. *Aust. J. Biol. Sci.*, (in press).
Peaker, M., 1983. *In* Mepham, T.B. (ed), *Biochemistry of Lactation*, Elsevier, Amsterdam-New York, 285-308.
Peel, C.J., Bauman, D.E., Gorwitt, R.C. and Sniffen, C.J., 1981. *J. Nutr., 11*, 1662-1671.
Pethick, D.W. and Lindsay, D.B., 1982. *Br. J. Nutr., 48*, 319-328.
Prior, R.L., 1982. *Fedn Proc., 41*, 117-122.
Reis, P.J., 1979. *In* Black, J.L., Reis, P.J. (eds), *Physiological and Environmental Limitations to Wool Growth*, University of New England Publishing Unit, Armidale, 223-242.
Remésy, C. and Demigné, C., 1976. *Ann. Rech. Vétér.*, 7, 329-341.
Robinson, J.J., Frazer, C. and Bennett, C.J., 1971. *J. Agric. Sci., Camb., 77*, 141-145.
Robinson, J.J., McDonald, I., McHattie, I. and Pennie, K., 1978. *J. Agric. Sci., Camb., 91*, 291-304.
Thordarson, G., McDowell, G.H., Forsyth, I.A. and Smith, S.V., 1983. *Proc. Nutr. Soc. Aust., 8*, 127.
Vernon, R.G., Clegg, R.A. and Flint, D.J., 1981. *Biochem. J., 200*, 307-314.
Weston, R.H., 1979. *Ann. Rech. Vet., 10*, 442-444.
Williams, A.J., Tyrrell, R.N., and Gilmour, A.R., 1978. *Aust. J. Exp. Agric. Anim. Husb., 18*, 52-57.
Wilson, S., MacRae, J.C. and Buttery, P.J., 1981. *Res. Vet. Sci., 30*, 205-212.
Wilson, S., MacRae, J.C. and Buttery, P.J., 1983. *Br. J. Nutr., 50*, 303-316.
Wynn, P.C., 1982. Ph.D. Thesis, University of Sydney.
Young, W.P., Creasy, R.K. and Rudolph, A.M., 1974. *J. Appl. Physiol., 37*, 620-621.

# ULTRASOUND IMAGING FOR LITTER SIZE DIAGNOSIS IN BREEDING FLOCKS

J.F. Wilkins, *Department of Agriculture, Agricultural Research Centre, R.M.B. 944, Tamworth, N.S.W. 2340.*
D.G. Fowler, *Department of Agriculture, Agricultural Research and Advisory Station, Glen Innes, N.S.W. 2370.*

## INTRODUCTION

Individual identification of ewes carrying various litter sizes in early pregnancy is very useful if not a basic requirement in many research programmes. Separation and differential management of ewes bearing and rearing different litter sizes is fundamental to many approaches of improving individual and overall productivity in the flock.

Many techniques have been attempted for pregnancy and multiple diagnosis and have been reviewed by Richardson (1972), Thwaites (1981) and Watt (1983). Many techniques are too expensive, impractical or inaccurate but some are quite useful for diagnosing pregnancy only. Few are capable of diagnosing litter size and of these real-time ultrasound imaging (R.U.I.) provides the best overall practical technique. This paper will discuss the basics of the technique and its uses in research and production.

## ASPECTS OF IMAGING TECHNIQUE

Since late 1970's real time imaging of high resolution has been possible with portable units. After initial screening to identify suitable machines, experiments were conducted to determine the accuracy of diagnosis at various stages of gestation (Fowler and Wilkins 1984a; Table 1).

**TABLE 1. Effects of stage of gestation (days) and litter size on accuracy (%) of scan prediction. (Sample sizes in parentheses).**

| | | | 1980 | | 1981 |
|---|---|---|---|---|---|
| Stage of gestation | | 40-47 | 56-68 | 83-96 | 54-70 |
| Litter size | 0 | 100 (38) | 100 (102) | 100 (48) | 100 (268) |
| | 1 | 95 (490) | 99 (924) | 99 (519) | 99 (1276) |
| | 2 | 67 (42) | 84 (96) | 82 (66) | 97 (1115) |
| | 3 & 4 | | | | 48 (48) |

Accuracy increased from 1980 to 1981 reflecting operator experience and improved technique. However diagnosis of pregnancy has always remained near 100% accuracy and diagnosis of multiple pregnancy in subsequent experiments (Wilkins and Fowler 1984a; Wilkins and Fowler unpublished) has been 95-99% accurate. Errors are nearly always underestimates. Very accurate diagnoses of zero, single, multiple can be made quickly over the range 45-100 days. Diagnoses in the last trimester take longer.

We find external linear array probes of frequency 3.0 or 3.5 MHz to be the most suitable to cover a wide range of stages of pregnancy. Probes of 5.0 MHz give better image clarity in shallow depth of scan but are less generally useful except at early stages. Preliminary results with internal probes (Wilkins and Fowler 1984b) have supported their use for very early pregnancy diagnoses (before 30 days) but accuracy with multiples at various stages needs evaluation.

### Large litter sizes

Exact prediction of litter sizes of 3 or more was poor in early experiments (Table 1). Large litter sizes were induced for several experiments in 1982. Accuracy of the prediction of various litter sizes (and the accuracy of detecting same) was examined from suitable data and the results are shown in Table 2. Most errors (22/25) were underestimates where litter size was 3 or more. We suggest repeated diagnoses will allow improvement on the above results if required (e.g. research flocks). It is not often required to identify 3 or more foetuses in commercial flocks.

**TABLE 2. Accuracy (%) of scan prediction and accuracy of detection for various litter sizes.**

| Litter size | 0 | 1 | 2 | 3 | 4 | 5 |
|---|---|---|---|---|---|---|
| Scan prediction | 100 | 97.7 | 97.8 | 86.4 | 60.0 | (1/1) |
| Detection | 100 | 99.7 | 95.9 | 81.4 | 75.0 | (1/6) |
| n | 64 | 607 | 320 | 86 | 16 | 6 |

### Early pregnancy

The fluid of the developing conceptus is detectable using R.U.I. before the embryo can be distinguished. In a small experiment involving 36 observations, pregnancy was confirmed by 27 days using an external probe (Fowler and Wilkins, unpublished). Data from this and other experiments using external probes (Wilkins *et al.* 1982; Wilkins *et al.* 1984) have shown that definable embryos were first found from 26-32 days but estimates of number were not reliable till around day 40.

Other reports (Botero-Herrera *et al.* 1983), using internal probes, have indicated 91-94% accuaracy of pregnancy diagnosis from 21 days. Our preliminary observations are in agreement (Wilkins and Fowler 1984b) but the accuracy of predicting numbers of embryos is yet to be established.

### Preparation

Various regimes of withholding feed and water have been examined (Wilkins and Fowler 1984a). Treatments tested experimentally have not significantly affected image clarity or diagnostic accuracy. However, we have found that gut distention can cause minor problems by displacing the uterus if feed conditions are very lush. Thus overnight fasting is a suitable and practical policy but by no means essential.

Preparation for external scanning must provide for good contact between the probe and the skin surface. Vegetable oil is commonly used as the coupling medium. To scan the abdominal area, the belly wool must be shorn for 5-10 cm above the udder. Scanning can be done using only the bare skin of the inguinal area and is highly accurate for pregnancy diagnosis. However, using this technique in an experiment involving 1500 observations, 22% of ewes carrying multiple foetuses were diagnosed single bearers compared to 3% when rescanned several days later after shearing the belly.

### Presentation

We have examined various alternative positions of presentation for scanning (Fowler and Wilkins 1984). The ewe held on her back gave the easiest and most accurate diagnoses. It was difficult to scan a standing ewe with external linear array probes. Using internal probes and an upright position could simplify sheep handling but this is not without problems such as orientation and depth of scan (Wilkins and Fowler 1984b) and its accuracy for litter size needs evaluation.

## APPLICATIONS

### Field monitoring/troubleshooting

In monitoring or troubleshooting exercises, scanning can provide a large amount of information for a relatively low input. For example, the proportions of zero, single and multiple bearing ewes describes the basic flock fertility and fecundity, indicating components subject to improvement. By estimating stages of pregnancy using foetal size we can infer the pattern of conception and predict the pattern of lambing. Using litter size pre-lambing and simple inspections post-lambing we can accurately measure lamb survival rates for different types of birth which would otherwise require a lot of work for lambing observations.

### Commercial flocks

Adoption of scanning depends on its cost and benefits to production in breeding flocks. Cost is largely dependent on the efficiency of the scanning operation. To date we have developed systems capable of diagnosing about 1000 ewes/day using 3 people and we expect this to be improved. The production benefits come from better control of nutritional and management requirements according to litter size throughout pregnancy and lactation. Several areas are likely to be responsive. The deficit in maternal wool production due to multiple bearing may be reduced. Maternal survival may be improved by reducing pregnancy toxaemia and parturition deaths. Lamb survival should be improved if birthweight can be controlled to reduce dystocia in singles and 'optimise' weights in multiples. Lamb growth may be enhanced by birthweight and improved lactation. Direct costs of supplementary feeding can be rationally allocated on requirement. Scanning also provides the commercial flock owner with greater opportunity to implement selection for fecundity and rearing ability. Management strategies to realise many of the above still require some large scale field development and/or testing.

### Stud flocks

The main advantages here are in simplifying or improving data collection. Accurate birth records require a lot of work and may be subject to large errors (Alexander *et al.* 1983). Knowledge of litter size prior to birth can minimise or eliminate such problems. Selection for fecundity is greatly streamlined particularly where individual pedigrees are not required. Stud flocks naturally benefit from improvements in general management and increased number of progeny.

### Research

Experiments involving production in breeding flocks are greatly assisted by being able to manipulate the proportions of ewes carrying various litter sizes within treatment groups. This may be particularly critical when cell size is small and be required early in pregnancy. Monitoring litter size is essential in reproductive wastage studies (Wilkins *et al.* 1984). Repeated observations of litter size on the same animals allows low levels of wastage to be detected using relatively few animals. Scanning can assist both accuracy and ease of collection of data for genetic research. Litter size diagnosis on large numbers of ewes greatly increases the scope of field research in particular. Many benefits and research applications are yet to be explored such as studies on *in utero* development and use in artificial breeding programmes.

ACKNOWLEDGEMENTS

We gratefully acknowledge the support of the Australian Meat Research Committee and the Rural Credits Development Fund.

REFERENCES

Alexander, G., Stevens, D. and Mottershead, B., 1983. *Aust. J. Exp. Agric. Anim. Husb., 23*, 361-368.

Botero-Herrera, O., Gonzalez-Stagnaro, C., Poulin, N. and Cognie, Y., 1983. *Proc. 34th Ann. Meeting of Federation Europeene de Zootechnie Madrid*. p. 702.

Fowler, D.G. and Wilkins, J.F., 1984a. *Livest. Prod. Sci., 11*, 437-450.

Fowler, D.G., and Wilkins, J.F., 1984b. *Proc. Aust. Soc. Anim. Prod., 15*, 681.

Richardson, C., 1972. *Vet. Rec., 90*, 264-275.

Thwaites, C.J., 1981. *Animal Breeding Abstr., 49*, 427-434.

Watt, B.R., 1983. *Univ. Syd. Post-grad. Comm. Vet. Sci. Proc. No. 67* (*Werribee*) pp. 207-225.

Wilkins, J.F. and Fowler, D.G., 1984a. *Proc. Aust. Soc. Anim. Prod., 15*, 769.

Wilkins, J.F. and Fowler, D.G., 1984b. *Proc. Aust. Soc. Reprod. Biol., 16*, 34.

Wilkins, J.F., Fowler, D.G., Bindon, B.M., Piper, L.R., Hall, D.G. and Fogarty, N.M., 1984. *Proc. Aust. Soc. Anim. Prod., 15*, 768.

Wilkins, J.F., Fowler, D.G., Piper, L.R. and Bindon, B.M., 1982. *Proc. Aust. Soc. Anim. Prod., 14*, 637.

# THE GLUCOSE TEST

R.A. Parr and I.P. Campbell, *Department of Agriculture, Animal Research Institute, Werribee, Victoria, 3030.*
J.L. Reeve, *Department of Agriculture, Rutherglen Research Institute, Rutherglen, Victoria, 3685.*

*Summary* The 'glucose test' consists of measuring the blood glucose concentration in ewes at 90-100 days of pregnancy and dividing the flock into those with high (i.e. above the median level) or low blood glucose concentrations. Most ewes (mean 88%, range 85-93%) with twin lambs and 90% of ewes which later suffered from pregnancy toxaemia had low blood glucose concentrations at the time of sampling. Moreover, 76% (range 68-83%) of non-pregnant ewes had relatively high blood glucose concentrations. In a study with high fecundity ewes it was found that most lamb deaths were associated with ewes having low blood glucose concentrations at the time of blood sampling. These ewes however were responsible for 2.5 times more productivity (live lamb weight at birth/ewe joined) than an equal number of high glucose ewes in the flock.

## INTRODUCTION

Insufficient feed in late pregnancy can result in pregnancy toxaemia (Reid 1968), restrictions in foetal growth (Mellor and Matheson 1979), and can be a predisposing factor in peri-natal lamb mortality (Underwood and Shier 1942). In contrast, over-feeding some ewes during this period may increase the incidence of dystocia (Curll *et al.* 1975).

Attempts to identify those ewes in the flock which require extra feed in late pregnancy have aimed at diagnosing foetal number in mid pregnancy. A mobile x-ray unit was developed by Rizzoli *et al.* (1976) and, with the advent of video fluoroscopy, instantaneous diagnosis of foetal number in large flocks became possible (Beach 1980). More recently, real time ultrasonic scanning of ewes has been developed (Fowler and Wilkins 1982) providing accurate diagnosis of multiple pregnancy. These approaches, however, give no indication of a ewe's nutritional status or how she is coping with the demands of pregnancy. Moreover, the energy requirements of a large ewe with a single foetus (according to ARC recommendations) are greater than those of a smaller ewe with twin foetuses. We have recently proposed an alternative approach to the identification of ewes requiring improved nutrition in late pregnancy (Parr *et al.* 1982). This method requires a single measurement of blood glucose concentration of each ewe at about day 90 of pregnancy. It does not aim at diagnosis of multiple pregnancy but identifies those ewes with relatively low blood glucose concentrations compared with other ewes.

## MATERIALS AND METHODS

### Experiment 1. Blood Glucose Profiles in Ewe Flocks

Four flocks of Border Leicester × Merino ewes (n=97–567) were joined to Dorset Horn rams between October and December. Ewes were yarded overnight when their mean stage of pregnancy was 90 days (range within flocks; 53-103 days). The following morning a blood sample (10 ml) was collected by jugular venipuncture from each ewe. Plasma samples were later measured for glucose concentration according to the method described by Parr *et al.* (1984b). Ewes were grouped into fortnightly mating intervals and the median glucose concentration was determined within each group. This enabled us to account for the effect of stage of pregnancy on plasma glucose concentration. Ewes with blood glucose concentrations below the median were designated as low-glucose ewes (Group L) and those above the median as high-glucose ewes (Group H). Each flock was kept intact and ewes were regularly inspected for signs of metabolic disorders. At lambing, three of the flocks were inspected daily and ewes which were dry or with single or multiple lambs were recorded.

### Experiment 2. Flock Glucose Profiles and Productivity

In a study of the potential application of blood glucose measurements in a Booroola Merino flock (Parr *et al.* 1984a), 120 mixed age Booroola Merino ewes were blood sampled (using the methodology described earlier) 95-100 days after a synchronized joining. The ewes were managed as one flock and, at lambing lambs were weighed and identified with ewes.

## RESULTS

### Experiment 1.

The incidence of pregnancy toxaemia was low, occurring in 10 ewes in total. With one exception all ewes with pregnancy toxaemia at day 120-130 of pregnancy had been designated as Group L ewes at the time of blood sampling (Table 1).

Ewes with multiple foetuses had a high probability of being designated Group L ewes at the time of blood sampling (mean 88%; range 85–93%). Conversely, non-pregnant ewes had a high probability of being included in Group H (mean 76%; range 68–83%). Ewes bearing single lambs were evenly distributed between Groups L and H in the 3 flocks where lambs were identified with ewes.

**TABLE 1. Distribution of ewes between groups of low and high concentrations of glucose.**

| Flock | No. of ewes(n) | Metabolic disorders | | Multiples | | Non-Pregnant | |
|---|---|---|---|---|---|---|---|
| | | Gp L | Gp H | Gp L | Gp H | Gp L | Gp H |
| 1 | 100 | 4 | 1 | 36 | 6 | 3 | 11 |
| 2 | 142 | — | — | 14 | 1 | 9 | 19 |
| 3 | 567 | 1 | — | 72 | 13 | 17 | 82 |
| 4 | 97 | 4 | — | Not Determined | | Not Determined | |

Experiment 2.

The overall pregnancy rate (EL/EJ) was 58% and for groups L and H was 83 and 33%. Most multiple bearing ewes (78%) were designated Group L at the time of blood sampling. Most non-pregnant ewes were in group H with 70% LB/EJ compared with 203% in Group L.

Of the ewes designated Group L, 25% lost one or more lambs at birth compared with 7% of the ewes in Group H. Lamb birth-weights within each litter size of Group L were lower than birthweights of lambs in Group H, though these differences failed to reach significance ($P>0.05$). Group L ewes had 2.5 times the level of total lamb weight and lamb weight per ewe joined than did Group H ewes.

## DISCUSSION

A single determination of blood glucose concentration from each ewe at mid pregnancy, in addition to providing a probability estimate of her pregnancy status (i.e., 0, 1 or more foetuses), reveals how a ewe is coping with pregnancy, exposing the balance between supply and demand of glucose. Using this simple approach a more equitable feeding and management programme could be devised. This would ensure more feed is supplied to those with a greater apparent need (i.e. those with low glucose) and less feed is wasted on non-pregnant ewes or those with less apparent need. This rationale is of particular importance in flocks with either natural or induced high fecundity. Two sub-flocks could be established, a low glucose flock of potentially high productivity and a high glucose flock of potentially lower productivity. Feed inputs could then more closely match the nutritional requirements of each sub-flock. In other studies (Campbell and Parr, unpublished data) we have found that blood glucose concentration is significantly correlated with condition score in late pregnancy. The relationship between blood glucose and live weight is less clear, particularly towards the end of gestation when the weight of the conceptus may have a confounding influence.

The trend of lower birthweights of lambs in Group L of the Booroola Merino flock was also seen as a significant reduction of Group L single and twin lamb birthweights in a prime lamb flock (Parr *et al.* 1984b). This effect however was not consistent in all flocks studied. When birthweights of Group L lambs are low it is apparent that foetal growth is restricted and lambs fail to reach their potential birthweight.

Ewes associated with a relatively low blood glucose concentration at mid-pregnancy are those which appear to have high potential production, but also to have most problems. With strategic management such problems as pregnancy toxaemia, low lamb birthweight and poor peri-natal survival may be reduced. Methods of flock management based on the glucose test are currently under investigation. Stocking rate adjustments or strategic use of high energy feed supplements, particularly during drought periods are likely applications of this work. With the development of low cost, simple 'on-the-spot' methods of blood glucose measurement based on portable diabetic instruments, the technique should find ready application in the sheep industry.

## REFERENCES

Beach, A.D. 1980. *Proc. Aust. Soc. Reprod. Biol., 12*, 90.
Curll, M.L., Davidson, J.L. and Freer, M., 1975. *Aust. J. Agric. Res., 26*, 553-565.
Fowler, D.G. and Wilkins, J.F., 1982. *Proc. Aust. Soc. Anim. Prod., 14*, 491-494.
Mellor, D.J. and Matheson, I.C., 1979. *Quart. J. of Exp. Physiol., 64*, 119-131.
Parr, R.A., Campbell, I.P. Cahill, L.P., Bindon, B.M. and Piper, L.R., 1984a. *Proc. Aust. Soc. Anim. Prod., 15*, 517-520.
Parr, R.A., Campbell, I.P. Reeve, J.L. and Chamley, W.A., 1982. *Proc. Aust. Soc. Reprod. Biol., 14*, 101.
Parr, R.A., Campbell, I.P. Reeve, J.L. and Chamley, W.A., 1984b. *Aust. J. Agric. Res.* (in press).
Reid, R.L., 1968. *Adv. in Vet. Sci., 12*, 163-238.
Rizzoli, D.J., Winfield, C.G., Howard, T.J., Englund, I.K.J. and Goding, J.R., 1976. *J. Agric. Sci., Camb., 87*, 671-677.
Underwood, E.J. and Shier, F.L., 1942. *J. Agric. W.A., 19*, 37.

# CONCENTRATION OF β-HYDROXYBUTYRATE IN PLASMA OF EWES IN LATE PREGNANCY AND EARLY LACTATION, AND SURVIVAL AND GROWTH OF LAMBS

J.Z. Foot, L.J. Cummins, S.A. Spiker and P.C. Flinn, *Department of Agriculture, Pastoral Research Institute, P.O. Box 180, Hamilton, Vic. 3300.*

*Summary* Concentrations of β-hydroxybutyrate in plasma of ewes reflect the balance between fat mobilisation (dependent on the amount of fat stores in the body and on the difference between the intake of and demand for nutrients) and the capacity to utilise the ketone bodies produced. Consequently they may be used as an index of the inadequacy of intake of nutrients by ewes to meet their requirements during periods of high demand. β-hydroxybutyrate in the plasma of 67 grazing Corriedale ewes with one to three lambs was measured in late pregnancy and early lactation. Results were related by multiple regression analysis to birth weights, survival and growth of lambs. Mortality was highest in the triplets (44%).

In late pregnancy concentrations of β-hydroxybutyrate in the plasma were positively related ($r = 0.63$, $P < 0.001$) to birth weights of lambs (g/kg liveweight of ewe). In early lactation concentrations of β-hydroxybutyrate were related to the number of lambs suckled ($r = 0.65$, $P < 0.001$) and the total growth rate of the lambs (g/d/ewe) ($r = 0.63$, $P < 0.001$). Values averaged $0.47 \pm 0.08$, $0.87 \pm 0.20$, $1.32 \pm 0.52$ and $2.23 \pm 0.41$ mM (mean ± SD) for ewes suckling 0, 1, 2 or 3 lambs. In pregnancy and lactation the time to and from lambing (days) was negatively correlated with concentrations of β-hydroxybutyrate in the plasma ($P < 0.05$ for days prelambing and $P < 0.01$ for days post-lambing).

The results indicate that concentrations of β-hydroxybutyrate in the plasma of ewes in late pregnancy and early lactation may be useful as an index of the extent to which the intake of the grazing animal fails to meet its requirements.

## INTRODUCTION

In late pregnancy and early lactation the nutritional requirements of ewes are very high. At the same time the penalties for not meeting requirements are likely to be more severe than at any other stage of the reproductive cycle. These penalties include reduction in survival of lambs, poor growth of lambs, poor growth of wool (with associated adverse effects on fibre strength) and, in severe cases, mortality of ewes.

Concentrations of β-hydroxybutyrate (β-OHB) in plasma have been shown to be a good index of severe undernourishment in sheep under field conditions (Russel 1977). This is because concentrations of β-OHB reflect the balance between fat mobilisation and the capacity of the animal to make use of the ketone bodies produced. The extent of fat mobilisation depends on the amount of fat stored in the body, and the difference between the intake of, and demand for, nutrients (particularly energy). Russell *et al.* (1977) found that ewes grazing pasture and having a concentration of β-OHB in the plasma of 0.71 mM were in energy balance. 'Moderate' undernourishment and 'severe' undernourishment gave rise to concentrations of 1.1 and 1.6 mM.

In this experiment we examined the relationships between concentrations of β-OHB in plasma from grazing ewes and the birth weights, survival and growth of their lambs. We wished to assess the value of concentrations of β-OHB for monitoring undernutrition in future work on the effects of nutrition on reproductive efficiency of grazing ewes.

## MATERIALS AND METHODS

### Animals

Ewes were taken from an experiment to assess the effects of Fecundin and nutrition at mating on fecundity and embryonic survival (Cummins *et al.* 1984). Sixty-seven mature Corriedale ewes were sampled from a total of 167. These included 5 barren ewes, 13 with single lambs, 34 with twins and 15 with triplets as determined by a Real-time Ultrasound Scanner.

### Treatments and management

The ewes grazed together until approximately 90 days of gestation when they were divided into two groups for treatment during late pregnancy.

*Nutritional treatments in late pregnancy* High plane of nutrition (H) ewes in this group included all those carrying three foetuses and half of those carrying twins. They grazed a perennial ryegrass/sub clover dominant pasture with an availability of about 1 t DM/ha in late winter, (approximately 100 kg DM/ewe). They were offered lupin grain as a supplement to pasture, initially 400 g/sheep/d increasing twice weekly to 600 and 750 g/sheep/d respectively for ewes with twins and triplets just before lambing. Low plane of nutrition (L): ewes included the remaining twin-bearing ewes, those with single foetuses and barren ewes. The pasture was of the same type as that grazed by ewes in the H treatment group, but with a slightly higher availability. The ewes were a supplement of lupin grain, starting at 200 g/sheep/d and increasing gradually to 400 g/sheep/d just before lambing.

*Lambing* Ewes with triplets lambed individually in small, sheltered, grass plots, while the other ewes lambed in small groups.

*Nutritional treatments in early lactation* After lambing all ewes grazed perennial ryegrass/subterranean clover pastures in the spring growth phase. Ewes on treatment H had 100 g/sheep/d lupin grain as a supplement to abundant pasture (just over 3 t DM/ha). Ewes on treatment L received no supplement and grazed a pasture which had an availability of just over 2 t DM/ha.

Ewes on treatment H in lactation were those with triplets and half of the twin-bearing ewes from each of the pregnancy nutritional treatment groups. Ewes on treatment L in lactation were those which were barren or which had a single lamb, and half of the twin-bearing ewes from each of the nutritional treatment groups during pregnancy.

Samples and measurements

Ewes were weighed and scored for body condition (Russel *et al.* 1969) when blood samples were taken for measuring concentrations of β-OHB in the plasma. In late pregnancy this was 6 to 25 days before lambing (17 ± 4.4) (mean ± SD) and in early lactation 16 to 35 days after lambing (24 ± 4.5).

Lambs were weighed at birth and marking, which coincided with the time of blood sampling and weighing of ewes. The β-OHB concentrations in plasma were measured by the method of Zivin and Snarr (1973).

Calculations and statistical analysis.

The average liveweights, body condition scores (CS) and concentrations of β-OHB in plasma of ewes in late pregnancy and early lactation were calculated for animals with 0, 1, 2 (H and L treatments) and 3 lambs. Average birth weights of lambs (g.kg liveweight of the ewe) and their rate of growth from birth to marking (g/d/ewe) were calculated.

The relationships between concentrations of β-OHB in the plasma of ewes (yl) and the birth weights of lambs (g/kg liveweight of the ewe) and days to lambing at sampling were analysed using stepwise multiple regression. The relationships between concentrations of β-OHB in the plasma of ewes in lactation (y2) and the rate of growth of lambs to marking (g/d) or number of lambs suckled, and days from lambing were also analysed by stepwise multiple regression.

The equation was of the form:-

$$y = a + b_1 x_1 + b_2 x_2$$

where a = the constant of the regression equation

$x_1$ and $x_2$ were the independent variables (birth weight of lambs and days to lambing for yl and rate of growth of lambs, or number of lambs suckled, and days from lambing y2)

$b_1$ and $b_2$ were the partial regression coefficients for $x_1$ and $x_2$

## RESULTS

The liveweights of ewes in late pregnancy were 66.7 ± 12.7, 66.7 ± 5.27, 68.4 ± 6.60, 68.4 ± 6.31 and 74.7 ± 3.17 kg (means ± SD) for barren ewes, ewes with one foetus, two foetuses (H and L treatments) and three foetuses live until term.

Birth weights of lambs (g/kg liveweight of the ewe) were 78 ± 23.8, 139 ± 18.3 and 159 ± 15.6 (means ± SD) from ewes with one, two or three lambs.

Growth rates of lambs from birth ot marking (g/d/ewe) were 355 + 62.8, 537 ± 94.6 and 586 ± 69 (means ± SD) for lambs form ewes suckling one, two or three offspring.

**TABLE 1. Concentration of β-OHB in plasma (mM) from ewes in late pregnancy and early lactation, condition score of ewes in late pregnancy and early lactation (mean ± SD).**

| Lambs born | Condition score in late pregnancy | Concentration of β-OHB pregnancy | Concentration of β-OHB lactation | Lambs suckled | Change in condition score | No. ewes |
|---|---|---|---|---|---|---|
| 0 | 3.1 ± 0.30 | 0.50 ± 0.202 | 0.48 ± 0.095 | 0 | +0.5 | 5 |
| 1 | 3.4 ± 0.21 | 0.85 ± 0.200 | 0.90 ± 0.222 | 1 | −0.2 ± 0.32 | 13 |
| 2 H(1) | 3.4 ± 0.21 | 1.09 ± 0.302 | 1.30 ± 0.520H | 2 | −0.8 ± 0.33 | 9 |
| | | | 1.32 ± 0.400L | 2 | −0.9 ± 0.15 | 7 |
| 2 L | 3.2 ± 0.34 | 1.20 ± 0.321 | 1.22 ± 0.379H | 2 | −0.6 ± 0.30 | 7 |
| | | | 1.45 ± 0.683L | 2 | −0.5 ± 0.25 | 8 |
| 3 | 3.2 ± 0.28 | 1.39 ± 0.325 | 0.46 ± 0.028 | 0 | −0.2 | 2 |
| | | | 0.78 ± 0.127 | 1 | −0.4 | 4 |
| | | | 1.26 ± 0.611 | 2 | −0.6 ± 0.25 | 6 |
| | | | 2.23 ± 0.410 | 3 | −1.1 | 3 |
| 3P(2) | 3.3 ± 0.24 | 0.99 ± 0.228 | | | | |

(1) H = 'high' nutritional treatment group. L = 'low' nutritional treatment group
(2) 3P = results for five ewes which had one or more foetuses dying before term

The CS of ewes in late pregnancy and concentrations of $\beta$-OHB in plasma of ewes in late pregnancy and early lactation are shown in Table 1, together with the numbers of lactating ewes included in the means. Three twin-bearing ewes were omitted from the results for lactation because they lost one or both lambs. They were all in different nutritional treatment groups. Results from all 67 ewes during pregnancy were included.

Five ewes with 3 foetuses (3P in Table 1) had one or more foetuses which had clearly died well before term. The lower concentration of 3-OHB in the plasma of these ewes reflects the lower foetal burden at the time of blood sampling. Of the 45 foetuses carried by triplet-bearing ewes, 10 died pre-partum and 10 died from 1-23 days after birth.

Concentrations of $\beta$-OHB in plasma of pregnant ewes was positively related to the number of lambs carried ($P < 0.001$) but was not significantly affected by nutritional treatment in late pregnancy. In lactation, concentrations of $\beta$-OHB were positively related to the number of lambs suckled ($P < 0.001$) and the concentrations of $\beta$-OHB were higher in plasma from ewes on treatment L than on treatment H ($P < 0.05$).

The regression coefficients for multiple regression equations relating concentrations of $\beta$-OHB in plasma to birthweight or rate of growth of lambs, or number of lambs suckled, and to the timing of blood sampling relative to parturition are shown in Table 2. All equations were highly significant ($P < 0.001$).

**TABLE 2. Multiple regression equations relating concentrations of $\beta$-OHB in plasma (y) to birthweight, rate of growth of lambs or number of lambs suckled and time from sampling.**

| Eq.No. | y[1] $b_1$[2] | $b_2$ | a constant | r.s.d.[5] | R |
|---|---|---|---|---|---|
| | $x_1$[3] = BWT | $x_2$[4] = D1 | | | |
| 1 | $y1 = 0.0052\, x_1$ | $- 0.019\, x_2$ | + 0.777 | 0.276 | 0.67 |
| | $x_1$ = LG | $x_2$ = D2 | | | |
| 2 | $y2 = 0.0016\, x_1$ | $- 0.046\, x_2$ | + 1.563 | 0.381 | 0.73 |
| | $x_1$ = LS | $x_2$ = D2 | | | |
| 3 | $y2 = 0.43\, x_1$ | $- 0.047\, x_2$ | + 1.580 | 0.365 | 0.75 |

(1) y1 concentration of $\beta$-OHB in plasma of ewes during pregnancy (mM)
y2 concentration of $\beta$-OHB in plasma of ewes during lactation (mM)
(2) $b_1$ partial regression coefficient for $x_1$
$b_2$ partial regression coefficient for $x_2$
(3) $x_1$ BWT; birth weight of lambs (g/kg weight of ewe)
LG; total rate of growth of lambs (g/d/ewe)
LS; number of lambs suckled/ewe
(4) $x_2$ D1, days to lambing at sampling in pregnant ewes
D2, days from lambing at sampling in lactating ewes
(5) r.s.d., standard error of the estimate.

Concentrations of $\beta$-OHB in plasma from ewes in pregnancy (yl) were positively related to birth weight of the lambs (g/kg liveweight of the ewe) ($P < 0.001$) and negatively to the number of days before lambing that the sample was taken ($P < 0.05$), (Equation 1).

In lactation, concentrations of $\beta$-OHB (y2) were positively related to rate of growth of lambs and negatively to the number of days from lambing ($P < 0.001$ in both cases), (Equation 2). Using the number of lambs suckled as $x_1$, instead of rate of growth of lambs, gave an equally good prediction of y2 (Equation 3).

## DISCUSSION

Concentrations of $\beta$-OHB in the plasma of ewes in this experiment reached levels which indicated 'moderate' and 'severe' undernutrition according to the definitions suggested by Russel *et al.* (1977).

Using equations 1 and 3 in Table 2 we calculated that in this experiment 75 kg ewes with three lambs, each with a birth weight of 4 kg, would have experienced moderate undernutrition from four weeks before lambing until five weeks after. Within this time they would have been severely undernourished for the first three to four weeks of lactation. Ewes suckling triplets lost more than 1 unit of CS in this period (Table 1).

Similar calculations may be made for ewes with two 5 kg lambs. Theoretically they would have been moderately undernourished from three weeks before lambing to four weeks after lambing. Within this period they would have been severely undernourished during the first two to three weeks of lactation.

The small effect of nutritional treatments may have been due to two factors. First, all ewes were in good body condition; each group had an average CS of between 3.2 and 3.4 in late pregnancy. Second, the so-called L nutritional treatment may not have limited the intake of nutrients of the ewes any more than the H nutritional treatment if the voluntary intake of the ewes rather than the availability of pasture was limiting.

Concentrations of β-OHB in plasma may be measured in samples taken from ewes immediately after they are gathered off pasture. This cannot be done when measuring concentration of glucose in plasma, so the use of glucose is restricted as an index of nutritional state in late pregnancy and early lactation.

It seems that measurements of β-OHB may be of value in studying nutritional stress in grazing ewes during late pregnancy and early lactation. Such studies will need to include measurements of the availability and quality of the pasture, intake of herbage by individual animals and reserves of body fat in individual animals using tritiated water space.

These studies will provide a good basis for decisions on nutritional management of grazing ewes during this critical phase of their reproductive cycle, and for assessing the economic consequences of such decisions.

## REFERENCES

Cummins, L.J., Spiker, S.A., Cook, C. and Cox, R.I. 1984. This volume, 326-328

Russel, A.J.F., 1977. *In* Lister, D. (Ed) *Blood Profiles in Animal Production*. Occasional Publication No. 1 British Society of Animal Production, p. 31-39.

Russel, A.J.F., Doney, J.M. and Gunn, R.G., 1969. *J. Agric. Sci., Camb.*, *72*, 451-454.

Russel, A.J.F., Maxwell, T.J., Sibbald, A.R. and McDonald, D., 1977. *J. Agric. Sci., Camb.*, *89*, 667-673.

Zivin, J.A. and Snarr, J.F., 1973. *Anal. Biochem.*, *52*, 456-461.

# MATERNAL BEHAVIOUR IN SHEEP AND ITS PHYSIOLOGICAL CONTROL

P. Poindron, *School of Agriculture (Animal Science), University of Western Australia, Nedlands, Western Australia 6009.*
P. Le Neindre, *Laboratoire de Production de Viande, INRA de Theix, 63122 Saint Genes, France.*
F. Levy, *Laboratoire de Comportement Animal, INRA de Nouzilly, 37380 Monnaie, France.*

## INTRODUCTION

Survival of lambs is an important parameter in the productivity of a flock. As most losses occur in the first two days of life, proper maternal behaviour at lambing is essential to minimize the detrimental effects of environmental factors. Although the ewe-lamb bond is usually established without major problems, there are situations in which maternal behaviour can become a limiting factor. This can be due to the failure of the ewe either to show maternal interest immediately after parturition, or to establish a strong bond with her litter, especially in the case of multiple births. Also the introduction of new techniques (e.g. oestrous synchrony, increased prolificacy) can lead to the appearance of problems that were absent of marginal in more traditional situations.

A good understanding of maternal behaviour and its mechanisms of control can prove to be helpful when assessing existing problems or when introducing new management techniques. The identification of factors controlling the establishment of the mother-young relationship can also lead to the development of new methods of fostering and provide sound guidelines for selection. This paper will review current knowledge concerning the mechanisms controlling maternal behaviour and its selective component. We will limit ourselves to the period around parturition since it is the most critical time for the survival of the lambs. Lastly, the practical implications of these findings and the prospects for research will be discussed.

## CONTROL OF THE ONSET OF MATERNAL INTEREST TOWARDS THE NEWBORN LAMB

### The sensitive period

Maternal behaviour towards the neonate is present in all ewes only at parturition (Fig. 1). Although maternal interest can be observed at times other than lambing, it is never seen in all females, and in most cases it is limited to a few hours before the birth of the lamb (Alexander 1960; Poindron and Le Neindre 1980). In addition, the attraction for the neonate fades rapidly after lambing if the mother is separated from her offspring for a few hours. After a 4-hour separation from birth about 50% of mothers would not accept their lamb and this proportion increases to about 80% after 24 h separation (Poindron and Le Neindre 1980). By contrast, if the lamb is kept with its mother for the first 24 to 48 h, then separation at a later time is of no consequence in maternal reacceptance of the young. Therefore the first few hours following parturition represent a critical phase in the establishment of the mother-young bond. During this 'sensitive period' the lamb allows her to remain maternal beyond the initial period of temporary receptivity. It is also clear that the interest of the ewe is not limited to one (her) lamb initially, but that any lamb, especially if newly born, will be accepted at this time (Poindron *et al.* 1980a).

### Physiological control of maternal interest

The rapid changes in the reactions of a ewe towards lambs at the time of parturition are closely related to physiological changes occurring during the birth process. So far two main factors controlling the onset of maternal behaviour have been identified.

Steroid hormones produced by the ovaries and the placenta, especially oestrogens, facilitate the onset of maternal behaviour. Treatments used to induce lactation, and which include the administration of progesterone and oestradiol, are also effective in inducing maternal behaviour in 50% of dry ewes (Le Neindre *et al.* 1979). This is not due to the induction of lactation itself, since a single injection of 20 mg of oestradiol is sufficient to lead to maternal receptivity within less than 24 h, a time too short for any milk production to be stimulated (Poindron and Le Neindre 1980). Further studies have shown that the action of oestrogens is not due to the increased secretion of prolactin following OB administration (Poindron *et al.* 1980b). Induction of lambing with 20 mg of oestradiol benzoate is associated with high levels of oestradiol in maternal blood for more than 24 h, as well as with a lengthening of the sensitive period (Poindron and Le Neindre 1980). By contrast, hormonal changes following induction of lambing by dexamethasone are similar to those observed in spontaneous parturition, as the drug acts mainly on the foetus, and the length of the sensitive period is not affected by this treatment. Control through oestrogens is also indicated by the occurrence of spontaneous maternal interest at various stages of the reproductive cycle. As seen in Figure 1, maternal interest is present at oestrus and during the last 10 days of pregnancy, both corresponding to periods of enhanced oestrogen secretion. On the other hand, during mid pregnancy when the progesterone concentration is high, no maternal behaviour is observed. This would indicate that in a normal situation oestrogens rather than progesterone are responsible for the onset of maternal receptivity, despite the fact that administration of exogenous progesterone can have some facilitatory effect in dry ewes (Poindron and Le Neindre 1980).

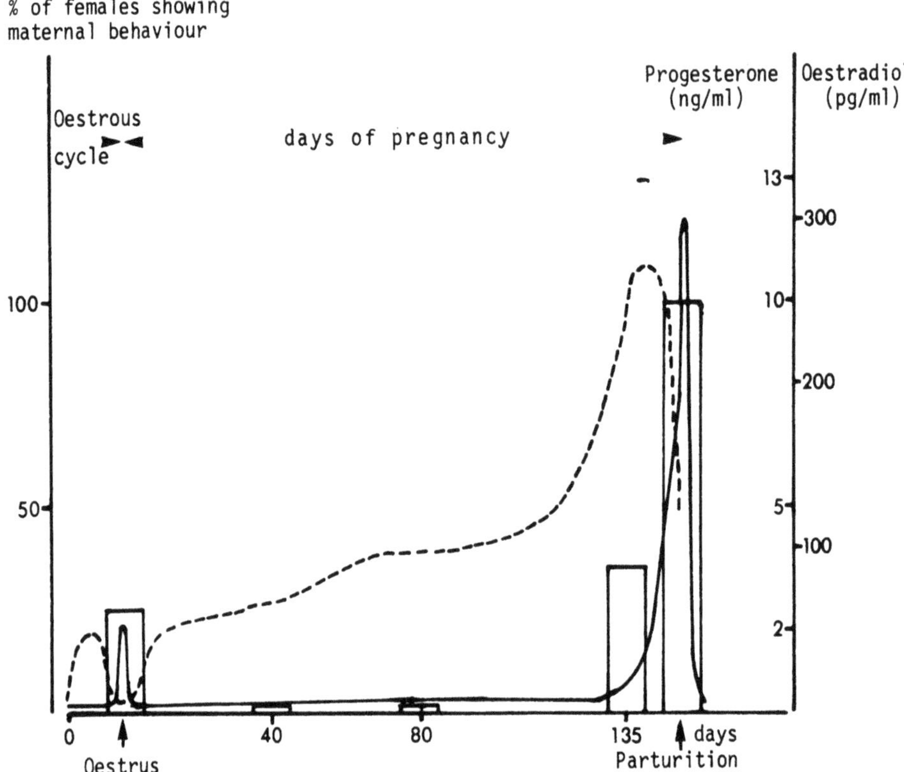


**FIGURE 1. Maternal responsiveness of ewes towards newborn lambs at various stages of the reproductive cycle, in relation to blood concentration of oestradiol and progesterone. (Adapted from Poindron and Le Neindre 1980).**

More recently we have shown that steriod hormones are not the only factor facilitating the acceptance of a lamb. Mechanical stimulation of the genital tract, either with the help of a vibrator in dry ewes, or with a rugby ball bladder in parturient females, has very clearcut positive effects on maternal behaviour (Keverne *et al.* 1983). While only 20% of dry ewes primed with a progestagen by vaginal sponge for 12 days and an injection of 25 mg of oestradiol after sponge withdrawal showed maternal behaviour, the proportion rose to 80% in ewes receiving, in addition, 5 min vaginal stimulation just before the behavioural test (Figure 2A). In a second experiment ewes were offered an alien newborn lamb in addition to their own about one hour after parturition; one group also underwent 5 min genital stimulation immediately beforehand (Figure 2C). Although both groups accepted the alien lamb, licking behaviour (especially of the alien) was enhanced by more than 50% in ewes receiving genital stimulation (about 15 min vs 8 ½ min). Another feature of genital stimulation in dry and post-parturient ewes was the immediate effect of the treatment.

Whereas hormones induced a response in a matter of hours, genital stimulation led to immediate acceptance of the lamb in most cases. The latter therefore appears to be a major factor in the onset of maternal behaviour at parturition.

Sensory cues involved in the development of maternal behaviour at parturition

The sensitive period represents a phase during which the ewe is particularly receptive to the neonate. The contact between the mother and her young allows the normal development and consolidation of maternal behaviour. Sensory cues from the lamb play an important role in this process, and odours from amniotic fluids and the lamb are essential in this respect.

At parturition ewes accept any newborn. However the behaviour of the mother is affected to a degree if the neonate is exchanged for a 12-h old lamb which is dry. Licking of the lamb is decreased in the latter case and aggressive behaviour is observed in about 50% of ewes. Also, ewes which have been deprived of their lamb for 12 h from birth more readily accept an alien newborn than their own 12-h old young (Poindron *et al.* 1980a). These results suggest that a newborn is more attractive than a 12 to 24 h-old lamb. Two major reasons can account for this difference in behaviour towards the lambs according to their age. There is growing evidence in goats (Gubernick 1981) that the mother actually labels her young through licking and feeding and such a

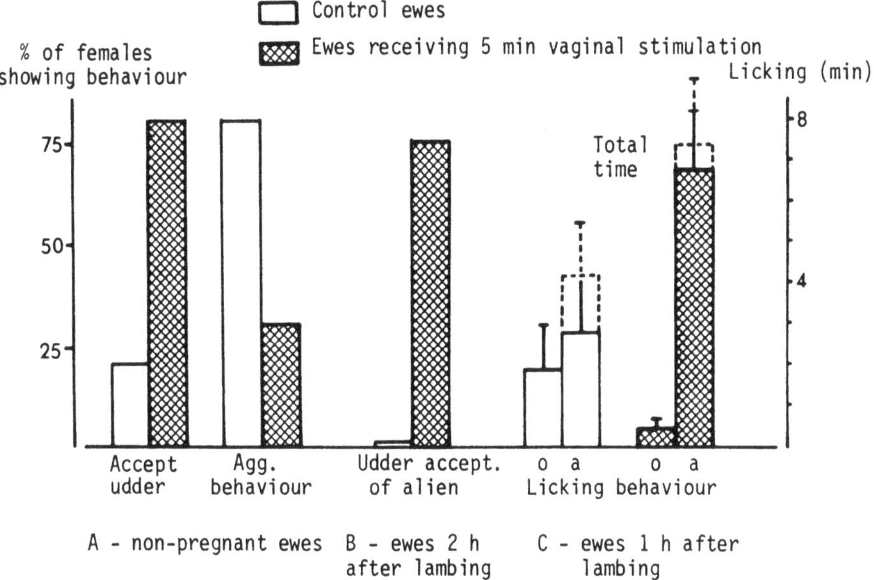


**FIGURE 2. Effects of genital stimulation on maternal behaviour in various situations.**

possibility might also exist in sheep (G. Alexander, pers. comm.). In this case ewes would reject a 12-h old alien lamb because it has already been labelled by another mother.

This point is discussed in more detail later in this paper. The second factor which could lead to a better acceptance of the newborn is the presence in its coat of amniotic fluid, whereas lambs 12 hours old are already dry. Amniotic fluid appears to be very attractive to parturient dams only (Levy *et al.* 1983). When offered the choice between food contaminated with amniotic fluids or water, in a test which excluded any interference with other cues from the lamb, ewes were strongly repelled by amniotic fluid at all stages of the reproductive cycle, except for a few hours around parturition, where it became highly attractive (Figure. 3). Therefore a parturient ewe is more likely to be responsive towards a newborn lamb wet with anmiotic fluid than towards a dry lamb. Further research has shown that wetting dry alien lambs with these fluids facilitates their acceptance by a parturient ewe, even if the fluids come from another female (Levy and Poindron 1984). It also appears that mechanical stimulation of the genital tract in the first hours after parturition delays the fading of the attraction of amniotic fluid with time after birth (F. Levy, unpublished).

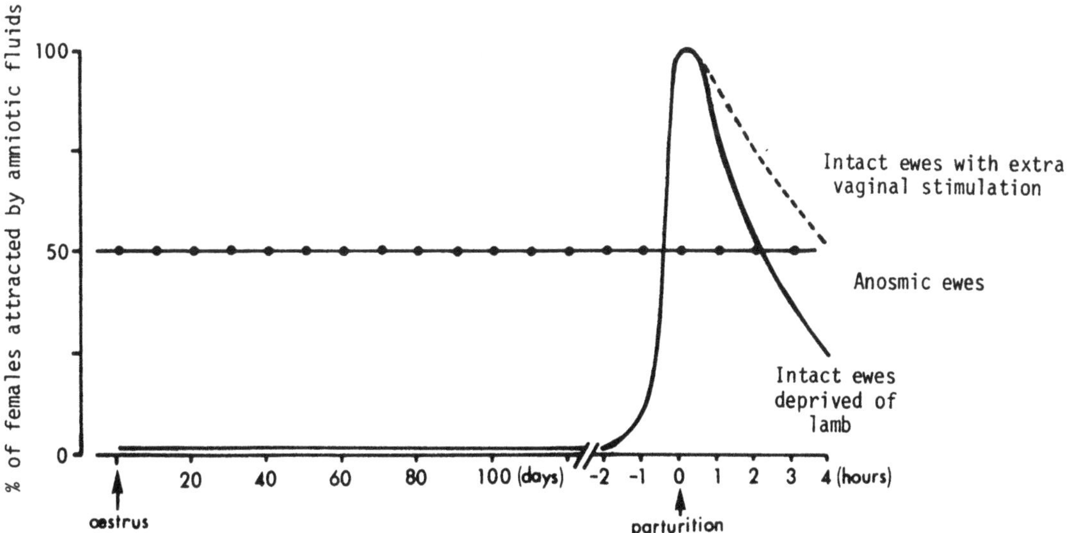


**FIGURE 3. Schematic representation of reactions towards amniotic fluids in intact and anosmic ewes. (Adapted from Levy *et al.* (1983) and F. Levy, unpublished data).**

Olfactory cues from the lamb are not only important to ensure a rapid response from the ewe as the lamb is born; they are also essential for the ewe to remain maternal once the endocrine and tactile factors controlling the onset of responsiveness have disappeared. This can be best illustrated by the response of ewes to partial separation from their lambs at birth. Ewes prevented from suckling their lambs for 12 h, or from suckling and licking them, still accept their young when reunited, provided that the lambs are kept very close to them (50 mm). When distance between the dam and her young is increased to 1 m, the proportion of ewes remaining maternal falls to 50% (vs 30% in mothers totally separated from their lambs; Poindron *et al.* 1980a). To obtain complete olfactory isolation we used airthight boxes with two transparent sides so that sight of the young was not impaired (P. Poindron, unpublished). Ventilation was provided by 200 mm rubber hoses and sounds from the boxes, though slightly modified, were clearly audible, due to their transmission by these hoses. Using these boxes we found that ewes which could see and hear but not smell their lambs for 8 h did not accept them more frequently than ewes totally separated from their lambs (about 30% in both groups). But when the air ventilating the box containing the lamb was returned to the ewe, acceptance occurred in about 60% of cases, regardless of whether the ewe could see her lamb or not.

These experiments all underline the importance of olfactory cues relative to other sensory information in the normal development of maternal behaviour after parturition. Thus it is possible that olfaction plays a major part in the sensitive period for the establishment of maternal drive, as well as in the later process of discrimination between lambs, although the exact nature of the relevant odours remains to be identified.

Maternal behaviour at parturition in maiden ewes

The results mentioned so far relate to multiparous ewes and do not apply directly to primiparous females. It is known that lamb mortality is higher in ewes lambing for the first time, partly because of disturbances in maternal behaviour (Alexander 1960; Shelley 1970), although the extent of the problem remains little documented in field conditions. In a study carried out indoors on three different breeds of sheep, we found that the proportion of primiparous ewes failing to accept their lamb within three hours of birth ranged from 8% (Romanov breed) to 29% (Ile de France), whereas all multiparous mothers accepted their young regardless of the breed (Poindron and Le Neindre 1984) (Figure 4). Similarly, for Merino ewes lambing in the field in March in south Western Australia, there was 20% lamb desertion in maiden vs 2% in females lambing for the second or third time (P. Poindron unpublished). These differences can be interpreted either as indicating an improvement in maternal behaviour in experienced ewes (Alexander *et al.* 1984) or as reflecting greater birth difficulties in younger ewes; both factors are probably involved.

Experiments such as hormonal induction of maternal behaviour in dry ewes, or testing maternal behaviour 10 days before lambing indicate that inexperienced ewes are not as responsive to hormones as experienced ewes (Le Neindre *et al.* 1979; Poindron and Le Neindre 1980). In both situations differences observed in relation to parity cannot be attributed to birth difficulties since ewes do not experience parturition at the time of test.

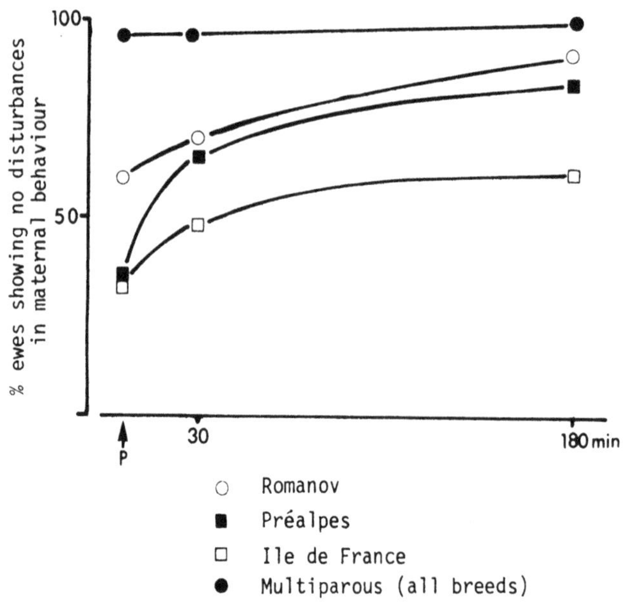


**FIGURE 4. Proportion of primiparous and multiparous ewes showing no disturbances of maternal behaviour in the three first hours after lambing in three different breeds. (Adapted from Poindron and Le Neindre 1984).**

On the other hand, birth difficulties can be a contributing factor in reduced maternal interest at parturition, as is well known in the case of severe dystocia. A similar phenomenon can occur in primiparous ewes, even in the absence of any obvious sign of dystocia. When comparing primiparous and multiparous ewes of three different breeds, lambed indoors, we found that the frequency of lamb desertion decreased as prolificacy increased (Ile de France 29% — prolificacy 1.5; Prealpes du Sud, 16% — prolificacy 1.6; Romanov 8% — prolificacy 2.4). This relationship was also associated with variations in the mean weights of the lambs according to the breed (IF: 3.9 kg; PA: 2.9 kg; R: 2.2 kg).

Furthermore, in primiparous IF ewes, the mean weights of lambs deserted by their dams were significantly higher than of those accepted by their mothers (4.24 kg vs 3.56 kg; $P < 0.01$) and the proportion of ewes rejecting their lambs was higher in single-bearing IF ewes than in multiple-bearing IF ewes (39% vs 23%; $P < 0.04$). However, these two last relationships were not found in the other two breeds. Therefore, supposing that birth weight is an acceptable indicator of the degree of birth easiness, these results suggest that rejection of the lamb due to birth difficulties is more likely in some breeds than in others.

Easiness of birth as indicated by length of labour produces a similar relationship (Alexander 1960; Shelley 1970). Our results also underline that analysis of differences, whether individual or between breeds, is best carried out on primiparous ewes, where interactions between the various factors are most likely to be observed.

Another interaction to bear in mind, especially under Australian conditions, is that which might exist between experience and nutrition, though such a relationship is not well documented. Under-nutrition during pregnancy is probably associated with poorer maternal care (Alexander 1960) but the effects of nutrition on maternal behaviour *per se*, especially in interaction with parity and maternal experience, remain largely unknown. There is indirect evidence from studies on white-tailed deer (Langenau and Lerg 1976). At a low diet mothers frequently rejected their young, although milk was present in the udder. Rejection of offspring also appeared more frequent in primiparous mothers than in does which had fawned the year before but numbers of animals in the study were too small to draw conclusions on this latter point.

Thus, maternal behaviour in inexperienced ewes is not elicited by physiological parameters as easily as in experienced mothers. In addition, external factors such as birth difficulty or nutrition are more likely to interfere with the normal development of maternal behaviour in maiden ewes, while experience in older animals helps them overcome disturbance by external factors.

## CONTROL OF THE SELECTIVITY OF MATERNAL BEHAVIOUR

### Evidence for selective behaviour and dynamics of establishment

It is well established that most ungulates preferentially suckle their own young. This is most obvious in sheep and goats, where mothers will reject, often with aggressive behaviour, any alien young trying to suck. Sucking of alien mothers by lambs can be observed, though this is usually restricted to over-crowded situations (Poindron and Le Neindre 1980). Selective behaviour at suckling is established very quickly after parturition and most ewes will reject alien lambs after a mother-young contact of only two hours, even if the alien young are neonates which have not been licked by their own dam (Poindron and Le Neindre 1980). This, of course, greatly limits the opportunity for fostering extra lambs.

Recognition at a distance is another aspect of discriminatory maternal behaviour in sheep but as it does not seem as closely associated with selective suckling as olfactory recognition, it will be treated in another section.

### Sensory cues involved in discriminatory behaviour of ewes at suckling

All experiments carried out suggest that acceptance at suckling depends primarily on olfactory cues. Ewes made anosmic before parturition, either by olfactory bulb removal or by destruction of the nasal mucosa by zinc sulphate, fail to establish selective behaviour at suckling. When olfactory cues are suppressed after the dam has established her selective behaviour, results are not as consistent, but selectivity is still clearly disturbed. Anosmia by zinc sulphate two weeks after lambing led to acceptance of alien lambs in about 50% of ewes, whereas the other half showed disturbance in their ability to suckle their own lambs. By contrast, recognition at a distance was not affected (Poindron and Le Neindre 1980). Washing lambs also affects their acceptance by the mother (Alexander and Stevens 1981). Thus anosmia appears suitable as a fostering method only if it is performed before parturition. It is possible that, although olfaction plays the primary role in selectivity at suckling, other cues come into play several weeks after lambing. This could explain part of the variation observed when anosmia is performed when the lamb is two weeks old.

The chemical components responsible for olfactory discrimination between individual lambs have never been identified, although such identification might allow the development of fostering techniques using substances to mask the odour of an alien lamb. Until recently it was thought that the lamb simply had some individual smell that the ewe learned to recognise. But more recent work carried out in sheep and in goats, suggest that other factors are also involved in the acceptance of young. As previously described, amniotic fluid plays a role in the attraction of the ewe towards a lamb. This substance appears to some extent to be 'anti-selective' in that it facilitates the adoption of alien lambs, especially if they are dry (Levy and Poindron 1984; F. Levy, unpublished).

Other studies in goats have led to the suggestion that, rather than the existence of individual kid odours as the basis for maternal discrimination between the young, does might label their own kids through licking and suckling, and reject kids which have been labelled by other dams. As a corollary unlicked kids fed artificially with a milk substitute would still be accepted since they have not been labelled (Gubernick 1981). Whether labelling occurs in sheep remains an open question. In many of our experiments there are situations where ewes clearly discriminated between their own lamb and an unlabelled alien neonate. Also, ewes unable to lick and feed their lambs were still able to discriminate as long as they could smell them (Poindron and Le Neindre 1980) and lambs fed with the colostrum of alien ewes were accepted by their dams, although the mothers could smell but not lick them before the test (P. Poindron, unpublished). On the other hand, in separate trials unlabelled lambs fed artificial milk were more readily accepted than labelled lambs and ewes unable to suckle their lamb did not establish selective behaviour (G. Alexander, pers. comm.). One difficulty in interpreting results on labelling in goats and sheep is that unlabelled young, as they are not licked, retain more amniotic fluid in their coat then young cleaned by their dams, and thus possibly become more attractive.

### Mechanisms controlling the establishment of selective suckling

Studies on the dynamics of the establishment of selective behaviour indicate that the ewe rapidly learns to recognise some olfactory characteristic of her lamb, and is thereafter unable to become attached to another lamb. However, it is clear that ewes experiencing multiple births can become attached to more than one lamb. Comparisons between single and twin-bearing ewes tend to suggest that the expulsion of the foetus is implicated in the control of selective behaviour. Single-bearing ewes begin to show signs of discrimination between an alien neonate and their own as early as 30 min after parturition (Poindron *et al.* 1980a), while twin-bearing ewes normally accept their second lamb, even if it is expelled much later than 30 min after the first one.

Furthermore, if the second born is exchanged for an alien newborn, ewes show signs of rejection in only 20% of cases (3/16, P. Poindron, unpublished) whereas this proportion is doubled if the alien neonate is added to a single-bearing ewe after she has been in contact with her own lamb for 30 min (6/14). Subsequent experiments have more clearly illustrated the role of genital stimulation in the establishment of selectivity. Post-parturient ewes, having already established selective behaviour, will accept an alien neonate two hours after parturition following mechanical stimulation of the genital tract for 5 min (9/12 cases — Figure 2B) (Keverne *et al.* 1983). Although the mechanism by which genital stimulation acts on the brain is unknown, it remains a potential technique for enhancing the success of fostering lambs onto ewes in the first hours following parturition.

### Mechanisms of discrimination at a distance

Olfactory recognition of the lamb appears limited to distances less than 0.25m. Beyond this, mothers rely on sight and hearing to recognise their young (Alexander and Shillito 1977a; Poindron and Le Neindre 1980). Recognition by sight, which depends mainly on cues from the head of the lamb (Alexander and Shillito 1977b), appears most efficient at distances shorter than 6 to 8 meters (Alexander and Shillito 1978).

No systematic study has been so far undertaken of the mechanisms of establishment of recognition by sight and hearing of the lamb by its mother. Whether it relies on a process of rapid acquisition as for olfactory recognition, or is acquired more slowly is not known, and its importance in the event of fostering later than the first days after lambing may have been overlooked. The only information available is that the ability of ewes to recognise their young is impaired by lambing in overcrowded situations indoors rather than in a paddock situation (Poindron and Schmidt 1984).

## CONCLUDING REMARKS

Recent progress in our understanding of the control of maternal behaviour offers immediate perspectives on its application to reduce lamb losses. However, these new techniques need to be adapted to the conditions of the Australian wool industry. In addition a long-term approach to bring lamb survival to an optimum implies further research on aspects of maternal behaviour inherent in Merino ewes under Australian conditions (e.g. primiparous ewes, interaction with nutrition, multiple-bearing ewes).

### Practical applications

Empirical (and successful) fostering techniques have been in use for a long time. However, there are two main drawbacks. Firstly, most procedures only allow the *replacement* of a dead lamb by an alien (twin or young deserted by its dam); the *addition* of an extra lamb remains very difficult. Secondly, fostering implies confinement of the foster ewe, usually impractical under the extensive conditions prevailing in most parts of Australia, except perhaps for stud breeders. On the other hand, there is a trend to use multiple birth as a means of increasing productivity, either by use of prolific strains such as Booroola or hormonal techniques to induce multiple ovulation. Although increasing the number of lambs born is quite easy, the maternal ability of Merino ewes to rear multiples reamins unchanged and is usually poor (Alexander *et al.* 1981). As a consequence the need for fostering is increased. Recent studies on the control of maternal behaviour could help to develop techniques more suitable for Australian conditions. Three possible methods are presented in Table. 1.

**TABLE 1. Possible fostering methods in sheep in relation to the physiological control of maternal behaviour**

| Type of technique | Type of ewes used | Advantages (+) and disadvantages (−) of technique |
|---|---|---|
| A Vaginal stimulation in parturient ewes | Parturient ewes with single lambs | +: cheapest and simplest. Suitable for flocks with a high degree of synchrony. −: need for high supervision at lambing. Foster ewes unknown in advance. period for fostering limited to a few hours, tend to produce 2 lambs reared per lambing ewe. |
| B Anosmia before lambing | Parturient ewes with single lambs | +: greater duration of fostering period than in A, characteristics of lambs of little importance in success of fostering. Highest rate of fostering success. −: single bearing ewes must be known before lambing, tends to produce 2 lambs reared per lambing ewe. |
| C Induction of lactation in non-pregnant ewes | Old dry ewes | +: easiest to manage on a large scale. Tends to give 1 lamb per ewe in the whole flock. −: induction of lactation is time consuming. most expensive method. need for pregnancy diagnosis to identify dry ewes at least 1 month before lambing. |

Other techniques are also available either as an alternative or in association with the methods presented here (tranquillizer, Tomlison *et al.* 1982; methods for masking and transferring the odour of the lamb, G. Alexander, pers. comm.). However, the adequacy and cost of the methods presented here remain to be assessed in farm situations. An attempt in this direction is currently being made at the University of Western Australia research farm, where non-pregnant ewes are being used to foster lambs from primiparous and multiple-bearing Merino ewes.

Future prospects of research

More research needs to be carried out on the onset of maternal responsiveness to increase the efficiency of methods available for inducing maternal behaviour in non-pregnant ewes. The validity of some of the results obtained needs to be verified for Merino ewes. For example, the efficiency of vaginal stimulation has never been tested in Merinos. In fact, breed and strain comparison concerning the control of maternal behaviour should be undertaken more systematically. In connection with the use of non-pregnant ewes to foster lambs, less time consuming methods to induce lactation need to be developed. Lastly, the association between anosmia and the induction of maternal behaviour in dry ewes may render this method more attractive, but this remains to be investigated.

Fostering techniques would also benefit from further work on the mechanisms controlling selective behaviour, especially when a lamb is to be presented to a ewe which has already established a selective bond with her own young. These studies should include investigations about the existence of labelling in sheep. Research on selective behaviour could also be important for understanding differences in the ability of mothers to form a strong bond with more than one lamb, one of the main problems encountered by multiple-bearing Merino ewes (Alexander *et al.* 1981). It cannot be ruled out that this is related to some variation in the degree of selectivity of ewes.

Considering the high level of lamb losses generally reported in maiden ewes, further studies on this topic appear necessary, especially under Australian conditions where the proportion of young ewes is usually high.

There is a clear necessity to identify the mechanisms leading to the absence of maternal behaviour in primiparous ewes, if long-term progress is to be achieved. In addition, maiden ewes seem to represent a suitable model of study interactions between physiological, genetic and environmental factors, since individual and breed differences are best observed in such animals. A point of particular interest is the study of interactions between maternal behaviour and nutrition, this latter factor being of great importance in Australia. These various studies would help to define criteria for selection on maternal behaviour, a field which has remained largely unexplored, although it could contribute in the long term to the efficient control of lamb losses.

Lastly, it must be kept in mind that lamb survival is also influenced by the young and its behaviour. Some work has been carried out (Alexander and Williams 1964, 1966; Stevens *et al.* 1984) but more research needs to be developed in this field. A point of particular interest already indicated will be to further assess the part the neonate plays in the occurrence of early cessation of contact with the dam, a major issue for the survival of multiple lambs.

## REFERENCES


Alexander, G., 1960. *Proc. Aust. Soc. Anim. Prod., 3*, 105-114.

Alexander, G., Kilgour, R., Stevens, D. and Bradley, L.R., 1984. *J. Appl. Ethol.* (in press).

Alexander, G. and Shillito, E.E., 1977a. *Appl. Anim. Ethol., 3*, 127-135.

Alexander, G. and Shillito, E.E., 1977b. *Appl. Anim. Ethol., 3*, 137-143.

Alexander, G. and Shillito-Walser, E.E., 1978. *Appl. Anim. Ethol., 4*, 81-85.

Alexander, G. and Stevens, D., 1981. *Appl. Anim. Ethol., 6*, 77-86.

Alexander, G., Stevens, D., Kilgour, R. and De Langen, H., 1982. *Appl. Anim. Ethol.* (in press).

Alexander, G. and Williams, D., 1964. *Science, 146*, 665-666.

Alexander, G. and Williams, D., 1966. *J. Agric. Sci., 67*, 181-189.

Gubernick, D.J., 1981. *Parental Care in Mammals*. Plenum Press, New York, 243-289.

Keverne, E.B., Levy, F., Poindron, P. and Lindsay, D.R., 1983. *Science*, 219, 81-83.

Langenau, E.E. and Lerg, J.M., 1976. *Appl. Anim. Ethol., 2*, 207-223.

Le Neindre, P., Poindron, P. and Delouis, C., 1979. *Physiol. Behav., 22*, 731-734.

Levy, F. and Poindron, P., 1984. *Biol. Behav.* (in press).

Levy, F., Poindron, P. and Le Neindre, P., 1983. *Physiol. Behav., 31*, 687-692.

Poindron, P. and Le Neindre, P., *1980. Advances in the Study of Behaviour, 11*, 75-119.

Poindron, P. and Le Neindre, P., 1984. *Genetique, Selection, Evolution*. (in press).

Poindron, P., Le Neindre, P., Raksanyi, I., Trillat, G. and Orgeur, P., 1980a. *Reprod. Nutr. Develop., 20*, 817-826.

Poindron, P., Orgeur, P., Le Neindre, P., Kann, G. and Raksanyi, I., 1980b. *Hormon. Behav., 14*, 173-177.

Stevens, D., Alexander, G., Mottershead, B.E. and Lynch, J.J., 1984. *Proc. Aust. Soc. Anim. Prod., 15*, 751.

Tomlison, K.A., Price, E.O. and Torell, D.T., 1982. *Appl Anim. Ethol., 8*, 109-117.

# CONSTRAINTS TO LAMB SURVIVAL

G. Alexander *CSIRO, Division of Animal Production, Prospect, NSW 2149.*

## INTRODUCTION

Failure to survive the perinatal period is a major form of reproductive inefficiency in sheep farming; not only is the opportunity of the ewe to reproduce lost for a year, but perinatal deaths reduce the volume of lamb for sale or reduce the number of surplus sheep for sale by the enterprise; they reduce the options for selection in breeding programmes and for culling to increase flock productivity, and they decrease the options to adjust flock structure for flexibility in management (Mullaney and Hyland 1967; Heath and Lee 1975). The opportunity cost of the 3-10% reduction in annual wool production due to pregnancy followed by lamb death, or about one bale per 1000 ewes lambing, is also appreciable (Corbett 1979).

This paper reviews and attempts to give biological and practical perspective to information on the causes of death, to discuss methods for reducing mortality and to suggest avenues for research. Precedence is given to Australasian work and to reviews, but unfortunately much useful Australian data has not yet been adequately analysed and published. Lamb mortality in Australia has been reviewed previously by Watson (1972) and Haughey (1983a).

Consideration of survival is particularly relevant at the present time, when methods for increasing ovulation rate are being promoted.

## MAGNITUDE OF LAMB MORTALITY AND ITS ASSESSMENT

Mortality of lambs can be assessed directly from carcass collection or indirectly by examination of ewes' udders to assess 'wet or dry', and of the breech for staining indicative of recent birth (Douglas 1965); losses on commercial properties are usually assessed by inference from the numbers of lambs present at 'lamb marking'. Losses can be grossly under-estimated if carcasses are not searched for carefully, or are removed by scavanging animals if there are many multiple births. An important new research method, real-time ultrasound scanning, allows accurate determination of total foetal number in a flock (Fowler and Wilkins 1982), and hence facilitates exact estimation of mortality.

Publications reporting levels of lamb mortality in Australia and elsewhere are too numerous to list. Rates vary from about 5% (McHugh and Edwards 1958) to more than 70% (Smith 1962). In a NSW survey covering 102 Merino breeders, 28% of flocks lost 15-24% of lambs, 18% lost 25-34% and 5% lost 35% or more (Plant *et al.* 1976), and this is probably representative of the national flock. Average levels of 10-20% are also experienced in NZ (Dalton *et al.* 1980)), the UK (Eales *et al.* 1983) and in other sheep raising countries (Cundiff *et al.* 1982). Lamb mortality in Australia as judged from lamb marking percentages has changed little during this century (Plant 1981).

## PATTERNS OF MORTALITY

Patterns of loss recur in the many flocks in which mortality has been closely studied.

### Age at Death

Deaths are invariably concentrated in the first few days after birth (Hight and Jury 1979), reflecting the problems of transition from a totally dependent intra-uterine existence to a less dependent extra-uterine life.

### Birth Weight and Mortality

Mortality in sheep and in other species is typically related to birth weight by a 'U'-shaped curve, the completeness of which will depend on the range of birth weights encountered (Mullaney 1969). Mortality is lowest at a weight between 3 and 5 kg, the optimum weight for survival, depending on breed and age of ewe (McMillan 1983).

### Litter Size and Mortality

Mortality usually increases with litter size (Watson 1972; Johnson *et al.* 1982), a relationship that is associated with a decline in birth weight as litter size increases. Mortalities of singles and twins in the same birth weight range are usually similar (Purser and Young 1964), but twin mortality can be above that of singles of the same weight (Stevens *et al.* 1982). Lambs born as triplets or quads have a lower chance of survival than twins or singles of the same birth weight (Hinch *et al.* 1983), probably because of competition and birth hypoxia.

### Age and Parity of Ewes

Age and parity are usually confounded. Mortality can be as high as 33% in primiparas (McMillan 1983; Dalton and Rae 1978), and is minimal over the next four or five pregnancies before rising again (Watson 1972).

Breed Differences

Differences in lamb mortality between various breeds and crosses, and even between strains of the same breed, running under comparable conditions have been reported by many workers (Dun and Hamilton 1965; George 1969; Dalton *et al.* 1980; Cundiff *et al.* 1982). Losses of lambs from crossbred ewes are usually lower than from pure breds (Dulton *et al.*1976; Ch'ang and Evans 1978), and dystocic losses in Australian Dorsets are notoriously high (George 1969). Breed differences may also influence the survival of multiples relative to singles (K. Haughey and B. McGuirk, pers. comm.), largely because of birth weight relationships and differences in milk supply.

Sex of Lamb

Male lambs consistently have a lower survival rate than female lambs of similar birth weight (Hight and Jury 1970).

STAGES OF VULNERABILITY

Lamb mortality is usefully considered against a biological background that emphasises developmental stages associated with deaths.

Prenatal Development

The stage is set for subsequent survival problems, even before implantation when the embryos must compete for the limited number of discrete attachment sites, which do not subsequently become available even if one foetus in a litter should die (Robinson 1981). The number and ultimate size of the placentomes supplying each foetus strongly influence the relative potential for foetal growth (Alexander 1974a; Mellor 1983; Bell 1984).

The stage of rapid foetal growth during the last two months or pregnancy is also vulnearable to growth retardation through inadequate nutrient or oxygen supply especially in polytocous pregnancy (Donald and Purser 1956; Mellor 1983; Bell 1984; Egan 1984).

Orientation of the foetus in preparation for birth is another critical stage, but mechanisms are unknown (Scanlon 1976).

Birth

The immediate perinatal period represents a most critical stage for survival. Birth is accomplished by the application of hydrostatic pressures with the potential to disrupt tissue (Haughey 1973a,b). Birth is also accompanied by a period of hypoxia and acidaemia during the transition from placental to pulmonary respiration; and those can be fatal if prolonged. Prematurity and associated lack of pulmonary surfactant, are not features of lamb mortality, as they are in man (Alexander 1976).

Post Natal Thermo-regulatory Adjustments

Circulatory and metabolic thermoregulatory adjustments are inevitable immediately after birth, when evaporation of fluids in the coat begins (Alexander 1974b). Increases in metabolic rate are achieved by activation of brown fat, which almost all newborn ungulates possess, and by shivering. Brown fat thermogenesis is activated before shivering, and so permits a significant increase in metabolism without interfering with the fine muscular movement necessary if the newborn lamb is to find the teats and suck (Alexander 1980). However, thermorgeulatory responses have clear limits (Alexander 1979), and under conditions of rapid heat loss, the lamb's responses may be inadequate; hypothermia results, often leading to death. Cold resistance of lambs varies widely, and is breed dependent; Merinos rank poorly (Slee 1981). Small lambs, hypoxic lambs, lambs that have experienced difficult birth, fasting lambs and lambs with sparce birthcoasts all have a poor cold resistance. However, the size of the energy reserves is not obviously related to cold resistance (Alexander and Bell 1975). Less is known about the ability of newborn lambs to resist hyperthermia (Alexander and Williams 1962, Smith 1965) although it can be an important cause of death (Smith 1964).

Behaviour of the Ewe and Lamb

Peripartum behaviour patterns have been much described for sheep including Merinos (Arnold and Morgan 1975; Alexander 1980). The physiological control the maternal patterns is described by Poindron *et al.* (1984). These patterns ensure that the ewe supplies only her own offspring with milk, the sole form of nourishment for young lambs; and inappropriate behaviours at this time are major causes of mortality.

Isolation seeking to faciliatate exclusive bonding, and birth site selection to minimize environmental hazards are often not followed to the best advantage (Stevens *et al.* 1981); indeed birth sites are often chosen 'unwisely' (Kilgour *et al.* 1983).

Grooming of the newborn is essential for the formation of a firm exclusive maternal bond, but inadequate grooming and poor bond formation are common with multiple births and difficult births, and where other parturient ewes participate in grooming (Kilgour *et al* 1983; Alexander 1980, 1984).

Failure of mothers to recognize their litter size readily, during the 24 h *post partum*, when lambs are learning to follow, can lead to separation and death of twin lambs, especially if the ewe leaves the birth site earlier than

4-6 h after birth (Alexander *et al.* 1983a). This behaviour is more common in Merinos than in other breeds and the incidence of resulting separations is dependent on lamb genotype (Stevens *et al.* 1984). Separation from multiples is also more common in ewes without experience of twin-care than in experienced ewes (Alexander *et al.* 1984). These behavioural aberrations can also lead to problems in recording litter size and pedigree (Alexander *et al.* 1983b).

Little is known about the nature of the guiding cues in lamb teat seeking (Vince 1984) and about whether olfaction is important as it is in newborn rabbits (Hudson and Distel 1983). The search for the teats is hampered by cold exposure (Alexander 1980) or birth injury (Haughey 1980), and by primiparas failing to co-operate with lambs attempting to suck for the first time; continued failure to find the teats diminishes the changes of successful sucking (Alexander 1980).

### Energy Reserves

The energy reserves of newborn lambs are meagre by comparison with reserves in many other species (Alexander 1974a). Total reserves are only about 4000 kJ. Since metabolic rates range from about 15W at thermoneutrality to about 100W representing the maximum thermogenic effort, or about 60W representing the rate substainable in the cold, starving lambs exhaust their reserves after about 3 days in warm conditions and after about 20 h or less in the cold (Alexander 1962). This explains the age distribution amongst neonatal deaths. There are wide variations both in energy reserves and maximum metabolic rate, that are only partly due to birth weight (Alexander 1962, 1974a).

## RELATIVE IMPORTANCE OF VARIOUS CAUSES OF DEATH

### Methods

Historically two approaches to assessing the causes of lamb deaths in Australasia have been used during a series of studies spanning 30 years.

*Autopsy approach* The autospy approach coupled with pathology was used in large scale surveys in NZ by McFarlane (1965), in eastern Australia by him and colleagues (Hughes *et al.* 1964; Haughey 1973a,b) and in WA by Dennis (1974). The procedure included carcass examination for evidence of autolysis indicative of prenatal death, examination of lungs for evidence of breathing, of feet for evidence of walking, of liver for evidence of birth trauma, of subcutaneous tissues for oedema indicative of dystocia, of thyroids for evidence of goitre, and of fat for evidence of depletion indicative of starvation. More recently, autopsies have included examination of the extremities for evidence of a yellowish oedema indicative of cold exposure (Haughey 1973c), and of brain and spinal cord for haemorrhages indicative of birth injury (Haughey 1973a,b). The carcass was also examined to determine whether any predator damage had occurred before or after death. Damage definitive of specific predators has been described by Rowley (1970). The autopsy method provides a basis for classifying lambs according to their time of death in relation to birth, and provides information about causes of death in a broad sense.

The surveys have shown that *antepartum* deaths usually comprise less than 2% of the total, with the remainder variably divided between *partum* and *post partum* deaths. Starvation is usually implicated in *post partum* deaths, hence the term 'starvation-mismothering syndrome', a term that rivals 'ideopathic' in explaining the circumstances of death. Other short-comings are the lack of criteria for reliably determining whether death was due to cold exposure as discussed later, or for assessing whether many of the lambs killed by predators soon after birth were already debilitated.

The autopsy approach has revealed a variety of minor specific causes of death including infection, specific nutrient deficiencies and congential abnormalties (Haughey 1983a). In the last 10 years autopsy findings have also focussed attention on a high incidence of central nervous system (CNS) birth injury that was previously unsuspected (Haughey 1973a,b).

*Direct observation* The other approach to assessing causes of death is based on direct observation and recording of behaviour, together with objective measurements such as lamb temperatures. This approach has been widely used in the CSIRO Animal Research Divisions and in State Departments of Agriculture (Alexander and Peterson 1961; Smith 1962; Alexander *et al.* 1967; Watson *et al.* 1968; Obst and Day 1968; Egan *et al.* 1972; Arnold and Morgan 1975; Luff 1980; Stevens *et al.* 1982). The approach is also based on a series of fundamental studies, already mentioned, on the physiology of thermoregulation, energy expenditure, placental and foetal growth and development, and mother/young behaviour in sheep (Alexander 1974a,b, 1978, 1979, 1980) that provided a framework for intelligent interpretation. The approach is labour intensive and can be applied to individual flocks only; it is not suitable for large scale surveys.

In recent years these two approaches have been used in conjunction (Plant 1977; Alexander *et al.* 1980; Haughey 1980), and together with a detailed investigation of husbandry procedures, were used by the 'Fertility Service' of the NSW Department of Agriculture in investigating flocks with poor reproductive performance (Luff 1980; Plant 1981). The service was able to identify and suggest remedies for the major causes of lamb deaths on mose properties where these exceeded 15%.

Causes of Death

Researchers largely agree on the catalogue of factors that contribute to lamb deaths, but there is controversy about their relative importance.

*Problems with parturition* Dystocia or difficult birth, due to foeto-pelvic disproportion or to malpresentation, has long been known as a cause of lamb death (Columella c. 64A.D.; George 1976). However, there is controversy about birth injury, which Haughey (1973a,b, 1983a) believes to be implicated in 56-80% of all lamb deaths, a claim based on his autopsy findings that a high proportion of dead lambs show haemorrhage in the CNS, grading from severe to barely perceptible, together with studies on experimentally delayed birth in ewes, and on survival and behaviour of their lambs (Haughey 1980). NZ workers (Gumbrell 1980; Duff *et al.* 1982) have also reported a high incidence of CNS lesions in lambs at autopsy. Other workers claim that the incidence of birth injury is lower, but there is little published data to support this (Alexander *et al.* 1980). In studies at Armidale NSW the incidence of difficult birth ranged from 4.1% of lambs born in Merinos to 34% in Dorsets (George 1975, 1976); about one third of all lambs deaths was attributed to difficult birth amongst the Dorsets, despite obstetrical assistance. In dystocic deaths the incidence of CNS haemorrhage is almost 100% (Haughey 1973a,b; Duff *et al.* 1982) which may help explain mechanisms, although there is no evidence the the haemorrhage is the direct cause of death; associated hypoxia and acidaemia could be more important. However, the important implication of Haughey's findings lies in the incidence of CNS haemorrhage in dead lambs in the starvation/mismothering/exposure category. Haughey (1983a) believes that birth injury has caused most of these lambs to fail to behave or thermoregulate normally, and that birth injury is the major primary cause of death in this category, although many of the 'injured' lambs would survive under fabourable conditions. This is a controversial issue. Duff *et al.* (1982) detected CNS lesions in only 32% of 'starved/ exposed' lambs in a study of 1414 lambing Romney ewes with an overall lamb loss of 16%; and Alexander *et al.* (1980) detected lesions in about 30% of lambs drying of chilling soon after birth, 20% of lambs dying of chilling subsequently and in only 10% of starved lambs, in a 5 year study covering 1153 fine wool Merinos with an overall loss of 19%. Duff *et al.* concluded that CNS birth injury was only a minor cause of starvation/ exposure deaths, because the injury was usually slight and because the proportion of injured lambs that had sucked before death was similar to that of uninjured lambs. However, the results of Alexander *et al.* showed that CNS haemorrhage decreased the cold resistance of lambs that had sucked. The incidence of CNS lesions in surviving lambs *in the field* has never been investigated, but in an experimental study involving slaughter of lambs (Haughey 1983a) only 21% of lambs with lesions died naturally. Difficult birth also interferes with normal maternal behaviour, and so could precipitate death from starvation even in the absence of CNS lesions. Variations in the incidence of CNS lesions amongst populations of dead lambs probably reflect breed or strain differences and differences in the seasonal incidence of other factors, such as weather or pasture scarcity that predispose to mortality.

Regardless of the controversy, problems of birth are clearly an important cause of mortality. Birth difficulty is related to birth weight (Gunn 1968; Duff *et al.* 1982) and to small pelvic size, at least in some breeds (Fogarty and Thompson 1974; McSporren and Fieldon 1979; Haughey and George 1982; Haughey 1984); both factors must be considered in devising strategies to improve survival.

*Cold exposure* Exposure, as a primary cause of lamb deaths, becomes apparent when large numbers of lambs die during or soon after periods of a few hours of low temperatures (less than about 5°C) with wind and rain, or after prolonged rain; deaths associated with 'bad' weather have been closely observed by Alexander *et al.* (1959, 1980), and Obst and Day (1968). However, death of individuals in these circumstances cannot be attributed with certainty to exposure as a primary cause, because lambs debiliated for other reasons, such as starvation, are highly susceptible to chilling; and conditions such as low birth weight, birth 'injury' and sparse birth coat predispose lambs to exposure, though they would survive under benign conditions. However, a relationship between birthcoat 'thickness' and survival is not a consistent finding in Australia (Mullaney 1966; Obst and Evans 1970).

There is no reliable autopsy criterion that death was due to exposure; the peripheral oedema described by Haughey (1973c) as cold injury is not seen in the many lambs that die rapidly soon after birth, (Alexander *et al.* 1980). Oedema is seen in lambs dying more slowly in 'bad' weather after having sucked; prolonged exposure seems to be required for severe oedema to develop, although Haughey (1973c) reported slight oedema in lambs exposed for as little as 4 h. Consequently, the presence of oedema at autopsy may be no more reliable than weather reports in assessing whether exposure has contributed to death, but lambs with oedema and with milk in the stomach are almost certain to have died from exposure (Alexander *et al.* 1980); their presence amongst lambs coming to autopsy would indicate the occurrence of weather severe enough to kill lambs.

It is not possible to estimate the proportion of Australasian lamb mortality that is primarily due to exposure.

*Starvation* The reasons for failure to suck successfully, in the large group of lambs in which death is attributed to starvation/mismothering, can be assessed from direct observation only, provided there is no obvious cause such as congenital abnormalities or severe birth injury. The reasons include failure of the ewe to bond with her lamb(s), 'accidental' separations after bonding (Alexander 1980), udder defects such as shearing injury or mastitis (Hayman *et al.* 1955; Quinlivan 1968) or delayed lactogenesis usually of nutritional orgin (McCance and Alexander 1959), competition with litter mates, and birth problems that affect behaviour of both the ewe and the lamb (Poindron *et al.* 1984). Problems specific to prolific sheep were reviewed by Holmes (1975).

*Size at birth* Extremely high or low birth weights are predisposing factors to death from birth injury, cold exposure or starvation. The highly susceptible birth weight groups for most breeds are below 3.0 and above 5.0 kg. These groups usually include only a small proportion of lambs, but account for a much higher proportion of the total mortality (14% and 32% respectively in the studies of Stevens *et al.* 1982). However, most dead lambs are in the medium birth weight group. In some breeds such as the Campbell Island feral sheep of NZ (Wilson and Orwin 1964), the Soay or Boreray in which birth weights are usually below 3 kg, the birth weight-mortality curves appear to be displaced to the left.

Birth weight of lambs is sensitive to a variety of influences including diseases and heat stress in pregnancy, prenatal nutrition, placental size, sex, litter size and parental characteristics (Bell, 1984), but there is little direct evidence to explain why the mortality of small lambs is high under field conditions. It is not due to infectious disease which accounts for less than 2% of lamb deaths in Australia (Hughes *et al.* 1971). Although the behaviour of small lambs has received little attention, there is relevant physiological and anatomical evidence that has been reviewed elsewhere (Alexander 1974a). Small lambs are disadvantaged in theromoregulation, in the relative magnitude of their energy reserves (Robinson 1981) and in the proportional size, and presumably function, of various tissues and organs.

*Minor causes* The catalogue covers a variety of causes that are minor on a national scale, but some of which can be catastrophic for a producer. They include congential malformations, which the autopsy based surveys show to account for only about 1% of deaths, various pre- and post-natal infectious diseases, deficiencies of minerals such as copper, iodine and selenium, plant toxins (Broadmeadow *et al.* 1984), and misadventure (Houston 1974). These causes have together accounted for less than 8% of deaths in large scale surveys; they have been adequately reviewed by Haughey (1983a).

*Predation* Predation by foxes, dingoes, feral dogs, feral pigs, eagles or ravens is often made the scapegoat by producers seeking to explain their losses. Their claims are based on the presence of potential predators and on the large proportion of carcasses showing mutilation. The majority of these lambs will have been dead or debilitated when mutilated (Dennis 1969). Many potential predators do not kill lambs (Alexander *et al.* 1967); but large losses due to ravens, foxes and pigs have been recorded (Smith 1964; Plant 1981). At autopsy, predation as the major cause of death is indicated by characteristic wounds, bleeding, replete adipose tissues and milk in the stomach, but in lambs dying soon after birth autopsy evidence could be equivocal.

Perspective

Starvation, cold exposure, problems with parturition, and size at birth are four, often interrelated factors that are involved singly or in combination in most lamb deaths, and which are themselves affected by a variety of factors, especially by the level of nutrition during pregnancy.

In my personal experience, when mortality levels are below about 15%, it is usually possible to attribute death confidently to a direct specific cause in a minority, say one third, of lambs. With the majority, most of the evidence will be circumstantial, and the relative importance of the various contributing factors will not be clear. With losses in excess of around 15%, major single factors may emerge (Plant 1977), such as dystocia in Dorsets (George 1976) or predation by feral pigs (Plant *et al.* 1978) or poor maternal care of twins by New England finewool Merinos (Alexander *et al.* 1980), or cold exposure (Obst and Day 1968). How much of the '15% core loss' in most flocks is due to birth-related problems remains to be seen.

The causes of lamb death are so complex and interrelated that overemphasis of any one cause by research or extension workers will be counter productive in reducing losses.

## STRATEGIES FOR IMPROVING SURVIVAL

Diagnosis of the causes of death in any one flock is necessary before appropriate remedial strategies can be recommended (Plant 1981). In the absence of factors such as disease, predation or shearing damage to udders, the major requirements for a high survival rate are clearly birth weight near the optimum for the breed, easy birth, protection from cold, and maximum contact between ewes and lambs during the first 12 h *post partum*. In fact, almost all lambs will survive if given adequate obstetrical assistance, warmth and food (Alexander *et al.* 1959; Alexander and Peterson 1961), but questions of economics, expected incidence of multiple births, likelihood of cold, wet and windy weather during lambing, and farmers' attitudes will determine the measures that are taken towards obtaining these objectives. Most Australian farmers seem reluctant to put a value on a newborn lamb; it would certainly be no more than a few dollars at present, whereas in Britain for example, farmers see each lamb as representing around 30 pounds income. However, these is an urgent need for practical field trials of various strategies for reducing mortality in Australia.

Nutrition during Pregnancy

Work on nutrition of the pregnant ewe has continued for many years (Robinson 1983; Bell 1984; Egan 1984 and these requirements are still debated, perhaps because of difficulty in agreeing on objectives such as control of metabolic diseases, permissible ewe weight loss and embryonic loss, optimal birth weight and vigour of lambs, maternal vigour for parturition amd adequate lactation. Fine tuning will no doubt continue, but at present recommendations are to avoid severe undernutrition at all stages of pregnancy and to feed well in the last third of pregnancy particularly ewes carrying litters.

Treatment of Ewes According to Litter Size

Ewes carrying multiples have a higher nutritional requirement than ewes with singles, and their requirement for supervision at lambing could be different. Ewes with singles are more likely to require obstetrical assistance than ewes with twins, and are less likely to be separated from lambs than ewes with twins; more ewes with twins may require intervention to rectify behavioural problems than ewes with singles.

Various methods for diagnosing litter size have been suggested and tried, but until recently none, including x-ray, has proven acceptable. Now, real-time ultrasound scanning shows promise if costs can be minimized (Fowler and Wilkins 1982). There is doubt whether survival of multiples can be improved nutritionally with pasture feeding (Johnson *et al.* 1982), but preferential treatment of ewes with multiples would be highly desirable when pasture conditions were poor (Langlands *et al.* 1974). Strategies for delivering supplements or drought rations without precipitating mismothering need to be developed; feeding out at night has shown promise.

Shearing and Crutching before Lambing

The practice of shearing ewes several weeks before lambing originated in NZ (Coop and Drake 1949) and is now practised in about half of the flocks in NSW. The rationale is that recently shorn ewes seek out shelter and lamb there. This does occur, but the practical value has been inconclusive in experimental studies (Alexander *et al.* 1980). There is also equivocal evidence that the cold exposure due to shearing pregnant ewes increases maternal nutrient supply to the conceptus, and can increase maternal appetite, with beneficial effects on birth weight and viability (Thompson *et al.* 1982), but in seasons of poor nutrition, lambing off-shears would be contra-indicated because under-nourished recently shorn ewes are cold susceptible.

Prelambing crutching is sometimes practiced for reasons of hygene, or in the belief that it facilitates the first suckling of the newborn lamb by the ewe, but there is little documented evidence about its benefits.

Selection of Lambing Paddock and Provision of Shelter

Shelter that reduces wind velocity across a lambing paddock can reduce heat loss from lambs sufficiently to increase survival significantly; for example from 83 to 91% in singles, and from 49 to 64% in multiples, in Merinos lambing in winter in New England (Alexander *et al.* 1980); and from 81 to 94% in singles in Merinos in western Victoria (Egan *et al.* 1972). In the New England study, rows of a sterile Phalaris hybrid (*P. aquatica* × *P. arundinaceae*) at 20 m intervals across the whole lambing paddock were used, but there are problems with the need to exclude grazing animals during establishment, and with farmer acceptance generally. A search for a less palatable hybrid is continuing. In the Victorian study, saved *P. aquatica* pasture with mown access paths was used. Trials with contrived shelter on known camping sites have been unsuccessful with woolly sheep largely due to the erratic use of the sheltered area. The use of taller windbreaks such as trees or shrubs to save lambs appears not to have been studied experimentally.

Stocking Rate and Paddock Size

There is almost no information on optimum stocking rates and paddock sizes for lambing, but mismothering appears to increase disproportionately with stocking rates above 18 per ha (Alexander *et al.* 1983), and ewes producing multiples may be best lambed in small paddocks with several watering points and good pasture, to minimize the chances of permanent separation of lambs from ewes. Flocks in which oestrus had been synchronized would be particularly prone to mismothering.

Husbandry Options at Lambing

There have been few if any comparisons of various husbandry options during lambing in which various key comparisons, such as drifting versus set stocking or shepherding versus no shepherding, have not been confounded with other factors, such as the presence or absence of fox predation (Moore *et al.* 1966) or the age at which lambs were mulesed and marked (Tyrrell and Giles 1974).

*No care* Many producers consider that any interference with the lambing flock is likely to cause mismothering that would outweigh advantages of saving cast or dystocic ewes, an assumption that does not hold with sheep used to the presence of a shepherd (Stevens *et al.* 1982). Proponents assume that natural selection will eliminate ewes with problems, and lead to development of a flock with 'easy-care' characteristics (Warren 1975).

*Minimal interference* The most common practice in Australia is to inspect the flock once or twice daily with the purpose of saving ewes in difficulty. However, this could disturb lambing ewes and cause further losses.

Fostering or bottle feeding of stray lambs is sometimes practised on a small scale with family labour; methods of fostering are currently being investigated (Alexander *et al.* 1984 and unpublished).

*Intensive shepherding outdoors* Intensive care of the lambing flock is common in many countries, but is rare in Australia. Ewes are inspected day and night, obstetrical assistance is provided, ewes stealing lambs are isolated, unsuckled lambs are identified and assisted to suck, often after stomach-tube feeding with cows' colostrum (Robinson 1981), chilled lambs may be revived with heat and intraperitoneal glucose injections (Eales *et al.* 1982), problem mothers are penned with their lambs, and fostering is practised with the object of having all lactating ewes rear 1 or 2 lambs.

*Drift lambing* With drift lambings, each day's crop of lambs and new mothers is separated from the pregnant ewes for ease of management and recording of pedigree (Mayoh 1969). The method is largely confined to research establishments. To minimize interference with bonding, the unlambed ewes should be moved on. A potential advantage is the possibility of drifting into a sheltered paddock if bad weather threatens. At Trangie NSW, mortalities were higher with set stocking without supervision than with drifting with supervision (Giles 1968), but the difference vanished when set stocking was confined to ewes that lambed within a day or two of each other (Tyrrell and Giles 1974).

*Indoor Lambing* Intensive lambing under shelter is widely practised in North America and Europe and has been practised In Australia with valuable stud animals or where deaths due to exposure are common (Blackburn and Rizzoli 1968). Artificial rearing of problem lambs can be incorporated. Indoor lambing is still used occasionally in Australia. To facilitate bonding, ewes and lambs are confined to small pens for 24 h, or longer with multiples. Lambing cubicles with a high entry step, leading from a common pen, facilitate bonding and identification (Gonyou and Stookey 1981). Indoor lambing is capital and labour intensive.

## Culling and Selection for Survival

*Culling* The possibility of improving reproductive performance by culling out ewes that do not rear lambs in early life has been often considered and sometimes practiced (Aitkens 1980; Purser and Young 1983).

On the basis of records from 3 Merino and 1 Dorset flock, covering 4 lambings, Haughey (1983a) claimed that 63% of failures to rear lambs were by ewes that repeatedly lost lambs. However, these data have since been subjected to rigorous analysis that indicates that rearing performance has a statistical repeatability of only 10% (Haughey, 1984), a level that is similar to that for other components of reproductive performance (Cundiff *et al.* 1982). A variety of genetic and non-genetic factors could contribute to repeated rearing failures; they include small pelvic size (Haughey, 1984) and foetal oversize associated with a prolonged gestation (Haughey 1983a), and udder malformation, shearing injuries and mastitis. A practical culling policy has yet to be devised.

*Genetic improvement* For several decades economic conditions in rural Australia have favoured the reduction of inputs, including labour and capital for lambing systems; the 'no care' option is probably the most common today. The use of sheep with a genetic propensity for lamb survival would therefore be an attractive option.

It is not surprising that natural selection should fail to make rapid improvements in survival, in view of the variety of factors that contribute to mortality and their fluctuating importance from year to year, and because any character important in survival is likely to be controlled by many genes. In additions, sheep classing may have acted against natural selection for survival. The evidence that survival is heritable was reviewed by Cundiff (1982), Piper (1982) and Fogarty (1984). The average heritability of the maternal component of survival (i.e. ewes' ability to rear lambs) was estimated as twice that of the lamb viability component, but standard errors were high and both values were low (0.08 and 0.04). Haughey (1983b) presented direct evidence that survival due to both factors combined is heritable, at least in one flock.

There is also evidence that some of the components of survival are heritable. For example, cold resistance of lambs is highly heritable (20-30%) (Slee pers. comm.), and correlates of cold resistance such as birthcoat and haemoglobin type are heritable (Schinckel 1955; Obst and Evans 1970). Birth coat thickness which contributes to cold resistance is also heritable (Purser and Karam 1967), and differences in survival rates of progeny between rams, may be genetically based through effects on birth weight (Knight *et al.* 1979). There are breed differences in the incidence of dystocia (Geoge 1975, 1976; Whitelaw and Watchorn 1975) and even between strains of the same breed such as Dorsets and Merino (Dun and Hamilton 1965; George 1980). Susceptibility to dystocia appears to be heritable (Smith 1977) and there are suggestions that pelvic size may be inherited (Haughey 1983a).

Selection programmes with Merino appear to have increased lamb survival at Trangie (Aitkins 1980) and Trundle (Donnelly 1982) in NSW and in Romneys in NZ (Parker 1974; Hight 1975; Quinlivan 1981) but it is not clear how much of the improvement is due to culling, to selection for increased litter size, and to improved lamb survival. Similarly, NZ claims about the development of 'easy care' lambing attributes (Warren 1975) especially by sheep that ran feral for generations, do not appear to be well documented (Kilgour and de Langen 1980). However, the 'Marshall' Romneys (Anderson and Marshall 1967) that had run on harsh hill country in the South Island have 95% survival of lambs, superior to that of other Romneys (Anon 1976), and feral sheep after 90 years in the subantarctic climate of Campbell Island are said to show a high and heritable cold resistance and high survival rate despite small body size (M.L. Bigham pers. comm.).

In view of the low apparent heritability of maternal and offspring components of lamb survival, it is 'appropriate to focus efforts on specific components of survival to develop more effective selection criteria' (Cundiff *et al.* 1982). These components should be scored on a quantitative rather than quantal basis (B. McGuirk, pers. comm.). Promising components include cold resistance of lambs, lamb vigour, ease of birth and mothering behaviour; it may be more difficult to select for an appropriately narrow range of birth weight, but factors such as placental size that control birth weight and with a possible genetic basis are being considered (Davis *et al.* 1981). Selection criteria for mothering ability would include birth site selection in respect to isolation and shelter, duration and intensity of grooming, period that ewe remains at the birth site, ability to recognize litter size and to keep multiples together, bond strength, and temperament that might reflect ease of disturbance during bond formation.

The Merino is a poor mother by repute (Whately *et al.* 1974) and ranks poorly when assessed subjectively, but the breed is adapted to difficult conditions (Squires 1981) and one must ask what other behavioural characteristics such as foraging or ranging behaviour would change if ewes were selected, for example, on the period for which they remain at the birth site.

## CONCLUSION

Loss of lambs remains a major source of reproductive inefficiency in the Australian sheep industry, despite research that has provided understanding of the wide variety of contributing factors and of the associated physiology and pathology. The major factors are birth weight extremes, problems with birth, starvation mostly from behavioural causes, and cold exposure, acting singly or in combination. High losses are usually associated with single causes, but there is a core level of about 15% that has not shown itself to be amenable to reduction without an input of labour and/or capital, that most producers regard as unacceptably high.

Since mortality in litters of more than one or two lambs commonly approach or exceed 50%, it cannot be taken for granted that stimulation of ovulation rate, currently being promoted, will satisfy producer expectations, unless ewes with an ability to rear multiples are used.

Use of genetically superior animals is an attractive option to improve lamb survival with low cost to the producer, but their development requires study of the genetics of the components of survival, particularly easy birth, superior maternal behaviour and good lamb vigour.

Other areas requiring research are birth weight determination and the role of placental function, and valid comparisons of various husbandry options for lambing. The contribution of birth problems, including birth injury, to death of lambs in the starvation/mismothering/exposure category requires investigation in a wide variety of flocks under a variety of conditions.

Stringent analysis and publication of the large body of unpublished data on lamb losses in Australia should have a high priority.

## REFERENCES

Reference to all relevant publications is not possible in the space allowed. Precedence has been given to reviews and to recent publications that cite earlier work.


Alexander, G., 1962. *Aust. J. Agric. Res., 13*, 144-164.

Alexander, G., 1974a. *In* 'Size at Birth', Ciba Foundation Symposium 27 (new series). Elsevier: Amsterdam, 215-245.

Alexander, G., 1974b. *In* Montheith J.L. & Mount, L.E. (eds) *20th Nottingham Easter School in Agric. Sci.*, Butterworths, London 173-203.

Alexander, G., 1976. *Supplement to Proc. combined meeting Endoc. Soc. Aust. and N.Z. Soc. of Endoc. 9*, S1.

Alexander, G., 1978. *In* Naftolin, F. (ed) *Abnormal Fetal Growth: Biological Bases and Consequences*. Dahlem Konferenzen, Berlin 149-164.

Alexander, G., 1979. *In* Robertshaw, D.(ed) *International Review of Physiology, 20*, 43-155. III. Univ. Park Press, Baltimore.

Alexander, G. 1980. *Reviews in Rural Science, IV*, 99-107.

Alexander, G., 1984. *Wool Technology and Sheep Breeding*, (in press).

Alexander, G. and Bell, A.W., 1975. *Boil. Neonate, 26*, 182-194.

Alexander, G., Kilgour, R., Stevens, D. and Bradley, L.R., 1984. *Appl. Anim. Behav. Sci., 12*, 363-372.

Alexander, G., Lynch, J.J., Mottershead, B.E., and Donnelly, J.B., 1980. *Proc. Aust. Soc. Anim. Prod, 13*, 329-336.

Alexander, G., Mann, T., Mulhearn, C.J., Rowley I.C.R., Williams, D. & Winn, D., 1967 *Aust. J. Exp. Agric. Anim. Husb, 7*, 329-336.

Alexander, G. & Peterson, J.E., 1961. *Aust. Vet. J., 37*, 371-381.

Alexander, G., Peterson, J.E. and Watson, R.H., 1959. *Aust. Vet. J., 35*, 433-441.

Alexander, G., Stevens, D. and Bradley, L.R., 1984. *Proc. Aust. Soc. Anim. Prod., 15*, 231-234.
Alexander, G., Stevens, D., Kilgour, R., De Langen, H., Mottershead, B.E. and Lynch, J.J., 1983a. *Appl. Anim. Ethol., 10*, 301-317.
Alexander, G., Stevens, D. and Mottershead, B., 1983b. *Aust. J. Exp. Agric. Anim. Husb., 23*, 361-368.
Alexander, G. and Williams D., 1962. *Aust. J. Agric. Res, 13*, 122-143.
Alexander, G. and Williams D., 1966. *J. Agric. Sci. Camb, 67*, 181-189.
Anderson, A.R. and Marshall, R.R., 1967. *Sheep Farming Annual, 30*, 35-46.
Avon, 1976. *Ann. Rep. Res. Div. NZ. Min. Ag. and Fish.* 1975-76, p 87.
Arnold, C.W. and Morgan, P.D., 1975. *Appl. Anim. Ethol, 2*, 25-46.
Atkins, K.D., 1980. *Proc. Aust. Soc. Anim. Prod., 13*, 174-176.
Bell, A.W., 1984. This volume, 144-152.
Blackburn, A.G. and Rizzoli, D. 1968. *J. Agric. Vic, 66*, 452-462.
Ch'ang, T.S. and Evans, R., 1978. *4th World Conference on Animal Production*, Buenos Aires 2, 673-681.
Broadmeadow, A.C.C., Cobon, D., Hopkins, P.S. and O'Sullivan, B.M., 1984. This volume, 216-219.
Columella, L.J.M., C.64 A.D. 'On Agriculture'. Transaltion from the Latin by E.S. Foster and E.H. Heffner (1968). Heinmann, London, p. 249.
Coop, I.E. and Drake, J.H., 1948. *Proc. N.Z. Soc. Anim. Prod, 9*, 122-129.
Corbett, J.L., 1979. *In* Black, J.L. and P.J. Reis (eds) *Proc. National Workshop 'Physiological and Environmental Limitations to Wool Growth, Leura 1978.* University of New England Press, 79-98.
Cundiff, L.V., Gregory, K.E. and Koch, R.M., 1982. *In, Proc. 3rd World Congr. on Genetics Appl. to Livestock Prod.* (*Madrid*) 5, 310-337.
Dalton, D.C., Bigham, M.L. and Sumner, R.M.W., 1976. *N.Z. J. Exp. Agric, 4*, 27-33.
Dalton, D.C., Knight, T.W., and Johnson, D.L., 1980. *N.Z. J. Agric. Sci., 23*, 167-173.
Dalton, D.C. and Rae, A.L., 1978. *Anim. Breeding. Abstr., 46*, 657-680.
Davis, S.R., Rattray, P.V., Petch, M.E. and Dugdnzich, D.M., 1981. *Proc. N.Z. Soc. Anim. Prod., 41*, 218-223.
Dennis, S.M., 1969. *Aust. Vet. J., 45*, 6-9.
Dennis, S.M., 1974. *Aust. Vet. J., 50*, 443-449.
Donald, A.P. and Purser, A.F., 1956. *J. Agric. Sci. (Camb), 48*, 245-249.
Donnelly, F.B., 1982. *Proc. Aust. Soc. Anim. Prod. 14*, 23-34.
Douglas, D.S., 1965. 'Identify those Dry Ewes', *Bull: A136 N.S.W. Dept. Agric.*
Duff, X.J., McCutcheon, S.N., and McDonald, M.F., 1982. *N.Z. Soc. Anim. Prod., 42*, 15-17.
Dun, R.B. and Hamilton, B.A., 1965. *Aust. J. Exp. Agric. Anim. Husb*, 5, 236-242.
Eales, F.A., Small, J. and Gilmour, J.S., 1982. *Vet. Rec., 110*, 121-123.
Eales, F.A., Small, J. and Gilmour, J.S., 1983. *In* Haresign, W. (ed) *35th Nottingham Easter School in Agric Sci.* Butterworths, London, 289-98.
Egan, A.R., 1984. This volume, 262-268.
Egan, J.K., McLaughlin, J.W., Thompson, R.L. and McIntyre, J.S., 1972. *Aust. J. Exp. Agric. Anim. Husb., 12*, 470-472.
Fogarty, N.M., 1984. This volume, 226-233.
Fogarty, N.M. and Thompson, J.M., 1974. *Aust. Vet. J., 50*, 502-506.
Fowler, D.G. and Wilkins, J.F., 1982. *Wool Tech. Sheep Breed. 30*, 111-117.
George, J.M., 1969. M. Rur. Sci. Thesis, University of New England.
George, J.M., 1975. *Aust. Vet. J., 51*, 262-265.
George, J.M., 1976. *Aust. Vet. J., 52*, 519-523.
George, J.M., 1980. *Proc. Aust. Soc. Anim. Prod., 13*, 494.
Giles, J.R., 1968. *Proc. Aust. Soc. Anim. Prod., 7*, 235-238.
Gonyou, H.W. and Stookey, J.M., 1981. *J. Anim. Sci., 51* (*Suppl.*), 128.
Gumbrell, R.C., 1980. *Proc. NZ Vet. Assoc. Sheep Soc., 10*, 94-96.
Gunn, R.G., 1968. *Anim. Prod., 10*, 213-215.
Haughey, K.G., 1973a. *Aust. Vet. J., 49*, 1-8.
Haughey, K.G., 1973b. *Aust. Vet. J., 49*, 9-15.
Haughey, K.G., 1973c. *Aust. Vet. J., 49*, 554-563.
Haughey, K.G., 1980. *Reviews in Rural Science IV*, 109-111.
Haughey, K.G., 1983a. *Refresher Course for Veterinarians Proc. No. 67.* The Post Graduate Committee in Veterinary Science, University of Sydney 135-147.
Haughey, K.G., 1983b. *Aust. Vet. J., 60*, 361-3.
Haughey, K.G. 1984. This volume, 210-212.
Haughey, K.G. and George, J.M., 1982. *Proc. Aust. Soc. Anim. Prod., 14*, 26-29.
Hayman, R.H., Turner, H.N. and Turner, E., 1955. *Aust. J. Agric. Res., 6*, 446-455.
Heath, J. and Lee, A., 1975. 'Lambing percentage - why worry?' *Dept. Agric. Hamilton, Vic.*

Hight, G.K., 1975. *In, Agricultural Research in the New Zealand Ministry of Agriculture and Fisheries.* Annual Report of the Agricultural Research Division, 1975-76.
Hight, G.K. and Jury, K. E., 1970. *N.Z. J. Agric. Res, 13*, 735-752.
Hinch, G.N., Kelly, R.W., Owens, J.L. and Crosbie, S.F., 1983. *N.Z. Soc. Anim. Prod., 43*, 29-32.
Holmes, R.J., 1975. *Proc. N.Z. Vet. Ass., Sheep Soc.,* 5, 67-75.
Houston, D.C., 1974. *Vet. Rec., 95*, 575.
Hudson, R. and Distel, H., 1983. *Behaviour, 85*, 260-275.
Hughes, K.L., Hartley, W.J., Haughey, K.G. and McFarlane, D., 1094. *Proc. Aust. Soc. Anim. Prod,* 5, 92-99.
Hughes, K.L., Haughey, J.G. and Hartley, W.J., 1971. *Aust. Vet. J., 47*, 472-476.
Johnson, D.L., Clarke, J.N., Maclean, K.S., Cox, E.H., Amyes, N.C. and Rattray, P.V., 1982. *Proc. N.Z. Soc. Anim. Prod, 42*, 13-14.
Kilgour, R. and De Langen, H., 1980. *Reviews in Rural Sc. IV*, 117-118.
Kilgour, R., Alexander, G. and Stevens, D., 1983. *Proc. 2nd Asia-Oceania Congress on Perinatology*, Auckland, Feb. 1982, 118-123.
Knight, T.W., Hight, G.K. and Winn, G.W., 1979. *N.Z. Soc Anim. Prod, 39*, 87-93.
Langlands, J.P., Donald, G.E. and Paul, D.R., 1984. *Aust. J. Exp. Agric. Anim. Husb., 24*, 57-64.
Luff, A., 1980. 'A Service for All Seasons.' *N.S.W. Department of Agriculture.*
Mayoh, M.G., 1969. *N.S.W. Dept. Agric., Div. Anim. Ind. Bull. A235.*
Mellor, D.J., 1983. *Br. Vet. J., 139*, 307-324.
McCance, I. and Alexander, G., 1959. *Aust. J. Agric. Res.*, 10, 697-719.
McHugh, F. and Edwards, M.S.H., 1958. *J. Agric. Vic., 56*, 425-438.
McFarlane, D., 1965. *N.Z. Vet. J., 13*, 116-135.
McMillan W.H., 1983. *Proc. N.Z. Soc. Anim. Prod., 43*, 33-36.
McSporran, K.D. and Fielden, E.D., 1979. *N.Z. Vet. J., 27*, 75-78.
Moore, R.W., McDonald, I.M. and Messenger, J.J., 1966. *Proc. Aust. Soc. Anim. Prod., 6*, 157-160.
Mullaney, P.D., 1966. *Aust. J. Exp. Agric. Anim. Husb., 6*, 84-87.
Mullaney, P.D., 1969. *Aust. J. Exp. Agric. Anim. Husb., 9*, 157-163.
Mullaney, P.D. and Hyland, P.G., 1967. *Aust. J. Exp. Agric. Anim. Husb, 7*, 304-307.
Obst, J.M. and Day, H.R., 1968. *Proc. Aust. Soc. Anim. Prod., 7*, 239-242.
Obst, J.M. and Evans, J.V., 1970. *Proc. Aust. Soc. Anim. Prod., 8*, 149-153.
Parker, A.G.H., 1974. *Proc. NZ Vet. Assoc. Sheep Soc.,* 5, 94-98.
Piper, L.R., 1982. *Proc. 2nd World Congress on Genetics Applied to Livestock Production (Madrid)* 5, 271-281.
Plant, J.W., 1977. *Proc. Ann. Conf. Aust. Vet. Assoc., 54*, 157-158.
Plant, J.W., 1981. *Refresher Course for Veterinarians Proc. No. 58*, Postgraduate Committee in Veterinary Science, University of Sydney, 411-440.
Plant, J.W., Ferguson, B.D., O'Halloran and Marchant, R., 1976. *Proc. Annu. Conf. Aust. Vet. Assoc., 53*, 189-190.
Plant, J.W., Marchant, R., Mitchell, T.D. and Giles, J.R., 1978. *Aust. Vet. J., 54*, 426-429.
Poindron, P., Le Neindre, P. and Levy, F., 1984. This volume 191-198.
Purser, A.F. and Karam, H.A., 1967. *Anim. Prod., 9*, 75-85.
Purser, A.F. and Young, G.B., 1964. *Anim. Prod., 6*, 321-329.
Purser, A.F. and Young, G.B., 1983. *Br. Vet. J., 139*, 296-306.
Quinlivan, T.D., 1968. *N.Z. Vet. J., 16*, 149-153.
Quinlivan, T.D., 1981. *Refresher Course for Veterinarians Proc. No. 58*, Postgraduate Committee in Veterinary Science, University of Sydney, 280-290.
Robinson, J.J., 1981. *Livest. Prod. Sci., 8*, 273-281.
Robinson, J.J., 1983. *In* Haresign, W. (ed) *35th Nottingham Easter School in Agric. Sci.* Butterworths, London, 111-131.
Rowley, I., 1970. *CSIRO Wild. Res., 15*, 79-123.
Scanlon, P.F., 1976. *J. Anim. Sci., 42*, 1217-1219.
Schinckel, P.G., 1955. *Aust. J. Agric. Res., 6*, 595-607.
Slee, J., 1981. *Livest. Prod. Sci., 8*, 419-429.
Smith, G.M., 1977. *J. Anim. Sci., 44*, 745-753.
Smith, I.D., 1962. *Aust. Vet. J., 38*, 500-507.
Smith, I.D., 1964. *Proc. Aust. Soc. Anim. Prod., 5*, 100-116.
Smith, I.D., 1965. *Aust. J. Exp. Agric. Anim. Husb.,* 5, 110-114.
Squires, V., 1981. 'Livestock Management in the Arid Zone'. Inkata Press, Melbourne.
Stevens, D., Alexander, G. and Lynch, J.J., 1981. *Appl. Anim. Ethol., 7*, 149-155.
Stevens, D., Alexander, G. and Lynch, J.J., 1982. *Appl. Anim. Ethol., 8*, 243-252.

Stevens, D., Alexander, G., Mottershead, B. and Lynch, J.J., 1984. *Proc. Aust. Soc. Anim. Prod., 15*, 751.
Thompson, G.E., Bassett, J.M., Sampson, D.E. and Slee, J., 1982. *Br. J. Nitr., 48*, 59-64.
Tyrrell, R.N. and Giles, J.R., 1974. *Aust. J. Exp. Agric. Anim. Husb, 14*, 600-603.
Vince, M.A., 1984. *Anim. Behav., 32*, 249-254.
Warren, H.D., 1975. *Proc. NZ Vet. Ass. Sheep Soc.,* 5, 76-78.
Watson, R.H., 1972. *World Rev. Anim. Prod, 8(2)*, 104-113.
Watson, R.H., Alexander, G., Cumming, I.A., Mcdonald, J.W., McLaughlin, J.W., Rizzoli, D.J. and Williams, D., 1968. *Proc. Aust. Soc. Anim. Prod., 7*, 243-249.
Whateley, J., Kilgour, R. and Dalton, D.C., 1974. *Proc. N.Z. Soc. Anim. Prod., 34*, 28-36.
Whitelaw, A. and Watchorn, P., 1975. *Vet. Res., 97*, 489-492.
Wilson, P.R. and Orwin, D.F.G., 1964. *N.Z. J. Sci., 7*, 460-490.

# CAN REARING ABILITY BE IMPROVED BY SELECTION?

K.G. Haughey, *Department of Veterinary Clinical Studies, University of Sydney, Camden, New South Wales, 2570.*

*Summary* In three flocks the mean repeatability of rearing performance was 0.10. Rearing success at two years of age was indicative of superior rearing performance at subsequent lambings. Differences in maternal pelvic dimensions partially explained repeatability in two flocks. In a third flock there was no apparent association between pelvic dimensions and rearing efficiency. In this flock, perinatal mortality differed significantly in two sub-flocks selectively bred from foundation ewes with high or low lifetime rearing efficiency. All the evidence indicates that improved lamb survival has to be seen mainly as a successful partnership between mother and offspring throughout pregnancy, parturition and lactation. Selection for rearing ability has unexploited potential for improving lamb survival.

## INTRODUCTION

On average 25% of pregnant ewes in New South Wales fail to rear progeny to weaning, with about 22% doing so before lamb-marking (Luff 1980), and an estimated 3% losing them between lamb-marking and weaning. Data supplied by the Australian Bureau of Statistics shows that lamb-marking percentage in New South Wales improved from 67% in 1951 to 76% in 1981. The strategies used to improve marking percentage directly, including selection for multiple births, improved nutrition of pregnant ewes, and provision of supervision and shelter for lambing ewes, are unlikely to be the sole reasons for the improvement as over the same period there has been extensive pasture improvement, greatly reduced rabbit populations and a marked shift in breed structure towards the more fecund Border Leicester × Merino ewe (personal communication, Australian Bureau of Statistics, 1982). Has the improvement been slow because we have been unable to sell our 'correct' solutions to the industry? Or is it because we have yet to discover other avenues for improving lamb survival? Studies of the lifetime rearing performance of ewes and preliminary results from selective breeding support the latter view.

## MATERIALS AND METHODS

### Lifetime rearing performance

We examined the rearing efficiency of ewes over four lambings from not more than five joinings in one Dorset Horn (flock 4) and three Merino flocks (flocks 1, 2 and 3). As similar data were not available for all flocks, rearing efficiency has been expressed two ways — rearing ability and rearing performance. *Rearing ability*, the number of occasions a ewe reared at least one lamb to weaning (flocks 1, 3 and 4), or to lamb-marking (flock 2), was used to describe the nett rearing efficiency of all flocks and to compare the pelvic dimensions of good and poor mothers in flocks 2, 3 and 4. Good mothers successfully reared on three or all occasions and poor mothers failed to rear a lamb on 2, 3 or all occasions. *Rearing performance* was expressed as the ratio, lambs weaned/lambs born (LW/LB) per ewe in flocks 1, 3 and 4 since these data were available. The conjugate (CD) and transverse (TD) pelvic diameters were determined after slaughter or radiographically at five or six years of age, and pelvic area (PA) was calculated as their product. The repeatability of rearing performance was estimated both as an intra-class correlation and as the regression of performance at lambings 2, 3 and 4 on rearing performance at first lambing at two years of age. The relationship between rearing performance and pelvic dimensions was also estimated by least squares methods.

### Selection for rearing ability

In 1975 we established two line-bred Merino selection flocks from ewes comprising flock 2. The high efficiency (HE) flock comprises descendants of foundation ewes which reared at least one lamb on three or all occasions.

The low efficiency (LE) flock comprises descendants of foundation ewes which failed to rear their progeny on 2, 3 or all occasions. The selection procedures have been described elsewhere (Haughey 1983). Since 1980 there have been sufficient breeding ewes to permit an early assessment of the response to selection. Obstetrical assistance was given only to ewes unable to deliver dead foetuses. Mortality rate, in the period birth to weaning, was calculated as the percentage of single or twin lambs dying, relative to the number born dead or alive. Age-specific mortality rate was calculated as the percentage of lambs dying during the period, relative to the number alive at the beginning of the period. The differences in mortality were evaluated by the Chi-square test.

## RESULTS

### Lifetime rearing efficiency

A mean of 26.3% (range — 15.2% to 30.3%) of ewes in the four flocks failed to rear progeny on 2, 3 or all occasions, and these accounted for a mean of 62.7% (range — 57.1% to 65.5%) of all rearing failures. Rearing performance in flocks 1, 3 and 4 was repeatable.

The intra-class repeatability estimates ranged from 0.05 to 0.17, with a mean of 0.10. Ewes rearing at least one lamb at two years of age in flocks 1, 3 and 4 outperformed those which failed to do so by a mean of 7.6% LW/LB (range — 2.8% to 14.7%) at subsequent lambings. Some or all pelvic dimensions of good mothers in flock 3 and 4 were significantly larger than those of poor mothers, whereas in flock 2 they were not significantly different (Table 1).

**TABLE 1. Differences in mean pelvic dimensions between ewes of good or poor rearing ability.**

| | Good Mothers | Poor Mothers | P |
|---|---|---|---|
| *Flock 2* | | | |
| Transverse diameter, cm | 8.27 | 8.09 | ns |
| Conjugate diameter, cm | 11.39 | 11.51 | ns |
| Area, $cm^2$ | 94.16 | 92.80 | ns |
| *Flock 3* | | | |
| Transverse diameter, cm | 8.24 | 8.10 | ns |
| Conjugate diameter, cm | 11.18 | 10.36 | * |
| Area, $cm^2$ | 92.32 | 84.32 | * |
| *Flock 4* | | | |
| Transverse diameter, cm | 9.40 | 9.28 | * |
| Conjugate diameter, cm | 10.07 | 9.91 | * |
| Area, $cm^2$ | 94.62 | 91.96 | * |

* $p < 0.05$

The rearing performance (LW/LB) of flocks 3 and 4 increased significantly by 16.3% ($p < 0.05$) and 6.3% ($p < 0.05$) per cm increase in CD respectively. Increasing PA increased LW/LB by 1.4% ($p < 0.05$) and 0.5% ($p < 0.05$) per $cm^2$, respectively. TD had no significant effect on LW/LB. Pelvic dimensions influenced the survival of single lambs more than that of twins. In flock 3 the increases in LW/LB per cm increase in CD were 18.6% ($p < 0.05$) and 9.2% ($p < 0.05$), respectively for singles and twins. In flock 4 the respective values were 12.8% ($p < 0.05$) and zero for singles and twins. There was a significant curvilinear ($p < 0.05$), as well as a linear, relationship between CD and PA and LW/LB with the slope of the regressions being steeper below the mean pelvic dimensions than above. Differences in pelvic dimensions only partially accounted for the repeatability of LW/LB as the estimates for flocks 3 and 4 were reduced from 0.15 to 0.05, and from 0.09 to 0.08, respectively, by differences in CD.

Selective breeding for rearing ability

Preliminary results from 1980 to 1983 inclusive indicate that the mean mortality rates, birth to weaning, of both single and twin lambs were significantly lower in the HE flock (Table 2).

**TABLE 2. Lamb mortality, birth to weaning, 1980-83 inclusive, relative to birth-type of lambs and rearing ability of maternal ancestors.**

| | Single Lambs | | Twin Lambs | |
|---|---|---|---|---|
| Flock | HE(1) | LE(2) | HE | LE |
| Total lambs born | 416 | 320 | 206 | 134 |
| Total deaths | 78 | 89 | 44 | 49 |
| Mean mortality, % | 18.8 | 27.8 | 21.4 | 36.6 |
| Significance, p < | ** | | ** | |
| Range of mortality between years, % | 12.1 −27.8 | 16.7 −32.9 | 15.4 −47.1 | nil −68.4 |

(1) High Efficiency flock
(2) Low Efficiency flock
** $p < 0.01$

The severe drought of 1980, 1981 and 1982 undoubtedly contributed to the high mortality rate. Comparison of age-specific mortality showed that most of the differences in mortality of single lambs were due to significantly higher mortality during or within three hours of parturition in the LE flock (HE v LE — 8.4% v 15.9%; $p < 0.01$). Mortality of LE twin lambs was higher during and within three hours of parturition (HE v LE — 3.4% v 9.7%; $p < 0.05$), and between 3 and 48 hours of age (HE v LE — 7.0% v 17.4%; $p < 0.01$).

## DISCUSSION

Our evidence indicates that rearing performance is repeatable and probably heritable. Repeatability was estimated at a mean of 0.10. The most likely explanation for the differences between the selection flocks in lamb mortality lies in the differences in rearing performances of maternal ancestors. Piper *et al.* (1982) estimated the repeatability and heritability of rearing performance at means of 0.09 and 0.10 respectively. The repeatable and apparently heritable nature of rearing performance implies that maternal and foetal factors are implicated in impaired lamb survival.

Maternal and foetal factors prejudicial to lamb survival include : teat and udder abnormalities (Moule 1954), aberrant maternal and neonatal behaviour (Alexander 1984), heritable low neonatal resistance to cold (Slee 1981) and maternal pelvic dimensions. As pelvic dimensions did not fully account for the repeatability of performance in flock 3 and 4, with a larger effect on the survival of singles compared to twin lambs, other intrinsic defects of the mother-offspring unit, including those specified, could have accounted for the balance. Rearing failure among the descendants of flock 2, in which pelvic dimensions were not associated with mothering ability, is often associated with a heritable foetal oversize (K.G. Haughey, unpublished data). Small maternal pelvic size is the maternal component of foeto-pelvic disproportion which predisposes to birth injury to the foetal central nervous system. Birth injury is the major component of perinatal death because it causes stillbirths and neonatal deaths due to depressed sucking and physical activity (Haughey 1973a,b, 1980, 1982). Our evidence suggests that selection for rearing ability offers unexploited potential for improving lamb survival.

Future research should concentrate on the identity and heritability of intrinsic factors promoting an effective partnership between mother and offspring including the nutritional and genetic determinants of foetal size, maternal pelvic dimensions, teat shape, the onset and level of lactation, maternal and neonatal behaviour and neonatal resistance to cold.

## ACKNOWLEDGEMENTS

The financial support of the Wool Research Trust Fund on the recommendation of the Australian Wool Corporation and the collaboration of Mr. J.M. George and Dr. B.J. McGuirk, CSIRO Division of Animal Production, are gratefully acknowledged.

## REFERENCES

Alexander, G., 1984. This volume, 199-209.
Haughey, K.G., 1973a. *Aust. Vet. J., 49*, 1-8.
Haughey, K.G., 1973b. *Aust. Vet. J., 49*, 9-15.
Haughey, K.G., 1980. *In* M. Wodzicka-Tomaszewska, T.N. Edey and J.J.Lynch (eds), *Reviews in Rural Science, No.IV*, University of New England, Armidale, p. 109-111.
Haughey, K.G., 1982. *Aust. Vet. J., 58*, 173-180.
Haughey, K.G., 1983. *Aust. Vet. J., 60*, 361-363.
Luff, A., 1980. *Final Report*, Sheep Fertility Service, New South Wales Department of Agriculture, p.1-45.
Moule, G.R., 1954. *Aust. Vet. J., 30*, 153-171.
Piper, L., Hanrahan, J.P., Evans, R. and Bindon, B.M., 1982. *Proc. Aust. Soc. Anim. Prod., 14*, 29-30.
Slee, J., 1981. *Livest. Prod. Sci., 8*, 419-429.

# INFLUENCE OF PLANE OF NUTRITION DURING LATE PREGNANCY AND LACTATION ON THE SURVIVAL AND GROWTH OF MERINO AND FIRST CROSS LAMBS

B.R. Beetson, *Department of Agriculture, Western Australia, 6151.*

*Summary* Ewes of differing fecundity (Merino, half Booroola cross Merino and quarter Booroola cross Merino) were joined with either Collinsville Merino, Poll Dorset or Border Leicester rams. Starting one month before lambing the ewes were supplemented with 1 kg/head/day (HP) or 0.5 kg/head/day (LP) of a 4:1 mixture of cereal and lupin grain for 106 days.

The crossbred lambs were 5 per cent heavier at birth yet had similar survival rates compared with the Merino lambs. By three and six months of age the liveweight difference had increased to 10 per cent.

The HP ewes consumed 60kg more grain than the LP ewes yet there were only small differences in birthweight and growth rates and no differences in survival of the lambs caused by the feed treatment. However the HP ewes were 9 kg heavier at weaning than the LP ewes.

## INTRODUCTION

Reproductive rates of Merino ewes in Western Australia can be increased by crossing with the highly fecund Booroola Merino but survival and growth rates of lambs from resulting multiple births are low (Beetson and Lewer 1983). Attempts to improve the survival and growth of lambs by increasing the level of supplementation during late pregnancy and early lactation and by using British breed rams as terminal sires are reported in this paper.

## MATERIALS AND METHODS

The flock consisted of 550 Merino (M), 515 half Booroola cross Merino (HB) and 483 quarter Booroola cross Merino (QB) ewes of mixed ages (3 to 6 years) on the Mt Barker Research Station in south Western Australia. In December 1982 the ewes were randomly drafted into three groups each with 183 M, 172 HB and 161 QB ewes, for joining with either Merino, Poll Dorset or Border Leicester rams in syndicates at one ram per 43 ewes for six weeks. The ewes were then run in age groups until one month prior to lambing when they were randomly allocated within sire joining groups to two feeding treatments. Ewes in each feeding treatment were allocated to one of 5 paddocks at a stocking rate of 9/ha. The pasture was typical of the dry residue of a mixture of subterranean clover and annual grasses described by Biddiscombe *et al.* (1980). Half the ewes received 520 g grain/head/day (low plane, LP) while the rest received 1090 g grain/head/day (high plane, HP). The supplement, a 4 to 1 mixture of cereal (oats:barley; 2:1) and sweet lupin grain, was fed for 106 days until the average age of the lambs was 10 weeks. The lambs were weaned at an average age of 16 weeks. The ewes were weighed before and after the feeding period whereas the lambs were weighed at birth, 16 and 24 weeks of age.

All data were anlaysed using appropriate statistical programmes from *Genstat* (Nelder 1977). Main effects tested were birth type, rearing rank, feed treatment and terminal sire breed. Final models adjusted for such factors as date of birth, sex of lambs, age and genotype of dam and any interactions if found to be significant ($P < 0.05$).

## RESULTS

All differences reported were significant at least at the 5 per cent level ($P < 0.05$).

### Birthweight

Lambs borne by HP ewes were heavier than those by LP ewes (3.46 vs 3.29 kg). Lambs sired by Merino rams were lighter than those by Poll Dorset and Border Leicester rams (3.28 vs 3.45 and 3.42 kg). Single, twin and triplet lambs weighed 3.96, 2.93 and 2.20 kg respectively.

### Liveweight at weaning

The HP lambs weighed 16.9 kg and the LP lambs, 15.0 kg at weaning. The response to feed treatment was greater among single than twin or triplet lambs. Table 1 shows the interaction between feed treatment and birth type ($P < 0.05$). Lambs reared as twins were 2.2 kg lighter than those reared as singles (14.3 vs 16.5 kg). Lambs sired by Poll Dorset and Border Leicester rams did not differ but both were heavier than those sired by Merinos (16.4 vs 15.0 kg).

### Liveweight at six months of age

The liveweights of the HP and LP lambs were 26.5 and 24.8 kg. Lambs sired by Merino rams were significantly lighter than those sired by Poll Dorset and Border Leicester rams (24.0 vs 26.8 and 26.2 kg). Single, twin and triplet born lambs weighed 26.5, 24.6 and 23.3 kg. Lambs reared as singles were 1.8 kg heavier than those reared as twins.

**TABLE 1. The influence of feed treatment and birth type on liveweight at weaning. Differences (HP-LP) are expressed as a percentage of the LP weight in brackets.**

| | Birth Type | | |
|---|---|---|---|
| Feed | Single | Twin | Triplet |
| HP | 17.9 (+16%) | 15.7 (+12%) | 14.1 (+9%) |
| LP | 15.4 | 14.1 | 13.0 |

Survival of lambs.

Factors significantly affecting mortalities while birth weight was left in the model were birth type and the interaction between sire breed and birth type (Table 2).

**TABLE 2. Number of lambs born and mortality of lambs (percent; unadjusted data) showing the interaction between birth type and terminal sire breed.**

| | Terminal Sire Breed | | | |
|---|---|---|---|---|
| | Merino | Poll Dorset | Border Leicester | Ave |
| *Birth Type* | | | | |
| Single | 294 ( 8) | 334 (11) | 315 (10) | 943 (10) |
| Twin | 223 (38) | 202 (29) | 180 (35) | 605 (34) |
| Triplet | 78 (64) | 57 (60) | 96 (52) | 231 (58) |
| Ave | 595 (26) | 593 (22) | 591 (25) | |

Triplet lambs sired by Border Leicester rams had a lower mortality rate; but when triplets were excluded from the data, this interaction was not significant.

Liveweights of the dams

The HP ewes were 9 kg heavier than the LP ewes after the feed treatments were terminated (Table 3).

**TABLE 3. Liveweights (kg) of dams at joining and at weaning showing effects due to feed treatment.**

| | Before feeding (1.11.1982) | After feeding (28.6.1983) |
|---|---|---|
| HP | 46.9 | 49.1 |
| LP | 46.9 | 40.1 |

## DISCUSSION

The high plane of nutrition resulted in lambs which were heavier than the LP lambs by 5 per cent at birth and 13 per cent at weaning (3 months of age). After weaning, both groups grazed for three months at the same plane of nutrition and the difference declined to 7 per cent.

Compensatory or 'catch-up' growth had halved the differences achieved during the feeding period. Moreover, the extra feed consumed by the HP ewes (amounting to 60 kg grain) had no effect on the survival of the lambs. The most pronounced effect of the feeding was on the liveweights of the ewes at weaning, suggesting that the grain replaced tissue reserves as the source of nutrients but had little effect on the total production of milk. This is consistent with the suggestion by MacRae (1983) that concentrated diets with low roughage content, resulting in elevated levels of blood glucose, stimulate insulin production which causes circulating nutrients to be partitioned away from the mammary gland and towards peripheral tissues.

The use of British breed rams as terminal sires had little influence on birthweight and survival of the lambs, contrary to suggestions by Hinch *et al.* (1983). However the crossbred lambs were, as expected (Nitter 1978), slightly heavier at three months and 10 per cent heavier at six months of age than the Merinos.

A surprising result was the interaction between feed treatment and birth type on liveweights at weaning: the higher plane of nutrition had more influence on the single lambs than the twins or triplets. This effect disappeared by six months of age.

At birth, twin and triplet lambs were 26 and 44 per cent lighter than singles but by 6 months of age 'catch-up' growth and reduced those differences to 7 and 12 per cent respectively. The poor survival rates of the multiple lambs was influenced neither by supplementary feed nor cross breeding in this experiment and is a problem needing further research.

REFERENCES

Beetson, B.R. and Lewer, R.P., 1983. *In* Land, R.B. (ed), *Proc The Genetics of Prolific Sheep*, Edinburgh. Butterworths, London.

Biddiscombe, E.F., Arnold, G.W., Galbraith, K.A. and Briegel, D.J ., 1980. *Agricultural Systems, 6*, 3.

Hinch, G.N., Kelly R.W., Owens, J.L. and Crosbie, S.F., 1983. *Proc. N.Z. Soc. Anim. Prod., 43*, 29.

Nelder, J.A. 1977. Genstat Manual. Statistics Department, Rothamstead Experimental Station, Harpenden, UK.

Nitter, G. 1978. *Anim. Breed. Abst., 46,131.*

Macrae, J.C., 1983. *In* Farrell, D.J., and Pran Vohra (eds) *Recent Advances in Animal Nutrition in Australia*. University of New England, Armidale, New South Wales, Australia.

# EFFECTS OF PLANT TOXINS ON PERI-NATAL VIABILITY IN SHEEP

A.C.C. Broadmeadow, D. Cobon, P.S. Hopkins and B.M. O'Sullivan, *Queensland Department of Primary Industries, Animal Research Institute, Yeerongpilly, QLD. 4105.*

*Summary*. Experiments have been conducted to establish whether sub-clinical levels of plant toxins fed to the pregnant ewe may be foetotoxic. *Tribulus terrestris* caused lowered teat seeking ability and high mortality in lambs whose dams were fed the plant at approximately 100 days gestation.

## INTRODUCTION

Despite satisfactory pregnancy rates in sheep in north west Queensland the average marking rate is only 40% (Entwistle 1972). Contributing factors have been shown to be heat stress and inadequate nutrition resulting in reduced birth weight (Hopkins *et al.* 1980). However, little work has been done on the effect of toxic plants. Many fodder plants contain toxins but few are consumed in quantities which present a problem of acute or chronic toxicity. Generally only these plants are regarded as toxic. The medical literature describes many compounds which present no significant problem to adults, but cause serious abnormalities with foetal development when consumed by pregnant women (Hays 1981). The effects of these foetotoxic agents may vary with dose and timing of exposure, and include death, mutation, malformation, behavioural defects and low birth weight (Stanley 1981). Lambs of low birth weight are less likely to survive to marking and survivors may grow less wool (Hopkins *et al.* 1980). In addition, observations by field officers of this Department have indicated that some lambs which appear normal at birth show very little desire to suck. We report preliminary results of studies on the effects of sub-clinical doses of plant toxins administered to the ewe during pregnancy on the peri-natal viability of lambs.

## MATERIALS AND METHODS

*Experiment 1.* Material from plants collected in western Queensland (Table 1) was extracted in 0.9% NaCl (50% W/V) and administered subcutaneously into newborn mice (10 males per treatment) twice daily for four days at a dose rate of $5 \times 10^{-5}$% live weight (V/W). Live weights and general condition were monitored for 14 days. Three sub-trials were conducted, two with plant extracts and one with oxalate (7.5 mg/ml) and nitrate (7.5 mg/ml) at the same dose rate as above.

Mice are born whilst developmentally immature compared with the sheep, and were therefore considered a suitable model in an initial screen for the sheep foetus in the last third of gestation.

*Experiment 2.* Ethanolic (70% V/V) extracts of plant material collected on 'Toorak' Research Station (Table 2) were evaporated under reduced pressure and administered intraperitoneally to pregnant ewes (5-8 per treatment) for 10 days (40 g DM equivalent per ewe per day) starting at approximately 100th day of gestation. Lamb deaths, birth weights and teat seeking times (Cobon and Broadmeadow 1984) were recorded.

*Experiment 3.* Rumen fistulated pregnant ewes (10 per treatment) were fed a standard ration. At approximately day 100 of gestation 300 g was replaced by 300 g hammermilled *Tribulus terrestris* or lucerne hay (*Medicago sativa*, controls) for 10 days. Levels of some serum enzymes during the feeding period were assayed using commercial kits (Boehringer). Lamb deaths, birth weights, growth rates and teat seeking abilities were recorded. All dead lambs were subjected to post mortem examination. Teat seeking ability was measured by an improved version of the method previously described. On four occasions individual times were compared with a normal group mean and scored as pass or fail.

## RESULTS

*Experiment 1.* The most striking effects of the treatments imposed were the very poor survival rates of the animals receiving *Tribulus* (40% and 0%) and *Desmodium* (0%). Intermediate effects were seen in *Bassia, Solanum* and *Atriplex*. Table 1 shows mean animal liveweights ± SEM at days 4 and 8 together with survival rates at days, 4, 8 and 15. *Tribulus, Desmodium, Solanum, Bassia* and *Atriplex* groups all showed additional signs of toxicity. These included sparse hair growth, extensive haematomas and scarring around the site of injection. These symptoms were not evident in the control, *Portulaca, Malvastrum, Rhynchosia* or nitrate treated animals. *Solanum* caused almost total inhibition of hair growth. Mice receiving oxalate were in poorer condition and showed lower growth rates than the controls but their survival rates were high.

**TABLE 1. Effect of plant extracts on live weights and survival rates in newborn mice**

| | Day 4 | | | Day 8 | | | Day 15 |
|---|---|---|---|---|---|---|---|
| | L.W. | SEM | SURV | L.W. | SEM | SURV | SURV |
| (g) | | (%) | (g) | | (%) | (%) | |
| Expt. 1.1 | | | | | | | |
| Control | 2.80 | .14 | 100 | 4.99 | .14 | 80 | 80 |
| *Portulaca oleracea* | 2.45 | .14 | 100 | 3.96 | .24 | 80 | 80 |
| *Rhynchosia minima* | 3.08 | .14 | 100 | 5.11 | .18 | 96 | 90 |
| *Solanum sp* | 2.63 | .07 | 100 | 3.86 | .10 | 100 | 80 |
| *Tribulus terrestris* | 3.60 | .42 | 40 | 6.45 | .66 | 40 | 40 |
| Expt.1.2 | | | | | | | |
| Control | 2.72 | .11 | 100 | 4.03 | .10 | 100 | 100 |
| *Portulaca oleracea* | 3.05 | .15 | 100 | 4.89 | .19 | 80 | 80 |
| *Rhynchosia minima* | 3.00 | .05 | 90 | 4.87 | .19 | 90 | 90 |
| *Bassia spp* | 2.63 | .08 | 90 | 3.07 | .24 | 70 | 50 |
| *Malvastrum spicatum* | 2.85 | .11 | 100 | 4.40 | .13 | 100 | 100 |
| *Atriplex sp* | 2.96 | .19 | 70 | 4.31 | .36 | 70 | 70 |
| *Solanum sp* | 2.24 | .19 | 80 | 3.93 | .44 | 60 | 60 |
| *Desmodium campylocaulon* | - | - | 0 | - | - | 0 | 0 |
| *Tribulus terrestris* | - | - | 0 | - | - | 0 | 0 |
| Expt.1.3 | | | | | | | |
| Control | 3.41 | .14 | 90 | 5.58 | .19 | 90 | 90 |
| Nitrate | 4.26 | .22 | 90 | 6.91 | .30 | 90 | 90 |
| Oxalate | 2.64 | .12 | 90 | 4.06 | .20 | 90 | 90 |
| Nitrate + Oxalate | 3.15 | .18 | 80 | 5.40 | .19 | 80 | 80 |

*Experiment 2.* All three plant extracts reduced birth weights and increased teat seeking times of lambs (Table 2). *Tribulus* produced the most pronounced effects with very low birth weights and teat seeking times four times those of the controls. Data on survival of lambs were inconclusive.

**TABLE 2. Effect of plant extracts on lamb viability**

| | Birth weight (kg) | SEM | Teat seeking time (s) | Survivors (%) |
|---|---|---|---|---|
| Control | 3.8 | .16 | 66 | 62 |
| *Hibiscus trionum* | 2.4 | .27 | 184 | 60 |
| *Abelmoschus ficulneus* | 2.3 | .27 | 152 | 100 |
| *Tribulus terrestris* | 1.9 | .32 | 243 | 50 |

*Experiment 3.* There were no significant differences ($P > 0.05$) in mean birth weights ($4.16 \pm 0.04$ kg and $4.27 \pm 0.23$ kg) or growth rates ($218 \pm 19$ g/day and $264 \pm 15$ g/day) of the surviving lambs between *Tribulus* and control groups respectively. Four of the ewes in the *Tribulus* group aborted. In three cases the lambs were mummified and were estimated to have died at 100-120 days gestation, while the fourth *Tribulus* ewe aborted an almost fully developed lamb close to term. A fifth ewe gave birth to twins which died on day 2 without having sucked. Both of the ewe's teats were functional and she showed strong mothering instinct. The surviving lambs recorded very poor performances in the teat seeking test (Table 3). All lambs of the control ewes survived.

Mean levels of S.G.O.T. and $\gamma$G.T. in the *Tribulus* group were slightly higher than in controls during the first half of the feeding period in the *Tribulus* group (Table 3) but differences were not statistically significant. Values were above the normal range (Anon 1982) in four of the animals, three of which aborted their lambs.

**TABLE 3. Effect of *Tribulus terrestris* in pregnant sheep**

| | Ewes, pregnant 100-110 d | | | | | | | | | Lambs | | | |
|---|---|---|---|---|---|---|---|---|---|---|---|---|---|
| | Serum enzyme values during feeding period (i.u./L) | | | | | | | | | Teat seeking ability test (% pass) | | | |
| | A.P.(1) | | | S.G.O.T.(2) | | | γ G.T.(3) | | | | | | |
| Day | 1 | 5 | 10 | 1 | 5 | 10 | 1 | 5 | 10 | 2 | 3 | 7 | 14 |
| Control | 35.7 | 50.3 | 47.1 | 54.9 | 54.1 | 49.6 | 46.1 | 43.8 | 40.2 | 57 | 71 | 71 | 71 |
| SEM | 3.6 | 5.2 | 4.8 | 4.5 | 6.5 | 3.5 | 4.8 | 4.3 | 2.8 | | | | |
| *Tribulus terrestris* | 33.0 | 38.1 | 37.8 | 68.8 | 63.5 | 45.5 | 61.8 | 58.0 | 52.9 | 14 | 14 | 28 | 42 |
| SEM | 3.5 | 4.0 | 3.6 | 14.2 | 8.7 | 2.7 | 12.4 | 10.4 | 8.3 | | | | |

(1) Alkaline phosphatase EC 3.1.3.1
(2) Aspartate amino transferase EC 2.6.1.1
(3) γGlutamyl transferase EC 2.3.2.2.

## DISCUSSION

Initial screening indicated that various plants contained compounds which were toxic to the newborn mouse. *Tribulus* and *Desmodium* appeared to be the most toxic of the plants tested. *Tribulus* is abundant in north west Queensland in some years and is readily eaten by sheep. Thus it was decided to study it as a possible model for investigating the effects of plant toxins on foetal development.

Experiment 3 showed that 300 g *Tribulus* fed daily for 10 days to pregnant ewes caused significant lamb losses. While direct losses through abortion and neo-natal death may be quantified, additional losses resulting from reduced teat seeking instinct still require elucidation. Birth weights did not show the depression seen in experiment 2. This may be an artifact of the small numbers in experiment 2 or could reflect the variable nature of this syndrome. It does however indicate that *Tribulus* may exert its effect via a mechanism other than by reducing birth weight.

At the level fed, *Tribulus* caused no visible symptoms of toxicity in the ewe, but marginal toxicity cannot be ruled out on the basis of the biochemical data. It is unclear whether the effect is due to the direct action of the toxin on the foetus or to the ewe's reaction to the toxin. The dose required to cause a response in the foetus is very much lower than that required to elicit a clinical response in the ewe. Recent work by Bourke (1983) in New South Wales required approximately 2.6 kg fresh *Tribulus terrestris* to be fed daily for 15 to 20 days to 28 kg Border Leicester weaners before the first symptoms were visible, and for more than 30 days to cause death. The total dose was therefore much more than the amount used in this trial. Sheep in north west Queensland may consume up to 300 g DM broadleaf forbs daily (including *Tribulus terrestris*) which represents 45% of their total intake (Lorimer 1976). The sole forb fed in this trial was *Tribulus* but it represented only 26% of total intake.

While the experiments described have concentrated on *Tribulus terrestris* other toxic plants may possibly elicit a similar response. Indeed different toxins may act synergistically. *Tribulus* has been shown to contain several toxins including nitrate, steroidal saponins and alkaloids. Its toxic effects have also been attributed to the mycotoxin sporidesmin derived from the saprophytic fungus *Pithomyces chartarum* (Kellerman *et al.* 1980). In these experiments no attempt was made to quantify the toxins as the aim was to determine if a problem existed. Results from experiment 1, supported by the work of Cobon and Broadmeadow (1984), suggest that nitrate is not responsible for the pre-natal and neo-natal losses at the level seen in experiment 3.

The results indicate that toxic plants ingested by the ewe may cause lamb losses via a pathway other than acute toxicity. This may contribute to the significant reproductive losses encountered under extensive grazing conditions. Field and laboratory studies are currently investigating the nature and extent of this process. The studies are to determine the susceptible gestational age and physiological effects in the sheep as well as the plant species involved and their critical stage(s) of growth. It may then be possible to formulate management strategies to combat the problem.

## ACKNOWLEDGEMENTS

We wish to thank Miss L. Volks for her technical assistance and the members of the Biochemistry and Pathology Branches of the Queensland Department of Primary Industries for performing the clinical assays and post-mortems.

REFERENCES

Anon, 1982. *Field Officers Manual*, Division of Animal Industry, Queenland Department of Primary Industries 4th Ed., 28-37.

Bourke, C.A., 1983. *Aust. Vet. J., 60*, 189.

Cobon, D. and Broadmeadow, A.C.C., 1984. *Proc. Aust. Soc. Anim. Prod., 15*, 666,

Entwistle, K., 1972. *Aust. Vet. J., 48*, 395-401.

Hays, D.P., 1981. *Drug Intelligence and Clinical Pharmacy, 15*, 444-458, 542-566, 639-650.

Hopkins, P.S., Nolan, C.J. and Pepper, P.M., 1980. *Aust. J. Agric. Res., 31*, 763-771.

Kellerman, T.S., Van Der Westhenzeu, G.C.A., Coetaer, J.A.W., Roux Celilia, Marasses, W.F.O., Minne, J.A., Bath, G.F. and Basson, P.A., 1980. *Onderstepoorte, Res. Vet. Sci., 47*, 231-261.

Lorimer, M.S., 1976. M. Agr. Sc. Thesis, Univ. of Qld.

Stanley, G.J., 1981. *Aust. Med. J., 1*, 688-693.

# THE EFFECT OF UDDER DAMAGE ON MILK YIELD, LAMB GROWTH AND SURVIVAL

D.J. Jordan, *Queensland Department of Primary Industries, Roma 4455.*
N. O'Dempsey, *Queensland Department of Primary Industries, Blackall 4472.*
R.G.A. Stephenson, *Queensland Department of Primary Industries, Yeerongpilly 4105.*
K. Wilson, *Queensland Department of Primary Industries, Barcaldine 4725.*

*Summary* Field studies and one pen study showed that ewes with one functional teat produced 15 to 42% less milk and reared 6 to 42 fewer lambs per 100 ewes than ewes with two functional teats. Lambs from the ewes with one functional teat grew up to 21% more slowly. The incidence of teat damage in commercial flocks ranged from 2 to 19% and increased by 2.2% per year of age.

## INTRODUCTION

Milk production of ewes influences early growth of lambs, as confirmed in studies on Merinos in north west Queensland (Stephenson *et al.* 1981). Moule (1954) and Hayman *et al.* (1955) reported udder damage in ewes of 6% and 21% respectively with resultant reduced lamb survival (43% and 12%) and growth (7% and 10%). Studies were conducted in western Queensland to determine the incidence and effects of udder damage on milk production, lamb growth and lamb survival.

## MATERIALS AND METHODS

Five experiments were conducted between 1979-1982 at Julia Creek, Charleville and Blackall — see Table 1 for details. In each experiment groups of ewes with either one or two functional teat(s) were tested pregnant by udder palpation or ballotment (Pratt and Hopkins 1975). Milk yield was assessed by the oxytocin method (McCance 1959) and lamb-marking percentages were recorded.

Ewes were run together in the four field experiments, and their lambs identified by the udder painting technique (Norton and Hales 1981). Ewes in the pen study were fed lucerne nuts. Lamb and ewe live weights were recorded at the end of the experiment.

Between 1979-1982, 5452 ewes from sixteen flocks in western Queensland were surveyed for udder damage. Details for each age group were recorded in two flocks; in the other fourteen flocks details were obtained for only one age group or for mixed ages.

Lamb-marking data were analysed using the chi-squared test, survey data by regression analysis and non-orthogonal analysis of variance, and all other data by analysis of variance using animals as replicates.

## RESULTS

Milk yields, lamb-marking percentages and lamb weights in the five experiments are presented in Table 1.

**TABLE 1. Ewe numbers, lamb age (wk), milk yield (mL $d^{-1}$), lamb-marking percentage and lamb live weight at marking in all experiments. Ewes with sound (S) and unsound (US) udders.**

| | Julia Creek 1979 | | | Pen 1980 | | | Charleville 1981 | | | Charleville 1982 | | | Blackall 1982 | |
|---|---|---|---|---|---|---|---|---|---|---|---|---|---|---|
| | S | | US | S | | US | S | | US | S | | US | S | US |
| No. of ewes | 44 | | 44 | 149 | | 135 | 97 | | 97 | 54 | | 56 | 129 | 141 |
| Lamb age | 6½ | | | 4–6 | | | 4–10 | | | 4–10 | | | 7–15 | |
| Milk yield | 420 | ** | 264 | 1139 | ** | 963 | 791 | * | 586 | 699 | ** | 403 | ND | ND |
| Marking % | 53.3 | | 42.1 | 91.3 | * | 81.6 | 88.7 | ** | 46.4 | 50 | | 42.9 | 26.4 | 20.6 |
| Lamb wt. | 9.1 | ** | 6.2 | 10.1 | ** | 8.8 | 15.0 | | 14.5 | 8.3 | | 7.7 | 15.7 | 15.2 |

*, ** between S and US values for the same location and year indicate the level of significant difference between them (* $P \leq 0.05$; ** $P \leq 0.01$). ND — not determined.

In the pen study, lambs of ewes with sound udders grew significantly faster than lambs of ewes with unsound udders (181 v 144 g $d^{-1}$, $P \leq 0.01$). At the end of the study ewes with sound udders were significantly lighter than ewes with unsound udders (38.0 v 39.8 kg, $P \leq 0.01$).

The survey of 5452 ewes in sixteen flocks showed that the mean (± S.E.) percentage incidence of ewes with unsound udders was 10.8 ± 1.20 (range 2 to 25% in various age groups) (Tables 2 and 3). In three flocks, incidence in weaners after their first shearing at between two and five months was 3.2, 7.0 and 10.7%. In the two flocks from which data were obtained for various age groups (Table 2), a significant ($P \leq 0.01$) linear relationship between percentage of ewes with unsound udders and age was obtained, with a regression coefficient (± S.E.) of 2.2 ± 0.45 for data pooled across sites. Differences between sites were indicated by significantly different ($P \leq 0.01$) intercepts (3.37 and 11.74 for Blackall and Barcaldine, respectively). The regression coefficient for age when all sixteen flocks were included was 1.87 ± 0.52 ($P \leq 0.01$).

**TABLE 2. Incidence of udder damage with age in two Merino flocks.**

| | Age of ewes at previous shearing (years) | | | | | | | |
|---|---|---|---|---|---|---|---|---|
| | 0.25 | 1 | 2 | 3 | 4 | 5 | 7 & 8 | Total |
| *Blackall 1982* | | | | | | | | |
| No. of ewes | – | 135 | 250 | 150 | 1216 | 150 | 100 | 2001 |
| % damaged | – | 5.9 | 7.2 | 10.0 | 12.3 | 14.7 | 20.0 | 11.6 |
| *Barcaldine 1982* | | | | | | | | |
| No. of ewes | 56 | – | 98 | 98 | 95 | 96 | – | 443 |
| % damaged | 10.7 | – | 14.3 | 25.5 | 20.0 | 19.8 | – | 18.7 |

**TABLE 3. Incidence of udder damage in fourteen Merino flocks.**

| Area and year | Age of ewes at previous shearing | No. of ewes | % damaged |
|---|---|---|---|
| Julia Creek 1982 | 1 to 7 years | 160 | 7.5 |
| Blackall 1980 | 3 to 7 years | 1277 | 9.7 |
| Blackall 1979 | 7 years | 100 | 3.0 |
| | 5 years | 100 | 4.0 |
| | 6,7 years | 100 | 9.0 |
| | 6,7 years | 111 | 17.1 |
| | 6 years | 100 | 8.0 |
| | 8 years | 84 | 9.5 |
| | 6,7,8 years | 120 | 13.3 |
| Charleville 1979 | 6,7 years | 150 | 9.3 |
| Roma 1982 | 2 to 6 years | 150 | 7.3 |
| | 2 to 6 years | 219 | 2.3 |
| | 5 months | 93 | 3.2 |
| Goondiwindi 1982 | 5 months | 244 | 7.0 |
| Total | | 3008 | 8.4 |

## DISCUSSION

The results suggest that a reduction in lamb growth and survival was due to the lower milk yield of ewes with unsound udders. This effect would be expected to be most significant under adverse nutritional conditions. However, while lamb-markings were higher in sound ewes than in unsound ewes in all our studies, differences were significant only when nutrition was good. The Julia Creek, Charleville 1982 and Blackall experiments were conducted under adverse seasonal conditions, resulting in low overall milk yields and large percentage reductions in ewes with damaged udders. Milk yields were generally less than the 500 mL $d^{-1}$ necessary for adequate survival rates (Stephenson 1982). Low numbers of surviving lambs coupled with the small group sizes at Julia Creek and Charleville would account for the failure of differences in lamb-marking percentages in those adverse years to reach significance.

The failure of differences in milk yield between sound and unsound ewes to be reflected in differences in live weight of lambs at marking may be a function of age at marking. Where lambs were marked at an early age, weight closely followed milk yield. The absence of a relationship at older marking ages could reflect some compensation in lambs once their rumens became functional and they grazed more. The higher milk yield in sound ewes would allow more lambs with low birth weights to survive in these groups, which would tend to reduce the weight differences between the groups at marking.

The milk yields of ewes with unsound udders measured more than half those of ewes with sound udders. This indicates partial compensation in the functional half of the udder.

Results of the survey of udder damage indicate that the problem remains as important as it was in the 1950s (Moule 1954; Hayman *et al.* 1955). The increase with age in the incidence of udder damage (2.2% per year) observed in the present study is similar to the value of 2.6% reported by Hayman *et al.* (1955).

Effects of management practices were suggested by variations in the incidence of udder damage (3 to 20% in ewes 7 years old) in the flocks surveyed. We observed that damage occurred mostly at shearing (especially with learner shearers), and was less when supervision was good and when belly wool was not removed from immediately in front of the udder. Most damage was caused by cuts that affected the patency of the teat canal. When healed, the teat appeared normal to the casual observer; however, milk could be expressed through the damaged teat in only 5% of cases.

The different intercepts in regressions for data in Table 2 may result from differences in age at first shearing. In the Barcaldine flock, sheep were first shorn at weaning as mixed sexes when teats were small and easily damaged. In the Blackall flock, first shearing was at 12 months when teats were better developed and more easily seen. The relatively high incidence of damage in young sheep in the survey suggests more care is required when shearing young than older sheep. Culling of maiden ewes with damaged udders where a high incidence is found at classing would be beneficial. The increase in incidence with age (2.2% per year) thereafter is relatively low.

Our data suggest that culling of ewes with damaged udders would be a useful management practice in many Queensland flocks, particularly those where the incidence of damaged udders is high. The high incidence in ewes older than 5½ years, together with their lower nett reproductive capacity (Rose 1972), makes it beneficial to cull at 5½ to 6½ years.

Improvement in shearing procedures, including better supervision of shearers, particularly learners, and retention of some belly wool immediately in front of the udder should assist in reducing the overall incidence of udder damage.

## ACKNOWLEDGEMENTS

We thank Mr. G. Roberts, Julia Creek, Mr J. Edwards, Cunnamulla and Mr D. Mayer, Brisbane for their assistance with the surveys and analyses of data. Financial assistance was given by the Australian Meat Research Council and by the Wool Research Trust Fund on a recommendation of the Australian Wool Corporation.

## REFERENCES

Hayman, R.H., Turner, H.N. and Turton, E., 1955. *Aust. J. Agric. Res., 6*, 446-455.
McCance, I., 1959. *Aust. J. Agric. Res., 10*, 839-853.
Moule, G.R., 1954. *Aust. Vet. J., 30*, 153-171.
Norton, B.W. and Hales, J.E., 1981. *J. Aust. Inst. Agric. Sci., 47*, 52-54.
Pratt, M.S. and Hopkins, P.S., 1975. *Aust. Vet. J., 51*, 378-380.
Rose, Mary, 1972. *Proc. Aust. Soc. Anim. Prod., 9*, 48-54.
Stephenson, R.G.A., Edwards, J.E. and Hopkins, P.S., 1981. *Aust. J. Agric. Res., 32*, 497-509.
Stephenson, R.G.A., 1982. *Qd. Agric. J., 108*, 329-332.

# REDUCTION OF THE EFFECTS OF HEAT STRESS ON LAMB BIRTH WEIGHT AND SURVIVAL BY PROVISION OF SHADE

R.G.A. Stephenson, *Queensland Department of Primary Industries, Animal Research Institute, Yeerongpilly, Q. 4105.*
G.R. Suter, *Queensland Department of Primary Industries, P.O. Box 282, Charleville, Q. 4470.*
A.S. Le Feuvre, *Queensland Department of Primary Industries, P.O. Box 231, Warwick, Q. 4370.*

*Summary* Three experiments were conducted to examine the benefit of providing shade in late pregnancy to reduce effects of heat stress on lamb birth weight and survival. In one pen experiment, birth weights of lambs in shadeless pens were significantly lower (average 19%) than those of lambs in shaded pens. In a second pen experiment, more deaths occurred in lambs born during high ambient temperatures (36°C) than in a period of moderate temperatures (28°C). In a paddock experiment, significantly more lambs (average 12%) were marked in each of 3 years in a paddock with shade trees than in a shadeless paddock. In all experiments survival rate was lower in lambs of lower birth weight. It was concluded that the provision of shade reduced the effect of maternal hyperthermia on birth weight, as well as the direct effect of high ambient temperature on lamb viability immediately after birth.

## INTRODUCTION

Reproductive wastage studies report persistently low lamb-marking percentages (c. 40%) in the semi-arid tropics which are characterised by a treeless environment with high summer temperatures (Moule 1954; Hopkins 1969). Many studies have identified the adverse effect of maternal hyperthermia during a summer gestation on lamb birth weight (e.g., Hopkins *et al.* 1980). Moule *et al.* (1956) and Rose (1978) both reported a direct relationship between lamb birth weight and survival. Hopkins and Pratt (1976) proposed provision of shade as beneficial in reducing effects of heat stress on fertility. The present experiments were designed to test the hypothesis that shade has potential also to improve lamb birth weights and survival.

This paper describes pen and paddock studies in which we examined the effect of shade and its interaction with level of nutrition in improving lamb survival.

## MATERIALS AND METHODS

The experiments were carried out at 'Toorak' Research Station, near Julia Creek in north west Queensland (long. 141°E; lat. 21°S). Summer is characterised by monthly maximum temperatures exceeding 35°C for six months of the year and solar radiation loads of 4 300 KJ $m^{-2}h^{-1}$, sufficient to induce hyperthermia in pregnant ewes and reduce birth weights of their lambs (Hopkins *et al.* 1980).

In pen Experiment 1, 80 pregnant ewes were divided into four treatment groups four weeks before expected parturition at the end of February. These groups were allocated to a factorial arrangement of two nutritional levels (500 and 1 200 g lucerne $ewe^{-1}d^{-1}$) × two shading treatments (a shaded v. an unshaded pen). A 24 hour surveillance was established to record parturition time, maternal behaviour immediately after parturition, lamb activity (times to rise to feed and to suck effectively), lamb birth weights and deaths over the first four days of life.

In pen Experiment 2, lamb birth weights and survival rates were recorded when 51 ewes lambed in shadeless pens during two periods of differing ambient temperatures. A basal ration of poor quality Flinders grass hay (0.9% N) was fed to ewes during the last 2-6 weeks of gestation. A period of five days of high maximum (36°C ± 0.9) daily temperatures occurred when 26 of the lambs were born. The remaining 25 lambs were born in the following six day period, when temperatures were moderate (28°C ± 1.1). Respiration rates greater than 100 $min^{-1}$ were observed in the ewes on days when ambient temperatures rose above 30°C.

In the paddock experiment, two adjoining 1 000 ha paddocks were used each year over three consecutive years to compare the lambing performance of ewes after a summer gestation. A plot of 66 Athol pine trees (*Tamarix aphylla*) had been established in the centre of one paddock three years earlier. No other shade was available in either paddock. On the basis of abdominal palpation (Pratt and Hopkins 1975), ewes expected to lamb over a three week period were allocated to both paddocks approximately four weeks before the onset of lambing. In year 1, 180 ewes, and in years 2 and 3, 200 ewes were placed in each paddock. Summer rainfall was above average (≥ 400 mm) for each of the three years, and ewes were stocked at the district average of 0.5 sheep $ha^{-1}$. An observation tower erected 400 metres from the shade plot was used by observers to monitor the amount of time ewes spent under the trees during the last four weeks of gestation. In the last year, a random sample of 15 ewes at day 135+ of gestation was removed from each paddock and lambed under surveillance in pens to record birth weights of lambs.

## RESULTS

### Pen Experiment 1

Lamb birth weights were significantly depressed when there was no shade and when nutrition was poor (Table 1). Under the experimental conditions imposed, shading had a greater effect than nutrition. Birth weights of the groups without shade in the high and low nutrition treatments were 23.3% ($P < 0.05$) and 15.4% ($P < 0.05$) lower than those of the same nutritional groups with shade provided. Values for the shaded and unshaded groups on poor nutrition were 13.4% ($P < 0.05$) and 4.3% (n.s.) lower than for the respective groups on the high plane of nutrition. Observed duration of parturition (25 ± 3 minutes) did not differ significantly between treatments. Three of the four ewes with the longest parturition times subsequently deserted their lambs. These three ewes were in the groups whose nutrition was poor; two were without shade and one was shaded. Lambs heavier than 2 kg from all groups were standing 9 ± 1 minutes after birth and were effectively sucking 24 ± 2 minutes after birth. Lambs lighter than 2 kg were born in all groups except the shaded, well-fed group. These lighter lambs were standing within 17 ± 3 minutes, but despite trying, all failed to suck effectively. Time of death varied from 24 to 48 hours after birth depending on shade treatment. Lamb mortality increased with severity of treatment (Table 1).

**TABLE 1. Birth weight (mean ± SEM) and mortality of lambs.**

| | High nutrition | | Low nutrition | |
|---|---|---|---|---|
| Attribute | Shade | No shade | Shade | No shade |
| Lamb birth weight (kg) | 3.0* ± 0.19 | 2.3 ± 0.15 | 2.6* ± 0.16 | 2.2 ± 0.17 |
| No. of lamb deaths | 2 | 3 | 5 | 9 |

* $P < 0.05$

### Pen Experiment 2

Most ewes lambed during the early morning and lambs were exposed to shadeless conditions during the immediate post-natal period of life. No significant differences in mean birth weight occurred between the high and moderate temperature periods. However, the birth weight of survivors was significantly higher than of non-survivors (Table 2). The survival rate of lambs born during moderate temperatures was better than of those born in the high temperature period but the difference was not significant (84 v. 65%). All dead lambs died within 24 to 36 hours of birth.

**TABLE 2. Birth weight (mean ± SEM) and survival of lambs during periods of different ambient temperatures at birth.**

| | Daily temperature | |
|---|---|---|
| Attribute | High | Moderate |
| Mean maximum (°C) | 36.0 ± 0.9 | 28.0 ± 1.1 |
| Lamb birth weight (kg) | | |
| All lambs | 2.94 ± 0.20 | 3.09 ± 0.12 |
| Survivors [1] (n) | 3.14 ± 0.14 (17) | 3.19 ± 0.16 (21) |
| Non-survivors (n) | 2.56 ± 0.14 (9) | 2.54 ± 0.15 (4) |

[1] Birth weight was significantly ($P < 0.05$) greater than for non-survivors during both temperature periods.

### Paddock Experiment

Significantly more lambs were marked each year in the paddock with the shade trees than in the shadeless paddock (Table 3). Lambs from ewes in year 3 that had the benefit of shade trees during gestation were heavier (12.9%, $P < 0.05$) at birth than lambs from ewes without shade. Most ewes in the shaded paddock spent approximately seven daylight hours in the shade to avoid the high temperatures.

**TABLE 3. Survival rate of lambs over three years in shaded and unshaded paddocks, and birth weight in year 3 (mean ± SEM).**

| Attribute | Shaded | Unshaded |
|---|---|---|
| Survival rate (%) Year 1 | 56 | 46 |
| 2 | 77* | 62 |
| 3 | 67* | 51 |
| Birth weight (kg) Year 3 | 2.72 ± 0.10* | 2.41 ± 0.13 |

* $P < 0.05$

## DISCUSSION

These results support previous work showing the depressing effect of heat stress on lamb birth weight and resulting viability and further implicate the direct effect of high ambient temperatures at birth. The early death (before 48 h) of lambs suggests dehydration due to hyperthermia (Smith 1961; Hopkins 1969). We propose that provision of shade has achieved a reduction in heat stress effects. The low nutritional plane in the pen experiment apparently exacerbated the effects of shadeless conditions on lamb viability. Nutrition of ewes during the paddock experiment was considered excellent.

It is concluded that conditions of heat stress typical of this environment may lower survival rates of autumn born lambs. Ewes lambing during hot, dry weather without shade will give birth to lambs with low chance of survival. Hopkins and Pratt (1976) highlighted some aspects of providing shade and we believe that industry will also benefit by improvement in lamb survival.

## ACKNOWLEDGEMENTS

This work was supported by a grant from the Wool Research Trust Fund on the recommendation of the Australian Wool Corporation.

## REFERENCES

Hopkins, P.S., 1969. M.V.Sc. Thesis, University of Queensland.
Hopkins, P.S., Nolan, C.J. and Pepper, P.M., 1980. *Aust. J. Agric. Res., 31*, 763-771.
Hopkins, P.S. and Pratt, M.S., 1976. *Proc. Aust. Soc. Anim. Prod., 11*, 153-157.
Moule, G.R., 1954. *Aust. Vet. J., 30*, 153-171.
Moule, G.R., Jackson, M.N.S. and Young, R.B., 1956. *Qld. Agric. J., 82*, 345-354, 399-401.
Pratt, M.S. and Hopkins, P.S., 1975. *Aust. Vet. J., 51*, 378-380.
Rose. M., 1978. M.Sc. Thesis, University of New South Wales.
Smith, I.D., 1961. *Aust. Vet. J., 37*, 205-210.

# BREEDING FOR REPRODUCTIVE PERFORMANCE

N.M. Fogarty, *Department of Agriculture, Agricultural Research Station, Cowra. N.S.W. 2794.*

## INTRODUCTION

The Australian sheep industry is diverse in both the environments utilized and types of production. The production systems vary from very extensive pastoral zone wool production to intensive lamb and wool production in higher rainfall and irrigation areas. The dominant production system is based on Merino sheep with wool the major product. However a high reproductive rate is important for all sheep breeders as considerable income is derived from sale of sheep, whether these be surplus sheep for restocking, slaughter, live sheep for export or prime lambs based on Merino crossbreds.

Biological and economic efficiency are both improved with higher levels of reproduction (Dickerson 1978). Increased reproductive rate spreads the high fixed energy input cost of maintaining the breeding ewe flock and replacements over more sale offspring. Flock reproductive rate also effects selection intensity and consequently the rate of genetic improvement in all traits under selection.

Breeding objectives for various types of sheep flocks have been extensively reviewed by Ponzoni (1982). For the Merino, as well as dam breeds used in the prime lamb industry, the trait, number of lambs weaned, contributed almost half the gain in economic units, with fleece weight, fibre diameter and liveweight together contributing the remainder (Stafford and Walkley 1979; Ponzoni 1982). Allowance for increased feed requirements that would accompany genetic changes in reproduction may reduce the contribution of number of lambs weaned to a large extent (Jones 1982). However, reproduction is an important trait in the breeding objective for a wide range of types of flocks and changing market trends (Ponzoni 1982), especially when surplus offspring are sold at weaning (Ponzoni and Walkley 1984).

Reproduction rate can be expresed in a number of ways, but generally refers to the number of lambs weaned (or marked) per 100 ewes joined. This is a complex trait comprising a multiplicative function of the component traits; fertility (lambed or not lambed), litter size (or fecundity) and lamb survival (individual and maternal rearing ability). In some intensive lamb production systems, a more complex definition (eg. weight of lamb weaned/ewe/year) may be used with the additional component traits of lamb weaning weight and lambings per ewe per year included. Lifetime reproduction which is influenced by age of first joining, age at puberty and ewe longevity may also be incorporated.

This paper reviews the opportunities for increasing reproduction and its components by genetic means with major emphasis given to Australian sheep. Deficiencies of knowledge and areas requiring further research are discussed.

## BETWEEN BREED VARIATION

There is considerable variation between breeds in their rates of reproduction, largely due to differences in ovulation rate, the main determinant of fecundity and to a lesser extent fertility at various seasons of the year or length of the breeding season and lamb survival. Fecundity ranges from around 1-1.2 in many Australian Merino flocks to in excess of 2.5 in breeds such as the Finnsheep and Romanov. These differences between breeds in fecundity are mainly an expression of the differences in ovulation rate. In the Booroola Merino mean litter size and ovulation rate are 2.5 and 4.2 which considerably exceed these values for other Merino strains (Piper *et al.* 1984).

Substantial heterosis or hybrid vigour is expressed for reproduction when crossbreeding is utilized (Nitter 1978). Maternal heterosis expressed by crossbred dams tends to be higher than individual heterosis for reproduction traits. Heterosis for the various components of reproduction are cumulative and can result in heterosis in excess of 40% for lambs weaned per ewe joined (McGuirk 1970; Fogarty *et al.* 1984). Ch'ang and Evans (1982) have demonstrated the importance of individual, maternal and paternal components of heterosis for reproduction. Development of new breeds by crossbreeding or infusion of new genes into an existing breed may be effective in obtaining desired attributes and utilizing heterosis, if recombination loss is not important. The Coopworth and Perendale in New Zealand and the Polypay in the U.S.A. are recent examples of newly developed breeds finding widespread acceptance.

## SELECTION WITHIN BREEDS

Selection of animals to be retained in the flock has a two-fold effect. Current flock performance is raised due to the superior lifetime production of retained animals, and the performance of the flock in future generations is affected by permanent genetic changes due to changes in gene frequencies.

Gains in current flock performance from selection are largely determined by the repeatability of the traits which are generally low and gains in current flock performance will be small. Changes in flock age structure necessitated by current flock selection will further dilute gains in total flock performance.

Gains in future generations can be predicted from:
$\Delta G = i\, h^2\, \delta_p / L$
where $\Delta G$ is the genetic response per year
$i$ is the standardised selection differential
$h^2$ is heritability
$\delta_p$ is phenotypic standard deviation
$L$ is generation interval in years.

Heritability estimates for the various reproduction traits are low, but are increased if mean performance over two or more joinings is used. Phenotypic variation for the traits is generally high. The selection intensity that can be applied and the generation interval are dependent on the level of net reproduction in the flock and the age structure in the ewe and ram flock.

Selection recommendations have generally involved selection for fecundity or twinning rather than complex traits or other component traits. This has been prompted by small predicted responses to selection for fertility (against barren ewes), ability to select ewes and rams at an early age and recommendations that environmental rather than genetic modifications be used to maximise lamb survival. However there is increasing evidence that genetic variation eixsts in all components and that selection for reproductive components other than fecundity should also be emphasised (Fogarty *et al.* 1982; Haughey and George 1982; Piper *et al.* 1982).

## REALISED RESPONSE

I am aware of five experiments in which response to selection for reproduction in sheep has been reported (Table 1). Selection has been aimed at increasing number of lambs born, but methods and specific criteria have varied between experiments. In summarising the results Land *et al.* (1983) noted the remarkably uniform rate of response of about 1.5% per year in all experiments except that of Mann *et al.* (1978) in which selection was only applied to rams and limited selection pressure was achieved. The responses reported are linear and close to predictions from the base populations, with no indication of any decline in response with time.

Four of the experiments illustrate that despite low heritability large total response can be achieved from selection for increased reproduction over a number of years. This results from the greater degree of variability in reproduction than other production traits and consequently the higher realised selection differentials that can be achieved in some instances. Initial screening of populations can achieve important increases in reproduction, such as 0.1 lambs born/ewe lambing in the experiment reported by Turner (1978).

**TABLE 1. Realised selection responses in number of lambs born/ewe lambing (Modified from Land et al. 1983).**

| Breed | Time (years) | Total response | Annual response | Predicted response | Reference |
|---|---|---|---|---|---|
| Merino | 10 | 0.22 | 0.022 | 0.015 | Atkins (1980a) |
| Merino | 14 | 0.38 | 0.02[1] | — | Turner (1978) |
| Merino | 10 | 0.02 | 0.002 | 0.004 | Mann *et al.* (1978) |
| Romney | 22 | 0.40 | 0.018 | 0.021 | Clarke (1972) |
| Galway | 15 | 0.35 | 0.023 | — | Hanrahan (1984) |

(1) Excluding initial effect of screening.

## COMPONENTS OF EWE REPRODUCTION

Reproduction is a complex trait comprising the components fertility, fecundity (which incorporates ovulation rate and embryo survival) and lamb survival (individual and maternal rearing ability). The components of ewe reproduction are under very different physiological control mechanisms and may well be controlled by different sets of genes. Thus it is important to examine in more detail the genetic variation in each component and the scope for genetic improvement.

### Fertility

Ewe fertility may be affected by age at puberty, breeding season and postpartum joining interval in accelerated lambing systems. Fertility of autumn joined adult ewes is generally close to the upper limit of 100% provided bodyweights are adequate and there is little scope for expression of any genetic variation. However, under an 8-monthly joining system fertility was an important contributor to variation in overall reproductive rate (Fogarty *et al.* 1982).

Considerable between breed variation and heterosis exists for age at puberty (see review by Nitter 1978). There is limited evidence of within breed genetic variation for age at puberty with heritability estimates ranging from 0.1 to 0.26 (Tierney 1979; Fogarty 1981; L.R. Piper and B.M. Bindon, unpublished data, cited by Piper,

1982). The number of oestrus cycles in the first breeding season, which is largely determined by age at puberty, appears to be moderately heritable (0.3) (Ch'ang and Rae 1970; Baker *et al.* 1979) and should respond to selection. Piper (1982) has pointed out that under the Australian environment Merino ewes generally do not reach sufficient liveweight to attain puberty in their first autumn-winter and selection response may not be effective in increasing ability to breed in their first year.

Sheep generally exhibit a seasonal pattern of oestrous activity which is triggered by decreasing daylength. There is considerable variation between and within breeds in the time of onset of oestrous activity and duration of the breeding season. In the extreme, may tropical breeds do not exhibit a distinct breeding season (Mason 1980). Onset and duration of the breeding season is effected by age, nutrition and postpartum interval (Ricordeau 1982) and can be modified by artificial day-length (Dunstan, 1977), use of the 'ram effect' (Oldham 1980) and exogenous hormones (Robinson 1967). Ovulation may occur without concurrent detection of oestrus which may result from 'silent heat' or inadequacies in methods for detection of oestrus. These factors contribute to the difficulties in assessing variation for length of the breeding season, although genetic variation appears to exist (Piper 1982). In a preliminary report Fahmy (1982) has shown some response to selection for early onset of the breeding season. A major gene has also been shown to suppress out of season breeding activity in Icelandic sheep (Dyrmundsson and Adalsteinsson 1980).

Variable and limited oestrous and ovulatory activity outside the autumn breeding season restricts flexibility of joining time, and limits reproductive rate at joinings outside the normal breeding season. This restriction limits the scope for more intensive accelerated lambing systems. There is limited evidence of genetic variation in length of the breeding season which needs to be comfirmed in Australian breeds by studies of between and within breed variation and by experimental selection.

### Ovulation rate and fecundity

Fecundity or number of lambs born per ewe lambing is largely a function of ovulation rate and embryonic survival. There is considerable variation between breeds, with fecundity ranging from 1.0 to about 2.5. Genetic variation for fecundity also exists within breeds and can be exploited by selection (Table 1).

Heritability estimates for ovulation rate range from $0.57 \pm 0.28$ for Galway ewes (Hanrahan 1980) to $.05 \pm .07$ for young Merino ewes (Piper *et al.* 1980). Heritability for ovulation rate is higher than that for litter size when both traits have been estimated in the same flock (Piper 1982). Hanrahan (1974) proposed that selection for increased ovulation rate would be more effective in increasing fecundity than direct selection for fecundity. Piper (1982) concluded that more recent data continued to support this proposal. Hanrahan (1984) has recently reported a regression of $0.14 \pm .06$ ($P < .02$) for daughter's fecundity on her dam's ovulation rate.

As ovulation rate increases there is a linear decline in embryonic survival which results in a curvilinear relationship between ovulation rate and fecundity. Hanrahan (1982) proposed a quadratic model for this relationship which is consistent with Finnsheep and Booroola Merino data and has a maxima at a litter size of about 3.0 with an ovulation rate of 6.1. Hanrahan (1982) concluded that there is no genetic variation for embryonic survival as this had not contributed to observed responses in selection for increased litter size. However Hanrahan (1984) has recently presented limited evidence that suggests improved embryo survival may have made a small contribution to response in litter size in his High Fertility line.

Ovulation rate is the major determinant of potential reproduction. However very few reports are available on the magnitude of genetic variation for this trait within the Merino and other Australian breeds. Accurate estimates of parameters are required, together with realised response to selection, to determine whether selection for ovulation rate will result in more rapid improvement than selection for fecundity. The question of genetic variation for embryonic survival requires further research.

### Lamb survival and maternal rearing ability

Annual losses of lambs from birth to marking in Australia are of the order of 20% or approximately 11 million lambs. Losses vary considerably but of all lambs born approximately 5% are born dead and 14% die 0 to 3 days after birth (Watson 1972). Large ewe breed differences in lamb survival have been reported in Australia (Iwan *et al.* 1971; Atkins 1980b) and overseas (Hight and Jury 1970). Heterosis for lamb survival (individual and maternal) has been demonstrated (5 to 25%), with the level of heterosis generally higher amongst primiparous ewes and with twin lambs (Hight and Jury 1970; McGuirk 1970; Iwan *et al.* 1971). Lambs of different breeds vary in resistance to heat loss under laboratory tests (Samson and Slee 1981; Slee 1981). The very high losses due to dystocia in the Dorset breed have been associated, at least in part, with small pelvic size (Fogarty and Thompson 1974; Haughey 1984). Pneumonia causes high lamb losses in the Border Leicester breed and there is evidence of genetic differences in susceptibility between flocks (Fogarty 1977).

In three Merino flocks Haughey and George (1982) showed that 25% of ewes failed to rear lambs at least twice from four lambings and these ewes accounted for 60% of total lamb losses. Lamb survival has been monitored in 15 flocks at Trangie that represent various Merino strains in the industry. The range for the 15 flocks in mean lamb survival from 1977 to 1981 was 66 to 85% for singles and 48 to 75% for multiples with flock differences being consistent over years. (I. Rogan, unpublished data). Since the 15 flocks have been bred and managed under the same conditions, these data strongly suggest there are genetic differences between them for lamb survival.

Variation in lamb survival amongst sire progeny groups is low, with heritability estimates for direct effects averaging about 0.04 (Cundiff *et al.* 1982). Because of the low heritability and extended generation interval required to progeny test sires, Piper *et al.* (1982) concluded there was little scope for improving lamb survival by direct selection over and above natural selection. It should be noted that inbreeding of the lamb and/or the ewe depresses lamb survival (Lax and Brown 1968), and thus non-additive genetic variation is important. The maternal component of lamb survival (ewe rearing ability) has a repeatability of 0.10 to 0.15 (Fogarty *et al.* 1982; Piper *et al.* 1982; Haughey 1984). These estimates suggest some improvement in current flock performance can be achieved by culling ewes with poor maternal rearing ability. In three flocks analysed by Haughey (1984) ewes that sucessfully reared lambs at two years of age reared on average 2.8 to 14.7% more lambs subsequently than those ewes that failed to rear a lamb at two years of age.

Estimates of heritability for maternal rearing ability at one lambing are about 0.1 for the Merino (Piper *et al.* 1982) and various purebreeds and crosses (Fogarty *et al.* 1982). As expected, the heritability estimates are somewhat higher when based on the mean of a number of lambing records (Fogarty *et al.* 1982). These parameter estimates are of a similar order to those for litter size and other measures of reproduction (Piper 1982). Thus response to selection would be expected, both in the current generation from culling ewes and in future generations from cumulative genetic gains.

The Trangie Merino Fertility selection flock has been subjected to selection for twinning, with ewes culled for failing to rear a lamb at any lambing. After approximately 10 years of selection, lamb survival to weaning of singles is 93 V 86% and of twins is 85 V 78% for the fertility and random flocks respectively (Atkins 1980a). Haughey (1983) has reported lamb mortality in Merino flocks selected for high and low efficiency of rearing of 19.5 V 32.3% for singles and 21.5 V 36.3% for twins respectively. Selection including maternal rearing ability in a commercial Merino flock has resulted in a 9% improvement in lamb survival (Donnelly 1982). These reports provide indirect evidence of genetic variation for maternal rearing ability, but parameter estimates and direct selection response need to be confirmed in adequately designed experiments.

Increases in potential reproduction rate due to genetic improvement, technology and/or management will only be beneficial if the extra lambs survive. Past recommendations have relied on environmental modification and lambing management to improve lamb survival and have generally ignored opportunities for genetic improvement. Recent evidence strongly suggests there is potential for genetic improvement of maternal rearing ability and research in this area should receive priority.

## CORRELATED RESPONSES

Selection for any one trait usually results in changes in other genetically correlated traits. The magnitude and sign of genetic correlations have important consequences for selection. These consequences include the use of correlated traits for indirect selection, changes in other important productive traits (complementary or antagonistic) following selection and development of optimum selection indexes to change combinations of important productive traits.

There are few estimates of genetic correlations (generally with large standard errors) amongst reproductive traits and between them and other productive traits. Fogarty *et al.* (1982) reported moderate negative genetic correlations (-.3 to -.4, not significant) between fecundity and other reproduction components fertility and lamb survival, and higher positive correlations between weight of lamb weaned per ewe joined and components, fertility, lamb survival and lamb weaning weight. Ch'ang and Rae (1972) reported moderate to high positive genetic correlations between liveweight and number of first year oestruses and reproductive traits in New Zealand Romneys, but Baker *et al.* (1982) using a larger data set have reported much lower genetic correlations for these traits.

Genetic correlations between reproduction and other traits in the Merino have been reviewed by Turner (1977). These indicated that the genetic correlation of reproduction with liveweight was small and positive, with wrinkle score, small and negative and with face cover, small and negative. Selection for high and low wrinkle score in Trangie Flocks resulted in a marked divergence in net reproductive rates between the flocks which was attributed to both ram and ewe effects (Atkins 1980a). Reported genetic correlations between reproduction and wool production are variable in size and magnitude, but tend to be small and negative (see Land *et al.* 1983).

## SELECTION OF RAMS

Rams do not directly express female reproductive traits and selection must be based on indirect traits in the ram or traits expressed by female relatives. For example in the Trangie Fertility flock, ram selection was based partly on birth type (Atkins 1980a), which is in effect selecting on the basis of the ram's fecundity.

A positive relationship between testicular growth in rams and fecundity of ewes between breeds has been reported (Land 1973; Hanrahan and Quirke 1977). Land (1973) proposed that testicular development in the male and ovulation rate in the female may be controlled by similar genes as similar hormones control reproduction in both sexes. Selection of rams for testicular development has produced divergent lines, with a realised heritability of about 0.4. Correlated changes in females included earlier onset of the breeding season in the high line (Land and Lee 1976), and a 15% divergence in ovulation rate (Land, cited by Bindon and Piper 1979) which is principally due to correlated changes in liveweight (Land *et al.* 1983). A correlation of .43 between female prolificacy and male testicular size in limited half-sib data (14 sires) has been reported by Ricordeau *et al.* (1979). While there appears to be between sire variation in testicular size, Purvis *et al.* (1984) reported no significant sire variation for ram serving capacity.

## INDIRECT SELECTION

Increased response to selection may be achieved by a better understanding of the relationships between components of reproduction and indirect selection traits that can be measured early in life in both sexes. In a theoretical study, Walkley and Smith (1980) compared expected response from direct selection for fecundity with indirect selection using three types of physiological traits, *viz.* male sex-limited, female sex-limited and a traits measurable in both sexes, as well as combined selection. Their results show there is usually scope for improvement in the rate of response with combined selection, which can be large if the heritability of the physiological trait and its genetic correlation with fecundity are high. They conclude by cautioning that realised response depends on the level of the various parameters, and lament the dearth of reliable estimates, especially for genetic correlations.

Various indirect traits have been advocated in selection for increasing reproductive rate (see Land 1974; Bindon and Piper 1979 Land and Carr 1979; Land *et al.* 1982, 1983). Selection for response to gonadotrophin releasing hormone (LH-RH) (Land *et al.* 1982) has produced direct responses but little or no change in prolificacy (Land, cited by Piper and Bindon 1984). It seems feasible that physiological traits may be found that are highly correlated with certain reproductive traits, such as ovulation rate. However, the search for such traits over the last ten years has been largely unsuccessful. The chance of a single physiological trait being highly genetically correlated with a complex reproductive trait, such as number of lambs weaned/ewe joined, seems remote. Such a complex trait encompasses the diverse components of ovulation rate, embryonic survival, lamb survival and maternal rearing ability, which are probably under very different physiological control systems.

## THE BOOROOLA MERINO

### Mode of action

The Booroola Merino is a member of the medium woolled non-Peppin strain. While it has wool and body characteristics typical of the strain (Piper and Bindon 1982a) it has a mean ovulation rate of 4.2 (range 1 to 10 and mean fecundity of 2.5 (range 1 to 7) (Piper *et al.* 1984). Piper and Bindon (1982 a,b) and Piper *et al.* (1984) have presented convincing evidence that a gene is segregating in the Booroola that has a major effect on ovulation rate. The putative gene (*F*) appears to act additively for ovulation rate and litter size (Piper *et al.* 1984).

### Crosses with other strains and breeds

Considerable research effort over the last few years has been devoted to evaluation of the Booroola (*F*) gene in various genetic backgrounds in Australia and New Zealand (see Piper *et al.* 1984). Many of these experiments were designed and implemented before it was realised that the Booroola effect was due to a single gene. Hence meaningful analysis of data will require the determination of *F* gene frequency and classification of ewes into the various genotype categories.

In comparison with purebred Medium Non-Peppin (MNP) and Collinsville Merino strains, Booroola cross ewes have shed 0.3 to 1.1 more ova and produced 20 to 30% more lambs weaned, despite lower lamb survival. Wool production and liveweight of Booroola cross ewes was slightly lower than Collinsville but similar to MNP Merino (Piper *et al.* 1979; McGuirk *et al.* 1984). In South Australia, Booroola × Bungaree Merino ewes produced 0.5-0.6 ($P < .05$) more lambs born per ewe lambing than Bungaree ewes (Ponzoni *et al.* 1984). Much higher lamb mortality amongst the Booroola cross ewes reduced the difference to 0.1-0.2 (n.s.) for lambs weaned. Booroola crosses with New Zealand Merinos have increased ovulation rate by 0.8 to 1.0 ova and fecundity by 0.5 to 0.6 lambs (Allison *et al.* 1977).

Booroola × Dorset (BD) ewes have produced 0.6 more lambs born per ewe lambing than Trangie Fertility × Dorset (TD) ewes over four years. This has resulted from a similar proportion of twin born lambs (42 V 46%) but considerably more triplet (22 V 4%) and quad (8 V 0%) births for BD than TD ewes. Birthweight of lambs from BD ewes was 0.4 to 0.6kg less (within birth category) than lambs from TD ewes, even though the lambs were of the same reciprocal cross genotype. This was associated with lower survival of lambs from BD ewes (Hall and Fogarty 1982; unpublished data). Booroola × Romney ewes have a significantly higher percentage showing oestrus in their first year (67 V 48%), with a higher ovulation rate (1.39 V 1.03) than Romneys and produced 0.8 to 0.9 more ova and 0.11 to 0.52 more lambs marked per adult ewe (Kelly *et al.* 1980).

The Booroola offers the opportunity for a quantum increase in fecundity in one generation compared with much slower improvement from within flock selection. However inability to readily identify carriers of the putative gene (homozygous or heterozygous), especially in rams, presently limits the controlled use of the Booroola in the sheep industry. The very high ovulation rate and fecundity, particularly amongst homozygous ewes and ewes with a background genotype for high fecundity, presents special problems for ewe and lamb management. Economically feasible management systems need to be developed to improve survival of the higher order multiple birth lambs under extensive Australian conditions.

## INPACT OF NEW TECHNOLOGY

Exogenous hormones have been used for many years to control breeding season and increase the level of fecundity in sheep flocks (Robinson 1967). Immunization of Merino ewes against oestrone, androstenedione or testosterone results in a 20% increase in lambs born through extra twins from an autumn joining (Cox *et al.* 1982). Responses achieved from these applications of technology tend to be greater amongst genotypes with higher reproductive potential. Genetic engineering offers potential for improvement of sheep flocks (Ward, 1982) and where appropriate it should be exploited to increase the rate of response to selection.

The extent to which increases in ovulation rate and fecundity are transformed into increased numbers of lambs weaned will depend on the managment of ewes during pregnancy and lambing. Appropriate management strategies need to be researched and developed to achieve high levels of reproduction (Jelbart and Dawe 1984). Real time ultrasound scanning to determine the number of foetuses in mid pregnancy (Fowler and Wilkins 1982) and determination of glucose status of ewes in mid-pregnancy (Parr *et al.* 1984) appear to be useful aids for refining the management of pregnant ewes and particularly for identifying ewes at risk.

The ability to freeze semen and embryos offers powerful research tools for geneticists. These techniques will allow more accurate assessment of response to selection than contemporary control flocks that are subject to genetic drift. Maternal effects can be measured directly and more efficient designs can be used to determine the importance of the various components of heterosis and recombination loss. The development of laparoscopic insemination (Killeen and Caffery 1982) has greatly improved conception rates with frozen semen and has made the use of frozen semen more feasible and practical.

## CONCLUSION

There is considerable opportunity for genetic improvment of reproductive rate in sheep. Research in Australia and overseas has demonstrated significant improvement in lambing rates by utilization of superior breeds, crossbreeding and within flock selection. All the components of net reproductive rate are important and should be included in selection, although the relative importance of each component will vary with the production system. Further research is required to estimate genetic parameters for length of the breeding season, ovulation rate, fecundity and maternal rearing ability in Australian breeds. These parameters are required to develop optimum selection programs that incorporate wool, liveweight and reproductive rate for the various production systems in the sheep industry.

Implementation of an effective selection program to improve reproductive is not without difficulty. Reproduction traits are directly expressed by ewes but selection of rams must be based on information from relatives, usually dams, or indirect traits. In most situations, progeny testing of rams increases the generation interval substantially because repeated measures of ewe progeny are usually required. Phenotypic expression of many traits is either binomial or as relatively few discrete levels, although the underlying variation may be normally distributed. Research effort to develop simple low cost techniques for identification of animals, collecting, recording and analysing reproductive data would be very useful to ram breeders.

It must be emphasised that selection is not a substitute for good management. Sheep flocks with high levels of fecundity require a high standard of management if they are to achieve commensurately high weaning percentages. Under intensive production systems, optimum ewe nutrition and lambing management strategies need to be developed to ensure high lamb survival and a continued high level of ewe reproduction, especially at out of season joinings and under accelerated lambing systems.

## REFERENCES


Allison, A.J., Stevenson, J.R. and Kelly, R.W., 1977. *Proc. N.Z. Soc. Anim. Prod., 37*, 230-234.

Atkins, K.D., 1980a. *Proc. Aust. Soc. Anim. Prod., 13*, 174-176.

Atkins, K.D., 1980b. *Aust. J. Expt. Agric. Anim. Husb., 20*, 288-295.

Baker, R.L., Clarke, J.N., Carter, A.H. and Diprose, G.D., 1979. *N.Z.J. Agric. Res., 22*, 9-21.

Baker, R.L., Clarke, J.N., Carter, A.H. and Diprose, G.D., 1982. *Proc. 2nd Wld. Congr. Genet. Appld. Livest. Prod.,* Madrid, *SY-5-9*, 519-524.

Bindon, B.M. and Piper, L.R., 1979. *In* Tomes, G.L., Robertson, D.E., Lightfoot, R.J. and Haresign, W. (ed), *Sheep Breeding*, 2nd Ed. Butterworths, London, 387-401.

Ch'ang, T.S. and Evans, R., 1982. *Proc. 2nd Wld. Congr. Genet. Appld. Livest. Prod.*, Madrid, *SY-6-D-33*, 796-801.

Ch'ang, T.S. and Rae, A.L., 1970. *Aust. J. Agric. Res., 21*, 115-129.
Ch'ang, T.S. and Rae, A.L., 1972. *Aust. J. Agric. Res., 23*, 149-165.
Clarke, J.N., 1972. *Proc. N.Z. Soc. Anim. Prod., 32*, 99-111.
Cox, R.I., Wilson, P.A., Scaramuzzi, R.J., Hoskinson, R.M., George, J.M. and Bindon, B.M., 1982. *Proc. Aust. Soc. Anim. Prod., 14*, 511-514.
Cundiff, L.V., Gregory, K.E. and Koch, R.M., 1982. *Proc. 2nd Wld. Congr. Genet. Appld. Livest. Prod.*, Madrid, *PS-V-3, 310-337.*
Dickerson, G.E., 1978. *Anim. Prod., 27*, 367-379.
Donnelly, F.B., 1982. *Proc. Aust. Soc. Anim. Prod., 14*, 30-32.
Dunstan, E.A., 1977. *Aust. J. Expt. Agric. Anim. Husb., 17*, 741-745.
Dyrumundsson, O.R. and Adalsteinsson, S., 1980. *J. Hered., 71*, 363-364.
Fahmy, M.H., 1982. *Proc. Wld. Congr. Sheep Beef Cattle Breed.*, New Zealand, *1*, 401-404.
Fogarty, N.M., 1977. *Proc. 3rd Int. Congr. Soc. Adv. Breed. Res. Asia Oceania,* Canberra, *7*, 8-11.
Fogarty, N.M., 1981. Ph.D. Diss., University of Nebraska, Lincoln, 176 pp.
Fogarty, N.M., Dickerson, G.E. and Young, L.D., 1982. *Proc. Aust. Soc. Anim. Prod., 14*, 435-438.
Fogarty, N.M., Dickerson, G.E. and Young L.D., 1984. *J. Anim. Sci., 58*, 301-311.
Fogarty, N.M. and Thompson, J.M., 1974. *Aust. Vet. J., 50*, 502-506.
Fowler, D.G. and Wilkins, J.F., 1982. *Proc. Aust. Soc. Anim. Prod., 14*, 491-494.
Hall, D.G. and Fogarty, N.M., 1982. *Proc. Aust. Soc. Anim. Prod., 14*, 651.
Hanrahan, J.P., 1974. *Proc. 1st Wld. Congr. Genet. Appld. Livest. Prod.*, Madrid, 1033-1038.
Hanrahan, J.P., 1980. *Proc. Aust. Soc. Anim. Prod., 13*, 405-408.
Hanrahan, J.P., 1982. *Proc. 2nd Wld. Congr. Genet. Appld. Livest. Prod.*, Madrid, *PS-V-2,* 294-309.
Hanrahan, J.P., 1984. *Proc. 2nd Wld. Congr. Sheep Beef Cattle Breed.*, South Africa (in press).
Hanrahan, J.P. and Quirke, J.F., 1977. *Anim. Prod., 24*, 148.
Haughey, K.G., 1983. *Aust, Vet. J., 60*, 361-363.
Haughey, K.G., 1984. This volume, 210-212.
Haughey, K.G., and George, J.M., 1982. *Proc. Aust. Soc. Anim. Prod., 14*, 26-29.
Hight, G.K., and Jury, K.E., 1970. *N.Z.J. Agric. Res., 13*, 641-659.
Iwan, L.G., Jefferies, B.C. and Turner, H.N., 1971. *Aust. J. Agric. Res., 22*, 521-535.
Jelbart, R.A. and Dawe, S.T., 1984. *Proc. Aust. Soc. Anim. Prod., 15*, 75-77.
Jones, L.P., 1982. *In* Barker, J.S.F., Hammond, K. and McClintock, A.E. (eds). *Future Developments in the Genetic Improvement of Animals*. Academic Press, Sydney, 119-136.
Kelly, R.W., Davis, G.H. and Allison, A.J., 1980. *Proc. Aust. Soc. Anim. Prod., 13*, 413-416.
Killeen, I.D. and Caffery, G.J., 1982. *Aust. Vet. J., 59*, 95.
Land, R.B., 1973. *Nature, 241*, 208-209.
Land, R.B., 1974. *Anim. Breed. Abstr., 42*, 155-158.
Land, R.B., Atkins, K.D. and Roberts, R.C., 1983. *In* Haresign, W. (ed). *Sheep Production*, Butterworths, London, 515-535.
Land, R.B. and Carr, W.R., 1979. *In* Shire, J.G.M. (ed). *Genetic Variation in Hormone Systems*, C.R.C. Press, Boca Raton, Fla., 89-112.
Land, R.B. and Lee, G.J., 1976. *Anim. Prod., 22*, 137.
Land, R.B., Gauld, I.K., Lee. G.J. and Webb, R., 1982. *In* Barker, J.S.F., Hammond, K. and McClintock, A.E. (eds). *Future Developments in the Genetic Improvement of Animals*, Academic Press, Sydney, 59-87.
Lax, J. and Brown, G.H., 1968. *Aust. J. Agric. Res., 19*, 433-442.
Mann, T.L.J., Taplin, D.E., and Brady, R.E., 1978. *Aust. J. Expt. Agric. Anim. Husb., 18*, 635-642.
Mason, I.L., 1980. *Prolific Tropical Sheep*,FAO Animal Production and Health Paper, No. 17, 124 pp.
McGuirk, B.J., 1970. M.Sc. Thesis. University of New South Wales, 234 pp.
McGuirk, B.J., Killeen, I.D., Piper, L.R., Bindon, B.M., Wilson R., Caffery, G. and Langford, C., 1984. *Proc. Aust. Soc. Anim. Prod., 15*, 464-467.
Nitter, G., 1978. *Anim. Breed. Abstr., 46*, 131-143.
Oldham, C.M., 1980. *Proc. Aust. Soc. Anim. Prod., 13*, 73-74.
Parr, R.A., Campbell, I.P., Cahill, L.P., Bindon, B.M. and Piper, L.R., 1984. *Proc. Aust. Soc. Anim. Prod., 15*, 517-520.
Piper, L.R., 1982. *Proc. 2nd Wld. Congr. Genet. Appld. Livest. Prod.*, Madrid, *PS-V*, 271-281.
Piper, L.R. and Bindon, B.M., 1982a. *In* Piper, L.R., Bindon, B.M. and Nethery, R.D. (eds). *The Booroola Merino*, CSIRO, Melbourne, 9-19.
Piper, L.R. and Bindon, B.M., 1982b. *Proc. Wld. Congr. Sheep Beef Cattle Breed.*, New Zealand, *1*, 395-400.
Piper, L.R. and Bindon, B.M., 1984. *Proc. 2nd Wld. Congr. Sheep Beef Cattle Breed.*, South Africa (in press).

Piper, L.R., Bindon, B.M., Atkins, K.D. and McGuirk, B.J., 1980. *Proc. Aust. Soc. Anim. Prod., 13*, 409-412.

Piper, L.R., Bindon, B.M. and Davis. G.H., 1984. *In* Land, R.B. and Robinson, D.W. (eds), *Genetics of Reproduction in Sheep*, Butterworths, London, (in press).

Piper, L.R., Bindon, B.M., Killeen, I.D. and McGuirk, B.J., 1979. *Proc. N.Z. Soc. Anim. Prod., 37*, 63-67.

Piper, L.R., Hanrahan, J.P., Evans, R. and Bindon, B.M., 1982. *Proc. Aust. Soc. Anim. Prod., 14*, 29-30.

Ponzoni, R.W., 1982. *Proc. 2nd. Wld. Congr. Appld. Livest. Prod.,* Madrid, *PS-VId-1*, 619-635.

Ponzoni, R.W. and Walkley, J.R.W., 1984. This volume, 378-381.

Ponzoni, R.W., Walker, S.K., Walkley, J.R.W. and Fleet, M.R., 1984. *In* Land, R.B. and Robinson, D.W. (eds). *Genetics of Reproduction in Sheep*, Butterworths, London, 127-138.

Purvis, I.W., Kilgour, R.J., Edey, T.N. and Piper, L.R., 1984. *Proc. Aust. Soc. Anim. Prod., 15*, 545-548.

Ricordeau, G., 1982. *Proc. 2nd. Wld. Congr. Genet. Appld. Livest. Prod.*, Madrid, *PS-V-4*, 338-347.

Ricordeau, G., Pelletier, J., Courot, M. and Thimonier, J., 1979. *Ann. Genet. Sel. Anim., 11*, 145-159.

Robinson, T.J. (ed). 1967. *The Control of the Ovarian Cycle in the Sheep*. Sydney University Press, Sydney, 258 pp.

Samson, D.E. and Slee, J., 1981. *Anim. Prod., 33*, 59-65.

Slee, J., 1981. *Livest. Prod. Sci., 8*, 419-429.

Stafford, J.E. and Walkley, J.R.W., 1979. *Proc. Aust. Assoc. Anim. Breed. Genet., 1*, 337-353.

Tierney, M.L., 1979. *In* Tomes, G.L., Robertson, D.E., Lightfoot R.J. and Haresign, W. (ed). *Sheep Breeding*, 2nd Ed. Butterworths, London, 379-386.

Turner, H.N., 1977. *Anim. Breed. Abstr., 45*, 9–31.

Turner, H.N., 1978. *Aust. J. Agric. Res., 29*, 327-350.

Walkley, J.R.W. and Smith, C., 1980. *J. Reprod. Fert., 59*, 83-88.

Ward, K.A., 1982. *In* Barker, J.S.F., Hammond, K. and McClintock, A.E. (eds). *Future Developments in the Genetic Improvement of Animals*. Academic Press, Sydney, 17-41.

Watson, R.H., 1972. *Wld. Rev. Anim. Prod., 8*, 104-113.

# LACTATION PERFORMANCE AND EARLY LAMB GROWTH IN MERINO SHEEP SELECTED FOR HIGH AND LOW WEANING WEIGHT

N.E. Heath, G.N. Hinch and C.J. Thwaites, *Department of Animal Science, University of New England, Armidale, N.S.W. 2351.*

*Summary* The lactation curve and peak milk yield of Merino ewes from lines selected for high (W+) and low (W−) weaning weight for 12 generations and that of random bred control ewes were examined. Marked differences in the shape of the lactation curve were apparent with peak yield being higher for W+ and lower for W− than for random controls. Part of these differences are attributable to differences in mature weight of the ewes and lamb sucking stimuli.

## INTRODUCTION

The lactation yield of a ewe is one of the major factors influencing the preweaning growth rate of her offspring and thus indirectly later productive traits and economic returns. Many of the factors contributing to variations in the lactation yield of ewes have been described (Boyazoglu and Treacher 1978), but we know very little about the relationship between ewe liveweight and milk yield or the effects of lamb size (Treacher 1978). The correlated response in lactation with selection for such factors as increased liveweight or preweaning growth rate needs to be quantified.

Pattie and Trimmer (1964) have shown substantial changes (10% in either direction) in the lactation performance of Merino ewes from divergent weaning weight selection lines. The present study was conducted to quantify the milk yield and lamb growth characteristics of divergent weaning weight selection lines described by Pattie (1965) after a further 20 years of selection. The nature of the lactation curve of the different selection lines and the relationships between liveweight and milk yield within and between these lines are examined.

## MATERIALS AND METHODS

### Animals

Merino ewes from the divergent weaning weight selection flocks described by Pattie (1965) were used. The ewes consisted of lines selected for 12 generations for high (W+) and low (W−) weaning weight and random bred controls (R).

Mixed aged ewes (3 to 8 years) rearing single lambs of their own genotype were selected at lambing (September 1983). The ewes were then milked at six day intervals beginning on day 4 ± 1 of lactation and continuing until day 40; thereafter, milking was at 12 day intervals until recording ceased at 76 days. The number of ewes in each group were W+, 29, W−, 16, R, 17. The ewes were machine milked using the oxytocin technique of McCance (1959) and milk production was recorded over a 4 hour period. Lamb liveweights were recorded at each milking.

A second group of random ewes rearing single lambs sired by weight + (R+) or weight — rams (R−) was milked at approximately weekly intervals in the same manner as described for the selection flocks.

The selection flock animals (W+, W−, R) grazed together on forage oats offered *ad libitum* throughout lactation. The R ewes rearing R+ or R− lambs grazed together on a phalaris/white clover pasture which had a lower dry matter availability than the forage oats.

### Analyses

Milk production data were analysed at each time of milking for effects of line and date of birth. Ewe age effects on milk production were also examined as each time. Ewe liveweights at the beginning of lactation and lamb growth rates were analysed by analysis of variance for line differences.

All data from R ewes rearing R+ and R− lambs were treated independently to the selection flock lines because of the potential differences between groups owing to the pre and post partum nutrition.

Multiple regression analysis was carried out to determine the influence of ewe liveweight at the onset of lactation and lamb birth weight on peak milk yield within and between selection lines. Ewe liveweight at the onset of lactation was taken to be representative of mature liveweight.

## RESULTS

The mean lactation curves for the five groups of ewes are presented in Figure 1. Differences in the selection lines were apparent both in the shape and amplitude of the curve; W− animals reached an earlier and lower peak than W+ and R ewes. Mean peak yield was 10% above R for the W+ and 25% below R and the W− ewes.

The W− line had significantly ($P < 0.05$) lower milk yield than R or W+ ewes after 22 days of lactation. There were no significant effects of ewe age on milk production at any time. Differences in the milk production of R ewes rearing R+ and R− lambs was significant ($P < 0.05$) only at the time of peak yield (14 days).

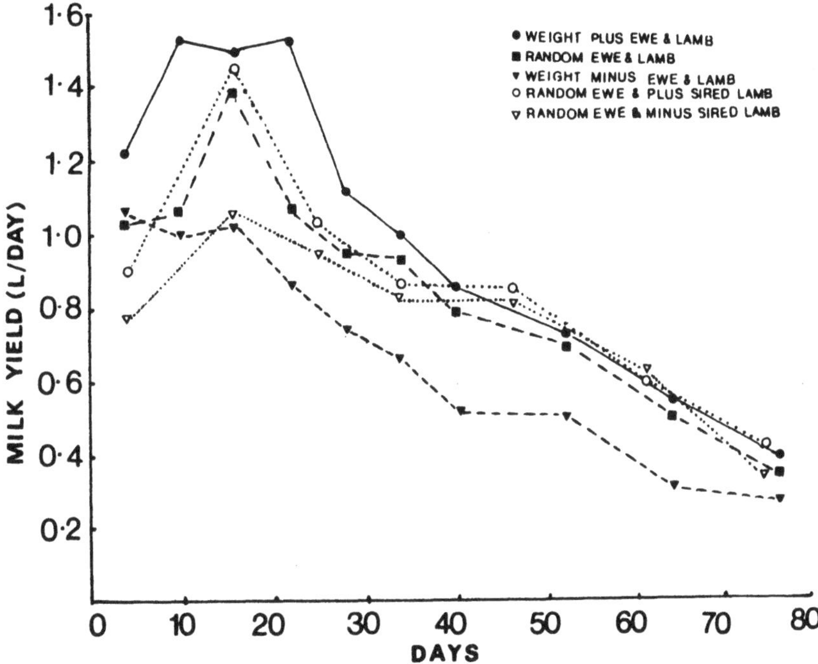


**FIGURE 1. Mean lactation curve of ewe groups.**

Based on the hypothesis of Taylor (1973) that peak milk yield is a function of metabolic liveweight, peak yield was predicted for the three ewe lines (Table 1). Only the peak yield of the W− ewes deviated markedly being 13% lower than predicted.

**TABLE 1. Mean peak lactation and predicted peak lactation of ewe liveweight (±SE) at the beginning of lactation, lamb birth weight (± SE), and lamb growth rate to 76 days (± SE).**

| Selection line | W+ | R | W− |
|---|---|---|---|
| Mean peak yield (l/d) | 1.53 | 1.39 | 1.05 |
| Predicted peak yield (l/d)[1] | 1.49 | 1.39 | 1.21 |
| Mean ewe liveweight (kg) | 46.8 ± 1.2 | 42.4 ± 1.5 | 35.2 ± 1.0 |
| Mean lamb birth weight (kg) | 4.6 ± 0.1 | 4.0 ± 0.1 | 3.2 ± 0.1 |
| Mean lamb growth rate (g/d) | 144.2 ± 8.4 | 136.3 ± 6.8 | 90.13 ± 5.0 |

[1]Predicted peak yield (l/d) : $0.09 \times W^{0.73}$ (Taylor 1973)

Multiple regression analyses showed that between and within lines 28.7 to 37.1% (mean 31.6%) of the variation in peak yield was associated with ewe weight and lamb birth weight. The relative importance of ewe weight and lamb birth weight varied between the different lines. For R ewes, ewe weight (51%) and lamb birth weight (49%) contributed approximately equal proportion of the variation in peak milk yield, while for W+ ewes a greater proportion was associated with lamb birth weight (75%) and for W− ewes a greater proportion with ewe weight (74%). For both the W+ and W− ewes, peak yield was significantly correlated with lamb growth rate ($r = 0.49$ and $0.56$, respectively; $P < .05$).

Lamb growth rates of the selection lines were ranked in the same order as peak milk yields, with W− lambs being significantly ($P < 0.001$) lower than W+ and R lambs (Table 1). Ewe liveweights differed significantly between lines ($P < 0.01$) at the beginning of lactation (Table 1).

## DISCUSSION

Differences in the lactation performance of the selection lines in this study are greater than those reported 20 years previously by Pattie and Trimmer (1964) for twin-rearing ewes and in this experiment marked differences in the shape of the lactation curve are also evident. The curves of the W+ and R ewes follow the pattern normally expected of single rearing ewes on adequate nutrition (Treacher 1978). These two lines also follow the predicted relationships between ewe metabolic liveweight and peak yield (Taylor 1973) which suggests that the increase in milk production of W+ ewes with selection for high weaning weight may largely be a function of ewe and/or lamb size change rather than a genetically correlated change in factors such as efficiency of utilization of nutrients for milk production.

However, the peak yield of W− animals was lower than predicted from their metabolic liveweight and the truncated lactation curve for these animals is characteristic of that of undernourished ewes or those in poor condition (Treacher 1978). This difference between lines may be partially a function of differences in appetite, with the W− ewes being unable to respond in terms of increased intake to the increased nutritient demands of lactation.

The slower growing W− lambs may also provide a poor sucking stimulus which would have major effects on peak milk yield and the shape of the initial portion (4-5 weeks) of the lactation curve (Langlands 1972). The possibility of differences in lamb sucking drive is indirectly supported by differences in the peak yield performance of R ewes rearing R+ and R− lambs; presumably the plus sired lambs stimulated a greater peak yield in ewes of the same potential. Such lamb effects were not evident in earlier studies of Pattie and Trimmer (1964) but this may have been due to the greater sucking stimulus imposed by twin rearing.

The multiple regression analyses also suggested that the correlated response in lactation performance associated with weaning weight selection may be attributable both to changes in ewe mature size and to lamb sucking stimuli. It would appear that for larger ewes with large lambs, the influence of the lamb is proportionately greater than for smaller ewes. This is consistent with the contention that the appetite of single W− lambs was too low to utilize the potential milk production of the W− ewes.

The differences in preweaning lamb growth rate and milk production of these lines are a function of both ewe size and lamb appetite. Further research is required to quantify these effects and clarify the underlying mechanisms which have led to the differences in weaning weights in these particular lines. It seems reasonable to assume that the hormone mechanisms which regulate glucose and fatty acid synthesis and hence influence milk synthesis, namely insulin, growth hormone and glucagon (Bines and Hart 1982), may have been altered during selection for growth. These selection lines are an ideal source of animals to examine the more fundamental endocrine mechanisms that control both lactation and lamb growth. Further examination of the mechanisms controlling these variables is intended along with continued studies of milk production and lamb growth relationships.

## ACKNOWLEDGEMENTS

This work is part of a joint project between the N.S.W. Department of Agriculture, Trangie and the Department of Animal Science, University of New England and is funded by the Australian Meat Research Committee.

## REFERENCES

Bines, J.A. and Hart, I.C., 1982. *J. Dairy Sci., 65*, 1375-1380.

Boyazoglu, J.G. and Treacher, T.T. 1978. *In, Milk Production in the Ewe*, European Assoc. for Animal Prod. Pub. *23*., p156.

Langlands, J.P., 1972. *Anim. Prod., 14*, 317-322.

McCance, I., 1959. *Aust. J. Agric. Res., 10*, 839-853.

Pattie, W.A. 1965. *Aust. J. Exp. Agric. Anim. Husb.,* 5. 361-367.

Pattie, W.A. and Trimmer, B., 1964. *Proc. Aust. Soc. Anim. Prod.,* 5, 156-159.

Taylor, St. C.S., 1973. *Proc. Brit. Anim. Prod., 2*, 15-26.

Treacher, T.T., 1978. *In, Milk Production in the Ewe,* European Assoc. for Anim. Prod. Pub. *23*, 31-38.

# OVULATION RATE AS A SELECTION CRITERION FOR IMPROVING LITTER SIZE IN MERINO SHEEP

L.R. Piper and B.M. Bindon, *CSIRO. Division of Animal Production, Armidale, N.S.W. Australia. 2350.*
K.D. Atkins and I.M. Rogan, *Department of Agriculture, Agricultural Research Centre, Trangie, N.S.W. Australia. 2823.*

*Summary* Ovulation rate has a number of theoretical advantages as a selection criterion to improve litter size in the sheep. In the Finnsheep and Galway breeds, the heritability of ovulation rate is higher than for litter size and it has been estimated that selection would be up to 3 times more effective if ovulation rate was used as the selection criterion. For the Merino flock is this study the heritability estimates for both traits are low (≅ 0.07) and not significantly different from each other. Selection for litter size should be more effective if ovulation rate is used as the selection criterion in the Merino, but the relative advantage will not be as great as for the Finnsheep and Galway breeds.

## INTRODUCTION

Hanrahan (1974,1980,1984) and others (Bindon and Piper 1976; Piper and Bindon 1984) have argued that selection for increased prolificacy (lambs born/ewe lambing, LB/EL) would be more effective if selection was based on ovulation rate (corpora lutea/ewe ovulating, OR) rather than prolificacy. The arguments in favour of OR include (i) reducing the generation interval by taking measurements before the first mating (ii) taking several measurements in one year to obtain the benefits of repeated records (higher heritability, higher realised selection intensity) without increasing the generation interval (iii) greater accuracy of measurement, (iv) lack of dependence on the male for expression of the trait, (v) little or no evidence for genetic variation in embryonic survival indicating that the genetic variation in LB/EL largely reflects genetic variation in OR and (vi) higher heritability of OR comapred with LB/EL in the Finnsheep (.45 v .10) and Galway (.57 v .06) breeds (Hanrahan 1980).

The relative efficiency of OR and LB/EL as selection criteria in any breed will depend on the relative heritabilities. Piper *et al.* (1980) published an initial estimate for OR in 18 month old Merino ewes which was low (0.05 ± 0.07) and little different from the value expected for LB/EL in Merino ewes of that age. In this paper we present preliminary estimates of the heritability of OR and LB/EL for Merino ewes aged 2 to 5 years from the same flock at the Trangie Agricultural Research Centre.

## MATERIALS AND METHODS

### Sheep and Environment

The flock studied consists of fifteen separate random-bred lines and has been described in detail by McGuirk *et al.* (1978). Line 15, which has a history of selection for increased numbers of lambs weaned (Atkins and Robards 1976), has not been included in these analyses. The remaining 14 lines were originally sampled from current industry sources and represent the major strains of Merino (fine, medium and strong wool) in NSW. Each line comprises 100 breeding ewes and is maintained as a seperate genetic entity by mating each year with 3 to 4 new rams from the appropriate parent flock.

### Experimental Procedures

Ewes born over the years 1975-1981 had their OR's assessed by mid-ventral laparoscopy two weeks prior to joining in February of each year. The initial observation was taken when the ewes were 15-16 months old; all ewes born in the flock and surviving to that age were examined. Following laparoscopy, approximately half (at random) of the 15-16 month old ewes in each line were retained as flock replacements and subsequent observations were taken annually until the ewes were 51-52 months of age. The study began in 1977 and, for the purpose of these analyses, the final records were taken in February 1983. Each year, with one exception, the ewes grazed with vasectomised rams for 3-4 months prior to laparoscopy. The exception was 1978 when the 1975 drop ewes were introduced to teasers only two days prior to laparoscopy. Those OR data have not been included in these analyses.

The flock lambed in a drift system (Dun and Eastoe 1970) in which inspections of lambing ewes were carried out twice daily to ascertain litter size at birth and pedigree records, and to assist with difficult births.

### Statistical Methods

The data for OR and LB/EL were analysed using least squares ANOVA methods. Separate analyses were carried out for each age of ewe using a model which included the effects of strain-line-year and sire within strain-line-year. Heritability estimates were derived from the sire variance component, and the total number of daughters, the number of degrees of freedom for sires and the weighted average number of daughters per sire for each analysis are given in Table 1.

**TABLE 1. Total number of daughters, degrees of freedom for sires (df Sires) and weighted average number of daughters/sire (k) for each age of ewe and trait analysis.**

| Trait | Statistic | Age 2 | 3 | 4 | 5 |
|---|---|---|---|---|---|
| OR | Total daughters | 2767 | 1327 | 1377 | 1073 |
| | df Sires | 184 | 130 | 129 | 107 |
| | k | 9.4 | 6.3 | 6.5 | 6.2 |
| LB/EL | Total daughters | 1300 | 1191 | 965 | 463 |
| | df Sires | 153 | 129 | 105 | 51 |
| | k | 5.1 | 5.6 | 5.6 | 5.6 |

RESULTS

The mean OR and LB/EL for each age of ewe are given in Table 2 and, as expected, the mean values for each trait increase with age. The mean LB/EL is also higher at each age than the mean OR. This may be due to the OR measurements being taken earlier in the breeding season than the OR at the time of conception, but could also result from nutrition/management changes occurring over the period prior to and during mating.

**TABLE 2. Least squares means (± SEM) and heritabilities (± SEM) of ovulation rate (OR) and prolificacy (LB/EL) for Merino ewes aged 2-5 years.**

| Trait and Statistic | Age 2 | 3 | 4 | 5 | all |
|---|---|---|---|---|---|
| Means | | | | | |
| OR | 1.18±.01 | 1.22±.01 | 1.34±.02 | 1.41±.02 | |
| LB/EL | 1.20±.01 | 1.30±.01 | 1.41±.02 | 1.51±.03 | |
| Heritabilities | | | | | |
| OR | 0.02±.04 | 0.06±.07 | 0.16±.08 | 0.02±.08 | 0.07±.03 |
| LB/EL | 0.03±.08 | 0.08±.08 | 0.01±.08 | 0.13±.14 | 0.06±.05 |

Half-sib heritability estimates for each trait are also given in Table 2. The estimates are generally low and, while pairs of estimates within traits may have differed significantly, there was no consistent pattern of change with age. The separate age estimates have therefore been combined and the pooled value for each trait is given in the final column of Table 2. The difference between the pooled estimates was not significant.

DISCUSSION

The similarity of the heritability estimates for OR and LB/EL found in this study may have resulted in part from the operation of scale effects. The OR measurements were taken 3-5 weeks earlier than the time of conception, and the mean OR in the OR data set was clearly less than in the LB/EL data set (Table 2). The difference in mean OR would be expected to slightly reduce the difference between the heritability estimates for the two traits. This limitation aside, the estimates for the two traits do not differ and this reduces the potential advantage that OR has as a selection criterion for LB/EL in the Merino. However, if there is little or no genetic variation in embryonic survival, which implies a genetic correlation between OR and LB/EL close to unity, then accuracy of measurement remains an important advantage favouring OR as a selection criterion. This, combined with its other advantages; lock of dependence on the male for the expression or OR, the ability to take several meaurements in one year and the opportunities for reducing the generation interval, means that in the Merino, OR will still be a more efficient selection criterion for improving LB/EL. However, the advantage will not be as great as in the case of the Finnsheep and Galway breeds.

ACKNOWLEDGEMENTS

Thanks are due to colleagues who assisted with the laparoscopies and to the staff at the Agricultural Research Centre and Mechelle Cheers, Yvonne Curtis and Robert Nethery for skilled technical assistance.

REFERENCES

Atkins, K.D. and Robards, G.E., 1976. *Aust. J. Exp. Agric. Anim. Husb. 16*, 315-320.

Bindon, B.M. and Piper, L.R., 1976. *In* G.J. Tomes., D.E. Robertson and R.J. Lightfoot (eds), *Sheep Breeding*, W.A.I.T., Perth, 357-371.

Dun, R.B. and Eastoe, R.D., 1970. *In, Science and the Merino Breeder.* (NSW Govt. Printer, Sydney) p.16-18.

Hanrahan, J.P., 1974. *Proc. 1st Wld. Congr. Genet. Appld. Livst. Prod. III*, 1033-1038.
Hanrahan, J.P., 1980. *Proc. Aust. Soc. Anim. Prod., 13*, 405-408.
Hanrahan, J.P., 1984. *Proc 2nd Wld. Congr. Sheep and Beef Cattle Breed.* (In press).
McGuirk, B.J., Atkins, K.D., Kowal,Eva, and Thornberry, K., 1978. *Wool Tech. and Sheep Breed., 26*, 17-24.
Piper, L.R. and Bindon, B.M., 1984. *Proc. 2nd Wld. Congr. Sheep and Beef Cattle Breed.* (In press).
Piper, L.R., Bindon, B.M., Atkins, K.D. and McGuirk, B.J., 1980. *Proc. Aust. Soc. Anim. Prod., 13*, 409-412.

# THE INHERITANCE OF REPRODUCTIVE PERFORMANCE

R.G. Beilharz and B.G. Luxford, *Agriculture and Forestry, The University of Melbourne, Parkville, Victoria, 3052.*

## INTRODUCTION

Why has mean lamb-marking percentage of Australian sheep flocks risen only from about 65 before 1930 to about 75 in 1970 (Ferguson 1973)? In the sciences of ecology and ethology it is accepted that species adapt to their environments so that the best possible use is made of the resources available therein (Krebs and Davies 1981). They do this because in every generation animals with lifetime strategies of reproduction optimal for the environment pass on the most genes to future generations (i.e. are most fit). Optimal lifetime strategies often vary between sexes. Only in environments recently improved can fitness (overall reproduction) rise. In all other situations the raising of one aspect of reproduction can only occur if other aspects fall (Goddard and Beilharz 1977). Geneticists agree. Robertson (1955) said 'I have taken it as axiomatic that in a population at equilibrium there will be no additive genetic variance in reproductive fitness. In the major components of fitness such as viability and fertility. . . there may be some utilisable genetic variation so that we can obtain some response to selection. But this must be at the expense of some other component of fitness, because, by definition we cannot increase overall fitness.' The fundamental question, therefore, about further genetic improvement of domestic animals in general, and Australian sheep in particular, is whether populations are at equilibrium, i.e. at a selection plateau for fitness (overall reproduction), in their environment.

Domestication introduces two new processes to the natural selection always acting on animal populations. Man alters the environment and he imposes demands on the animals such as requiring certain production levels before animals may breed. However, both these processes do no more and no less than alter the requirements for reproductive success for the animals in the population (Dickerson 1955). It may be that some populations of domestic animals are in various stages of disequilibrium. If so these populations will still have additive variance left for fitness and will be currently responding to the ever present selection for higher overall reproduction, as shown by an increasing number of lambs weaned per ewe joined in a control flock on a research station (Turner *et al.* 1972).

Animal breeding scientists employ selection objectives with lifetime reproduction as a major component (e.g. Ponzoni 1982). This implies the expectation that reproduction can be improved. However, in seeking selection criteria, attention has been focused more on the genetic variation in the measureable components of the phenotype than upon the selection objective itself (Dickerson 1982). While significant progress can be made in many of these component traits (Piper 1982), it is less clear that this improvement can be translated into an increase in measures that more closely resemble the selection objective, e.g. total weight of lambs weaned over a ewe's lifetime.

The low level and virtual absence of improvement in reproductive performance of Australian sheep strongly suggests that most Australian sheep populations are in equilibrium with, and limited by, the environment in which they are kept. In this paper we examine the consequences of this hypothesis for the breeding of sheep in practice.

## THEORY

An alternative hypothesis proposed for the lack of progress is that breeders have unconsciously selected against reproduction because they have favoured single-born animals over twins in their selection on production (Turner and Young 1969). However, there is evidence that selection on size (e.g. weaning weight) is associated with effective indirect selection for lifetime reproduction (Barlow and Hodges 1976; Luxford *et al.* 1982). Other evidence which seems to contradict the equilibrium hypothesis is that successful selection experiments have been carried out for litter size in Merinos (Turner 1978; Atkins 1980). But, were the environments in these studies representative of those of normal Australian sheep? Environmental level significantly affects genetic parameters found. This is well illustrated by Vinson (1982) who, in dairy cattle, found the relationship of milk yield and fertility to be, in general, stronly negative. However, the magnitude and even the sign of the measure was variable, with estimates from cattle on research stations consistently more positive due to their 'better' management level. We believe that a different way of visualising reproduction may explain its nature and its inheritance better.

Every animal develops and grows, matures, reproduces and dies. All aspects of such a lifetime pattern of growth, development and reproduction are relevant to how well the animal reproduces itself. Selection may act at all stages of such lifetime patterns. We visualise the instructions in the genes initiating and guiding the development of the individual within the disease and predation constraints of its particular environment (i.e. nutrution, temperature, etc.). Every stage of life influences subsequent stages. For example, early starvation, or genes that slow down early growth, will each delay puberty. There are thus definite, developmentally specifiable reasons why correlations between traits occur. Clearly, natural selection must favour in every population those individuals with lifetime patterns that best fit the prevailing environmental conditions.

Among domestic animals it is the reproductive performance of individual females that sets the limit to the rate of reproduction possible from the flock or herd. We expect that individual females will have lifetime strategies in which the climatic constraints and resources available to them from the environment will be optimally positioned so as to produce the maximum number of surviving offspring. Temperate species breed so as to avoid extreme weather killing the young and to provide feed at the right time. Female reproductive effort also becomes allocated optimally along a lifetime, the length of which is appropriate for the environment and production system (Stearns 1976). For extensively kept domestic animals, relatively long lifetimes and a slow start to breeding seem to have been best in the past.

There is much ecological evidence that reproduction is limited by the environment. A well-known example is the demonstration by Lack (1954) that clutch size in birds is regulated by the limiting ability of the parents to feed the nestlings.

In a preliminary experiment (Luxford and Beilharz 1982), we showed that selection in mice at first parity for either (a) total weight of litter at weaning, or (b) number of live young at birth, greatly reduced 'lifetime' reproduction (9 months of continuous breeding). The weights of mature (9-week) progeny produced over the 'lifetime', per dam mated, were: for line (a) — 535 gm; for line (b) — 650 gm; for the unselected control — 838 gm. Other selection experiments in mice (e.g. Wallinga and Bakker 1978) have measured 'lifetime performance' but on traits such as numbers of offspring born, rather than on mature young as in our case. Even so, under continuous mating Wallinga and Bakker's results were like ours, with the unselected line having better lifetime performance than the selected lines. We believe that herein lies the clue to distinguishing between the different hypotheses. If reproductive performance is measured on only small parts of total reproduction, one may fail to see a depression of lifetime reproduction which Goddard and Beilharz (1977) claim must follow selection of any component of reproduction.

Research on reproduction in sheep seems to have looked always at small components of reproduction. A sample of papers (Dun and Grewal 1963; Drinan 1968; Atkins and Robards 1976) shows that measurement of reproduction was done using a restricted component of lifetime reproduction which was assumed to reflect lifetime reproduction. Twins were usually better than singles. Such measurements and even statements merely referring to such measurements, were later quoted as evidence that twins have a higher lifetime reproductive performance. Clearly, none of these papers has studied lifetime reproduction of ewes. The data in these papers are, however, consistent with our mouse data, which show that the components of reproduction that were selected responded to selection. But the mouse data went on to demonstrate a drop in lifetime reproduction consistent with the equilibrium hypothesis. We suggest this is likely to be the case also with sheep and, as there seem to be few data on lifetime performance of sheep, further research is required. Resolution of this point is extremely important to define what are appropriate selection objectives for sheep breeding.

## THE CONSEQUENCES IN PRACTICE

The order of traits in a developmental sequence is critically important for understanding whether reproduction can be improved. Thus, if lactation places greatest demands on ewes in any reproductive episode, it is likely to be the limiting aspect of each breeding period. It follows that raising ovulation rate will not by itself lead to higher numbers of lambs weaned. In fact, the limiting nature of lactation will usually impose selection for an intermediate ovulation rate, just like clutch size in birds is held to an appropriate level by the feeding ability of the parents. The limiting effect of the environment seems generally accepted when extreme variants like Booroola genes for raising littersize can only be utilised successfully if the environment is substantially improved (Baillie *et al.* 1980). The same considerations apply to the use of technologies like Fecundin® to raise litter sizes. We maintain that in most situations even the relatively small increases resulting from selection of litter size or ovulation rate will need appropriate (even if small) improvements in the environment if they are not to have deleterious side effects. This leads us to the question of what can be done to improve reproduction genetically.

The first question to be answered is: Is our population at the selection limit for overall reproduction? If so, we would expect to find negative genetic correlations between components of reproduction. It is probably best to define a measure of reproduction for ewes as close as possible to the selection objective of overall reproduction, and to estimate its heritability. The estimate should be made in the environment and production system of our commercial animals. If the estimated heritability is greater than 0 we should still be able to make progress.

If our population is at a selection limit for overall reproduction, the next question becomes: Is it feasible (economically) to alter the environment or to redefine the production system? An economic analysis will provide the answer. We might be able to remove extremes of cold stress by providing shelter belts. Or it might be possible to supply extra feed at critical stages. In intensive industries for producing prime lambs it may be possible to artificially rear all young, or at least the excess above singles. In intensive systems a long breeding life might be sacrificed. Whenever the production system is redefined in this way there is room for further progress as the animals adapt to the altered situation.

For many of Australia's Merino flocks it may not be economically feasible to change the production system.

We can then ignore reproduction and select only on other traits such as fleece weight. This will eventually affect reproduction adversely. We need to find the appropriate economic balance between dropping reproduction and improvements in other production traits. If we do select for reproduction, the simplest possible scheme for indirect selection of fertility would lead to greatest producer adoption. As Luxford *et al.* (1982) found, selection on uncorrected weaning weight at any particular lambing produced similar positive secondary selection differentials for lifetime reproductive performance as selection on twinning. Further work to elucidate the genetic relationship between characters such as weaning weight and lifetime reproductive performance is required.

In conclusion, this paper highlights the need to alter the focus of attention away from simply estimating the genetic variation of the more easily measured components of the phenotype to a study of the selection objective itself, including its relationships with environmental parameters.

## REFERENCES

Atkins, K.D., 1980. *Proc. Aust. Soc. Anim. Prod., 13*, 174-176.

Atkins, K.D., and Robards, G.D., 1976. *Aust. J. Exp. Agric. Anim. Husb., 16*, 315-320.

Ballie, B.G., Bennett, N.W. and Butt, J.A., 1980. *Proc. Aust. Soc. Anim. Prod., 13*, 176-179.

Barlow, R. and Hodges, C.J., 1976. *Aust. J. Exp. Agric. Anim. Husb., 16*, 321-324.

Dickerson, G.E., 1955. *Cold Spring Habor Symp. on Quant. Biol., 20*, 213-224.

Dickerson, G.E., 1982. *Proc. 2nd World Cong. Genet. Appl. Livestock Prod.,* 5, 15-20.

Drinan, J.P., 1968. *Proc. Aust. Soc. Anim. Prod., 7*, 208-211.

Dun, R.B., and Grewal, R.S., 1963. *Aust. J. Exp. Agric. Anim. Husb., 3*, 235-242.

Ferguson, K.A., 1973. *In* Alexander, G. and Williams, O.B. (eds.) *The Pastoral Industries of Australia.* Sydney Univ. Press, Sydney. 521-545.

Goddard, M.E. and Beilharz, R.G., 1977. *Proc. 3rd Int. Cong. S.A.B.R.A.O., Animal Breeding Papers*, 4-19 to 4-21.

Krebs, J.R. and Davies, N.B., 1981. *An Introduction to Behavioural Ecology.* Blackwell, Oxford. 292 pp.

Lack, D., 1954. *The Natural Regulation of Animal Numbers.* Oxford Univ. Press, London. 343 pp.

Luxford, B.G. and Beilharz, R.H., 1982. *Proc. 2nd World Cong. Genet. Appl. Livestock Prod., 7*, 479-482.

Luxford, B.G., Beilharz, R.G. and Thompson, R., 1982. *Proc. 2nd World Cong. Genet. Appl. Livestock Prod., 8*, 630-635.

Piper, L., 1982. *Proc. 2nd World Cong. Genet. Appl. Livestock Prod.,* 5, 271-281.

Ponzoni, R.W., 1982. *Proc. 2nd World Cong. Genet. Appl. Livestock Prod.,* 5, 619-634.

Robertson, A., 1955. *Cold Spring Harbor Symp. on Quant. Biol., 20*, 225-229.

Stearns, S.C., 1976. *Quant. Rev. of Biol., 51*, 3-47.

Turner, H.N., 1978. *Aust. J. Agric. Res., 29*, 327-350.

Turner, H.N., McKay, E. and Guinane, F., 1972. *Aust. J. Agric. Res., 23*, 131-148.

Turner, H.N. and Young, S.S.Y., 1969. *Quantitative Genetics in Sheep Breeding.* MacMillan, Melbourne 332 pp.

Vinson, W.E., 1982. *Proc. 2nd World Cong. Genet. Appl. Livestock Prod.,* 5, 370-386.

Wallinga, J.H. and Bakker, H., 1978. *J. Anim. Sci., 46*, 1563-1571.

# REPRODUCTIVE PERFORMANCE OF CROSSBRED EWES DERIVED FROM BOOROOLA AND CONTROL MERINOS AND JOINED TO RAMS OF TWO TERMINAL SIRE BREEDS

B.M. Bindon, L.R. Piper *CSIRO, Division of Animal Production, Armidale, N.S.W. 2350.*
T.S. Ch'ang *CSIRO, Division of Animal Production, Prospect, N.S.W. 2149.*

*Summary* The reproductive performance and monetary value of lamb production was compared for ewes derived from crosses of Border Leicester and Dorset Horn rams with Booroola or random-bred control Merino ewes. The ewes were evaluated by joining with either Suffolk or a new crossbred sire line. In the two years of the study Booroola cross ewes weaned annually 27% more lambs and these returned 24% more dollars per ewe joined than crossbred ewes derived from control merinos. These advantages resulted from the increased prolificacy of Booroola crosses. Ewes joined to crossbred rams had significantly better lamb production primarily as a result of increased fertility.

## INTRODUCTION

The exceptional prolificacy of the Booroola Merino can be used to bring about rapid and permanent increases in the reproductive performance of Merino flocks. Ewe progeny of Booroola rams and medium non-Peppin (MNP) or Collinsville ewes weaned respectively 16% (Piper and Bindon 1982) and 29% (McGuirk *et al.* 1984) more lambs than did the progeny of MNP or Collinsville rams. The Booroola should also be of potential benefit to Australia's prime lamb industry, and the present investigation was designed to assess the performance of Booroola half-bred F1 ewes as dams for prime lamb production when joined to rams of two terminal sire breeds.

## MATERIALS AND METHODS

### Animals

The CSIRO Booroola (B) and randomly bred Control (C) flocks have been described in detail elsewhere (Turner 1978; Piper and Bindon 1982). The four ewe genotypes evaluated in this paper were generated by single-sire joining of mixed-age B and C Merino ewes in autumn with four new Border Leicester (BL) and Dorset Horn (DH) stud rams in each of the years 1976-1980. The ewe progeny (BLxC, BLxB, DHxC and DHxB) were reared and run as one flock and their reproductive performance as mixed age ewes was assessed by single sire joining in the autumn of 1982 and 1983 with either five Suffolk rams or with five rams of a new synthetic sire (SIROMT) line derived from crosses of the DH, Corriedale (Co) and Cheviot (Ch) breeds. These rams were aged 18-months at joining and new rams were used each year.

### Management and Observations

At Armidale NSW, in early March 1982 and 1983, the ewes of the four genotypes were allocated at random, within age and genotype to each of the single-sire joining groups. Daily service records were taken during the 35 day joining period in 0.2ha paddocks. Ovulation rate was assessed by laparoscopy, using local anaesthesia between three and ten days following a ewe's first service. Ewes were shorn in November and lambed in their sire groups in August/September. Lambing observations were restricted to obtaining pedigree, litter size and birth weight information. Ewes having lambing difficulty at times of daily inspection were assisted but no other special management procedures were used.

### Assessment of commercial lamb value

The objective of the study was to market lambs from the experiment using standard commercial practices for the district, employing the services of a livestock agent and the auction system. In 1982, as a drought contingency, all lambs were early weaned at an average age of 90 days and all were sold 30 days later as store lambs. In 1983 the lambs remained with their dams until sale, the date of which was decided by the agent on the basis of visual assessment of the lamb's suitability for market. The lambs were sold by auction in three drafts at approximate ages of 110, 125 and 155 days. In both years, lamb liveweight was recorded on the day of sale. No carcase data were recorded.

### Definition of traits and statistical procedures

Reproductive performance was evaluated in this study on the basis of *reproductive rate*, defined as the number of lambs weaned per ewe joined and present at lambing (i.e. LW/EJ). This is comprised of *fertility*, defined as the proportion of ewes joined that lambed (i.e. EL/EJ), *prolificacy*, defined as litter size of the lambing ewes (i.e. LB/EL) and *survival*, defined as the proportion of lambs born which survived to sale (i.e. LW/LB). *Ovulation rate* is defined here as the number of ovulations per ewe ovulating, based on counts of corpora lutea. Lamb production traits were *lamb and litter liveweight* (kg) and *lamb and litter value* ($) per ewe joined. For these traits the genetic contrasts contributing to the ewe's genotype (i.e. B vs C Merino ewes, BL vs DH rams) or arising from the ram to which she was joined (i.e. Suffolk vs SIROMT) were evaluated by analysis of variance using least squares methods to adjust the data for non-random fixed effects such as year of measurement and age of ewe (see Table 1 and 4). Distributions of ovulation rate and litter size for the four ewe genotypes were prepared from unadjusted values for all ewes in the two years of the investigation.

## RESULTS

### Comparison of ewe genotypes derived from Booroola and Control Merinos

Table 1 summarizes the least squares means for each of the four ewe genotypes for liveweight, ewe reproductive performance and lamb production, and also indicates the statistical significance of particular contrasts. The maximum number of records contributing to each mean is similar to that shown in brackets for pre-joining liveweight.

**TABLE 1. Least squares means (± SE) for pre-joining liveweight, the components of reproductive performance and lamb production traits for cross bred ewes derived from Booroola and control Merinos.**

| Ewe's sire breed | BL | | DH | | Contrast | |
|---|---|---|---|---|---|---|
| Ewe's dam breed | B | C | B | C | BvC | BLvDH |
| Liveweight (kg) | 48.3 ± .4 (182) | 52.0 ± .5 (108) | 48.4 ± .5 (129) | 49.8 ± .6 (106) | *** | NS |
| Fertility (EL/EJ) | .89 ± .02 | .83 ± .03 | .90 ± .03 | .87 ± .03 | NS | NS |
| Ovulation rate | 3.13 ± .07 | 2.02 ± .10 | 2.77 ± .09 | 1.71 ± .10 | *** | NS |
| Prolificacy (LB/EL) | 2.43 ± .06 | 1.66 ± .09 | 2.21 ± .07 | 1.51 ± .08 | *** | * |
| Lmb survival (LW/LB) | .72 ± .02 | .91 ± .03 | .71 ± .03 | .82 ± .04 | *** | NS |
| Reprod. rate (LW/EJ) | 1.51 ± .06 | 1.25 ± .07 | 1.42 ± .07 | 1.04 ± .08 | *** | * |
| Weight/lmb (kg) | 27.0 ± .2 | 28.9 ± .3 | 27.2 ± .3 | 29.3 ± .4 | *** | NS |
| Value/lmb ($) | 10.80 ± .13 | 11.62 ± .19 | 10.90 ± .17 | 11.74 ± .21 | *** | NS |
| Litter wt/EJ(kg) | 40.85 ± 1.63 | 35.43 ± 2.04 | 38.55 ± 1.93 | 30.25 ± 2.13 | *** | * |
| Litter value/EJ($) | 16.56 ± .80 | 14.05 ± 1.00 | 15.73 ± .95 | 11.92 ± 1.04 | *** | NS |

Statistical model included ewe and terminal sire genotype, year of measurement, age of ewe and all first-order interactions.
BvC = (BLxB + DHxB) vs (BLxC + DHxC); BLvDH = (BLxB + BLxC) vs (DHxB +DHxC).
* $P < 0.05$ *** $P < 0.001$

On average, B cross ewes had significantly lower pre-joining liveweights, similar fertility, higher ovulation rates, reduced lamb survival but higher overall reproductive rate than ewes derived from C Merinos.

Although individual lambs produced by B cross ewes were significantly smaller and had lower monetary value than lambs from C cross ewes, the total weight and value of litters from B cross ewes were significantly greater. Table 1 also shows that BL cross ewes are on average more productive than comparable DH cross animals. There were significant year differences for most traits but there were no significant year × main effect interactions.

### Distribution of ovulation rate and litter size

Table 2 shows that most C cross ewes had ovulation rates and litter sizes of one or two. By contrast, approximately 55% of B cross ewes had ovulation rates greater than two with the result that approximately 40% of BL × B and 30% of DH × B ewes respectively produced litters of three or more.

**TABLE 2. Distribution (%) of ovulation rate and litter size for mixed-age cross bred ewes derived from Booroola or Control Merino ewes.**

| Ewe genotype | No. of records | Trait | Ovulation rate or litter size: 1 | 2 | 3 | 4 | 5 | 6 | 7 |
|---|---|---|---|---|---|---|---|---|---|
| BLxB | 182 | O.R.[1] | 7.7 | 31.3 | 26.4 | 20.3 | 11.0 | 2.7 | 0.5 |
| BLxB | 171 | L.S. | 17.0 | 42.7 | 29.2 | 7.6 | 2.9 | 0.6 | — |
| BLxC | 108 | O.R. | 19.4 | 66.7 | 13.9 | — | — | — | — |
| BLxC | 98 | L.S. | 38.8 | 60.2 | 1.0 | — | — | — | — |
| DHxB | 129 | O.R. | 9.3 | 37.2 | 37.2 | 8.5 | 5.4 | 2.3 | — |
| DHxB | 115 | L.S. | 23.5 | 45.2 | 25.2 | 3.9 | 0.8 | — | — |
| DHxC | 106 | O.R. | 38.7 | 57.5 | 3.8 | — | — | — | — |
| DHxC | 96 | L.S. | 51.0 | 49.0 | — | — | — | — | — |

(1) O.R. = Ovulation Rate; L.S. = Litter Size.

Comparision of terminal sires

The effects of terminal sire genotype on ewe reproductive performance and lamb production traits are summarized in Table 3 which shows the least squares means after adjustment for years and ewe genotype.

**TABLE 3. Least squares means (±SE) for the reproductive performance and lamb production traits of ewes joined to Suffolk or SIROMT rams.**

| Terminal sire breed | Suffolk | SIROMT | Significance of difference |
|---|---|---|---|
| Fertility | 0.82 ± .02 | 0.92 ± .02 | *** |
| Ovulation rate | 2.42 ± .06 | 2.38 ± .06 | NS |
| Prolificacy | 1.94 ± .06 | 1.96 ± .05 | NS |
| Lamb survival | 0.76 ± .02 | 0.82 ± .02 | NS |
| Reprod. rate | 1.15 ± .05 | 1.43 ± .05 | ** |
| Weight/lamb sold (kg) | 28. 4 ± .2 | 27.8 ± .2 | *** |
| Value/lamb ($) | 11.29 ± .13 | 11.24 ± .12 | NS |
| Litter wt sold/EJ (kg) | 32.64 ± 1.38 | 39.88 ± 1.37 | *** |
| Litter value/EJ ($) | 12.58 ± .68 | 16.51 ± .67 | *** |

** $P < 0.01$ *** $P < 0.001$

Ewes joined to SIROMT rams were significatly more productive: this resulted from their significantly higher fertility and slightly better lamb survival, leading to a significant increase (20%) in reproductive rate. Although the lambs sired by SIROMT rams were smaller (2.6%) at sale than progeny of Suffolk sires, their average sale price was similar. The combined effects of reproductive rate and lamb value resulted in 31% advantage in the monetary return per ewe joined in favour of SIROMT sires (Table 3).

Lamb survival in relation to sire breed, litter size and birth weight

Lambs sired by SIROMT rams had slightly higher (6%) survival than progeny of Suffolk rams. The terminal sire breeds were ranked similarly in both years. This difference apprached significance for the two years; data (Table 3), but was more evident in 1983 when SIROMT-sired lambs had a 9% advantage in survival ($P < 0.01$). When lamb survival is related to litter size at birth for both years of the investigation (Table 4) it is seen that the survival advantage of SIROMT-sired lambs is confined to multiple births and is more pronounced as litter size increases. This sire × rearing rank interaction was not significant. The difference in lamb survival between the terminal sire breeds cannot be explained by differences in lamb birth weights, which were similar for Suffolk and SIROMT progeny in all litter size classes. Overall lamb survival was greatly influenced by litter size at birth (Table 4) and this is a reflection of the reduced birthweights of lambs born in litters of four or more. A notable result, however, is the relatively good survival (79%) of lambs born as triplets. This was an important source of the productivity advantage of B cross ewes.

Age of lambs at sale

In 1982 all lambs were sold on the one day. In 1983 lambs produced by B and C ewes were sold at average ages of 125 ± 1.0 and 116 ± 0.2 days respectively.Progeny of SIROMT and Suffolk sires were sold at average ages of 126 ± 1.2 and 116 ± 1.4 days respectively. These differences are statistically significant but are of only minor agricultural importance.

**TABLE 4. Least squares means (± SE) for lamb survival (%) and lamb birthweight (kg) in litter size classes sired by Suffolk and SIROMT sires in 1982 and 1983.**

| Sire breed | Trait | Litter size at birth: 1 | 2 | 3 | 4 | 5+6 |
|---|---|---|---|---|---|---|
| Suffolk | Survival | 89.8 ± 4.8 | 82.3 ± 2.7 | 73.7 ± 12.4 | 44.8 ± 7.9 | 28.0 ± 16.0 |
| | Birthweight | 4.93 ± .09 | 3.87 ± .05 | 3.27 ± .22 | 2.77 ± .14 | 2.43 ± .29 |
| | No. of lambs | 67 | 214 | 129 | 36 | 11 |
| SIROMT | Survival | 91.0 ± 4.7 | 88.1 ± 2.5 | 82.5 ± 11.6 | 59.1 ± 8.1 | 42.6 ± 10.7 |
| | Birthweight | 4.87 ± .08 | 3.88 ± .05 | 3.36 ± .21 | 2.66 ± .14 | 2.50 ± .19 |
| | No. of lambs | 75 | 248 | 111 | 40 | 25 |

Statistical model as for data in Table 1 but with the addition of terms for litter size of lambs and all first order interactions with other main effects.

## DISCUSSION

The main objective of this study was to evaluate F1 British breed × Booroola ewes in comparison with F1 BL × Merino and DH × Merino which are respresentative of prime lamb dams in the sheep industry. The results (Table 1) clearly show that despite their lower pre-joining liveweights, the superior prolificacy of Booroola cross ewes more than compensated for the reduced survival of lambs born in litters larger than two and this led to a significant advantage (27%) in reproduction rate of BL × B and DH × B ewes relative to the C Merino crosses. This advantage was maintained when lamb growth, sale age and sale vaule were considered with the Booroola cross ewes showing a 24% advantage in monetary return per ewe joined. This advantage may be an underestimate, since the drought contingencies of 1982 (i.e early weaning and premature sale of lambs as store animals) imposed differential penalities on ewes with higher reproductive rate. However the conditions experienced in the two years of the investigation were a realistic version of what happens in the lamb industry, so the results shown in Table 1 are probably meaningful. Only about 70% of B cross ewes evaluated here were carriers (*F*+) of the *F* gene. It is not known if further advantages would accrue if all crossbred ewes were *F*+. Large scale generation of *F*+ (e.g. BLxB) ewes could be obtained by crossing *FF* Border Leicester rams (i.e. homozygous for the F gene) with industry Merinos. For the present, *F*+ crossbred ewes such as those described here would only become available to the industry by crosses of British breed rams with the relatively small population of Merino ewes carrying the *F* gene.

The results reported here demonstrate some real advantages in using terminal sires developed from a mixture of breeds (i.e. SIROMT). Most of the advantage came from their better fertility and it is stressed that this was assessed under single-sire joining conditions with all rams used at age 1.5 years. Under practical management conditions, syndicate mating may eliminate some differences between sires in fertility.

ACKNOWLEDGEMENTS

The authors gratefully acknowledge the technical assistance of Y.M. Curtis, M.A. Cheers and R.D. Nethery, together with staff of the Arding Field Station, Armidale, and CSIRO data base, Prospect. The work was financed by the Australian Meat Research Committee and the Australian Wool Corporation.

REFERENCES

McGuirk, B.J., Killeen, I.D., Piper, L.R., Bindon, B.M., Wilson, R., Caffery, G. and Langford, C., 1984. *Proc. Aust. Soc. Anim. Prod., 15*, 464-466.

Piper, L.R. and Bindon, B.M., 1982. *In* L.R. Piper, B.M. Bindon and R.D. Nethery (eds). *The Booroola Merino,* CSIRO, Melbourne, 9-19.

Turner, H.N., 1978. *Aust. J. Agric. Res.,* 29. 327-350.

# PADDOCK MATING PERFORMANCE, MATING BEHAVIOUR AND TESTICULAR SIZE IN GENETICALLY LARGE AND SMALL MERINO RAMS

G.P. Davis, C.J. Thwaites, G.N. Hinch, *Department of Animal Science, University of New England, Armidale, N.S.W. 2351.*
B.P. Kinghorn, *N.S.W. Department of Agriculture, Trangie Agriculatural Research Centre, Trangie, N.S.W. 2823.*

*Summary* Male reproductive parameters have been studied in rams from lines selected for and against weaning weight. These lines differ in mature size and various components of flock fertility such as, return to service rate and lambing percentage, when compared with randomly selected controls.

Eighteen-month-old rams from these selection lines and the randomly bred control were studied from June 1983 to June 1984 at Armidale. Liveweight and testicular diameter were recorded monthly and serving capacity was tested, in June 1983 and in June 1984.

Serving capacity results indicated a higher number of serves by rams selected for weaning weight in 1983 but by 1984 there were no differences in any parameters. Testicular size was significantly different between lines when rams were 30-36 months of age but there was no difference at 21-30 months of age. Liveweight was always significantly different between lines.

Relevance of the results to total flock reproductive performance, and areas of futher research are discussed.

## INTRODUCTION

Selection of Medium-Peppin merinos for (W+) and against (W−) weaning weight at Trangie Agricultural Research Centre since 1951 has resulted in animals of significantly different mature size (Thompson 1983). Examination of flock records of the two selection lines and a randomly selected control (R) indicates that since 1975 there have been significant differences in reproductive performance, with the percentage of lambs weaned per ewe joined in the W+, R and W− lines averaging 80.0, 84.0, and 49.8 respectively ($P < 0.01$).

As part of a wider study of ewe, ram and lamb factors which may have contributed to the above differences in flock performance, the current paper examines the possibility that selection for and against weaning weight has adversely affected ram reproductive performance. Three approaches have been followed. Mating records have been analysed in an attempt to identify any specific deficiencies in ram performance under field conditions. Serving capacity tests, which are known to be significantly correlated with the proportion of ewes marked in the first cycle and the proportion of those which are inseminated (Kilgour and Wilkins 1980) were used to directly assess ram mating behaviour. Thirdly, possible between line differences in testicular size were examined in the light of knowledge that this factor is positively related to the proportion of ewes lambing (Gherardi *et al.* 1980), and that there is a positive genetic correlation between liveweight at weaning and testicular diameter at various subsequent ages (I. Purvis unpublished data). A realised response in testicular size at 18 months of age proportional to that observed in liveweight in these lines would be expected to have noticeably reduced testicular size and perhaps fertility in the W− rams.

## MATERIALS AND METHODS

### Animals and measurements

*Paddock mating* The 1981-1983 flock reproductive records collected during 6-week single-sire joinings of the W+, R and W− lines at Trangie Agricultural Research Centre, NSW, were used to estimate the following parameters, in each case as a percentage of the ewes joined: Ewes marked (% MKD), Ewes marked in the first 16 days of joining (% M1C), Ewes marked in the first 16 days of joining which returned to service (% RET), and, Ewes lambed (% LMB). All joinings took place in February-March and involved 15-40 ewes per ram. In 1983 a number of additional R ewes were also mated to W+ and W− rams.

*Serving capacity and mating behaviour* For the estimation of serving capacity and testicular size, thirty-three (10 W+, 13 R, 10 W−) rams born in July and August 1981 were available. The rams arrived in Armidale in two groups, the first in February 1983 and the second in March 1983. All rams had previous heterosexual experience through being joined (N = 21) or through semen collection trials (N = 12).

Serving capacity tests (Kilgour 1980), involving 2 × 20 minute and 2 × 1 hour pen tests were carried out at the beginning (June 1983) and end (June 1984) of the experimental period. During each such test observations were made on the number of serves (S), mounts (M) and the average time to first service (TFS), and the number of mounts required to achieve each service (MSR) was calculated.

*Testicular diameter and liveweight* Monthly measurements of fasted liveweight (LW) and testicular diameter (TDM = the sum of both testes corrected for skin thickness, Land and Sales 1977) were carried out over the experimental period (June 1983 – June 1984).

Six rams (3 W+, 3 R) died from flystrike or misadventure during the course of the study.

Statistical

The 1981-1983 paddock mating data from these three selection lines was analysed by fitting the model:

$$Y_{ijk} = u + G_i + A_j + GA_{ij} + e_{ijk} \quad (1)$$

where Y is the observation for N = 59 rams, u an overall mean, G the line effect, A the year effect, GA a line × year interaction and e error. The data involving R ewes joined with W+ and W− rams (N = 11) in 1983 was analysed separately by fitting the model:

$$Yi = u + G_i + e_i \quad (2)$$

where Y, u, G, and e are as in (1).

The serving capacity data for the 20 rams tested in both years was subjected to repeated measures analysis which revealed the only significant effect was due to year. The data for each year was then tested separately by fitting a line effect model as in (2). Testicular diameter and liveweight data were split into seasons and examined by fitting the model:

$$Y_{ijk} = u + G_j + RG_{ij} + M_k + MG_{jk} + RM_{ijk} \quad (3)$$

where Y,u and G are as before and RG is an error due to rams within lines, M is an effect due to month within each season, MG is an interaction between month and line and RM is residual error.

## RESULTS

### Paddock mating

The results in Table 1(a) show there were significantly more returns to service in ewes joined to W− rams than in those joined to either R or W+ rams. The percentage of ewes lambing in the W+ line is both significantly smaller than that for the R line and significantly larger than that for the W− line. There were no significant differences in any of the parameters studied when W+ and W− rams were mated to R ewes (Table 1 (b))

**TABLE 1. Least square means (s.e.) for ewes marked (% MKD), ewes marked in the first 16 days of joining (% M1C), ewes marked in the first cycle that returned to service in the rest of joining (% RET) and ewes that lambed (% LMB) expressed as percentages of ewes joined for each line (a) for rams mated to ewes of their own line, and (b) for W+ and W− rams mated to R ewes (N = number of rams).**

| Line | N | % MKD | % M1C | % RET | % LMB |
|---|---|---|---|---|---|
| a) | | | | | |
| W+ | 20 | 85.2 (2.3)[a] | 56.8 (3.0)[a] | 19.9 (2.6)[a] | 68.3 (3.5)[a] |
| R | 18 | 91.4 (3.1)[a] | 65.1 (4.1)[a] | 21.3 (3.5)[a] | 76.7 (4.7)[b] |
| W− | 21 | 88.2 (3.5)[a] | 63.1 (4.6)[a] | 32.4 (3.8)[b] | 55.5 (5.2)[c] |
| b) | | | | | |
| W+ | 5 | 84.1 (5.8)[a] | 44.0 (8.6)[a] | 18.0 (6.7)[a] | 58.6 (7.4)[a] |
| W− | 6 | 76.2 (5.3)[a] | 42.9 (7.9)[a] | 20.3 (6.1)[a] | 52.6 (6.7)[a] |

(Means with different superscripts within columns within (a) or (b) are significantly different $P < 0.05$).

### Serving capacity and mating behaviour

The only significant difference between lines in 1983 was for a larger number of services to be performed by the W+ rams (Table 2(a)). No significant differences were observed in 1984 (Table 2(b)).

**TABLE 2. Least square means (s.e.) for the total number of serves (S), mounts (M), the average number of mounts which resulted in a service (MSR), and the average time to first service in seconds (TFS) for 2 × 1-hour tests in June 1983 and 1984.**

| | M | S | MRS | TFS |
|---|---|---|---|---|
| a) 1983 | | | | |
| W+ | 54.4 (12.8)[a] | 15.4 (1.5)[a] | 5.7 (2.7)[a] | 141.1 (110.4)[a] |
| R | 59.3 (10.1)[a] | 10.8 (1.4)[b] | 5.5 (2.5)[a] | 227.9 (102.1)[a] |
| W− | 39.4 (10.8)[a] | 10.4 (1.4)[b] | 8.2 (2.5)[a] | 266.9 (103.2)[a] |
| b) 1984 | | | | |
| W+ | 78.6 (12.8)[a] | 11.0 (1.4)[a] | 5.1 (1.4)[a] | 270.3 (57.0)[a] |
| R | 65.4 (10.1)[a] | 10.8 (1.1)[a] | 5.1 (1.1)[a] | 102.3 (44.2)[a] |
| W− | 61.3 (10.3)[a] | 11.1 (1.1)[a] | 4.5 (1.1)[a] | 120.9 (44.2)[a] |

(Means with different superscripts within columns within (a) or (b) are significantly different $P < 0.05$).

Testicular diameter and liveweight

The fleece-free liveweights presented in Table 3 were significantly different between lines over all seasons. The same Table demonstrates that testicular diameters were significantly different between lines only in the Autumn and Winter of 1984. In both these seasons the W+ rams had significantly larger testes than the R and W− rams.

**TABLE 3. Least square means (s.e.) for liveweight in kilograms (LW) and testicular diameter in millimeters (TDM) over seasons.**

| | 1983 | | | 1984 | |
|---|---|---|---|---|---|
| | Winter | Spring | Summer | Autumn | Winter |
| a) LW | | | | | |
| W+ | 65.4 (1.1)[a] | 73.7 (1.1)[a] | 60.1 (1.3)[a] | 60.3 (1.2)[a] | 63.4 (1.8)[a] |
| R | 55.6 (1.0)[b] | 64.7 (1.0)[b] | 53.1 (1.0)[b] | 52.0 (1.0)[b] | 55.8 (1.5)[b] |
| W− | 46.9 (1.1)[c] | 52.7 (1.1)[c] | 43.2 (1.1)[c] | 42.7 (1.0)[c] | 45.0 (1.5)[c] |
| b) TDM | | | | | |
| W+ | 835 (31)[a] | 882 (16)[a] | 976 (21)[a] | 1003 (29)[a] | 865 (36)[a] |
| R | 796 (26)[a] | 869 (13)[a] | 940 (16)[a] | 899 (22)[b] | 772 (28)[b] |
| W− | 765 (30)[a] | 855 (14)[a] | 888 (16)[a] | 838 (22)[b] | 705 (28)[b] |

(Means with different superscripts within columns within (a) or (b) are significantly different $P < 0.05$).

## DISCUSSION

The results of the analysis of the paddock mating data, serving capacity and mating behaviour parameters, and testicular diameter and liveweight measurements combined indicate that the differences in flock reproductive performance between the W+, R and W− lines are unlikely to be a direct result of these components of ram fertility.

The significant differences between lines in lambing percentage (% LMB) when no difference in marking percentage (% MKD) exists (Table 1(a)) could be a result of either insemination failure, failure to conceive or early embryonic death. The higher return rate (% RET) in the W− line (Table 1(a)) suggests that reproductive wastage at an early stage is more prevalent in this than either the W+ or R lines. Insemination failure has been suggested as a major cause of reproductive wastage in some flocks (Killeen 1974, Cahill *et al.* 1974), and serving capacity may contribute to this (Kilgour and Wilkins 1980). Although insemination failure may have contributed to differences in flock fertility between lines in the current study, those differences are more likely to have been due to the ewe than the ram since they were not apparent when W+ and W− rams were joined to groups of R ewes (Table 1(b)). The latter result was however estimated on only a small number of rams and it is also possible that W+ and W− rams may behave differently towards ewes of their own line (and hence size) than to ewes of a different, intermediate size.

The only significant result to emerge from the pen tests of mating behaviour was that W+ rams achieved 50% more services than R and W− rams in 1983. Published results from studies involving the testing of mature, heterosexually experienced rams under first pen and then paddock mating conditions indicate a positive correlation between the number of services recorded in pens and the number of ewes marked in the first cycle of joining (Kilgour 1980). The current results do not follow this pattern in that the percentage of ewes marked in the first cycle was actually lowest in the W+ line (Table 1(a)). It should be noted, however that this result may be a ewe effect, and also that in the current work not all rams contributed to both pen and paddock mating results. Furthermore the rams were only 18 months old and heterosexually inexperienced when paddock mating took place. Further research is required to examine the relationship between pen test parameters and paddock mating performance in young heterosexually inexperienced rams from these lines.

Although selection has resulted in large divergences in the liveweight of these lines, there has apparently been no realised response in the testicular size of 21-30 months old rams (Table 3), the only significant difference between which occurred in Autumn and Winter 1984. While between line divergence in testicular size may occur in older rams, it was not apparent in the current work nor when the rams were first mated at 18 months of age (G.P. Davis unpublished data). Testicular size is highly correlated with sperm production, with each gram of testicular tissue producing about $20 \times 10^6$ sperm per day (Knight 1973), so that testicular factors are unlikely to have contributed to the differences observed between lines in flock fertility in the current work, where all rams grazed the same pastures before and during mating. The fact that W+ and W− rams performed similarly when joined to R ewes (Table 1(b)) also contraindicates differences between them in semen quality.

## ACKNOWLEDGEMENTS

The technical assistance of Mrs. S. Swan and Mr. B. Johnston and the financial support of the Australian Meat Research Committee is gratefully acknowledged.

## REFERENCES

Cahill, L.P., Kearins, R.D., Blockey, M.A.de B., Restall, B.J., 1974. *Aust. J. Exp. Agric. Anim. Husb., 14*, 723-725.
Gherardi, P.B., Lindsay, R.D., Oldham, C.M., 1980. *Proc. Aust. Soc. Anim. Prod., 13*, 48-50.
Kilgour, R.J., 1980. *In* Wodzicka-Tomaszewska, M., Edey, T.N., Lynch, J.J. (eds), *Reviews in Rural Science IV* UNE Armidale, N.S.W.
Kilgour, R.J., Wilkins, J.F., 1980. *Aust. J. Exp. Agric. Anim. Husb., 20*, 626-666.
Killeen, I.D., 1974. *Aust. J. Exp. Agric. Anim. Husb., 14*, 723-725.
Knight, T.W., 1973. Ph.D. Thesis, University of Western Australia, Nedlands, Western Australia.
Land, R.B. and Sales, D.I., 1977. *Anim. Prod., 24*, 83-90.
Thompson, J.M., 1983. PhD. thesis, University of Sydney, Sydney, N.S.W.

# BREEDING FOR REPRODUCTIVE RATE IN THE MERINO INDUSTRY AND THE CHOICE OF SELECTION CRITERIA

L.G. Butler and R.P. Lewer, *Animal Breeding and Research Institute, Katanning, Western Australia, 6317.*

*Summary* A nucleus breeding scheme aimed at improving reproductive rate (with associated control flock) has been established by the W.A. Department of Agriculture. Foundation nucleus ewes were stud maiden ewes indentified by laparoscopy as multiple ovulating. Selection is based primarily on ovulation rate and 17 per cent of all stud flocks in Western Australia are participating. Twenty six per cent of these use frozen semen from selected rams on fertility families within participating stud flocks.

## INTRODUCTION

In 1980, the Western Australian Department of Agriculture initiated a nucleus breeding scheme in co-operation with Stud Merino breeders. Its aim is to assist breeders to improve reproductive rate and to pass the genetic gains to the commercial sheep industry. Reproductive rate can be increased by selection as reported by Turner (1969) for the Merino and in other breeds by Hanrahan (1984). This paper briefly describes the project, outlines the estimated genetic gains and discusses the choice of selection criteria.

## MATERIALS AND METHODS

This project, described by Butler and Morrow (1984) has been designed along established nucleus breeding guidelines (Jackson and Turner 1972). In 1980 and 1981, laparoscopy was used to screen, during their first mating, multiple ovulating maiden ewes from the participating Merino studs into a registered nucleus flock. Nucleus ewes are joined with rams selected from those born within the flock. A control flock was established to allow estimation of genetic progress and the effect of screening.

'Fertility families' are established within the participating stud flocks, in most cases by identifying multiple bearing ewes using ultrasonic scanning (Fowler and Wilkins 1982). These ewes are inseminated with frozen-thawed semen from nucleus sires.

## RESULTS

A total of 76 studs (about 17% of Merino studs in W.A.) are participating in the scheme. The 12,000 maiden ewes screened over two seasons had an ovulation rate average of 1.08 eggs shed per ewe ovulating (Butler 1984). A total of 450 multiple ovulating ewes were selected as founders, and 600 ewes are now joined annually in the AB&RI Fertility Flock. In 1983 over 900 stud ewes were inseminated in 16 stud fertility families and this will be expanded to 1350 ewes on 20 studs in 1984.

Among the maiden ewes screened in year 1, about 8% of those that were ovulating had a twin ovulation, corresponding to a selection differential of about 1.86 standard deviations. These ewes had been mated to unselected rams on their home farms. Assuming a heritability of 0.05 (Piper *et al.* 1980) for ovulation rate in maiden ewes, the predicted response to screening was 1.3%. Rams born in year 1 from the 40% of screened ewes which repeated their twin ovulation performance immediately prior to their second mating were used in year 2. The predicted selection response assuming a repeatability of 0.2 (Lewer and Allison 1980) was 2.5%. When these rams are used in the studs' own 'fertility families', the predicted advantage of the progeny is about 2% over the contemporary stud progeny.

## DISCUSSION

The participation rate of 17% of W.A. Merino studs illustrates the level of industry acceptance of the project. Although at this early stage only 26% of participating studs are actively using semen, much higher levels of participation are expected in the future.

When this project started, ovulation rate was the only practical method by which foundation nucleus stock could be identified. Despite the low estimate of heritability for ovulation rate in maiden ewes found by Piper *et al.* (1980) the predicted genetic gains are positive and should lead to genetic improvement.

The breeding objective here is to increase reproductive rate rapidly while maintaining wool production and bodyweight. Direct selection is the most effective method of improving reproductive efficiency. Turner (1969) concluded that litter size was the most profitable selection criterion for raising reproductive rate. We believe that selection for ovulation rate would be even more effective for three reasons.

First, Hanrahan (1980) argues that the genetic correlation between ovulation rate and litter size is unity. Second, there are fewer random 'environmental' factors affecting the expression of ovulation rate than of litter size. The heritability of ovulation rate therefore should be greater than for litter size (Hanrahan 1984). However, this has not yet been established for the Australian Merino (Piper *et al.* 1980). Third, ovulation rate is likely to be measured with greater accuracy than litter size after birth (Alexander *et al.* 1983).

Given these assumptions, it seems logical to select on 'potential reproductive rate' based on ovulation rate and litter size at birth, in combination with some estimate of rearing ability, mothering ability or milk production if such characters are inherited. The most efficient method of selection is usually an index, but to construct an index we have to have estimates of the relationships between the characters. None yet exists for ovulation rate. One of the objectives of this project is to provide data on which estimates of relationship between ovulation rate, wool, and liveweight can be made. Meanwhile the best programme appears to be to select rams on the basis of the ovulation rate of their mothers and then to cull those below average for the flock on wool weight and liveweight.

## ACKNOWLEDGEMENTS

This work was supported by a grant from the Wool Research Trust Fund on the recommendation of the Australian Wool Corporation.

## REFERENCES

Alexander, G., Stevens, D. and Mottershead, B., 1983. *Aust. J. Exp. Agric. Anim. Husb., 23*, 361-368.
Butler, L.G., 1984. *Proc. Aust. Assoc. Anim. Breed. Genetics, 4*, 186-187.
Butler, L.G. and Morrow, G.F., 1984. *Proc. Aust. Assoc. Anim. Breed. Genetics, 4*, 184-185.
Fowler, D.G. and Wilkins, J.F., 1982. *Proc. Aust. Soc. Anim. Prod., 14*, 491-494.
Hanrahan, J.P., 1980. *Proc. Aust. Soc. Anim. Prod., 13*, 405-408.
Hanrahan, J.P., 1984. *Proc. 2nd World Cong. on Sheep and Cattle Breed., Pretoria* (in press).
Jackson, N., and Turner, H,N. 1972. *Proc. Aust. Soc. Anim. Prod., 9*, 55-63.
Lewer, R.P., and Allison, A.J., 1980. *Proc. N.Z. Soc. Anim. Prod., 40*, 248-253.
Piper, L.R., Bindon, B.M., Atkins, K.D. and McGuirk, B.J., 1980. *Proc. Aust. Soc. Anim. Prod., 13*, 409-412.
Turner, H.N., 1969. *Anim. Breed. Abs., 37*, 545-565.

# FEMALE REPRODUCTIVE CHARACTERISTICS IN HIGH FECUNDITY MERINO CROSSES WITH THE SOUTH AUSTRALIAN MERINO

J.R.W. Walkley, R.W. Ponzoni and S.K. Walker, *Department of Agriculture, Box 1671, G.P.O., Adelaide, S.A. 5000.*

*Summary* Reproductive characteristics of female progeny of high fecundity Merinos crossed with the South Australian Strong Wool Merino were studied.

There were differences between the strains in ovulation rate and litter size but no significant differences in number of lambs weaned.

An assessment of the effect of the 'F' gene on aspects of reproductive traits indicated that reproductive wastage between ovulation and birth may negate the superiority in ovulation rate conferred by the 'F' gene.

## INTRODUCTION

Reproductive rate is an economically important trait in sheep and may be improved by crossing existing flocks with breeds or strains already superior in this character. Three Merino strains — the Booroola (Bo), Trangie Fertility (TF) and 'T' Flock — have a high incidence of multiple births (McGuirk 1976) and hence the potential to improve reproductive rate in crosses with other Merino strains. The aim of this study was to evaluate the reproductive performance of Bo and TF crosses with South Australian Strong Wool Merinos (SAM) of the Bungaree (Bu) family group. Results on the effect of the Bo 'F' gene (Piper and Bindon 1982) on female reproductive traits are also presented.

## MATERIALS AND METHODS

### Location and Environment

The experiment was conducted at Minnipa Research Centre. For a description of the Minnipa environment and details of flock management see Walker *et al.* (1984).

### Comparisons of High Fecundity Cross Merino Ewes

Bu ewes at Minnipa were mated with Bu, Bo and TF rams thus generating Bu, ½Bo and ½TF progeny. Quarter Bo (¼Bo) ewes have also been generated. Single sire matings were used to produce experimental progeny. A minimum of five rams of each strain were used each year. The number of ewes of each age and strain for which records were available for analysis are shown in Table 1. The results reported are restricted to ovulation rate (OR) observed in the cycle prior to joining, ewe fertility ($E_{PJ}$), litter size ($L_{BP}$), lamb mortality between birth and weaning ($L_{DB}$) and number of lambs weaned ($L_{WJ}$). For further details on experimental design and observations on wool and liveweight traits see Ponzoni *et al.* (1982) and Ponzoni *et al.* (1984a). Experimental ewes were syndicate mated to Bu rams.

**TABLE 1. Number of ewes of each age, strain and year of birth.**

| Strain | Ewe age and year of birth | | | | | |
|---|---|---|---|---|---|---|
| | 2 years | | | 3 years | | 4 years |
| | 1978 | 1979 | 1980 | 1978 | 1979 | 1978 |
| Bu | 43 | 42 | 8 | 43 | 37 | 43 |
| ½Bo | 92 | 37 | 58 | 89 | 36 | 89 |
| ½TF | — | 67 | 85 | — | 61 | — |
| ¼Bo | — | — | 21 | — | — | — |

### The Effect of the 'F' Gene

Because the Bo sires were mated to Bu ewes (all ++) the expected frequency of animals carrying the 'F' gene in their ½Bo progeny groups is 1.0 (progeny of FF sires), 0.5 (progeny of F+ sires) or 0.0 (progeny of ++ sires). Bo sires were classified as FF, F+ and ++ on the basis of their daughters' ovulation rate observed between 9 and 66 months. Daughters which had $\geq$3 ovulations were defined as carriers of the gene. Then the proportion of daughters with the gene was calculated for each sire. Confidence intervals (90%) were estimated for this proportion (Snedecor and Cochran 1967) and sires were classified as FF, F+ or ++ if the confidence interval included the values 1.0, 0.5 or 0.0 respectively. In this way the effect of the 'F' gene on female reproductive traits (OR, $E_{PJ}$, $L_{BP}$, $L_{DB}$ and $L_{WJ}$) was assessed. Twelve of the 14 Bo sires used at Minnipa also had daughters in an experiment at Cape Borda Research Farm which meant a total of 957 records were available for sire classification. On average, approximately 68 records per sire were available with a range of 35 to 128.

Statistical Procedures

The data were analysed by least-squares analysis of variance using the procedure GLM in SAS (Freund and Littell 1981). The models fitted for the strain comparison are described by Ponzoni *et al.* (1984a).

When assessing the effect of the 'F' gene, only ½Bo ewes were included in the analysis. In this case, sire genotype (FF, F+ or ++) was included as a source of variation in addition to the other effects described by Ponzoni *et al.* (1984a).

## RESULTS

Reproductive Performance of Merino Strains

The least-squares means for reproductive characters of Bu, ½Bo, ½TF and ¼Bo experimental ewes are presented in Table 2. Note that the expected frequency of carrier animals among ½Bo is 60.7% and the expected frequency of the F gene 30.4% (Ponzoni *et al.* 1984a).

There were significant differences in OR at each age. The ½Bo ewes were always superior to the other 'strains' in OR. At 18 months of age, ½ and ¼Bo were equivalent. While ½TF had higher OR values than Bu the differences were only significant at 2.5 years.

The strains did not differ significantly in $E_{PJ}$. Significant differences between strains in $L_{BP}$ occurred at each ewe age. Litter size was always highest in the ½Bo. At 3 years of age the Bu and ½TF had the same litter size, but ½TF was superior at 2 years of age. One quarter Bo was intermediate between ½Bo and ½TF but differences were not significant. There were no significant differences among the strains in $L_{DB}$ and $L_{WJ}$.

The Effect of the 'F' Gene

The least-squares means of reproductive characteristics of the female progeny of 6 FF, 6 F+ and 2 ++ Bo sires are presented in Table 3. At 1.5 and 2.5 years of age there were significant differences ($P<0.05$) in OR. Sire genotypes ranked $++ < F+ < FF$ at 1.5 years while at 2.5 years the ranking was $++ = F+ < FF$. The fertility ($E_{PJ}$) of progeny of the Bo sire genotypes did not differ significantly at either age. Significant differences in $L_{BP}$ were observed at 2 years of age but not at 3; at the younger age, the sire genotypes ranked $FF > F+ = ++$. No significant differences were observed for $L_{DB}$ or $L_{WJ}$ at either age. Because the values shown in Tables 2 and 3 are least-squares means, sometimes based on slightly different numbers of observations, the product $E_{PJ} \times L_{BP} \times (1 - L_{DB})$ is not always exactly equal to $L_{WJ}$.

## DISCUSSION

Differences Among Strains

There were differences between the strains in the biologically important characters OR and $L_{BP}$ with the ½Bo clearly superior to the other strains. The superiority of the ½TF over Bu was of a smaller magnitude than for the ½Bo. The strains ranked in the same order for $L_{BP}$ as for OR, but the differences were smaller. This observation suggests that fertilization failure and/or embryonic mortality may be higher in the more fecund crosses; the relative contribution of these sources of reproductive wastage is however unknown. The ranking of the strains in $L_{WJ}$ (½Bo > ½TF > Bu) was consistent in all age groups. One-half Bo and ½TF weaned approximately 0.15 and 0.1 more lambs per ewe joined respectively than Bu ewes. Reproductive wastage between ovulation and birth and loss of lambs between birth and weaning are two major problems in the high fecundity crosses, both of which require further study, particularly the development of management practices directed towards the improvement of survival of lambs born in large litters.

Bu ewes cut more clean wool of greater average fibre diameter and have heavier liveweights than the fecund crosses. Unless these deficiencies in high fecundity crosses with South Australian Merinos can be remedied, their commercial use is debatable (Ponzoni *et al.* 1984a). Possible methods of correcting the current deficiencies are:

(i) developing a South Australian Merino which has the 'F' gene of the Bo, and,
(ii) constructing a synthetic (e.g. ½Bu:¼TF:¼'T' Flock) combining the desirable features of the strains and which would respond more readily to direct selection because of greater genetic variation.

The Effect of the 'F' Gene

The outstanding reproductive performance of the Bo Merino is due to a gene (or single segregating unit) 'F' affecting OR (Piper and Bindon 1982). Our work indicates however that the presence of the 'F' gene does not affect ewe fertility and that due to the greater reproductive wastage among ewes carrying the 'F' gene, differences between carrier and non-carrier ewes in the number of lambs weaned may be small or non-existent.

Three sire categories (FF, F+ and ++) were identified on the basis of the proportion of daughters with ≥3 ovulations when all ovulation records available were used. The incidence of daughters with ≥3 ovulations was 0.00, 0.30 — 0.46 and 0.82 — 0.96 for ++, F+ and FF sires respectively (Walker, S.K., Ponzoni, R.W., Walkley, J.R.W. and Morbey, A.S.C. unpublished data).

Because OR was recorded in the cycle prior to joining some inconsistency (see Table 3) between OR and

**TABLE 2. Least-squares means for number of corpora lutea per ewe ovulating (OR), number of ewes lambing per ewe joined ($E_{PJ}$), number of lambs born per ewe lambing ($L_{BP}$), number of lambs dying between birth and weaning per lamb born ($L_{DB}$) and number of lambs weaned per ewe joined ($L_{WJ}$).**

| Strain | OR | | | $E_{PJ}$ | | | $L_{BP}$ | | | $L_{DB}$ | | | $L_{WJ}$ | | |
|---|---|---|---|---|---|---|---|---|---|---|---|---|---|---|---|
| | 1.5 Years | 2.5 Years | 3.5 Years | 2 Years | 3 Years | 4 Years | 2 Years | 3 Years | 4 Years | 2 Years | 3 Years | 4 Years | 2 Years | 3 Years | 4 Years |
| Bu[(2)] | $1.15^{a(1)}$ (93)[(3)] | $1.30^{a}$ (80) | $1.33^{a}$ (43) | 0.66 | 0.84 | 0.76 | $1.00^{a}$ | $1.18^{a}$ | $1.20^{a}$ | 0.12 | 0.28 | 0.25 | 0.59 | 0.69 | 0.63 |
| ½Bo | $1.99^{b}$ (187) | $2.04^{b}$ (125) | $2.59^{b}$ (89) | 0.75 | 0.84 | 0.75 | $1.50^{b}$ | $1.53^{b}$ | $1.82^{b}$ | 0.35 | 0.35 | 0.38 | 0.74 | 0.86 | 0.75 |
| ½TF | $1.33^{a}$ (152) | $1.74^{c}$ (61) | | 0.73 | 0.92 | | $1.17^{cd}$ | $1.18^{a}$ | | 0.23 | 0.23 | | 0.64 | 0.83 | |
| ¼Bo | $1.84^{b}$ (21) | | | 0.74 | | | $1.30^{bd}$ | | | 0.28 | | | 0.69 | | |

(1) Means with no superscript or with a common superscript within each age group do not differ significantly ($P>0.05$) from each other.
(2) Bu = Bungaree; ½Bo = ½Booroola:½Bungaree; ½TF = ½Trangie Fertility:½Bungaree; ¼Bo = ¼Booroola:¾Bungaree
(3) In brackets total number of ewes with information to estimate least-squares means. Number of ewes contributing to OR, $E_{PJ}$, $L_{BP}$ and $L_{WJ}$ varied slightly depending on records available.

**TABLE 3. Least-squares means for number of corpora lutea per ewe ovulating (OR), number of ewes lambing per ewe joined ($E_{PJ}$), number of lambs born per ewe lambing ($L_{BP}$), number of lambs dying between birth and weaning per lamb born ($L_{DB}$) and number of lambs weaned per ewe joined ($L_{WJ}$) for progeny of ++, F+, and FF Booroola rams.**

| Bo sire genotype | OR | | $E_{PJ}$ | | $L_{BP}$ | | $L_{DB}$ | | $L_{WJ}$ | |
|---|---|---|---|---|---|---|---|---|---|---|
| | 1.5 Years | 2.5 Years | 2 Years | 3 Years | 2 Years | 3 Years | 2 Years | 3 Years | 2 Years | 3 Years |
| ++ | $1.04^{a(1)}$ (21)[(2)] | $1.26^{a}$ (11) | 0.65 | 0.82 | $1.16^{a}$ | 1.71 | 0.31 | 0.27 | 0.54 | 1.01 |
| F+ | $1.76^{b}$ (80) | $1.77^{a}$ (53) | 0.75 | 0.83 | $1.40^{a}$ | 1.54 | 0.33 | 0.29 | 0.72 | 0.90 |
| FF | $2.50^{c}$ (82) | $2.48^{b}$ (58) | 0.76 | 0.83 | $1.70^{b}$ | 1.59 | 0.36 | 0.40 | 0.81 | 0.84 |

(1) Means with no superscript or with a common superscript within each age group do not differ significantly ($P>0.05$) from each other.
(2) In brackets total number of ewes with information to estimate least-squares means. Number of ewes contributing to OR, $E_{PJ}$, $L_{BP}$ and $L_{WJ}$ varied slightly depending on records available. Numbers in tables 2 and 3 do not correspond exactly because for a few animals sire strain was known but individual sire identity was not.

$L_{BP}$ (or $L_{WJ}$) for daughters of ++ sires is not unexpected. Furthermore, note the limited (11) number of daughters available at 2.5 years of age.

Although the number of sires involved in our study is limited, the results indicate that reproductive wastage between ovulation and birth, and between birth and weaning, could be high enough to negate the superiority in terms of OR conferred by the 'F' gene. The practical significance of the biological differences observed among the Merino strains for reproductive traits is unclear because wool production is depressed in the high fecundity crosses (Ponzoni *et al.* 1984a). In the short term, the incorporation of the 'F' gene to SAM seems to offer the best prospects as preliminary analyses (Ponzoni *et al.* 1984b) suggest that the gene has no undesirable effects *per se* on wool or liveweight traits. However, it is not yet known whether, from an industry point of view, commercial ewes should have one or two copies of the gene (i.e. F+ or FF). The future utilisation of non-Bo fecund strains in South Australia is even less clear, though the development of a synthetic strain is an attractive proposition.

## ACKNOWLEDGEMENTS

The assistance of the staff at Minnipa Research Centre is acknowledged. CSIRO and the New South Wales Department of Agriculture co-operated in this work by providing Bo and TF rams. This work was supported by a grant from the Wool Research Trust Fund on the recommendation of the Australian Wool Corporation.

## REFERENCES

Freund, R.J. and Littell, R.C., 1981. SAS for linear models. A guide to the ANOVA and GLM procedures. SAS Institute Inc. N.C. U.S.A.

McGuirk, B.J., 1976. *Proc. Aust. Soc. Anim. Prod., 11*, 93-100.

Piper, L.R. and Bindon, B.M., 1982. *In* Barton R.A. and Smith W.C. (eds) *Proc. Wld. Cong. Sheep and Beef Cattle Breed., Vol. 1*, 395-400.

Ponzoni, R.W., Fleet, M.F., Walkley, J.R.W. and Walker, S.K. 1984b. *Anim. Prod.* (In Press).

Ponzoni, R.W., Walker, S.K. and Walkley, J.R.W., 1982. *In* Piper, L.R., Bindon, B.M. and Nethery, R.D. (eds) *The Booroola Merino* CSIRO: Australia, 51-59.

Ponzoni, R.W., Walker, S.K., Walkley, J.R.W. and Fleet, M.R., 1984a. *In* Land, R.B. and Robinson, D.W. (eds) *Genetics of Reproduction in Sheep*: Butterworth, London, 127-138.

Snedecor, G.W. and Cochran, W.G. 1967. Statistical Methods, 6th Edition. The Iowa State University Press, Ames, Iowa, U.S.A., 210-211.

Walker, S.K., Ponzoni, R.W., Walkley, J.R.W. and Morbey, A.S.C., 1984. *Anim. Reprod. Sci.* (In Press).

# TESTICULAR GROWTH AND THE EXPRESSION OF THE 'F' GENE IN YOUNG MERINO RAMS

C.M. Oldham and S.J. Gray, *School of Agriculture (Animal Science), University of Western Australia, Nedlands, Western Australia 6009.*
M.J. Carrick, *WA Department of Agriculture, Jarrah Road, South Perth, Western Australia, 6151.*

*Summary* The hypothesis that early testicular growth could be used to identify rams carrying the 'F' gene was tested in a flock in which it was established that the 'F' gene was segregating. No aspect of testicular growth was characteristic of the progeny of the half Booroola sires known to be carriers of the 'F' gene.

## INTRODUCTION

The high ovulation rate of the Booroola Merino is largely due to a dominant gene ('F') gene (Davis *et al.* 1982; Piper and Bindon 1982). Adult ewes with the 'F' gene often have an ovulation rate greater than three but no phenotypic expression for the 'F' gene has been identified in rams. Bindon and Piper (1976) found no difference in the testicular growth of young Booroola rams compared with control Merino rams. At this time the question of the 'F' gene had not arisen and thus the possibility of segregation within Booroola and Booroola cross rams was not investigated. Beetson (1982) measured the testicular diameter of Booroola × Collinsville rams at 23 weeks of age and found a strong genetic relationship with the ovulation rate of half sibs. We reasoned that a close study of the pattern of testicular growth of young rams would provide a range of criteria that could be used to identify individual rams as carriers of the 'F' gene.

## MATERIALS AND METHODS

The progeny of twelve 1/2 Booroola : 1/2 AMS (Australian Merino Society) rams comprising 187 females and 179 males born during a 10-day period in March 1982 at "Allandale Farm" Wundowie, Western Australia, formed the basis for this study.

Ovulation rates of females were determined by laparoscopy at 15 to 18 months of age. Testicular volume rams, estimated by comparative palpation of the right testicle (Oldham *et al.* 1978), and unfasted liveweight were recorded every 14 days from 14 to 32 weeks of age.

Testicular volume was plotted against time for every individual, and the time at which testicular growth changed from around 5 ml per 14 day period to around 15 ml per period was chosen arbitrarily as the age of onset of puberty. On the basis of ovulation rates of female progeny (Table 1) sires were classified as being either of F+ or ++ genotype.

The model fitted by oridinary least squares (Harvey 1960, 1982) included sire genotype as a fixed effect and sires as a nested random effect either with or without liveweight as a partial regression, fitted within sire genotype sub-classes. The liveweight in each case was coincident with the testicular measurement.

## RESULTS

The ovulation rates of daughters of 15 to 18 months of age were used to estimate the breeding value of each sire. These breeding values fell into two distinct classes (Table 1) and were thus used to classify five sires as F+ genotype and the remaining seven as ++ genotype.

The pattern of liveweight and testicular volume change between 14 and 32 weeks was virtually identical in the two sire genotype groups (Table 2). When liveweight was fitted as a partial regression, testicular growth was similarly identical in each genotype group.

The mean age at onset of puberty (20 weeks) was similar in both sire genotypes and furthermore this arbitrary measure was extremely variable within sire groups of each genotype (Table 2). Heritability estimates for testicular volume, using the nested model with individual sub-class partial regression on liveweight, ranged from 0.26 to 0.65 (Table 2). At least one and sometimes both of the nested sire groups were significant effects in this model at each of the ten time periods.

## DISCUSSION

Measurement of testicular growth from 14 to 32 weeks of age, whether adjusted for liveweight or not, did not provide any indication of phenotypic expression of the 'F' gene in males. The study included the period of rapid testicular growth which might be called the pubertal spurt of growth (Table 2). This growth spurt happened at the same time in both sire genotype groups.

While the 'F' gene had no effect on any measure of testicular growth, significent genetic variation is present at each of the ages measured. Thus it now appears very likely that the realised positive genetic correlation between testicular dimensions and ovulation rate reported by Knight (1984) and the positive genetic correlation of Beetson (1982) represent potentially usable genetic relationship in the absence or presence of the

'F' gene. Only heterozygote progeny were investigated in this study. It is possible that the 'F' gene affects testicular growth but is completely recessive.

**TABLE 1. The distribution of ovulation rate of the 15 to 18 month old progeny[1] of 12 individual 1/2 Booroola: 1/2 AMS rams or pooled progeny of backup AMS rams.**

| Sire | n | No. of corpora lutea 0 | 1 | 2 | 3 | 4 | Mean ov. rate[3] | (BV[2]) (ov. rate) |
|---|---|---|---|---|---|---|---|---|
| 1/2 Booroola | | | | | | | | |
| 1 | 72 | 6 | 56 | 10 | | | 1.15 | −0.28 |
| 2 | 57 | 15 | 40 | 2 | | | 1.05 | −0.38 |
| 3 | 9 | 0 | 1 | 4 | 4 | | 2.33 | +0.90[4] |
| 4 | 29 | 0 | 15 | 9 | 4 | 1 | 1.69 | +0.26[4] |
| 5 | 46 | 0 | 20 | 13 | 11 | 2 | 1.89 | +0.46[4] |
| 6 | 30 | 1 | 15 | 8 | 5 | 1 | 1.72 | +0.29[4] |
| 7 | 16 | 0 | 13 | 3 | | | 1.18 | −0.25 |
| 8 | 45 | 4 | 28 | 3 | | | 1.07 | −0.36 |
| 9 | 70 | 6 | 37 | 18 | 9 | | 1.56 | +0.13[4] |
| 10 | 20 | 2 | 16 | 2 | | | 1.11 | −0.32 |
| 11 | 39 | 9 | 29 | 1 | | | 1.03 | −0.40 |
| 12 | 30 | 2 | 19 | 9 | | | 1.32 | −0.11 |
| AMS | 187 | 42 | 133 | 19 | 1 | | 1.14 | |

[1] 1/4 Booroola, 3/4 AMS;

[2] Breeding value $= \left[\dfrac{1/2\, nh^2}{1 + (n-1)1/4\, h^2}\right] \circ (y - \bar{y})$ Where y = mean of half sib group; $\bar{y}$ = overall mean; n = number of individuals in half sib group; $h^2$ = heritability of character.

[3] Ovulation rates of ewes ovulating.

[4] Sires heterozygous (F+) for the 'F' gene.

**TABLE 2. The mean liveweight (± SEM) and mean volume of the right testicle (± SEM) for the rams of the two sires genotypes F+ and ++, measured every 14 days between 14 and 32 weeks of age. Also shown are the heritabilities (± SE) from pooled sire effects.**

| Age (weeks) | Liveweight (kg) | | Testicular Volume (ml) | | Heritability of testicular volume |
|---|---|---|---|---|---|
| | F+ | ++ | F+ | ++ | |
| 14 | 16 (.3) | 16 (.2) | 6 (0.3) | 6 (0.2) | 0.57 (0.306) |
| 16 | 21 (.4) | 21 (.3) | 11 (0.7) | 11 (0.5) | 0.36 (0.250) |
| 18 | 23 (.4) | 23 (.3) | 18 (1.4) | 17 (1.0) | 0.28 (0.226) |
| 20 | 24 (.5) | 24 (.4) | 27 (2.3) | 26 (1.6) | 0.42 (0.267) |
| 22 | 29 (1.2) | 29 (.8) | 45 (3.9) | 45 (2.8) | 0.26 (0.218) |
| 24 | 30 (.6) | 30 (.4) | 64 (4.7) | 65 (3.3) | 0.32 (0.239) |
| 26 | 33 (.6) | 32 (.4) | 74 (4.8) | 73 (3.4) | 0.38 (0.256) |
| 28 | 34 (.6) | 34 (.4) | 84 (4.3) | 83 (3.1) | 0.65 (0.332) |
| 30 | 36 (.6) | 35 (.4) | 86 (4.2) | 89 (3.0) | 0.41 (0.267) |
| 32 | 38 (.6) | 38 (.4) | 94 (4.3) | 97 (3.0) | 0.60 (0.319) |

ACKNOWLEDGEMENTS

We are grateful to the AMRC for financial support and to the participating farmers for their enthusiastic cooperation.

REFERENCES

Beetson, B.R., 1982. *In* Piper, L.R., Bindon, B.M. and Nethery, R.D. (eds) *The Booroola Merino*. CSIRO, Melbourne, 41-50.

Bindon, B.M. and Piper, L.R., 1976. *In* Tomes, G.J., Robertson, D.E. and Lightfoot, R.J. (eds) *Sheep Breeding*, WAIT Press, Perth, 357-371.

Davis, G.H., Montgomery, G.W., Allison, A.J., Kelly, R.W. and Bray, A.R., 1982. *N.Z. J. Agric. Res., 25*, 525-529.

Harvey, W.R. 1960. *USDA, Agric. Res. Publ. 20*, 1-157.
Harvey, W.R., 1982. *J. Anim. Sci., 54*, 1279-1285.
Knight, T.W., 1984. *N.Z. J. Agric. Res., 27*, 179-187.
Oldham, C.M., Adams, N.R., Gherardi, P.B., Lindsay, D.R. and Mackintosh, J.B., 1978. *Aust. J. Agric. Res., 29*, 173-179.
Piper, L.R. and Bindon, B.M., 1982. *In* Barton, R.A. and Smith, W.C. (ed.). *Proc. Wld. Congr. on Sheep and Beef Cattle Breeding*, Dunmore Press Ltd., Palmerston North, N.Z., 395-400.

# PROGENY TESTING FOR THE 'F' GENE USING PREPUBERTAL EWE LAMBS

C.M. Oldham, S.J. Gray and P. Poindron, *School of Agriculture (Animal Science), University of Western Australia, Nedlands, Western Australia 6009.*
B.M. Bindon, *CSIRO Division of Animal Production, Armidale, New South Wales 2350.*

*Summary* Two experiments were designed to test the hypothesis that prepubertal (approx 5 months old) ewes, daughters of rams carrying the 'F' gene for high ovulation rate and litter size, would have a higher ovulation rate in response to an injection of PMSG (500 iu) than flockmates not carrying the gene. In the first experiment some sire groups of ⅛ Booroola lambs clearly had a greater incidence of three or more ovulations compared with Merino flockmates. In the second experiment a similar superiority in ovulation rate in response to PMSG was associated with a known presence of the 'F' gene in Booroola × Border Leicester ewe lambs compared with Collinsville × Border Leicester controls. The combined results suggest that the ovulatory response of 5 month old ewe lambs to a single intramuscular injection of PMSG can be used to evaluate sires with respect to the 'F' gene.

## INTRODUCTION

It is generally agreed that the high fecundity of the 'Booroola' strain of Merino is associated with a major gene, ('F') or single segregating unit. Adult ewes carrying the 'F' gene have at least one record of ovulation rate and/or litter size of three or more during their lifetime (Piper and Bindon 1982). In the absence of phenotypic expression of the gene in rams, it is necessary to use the ovulation rate of daughters to select rams carrying the gene. As ewes may not ovulate spontaneously until they are 8 to 15 months of age, evaluation of sires may take as long as two years.

Recently it has been shown that the ovulation rate of adult ewes carrying the 'F' gene is more sensitive to PMSG than that of non carriers (Kelly *et al.* 1983). PMSG alone will induce ovulation in prepubertal ewes greater than four months of age (Mansour 1959). We hypothesised that differential response in the ovulation rate of ewe lambs five moths of age induced by PMSG, could be used to identify sires carrying the 'F' gene, thus allowing selection of sires within 10 months of their being joined with ewes.

## MATERIALS AND METHODS

### Experiment 1

As part of a cooperative research project between the University of Western Australia and the Australian Merino Society (AMS) 12 rams, Booroola × AMS Merino, selected for fleece weight and live weight, were joined with AMS ewes. Using the method of Piper and Bindon (1982) based on the spontaneous ovulation rate of daughters, 5 of the 12 rams were carriers of the 'F' gene. Subsequently 12 sons (¼ Booroola : ¾ AMS) of the five rams heterozygous for the 'F' gene (F+) were each joined with 50 AMS ewes, two rams per farm on six farms in February 1983. The rams were joined for four to six weeks in parallel with the farmer's normal Merino joining period. The progeny of each ¼ Booroola ram were individually identified at lambing and then run with the remainder of the Merino flock from weaning. In January 1984 all the 5 to 6 month old ewe progeny of the ¼ Booroola rams and 20 AMS Merino flockmates were injected intramuscularly with 500 iu of whole pregnant mare's serum, standardised against PMSG (Folligon; Intervet) in a rat ovarian growth bioassay. Ovulation rates were determined by laparoscopy three to four days later. At laparoscopy all ewes were weighed after an overnight fast.

The live weight and ovulation rate data based on farms and sire type within farms were treated by analysis of deviance (Nelder and Wedderburn 1972) using the statistical computer programme GENSTAT. The hypothesis in experiment 1, that the incidence of $\geq 3$ corpora lutea was greater in the progeny of some ¼ Booroola rams than in the pooled data for AMS rams was tested using the G test (Sokal and Rohlf 1981).

### Experiment 2

In response to the results obtained in the first experiment a second group of 6 month old prepubertal ewes whose dams' genotype was known to be 'FF', 'F+' or '++' were injected with 500 iu of PMSG (Folligon; Intervet), weighted, and their ovulation rate recorded by laparoscopy three days later. They were all sired by Border Leicester rams and their dams were Booroola Merino of 'FF' or 'F+' genotype, or Collinsville Merino '++'. In experiment 2 the same statistical method tested the difference between the incidence of $\geq 3$ corpora lutea in the progeny from dams with or without the 'F' gene.

## RESULTS

In the first experiment there were significant differences among farms in the live weight of sire groups independent of sire type (lowest 22 kg and highest 28 kg) but live weight was unrelated to the ovulatory response of sire groups (lambs ovulating and ovulation rate of lambs ovulating) to the injection of whole pregnant mares' serum. There were no differences in the ovulatory response between the six farm groups of

AMS lambs so the results for these ewes were pooled (mean ovulation rate of ewes ovulating = 1.62, range of 1.50 to 1.70; Table 1).

**TABLE 1. The live weight, distribution of ovulation rate, and mean ovulation rate of ewe lambs treated with 500 iu of whole pregnant mares' serum (Experiment 1). In the table, progeny of AMS rams are pooled while the progeny of ¼ Booroola rams are grouped after comparison with the pooled distribution of ovulation rate for the AMS lambs.**

| Hypothesised genotype | n | Live weight (kg) | No. corpora lutea 0 | 1 | 2 | 3 | 4 | 5 | 6 | 9 | Mean OR (1) |
|---|---|---|---|---|---|---|---|---|---|---|---|
| AMS | 113 | 25 | 27 | 43 | 35 | 6 | 2 | | | | 1.62 |
| ⅛ Booroola(2) *minus* 'F' gene | 109 | 24 | 30 | 40 | 31 | 4 | 2 | | | | 1.58 |
| ⅛ Booroola(3) *plus* 'F' gene | 84 | 25 | 17 | 16 | 19 | 18 | 8 | 3 | 2 | 1 | 2.63 |

(1) mean ovulation rate of ewes ovulating
(2) pooled result for the progeny of 7 sires
(3) pooled result for the progeny of 5 sires

The results for those groups of ⅛ Booroola lambs in which the incidence of ≥ 3 corpora lutea among lambs ovulating was greater ($P < 0.05$ to $P < 0.001$) than among the AMS lambs were pooled, and are shown in Table 1 as ⅛ Booroola progeny among which it is hypothesised that the 'F' gene was segregating. The progeny of the remaining seven Booroola cross sires are shown in Table 1 as ⅛ Booroola minus the 'F' gene.

In the second experiment the lambs from ewes known to carry the 'F' gene were lighter than the flockmates from ewes not carrying the 'F' gene but the incidence of ≥ 3 corpora lutea was greater ($P < 0.05$; Table 2).

**TABLE 2. The live weight, distribution of ovulation, and mean ovulation rate of ewe lambs treated with 500 iu PMSG (Experiment 2) grouped according to the genotype of their dam.**

| Dam genotype | n | Live weight ± SE (kg) | No. corpora lutea 0 | 1 | 2 | 3 | 4 | Mean OR (1) |
|---|---|---|---|---|---|---|---|---|
| ++ | 29 | 33.1 ± 0.67 | 2 | 17 | 8 | 2 | | 1.44 |
| FF orF+ | 15 | 31.4 ± 1.13 | 1 | 6 | 3 | 3 | 2 | 2.53 |

(1) Mean ovulation rate of ewes ovulating

## DISCUSSION

It is clear that a single intramuscular injection of 500 iu PMSG will induce ovulation in greater than 70% of 5 to 6 month old Merino ewe lambs. In Experiment 1, daughters of some Booroola cross sires showed a dramatically increased incidence of ≥ 3 ovulations. When viewed with experiment 2 which showed a similar phenomenon associated with 'F' genotypes it seems highly probable that the ovulation rate of 5 month old daughters in response to PMSG can be used to identify sires carrying the 'F' gene.

Subsequent experiments in New Zealand on ewe lambs of known genotype (FF of F+) suggest that this method may also distinguish between sires homozygous versus heterozygous for the 'F' gene (G. Davis, pers. comm.).

## ACKNOWLEDGEMENTS

We are grateful to the participating farmers and to Jenny Ford, Peter Moore, Dave Suckling, Steve Cairns and David Turner for technical assistance. To Glen De'ath for statistical help and to the AMRC for financial support.

## REFERENCES

Kelly, R.W., Owens, J.L., Crospie, S.F., McNatty, K.P. and Hudson, N., 1983. *Anim. Reprod. Sci., 6*, 199-207.

Mansour, A.M., 1959. *J. Agric. Sci. Camb., 52*, 87-93.

Nelder, J.R. and Wedderburn, R.W.M., 1972. *J. Roy. Stat. Soc., 135*, 87-93.

Piper, L.R. and Bindon, B.M., 1982. *In* Barton, R.A. and Smith, W.C. (eds), *Proc. World Congress on Sheep and Beef Cattle Breeding, Vol. 1: Technical.* Dunmore Press Ltd., Palmerston North, N.Z., 395-400.

Sokal, R.R. and Rohlf, F.J., 1981. *In* Sokal, R.R. and Rohlf, F.J. (eds), Biometry, 2nd edition. W.H. Freeman and Co., San Francisco, Chapter 17.

# NUTRITION FOR REPRODUCTION

A.R. Egan, *Animal Production Section, School of Agriculture and Forestry, The University of Melbourne.*

## INTRODUCTION

Reproduction in animals is the summation of a series of physiological events, each calling for, among other things, some specific level and composition of nutritional support. The processes span a long period each year, and, with grazing sheep in the Australian environments, nutritional conditions change predictably but with variability between years. Because of this, a set nutritional management programme will be sucessful only when averaged over a number of years. Animals faced with unlimited, high quality feed experience reproductive problems only through effects of high lamb birth weight and excessive fat deposition in the abdominal, renal and pelvic fat bodies. Where nutritional conditions fluctuate, successful management consists of have animals on good feed at the most critical periods of high demand in reproductive activity: late pregnancy and lactation. Factorial estimates (ARC 1980) give broad guidance in matching the needs of late pregnancy and lactation with seasons when supply is most likely to be adequate. This provides a form of 'coarse turning' although it fails to provide good nutrition for all evetns in all seasons. Other key events occurring in periods set six moths distant from those obvious high demand periods will, by design, fall at less favourable times of the year. The nutritionist and reproductive physiologist aim to identify those key events, and determine the penalties for under-nutrition or malnutrition at those times; and to determine the level, and timing of extra nutritional input to remove that penalty. Supplies of the appropriate types of feed can be inadequate at almost any time in the Australian environment. Statistical approaches can be used to formulate 'minimum risk' policies, but to follow these in every year may disavantage overall production. It is essential therefore to establish the magnitude of the penalty for moving to a plan with higher risk, but offering higher productivity in 'good seasons'.

The penalty needs to be assessed on two bases — effects on cash flow, and effects on long term productivity and profitability. If there are indirect and delayed penalties apart from effects on numbers of lambs produced in the current season, these also need to be identified and evaluated. Further, the benefits from tactical supplementary feeding need to be assessed: the costs may be offset against penalties avoided if we can put values on these.

At present the biological data allowing assessment of direct and delayed penalties for under-nutrition and the dose-response benefits of small supplementary inputs at key times are limited, although computer modelling offers the immense potential for running 'trial balances'. This review aims to identify periods of nutritional risk, the nature, and, where possible, and the magnitude of the benefit of nutritional support in those times.

## MEETING THE NUTRITIONAL REQUIREMENTS OF REPRODUCTIVE EVENTS

Broad patterns of requirements for metabolizable energy and intestinally digestible protein for pregnancy and lactation in ewes are well outlined by the ARC (1980), and by Robinson and Orskov (1975). These sources can be readily consulted and it is not intended to review the factorial and empirical methods whereby they have been derived. Tabulated information such as this is not always helpful or applicable to the Australian sheep industries. If ewes were hand-fed using these recommendations, reproductive performance of Australian sheep would probably be greatly improved. However, tables may overstate needs, in that variation in supply from these levels may have little adverse effect if under- and over-supply occur at the right times. Grazing intake is a function of availability, of digestibility, and of composition of plant species and parts selected by the animal (Arnold 1970). Measure of pasture allowance offer some potential (Rattray and Jagusch 1978) for control of availability and selection, but management at this level cannot solve problems of low voluntary intake, low digestibility and low protein content. These 'problems' may be overcome, in part, by exploiting the 'elasticity of requirements' associated with the ewe's content and capacity to use endogenous reserves. The major task of the sheep nutritionist is to know how much that elasticity may be relied upon, and where, if at all, the penalties for such a strategy will be expressed. Special metabolic characteristics which must be considered are related to the need within the total energy metabolism (exogenous and endogenous) to balance gluconeogenic precursors and amino acids in particular. The detection of adverse nutritional status, the prediction of the magnitude of the penalty and its distribution across the flock call for a wide range of approaches. Inadequacies in specific mineral nutrients will also have dramatic effects if the exogenous and endogenous supplies fail to support metabolic events in key organs during reproductive events.

Association of certain soil types, seasonal conditions and pasture growth patterns with mineral-responsive reproductive disorders is a well-developed field of specialisation in agronomy and grazing animal nutrition (Egan 1974). This review will not develop this aspect of nutrition for reproduction, except where more recent developments point to particular principles.

## NUTRITIONAL HISTORY, LIVE WEIGHT AND REPRODUCTIVE PERFORMANCE

There is no question that reproductive performance is related to some function of live weight, body condition or nutritional status; or to the direction and rate of change of weight of some body components. The 'dynamic' and 'static' effects of live weight on proportion of mated ewes lambing and presenting multiple offspring have been well explored (Underwood and Shier 1941; Moule 1962; Coop 1966; Killeen 1967; Edey 1968; Fletcher 1971; Fletcher 1974; Morley *et al.* 1978). There remains the need to translate this into specific mechanisms, currently thought to be related to level of energy and/or protein 'reserves' or to circulating metabolite levels reflecting these.

There is also no question that there are some times in the reproductive cycle when a shortfall in intake of nutrient can be met by mobilization of body reserves, and hence loss of weight, without any penalty. At other times a severe penalty will be incurred. Simply stated, effects of nutrition at any time are modulated by previous nutritional history. Our hypothesis is that availability of metabolites from both endogenous and exogenous sources can satisfy the specific needs of the foetus and of milk production, but that some function of the endogenous component, if this is too great, may also be translated into a metabolic pattern which can inhibit, or fail to stimulate, key reproductive events. This will adversely affect current or future reproductive performance.

In lactation it is difficult to feed enough to a ewe with twins or triplets to support high lamb growth rate without the ewe suffering some live weight loss. A 6-10 kg gain in live weight by the ewe during late pregnancy can reduce the maternal handicap on such lambs (File 1981). Some loss of live weight (or condition) at this time is permissible for ewes with single lambs, provided 'reserves' are not at a low level already. However, excessive rates of loss lead to poor lactation and possibly a delay in the subsequent return to oestrus. Effective mating will be delayed if the period of under-nutrition is extended through lactation and the post-weaning period (Cahill *et al.* 1984). It remains for us to define the relationships between commencing levels of body reserves, the rate of their loss, and the patterns of absorption of specific nutrients which are consistent with early return to oestrus and high ovulation rate. There is some evidence that even recovery in body condition over six months of access to good quality feed will not restore the ewe undernourished during late pregnancy to the same level of reproductive performance expressed by ewes held at a good live weight in pregnancy (Fletcher 1971).

In late pregnancy, setting a level and quality of nutrient intake to support the growth of the foetus without ewe live weight loss may be less important than setting nutritional conditions to ensure balance in utilization of endogenous and exogenous substrate. Fat can be mobilized and utilized without the immediate penalty of ketosis if sufficient gluconeogenic precursor is available (Russell *et al.* 1977; Egan and MacRae in preparations). In this sense the appearance of a high level of ketone bodies is seen as the result of poor fat catabolism rather than a high rate of fat mobilization. There is also, however, the need to preserve the nutritional status essential to the development of the mammary gland. It may be that nutrition in mid to late pregnancy should be viewed more from the point of view of 'priming the lactational pump' rather than 'feeding the foetus'.

Much of this is contrary to the way in which statements of nutritional requirements are generated on a factorial basis. Those approaches address the need for support of the current rates of tissue accretion rather than the inputs to support a high potential for subsequent stages of reproductive performance, lamb survival and growth. They assume a continuity of feeding based on adequacy at all times, without controlled or uncontrolled periods of deficit and excess.

Despite strong research efforts there remain major areas of uncertainty in our understanding of the interaction between history of nutrition (expressed in magnitude and composition of body reserves), the capacity to mobilize reserves, and tolerances in integration of the food consumed with rates of mobilization from endogenous sources, without loss to a subsequent function. Consequently we need:

(1) redefinition of priority targets, expressed as weight, rate of weight change, levels of energy and protein reserves;
(2) definition of the nutritional conditions to reach those targets;
(3) a knowledge of the biological and economic penalties for not reaching the targets. Both immediate and delayed penalties must be considered;
(4) metabolic indices of the existing status of the ewe in relation to targets.

It would be foolish to represent the aim of management as being to manipulate the status of the ewe to achieve the perfect nutritional/body condition interactions. The inputs necessary and the alternative patterns of nutritional management will be influenced by availability of herbage and the timing of all operations to suit local and seasonal conditions. However, there is a need to establish the conditions under which a higher level of nutrition or a different source of nutrition will benefit key events in the reproductive cycle.

## NUTRITION AND PRODUCTION OBJECTIVES

Feeding strategies and tolerances for weight loss in ewes will differ depending on whether the system produces lambs for replacment in a wool producing flock or prime lambs for meat. In a system of lamb replacement, the great capacity to recover from the severe under-nutrition, provided they survive, can be utilized. The penalties apart from lamb and, at the extreme, ewe losses, are time to reach maturity, and to enter the breeding flock (Allden 1970). In the prime lamb system, under-nutrition of ewes becomes a problem as soon as the rate of growth of lambs falls. These take longer to reach market weight and often compound the problem by carrying over into nutritional conditions less favourable for growth. Early supplementation of the ewe may overcome both these sets of problems, but at a cost. Options here include creep feeding, supplementation of lambs directly, or early subdivision of the ewe/lamb flock according to lamb weight and special nutritional management to promote faster growth in the light-weight lambs.

Wool growth is affected by the nutritional demands or stress of pregnancy, parturition and lactation. Tenderness in wool at this time is a major problem particularly with autumn-lambing ewes on pastures of decreasing nutritional value (Ralph 1984). The position of the weakness is often set at a less critical position in the fibre by shearing as close to the time of lambing as possible. There are climatic or other reasons for setting shearing at other times, and extra nutritional inputs may be called for to reduce stress and to support wool growth at this time. Since there are other constraints on shearing and lambing times, assessment of the penalties to wool growth and quality effects arising from nutritional conditions, and a knowledge of appropriate nutritional inputs to remove or minimize these penalties is essential. Ralph (1984) reported good results in improving fibre strength by supplementation with oats or oat-lupin mixtures. Defining points at which major wool quality faults are introduced by under-nutrition calls for knowledge of the sensitivity at key events. So also does improvement from satisfactory to superlative reproductive performance.

Selecting ewes for high fecundity, or increasing ovulation rate by hormonal manipulation, increases the nutritional demand and magnifies the vulnerability to nutritional stress. For these conditions the view of a 'body stores — supplementation' strategy must be tempered by recognition of the importance of current intake of gluconeogenic substrate and amino acids as well as metabolisable energy (MacRae and Egan 1983).

Producing more lambs per ewe and lambing more than once a year (e.g. Hall 1984) impose additional nutritional demands, or shift the nutritional demand to more difficult times of the year. White *et al.* (1982) drew up curves of the biological and contemporary economic effects of altered lambing time in Victoria. With possible cumulative effects of a shorter lambing interval we now need to know whether each curve will show similar benefits for equal inputs of extra feed at late pregnancy and early lactation; or whether each lambing time curve calls for different strategies in the choice of patterns of use of available pasture and supplements.

## THE PUBERTAL EWE

In a system sensibly stocked to a constant size of ewe flock consistent with grazing pressure and pasture growth curves, mating maiden ewes as early as possible increases opportunity for later culling on performance, and reduces the age profile in the flock. It reduces the size of the replacement component of the ewe flock and hence the maintenance costs of the breeding nucleus by up to one year per head over the lifetime of the ewe. Mating early in life may make the ewes more vulnerable to poor nutrition if there is a penalty in growth and recovery before the next mating, but if growth and pregnancy can be simultaneously supported the net advantage may be great. Season of birth, nutritional level, live weight and social interactions are all implicated in determining time of puberty in the lamb (Dyrmundsen 1981). Relationships between puberty and weight have been examined for most Australian breeds of sheep. Autumn born lambs may reaçh the target weight but still have delayed puberty because of seasonal anoestrus. In a flock of 10-11 months old Merinos, ovular ewes were heavier than their anovular flock mates (Murtagh *et al.* 1984). More rapid growth rate to six months of age, raising lean body mass or nutrient status, can have a positive influence on reproductive performance. The penalty for late achievement of weight targets can be an eight to ten month hiatus until the subsequent flock mating time. Even at 18 months, slow growth due to cyclic seasonal feed conditions may cause deferral of mating for another year, and animals culled on this basis may represent an unnecessary reduction in rate of genetic progress in other areas.

## PRE-JOINING NUTRITION

Since early work of Underwood and Shier (1941) the challenge to identify the factors involved in the entry into full cyclic behaviour in the flock, with increased numbers of multiple ovulation, has been pursued. This has been associated with a rising plane of nutrition, increasing live weight, or the achievement of a heavy live weight (Coop 1966). Because of the statistical problems in this type of work, inconclusive results are often encountered, and the principal nutritional and physiological factors in body condition have not been identified (Lindsay 1976). Positive responses to very short term provision of supplements, particularly lupins, have been reported (Knight *et al.* 1975; Marshall 1978) and there are both postive and negative reports on effectiveness of very short-term supplementation with barley, oats, beans, or peas in eliciting an increase in ovulation rate or lambs born per ewe lambing (Reeve *et al.* 1979). The reasons to which these differences are ascribed have generally focussed on either the condition or weight or body composition of the ewe at the time of mating, or on the components of the supplement — energy, protein, gluconeogenic precursor yield (Rowe *et al.* pers. comm.) or minerals (Lindsay 1976). The effects of inputs of nutrients which improve ovulation rate are probably mediated through a hormonal response to that change in nutritional status (Brien *et al.* 1976.) The mechanism remains elusive, but the most popular interpretation seems to centre on a protein effect on ovulation rate which may have a 'para-pharmacological' aspect.

## NUTRITION DURING PREGNANCY

The pregnant ewe has immense capabilities for sustaining the fertilized ovum and the developing embryo and foetus. While embryonic loss is an identifiable contributor to reproductive wastage, the role in this of undernutrition is not seen as a major one (Braden 1971). Nutritional deprivation must be severe to result in losses at an early stage of pregnancy, but can result in impairment of rate of foetal growth (Everett 1964). Specific deficiencies in trace elements may be of more concern and the effects of transient inadequacy of copper, zinc, iodine and possibly manganese and selenium need more consideration. These more specific nutrients are discussed later.

In the last trimester of pregnancy, the foetus grows at an increasing rate and greatly increases its demand. Nutritional requirements are developed on the basis of factorial estimates of energy and protein needs at this stage, assuming that the ewe remains at the same live weight. However, inadequate energy intake can result in mobilization of body reserves with no penalty in the survival of ewe and lamb. With high rates of fat mobilization in the ewe with advancing pregnancy, ketosis can occur. There are two problems:

(1) Continuing chronic under-nutrition can result in early depletion of reserves; and also reduce the effective development of mammary tissue for the subsequent lactation. Ketone bodies rise and glucose falls in concentration in the blood, the magnitude increasing with numbers of foetuses (Parr 1984). In this case total energy intake must be raised to control loss of live weight. The problem is greater in twin-bearing ewes and use of indices such as blood glucose level (Parr 1984) or ketone level (Foot *et al.* 1984) can allow separation for differential nutrition of twin-bearing ewes.

(2) Acute under-nutrition of previously well-nourished ewes. This can arise in adverse weather or changes in paddock. The capacity for mobilization is high, but the balance of endogenous substrates presented is inappropriate. Ketosis appears, often without an initial fall in blood glucose level, and appetite fails. The animal responds to inputs of glucose only in the first day or so of appearance of the disorder: thereafter, in order to restore appetite infusion of glucose and branched-chain amino acids is necessary (A.R. Egan, unpublished data).

Separation of a flock on the basis of low glucose level or of rising ketone level works well in case (1) but not in case (2). It seems sensible therefore to hold back the condition of ewes in the first two thirds of pregnancy, and allow a rising plane of nutrition in late pregnancy. High levels of energy stores in ewes are not necessarily helpful at this stage.

Supplementation with a high protein concentrate may aid in reducing the rate of appearance of ketones in the plasma if ewes are faced with inadequate quantities of feed, accompanied by live weight loss and fat mobilization (Egan, Campbell, Schlink, Dixon and Hart, unpublished data).

Birth weight of lambs and their survival are improved by such supplements, even when ewes lose live weight (*concepta*-free) in the last trimester. There is, however, a second consequence of nutritional state in mid-to late-pregnancy — the level of lactation may be lower in ewes losing weight at this stage, which would greatly influence growth rate of twin lambs in particular. File (1981) suggests that twin-bearing ewes must gain 200g/d in the last trimester to support lamb growth rates equal to those of singles. Dove *et al.* (1984) provided supplements of oat grain or sunflower meal (500g/d) to crossbred ewes in late pregnancy or early lactation, or both. With maternal weight at lambing increasing by 5kg in supplemented groups, there was no effect on lamb birth weight, but a marked increase in lamb growth rates up to weaning. With single lambs there was little difference between the two supplements or the various lengths or times of supplementation. With twins, supplementation prior to parturition had a beneficial effect greater than that of *post-partum* supplementation.

In current studies on short periods of supplementation of Corriedale and Merino ewes in late pregnancy, early lactation, or both (Egan, Campbell, Schlink, Dixon and Hart, unpublished) there is evidence of a benefit in wool production and particularly minimum fibre diameter, and in rate of growth of lambs with lupin grain fed at 500g/d for 20-30 days in late pregnancy. There was less benefit from the same amount of lupin fed in early lactation and though loss of live weight by the ewes was less over the lactation period, lamb growth rate was slower, and tenderness faults were more common in the ewes' wool. Supplementation over both periods did not greatly improve lamb growth rates. The patterns of body protein and fat deposition and mobilization established under these conditions are currently being examined. Dose response curves for supplements have not yet been established.

## CHANGING GEARS AT PARTURITION AND LACTATION

The lactating ewe can metabolize endogenous fat rapidly with what appears to be little metabolic abnormality (Annison 1984; Egan and MacRae, unpublished). The provision of substrate for protein and lactose synthesis, rather than feeding to prevent live weight loss, is a reasonable strategy in ewes in good condition at parturition. However, for ewes in poor condition at parturition, feeding at levels recommended by the ARC appropriate for a high milk production often results in live weight gain and a lower-than-expected milk production and live weight gain in the lamb.

This suggests a difference in partition of use of nutrients providing for ewe and lamb survival, rather than rapid lamb growth with possible adverse consequences for the ewe. The prolific ewe is particularly vulnerable, but feeding *post-partum* to improve milk production can be less effective than earlier feeding to ensure good condition at commencement of lactation.

Ewes with multiple lambs inevitably use body reserves at peak lactation supporting a high rate of lamb growth. The source of substrate, endogenous and exogenous, and the partition of metabolites into body tissue, wool and milk, are markedly affected by the availability of extra feed and also by the pattern of nutrients absorbed. A clear statement relating performance in these respects to rate of deposition or mobilization of fat and protein awaits the outcome of current work.

## LAMB SURVIVAL

Causes of perinatal lamb mortality include among others, those arising directly and indirectly from nutritional status of the ewe. Data from surveys in many regions covering northern and inland semi-arid to southern and elevated cold wet environments suggest losses of about 11 million lambs in Australia annually (McGuirk 1982). It is not unusual to find that 20-25% of pregnant ewes fail to rear a lamb to marketing (Anon. 1980). Low birth weight of lambs, mismothering and lamb starvation, vulnerability to predators, and susceptibility to exposure all have nutritional components (Alexander 1980) and will be reduced with a nutritional regime for the ewe which, on empirical observation, supports a live weight gain of 6-8kg in late pregnancy and lamb birth weight of 3.5 — 4.5kg (Ferguson 1982). *Post-partum* improvement in nutrition does not compensate for adverse nutritional conditions *pre-partum*. Peak lactation yield may be improved but the early postnatal period is critical. Ewes with twins or triplets will need more nutrients, and the need to identify and specially manage these ewes is recognized. The means are discussed elsewhere in this volume (Parr 1984; Foot *et al.* 1984).

## POST-LACTATIONAL MANAGEMENT

There is evidence (Fletcher 1974; Cahill *et al.* 1984) that the nutritional conditions during late pregnancy and early lactation have 'carry-over' effects on the reproductive performance of the ewe at the subsequent mating six months later. The way in which this relates to timing in recovery of live weight, body fat and protein stores is unclear but is receiving attention by our group. This may influence the effectiveness of 'flushing' when the next cycle commences. Falling body condition after weaning, and maintenance of low live weight up to mating (the traditional pattern for dry ewes in areas of seasonal pasture production) may have consequences which cannot be overcome by short-term feeding (Cahill *et al.* 1984) or which limit the capacity for response to surges of supplementation. It is essential that these patterns be examined in controlled experiments to identify *metabolic conditions* in which reproductive performance is adversely affected. Effort in this direction has focussed on times close to mating. Even the experimental methods of generating different body conditions and weights over short periods prior to a re-alimentation and flushing may not be the appropriate model to assess effects of change in live weight in the complete year-round reproduction cycle.

## NUTRITION AND REDUCED LAMBING INTERVAL

In systems where reduced lambing interval is sought (rebreeding after drought, changes in lambing time and three lambings in two years) the most servere delayed penalties of low live weight or prolonged anoestrus associated with lactation are apparent. Success in these management strategies calls for knowledge of interactions between hormones, body condition and current nutrition in setting the stage for ovulation, behavioural oestrus, implantation, foetal growth and next lactation. Hunter and van Aarde (1973) studied the influence of lactation, level of nutrition and presence of rams on the incidence of oestrus in mutton Merinos lambing in November, March/April or July in South Africa. While season of lambing had important overall effects, within any one time the level of nutrition determined when ewes recommenced cyclic oestrous activity.

Attempts to reduce lambing interval, either to lambing on an eight-month basis (three lambings in two years) or a six-month basis (twice-a-year lambing) produce major problems (Bourke 1964; Hulet 1978; Geytenbeek, pers. comm.). In particular, the lactation/suckling effects on oestrous activity are inescapable under lamb-rearing conditions. Geytenbeek, Egan and Findlay (unpublished) have examined some nutritional effects on *post-partum* anoestrus in Dorset × Merino ewes. In a control group grazing autumn-winter pasture (Mediterranean-type environment north of Adelaide) at high stocking rate, 50% of ewes exhibited first *post-partum* oestrus over a period from 40 to 120 days after lambing with a modal value of 68 days. Of the parallel group of ewes given 500g lupins/head.day, 85% exhibited oestrus, scattered over a period of 45-90 days with a modal value of 60 days. Nutrition during only 10 days (at 20-30 days post-partum) was as effective as extended feeding over the 30 days *post-partum*, and more effective than sustained feeding. In a separate experiment, faba leans had no effect, but the lupins gave a similar response. The sudden withdrawal of lupins may play a role i.e. a *decrease* in inputs of nutrients following a sudden increase may initiate signals for re-entry into reproductive activity. The observations need further examination, but the mechanism is likely to involve an endocrine-nutrition interaction which needs to be examined. The possible interaction between nutrition, progesterone conditioning and the 'ram effect' needs full investigation (Geytenbeek *et al.* 1984) if early resumption of cyclic activity in the lactating ewe is to be exploited. Weaning of lambs can decrease the interval to first *post-partum* oestrus (Hunter and van Arde 1973; Cognie *et al.* 1975) but nutritional and physiological strategies permitting mating during early or mid lactation would allow more options. Exogenous hormone treatment can be effective, but the mechanisms may be more sophisticated or more complicated than merely over-riding the endocrinological effects of suckling and lactation for a period of time.

## SPECIFIC NUTRITIONAL FACTORS

The discussion has centred so far on energy, protein, gluconeogenesis and body stores of fat and protein but other nutrients are important in specific circumstances. Deficiencies in most of these are readily diagnosed, and treatment is not difficult although cost benefits have not been well worked out. Such deficiencies are though of as all or nothing and usually treatment of flocks in undertaken only if the expense is low. Two recent considerations need to be mentioned:

*Cobalt* The Vitamin $B_{12}$ requirements of the ewe may be satisfied from pastures of cobalt content 0.06ppm. The ewe and foetus may suffer no loss of production, and *post-partum*, the ewe may sustain an adequate level of milk production. However, it is clear from the work of Quirk (1984) in Queensland that the suckled lamb receives insufficient Vitamin $B_{12}$. As the lamb adopts the ruminant habit it remains vulnerable because its requirements are high, it has exhausted any tissue reserves, and neither milk nor pasture provide adequate sources. Formimino-glutamic acid excretion in the urine of lambs is suggested by Quirk (1984) to be the easiest and most senstive index of cobalt insufficiency affecting the most vulnerable section of the flock, and cobalt treatment of the ewe is effective;

*Zinc* Egan (1972) and Masters (1981) in two series of experiments found evidence of response to zinc supplementation. The effect was on returns to service (Egan) and lambs born per ewe mated (Egan, Masters). The effect was not seen in other similar environments, but the possibility of marginal zinc deficiency affecting reproductive performance in areas where permanent pastures on sand or podzolic soils may not be fertilized with zinc must be kept in view. Slow-release zinc devices in the ewe during spring may support earlier oestrus and higher ovulation rates in mid summer, as judged by time of lambing and twinning rate (Egan, in prep.).

## THE RAM

In the male, reproduction is associated with establishment of potential to produce spermatozoa, and sustaining that potential with nutritional inputs to support the single physiological event of serving the ewe in oestrus. Good body condition at commencement of mating and attention to likely Vitamin A and D inadequacies or mineral deficiencies are imperative. Because the numbers of rams are small and their performance is important, there is little excuse for adverse reproductive performance in the flock to arise from nutritional causes.

## CONCLUSIONS

The nutritional considerations in reproductive performance focus heavily on the ewe flock and on the nutrition of the pre-weaned lamb. Quantitative nutrition based on estimates of requirements derived in pen experiments or based on factorial calculations excluding ewe live weight gain or loss and the composition of substances mobilized possibly preclude some nutritional strategies essential in the major Australian sheep production environments. Minimum inputs must be those which permit live weight loss at rates which do not result in acute metabolic problems for ewe and foetus or adversely affect wool quality (tenderness) within the staple length, or do not have carry-over effects on a subsequent performance. Loss of live weight implies an earlier gain in live weight, and cyclically a subsequent live weight gain. The times for gain are concurrent with late pregnancy.

Weight losses in late pregnancy and early lactation can be accepted, but demand the control of rate of fat mobilization in late pregnancy, and early lactation. The carry-over effects of selected paths of gain or loss of live weight are seen as factors influencing the response to increased plane of nutrition or 'surge' feeding of supplements prior to mating.

While ovulation rate can be increased by lupin feeding for example, this depends on the basal state from which the control animals operate. The interplay of nutrients absorbed with capacity to mobilize from labile reserves, and the way in which the endogenous endocrine balances interact to influence key events of follicle development, ovulation, expression of oestrus, mammary gland development and lactation level must be understood before responses to exogenous hormones can be predicted.

## REFERENCES

Agricultural Research Council, 1980. Feeding Requirements of Runiment Livestock. Agricultural Research Council (London).

Alexander, G., 1980. *In* Wodzicka-Tomaszewska, M., Edey, T.N. and Lynch, J.J. (eds) *Reviews in Rural Science, 4.* University of New England Publ. Unit, Armidale, 99-105.

Allden, W.G., 1975. *Nutr. Abst. Rev., 40,* 1167-1184.

Annison, E.F., Gooden, J.M., Hough, G.M. and McDowell, G.H. 1984. This volume, 174-181.

Anon., 1980. Final Report of Sheep Fertility Service, Wagga Wagga, New South Wales Department of Agriculture.

Arnold, G.W., 1970. *In* Phillipson, A.T. (ed) *Physiology of Digestion and Metabolism in the Ruminant*, Oriel Press. Newcastle upon Tyne. 264-276.

Bourke, M.E., 1964. *Proc. Aust. Soc. Anim. Prod.,5*, 129-134.

Braden, A.W.H., 1971. *Aust. J. Exp. Agric. Anim. Husb.,11*, 375-378.

Brien, F.D., Baxter, R.W., Findlay, J.K. and Cumming, I.A., 1976. *Proc. Aust. Soc. Anim. Prod., 11*, 237.

Cahill, L.P. Anderson, G.A. and Davis, I.F., 1984. *Proc. Aust. Soc. Anim. Prod.,15*, 278-281.

Cognie, Y., Hernandex-Barret, M. and Sammande, J., 1975. *Ann. Biol. Anim. Bioch. Biophys.,15*, 329-343.

Coop, I.E.,1966. *J. Agric. Sci. (Camb.),67*, 305-323.

Dove, H., Freer, M., Axelsen, A. and Downes, R.W., 1984. *Proc. Aust. Soc. Anim. Prod.,15*, 329-332.

Dyrmundsen, O.R., 1981. *Livestock Production Science,8*, 55-65.

Edey, T.N., 1968. *Proc. Aust. Soc. Anim. Prod., 7*, 188-191.

Egan, A.R., 1972. *Aust. J. Exp. Agric. Anim. Husb., 12*, 131-135.

Egan, A.R., 1974. *In* Nicholas, D.J.D. and Egan, A.R. (eds). *Trace Elements in Soil-Plant-Animal Systems.* Academic Press, London.

Everett, G.C., 1964. *Nature, 201*, 1341-1343.

Ferguson, B.D., 1982. *Proc. Aust. Soc. Anim. Prod., 14*, 23-34.

File, G.C. 1981. *Wool Technol. Sheep Breed., 29*, 7-11.

Fletcher, I.C., 1971. *Aust. J. Agric. Res., 22*, 321-330.

Fletcher, I.C., 1974. *Proc. Aust. Soc. Anim. Prod., 10*, 261-264.

Foot, J.Z., Cummins, L.J., Spiker, S.A. and Flinn, P.C., 1984. This volume, 187-190.

Geytenbeek, P.E., Oldham, C.M. and Gray, S.J., 1984. *Proc. Aust. Soc. Anim. Prod., 15*, 353-356.

Hall, D.G., 1984. *Proc. Aust. Soc. Anim. Prod., 15*, 66-79.

Hulet, C.V., 1978. US Dept. Agriculture, North Carolina, Reg. Publ. 248.

Hunter, G.L. and Van Aarde, I.M.R., 1973. *J. Reprod. Fert., 32*, 1-8.

Killeen, I.D., 1967. *Aust. J. Exp. Agric. Anim. Husb., 7*, 126-136.

Knight, T.W., Oldham, C.M. and Lindsay, D.R., 1975. *Aust. J. Agric. Res., 25*, 567-571.

Lindsay, D.R.,1976. *Proc. Aust. Soc. Anim. Prod., 11*, 217-224.

McGuirk, B.J., 1982. *In* B.J. McGuirk, *Proc. Aust. Soc. Anim. Prod., 14*, 23-34.

MacRae, J.C. and Egan, A.R., 1983. *Brit. J. Nutr., 49*, 385-393.

Marshall, T., 1978. *In* D.R. Lindsay (ed) *Sheep ferility, Proc. Aust. Soc. Amin. Prod. W.A.* p.21.

Masters, D.G.,1981. *In* J. McC. Howell, J.W. Gawthorn and C.L. White (eds) *Trace Element Metabolism in Man and Animals*, Aust. Acad. Sci., Canberra, 331-336.

Morley, F.W., White, D.H., Kenney P.A. and Davis, I. F., 1978. *Agric. Systems, 3*, 27.

Moule G.R., 1982. *Proc. Aust. Soc. Amin. Prod., 4*, 195-200.

Murtagh, J.J., Gray, S.J., Lindsay, D.R. and Oldham, C.M., 1984. *Proc. Aust. Soc. Anim. Prod., 15*, 490-493.

Parr, R., 1984. M. Agr. Sc. Thesis, University of Melbourne.

Quirk, M., 1984. Master of Agricultural Science Thesis, University of Queensland.

Ralph, I.G., 1984. *Proc. Aust. Soc. Anim. Prod., 15*, 549-552.

Rattray, P.V., and Jagusch, K.T., 1978. *Proc. N.Z. Soc. Anim. Prod., 38*, 121-126.

Reeve, J.L., Kenney, P.S., Baxter, R.W. and Cumming, I.A., 1976. *J. Reprod. Fert., 46*, 519-520.

Robinson, T.J. and Orskov, E.R., 1975. *Wld. Rev. Anim. Prod., 11*, 63-76.

Russell, A.J.F., Maxwell, T.J., Sibbald, A.R. and McDonald, D., 1977. *J. Agric. Sci. (Camb.), 89*, 667-673.

Underwood, E.J. and Shier, F.L., 1941. *Journal of the Department of Agriculture, W.A., 18* (2nd series) 13-16.

White, D.H., Bowman, P.J. and Morley, F.H.W., 1982. *Proc. Aust. Soc. Anim. Prod., 14*, 35-46.

# PARTITIONING OF ENERGY IN YOUNG LACTATING SHEEP

J.Z. Foot, P.G. Heazlewood and K. Joseph, *Department of Agriculture, Pastoral Research Institute P.O. Box 180, Hamilton, Vic. 3300.*

*Summary* The partitioning of nutrient energy in young lactating ewes was studied in an experiment with housed two-year-old ewes of three genotypes: large Merinos (LM), average liveweight 50 kg; small Merinos (SM) 44 kg, and Corriedales (C) 56 kg. The ewes were in moderate condition at lambing (18 to 24% fat in the total body). Five ewes of each genotype were given a metabolisable energy (ME) allowance estimated to support 0.75 l (L treatment) and five were given an allowance estimated to support 1.5 l milk/d (H treatment). In the first five weeks of lactation, there was no significant difference in production of milk and growth of lambs between nutritional treatments. Corriedale lambs from ewes on the L treatment grew at 0.28 kg/d, significantly faster ($P < 0.05$) than other treatments.

Ewes on treatment L lost 5.0 kg liveweight and 3.0 kg body fat (measured by tritiated water dilution) equivalent to 5.3 MJ ME/day of feed. These losses were significantly greater ($P < 0.05$) than those in ewes on treatment H which lost 0.6 kg ($P < 0.01$) liveweight and 1.3 kg fat ($P < 0.05$), equivalent to 2.3 MJ ME/day.

Ewes on both treatments had growth rates of wool which were 25 to 40% lower during lactation than they were prior to lambing. The decline in growth rate of wool did not differ significantly between nutritional treatments, but there was a significantly greater decline, 41%, in wool growth of LM than of C, 25%, on the H treatment ($P < 0.05$).

Further research is needed on partitioning of nutrients in young ewes which are poorer in body condition as with less body energy to mobilise for maintenance of lactation they would be more vulnerable to nutritional stress.

## INTRODUCTION

Ewes lambing for the first time at two years of age or less have rarely reached mature body size at parturition. During late pregnancy and early lactation nutrient intake of these ewes may be insufficient to allow maternal body growth to take place at the same time as foetal growth, milk production (hence lamb growth) and wool production.

Reproductive performance is usually lower in two-year-old ewes than it is in older ewes; frequently a major contributing factor to this is the lower survival rate in lambs born to the younger ewes. For instance this autumn (1984) mortality of single-born lambs from our Merino flock in south-western Victoria was 17% in those from two-year-old ewes and 3% in those from mature ewes. The higher mortality has been related to the lower birthweights of lambs from young ewes (Dalton *et al.* 1980), but this is not invariably the case and it can be caused by inadequate production of milk (J.Z. Foot, unpublished data). Poor nutrition probably has a greater adverse effect on the reproductive performance of two-year-old ewes than on that of older ewes. However there is little information of the way in which nutrient energy is partitioned in young ewes, and of responses in their production to suboptimal nutrition.

In this experiment we concentrated on the five weeks post-lambing in two-year-old ewes with their first lambs. Changes in liveweight and body composition, yield and composition of milk, growth rates of lambs and of wool were measured in ewes of three genotypes given two levels of nutrition.

## MATERIALS AND METHODS

### Management and experimental design

Thirty 2-year-old ewes, all due to lamb within a period of a few days were brought into individual pens in an animal house about ten days before lambing in September. Half the sheep received amounts of feed calculated to provide sufficient metabolizable energy (ME) to cover the requirements for maintenance plus 0.75 l milk/d (treatment L), the others received maintenance plus an allowance for 1.5 l milk (treatment H). Amounts of feed were calculated allowing 0.4 MJ ME/$kg^{0.75}$ daily for maintenance, based on ewe liveweight one to four days after lambing, and 7.3 MJ ME/l milk (Robinson 1978).

### Animals and feed

The ewes were of three genotypes: small fine Western District (Victoria) Merinos (SM), large strong wool South Australian Merinos (LM) and Corriedales (C). Each treatment group consisted of five animals from each genotype. The diets comprised fixed amounts (fresh weight) of lupin grain (0.6 kg), lucerne hay (0.4 kg) and grass hay (0.1 kg) with adjustments for the liveweights of individual animals being made by varying the amount of oaten grain given to each ewe.

### Procedures and measurements

The ewes were weighed weekly and subjective assessments of body condition (CS) were made (Russel *et al.* 1969). Indirect estimates of body fat were made three to four days after lambing and again about five weeks later. Tritiated water space and body weight were used to estimate fat (Foot *et al.* 1979). The gross energy of body fat was taken as 40 MJ (Agricultural Research Council 1980). The potential value of mobilised fat as a substitute for dietary ME was calculated (France *et al.* 1983).

Output of milk was measured once a week, using 5 i.u. of oxytocin and machine milking. Milk was analysed for total dry matter, protein and fat. The energy content of milk produced was calculated from the fat content and stage of lactation (Brett *et al.* 1972).

Lambs were weighed at birth and weekly thereafter. Growth rates of lambs were calculated for the period between the first and final estimates of body fat of ewes.

Growth of wool in the ewes was measured using dyebands. The rates of growth from shearing in August the previous year until September, just before lambing, were compared with the rates of growth then on until shearing in late November. The latter period spanned nine to eleven weeks of lactation.

## RESULTS

The mean ME content of feed given to the ewes was 18.5 MJ/d for those on treatment H and 13.0 MJ/d for those on treatment L. Birthweights of lambs and liveweights and estimated body weights of body fat for the three genotypes one to four days after lambing are shown in Table 1. Liveweights of ewes on the two nutritional treatments were similar at this stage, averaging 49.5 kg for treatment H and 50.6 kg for treatment L. The LM and C ewes were approximately 10 kg lighter than four-year-old ewes at CS 3.5 of the same genotypes, the SM were approximately 6 kg lighter than four-year-old ewes. By the fifth and sixth week of lactation liveweights of the ewes in the treatment groups differed significantly ($P < 0.01$) and were $48.7 \pm 6.59$ for the H and $45.6 \pm 6.31$ (mean $\pm$ SD) for the L group.

**TABLE 1. Birthweights of lambs and liveweight and body fat of ewes after lambing (mean ± SD).**

| | Birthweight of lambs (kg) | Liveweight (kg) | Condition score | Body fat (kg) |
|---|---|---|---|---|
| Large Merinos | $4.7 \pm 0.63$ | $50.0 \pm 4.70$ | 2.9 | $9.1 \pm 2.05$ |
| Small Merinos | $4.2 \pm 0.70$ | $44.3 \pm 3.82$ | 2.8 | $8.2 \pm 1.83$ |
| Corriedales(1) | $5.1 \pm 0.53$ | $55.7 \pm 5.27$ | 3.2 | $13.3 \pm 2.84$ |
| LSD ($P < 0.01$) | 0.8 | 4.7 | | 2.5 |

(1) Includes one pair of twins, all other lambs single-born.

Changes in liveweight, body fat and CS during the first five weeks of lactation are shown in Table 2, together with results for yields and fat content of milk, and rates of growth of lambs and greasy fleece.

**TABLE 2. Effects of breed and of feed allowance in early lactation (c. 5 weeks) on changes in liveweight, body fat, condition scores and wool growth in ewes of three genotypes, and on their yield of milk energy and the growth rates of their lambs.**

| Nutritional treatment | H | | | L | | |
|---|---|---|---|---|---|---|
| Breed | LM | SM | C(1) | LM | SM | C |
| *Ewe* | | | | | | |
| Change in liveweight (g/d) | −85[ab(2)] | +33[a] | −34[ab] | −182[c] | −166[c] | −157[bc] |
| Change in CS | −0.4[a] | −0.4[a] | −0.3[a] | −1.1[b] | −0.5[a] | −0.6[a] |
| Change in body fat (g/d) | −25[a] | −52[ab] | −55[ab] | −126[c] | −86[bc] | −88[bc] |
| ME equivalent of fat loss (MJ/d) | 1.3[a] | 2.8[ab] | 2.9[ab] | 6.7[c] | 4.6[bc] | 4.7[bc] |
| *Milk* | | | | | | |
| Yield (kg/d) | 1.6 | 1.7 | 1.5 | 1.4 | 1.4 | 1.7 |
| Fat (%) | 6.6 | 6.0 | 7.6 | 7.2 | 6.9 | 8.4 |
| Energy in milk (MJ/d) | 6.9 | 7.0 | 7.3 | 6.6 | 6.3 | 8.1 |
| *Lamb* | | | | | | |
| Rate of growth (kg/d) | 0.22[a] | 0.22[a] | 0.23[a] | 0.23[a] | 0.21[a] | 0.28[b] |
| *Growth of greasy wool* | | | | | | |
| During lactation (g/d) | 9.0[a] | 8.4[a] | 11.7[b] | 9.7[ab] | 7.3[a] | 7.8[a] |
| Decline in lactation (%) | 41[a] | 28[ab] | 25[b] | 38[ab] | 30[ab] | 38[ab] |

(1) Results from one Corriedale ewe with twin lambs were omitted from the analysis.
(2) Means followed by different superscripts differ significantly ($P < 0.05$) (Duncan's new multiple range test).

There were no significant effects of nutritional treatments on milk, growth rates of lambs or growth of wool. However rates of loss of liveweight, body fat and CS were significantly greater in ewes on treatment L than in those on treatment H. Total losses in liveweight (5.0 kg) and in body fat (3.0 kg) in ewes on treatment L were significantly greater ($P < 0.01$) than they were in ewes on treatment H (0.06 kg liveweight and 1.3 kg fat).

The genotype of the ewe had significant effects in a few cases. LM ewes on treatment L tended to experience greater losses of liveweight, body fat and CS than the other two genotypes; for CS this difference was significant ($P < 0.05$). The non-significantly higher energy output in milk from C ewes on treatment L resulted in their lambs growing significantly faster ($P < 0.05$) than all the others. The depression in the rate of growth of wool in the 9 to 11-week period of lactation, as a proportion of the rate of growth between the previous shearing and lambing, was less in C ewes on treatment H than LM ewes on treatment H ($P < 0.05$).

## DISCUSSION

Some of the depression in growth of wool during lactation may have been due to the effects of housing and change in diet. However the rate of wool growth remained the same when ewes were returned to pasture as it had been when they were in the animal house. Obviously dry sheep should have been included to allow the effects of changes in nutrition and management to be distinguished from those of lactation.

The ewes in this experiment were not fully grown at the start of lactation. Nevertheless their body reserves and the protein content of their diet were sufficient to allow body fat to be mobilised. This prevented the undernutrition imposed by the lower ME intake of ewes in treatment L from reducing milk yield in the first five weeks of lactation. Many ewes lambing for the first time are much leaner than these were. Therefore future research is necessary to see if ewes with lower body reserves would be able to respond in the same way as the ewes in this experiment to the demands of lactation when availability of feed is restricted.

## REFERENCES

Agricultural Research Council, 1980. *The Nutrient Requirements of Ruminant Livestock*. Commonwealth Agricultural Bureaux, Slough.

Brett, D.J., Corbett, J.L. and Inskip, M.W., 1972. *Proc. Aust. Soc. Anim. Prod., 9*, 286-291.

Dalton, D.C., Knight, T.W. and Johnson, D.L., 1980. *N.Z. J. Agric. Res. 23*, 167-173.

Foot, J.Z., Skedd, E. and Macfarlane, D.N., 1979. *J. Agric. Sci., Camb., 92*, 69-81.

France, J., Neal, H.D. St. C., Probert, D.W. and Pollett, G.E., 1983. *Agric. Systems, 10*, 213-244.

Robinson, J.J., 1978. *In* Boyazoglu, J.G. and Treacher, T.T. (eds), *Milk Production in the ewe*. E.A.A.P. Publication No. 23, 53-65.

Russel, A.J.F., Doney, J.M. and Gunn, R.G., 1969. *J. Agric. Sci., Camb. 72*, 451-454.

# LONG-TERM EFFECTS OF NUTRITION OF EWE LAMBS IN THE NEONATAL PERIOD

A.H. Williams, *Department of Agriculture, Animal Research Institute, Werribee*, Victoria 3030.

*Summary* The long-term consequences on 136 ewe offspring of reducing the feed intake of their dams over 8 weeks beginning either 2 weeks before or 6 weeks after lambing was studied. Growth was retarded in lambs born to underfed mothers. These lambs took up to three years to reach liveweights similar to lambs born of well-fed mothers and reached puberty on average 35 days later and 6 kg lighter than control lambs. The ovulation rates of lambs born to underfed dams were depressed over the first 3 years of life. Clean wool production was only affected at their first shearing. It is concluded that good nutrition during the period 2 weeks before to 14 weeks after lambing is important, especially to ensure development of the reproductive potential in ewe lambs.

## INTRODUCTION

It is possible that some management practices when applied to very young ewe lambs might adversely affect their reproductive performance when they are adults. In this context, the feeding of ewe lambs which are born either during, or immediately before a drought is an important area requiring investigation. Previous studies have already reported that underfeeding of ewe lambs results in a decreased incidence of twinning throughout the reproductive life-span of the adult ewe (Reardon and Lambourne 1966; Gunn 1977). The present experiments studied the effects of nutrition during the period from birth to weaning on growth, productivity and reproductive potential of lambs over the first three years of life.

## MATERIALS AND METHODS

A flock of 406 Border Leicester × Merino ewes mated to Dorset Horn rams were ranked on liveweight and allocated at random from within liveweight categories into 4 equal sub-flocks. The experiment was a 2 × 2 factorial design. There were two levels of nutrition: 1. Low protein-low energy (LL, pasture hay designed to maintain 35 kg fleece free ewe liveweight). 2. High portein-high energy (HH, pasture *ad. lib.*, + 500 g lupin grain/ewe/day + cereal grain *ad. lib.*). These rations were offered for 8 weeks beginning either two weeks before (Period 1; 15/4/80 to Period 2; 10/6/80 period) or six weeks after mean expected lambing date (Period 2; 9/6/80 to 4/8/80 period). At other times the ewes were offered pasture *ab lib.* + oats at twice the level designed to maintain 35 kg fleece free ewe liveweight.

The ewe lambs (n = 136) born as singles to these ewes were weaned on 4/8/80 (at mean age 14 weeks) and were grazed on adequate levels of pastures to achieve liveweight gains thereafter. The liveweights of the young ewes were measured at regular intervals throughout the period of observation. Wool growth, clean wool yield and fibre diameter were measured at their 1st and 2nd shearing. Wool growth alone was measured at the 3rd shearing. The young ewes were run with vasectomized rams from 3/1/81 (8-9 months of age) and oestrus was recorded. Once the majority of the ewe weaners had shown first oestrus (mid autumn) the ewes were examined by laparoscopy to determine ovulation rate over 3 successive oestrous cycles. A sample of 15 ewes from each group was then given 800 i.u. PMSG (Folligon, Intervet) on day 12 of the next cycle. Ovulation rate at the next oestrus was recorded. The observations of ovulation rate were repeated in 1982 and 1983. PMSG response was repeated in 1983.

## RESULTS

The growth rate of the ewe lambs was restricted during the period of suboptimal nutrition. The effect was most severe in the lambs that were underfed in period 2; these lambs did not gain any weight during the treatment period (Table 1). Regardless of whether the dams were restricted at period 1 or period 2, the ewe lambs gained liveweight at similar rates to those from well-fed dams when the dams were returned to the standard level of nutrition. Thus ewe lambs in the LL treatment remained significantly ($P < 0.01$) lighter than ewe lambs in the HH treatment throughout the three year observation period (Table 1).

The maiden ewes which underwent LL nutrition treatments at periods 1 and 2 exhibited their first oestrus 4.4 weeks later ($P < 0.01$) and 6 kg lighter than the HH maiden ewes. Ovulation rate in LL ewes was significantly ($P < 0.05$) lower than in HH ewes up to 3 years of age. Similarly response to PMSG was significantly ($P < 0.05$) lower in LL than in HH ewes. Greasy fleece weight and fibre diameter did not differ between groups ($P > 0.05$) at any age. However, significantly ($P < 0.05$) less clean wool was cut from LL lambs than HH lambs at one year old.

## DISCUSSION

Undernutrition of ewe lambs for 8 week periods commencing either two weeks before birth or six weeks after birth resulted in lower lamb liveweights at weaning compared with lambs born of well fed dams. This effect persisted up to three years of age. Underfeeding early in life also delayed onset of puberty and depressed ovulation rates during the first 3 breeding seasons. The results are consistent with the observation of Readon and Lambourne (1966) and Gunn (1977) that the fecundity of adult ewes is influenced by their nutritional history through early life. These results are also consistent with those of Langlands *et al.* (1984b) who showed that grazing sheep at different stocking rates from birth to 15 months of age affected age and liveweight at first oestrus and fecundity as adults.

**TABLE 1. Effect of nutrition of dams on the growth of daughters' onset of puberty, ovulation rate, response to PMSG and clean fleece production.**

| Nutrition | | LL | | | HH | | |
|---|---|---|---|---|---|---|---|
| Period | | 1 | 2 | Mean | 1 | 2 | Mean |
| *Before Puberty* | | | | | | | |
| Liveweight (kg) | | | | | | | |
| Age (weeks) | 0 | 3.6 | 3.6 | 3.6 | 3.6 | 3.6 | 3.6 |
| | 7 | 8.2 | 13.5 | 11.4 | 16.1 | 13.2 | 14.6 |
| | 14 | 13.7 | 13.5 | 13.6 | 22.6 | 22.5 | 22.5 |
| | 34 | 23.3 | 23.0 | 23.1 | 29.4 | 29.4 | 29.4 |
| *Puberty* | | | | | | | |
| Age (weeks) | | 49.9 | 50.4 | 50.1 | 45.7 | 45.7 | 45.7 |
| Lwt (kg) | | 34.6 | 35.3 | 35.0 | 41.0 | 40.8 | 40.9 |
| *After Puberty* | | | | | | | |
| Age (weeks) | | | | | | | |
| 54-58 | OR | 1.01 | 1.00 | 1.00 | 1.27 | 1.14 | 1.21 |
| | LWT | 35.3 | 35.4 | 35.3 | 41.5 | 41.0 | 41.3 |
| 97-108 | OR | 1.45 | 1.49 | 1.47 | 1.66 | 1.61 | 1.64 |
| | LWT | 54.7 | 55.4 | 55.1 | 59.1 | 59.6 | 59.3 |
| 147-154 | OR | 1.46 | 1.37 | 1.41 | 1.53 | 1.54 | 1.54 |
| | LWT | 59.0 | 59.4 | 59.3 | 61.9 | 62.2 | 62.0 |
| *Ovulation rate followng PMSG (800 i.u.)* | | | | | | | |
| 60 weeks | | 2.58 | 2.00 | 2.29 | 2.89 | 3.40 | 3.15 |
| 108 weeks | | 1.71 | 2.33 | 2.03 | 2.94 | 2.53 | 2.80 |
| *Clean Fleece Weight (kg)* | | | | | | | |
| 75 weeks | | 2.20 | 2.25 | 2.23 | 2.38 | 2.41 | 2.39 |
| 123 weeks | | 2.40 | 2.41 | 2.41 | 2.46 | 2.53 | 2.49 |

The increases of 3.4, 4.0 and 4.8% in ovulation rate per kg liveweight between the LL and HH groups at 1, 2 and 3 years of age respectively are higher than the 2.0-2.5% quoted by Morley *et al.* (1978) for BL × Merino ewes. This indicates that the early nutrition of these ewes may have affected ovulation rate by a process which acts in addition to the liveweight effect on ovulation rate. The long term influence of poor nutrition during the early post natal period upon ovarian function perhaps arises from an effect on ovarian follicular pools since the first wave of antral follicular growth takes place during the early post natal period (Kennedy *et al.* 1974; Trounson *et al.* 1974).

The time taken to compensate for a growth handicap may be influenced by the stage of growth at which nutritional deprivation occurs. Allden (1970, 1979) suggested that sheep underfed after six months of age rapidly overcome growth impairments but those restricted at an earlier age recovered more slowly.

In the present study, wool production of the ewes as adults was only affected during their first year by the early undernutrition. This contrasts with the work of Schinckel and Short (1961), who imposed severe nutritional deprivation but agrees with data from other work on less severe stress (Allden 1970; Langlands *et al.* 1984a).

This study supports the conclusion that early growth pattern can affect adult body size, and that ovulation rate can be influenced by the pattern of early growth.

REFERENCES

Allden, W.G., 1970. *Nutr. Abstr. Rev., 40*, 1167-1184.
Allden, W.G., 1979. *Aust. J. Agric. Res., 30*, 939-948.
Gunn, R.G., 1977. *Anim. Prod., 25*, 155-164.
Kennedy, J.P., Worthington, C.A. and Cole, E.R., 1974. *J. Reprod. Fert., 36*, 275-282.
Langlands, J.P., Donald, G.E. and Paull, D.R., 1984a. *Aust. J. Exp. Agric. Anim. Husb., 24*, 34-46.
Langlands, J.P., Donald, G.E. and Paull, D.R., 1984b. *Aust. J. Exp. Agric. Anim. Husb., 24*, 47-56.
Morley, F.H.W., White, D.H., Kenney, P.A. and Davis, I.F., 1978. *Agricultural Systems, 3*, 27-45.
Reardon, T.F. and Lambourne, L.J., 1966. *Proc. Aust. Soc. Anim. Prod., 6*, 106-108.
Schinckel, P.G. and Short, B.F., 1961. *Aust. J. Agric. Res., 12*, 176-202.
Trounson, A.O., Chamley, W.A., Kennedy, J.P. and Tassell, R., 1974. *Aust. J. Biol. Sci., 27*, 293-299.

# THE MINIMUM PERIOD OF INTAKE OF LUPIN GRAIN REQUIRED BY EWES TO INCREASE THEIR OVULATION RATE WHEN GRAZING DRY SUMMER PASTURE

C.M. Oldham and D.R. Lindsay, *School of Agriculture (Animal Science), University of Western Australia, Nedlands, Western Australia 6009.*

*Summary* An experiment on a flock of 1,895 adult Merino ewes grazing dry summer pasture showed that an average intake of 750 grams of lupin grain/head/day for only six days was sufficient to stimulate an extra 20 to 30 corpora lutea/100 ewes ovulating compared with unsupplemented flockmates. The intake of lupins by the supplemented ewes did not induce a measurable difference in the live weight between ewes in the two flocks.

## INTRODUCTION

Relatively short term nutritive treatments can change the reproductive performance of Merino ewes in south Western Australia (Knight *et al.* 1975; Lightfoot and Marshall 1975). Cereal grain before and during joining did not improve the reproductive performance of ewes (Marshall 1974) but whole lupin grain fed to ewes as a supplement just before and during joining increased their ovulation rate. In many of the experiments described by the authors above, inceases in ovulation rate were not associated with complementary differences or increases in the live weight of the ewes. This finding appears to be at odds with the conclusion of Morley *et al.* (1978) who, after reviewing the results of a large number of published experiments concerning nutrition and ovulation rate, concluded: 'Since ovulation rate was much more closely related to live weight than to live weight change over a wide range of genotypes, the ovarian response to the current plane of nutrition was less than to the *body reserve status* of the ewe. This suggests that the value of *flushing* ewes over a brief period before joining is unlikely to improve lambing percentages greatly, unless ewes are initially in poor condition'. However it was hypothesised by Knight *et al.* (1975) that the increased ovulation rate of ewes in response to a supplement high in crude protein in south Western Australia was a direct function of the very low crude protein content of the pasture available to ewes in spring and summer.

In sows and gilts it has been shown that increased nutrient intake in the last six days before oestrus, can induce an increase in their ovulation rate (Zimmerman *et al.* 1960). In sheep, metabolic rate responds rapidly to changes in nitrogen intake (Egan 1964). We reasoned that the speed of the response in ovulation rate of ewes grazing a base diet low in crude protein to increased protein intake may parallel that observed in sows and gilts. This paper describes an experiment designed to test the hypothesis that the ovulation rate of Merino ewes grazing pasture low in crude protein would increase within the space of one oestrous cycle if their diet were supplemented with lupins; that is, before correlated increases in their body weight are measurable. Ovulation rate may respond to specific nutrition before live weight but prolonged supplementation will eventually result in live weight gain.

## MATERIALS AND METHODS

### Animals

The experiment was done on a uniform flock of 1,895 three and a half year old Merino ewes which were in 'forward store' to 'fat' condition and owned by W. and P. Emmott of Dowerin, 218 km NE of Perth, Western Australia.

### Supplementation

Supplemented ewes received whole lupin grain at the rate of 750 g head/day from the day of joining until 24 hours after they were seen to have been marked by harnessed entire rams.

### Ovulation rate

The number of ovulations per ewe was assessed by counting the number of corpora lutea (CL) on their ovaries at laparoscopy (Oldham *et al.* 1976) approximately six days after the ewes were first marked by rams. When less than 40 ewes were marked per nutrition treatment per day, the ovaries of all ewes were examined. When more than 40 ewes were marked the ovaries of a random sample of approximately 40 ewes were examined.

### Experimental Procedure

In an attempt to ensure random oestrous activity among ewes during the experiment vasectomised rams were run with the flock between 20th December 1975 and 31st January 1976. To train ewes to eat lupins during the experiment they were fed 250 g of lupin/head/day on the ground for seven days beginning six weeks before the experiment. At the end of this time all ewes actively completed for the grain when it was fed. On 31st January the flock was serially drafted into two and ear-tagged:

flock 1 — 927 ewes with white tags to receive the lupin supplement,
flock 2 — 968 ewes with orange tags to act as unsupplemented controls.

Each flock was joined with 2.4 per cent of harnessed mature Merino rams in a 250 ha paddock. Both paddocks were adjacent to the same complex of small holding paddocks and sheep handling yards. The feed available in each was dry subterranean clover and annual ryegrass. The amount of feed available was judged by the farmer to be adequate to carry the ewes, without supplementation, until lambing.

Ewes in flock 1 were fed their first full lupin supplement at approximately 1000 h on 31st January and thereafter at the same time daily for 14 days. The supplement was fed in a long trail on hard, bare ground in one of the holding yards. During the first three days the ewes were not released from the holding yard until the last of the lupin was eaten (2-3 hours). Thereafter the gate was left open after feeding and the ewes allowed to drift back to the paddock. All ewes readily ate the lupin supplement. No lupin supplement was fed on days 15, 16 and 17 of the experiment. At 0400 h each morning the two flocks were mustered and the ewes marked by the rams in the previous 24 hours were drafted off, weighed and branded with a number (1-17) appropriate to the day of the experiment. The ewes were then returned to their original flocks. Flock 1 was then fed the lupin supplement and flock 2 returned to its paddock.

The ewes branded the previous day were drafted into a third flock (flock 3) to await laparoscopic examination of their ovaries five to six days later. Ewes destined for laparoscopy on any particular day were drafted from flock 3 in the morning, held in the yards off feed and water during the day and examined that night between 1700 and 2400 h. After laparoscopy all ewes were bulked into a fourth flock.

## RESULTS

The advantage in ovulation rate in favour of the ewes supplemented with lupin grain, on successive days after the start of feeding, is shown in Figure 1. The first significant difference in the proportion of ewes with twin ovulations was observed on the sixth day after feeding commenced ($x^2 = 13.6$, $P < 0.001$). A difference ($P < 0.05$) was maintained each day in favour of the ewes receiving the lupin supplement as long as supplementation continued. On day 3 the 16% difference in ovulation rate was not significant. On that day only 18 ewes were marked in the supplemented flock and 30 in the control flock. On day 4 comparable figures were 29 vs 23 ewes and on day 17, 22 vs 33 ewes. On all other days of the experiment the ovaries of more than 30 ewes from each flock were examined to determine ovulation rate. The number of ewes remaining unmarked by the rams did not permit the experiment to be continued past day 17 but ewes reverted to an ovulation rate similar to that of the control ewes within three days of withdrawal of the lupin supplement.

The mean live weight, at oestrus, of the ewes laparoscoped each day is also shown in Figure 1. Differences in ovulation rate in favour of the ewes fed lupin were not related to differences in live weight between the two groups of ewes. In fact, the intake of lupins by the supplemented ewes did not induce a significant measurable difference in their live weight over the ewes in the control group. The 261 ewes with twin ovulations were heavier by 2 kg than the 972 ewes with single ovulations ($55.2 \pm 1.21$ vs $53.2 \pm 0.85$; mean kg $\pm$ SE). However, the relationship was independent of lupin supplementation. The mean live weight of the 432 single ovulating ewes fed lupin was 53.5 kg vs 52.9 kg for the 540 single-ovulating control ewes. Similarly, the mean live weight of the 181 twin ovulating ewes fed lupin was 55.1 kg vs 55.5 kg for the 80 twin-ovulating control ewes.

## DISCUSSION

The intake of 750 g of lupin grain/head by Merino ewes in 'forward store' to 'fat' condition, grazing a dry pasture of sub clover and annual rye grass, induced a significant increase in their ovulation rate when feeding commenced six days before ovulation, or around day 10 of their oestrous cycle. Since this experiment was completed our result has been confirmed by Lightfoot *et al.* (1976) who reported that the first significant difference in the incidence of twin births in a flock of Merino ewes fed a lupin supplement during joining compared with unsupplemented flockmates was seen in ewes mated on day 7 of joining. In most published experiments designed to elucidate the influence of and/or interactions between supplementary nutrition, live weight and ovulation rate of ewes, the ovulation rate of the ewes was not assessed before, and the effects of treatments were not evaluated until three to six weeks after nutritional treatments began. In the light of the speed of the response of ovulation rate to lupin intake of our ewes the results of such experiments are extremely difficult to evaluate as they inevitably confound possible effects associated with specific nutrition with those associated with the live weight of the ewes. The results support the hypothesis of Knight *et al.* (1975) that the quality (crude protein content) of their diet may be an overriding factor suppressing the ovulation rate of ewes in south Western Australia from more closely approaching their inherent potential. This hypothesis is further supported by the fact that Knight (1979) reproduced the response in ovulation rate to lupin intake in both fat and thin Romney ewes in New Zealand where the lupin and control diets contained 21 per cent and 11 per cent crude protein respectively.

The apparent failure of the ewes fed lupins to gain live weight during the experiment was probably due to several factors. They were initially in good condition they may have substituted lupin for a large part of their intake of pasture, and they were fed the supplement for only 14 days.

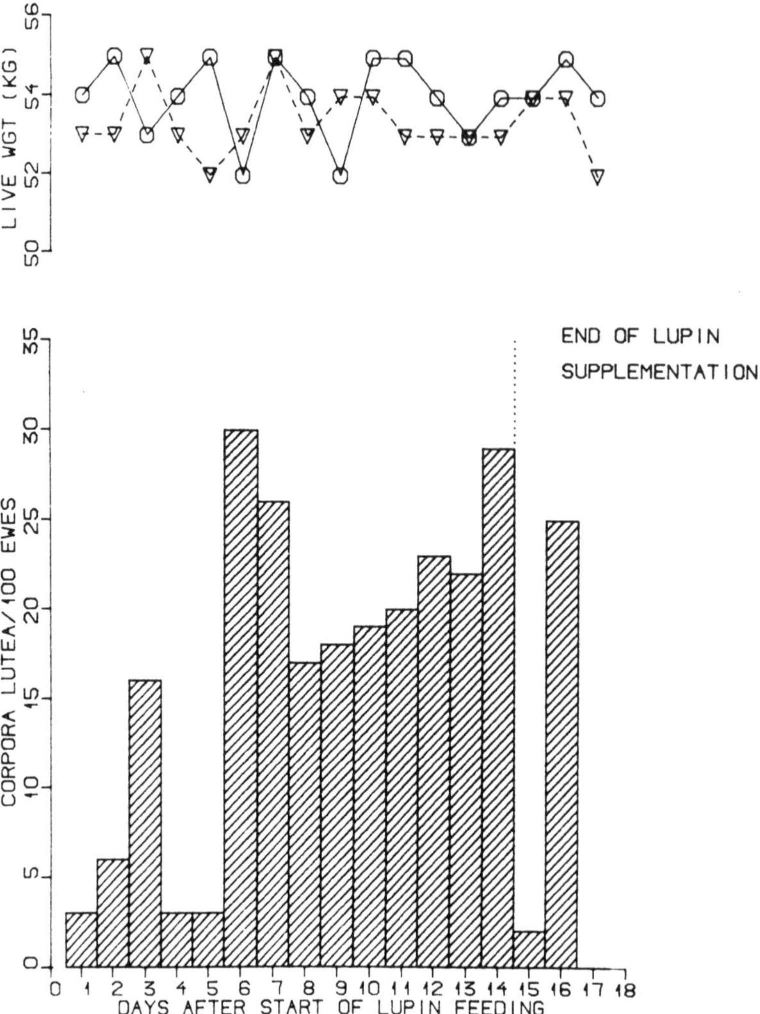


**FIGURE 1. The advantage in favour of ewes supplemented with lupin grain over unsupplemented flockmates (histograms, in CL/100 ewes) on successive days after feeding commenced. Also shown is the mean live weight (kg) at oestrus of the ewes whose ovaries were observed. (o, + lupins; △, controls).**

ACKNOWLEDGEMENTS

This work was supported by the Australian Mean Research Committee. We wish to thank W and P Emmott for the use of their ewes, and P Gherardi, P Chidzey and S Abbott for technical assistance.

REFERENCES

Egan, A.R., 1964. PhD Thesis, University of Western Australia.

Knight, T.W., 1979. *Proc. Aust. Soc. Reprod. Biol. 11* (Abstr. 42).

Knight, T.W., Oldham, C.M. and Lindsay, D.R., 1975. *Aust J. Agric. Res., 26*, 567-575.

Lightfoot, R.J. and Marshall, T., 1975. *Aust. Soc. Reprod. Biol., 7* (Abstr. 36).

Lightfoot, R.J., Marshall, T. and Croker, K.P., 1976. *Proc. Aust. Soc. Anim. Prod., 11*, 5.

Marshall, T., 1974. *In* Lindsay, D.R. (ed) *Sheep Fertility. Proceedings of a Seminar*, University of Western Australia, 31-36.

Morley, F.H.W., White, D.H., Kenney, P.A. and Davis, I.F., 1978. *Agricultural Systems, 3*, 27-45.

Oldham, C.M., Knight, T.W. and Lindsay, D.R., 1976. *Aust. J. Exp. Agric. Anim. Husb., 16*, 24-27.

Zimmerman, D.R., Spies, H.G., Self, H.L. and Casida, L.E., 1960. *J. Anim. Sci., 19*, 295-300.

# OVULATION RATES IN EWES: THE ROLE OF ENERGY-YIELDING SUBSTRATES

E. Teleni, J.B. Rowe and K.P. Croker, *Sheep and Wool Branch, Department of Agriculture, South Perth, 6151, Western Australia.*

*Summary* The effect of intravenous infusion of energy-yielding substrates (glucose and acetate) on ovulation rates (OR) in ewes was studied. A total of 177 mature Merino ewes were divided into three treatment groups of between 50 and 69. Increases in OR (0.39) in response to 564 g lupin seed/day, fed as a supplement for nine days before ovulation, were similar to those achieved (0.32) when a solution of glucose and acetate (providing similar quantities of these metabolites as supplied by the lupin supplement) was infused intravenously for the same period. It is concluded that the increase in OR for ewes fed the lupin seed supplement might be explained principally by the increased supply of energy-yielding substrates.

## INTRODUCTION

The influences of nutritional stimuli on the reproductive performances of ewes are well documented but the effects of dietary constituents, in particular protein and energy, on ovulation rate (OR) are unclear. Studies in Western Australia have shown that supplementary feeding of ewes with sweet lupin seed (Lightfoot and Marshall 1974; Knight *et al.* 1975) can produce dramatic increases in OR. The unique feature of this response is that ewes respond to lupin feeding in a little as six to seven days after supplementation starts (Lightfoot *et al.* 1976; Oldham 1980) and before any measurable change in live weight can be detected. The response appears therefore to be different (Lindsay 1976) from that accounted for in the 'static' and 'dynamic' live weight theories (Coop 1966).

It is not clear what component of lupins is primarily responsible for the increase in OR but it has been suggested (Lindsay 1976) that it is the protein which is present at a high concentration (approximately 30%). Feeding studies in our laboratory (Rowe *et al.* 1983) showed that estimated supplementary metabolizable energy (ME) intake was more strongly related ($r^2 = 0.87$) to OR than was the estimated amount of protein available at the small intestine ($r^2 = 0.40$). We report here results of studies in which we have explored further the hypothesis that the energy-yielding substrates may be equally, or more important, than protein in influencing OR.

## MATERIALS AND METHODS

Oestrus in 177 mature Merino ewes, individually penned and fed, was synchronized using progestagen-impregnated sponges (Repromap, Upjohn, NSW) inserted in the vagina for 13 days. Three nutritional treatments: (i) basal fed at estimated maintenance level (B); (ii) B + 750 g lupin seed/day; and (iii) B + intravenous infusion of glucose (525.1 mmole/day) and acetate (1122.2 mmole/day) were given for a period of nine days before the estimated time of ovulation. The rates of infusion of the two energy-yielding substrates were based on estimates of the entry rates of glucose and acetate in ewes fed B or B + 750 g lupin seed/day (Teleni *et al.* 1984) and estimates of the effect of different levels of glucose, infused intravenously, on endogenous glucose entry rates in ewes fed B (E. Teleni, unpublished data). All ewes were examined by laparoscopy to determine OR three days after the estimated time of ovulation.

## RESULTS

Table 1 shows that intravenous infusion of glucose + acetate into ewes fed a maintenance diet is as effective in increasing OR as is feeding a daily supplement of 564 g lupins. The mean increases in OR's were 31% and 26% for the lupin and infusion groups respectively, but the OR's in these two groups were not significantly different.

## DISCUSSION

The total energy available to the ewes was higher in the lupin group than in the infused group since the infusion did not account for lipids, $\beta$-hydroxybutyrate or propionate not converted to glucose. In addition, the high entry rate of urea in lupin fed ewes (E. Teleni, unpublished observation) suggests that at least a proportion of this urea is from amino acid deamination for direct energy production and for gluconeogenesis. It appears possible that the nutritional stimulus on OR in lupin feeding may be mediated via glucose or acetate or both. The strong relationship between glucose entry rate and intake of ME (Judson and Leng 1968) is consistent with a suggestion that glucose is an important substrate. Whether the stimulatory effect of glucose and/or acetate on OR is through induction of a hormonal environment conducive to twin ovulation or through a direct, non-hormonal effect is, a matter for speculation.

**TABLE 1. Live weight (kg), ME intake (MJ/ewe/day), protein intake (g/day) and ovulation rates in ewes given a basal maintenance diet (B), B + lupin and B + infusion of glucose + acetate.**

| | B | B + lupin | B + (glucose + acetate) |
|---|---|---|---|
| Number of ewes | 69 | 58 | 50 |
| Live weight (Mean ± SEM)[1] | 53.7 ± 0.8 | 52.7 ± 0.6 | 53.2 ± 0.7 |
| ME intake (Mean ± SEM) | 6.07 ± 0.06 | 12.21 ± 0.20[2] | 7.97 ± 0.84[3] |
| Protein intake (Mean ± SEM) | 80 ± 0.8 | 246 ± 5.0 | 72 ± 1.6 |
| Ewes ovulating: single | 44 | 20 | 21 |
| twin | 16 | 35 | 28 |
| OR/ewe ovulating | 1.25 | 1.64** | 1.57* |

* $P < 0.05$ ** $P < 0.01$. Significantly different from basal, using $\chi^2$.
(1) Live weight on day of laparoscopy examination.
(2) Includes a mean daily lupin intake of 564 g (6.82 MJ).
(3) Includes a mean daily energy infusion of 2.49 MJ.

To utilise this understanding of the nutrient requirements for improved OR under commercial production systems there are two major areas where additional information is required. First, the effects of combinations of nutrient(s) (acetate/glucose) must be examined more closely, together with the development of an hypothesis for the specific mode by which these nutrients affect OR. Second, there is substantial variation in the response in OR of different flocks to supplementary feeding immediately prior to ovulation. The reason for this variability must be understood before any treatment or nutritional management programme could be advocated for on-farm use.

ACKNOWLEDGEMENTS
We thank P.J. Murray, W.R. King and R. Stewart for technical assistance and Messrs M.A. Johns, T.J. Johnson and Dr R.W. Kelly for help with laparoscopies. The Western Australian Royal Agricultural Society provided animal house facilities. The project was supported by the Australian Wool Corporation.

REFERENCES

Coop, I.E., 1966. *J. Agric. Sci., Camb., 67*, 305-323.
Judson, G.J. and Leng, R.A., 1968. *Proc. Aust. Soc. Anim. Prod., 7*, 354-358.
Knight, T.W., Oldham, C.M. and Lindsay, D.R., 1975. *Aust. J. Agric. Res., 26*, 567-575.
Lightfoot, R.J. and Marshall, T., 1974. *J. Agric. (West. Aust.), 15*, 31-32.
Lightfoot, R.J., Marshall, T. and Croker, K.P., 1976. *Proc. Aust. Soc. Anim. Prod., 11*, 5p.
Lindsay, D.R., 1979. *Proc. Aust. Soc. Anim. Prod., 11*, 217-224.
Oldham, C.M., 1980. *PhD. Thesis*, University Western Australia.
Rowe, J., Murray, P., Croker, K. and Johns, M., 1983. *In* Oldham, C.M., Paterson, A.M. and Pearce, D.T. (eds), *Reproduction in Farm Animals. Australian Society of Animal Production, WA Branch*, 133-139.
Teleni, E., Pethick, D.W., Murray, P.J. and King, W.R., 1984. *Proc. Nutr. Soc. Aust., 9*, (in press).

# THE ROLE OF EMBRYO GENE TRANSFER IN SHEEP BREEDING PROGRAMMES

K.A. Ward, J.D. Murray, C.D. Nancarrow, M.P. Boland and R. Sutton, *CSIRO Division of Animal Production, Box 239 PO, Blacktown, N.S.W. 2148.*

## INTRODUCTION

The twentieth century will be recorded as the age when man developed an unprecedented ability to intervene in his environment. The technological advances which have contributed to this are numerous, but, recently, a new member has been added to this list, namely the ability to manipulate biological systems by means of the powerful techniques of recombinant DNA and gene cloning. This methodology is still in its infancy, and its potential to influence the genetic expression of higher organisms remains speculative. Nevertheless, it is possible that its application to domestic livestock may result in substantial phenotypic changes to these animals. These possible changes form the subject of this paper, particularly as they relate to the improvement in the genetic properties of the sheep breeds of economic importance to the Australian farming industry.

The sheep has been one of man's oldest domesticated animals. It probably originated in Europe and Asia in the Pleistocene Era, about one million years ago, and our modern breeds can be traced back to two ancestors, the mouflon of Europe (*O. musimon*) and the Asiatic urial (*O. vignei*) (Rice and Andrews 1951). The genetic gains which have been achieved by selective breeding techniques are obvious when one considers that the original animals possessed a hairy coat, with wool as an underfur. While some breeds still maintain this property, our important modern wool-producing breeds have been selected to reduce hair growth to a minimum, with the result that the woolly areas of the body are almost totally devoid of hair.

The wool-producing potential of sheep is a major factor in their agricultural value but, in addition, the animals provide an important source of meat. Selective breeding of the animals has therefore had to follow two distinct, and at times contrary, selection pathways. The result is that today we have numerous breeds of sheep which vary considerably in their fecundity, wool-producing qualities and meat production. The genetic control exerted over many of these phenotypic traits is very complex, and certainly results from the concerted action of a large number of genes. The movement of these gene complexes between different breeds is at best a slow process by conventional breeding techniques, and in certain situations appears to resist all such attempts. While the difficulties should not be underestimated, it is possible that, as the technology continues to develop, direct gene transfer using recombinant DNA techniques might be able to aid in the movement of some of these gene complexes between breeds, thus providing a useful additional tool for the sheep breeder.

In this paper, we will review the current status of *in vitro* gene transfer in animals, and then discuss the areas where the techniques might prove to be of value for improving the productivity of sheep, and the problems which may be encountered. No attempt will be made to describe the procedures involved in the preparation of recombinant DNA or the isolation and characterization of gene sequences, since these are discussed in great detail in several reviews (Wu 1979; Maniatis *et al.* 1982).

## THE CURRENT STATUS OF GENE TRANSFER IN ANIMALS

The ability of mammalian cells to incorporate foreign DNA into their genome was first demonstrated in cell culture. Initially, adenovirus DNA fragments were mixed with calcium phosphate and the resulting precipitate mixed with cultured cells. These cells were subsequently shown to be transformed by the adenovirus DNA-encoded functions (Graham *et al.* 1973,1974). The mechanism of this integration of foreign DNA is unclear, but the observation indicates that mammalian cells possess a repertoire of DNA processing enzymes which can cut the chromosomal DNA and ligate into it the foreign fragments of genetic information. The potential value of this technique was obvious, and was soon extended by several different laboratories. Wigler *et al.* (1977) used the technique to introduce DNA into cells such that the cells which had incorporated the DNA could be selected on defined growth medium. It was subsequently shown that cells which were able to take up this selectable marker DNA were also able to incorporate other DNA fragments at the same time (Wigler *et al.* 1979). This phenomenon was called co-transformation, and provides a powerful method for the introduction of recombinant DNA into cell cultures. It has been used by a number of laboratories, and the number of selectable marker DNA species has been extended to include the hypoxanthine-guanine phosphoribosyltransferase gene (Graf *et al.* 1979; Willecke *et al.* 1979), the bacterial xanthineguanine phosphoribosyltransferase gene (Mulligan and Berg 1980), the adenine phosphoribosyl transferase gene (Wigler *et al.* 1979), and a mutant dihydrofolatereductase gene (Wigler *et al.* 1980). The demonstration that mammalian cells had the ability to incorporate foreign DNA stimulated efforts to devise a method for the transfer of such DNA to mammalian embryos. The efficiency with which cells incorporate DNA by the calcium phosphate precipitate method varies from one in $10^5$ to one in $10^6$, which precludes the use of the method for embryo transfer studies. However, one possible route was direct microinjection. The fact that mouse embryos could incorporate foreign DNA was suggested by the experiments of Jaenisch and Mintz (1974), where intact SV40 DNA was microinjected into the blastocoel cavities of mouse blastocysts and subsequently shown to be integrated into the mouse embryo

DNA. The potential of microinjection was further enhanced by the demonstration in cell culture that very high transformation efficiencies could be achieved by the direct microinjection of DNA into cells (Anderson 1980; Capecchi *et al.* 1980).

The first successful integration of recombinant DNA into embryos by this technique was achieved by Gordon *et al.* (1980), who reported the incorporation into mice of a recombinant plasmid consisting of the Herpes simplex virus thymidine kinase gene, the SV40 virus origin of replication and the plasmid pBR322. Almost simultaneously, three other laboratories also reported the successful transfer of recombinant DNA to mice. Costantini and Lacy (1981) successfully transferred the rabbit $\beta$-globin gene, and showed that the mice containing the sequences were able to transmit the DNA to their progeny. In addition, the site of integration of the DNA was determined in the case of one animal, where it was found to be entirely located at one chromosomal site on one homologue of chromosome 1. Wagner *et al.* (1981) also successfully transferred the rabbit $\beta$-globin gene, and provided immunological evidence to support the view that the gene was also expressing its $\beta$-globin product in the mice. Wagner *et al.* (1981) reported the transfer of a human $\beta$-globin gene and the Herpes thymidine kinase gene, and presented enzymic analyses which indicated the presence of a functional viral thymidine kinase enzyme present in homogenates of the mice foetuses. Gordon and Ruddle (1981) reported the stable germ line transmission of the Herpes thymidine kinase gene and also of a construct containing the human leukocyte interferon gene. The integration of a second human gene, the human insulin gene, was also reported by Burki and Ullrich (1982).

The creation of transgenic mice fulfilled the promise offered by the new molecular biology techniques, namely that it is possible to transfer DNA between organisms regardless of the biological source of the DNA. In these early papers, the efficiency with which transgenic mice were produced was of the order of 0.2% to 0.5% of the initial embryos injected. This efficiency, however, was sufficient for the technique to be used for studies of the control of gene function. Brinster *et al.* (1981) showed that a fusion plasmid constructed from the Herpes simplex thymidine kinase gene and the promotor/regulatory region of the mouse metallothionein-I gene could be expressed at high level in transgenic mice. The work was extended in a later study (Brinster *et al.* 1982) in which a series of deletion mutants of the metallothionein promotor sequence were made and their effects studied on the expression of the Herpes thymidine kinase gene to which they were fused. These experiments showed that the heavy metal regulation of the thymidine kinase gene, controlled by the metallothionein promotor sequence, seemed to be functioning in a normal manner. However, it became apparent that the regulation of the integrated genes varied from generation to generation. Progeny of mice containing the thymidine kinase/metallothionein in promotor fusion gene frequently expressed the thymidine kinase gene product at a level which was very different from that of the parental animals (Palmiter *et al.* 1982). The explanation for this phenomenon is not understood, but may be the result of a reorganization of the chromatin structure in the germ line of the parental animals, such that the introduced gene is in a different configuration in the progeny. Such a possibility must involve only subtle changes, however, since neither the physical location of the gene on the host chromosome nor the actual nucleotide sequence of the gene seems to change during this period.

The preceding experiments showed that the introduced genes in transgenic mice were functional. The actual alteration of the phenotype of an animal by the introduction of recombinant DNA was shown by Palmiter *et al.* (1982). In this series of experiments, the rat growth hormone gene was fused to the same mouse metallothionein-I regulatory/promotor sequence that had been used in their earlier studies of the Herpes simplex thymidine kinase gene. The introduction of the growth hormone fusion gene to transgenic mice resulted in the animals growing faster and for a longer time than litter mates which did not possess the fusion gene. In different animals, the level of growth hormone circulating in the blood stream varied considerably, and did not show a strong correlation with the rate of growth of the animals, which suggested that the hormone was exerting its effect through some secondary hormone mechanism. Nevertheless, it is clear that by increasing the level of the hormone, growth was strongly stimulated. The commercial implications of the work were fully appreciated by the research workers, who stated:

"The implicit possibility is to use this technology to stimulate rapid growth of commercially valuable animals". More recently, these workers have shown that a similar fusion gene constructed from the mouse metallothionein promotor sequence and a human growth hormone sequence has the same effect on the phenotype of transgenic mice (Palmiter *et al.* 1983). This indicates that the affinity of the rat growth hormone and human growth hormone molecules for mouse receptor molecules is functionally the same, but more importantly, further bolsters the view that functional genes from significantly different sources of DNA can operate in transgenic animals, and follow apparently normal patterns of inheritance.

Since the studies of Palmiter, Brinster and their colleagues, research into the effects of transgenic animals has followed a number of different paths. Some laboratories have continued intensive research into the regulation of the introduced genes in their new genomic environment. For example, Lacy *et al.* (1983) have shown that the foreign DNA sequences can integrate in a large number of different chromosomal sites, and suggest that these sites might impose different levels of expression on the genes because of the influence of neighbouring sequences. McKnight *et al.* (1983) have shown that the chicken transferrin gene is readily expressed in transgenic mice, and obtain maximal expression in the mouse liver, suggesting that tissue specific controls which operate in the chicken (and presumably still carried by the chicken gene) are able to operate in the mouse.

A second experimental use of transgenic mice is to study the effect of the foreign genes on normal development. This is made possible by the fact that the insertion of the foreign DNA into the mouse chromosomes is random, and that sometimes this random insertion disrupts a gene which plays an important role during development (Jaenisch *et al.* 1983; Schnieke *et al.* 1983; Wagner *et al.* 1983). Such animals are usually recognised by the fact that homozygotes are not viable. It is the aim of such work to look for the disruption of genes which are important to normal embryogenesis. A third pathway of research into the formation of transgenic animals in its application to domestic livestock. In this paper, only the application to sheep will be considered in detail. For a wider consideration of the applications of the technology to domestic animal production, the reader is referred to the following recent papers (Jones 1980; Ward 1982; Smith and Hespell 1983; Franklin 1984; Ward *et al.* 1984).

The broad principles of gene transfer which have been established in mice probably also apply to larger animals such as sheep, cattle and pigs. Thus, techniques for the production of single-cell sheep embryos and their re-implantation in surrogate ewes have been established (Nancarrow *et al.* 1984), and these differ from the mouse techniques in detail only. One technical problem which has been encountered in the sheep work is the difficulty experienced in seeing the pronuclei of the embryos, owing to the presence of yolky cytoplasmic inclusions. Since it is essential that the pronuclei be clearly visualized for successful injections to be accomplished, this problem has received considerable attention. Recently techniques have been established which allow the visualization of sheep pronuclei in about 40% of embryos observed. Such embryos appear to have normal viability when treated and then implanted in surrogate ewes without first being subjected to microinjection procedures (J.D. Murray, unpublished data). However, it has yet to be shown that these teated embryos can withstand the trauma of injection and then continue normal development. While it is probable that further technical difficulties will yet be encountered in the application of the mouse techniques to larger animals, it appears at present that no major limitation exists to prevent such transfers from succeeding.

## GENES WHICH MIGHT INFLUENCE DOMESTIC ANIMAL PRODUCTION

The probability that gene transfer in large animals will succeed makes it important to consider carefully the type of genes which should be used in such transfer experiments. These fall into a number of different categories. First, there are genes which are likely to affect whole body growth directly. Second, there are genes which are concerned with selected animal products, such as wool and milk. Finally there are those genes which might influence other complex physiological processes such as reproduction, nutritional performance, or thermal regulation.

The choice of DNA sequence to be used in any of these categories must be carefully evaluated under at least four criteria. Firstly, the phenotypic response desired in the animal must be the result of a small number of genes, and preferably a single gene. Although large multi-gene families can be isolated, it is not desirable to have to consider modifying such gene complexes with foreign regulator or promotor elements and then reintroducing them to recipient embryos. This requirement imposes a restriction on the potential application of this technology, because most phenotypic characteristics which are used as criteria for selection in animal breeding experiments involve multi-gene interactions, most of which are very poorly understood. Secondly, the gene sequence which is to be used in the transfer experiment must be able to be identified and isolated. With current techniques, this means that an identifiable gene product must be available, from which a nucleic acid probe can be constructed to identify the gene sequence itself. Such probes are usually complementary DNA (cDNA) molecules prepared from messenger RNA preparations containing the sequence of interest. Thirdly, the expression of the recombinant DNA gene in its new host genome should be independant of the temporal constrictions of mammalian development. Very little information is yet available concerning the regulation of gene sequences during development, and hence it is not practicable to consider modifying genes to include such temporal developmental control. Finally, the expression of the recombinant DNA gene must not interfere with the operation of genes essential for the survival of the animal. This requirement is not easily evaluated, and to some extent can only be assessed by intensive investigation of the new, genetically-modified animal. However, the introduction of sequences which are designed to alter specific production characteristics, but at the expense of essential cellular cofactor, growth factor or enzyme substrate concentrations, would be unlikely to enhance animal productivity. In view of the severe restrictions which exist concerning the choice of DNA which might be considered for gene transfer experiments, it is clearly of great importance for physiologists and molecular biologists to collaborate closely with one another.

Whole body growth and the production of selected animal products can be modified either by changing the level of a growth factor or hormone critical to the particular growth process, or by removing a nutritional block either by repairing a missing enzyme pathway or by providing the animal with an enzyme system it does not normally possess. The most obvious gene to influence the growth performance of domestic animals is the growth hormone gene, since it has been shown to have this desirable effect on mice (Palmiter *et al.* 1982). This gene has a number of advantages for gene transfer, given our current state of knowledge concerning the control and integration of DNA introduced into a foreign genomic environment. The most important of these is that the gene can operate in all the tissues of the animal and still have its effect. This is not the case with many genes, which must be expressed specifically in one or a few tissues. Secondly, only a single gene is involved in order to obtain the desired phenotypic response. Thirdly, its modification by fusion to a strong regulator or promotor gene sequence, the mouse metallothionein regulator sequence, has been well-documented by Palmiter's laboratory. Currently, the growth hormone sequence is known to be in use in sheep transfer experiments at CSIRO Division of Animal Production, goat and sheep transfers at the University of Pennsylvania (R.L. Brinster, *pers comm*), pig transfers at the University of Adelaide (J. Brooker, *pers comm*) and sheep transfers at the University of Melbourne (M. Brandon, *per comm*). In view of the popularity of this particular sequence both from a scientific and a commercial point of view, it is highly likely that other laboratories are also carrying out similar research.

The approach our group is taking is to isolate the gene sequence for the sheep growth hormone and modify it with a sheep metallothionein in regulatory sequence. The initial stages of this work have been completed, such that a sheep growth hormone complementary DNA (cDNA) probe has recently been characterised and used to isolate a genomic DNA sequence from a library of cloned sheep genomic DNA. Characterisation of this gene is currently proceeding in preparation for its fusion with a sheep metallothionein-I regulatory sequence, which has also recently been isolated and characterised (J. Mercer, unpublished data). After evaluation, it is hoped that this fusion gene will be transferred to sheep.

The effects of such a growth hormone-metallothionein regulator fusion gene on the sheep phenotype cannot be predicted. If the animals follow the pattern established by the mouse experiments, a substantial increase in growth rate might be expected, leading to a larger animal or to an animal which achieves a defined market size in a shorter time. It is also possible that such animals may utilise feedstuffs more efficiently, and their carcass may contain less fat and more muscle. However, substantial physiological testing of such animals will be necessary in order to evaluate their potential. Such testing must include evaluation of reproductive performance, growth rates, nutritional requirements and general behaviour patterns.

The application of gene transfer to increasing the production of a useful animal product such as wool or milk involves the identification of a step in the process which is rate limiting and which can be made non-limiting by the introduction of a modified gene. The rate-limiting step may be the concentration of a hormone, or the availability of a particular substrate, or the activity of an enzymic pathway. In the production of wool, such a rate-limiting process can be identified. Wool growth is limited by the availability of the amino acid cysteine, and when this anino acid is supplied to the sheep in such a way that it avoids degradation in the rumen, wool growth can be increased substantially (Reis and Schinkel 1963; Reis *et al.* 1973 a,b). Cysteine is an essential amino acid for all mammals, including sheep, and hence must be supplied by the diet of the animal. However, it can be synthesised by bacteria and other auxotrophs. The basis of the biosythetic pathway is depicted below, and involves the activation of serine by acetyl-Coenzyme A to form o-acetylserine (reaction 1), which then is reacted with hydrogen sulphide to form cysteine (reaction 2).

$$\text{(1)} \quad \text{serine} + \text{acetyl-CoA} \xrightarrow[\text{serine transacetylase}]{} \text{o-acetylserine} + \text{CoA}$$

$$\text{(2)} \quad \text{o-acetylserine} + H_2S \xrightarrow[\text{o-acetylserine sulphydrylase}]{} \text{cysteine} + \text{acetate}$$

The hypothesis is that the sheep contains hydrogen sulphide in its digestive tract, particularly in the rumen. The sulphide ions are able to diffuse into the cellular epithelia which line the tract so that a source of hydrogen sulphide exists within these cells. The other substrate necessary for the pathway to function is serine, and since this is a non-essential amino acid, it is readily available to the digestive tract epithelia. Thus, if the two enzymes which catalyze the serine to cysteine conversion could be introduced into the digestive tract epithelia, it is possible that cysteine could be synthesised by the sheep.

The aim is to isolate from the bacterium *E.coli* the genes which encode the two enzymes which catalyze this process. The serine transacetylase is encoded by the *E.coli* cysE gene, and o-acetyl serine sulphydrylase is encoded by the *E.coli* cysK gene (Umbarger 1978). The approach that has been adopted for this purpose is to construct transducing phage which contain the genes of interest, and then to isolate from these the relevant pieces of DNA and subclone them into bacterial plasmids for structural and functional characterisation. In the case of the cysE gene, considerable progress has been made. The transducing phage $\lambda$dCysE$^+$ has been constructed (Ingle and Loughlin 1980) and shown to carry the cysE gene. Fragments of this phage have been subcloned into the plasmid pMB9 and two plasmids, pDC1 and pDC2, have been shown to contain the cysE gene from the bacteriophage DNA (J.E. Cronan unpublished data). Characterisation of this DNA is currently in progress. This involves the determination of its nucleotide sequence, in order to look for an open reading frame which might specify the sequence of the gene, and a series of complementation studies with deletion mutants of the plasmid inserts, which, when tested against the cysE$^-$ bacterial strains, are hoped to define the functional gene within the plasmid insert.

Progress with the cysK gene, which encodes the sulphydrylase enzyme, is slower, owing to the lack of suitable bacterial mutants to test for the cysK$^-$ lesion. However, a transducing phage containing the cysA gene of the cysteine pathway has been isolated (R.E. Loughlin unpublished data), and genetic studies have shown that the cysA and cysK genes are very closely linked on the *E. coli* chromosome. It is therefore possible that the cysA-containing bacteriophage also carries the cysK gene, and this will be tested as soon as suitable bacterial mutant strains are constructed.

An advantage associated with this system is that tissue-specific expression of the two genes in the sheep may not be necessary. This is because the entire enzymic pathway can only operate in the presence of hydrogen sulphide, and this is only available in the animal's digestive tract. It is possible that the universal operation of the first enzyme in the pathway, the serine transacetylase, might cause some biochemical disturbance by depleting available levels of the important cellular cofactor, coenzyme A. Furthermore, the energy cost involved in the futile production of the enzymes in all tissues of the animal might prove to negate any positive benefits derived from the increased cysteine supply. However, there is no way of predicting these, and answers will only be known when animals are produced which contain the introduced, functional genes.

Increased wool growth might also be achieved by the introduction of modified wool keratin gene sequences. A substantial body of knowledge is now accumulating which defines the organization of the wool keratin genes within the sheep genome (Ward *et al.* 1982; Powell *et al.* 1983; Rogers 1984), and current work is aimed at the definition of the sequences which control their expression. When such knowledge is available, it would be feasible to modify a keratin gene to allow its overexpression, and to then introduce this gene to a sheep. In order for this to be useful in the production of wool, however the modified gene must be expressed tissue-specifically, because keratin proteins are terminal differentiation products which are lethal to the cells which produce them. Assuming that expression specific to wool follicles can be accomplished, it remains necessary to show that an increase in the level of a specific keratin mRNA would increase the rate of wool production. In the light of current knowledge, this seems unlikely, since it is known that cysteine levels are already rate-limiting. However, if the cysteine block can be overcome by the introduction of the serine-to-cysteine pathway described above, it is then possible that keratin mRNA levels might become rate-limiting. A detailed physiological and biochemical study would need to be undertaken in order to ascertain whether this is correct, utilising similiar experimental protocols to those devised to establish the cysteine limitation which currently exists.

The application of gene transfer to improving some of the more complex physiological functions of sheep, such as reproduction itself, for example, is a formidable task. This is because such functions involve the interaction of a number of genes, some of which have probably not yet been identified in terms of their gene products. However, there are some areas where an impact might be made even with our present knowledge, and these areas present a fascinating environment for the interaction of physiologist and molecular biologist.

It is quite feasible to isolate some of the genes which encode hormones important to the reproductive process, such as follicle stimulating hormone (FSH) or luteinising hormone (LH). However, although their isolation would be a relatively simple task, a far greater problem would be their modification to give a controllable change in reproductive performance. One of the reasons for this is that the levels of the hormones fluctuate in a cyclic fashion during the reproductive cycle itself. If the hormone FSH is used as an example, the peaks of hormonal concentration which occur are due mainly to a release of the hormone from cellular repositories, rather than rapid synthesis of the hormone, although some changes in synthetic rate probably contribute to the changes in hormonal concentration. It would therefore not be of any value of modify the hormone-encoding gene if this resulted in its being produced in a high level which by-passed the cellular storage depots, since this would probably result in a chronic high level of the hormone, with almost certain infertility the result. It is possible that deregulation of the gene might be achieved such that much higher storage levels of the hormone could be achieved, possible resulting in higher peak concentrations of the hormone in the circulation. A more productive approach, however, might be the modification of the structural portion of the gene in such a way that the half-life of the hormone was altered. This would involve a knowledge of the degradative pathway of the hormone, together with sites on the molecule which, if altered, would influence this degradation without altering the affinity of the molecule for its cellular receptors.

An alternative approach to the above is provided by the property of many of the reproductive hormones, namely that their release to the circulation is controlled by a releasing hormone. This provides a second target for modification, although precise information would also be required concerning the appropriate level of releasing hormone. Again, it would not be feasible to allow unrestricted synthesis of the releasing hormone. Hence, any gene modification must provide for precise regulation. One such hormone which appears worthy of serious consideration is inhibin. This hormone has been shown to inhibit the release of FSH and LH (Blanc 1980), although much of the information concerning it has relied on partially-purified aqueous tissue extracts.

Recently, however, a peptide has been isolated from human seminal plasma which has considerable inhibin-like activity when tested against 21-day old mouse pituitaries. A partial amino acid sequence for this peptide has been obtained, and a chemically-synthesised peptide matching the deduced sequence shown to possess similar inhibin-like activity. (Ramasharma *et al.* 1984). Since the human analogue of inhibin is active against mouse pituitaries, it is possible that considerable conservation of inhibin structure exists across species. If this is so, the human sequence may provide sufficient information for the synthesis of a DNA probe with which to identify the ovine gene in a cloned genomic library. Such a gene is a useful candidate for modification in order to alter FSH release during the reproductive cycle of the sheep. For example, it has recently been suggested that the elevated prolificacy of the Booroola Merino sheep is the result of elevated levels of FSH, possibly induced by a deficiency in inhibin (Cummins *et al.* 1983; Bindon 1984). If this is confirmed, it would support the view that reproductive performance might be altered by genetic engineering of the inhibin gene, provided that the endogenous inhibin gene activity could be partially or completely supressed. In order to achieve this, detailed information concerning the mode of action of the hormone is required, particularly with regard to its receptor-binding characteristics.

## CONCLUSION

The modification of hormone-encoding genes to influence reproductive performance in sheep provides only one of many examples of the way in which genetic engineering might influence the sheep industry. It is important to remember, however, that the technique cannot yet be considered a replacement for conventional breeding techniques, because of the importance of multiple-gene interactions in the conventional selection process. Genetic engineering should be regarded as a way of making changes in highly-selected areas of the genome. These changes could be the alteration of key hormonal concentrations, such as growth hormone, or the re-introduction of an enzymatic pathway lost during the course of evolution, the example discussed being the conversion of serine to cysteine. Animals produced by such experiments would then be considered as breeding stock to be introduced into the conventional animal breeding programmes if their phenotypes were judged to provide useful characteristics.

It is self-evident that, before any genetically-engineered animals were released to the domestic market, considerable physiological evaluation of their performance is required. This involves studies of their reproductive and nutritional performance, monitoring of their behavioural characteristics, and the stability of transfer between generations of the introduced genetic information. Such evaluation studies necessarily involve a considerable number of years of research. Even if such sheep were available at present, it would be at least 5-10 years before such animals could be released.

At the current stage of the research, the most important factor in maintaining its momentum is to develop sufficient dialogue between scientists who understand the complex physiology of our domestic animals and scientists with the skills required for the isolation and modification of genes. It is within our grasp to consider the reconstruction of many enzymic pathways which have been rendered non-functional during the course of evolution. The most important of these for improving animal production must be identifed, together with methods for identifying the gene products associated with the pathways. When this is accomplished, it is certain that the sheep which form the basis of the major breeding programmes will be considerably different in genetic potential, and probably in phenotype, from those in use today.

## REFERENCES


Anderson, W.F., Killos, L., Saunders-Haigh, L., Kretschmer, P.J. and Diacumakos, E.G., 1980. *Proc. Natl. Acad. Sci. USA, 77*, 5399-5403.

Bindon, B.M., 1984. *Aust. J. Biol. Sci., 37*, 163-189.

Blanc, M.R., 1980. *Reprod. Nutr. Dev., 20*, 573-586.

Brinster, R.L., Chen, H.Y., Trumbauer, M., Senear, A.W., Warren, R. and Palmiter, R.D., 1981. *Cell, 27*, 223-231.

Brinster, R.L., Chen, H.Y., Warren, R., Sarthy, A. and Palmiter, R.D., 1982. *Nature, 296*, 39-42.

Burki K. and Ullrich, A., 1982. *The EMBO Journal, 1*, 127-131.

Capecchi, M., 1980. *Cell, 22*, 479-488.

Cummins, L.J., O'Shea, T., Bindon, B.M., Lee, V.W.K. and Findlay, J.K., 1984. *J. Reprod. Fert., 67*, 1-7.

Costantini, F. and Lacy, E., 1981. *Nature, 294*, 92-94.

Franklin, I.R., 1984. *Proceedings of 2nd World Congress on Sheep and Beef Cattle Breeding, Pretoria, South Africa*. (in press).

Gordon, J.W. and Ruddle, F.H., 1981. *Science, 214*, 1244-1246.

Gordon, J.W., Scangos, G.A., Plotkin, D.J. Barbosa, J.A. and Ruddle, F.H., 1980. *Proc. Natl. Acad. Sci. USA, 77*, 7380-7384.

Graf, L.H., Urlaub, G. and Chasin, L., 1979. *Som. Cell Genet., 5*, 1031-1044.

Graham, F.L., Abrahams, P.J., Mulder, C., Meijneker, H.L., Warnaar, S., De Vries, F.A.J., Fiers, W. and Van Der Eb, A.J., 1974. *Cold Spring Harbor Symp. Quant. Biol., 39*, 637-650.

Graham, F.L. and Van Der Eb, A.J., 1973. *Virology, 54*, 536-539.

Ingle, C.P. and Loughlin, R.E., 1980. *Aust. J. Biol. Sci.*, *33*, 605-612.
Jaenisch, R., Harbers, K., Schnieke, A., Lohler, J., Chumakov, I., Jahner, D., Grotkopp, D., and Hoffmann, E., 1983. *Cell*, *32*, 209-216.
Jaenisch, R. and Mintz, B., 1974. *Proc. Natl. Acad. Sci. USA*, *71*, 1250-1254.
Jones, K.W., 1980. *Br. Med. Bull.*, *36*, 173-180.
Lacy, E., Roberts, S., Evans, E.P., Burtenshaw, M.D. and Costantini, F.D., 1983. *Cell*, *34*, 343-358.
Lo, C.W., 1983. *Mol. Cell. Biol.*, *3*, 1803-1814.
Maitland, H. and McDougall, J., 1977. *Cell*, *11*, 233-241.
McKnight, G.S., Hammer, R.E., Kuenzel, E.A. and Brinster, R.L., 1983. *Cell*, *34*, 335-341.
Maniatis, T., Fritsch, E.F. and Sambrook, J., 1982. *Molecular cloning. A laboratory manual.* Cold Spring Harbor Laboratory, USA.
Minson, A.C., Wildy, P., Buchan, A. and Darby, G., 1978. *Cell*, *13*, 581-587.
Mulligan, R.C. and Berg, P., 1980. *Science*, *209*, 1422-1427.
Nancarrow, C.D., Murray, J.D., Boland, M.P., Sutton, R. and Hazelton, I.G. This volume, 286-288.
Palmiter, R.D., Brinster, R.L., Hammer, R.E., Trumbauer, M.E., Rosenfeld, M.G., Birnberg, N.C. and Evans, R.M., 1982. *Nature*, *300*, 611-615.
Palmiter, R.D., Chen, H.Y. and Brinster, R.L., 1982. *Cell*, *29*, 701-710.
Palmiter, R.D., Norstedt, G., Gelinas, R.E., Hammer, R.E. and Brinster, R.L., 1983. *Science*, *222*, 809-814.
Powell, B.C., Sleigh, M.J., Ward, K.A. and Rogers, G.E., 1983. *Nucl. Acids Res.*, *11*, 5327-5346.
Ramasharma, K., Sairam, M.R., Seidah, N.G., Chretien, M., Manjunath, P. and Schiller, P.W., 1984. *Science*, *223*, 1199-1202.
Reis, P.J. and Schinkel, P.G., 1963. *Aust. J. Biol. Sci.*, *16*, 218-230.
Reis, P.J., Tunks, D.A. and Downes, A.M., 1973a. *Aust. J. Biol. Sci.*, *26*, 249-258.
Reis, P.J., Tunks, D.A. and Sharry, L.F., 1973b. *Aust. J. biol. Sci.*, *26*, 635-644.
Rogers, G.E., 1984. *Ann. N.Y. Acad. Sci.* (*in press*).
Schnieke, A., Harbers, K. and Jaenisch, R., 1983. *Nature*, *304*, 315-320.
Smith, C.J. and Hespell, R.B., 1983. *J. Dairy Sci.*, *66*, 1536-1546.
Stewart, T.A., Wagner, E.F. and Mintz, B., 1982. *Science*, *217*, 1046-1048.
Wagner, E.F., Covarrubias, L., Stewart, T.A. and Mintz, B., 1983. *Cell*, *35*, 647-655.
Wagner, E.F., Stewart, T.A. and Mintz, B., 1981. *Proc. Natl. Acad. Sci. USA*, *78*, 5016-5020.
Wagner, T.E., Hoppe, P.C., Jollick, J.D., Scholl, D.R., Hodinka, R.L. and Gault, J.B., 1981. *Proc. Natl. Acad. Sci. USA*, *78*, 6376-6380.
Ward, K.A. 1982. *In* Barker, J.S.F., Hammond, K. and McClintock, A.E. (eds) *Future Developments in the Genetic Improvement of Animals*, Academic Press, Australia, 17-39.
Ward, K.A., Murray, J.D., Nancarrow, C.D., Boland, M.P. and Sutton, R., 1984. *N.S.W. Veterinary Proceedings*, *20*, (in press).
Ward, K.A., Sleigh, M.J., Powell, B.C. and Rogers, G.E., 1982. *Proc. of 2nd. World Congr. Genet. Appl. Livestock Product.*, *6*, 146-156.
Wigler, M., Pellicer, A., Silverstein, S. and Axel, R., 1978. *Cell*, *14*, 725-731.
Wigler, M., Pellicer, A., Silverstein, S., Axel, R., Urlaub, G. and Chasin, L., 1979. *Proc. Natl. Acad. Sci. USA*, *76*, 1373-1376.
Wigler, M., Perucho, M., Kurtz, D., Dana, S., Pellicer, A., Axel, R. and Silverstein, S., 1980. *Proc. Natl. Acad. Sci. USA*, *77*, 3567-3570.
Wigler, M., Silverstein, S., Lee, L.-S., Pellicer, A., Cheng, T., and Axel, R., 1977. *Cell*, *11*, 223-232.
Wigler, M., Sweet, R., Sim, C.K., Wold, B. Pellicer, A., Lacy, E., Maniatis, T., Silverstein, S. and Axel, R., 1979. *Cell*, *16*, 777-785.

# EFFECT OF GONADOTROPHIN RELEASING HORMONE IN THE PRODUCTION OF SINGLE-CELL EMBRYOS FOR PRONUCLEAR INJECTION OF FOREIGN GENES

C.D. Nancarrow, J.D. Murray, M.P. Boland, R. Sutton and I.G. Hazelton *CSIRO, Division of Animal Production, P.O. Box 239, Blacktown NSW, Australia, 2148.*

*Summary* Various protocols for the production of single-cell embryos for pronuclear injection were examined. In order to recover these embryos in the morning, intravaginal progestagen sponges were removed at 19:00 h, PMSG having been injected 48 h earlier. Insemination was carried out surgically 20-24 h after sponge removal and was accompanied by two GnRH injections. When eggs were collected 62 h after sponge removal only 6% had cleaved and 18% of fertilized eggs had visible pronuclei. GnRH increased both ovulation rate and fertilization rate; embryos collected per ewe increased from 1.9 to 4.2

## INTRODUCTION

The development of techniques for splicing and recombining selected pieces of DNA has enabled single genes and the regions controlling their expression to be isolated and studied. By combining these genetic engineering techniques with those of micromanipulation and transfer of mammalian embryos, it is now possible to alter directly single gene characteristics of individual animals (Palmiter *et al.* 1982). The adoption of these principles to domestic livestock could significantly improve the genetic base from which stud breeders may then select superior animals (Ward *et al.* 1984).

The aim of this work was to establish protocols for superovulation of sheep and subsequent egg recovery, which provided sufficient single-cell embryos for pronuclear microinjection at appropriate times of the day.

## MATERIALS AND METHODS

Synchronization of oestrus in Merino and Merino cross ewes was achieved by treatment for 14 days with intravaginal progestagen sponges (Repromap, Upjohn Pty Ltd). All ewes received 1500 i.u. PMSG (Folligon, Intervet Australia Pty Ltd) by intramuscular injection 48 h before removal of the sponges. Treatment protocols for 13 experimental groups of sheep are shown in Table 1. These were based on our observations that many ewes treated with PMSG 48 h before progestagen withdrawal were marked by rams within 24 h of sponge removal. With ovulation occurring 27-30 h after onset of oestrus and the first cleavage within another 20-24 h (Braden and Baker 1973), the range of time to recover single-cell embryos from these sheep would be between 54 and 78 h after sponge removal.

In the first experiment embryos were collected at 54 (Group 1) or 64 h (Group 2) after sponge removal and within these treatments, the efficiency of fertilization following natural service or surgical insemination was compared (Table 2). Half of the ewes were served daily by 2 fertile rams at 08:00, 13:00 and 17:00 until the end of heat and the remainder were artificially inseminated 24 h after sponge removal by depositing 50 $\mu$l fresh pooled diluted semen, obtained from another 4 rams, into the tip of each uterine horn (Boland and Gordon 1978). This latter group was exposed to harnessed vasectomized rams until the time of insemination. Oviducts were flushed with 3 ml of phosphate-buffered saline (PBS) containing 5% sterile sheep serum. Eggs were examined and classified as fertilized if sperm were noted on the zona pellucida or cleavage had occurred.

In the second and third experiments (Groups 3 to 6 and 7 to 13 respectively) two other times of collection (56 and 62 h) were tested (Table 1). It had previously been noted that many eggs retained cumulus cells and these two experiments were designed to alleviate this problem by the use of gonadotrophin releasing hormone (GnRH) to synchronize ovulation. Two injections of GnRH ( 25 $\mu$g in 1 ml PBS containing 0.1% bovine serum albumin) were spaced 90 min apart and given around the time of insemination (Foster 1978). Some of the embryos were examined with Nomarski DIC optics for the visualization of pronuclei.

The data were treated to Chi-square analysis (Experiment 1) or analysis of variance.

## RESULTS

None of the 145 embryos collected from groups 1, 3 and 6 at 54-56 h after sponge removal had cleaved, whereas around 6% of 323 collected at 62 h (groups 4, 5, 7-13) and 23% of 52 collected at 64 h were multicellular (Table 1).

The mean ($\pm$ S.E.M.) ovulation rate for ewes in Experiment 1 was 7.7 $\pm$ 1.2. Data for treatment groups 1 and 2 are shown in Table 2. All ewes in the natural service group were mated within 48 h of sponge removal. Half of the other ewes were marked at the time of insemination and these had a higher fertilization rate than those unmarked (98% vs 86%; $P < 0.05$). There was no difference between the differently bred groups in the rate of recovery of eggs. However surgical insemination resulted in a higher fertilization rate than did natural mating (89% vs 53%; $P < 0.01$); average recovery of fertilized eggs per donor was 4.2 and 2.2, respectively.

**TABLE 1. Protocols used for synchronization of oestrus and superovulation of sheep to produce single-cell embryos.**

| Group (Total number of ewes) | Sponge removal Day 1 | GnRH Day 2 | Artificial insemination Day 2 | Egg recovery Day 3 or 4* | Time from sponge removal to recovery (h) | % of fertilized eggs cleaved |
|---|---|---|---|---|---|---|
| | Timing of events (day number) | | | | | |
| 1(20) | 08:30 | - | 08:30[1] | 14:30 | 54 | 0 |
| 2(20) | 17:00 | - | 17:00 | 09:00* | 64 | 23.1 |
| 3,6(24) | 08:00 | 07:30<br>09:00 | 08:00 | 16:00 | 56 | 0 |
| 4,5(24) | 19:00 | 16:30<br>18:00 | 17:00 | 09:00* | 62 | 6.7 |
| 7-13(90) | 19:00 | 16:30<br>18:00 | 15:30 | 09:00* | 62 | 6.2 |

[1] Half of the ewes in each of Groups 1 and 2 were mated to fertile rams, the other half artificially inseminated.

**TABLE 2. Ovarian response and supply of zygotes as influenced by time of progestagen withdrawal and method of insemination (Groups 1 and 2).**

| Method of insemination | Natural | | Artificial | |
|---|---|---|---|---|
| Time of sponge removal | 08:30 | 17:00 | 08:30 | 17:00 |
| Group | 1 | 2 | 1 | 2 |
| No. of ewes | 10 | 10 | 10 | 10 |
| No. ovulated | 9 | 10 | 9 | 10 |
| Mean (±s.e.m.) ovulation rate/ewe treated | 6.2±1.2 | 8.1±1.9 | 11.3±2.1 | 4.9±1.0 |
| No. (%) of eggs recovered | 42 (67.7) | 43 (53.1) | 61 (54.0) | 35 (71.4) |
| Fertilization rate (%) | 50 | 55.8 | 98.2 | 80.0 |
| No. of fertilized eggs per ewe | 1.9 | 2.4 | 5.6 | 2.8 |

In Experiments 2 and 3, 138 ewes were treated and recorded 1196 ovulations yielding 550 eggs of which 393 were fertilized. Injections of GnRH increased the ovulation rate over control ewes and this was reflected in the number of eggs both recovered and fertilized per ewe (Table 3). There was also a tendency for fertilization rate to be higher than in controls ($P = 0.060$). No eggs recovered from GnRH treated groups were enveloped in cumulus cells. Pronuclei were seen in 34 (18%) of 185 fertilized eggs examined.

**TABLE 3. Effect of gonadotrophin releasing hormone on ovarian response and supply of zygotes in sheep treated with PMSG (Mean ± SEM) (Groups 3 to 13).**

| Parameter | Gonadotrophin releasing hormone − | Gonadotrophin releasing hormone + | Analysis of variance probability |
|---|---|---|---|
| Ovulation rate | 7.4 ± 0.8 | 9.9 ± 0.7 | 0.027 |
| Recovery: | | | |
| number (eggs/ewe) | 3.2±0.6 | 5.0±0.6 | 0.044 |
| rate(%) | 45.5±6.9 | 49.7±4.1 | 0.606 |
| Fertilization: | | | |
| number (eggs/ewe) | 1.9±0.5 | 4.2±0.8 | 0.032 |
| rate (%) | 49.8±8.4 | 75.7±9.7 | 0.060 |

## DISCUSSION

Suitable numbers of single-cell embryos were obtained from all groups of sheep. However, when the progestagen was withdrawn in the morning it was necessary to recover eggs at an inconvenient time in the afternoon; withdrawal in the evening allowed adequate time for micromanipulation and embryo transfer after morning recoveries.

The use of GnRH improved the yield of fertilized eggs twofold. This was due to unaccountable increases in both ovulation and fertilization rates. GnRH caused release of LH in all 5 ewes within 2 h of beginning of treatment whereas LH peaks in 5 control ewes occurred over a 22 h period (C.D. Nancarrow *et al.* unpublished observations).

Transfer of genes remains to be successfully carried out and this depends on processing sufficient embryos to overcome the problems of survival of injected embryos, gene incorporation and expression. At the same time, other techniques will be introduced such as non-destructive methods to enhance pronuclear visibility, insemination via endoscopy, and *in vitro* fertilization, which should increase the efficiency of production of single-cell embryos for injection.

## ACKNOWLEDGMENTS

We thank Mr. J.T.A. Marshall, Miss A. Doff and Miss G. Moreta for their excellent technical help, and the Reserve Bank of Australia for providing a fellowship to M.P.B.

## REFERENCES

Boland, M.P., and Gordon, I., 1978. *Ir. Vet. J., 32*, 123-125.

Braden, A.W.H. and Baker A.A. 1973. *In* Alexander, G. and Williams, O.B. (eds), *The Pastoral Industries of Australia*, Sydney University Press, Sydney, 269-302.

Foster, J.P., 1978. *In* Crighton, D.B., Haynes, N.B., Foxcroft, G.R. and Lamming, G.E. (eds), *Control of Ovulation*, Butterworths, London, 101-116.

Palmiter, R.D., Brinster, R.L., Hammer, R.E., Trumbauer, M.E., Rosenfeld, M.G., Birnberg, N.C. and Evans, R.M., 1982. *Nature, 300*, 611-615.

Ward, K.A., Murray, J.D., Nancarrow, C.D., Boland, M.P., and Sutton, R., 1984. *This Volume,* 279-285.

# CULTURE OF PREIMPLANTATION SHEEP AND GOAT EMBRYOS

P. Quinn, G.M. Warnes, S.K. Walker and R.F. Seamark, *Department of Obstetrics and Gynaecology, University of Adelaide, The Queen Elizabeth Hospital, Woodville, S.A. 5011.*

*Summary* Sheep and goat embryos from the 1-cell to 16-cell stage were cultured in sheep oviduct fluid (SOF) medium containing 10% homologous heat inactivated serum. This medium allowed over 75% of the embryos to develop to the expanded blastocyst stage. When goats were cultured further, 70% of the blastocysts hatched from their zona pellucida. Development in vitro was slower than that *in vivo*. The capacity of an embryo to develop in culture has proved invaluable in optimizing conditions for transport and storage of both sheep and goat embryos.

## INTRODUCTION

There are few reports in the literature of extended culture of early preimplantation sheep or goat embryos. One of the first attempts was that of Moore (1970) who cultured 2 to 8-cell sheep embryos for up to 48h in a balanced salt solution. A significant improvement in the cleavage rates and viability of cultured sheep and cattle embryos was obtained by Tervit *et al.* (1972) who formulated a medium (synthetic oviduct fluid, SOF) based on the biochemical analysis of sheep oviduct fluid. These authors reported that 1- to 8-cell sheep embryos developed to the morula to early blastocycst stage during 3 to 6 days of culture in SOF medium. Several groups of workers have found that Dulbecco's phosphate buffered saline (PBS) containing homologous serum is adequate for maintaining day 5 to later stage sheep and goat embryos for up to 2 days of culture (Moore and Spry 1972; Bilton and Moore 1976; Willadsen 1977). However, to assess the developmental potential of earlier stages *in vitro*, a bicarbonate buffered medium such as SOF is required (Tervit and Gould 1978).

The studies reported here were undertaken to assess the developmental capacity of preimplantation sheep and goat embryos during extended (>2 days) periods of culture in SOF medium. The culture technique was used to analyse the effect of storage at 4°C for 2 days on the viability of goat embryos.

## MATERIALS AND METHODS

Goat and sheep embryos at the 1-cell to early morula stage (mostly 4 to 8-cell stage) were collected from superovulated adult animals using procedures described elsewhere (goats; Armstrong *et al.* 1983: sheep; Walker *et al.* 1984). During collection, assessment, transfer to recipients or transport to the laboratory, the embryos were handled at 25-30°C in Hepes buffered SOF medium containing 10% heat-inactivated homologous serum. The embryos were cultured at 37°C in an humidified atmosphere of 5%$CO_2$:5%$O_2$:90%$N_2$ in 30-50ul drops of bicarbonate-buffered SOF medium containing 10% homologous serum under paraffin oil. The composition of the SOF-medium has been described by Tervit *et al.* (1972) and the procedure of preparation of medium used in our laboratory has been reported previously (Quinn *et al.* 1984). The embryos were assessed at x80 magnification under a dissecting microscope at least once a day during culture and the numbers reaching the expanded blastocyst stage recorded.

## RESULTS

The development of preimplantation goat and sheep embryos *in vitro* is given in Table 1. Over 75% of the embryos developed to expanded blastocysts after culture for up to 12 days. The most frequent protocol followed was to collect 4-8 cell embryos on the third day after the onset of oestrus (= day 0). These embryos began to hatch from the zona pellucida after 5 days of culture. This rate of development *in vitro* of both goat and sheep embryos was delayed by approximately 2 days in comparison to embryos developing entirely *in vivo*. For instance, after 4-5 days of culture, embryos collected on day 3 at the 4-8 cell stage had only reached the morula to early blastocyst stage whereas *in vivo*, the embryos would be expanded and/or hatched blastocysts. When 57 of the goat embryos which had reached the expanded blastocyst stage were cultured further, 40 (70%) hatched completely from the zona pellucida and in a few instances, the inner cell mass began to differentiate into primary embryonic layers (yolk sac, amnion) by day 13 (10th day of culture).

In most instances, embryos developing to the blastocyst stage were frozen in liquid nitrogen using the techniques described by Quinn *et al.* (1982). Few (<10%) of these embryos survived the freeze-thaw treatment as judged by their subsequent development *in vitro*. In one instance, however, twin kids were born upon transfer of frozen-thawed goat blastocysts to a synchronized recipient.

Eighty percent (16/20) of goat embryos at the 4-8 cell stage stored in Hepes-buffered SOF at 4°C for 2 days developed upon culture. There was no difference in the number of blastocysts developing in culture between media containing 10 or 20% goat serum (8/11 and 8/9 respectively). Subsequently, this protocol has been used as a back-up system for 3 seasons in a commercial goat embryo transfer program and 7 of 10 recipients which each received 2 stored eggs became pregnant and 11 kids were born.

**TABLE 1. Development of preimplantation goat and sheep embryos to expanded blastocysts *in vitro*.**

| Day of Collection | Stage | No. replicates | No. Embryos cultured | Days cultured | No. embryos developing | % |
|---|---|---|---|---|---|---|
| a) Goat | | | | | | |
| 2 | 1 cell | 3 | 46 | 7-12 | 36 | 78 |
| 2 | 2 cell | 1 | 4 | 10 | 3 | 75 |
| 2 | 4 cell | 1 | 4 | 9 | 3 | 75 |
| 3 | 2 cell | 1 | 4 | 12 | 3 | 75 |
| 3 | 4 & 8 cell | 5 | 62 | 6-10 | 54 | 87 |
| 3 | 8 cell | 1 | 6 | 6 | 6 | 100 |
| 4 | 4 & 8 cell | 1 | 10 | 5 | 8 | 80 |
| 4 | 8 & 16 cell | 4 | 15 | 4-7 | 14 | 93 |
| b) Sheep | | | | | | |
| 4 | 8-16 cell | 4 | 26 | 3 | 21 | 81 |
| 5 | morula | 1 | 41 | 2 | 38 | 93 |
| 6 | morula | 1 | 8 | 1 | 7 | 88 |

## DISCUSSION

The results obtained in this study show that a high proportion (>75%) of preimplantation goat and sheep embryos can be cultured to expanded blastocysts and in the case of goats at least, 70% of these blastocysts subsequently hatch from the zona pellucida. In addition, with the goat, there does not appear to be any stage-specific block to development to the blastocyst stage, as has sometimes been postulated, as nearly 80% of 1-cell embryos developed to expanded blastocysts *in vitro*. The inclusion of homologous serum as the protein component in the medium is probably responsible for the high (70%) proportion of goat embryos hatching from the zona pellucida.

The development of both goat and sheep embryos *in vitro* was slower than that *in vivo*. This would make the transfer to recipients of embryos cultured for longer than 2 days more difficult as it would be harder to synchronize the developmental age of the embryos with the recipient. For instance, should the synchronization be based on chronological age or morphological appearance? The latter would probably be better but further studies are needed to clarify this point.

One application of culture *in vitro* of preimplantation embryos is the assessment of various treatments, such as storage or manipulation, on the subsequent viability of the embryos. It was shown for example that 8-cell goat embryos could be stored for 48h at 4°C with little detrimental effect on their subsequent development *in vitro* and this technique has been applied to the field situation with the satisfactory results reported. The results reported here comfirm the observations of Tervit *et al.* (1972) that SOF medium is suitable for the culture of sheep embryos. It also appears to be adequate for goat embryos and the Hepes-buffered version is adequate for short term storage and transport of goat and sheep embryos. Therefore, there seems to be little point in trying to optimize further the culture media requirements for sheep and goat embryos. Studies should now be undertaken to optimize conditions for cryopreservation of sheep and goat embryos and manipulation *eg* embryo splitting, micro-injection of cells and macromolecules. The subsequent *in vitro* development of the treated embryos should give a good indication of their *in vivo* development upon transfer.

## REFERENCES

Armstrong, D.T., Pfitzner, A.P., Warnes, G.M. and Seamark, R.F., 1983. *J. Reprod. Fert.*, *67*, 403-410.
Bilton, R.J. and Moore, N.W., 1976. *Aust. J. Biol. Sci.*, *29*, 125-129.
Moore, N.W., 1970. *Aust. J. Biol. Sci.*, *23*, 721-724.
Moore, N.W., and Spry, G.A., 1972. *J. Reprod. Fert.*, *28*, 139.
Quinn, P., Barros, C. and Whittingham, D.G., 1982. *J. Reprod. Fert.*, *66*, 161-168.
Quinn, P., Warnes, G.M., Kerin, J.F. and Kirby C., 1984. *Fertil. Steril.*, *41*, 202-209.
Tervit, H.R., and Gould, P.G., 1978. *Theriogenology*, *9*, 251-258.
Tervit, H.R., Whittingham, D.G. and Rowson, L.E.A., 1972. *J. Reprod. Fert.*, *30*, 493-497.
Walker, S.K., Smith, D.H., Little, D.L., Warnes. G.M., Quinn, P., and Seamark, R.F., 1984. This volume, 306-309.
Willadsen, S.M., 1977. *Ciba Fdtn. Symposium, 52*, 175-189.

# CURRENT PROBLEMS AND FUTURE POTENTIAL OF ARTIFICIAL INSEMINATION PROGRAMMES

W.M.C. Maxwell, *Animal Breeding and Research Institute, Katanning, Western Australia, 6317.*

## INTRODUCTION

Over the past 10-15 years interest in commercial use of artificial insemination (AI) in sheep has steadily increased in Australia, particularly in Western Australia. This has been characterised by a movement away from on-farm AI using 'home-grown' rams, to the use of 'top' sires in studs or commercial artificial breeding (AB) centres, or in nucleus type group breeding schemes. The traditional methods for on-farm AI using freshly collected semen have limitations for use with rams which may be located at a remote site, or may be unavailable at the time of the AI programme. Until recently, the lack of practical methods for semen storage have made it necessary either to transport rams to the site of insemination, or take ewes to an AI centre. This is expensive and limits the number of ewes which can be inseminated. The stress of transporting both ewes and rams can also affect fertility. Synchronisation of oestrus is another development important to efficient AB programmes because of the need to inseminate large numbers of ewes with semen from specific rams.

This paper will outline developments in the areas of semen storage, synchronisation of oestrus and AI technique, and examine how they might be applied in the future.

## STORAGE OF RAM SEMEN

Successful storage of semen allows the more widespread use of individual sires and the preservation of genetic material for future use. During the ninteenth century several studies were made on various media and methods for storage of spermatozoa. However, it was the work of Ivanov (1907) on synthetic media for the dilution of spermatozoa which led to the large scale development of AI in farm animals (horses, cattle and sheep), mainly in the Soviet Union. Artificial insemination in cattle and sheep was based largely on phosphate buffered diluents (Kuznecova *et al.* 1932), with semen diluted and held at 30°C for up to 1 h before use.

During the 1930's a number of Soviet workers investigated the value of various lipids, including egg yolk and milk, in protecting spermatozoa against cold shock (reviewed by Milovanov 1962). Lardy and Phillips (1939) applied these findings in the development of the egg yolk-phosphate diluent for the preservation, transportation and insemination of bull semen. Soon afterwards, another generally accepted storage medium, the egg yolk-citrate diluent was elaborated by Willett and Salisbury (1942). In addition to preserving the viability of spermatozoa, 'egg yolk' diluents were found to protect the spermatozoa from cold shock, and also to provide a sutiable medium in which the spermatozoa could be cooled and stored at 5-10°C (Chang and Walton 1940; Easley *et al.* 1942). This storage method is known as 'chilled' or 'liquid storage', and is used for short-term preservation of semen. Longer-term storage requires freezing of semen.

### Fresh and chilled semen

After 1950 a large number or reports were published on the use of fresh and chilled-stored ram semen. The majority of workers used cow milk (whole or skimmed), egg yolk-citrate or egg yolk-phosphate diluents for the storage of semen at 2-5°C or under anaerobic conditions at ambient temperatures. Many reports of these techniques were from Soviet sources and these have been reviewed by Maxwell (1978). These workers investigated the major problem that although the motility of spermatozoa could be maintained during several days of storage, the lambing rate following normal cervical insemination declined after 24 h of semen storage (Salamon and Robinson 1962; Rabocev 1965, 1966; Smagulov and Martynov 1966; Lopyrin and Rabocev 1968; Jones *et al.* 1969; Lapwood *et al.* 1972; Firth 1976; Watson and Martin 1976; Colas and Courot 1976). The factors considered to account for this decline in fertility were an impairment of fertilising capacity of spermatozoa during storage, a decrease in the viability of spermatozoa in the female reproductive tract, and embryonic loss (Lopyrin 1971). Using uterine insemination following midventral laparotomy, Salamon *et al.* (1979) showed that rates of fertilisation of ova could exceed 90% after storing semen for up to four days at 5°C, but from 4 to 10 days fertilising capacity progressively declined (Table 1). The authors reported lambing rates of 40.0, 25.0 and 5.0% for semen chilled-stored for 6, 8 and 9 days respectively. Decreased fertilisation of ova with storage beyond four days was not associated with an increase in early embryonic loss (Salamon *et al.* 1979). The development of more practical methods for intrauterine insemination of sheep may make possible the extension of the useful storage period for chilled semen up to 8-9 days.

Fertility following cervical insemination with fresh and chilled-stored semen depends on the dilution rate of the semen, the number of spermatozoa in the inseminate dose and the number of inseminations per oestrus. At a synchronised oestrus, fertility is best when a low volume of highly concentrated spermatozoa is delivered into the female tract (Allison and Robinson 1971). Best results with ewes in natural oestrus occur when the inseminate dose contains 100 to 125 million normal motile spermatozoa (Salamon 1962). Higher doses do not improve fertility, but double insemination is required with chilled-stored semen at a natural oestrus (Salamon *et al.* 1979, Table 2). This is not the case for ewes in synchronised oestrus due to decreased sperm transport after hormonal treatments (Quinlivan and Robinson 1969; Colas *et al.* 1973). Colas *et al.* (1973) reported that similar fertility could be achieved with either single or double insemination with the same total sperm dose.

**TABLE 1. Fertilisation after surgical insemination with fresh and chilled-stored ram semen (adapted from Salamon *et al.* 1979).**

| Period of semen storage (days) | No. of ewes yielding eggs | Ewes yielding fertilised eggs No. | % |
|---|---|---|---|
| 0 | 34 | 32 | 94.1 |
| 2 | 16 | 16 | 100.0 |
| 4 | 22 | 21 | 95.5 |
| 6 | 20 | 15 | 75.0 |
| 8 | 17 | 9 | 52.9 |
| 9 | 18 | 2 | 11.1 |
| 10 | 15 | 1 | 6.7 |

**TABLE 2. Lambing after single and double cervical inseminations with fresh and chilled-stored semen (adapted from Salamon *et al.* 1979).**

| Age of semen (days) | No. ewes lambed/inseminated (%) | |
|---|---|---|
| | Single insemination | Double insemination |
| 0 | 45/78 (57.7) | 52/78 (66.7) |
| 1 | 14/46 (30.4) | 25/44 (56.8) |
| 2 | 11/41 (26.8) | 13/28 (46.4) |
| 3 | 2/43 ( 4.7) | 22/53 (41.5) |

Frozen semen

The most significant discovery on the freezing of mammalian semen was made by Polge *et al.* (1949) who, in their studies on the low temperature storage of fowl spermatozoa, brought to prominence the cryoprotective properties of glycerol. The authors reported that the motility of fowl spermatozoa was fully resumed after thawing semen frozen to −79°C in Ringer's solution containing 40% glycerol. After the discovery of this protective effect of glycerol, the techniques for freezing spermatozoa were further developed, and workers in many countries contributed to a rapid expansion of AI in cattle using frozen semen. The achievements made on deep freezing of bull semen also initiated the development of freezing methods for the semen of other mammalian species.

During the 1960's and early 1970's there were many reports of successful AI of sheep with frozen ram semen (Salamon 1967; Salamon and Lightfoot 1967; Platov 1968; Loginova and Zeltobrjuk 1968; Fraser 1968; Mattner *et al.* 1969; Lightfoot and Salamon 1970; Salamon and Lightfoot 1970 a,b; Colas and Brice 1970; Salamon 1971, 1972). AI with frozen semen has been performed on an experimental scale in many countries since these first successful reports.

The freezing procedure developed by Salamon and co-workers in Sydney (Salamon and Visser 1972) involved the dilution of freshly collected ejaculates with a Tris-based medium containing the cryoprotective agent glycerol, slow cooling of the diluted semen to 5°C, followed by rapid freezing to −79°C by pelleting on solid carbon dioxide (dry ice). The pellets were stored in liquid nitrogen (−196°C). Salamon and Visser (1974) reported no difference between the fertility of semen frozen-stored for two weeks and that of semen frozen-stored for five years by the above method. Subsequent fertility tests on a semen bank laid down in 1968 have indicated no decline in fertility if the semen is maintained at the temperature of liquid nitrogen (Table 3).

**TABLE 3. Lambing and pregnancy results after cervical insemination with semen stored for 3, 5, 7, and 11 years (adapted from Salamon 1980).**

| Test No. | Period of storage | No. of ewes inseminated | lambed | % lambed |
|---|---|---|---|---|
| 1 | 3 years | 172 | 91 | 52.9 |
| 2 | 5 years | 70 | 37 | 52.9 |
| 3 | 7 years | 143 | 74 | 51.7 |
| 4 | 11 years | 159 | 88 | 55.3 |
| | Fresh semen (control) | 160 | 105 | 65.6 |

Another procedure for freezing ram semen was developed independently in France by Colas (1975a). Freshly collected semen was diluted with a lactose-egg yolk solution, cooled to 4°C, rediluted with skim milk containing glycerol, and packaged in straws before freezing in liquid nitrogen vapour. As before, the frozen straws were stored in liquid nitrogen (−196°C).

Many attempts have been made since the development of these methods to solve the special problems associated with the freezing procedures so that consistent results could be obtained in commercial sheep breeding programs (reviewed by Graham *et al.* 1978). Despite intensive laboratory studies and considerable progress, the procedure of freezing and thawing ram semen inevitably reduces the viability of spermatozoa. Fertility following cervical insemination with frozen semen tends to be considerably lower than after using fresh or chilled semen (Maxwell *et al.* 1980). The major obstacles to fertility are the establishment of a large enough population of viable spermatozoa in the cervix, which may necessitate centrifugation or double inseminations (Salamon 1977), and imparied sperm transport from the cervix through the genital tract of the ewe (Lightfoot and Salamon 1970a). These problems have been partially overcome by increasing the depth of cervical insemination (Graham *et al.* 1978), and more recently using intrauterine insemination by laparoscopy (Killeen *et al.* 1982).

## SYNCHRONISATION OF OESTRUS

During the normal breeding season (generally late January to April/May in Australia) about 6-8% of ewes in a flock can be expected in oestrus each day. In order to detect each ewe in oestrus during a traditional AI program, drafting of the flock and subsequent insemination is carried out daily for a period of at least 19 days (Salamon 1976). This practise is time consuming, does not allow best utilisation of the manpower available, and raises difficulties in management of flocks. Various forms of synchronisation of oestrus have been developed to overcome these shortcomings.

The principles upon which oestrus synchronisation has been based during the normal breeding season are either the shortening of the luteal phase of the cycle by an injection of prostaglandin, or the suppression of the cycle in all ewes by treatment with a progestagen-impregnated intravaginal sponge for 12-14 days (Robinson 1965).

In order to synchronise the whole flock with prostaglandin two injections are necessary, spaced 8-14 days apart. Not all ewes express oestrus, and fertility is generally poor following AI at the first oestrus after synchronisation with prostaglandins (Fairnie and Wales 1982).

Upon cessation of treatment with progestagen sponges, a common start is given to the ewes for resumption of cycling, and most (85-95%) will express oestrus over 2-3 days. Fertility following insemination in this first oestrus after sponge withdrawal is generally lower than in untreated ewes (Robinson and Moore 1967). This problem can be overcome in two ways.

First, ewes may be treated with a single intramuscular injection of PMSG (Pregnant Mares Serum Gonadotrophin) at the time of sponge removal to provide a better synchronisation of oestrus and ovulation (Robinson and Smith 1967). Using this method, it is not necessary to detect oestrus, but ewes should be joined with 10% teaser rams at sponge removal to help stimulate ovulation, particularly when the technique is used in the non-breeding period (Pearce *et al.* 1983). The majority of ewes will be in oestrus within 48 hr and ovulate at a median time of 56 hr (54.6 - 57.1, 95% fiducial limits, W.M.C. Maxwell 1984 unpublished). A single fixed time insemination should be performed on all ewes at 55-56 hr (cervical insemination, Colas 1975b) or 60-72 hr after sponge removal (laparoscopic insemination, W.M.C. Maxwell 1984 unpublished). The above synchronisation technique is effective both during spring (non-breeding season) and autumn (breeding season, Colas and Brice 1976; Pearce *et al.* 1983). The method allows a precise planning of an AI programme so that the date of the programme, the exact number of ewes to be inseminated each day, and the time of insemination can be controlled.

The second way of utilising progestagen sponges during the normal breeding season (avoiding the expense of using PMSG) is to inseminate the ewes at the second oestrus following sponge removal (Robinson 1976, Salamon 1976). The ewes in the second oestrus are still well synchronised (80-90% in oestrus over 6-7 days), and the lambing to a single cervical insemination with fresh-diluted semen is normal (50-60%). As for natural oestrus, harnessed teaser rams are run with the ewes from 16-17 days after sponge removal, and marked ewes drafted off once or twice daily.

In south Western Australia the time for most commercial AI programs is spring and summer (Oldham 1980). Much of this time is the natural non-breeding season for many ewes. During this season, ewes which are pre-conditioned by a period of isolation from rams, will respond to the sudden presence of rams by recommencing sexual activity. This response to sudden exposure to rams is called the 'ram effect' (see Pearce and Oldham 1984).

The peak of induced oestrous activity has been utilised for AI programmes by running 'teaser' rams with ewes for 14 days before the commencement of the programme (Fairnie 1976). This ensures that all ewes are cycling and expressing sexual activity over a period of 7-10 days. If the ewes receive a small dose of progesterone (10 mg intramuscular) at the time the teasers are introduced, short cycles (and ineffective ovulations) are eliminated, and the synchrony of the first oestrus is much better (3-4 days, Cognie *et al.* 1982).

Using the 'ram effect' plus a progesterone injection has distinct advantages as a synchronisation method. It is inexpensive and requires a minimum of labour, but ensures that a large proportion of the flock are inseminated in a short time. This technique is gradually being utilized by some of the larger group breeding schemes (e.g. the Australian Merino Society), who have traditionally used the 'ram effect' alone for synchronisation of oestrus for AI in spring (Corke 1980, Cognie *et al.* 1982).

## ARTIFICIAL INSEMINATION TECHNIQUES

Practical systems for AI of sheep have been available for many years (reviewed by Emmens and Robinson 1962). The traditional method is to draft the ewe flock once or twice daily, separate ewes exhibiting a natural oestrus as identified by teaser rams, and then inseminate these ewes with freshly collected and diluted semen. The ewes are generally inseminated by suspension of the hindquarters over a elevated rail and deposition of semen within the first fold of the cervix using a plastic pipette, speculum and headlamp (the 'cervical' or 'over-the-rail' insemination method, Salamon 1976).

Two other methods for sheep have recently come into common usage in Australia. First is the vaginal technique, commonly termed the 'shot-in-the-dark' or SID method (Fairnie and Wales 1982). For this simplified technique, the ewe is restrained in a standing position against the side of a pen or race. The inseminate (usually a large volume of fresh undiluted semen) is deposited in the vagina using a straight plastic pipette without the use of a speculum or any attempt to locate the entrance to the cervix. The Australian Merino Society inseminate some 100,000 ewes annually in Australia (Fairnie and Wales 1982), claiming wide usage of the SID method with satisfactory fertility. However, the only published information on the method reported poor fertility following vaginal insemination of sheep (Kerton *et al.* 1984).

The second sheep AI method in current usage is that of intrauterine insemination. During normal insemination it is rarely possible to pass the inseminating pipette through the cervix of the ewe. The cervix constitutes an initial barrier to the ascent of spermatozoa, and a relatively small proportion of cells deposited reach the site of fertilisation. Lopyrin and Loginova (1958) reported that frozen-thawed ram spermatozoa penetrated more slowly to the cranial cervix than did those from fresh semen, and Salamon and Lightfoot (1967) found very low numbers of spermatozoa in the Fallopian tubes and a low fertilisation rate following normal cervical insemination with frozen semen. Lightfoot and Salamon (1970a) ascribed the problem to a decreased rate of sperm transport through the genital tract of the ewe. Further, they demonstrated a failure of the establishment and maintenance of the cervical population of spermatozoa which could be partially overcome by use of inseminates with a high concentration of motile spermatozoa.

A number of attempts have been made to overcome the problem of transport of frozen-thawed ram spermatozoa by bypassing the cervix, and placing semen directly in the uterus. Until recently, these intrauterine insemination techniques have not proved very practical as they have relied either on deposition of semen in the uterine horns after major surgery (Lightfoot and Salamon 1970a,b) or a forceful entry to the uterus through the cervix (Andersen *et al.* 1973; Fukui and Roberts 1976).

An alternative route for the deposition of semen closer to the site of fertilisation in animals is the intraperitoneal method. This AI technique involves the blind deposition of semen directly into the peritoneal cavity in the general area of the broad ligament. It has proved successful in several avian and mammalian species (reviewed by Vanden Bosch and Hafez 1975), and more recently in sheep (Negogatikov *et al.* 1981). Although the fertility reported by these workers was promising, the temporal relationship between successful insemination and oestrus/ovulation remains unclear, and there appears to be a high rate of implantation failure following intraperitoneal insemination in some mammals (Vanden Bosch and Hafez 1975). Very little is known about intraperitioneal insemination in sheep, and further research is required to determine whether this method could be applied for use with frozen semen.

A method developed by Killeen and Caffery (1982) has given more promising results. This utilises a rapid laparoscopic location of the uterus, and direct injection of semen into the uterine horns from a fine pipette through a separate canula (Killeen *et al.* 1982). Recent work has confirmed that use of the technique with frozen semen is as effective as cervical insemination of fresh semen (Maxwell *et al.* 1984a) or natural mating over one synchronised oestrous cycle (Maxwell and Butler 1984). Moreover the technique has proved more efficient for use with frozen semen than cervical insemination because of the relatively small dose of semen used (0.04 ml containing 40 million total spermatozoa for laparoscopic insemination cf. 0.2 ml containing 200 million total spermatozoa for cervical insemination; Maxwell *et al.* 1984b). Using progestagen sponges and PMSG for synchronisation of oestrus , a relatively wide range of time is available over which ewes can be successfully inseminated with frozen-thawed semen by laparoscopy (Table 4). However embryonic loss is high if semen is deposited too early (< 48 hr after sponge removal; Maxwell *et al.* 1984).

AI with frozen semen by laparoscopy was practised over a wide area of south Western Australia during the 1983/84 breeding season in association with the Animal Breeding and Research Institute's (AB&RI) cooperative breeding projects. A summary of the results from the Fertility Breeding Project is presented in Table 5. In this AI program 837 ewes were inseminated on 17 Merino studs. All ewes received 0.02 ml frozen-thawed semen in each uterine horn 60-66 hr after sponge removal/PMSG injection. There was no variation in the percentage of ewes lambing within the four geographical regions covered. However, lambing percentage was lower in the Great Southern than in the Eastern Wheatbelt ($p < 0.005$), where it was in turn lower than the Midlands and Central Wheatbelt ($p < 0.05$). These results suggest that research may be required on the reasons for variability in fertility following AI of sheep in different geogrphical and/or climatic regions.

**TABLE 4. Lambing after insemination with fresh and frozen semen (W.M.C. Maxwell 1984, unpublished data).**

| Semen (type of insemination) | Insemination time after sponge removal (hr) | No. of ewes inseminated | No. of ewes lambed (%) |
|---|---|---|---|
| Fresh control (cervical) | 55 | 89 | 52 (58.4) |
| Frozen (intrauterine) | 24 | 34 | 3 ( 8.8) |
| | 36 | 34 | 6 (17.6) |
| | 48 | 86 | 42 (48.8) |
| | 60 | 81 | 46 (56.8) |
| | 72 | 54 | 36 (66.7) |

**TABLE 5. Lambing following AI by laparoscopy of stud Merino ewes with frozen semen in four geographical regions of Western Australia during November 1983 (L.G. Butler, W.M.C. Maxwell and G.F. Morrow 1984, unpublished data).**

| Geographical region | No. ewes inseminated (no. of studs) | No. ewes lambed (%) |
|---|---|---|
| Great Southern | 452 (7) | 187 (41.4) |
| Eastern Wheatbelt | 155 (5) | 88 (56.8) |
| Midlands & Central Wheatbelt | 175 (4) | 119 (68.0) |
| Experance Sandplain | 55 (1) | 30 (54.5) |
| Total and mean | 837 | 424 (50.7) |

Conception rates between 50 and 70% have been reported after using intrauterine insemination with fresh semen (Killeen and Caffery 1982, Kerton *et al.* 1984). However the real value of the laparoscopic insemination technique remains to be assessed with stored semen. Considerable research is still required to determine the best site of insemination, optimum dose of spermatozoa in the inseminate, whether embryonic loss is increased after insemination by laparoscopy, and the use of the technique for chilled-stored semen.

## FUTURE POTENTIAL FOR AI PROGRAMMES IN AUSTRALIA

The use of AI in sheep has become more common in the Australian stud Merino industry and the larger group breeding schemes. This was achieved initially through the use of fresh semen on-farm, and more recently frozen semen from licensed artificial breeding (AB) centres, which is processed and sold according to regulations controlled by the various state governments. Some 200,000 sheep were artificially inseminated with fresh semen throughout Australia during the 1983/84 breeding season (estimate from survey of the A.M.S., other group breeding schemes and AB centres), and in Western Australia alone over 22,000 ewes were inseminated commercially with frozen semen by laparoscopy (estimate from survey of licensed AB centres).

Although the ewe population of W.A. is 14.5 million (Aust. Bureau of Statistics 1984), the 22,000 ewes inseminated with frozen semen represent a significant proportion of the ewes in ram-breeding flocks; some 5% of the stud ewe population of approximately 400,000 (Stud Merino Breeders Association of W.A.). The small number of AI sires currently in use may therefore have considerable influence on the sheep industry in W.A.

Breeders may have different objectives for using AI, but the overall aim will be to improve their incomes. With careful selection of genetically superior 'sires-to-breed-sires', benefits will doubtless flow to the flock ram buyer. The various objectives of AI may be grouped according to whether they are for short or long term purposes.

Short term objectives:

(a) To effect a rapid change to a more desirable genotype.
(b) To control disease, particularly those transmitted venereally.
(c) To increase the usage of famous or fashionable sires.

Unfortunately, much of the AI which is currently practised is concerned with the latter objective.

Long term objectives:

(a) To achieve long term genetic gain.

(b) To identify accurately genetically superior sires.

Once identified on the basis of breeding values for productive traits, superior genetic material can be rapidly disseminated through the use of AI. Figure 1 outlines the potential influence of one ram on the industry using current technology by comparing the number of lambs produced per year from natural mating, AI with fresh semen and AI with frozen semen. The average lamb marking percentage in south Western Australia is 65% (Aust. Bureau of Statistics 1984), so the average naturally mated ram leaves 22 lambs per year. If AI with fresh semen is used, and a lamb marking percentage of 50% is obtained, some 500 lambs may be produced. Frozen semen may be collected and stored from a sire for most of the year. The example in Figure 1 presents conservative estimates of 8 months semen collection from one ram at a rate of 9 ejaculates per week. The AI ram in this case can leave up to 12,000 lambs per year.

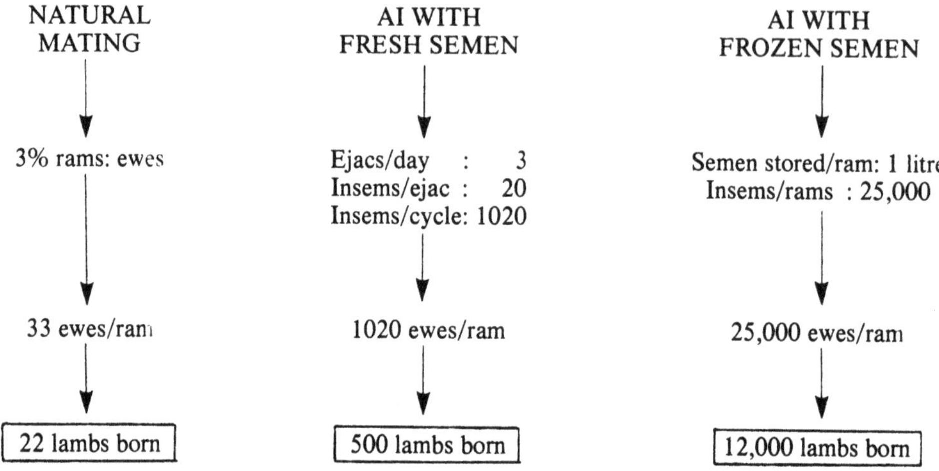


**FIGURE 1. The potential number of lambs sired by a ram each year using natural mating, AI with fresh semen, and AI by laparoscopy with frozen semen (data applies to south Western Australia).**

Examples of the potential use of AI for long-term genetic gain are the nucleus group breeding schemes (e.g. the A.M.S. and the Fertility Breeding Project; Butler and Morrow 1984). In such schemes, the ewes of contributing members are screened for productive characteristics and the best 5 or 10% formed into a nucleus flock. Selection is carried out within the nucleus, and genes can be returned to the contributing flocks by rams or semen. Each year the A.M.S. transports a small number of highly-selected rams to contributors all over Australia where they are used for AI programmes using fresh semen, generally without synchronisation of oestrus. Each ram provides enough semen to inseminate an average of 5,000 ewes (Fairnie and Wales 1982).

AI with frozen semen at a fully synchronised oestrus is used within the Fertility Breeding Project because the number of visually acceptable rams with high breeding values for reproductive rate is not sufficient to supply the many widely located breeders requiring rams in a limited breeding season (mainly November). Frozen semen is the only technique which allows the same ram to cover ewes over a wide geographical area in a short period.

An example of a project designed to identify genetically superior sires is the Stud Merino Breeders Association of W.A.'s sire referencing scheme (SRS) for sheep. SRS aim to provide a method for accurate identification of rams of high breeding value across studs. The procedure is to progeny test one or more reference sires in a number of participating studs by AI with frozen semen. This establishes a genetic link between flocks, and allows breeders to compare their rams with reference rams and thus indirectly with other 'outside' rams in the scheme. By comparing contemporarily tested home bred sires with the reference sire, it is possible to identify outstanding rams which may have been otherwise undiscovered. AI is an integral part of the system, enabling a small number of rams to be used in a large number of flocks.

## CONCLUSIONS

Techniques for AI, semen storage and synchronisation of oestrus in sheep are well advanced. Since results with frozen semen are still variable, further research on its use in commercial AI programmes is required. The recent success with laparoscopy points to the possibility of other insemination routes (e.g. deep cervical and intraperitoneal). Research is also needed on the use of vaginal ('SID') and intrauterine AI with fresh and chilled semen at a natural or synchronised oestrus.

AI of sheep is becoming more common in Australia. Its future benefit to the industry now depends on accurate identification of genetically superior sires, and their use in programmes designed to achieve long term genetic gain.

## REFERENCES


Allison, A.J. and Robinson, T.J., 1971. *Aust. J. Biol. Sci., 24*, 1001-1008.

Andersen, K., Aamdal, J. and Fougner, J.A., 1970. *Zuchthyg., 8*, 113-118.

Australian Bureau of Statistics, 1984. *Livestock and Livestock Products, Western Australia, Season 1982-83*. Catalogue No. 7221.5

Butler, L.G. and Morrow, G.F., 1984. *Proc. 4th Conf. Aust. Ass. Anim. Breed. Genetics, Adelaide*, 184-185.

Chang, M.C. and Walton, A., 1940. *Proc. Roy. Soc., B, 129*, 517-521.

Cognie, Y., Gray, S.J., Lindsay, D.R., Oldham, C.M., Pearce, D.T. and Signoret, J.P., 1982. *Proc. Aust. Soc. Anim. Prod., 14*, 519-522.

Colas, G., 1975a. *J. Reprod. Fert., 42*, 277-285.

Colas, G., 1975b. *Ann. Biol. Anim. Bioch. Biophys., 15*, 317-327.

Colas, G. and Brice, G., 1970. *Ann. Zootech., 19*, 353-357.

Colas, G. and Brice, G., 1976. *Proc. 8th Int. Congr. Anim. Reprod. Artif. Insem., Cracow, IV*, 977-980.

Colas, G. and Courot, M., 1976. *In* Tomes, G.J., Robertson, D.E. and Lightfoot, R.J. (eds) *Sheep Breeding*. Western Australian Institute of Technology, Perth, 455-466.

Colas, G., Thimonier, J., Courot, M. and Ortavant, R., 1973. *Ann. Zootech., 22*, 441-451.

Corke, D.G., 1980. *Proc. Aust. Soc. Anim. Prod. 13*, 81-82.

Easley, G.T., Mayer, D.T. and Bogart, R., 1942. *Amer. J. Vet. Res., 3*, 358-366.

Emmens, C.W. and Robinson, T.J., 1962. *In* Maule, J.P. (ed) *The Semen of Animals and Artificial Insemination*. Commomwealth Agricultural Bureaux, Farnham Royal, U.K., 205-251.

Fairnie, I.J., 1976. *In* Tomes, G.J., Robertson, D.E. and Lightfoot, R.J. (eds) *Sheep Breeding*. Western Australian Institute of Technology, Perth, 500-508.

Fairnie, I.J. and Wales, R.G., 1982. *In* Barton, R.A. and Smith, W.C. (eds) *Proceedings of the World Congress on Sheep and Beef Cattle Breeding, II*, The Dunmore Press Limited, Palmerston North, New Zealand, 311-320.

Firth, J.H., 1976. *In* Tomes, G.J., Robertson, D.E. and Lightfoot, R.J. (eds) *Sheep Breeding*. Western Australian Insititue of Technology, Perth, 475-481.

Fraser, A.F., 1968. *Proc. 6th Int. Congr. Anim. Reprod. Artif. Insem., Paris, 2*, 1033-1035.

Fukui, Y. and Roberts, E.M., 1976. *In* Tomes, G.J., Robertson, D.E. and Lightfoot, R.J. (eds) *Sheep Breeding*. Western Australian Institute of Technology, Perth, 482-494.

Graham, E.F., Crabo, B.G. and Pace, M.M., 1978. *J. Anim. Sci., 47* (Suppl. 2), 80-103.

Ivanov, E.E., 1907. *Archives de Sciences Biologiques, Inst. Imp. Med. Exp., St. Petersbourg, 12*, 377-403.

Jones, R.C., Martin. I.C.A., and Lapwood, K.R., 1969. *Aust J. Agric. Res., 20*, 141-150.

Kerton, D.J., Mcphee, S.R., Davis I.F., White, M.B., Banfield, J.C. and Cahill, L.P., 1984. *Proc. Aust. Soc. Anim. Prod., 15*, 701.

Killeen, I.D. and Caffery, G.J ., 1982. *Aust. Vet. J., 59*, 95.

Killeen, I.D., Caffery, G.J. and Holt, N., 1982. *Proc. Aust. Soc Reprod. Biol. Ann. Conf., Sydney*, 104.

Kuznecova, N., Milovanov, V.K., Neumann, O., Nagaev, V. and Skatkin, P., 1932. *Artifical Insemination oj Cattle and Sheep*. Selkhozgiz, Moscow. (In Russian).

Lapwood, K.R., Martin, I.C.A. and Entwistle, K.W., 1972. *Aust. J. Agric. Res., 23*, 457-466.

Lardy, H.A. and Phillips, P.H., 1939. *Proc. Amer. Soc. Anim. Prod., 32nd Ann. Meet.*, 219.

Lightfoot, R.J. and Salamon, S., 1970a. *J. Reprod. Fert., 22*, 385-398.

Lightfoot, R.J. and Salamon, S., 1970b. *J. Reprod. Fert., 22*, 399-408.

Loginova, N.V. and Zeltobrjuk, N.H., 1968. *Ovtsevodstvo, 14(9)*, 22-25.

Lopyrin, A.I., 1971. *Biology of Reproduction in Sheep*. Kolos, Moscow. (In Russian).

Lopyrin, A.I. and Loginova, N.V., 1958. *Ovtsevodstvo, 4(8)*, 31-33.

Lopyrin, A.I. and Rabocev, V.K., 1968. *Ovtsevodstvo, 14(7)*, 20-22.

Mattner, P.E., Entwistle, K.W. and Martin, I.C.A., 1969. *Aust. J. Biol. Sci., 22*, 181-187.

Maxwell, W.M.C., 1978. Ph D. thesis, University of Sydney, Sydney, N.S.W., 290 pp.

Maxwell, W.M.C. and Butler, L.G., 1984. *Proc. 4th Conf. Aust. Ass. Anim. Breed Genetics, Adelaide*, 192-193.

Maxwell, W.M.C., Butler, L.G. and Wilson, H.R., 1984a. *J. Agric. Sci., Camb., 102*, 233-235.

Maxwell, W.M.C., Curnock, R.M., Logue, D.N. and Reed, H.C.B., 1980. *Theriogenology, 14*, 83-89.

Maxwell, W.M.C., Wilson, H.R. and Butler, L.G., 1984b. *Proc. Aust. Soc. Anim. Prod. 15*, 448-451.

Milovanov, V.K., 1962. *Biology of Reproduction and Artificial Inseminatiuon of Animals*. Seljhozizdat, Moscow. (In Russian).
Negogatikov, G., Zhirmokleev, V. and Zarudnev, S., 1981. *Zhivotnovodstvo, 1*, 54-56.
Oldham, C.M., 1980. Ph D. thesis, University of Western Australia, Perth, 186 pp.
Pearce, D.T. and Oldham, C.M., 1984. This volume, 26-34.
Pearce, D.T., Oldham, C.M., Gray, S.G., Lindsay, D.R. and Wilson, H.R., 1983. *In* C.M. Oldham, A.M. Paterson and D.T. Pearce (eds) *Proceedings of a Seminar on Reproduction in Farm animals*. The Australian Society of Animal Production (WA Branch), 159-166.
Platov, E.M., 1968. *Ovtsevodstvo, 14(8)*, 14-18.
Polge, C., Smith, A.U. and Parkes, A.S., 1949. *Nature, London, 164*, 666.
Quinlivan, T.D. and Robinson, T.J., 1969. *J. Reprod. Fert., 19*, 73-86.
Rabocev, V.K., 1965. *Ovtsevodstvo, 11(9)*, 14-16.
Rabocev, V.K., 1966. *Ovtsevodstvo, 12(9)*, 33-36.
Robinson, T.J., 1965. *Nature, London, 206*, 39-41.
Robinson, T.J., 1976. *In* Tomes, G.J., Robertson, D.E. and Lightfoot, R.J. (eds) *Sheep Breeding*. Western Australian Institute of Technology, Perth, 509-523.
Robinson, T.J. and Moore, N.W., 1967. *In* Robinson, T.J. (ed) *The Control of the Ovarian Cycle in the Sheep*, Sydney University Press, Sydney, 116-132.
Robinson, T.J. and Smith, J.F., 1967: *In* Robinson, T.J. (ed) *The Control of the Ovarian Cycle in the Sheep*, Sydney University Press, Sydney, 144-157.
Salamon, S., 1962. *Aust. J. Agric. Res., 13*, 1137-1150.
Salamon, S., 1967. *Aust. J. Exp. Agric. Anim. Husb., 7*, 559-561.
Salamon, S., 1971. *Aust. J. Biol. Sci., 24*, 183-185.
Salamon, S., 1972. *Proc. 7th Int. Congr. Anim. Reprod. Artif. Insem., M*ünchen, II, 1493-1495.
Salamon, S., 1976. *Artificial Insemination of Sheep*. Publicity Press Ltd., Chippendale, N.S.W., 104 pp.
Salamon, S., 1977. *Aust. J. Agric. Res., 28*, 477-479.
Salamon, S., 1980. *Proc. 9th Int. Congr. Anim. Reprod. Artif. Insem., Madrid, V*, 420-421.
Salamon, S. and Lightfoot, R.J., 1967. *Aust. J. Agric. Res., 18*, 959-972.
Salamon, S., and Lightfoot, R.J., 1970. *J. Reprod. Fert., 22*, 409-423.
Salamon, S., Maxwell, W.M.C. and Firth, J.H., 1979. *Animal Reproduction Science, 2*, 373-385.
Salamon, S., and Robinson T.J., 1962. *Aust. J. Agric. Res., 13*, 271-281.
Salamon, S., and Visser, D., 1972. *Aust. J. Biol. Sci., 25*, 605-618.
Salamon, S., and Visser, D., 1974. *J. Reprod. Fert., 37*, 433-435.
Smagulov, S.B. and Martynov, J.V., 1966. *Ovtsevodstvo, 12(8)*, 30-31.
Vanden Bosch, G.C. and Hafez, E.S.E., 1975. *The Journal of Reproductive Medicine, 14*, 117-122.
Watson, P.F. and Martin, I.C.A., 1976. *Theriogenology, 6*, 553-558.
Willett, E.L. and Salisbury, G.W., 1942. *Cornell University Agric. Exp. Sta. Memorandum* No. 249.

# PRESELECTION OF SEX OF LAMBS BY LAYERING SPERMATOZOA ON PROTEIN COLUMNS

I.G. White and G. Mendoza, *Department of Veterinary Physiology, University of Sydney, Sydney, N.S.W. 2006.*
W.M.C. Maxwell,[1] *Department of Animal Husbandry, University of Sydney, Sydney, N.S.W. 2006.*

[1] Present address: Animal Breeding and Research Institue, Katanning, W.A. 6317.

*Summary* The possibility of producing a preponderance of either male or female lambs, as required, would seem to have obvious advantages to the sheep industry. Preliminary studies indicate that it may be possible to separate the male and female producing spermatozoa of the ram by diluting semen on a column of bovine serum albumin (BSA) and allowing the spermatozoa to swim into it. In this trial, involving 87 ewes, spermatozoa from the top of the BSA column produced 36.4% male and 63.6% female lambs, while spermatozoa from the bottom of the column produced 75.0% male and 25.0% female offspring.

## INTRODUCTION

Altering the sex ratio in the human population could have important social implications (Westoff and Rindfuss 1974) and application of the technique to farm animals should prove beneficial in the selection of superior breeding stock. Varying the proportion of X and Y chromosome bearing spermatozoa in the semen is one possible way of altering the sex ratio and attempts to separate the two types of spermatozoa have been reviewed by Beatty (1974). Although these results have not been very consistent, a recent technique which also enhances semen quality, offers a new approach to sex preselection. It is based on the ability of spermatozoa bearing the Y chromosome to outdistance slower swimming spermatozoa bearing the X chromosome in a viscous medium containing protein. In this way it has been possible to isolate up to 85% of Y chromosome bearing human spermatozoa of which 90-98% were progressively motile (Ericsson *et al.* 1973). Attempts have been made to separate the spermatozoa of farm animals such as the stallion and pig using a modification of this technique (Goodeaux and Kreider 1978; Dixon *et al.* 1980). However, Dixon *et al.* (1980) were not able to modify piglet sex ratio. Maxwell *et al.* (1984) obtained an improved post-thawing survival of ram spermatozoa when diluted semen was layered on a column of 6% bovine serum albumin (BSA) in Tris diluent, and the bottom layer of the column isolated for freezing after two hours holding time. The aim of the present experiment was to assess the sex ratio of lambs from synchronised ewes which were inseminated with ram spermatozoa separated on such protein (BSA) columns.

## MATERIALS AND METHODS

Three Merino rams were electroejaculated two or three times each week for six weeks during September and October. Semen samples were assessed for motility of spermatozoa on a 0-5 scale and pooled if the score was four or higher.The pooled semen was extended at room temperature to a concentration of $200 \times 10^6$ spermatozoa per ml with diluent A: 300 mM fructose and 95mM citric acid. In test tubes (10 cm long, 1.4 cm i.d.) the diluted semen (2 ml) was layered onto 6 ml of diluent B: 360 mM Tris, 33.3 mM glucose, 113.7 mM citric acid and 6% (w/v) BSA. After 2 hours holding at room temperature, the top (2 ml) and bottom (6 ml) portions of the column were collected in 10 ml centrifuge tubes and spun at 1000 g for 15 minutes; the supernatants were discarded and the spermatozoa from appropriate portions pooled and reconsitiued with diluent A. The suspensions of spermatozoa were recentrifuged, the supernatants discarded and the spermatozoa diluted 1:1 with diluent C: 360 mM Tris, 33 mM glucose, 113 mM citric acid, 18% (v/v) egg yolk and 6% (v/v) glycerol. The diluted spermatozoa were cooled to 5°C in 1½ hours, pellet frozen on dry ice and stored in liquid nitrogen until insemination (Visser and Salamon 1973).

Eighty seven Merino ewes were treated with intravaginal sponges (30 mg Chronogest, Intervet Australia Pty Ltd, Robinson 1965) for 12 days. At sponge removal each ewe received an intramuscular injection of 400 i.u. PMSG (Gravamed, Beresford Laboratories). The ewes were divided into two groups (a and b) and received two cervical inseminations (50 and 60 hours after sponge removal) with $100 \times 10^6$ motile spermatozoa from top (group a) or bottom (group b) portions of the BSA column.

## RESULTS

Lambing data for ewes inseminated with spermatozoa from the top and bottom portions of the BSA columns are summarised in Table 1. Chi square comparisons of the numbers of male lambs after inseminating spermatozoa from the top and bottom of the BSA column showed a significant difference ($P < 0.01$), as did comparisions of the number of female lambs ($p < 0.01$). There was no significant difference in fertility of ewes inseminated with sperm from the top and bottom portions of the columns.

**TABLE 1. Lambing following insemination of spermatozoa from the top and bottom of BSA columns.**

| BSA column portion | No. ewes treated | No. lambed (%) | No. males (%) | No. females (%) |
|---|---|---|---|---|
| Top | 44 | 21 (47.7) | 8 (36.4) | 14 (63.3) |
| Bottom | 43 | 25 (58.1) | 18 (75.0) | 6 (25.0) |

## DISCUSSION

The results of this experiment showed that ewes inseminated with spermatozoa from the bottom portion of the BSA column produced more male, and correspondigly fewer female, lambs than did spermatozoa from the top portion. Thus the Ericsson *et al.* (1973) technique modified for ram spermatozoa, raises the possibility of separating X and Y chromosome bearing spermatozoa populations in the sheep. Unfortunately, it is not possible to identify these two types of sperm in the semen of the ram or other domestic species although the Y chromosome of human sperm can be stained with quinacrine (Barlow and Vosa 1970). The additional beneficial effects of this separation procedure observed in other species are increased sperm motility, reduced seminal debris and perhaps reduced sperm with abnormal morphology (Ericsson and Glass 1982).

Insemination of spermatozoa separated on albumin columns has apparently resulted in a significant shift in the sex ratio of man, cattle and rabbits (Ericsson and Glass 1982). The results reported here suggest that this is also true for the sheep, and while caution is necessary in evaluating an experiment based on relatively small numbers of ewes, the technique seems sufficiently promising to warrant more extensive artificial insemination trials.

## ACKNOWLEDGEMENTS

The authors acknowledge the valuable help of Dr S. Salamon, Department of Animal Husbandry, University of Sydney, Dr R.J. Ericsson, Gametrics Limited, California and Mrs B. Maxwell.

## REFERENCES

Barlow, P. and Vosa, C.G., 1970. *Nature, London, 226*, 961-962.

Beatty, R.A., 1974. *Bibliography of Reproduction, 23*, 120-129.

Dixon, R.E., Songy, E.A., Thrasher, D.M. and Kreider, J.L., 1980. *Theriogenology, 13*, 437-444.

Ericsson, R.J. and Glass, R.H., 1982. *In* Amann, R.P. and Seidel, G.E., *Prospects for Sexing Mammalian Sperm*, Colorado Association of University Press, Boulder, 201-211.

Ericsson, R.J., Langevin, C.N. and Nishino, H., 1973. *Nature, London, 246*, 421-424.

Goodeaux, S.D. and Krieder, J.L., 1978. *Theriogenology, 10*, 405-414.

Maxwell, W.M.C., Mendoza, G. and White, I.G. 1984. *Theriogenology, 21*, 601-607.

Robinson, T.J., 1965. *Nature, London, 206*, 39-41.

Visser, D. and Salamon, S., 1973. *Aust. J. Biol. Sci., 26*, 513-516.

Westoff, C.F. and Rindfuss, R.R., 1974. *Science, 184*, 633-636.

# SEMEN DEPOSITION AND FERTILITY IN OVINE ARTIFICIAL BREEDING PROGRAMMES

M.D. Rival, P.J. Chenoweth, and L.I. McMicking, *Pastoral Veterinary Centre, University of Queensland, Goondiwindi 4390.*

*Summary* Comparison of 'over-the-rail' and 'shot-in-the-dark' insemination techniques in naturally cycling ewes gave a significant ($P < 0.005$) advantage in non-return-to-service rates for the former when using fresh extended semen (150 and $300 \times 10^6$ motile spermatozoa in 0.1 ml). Non-return-to-service rates following the insemination of naturally cycling ewes by laparoscopy ($50 \times 10^6$ motile spermatozoa; frozen-thawed) were equivalent to those following both 'over-the-rail' insemination ($100 \times 10^6$ motile spermatozoa; fresh, extended) and natural service (2% rams). Non-return-to-service rates following the insemination of fresh extended semen by laparoscopy ($50 \times 10^6$ motile spermatozoa) did not differ significantly from either 'over-the-rail' insemination ($200 \times 10^6$ motile spermatozoa) or natural service (5% rams) in oestrous-induced ewes. The use of a device which expels semen with the aid of compressed $CO_2$ was compared with the standard 'over-the-rail' insemination procedure at two dose rates (50 and $100 \times 10^6$ motile spermatozoa). Neither method nor dose rate had any significant effect on non-return-to-service rates.

## INTRODUCTION

Ovine artificial insemination (AI) programmes offer studbreeders a means of achieving rapid genetic improvement by using fewer and younger rams, by facilitating performance testing of sires, and by enabling them to share expensive rams or to introduce diverse genetic material (Fairnie and Wales 1982; Hewett 1982).

Nevertheless, AI programmes employing fresh semen have limited application due to the small number of ewes which can be allocated to individual sires. In general, approximately 30 ewes can be allotted per sire per day (Fairnie and Wales 1982) assuming a standard inseminate of $120 \times 10^6$ motile spermatozoa (Salamon 1962) from rams producing $4000 \times 10^6$ spermatozoa per day (Cameron *et al.* 1984). In addition, spare rams are required to compensate for possible low fertility during the AI programme (Hewett 1982). With frozen-thawed semen, deposition in the ewe's cervix results in variable fertility (Fukui and Roberts 1976; Maxwell *et al.* 1984) and although AI by laparoscopy (AI/LAP) presents a promising alternative (Killeen *et al.* 1982; Maxwell *et al.* 1984) it is time-consuming and requires expensive equipment.

Site of semen deposition is the major factor affecting efficiency, labour input, and ram:ewe ratios in sheep AI programmes. The purpose of this work was to investigate the performance of conventional AI methods (i.e. deposition site and insemination dose) and to make a preliminary assessment of a new concept in AI of ewes.

## MATERIALS AND METHODS

### Animals

A total of 3305 adult Merino ewes (South Australian strain) were randomly allocated to the different AI trials and control groups. These trials were conducted during two consecutive Autumn breeding seasons (1982-83) except for the AI/LAP trial which was conducted in December 1983. Numbers and treatments of ewes are shown in Table 1.

**TABLE 1. Experimental groups and treatments.**

| Trial | *Number of ewes* | | Oestrous Manipulations | Breeding Period |
|---|---|---|---|---|
| | AI Group | Controls | | |
| SID/OTR | 372 | 356 | naturally cycling | Autumn 1982 |
| AI/LAP(fresh)/OTR | 256 | 141 | oestrous-induced | December 1983 |
| AI/LAP(frozen)/OTR | 919 | 63 | naturally cycling | Autumn 1983 |
| $CO_2$ assisted/OTR | 1198 | - | naturally cycling | Autumn 1983 |

Teaser wethers, prepared with three weekly parenteral injections of 150mg testosterone (Banrot, Cooper Wellcome; Croker *et al.* 1982), were equipped with mating harnesses and placed with ewes at ratios of 2% (natural oestrus) and 5% (induced oestrus). Oestrous detection for AI commenced on completion of the injection regime.

Rams for AI were screened for semen quality, trained to use the artificial vagina, and placed in sheds for eight weeks prior to use.

### Synchronisation of oestrus

Ewes in AI/LAP investigations on fresh semen deposition were prepared with a 12 day intravaginal progestagen-impregnated sponge treatment (Chronogest 30mg, Intervet) immediately followed by a subcutaneous injection of PMSG (Pregnecol 400 i.u., Livestock Laboratories). AI, or mating, occurred at the first oestrus after sponge removal. Ewes in all other trials were inseminated, or served, at natural oestrus.

Semen collection and handling

Semen destined for immediate use was assessed for concentration and was required to pass minimum standard for mass activity (++) and individual motility (70%) before extension in UHT milk diluent (Kerton *et al.* 1984) to give an inseminate volume of 0.1 ml. For AI/LAP, spermatozoal motility was regularly monitored. Extended semen was maintained at 33°C in a water-bath until used. Semen for freezing was processed as described by Visser and Salamon (1973) to give a final inseminate containing no less than $50 \times 10^6$ motile spermatozoa in .05ml.

Insemination

Frozen semen pellets were individually thawed at 37°C and used within 15 minutes. Fresh extended semen was used within 60 minutes of collection. In all trials, observations for oestrus were conducted at 7.00 a.m. and 6.00 p.m. and insemination occurred eight to 14 hours after oestrous ewes were segregated.

'Shot-in-the-dark' (SID) and 'over-the-rail' (OTR) techniques employed were as previously described (Hewett 1982; Kerton *et al.* 1984). AI/LAP was performed in the manner described by Maxwell *et al.* (1984). Each uterine horn received .025 ml of semen containing $25 \times 10^6$ spermatozoa. For $CO_2$ assisted insemination, the AI pipette was inserted in the cervical opening with the aid of a speculum and headlamp. This pipette was modified to include a collar near its tip which formed a seal with the cervix to minimise backflow. A hand-held device containing a compressed $CO_2$ bulb ('Sparklet') was connected to the pipette via a three-way tap. This tap permitted semen to be drawn into the pipette prior to expulsion which was brought about by a one second burst of approximately 100ml of $CO_2$. This procedure was compared with OTR at two dose rates of fresh extended semen (50 and $100 \times 10^6$ motile spermatozoa).

Evaluation and analysis

Treatments were evaluated by non-return-to-service rates (NR) at the subsequent oestrous cycle obtained by using entire rams equipped with mating harnesses. Service marks were checked three times over 21 days. Previous trials in this environment have shown high correlations between NR and lambing rates (Rival and McMicking, unpublished data), so additional management procedures to facilitate the collection of lambing data were not considered. Data were subject to chi square analysis.

## RESULTS

SID v OTR

The SID technique of insemination produced non-return rates which were 15% lower than the OTR technique or natural service, regardless of the number of spermatozoa used. The numbers of spermatozoa did not affect the NR rate when SID was used but with the OTR method $300 \times 10^6$ spermatozoa were superior to $150 \times 10^6$ spermatozoa per dose ($P < 0.05$). Overall, there was no difference in the return rates (50%) between OTR insemination and natural service with 2% rams.

AI by laparoscopy

AI by laparoscopy using fresh extended semen and $50 \times 10^6$ motile spermatozoa gave a non-return rate of 52%. This was identical to the non-return rate using either OTR with $200 \times 10^6$ motile spermatozoa or natural service with 5% rams.

Frozen-thawed semen deposited by laparoscopy in a dose of $50 \times 10^6$ motile spermatozoa gave a non-return rate of 60%, which was not significantly different from 67% using OTR and $100 \times 10^6$ fresh spermatozoa, or 56% with natural service and 2% rams.

$CO_2$-assisted insemination

Using $CO_2$-assisted insemination and OTR at doses of either 50 or $100 \times 10^6$ motile spermatozoa we obtained non-return rates of 63%.

## DISCUSSION

Our results show that the SID method is inferior to the more conventional system of OTR when insemination was at 14 h after the detection of oestrus. This agrees with the conclusions of Kerton *et al.* (1984) who inseminated ewes at a fixed time after a synchronisation treatment. We were not able to reproduce the finding of Fairnie and Wales (1982) or Hewett (1982) who found no difference between the two techniques. Hewett (1982) in particular, claimed that the SID method was preferable to OTR for maiden ewes because the SID method gave better rates of pregnancy. In other trials, we have not found any difference between maidens and adult ewes using the OTR method (Rival, unpublished data) but we always use a smaller speculum with maiden ewes.

AI following laparoscopy, even with frozen semen is equally as successful as the OTR method using fresh semen but has the advantage of using less spermatozoa per dose. The advantage has also been shown by Killeen and Caffery (1982), Killeen *et al.* (1982), Kerton *et al.* (1984) and Maxwell *et al.* (1984).

These results show that AI can be used successfully to extend the influence of individual rams over natural service and among AI techniques, laparoscopy allows more ewes to be inseminated than the conventional techniques.

The method of propelling semen using $CO_2$ proved to be no better than the normal method of insemination in this series of experiments even when doses were relatively low. The success of doses as small as $50 \times 10^6$ spermatozoa is surprising in view of the work of Salamon (1962) who showed a linear response in fertility to numbers of spermatozoa per inseminate from $25 \times 10^6$ to $125 \times 10^6$. However Entwistle and Martin (1972) obtained maximal fertility with only $50 \times 10^6$ spermatozoa so that we may not have had sufficiently few spermatozoa to emphasise differences between the two techniques. In any case the use of $CO_2$ was not detrimental to NR rates which suggests that further work on this method is warranted.

## ACKNOWLEDGEMENTS

The authors are indebted to Mr. E.G. Sylvester, 'Undabri Merino Stud', Goondiwindi, for supplying trial sheep, facilities, and labour. Work was supported by University of Queensland research funds.

## REFERENCES

Cameron, A.W.N., Fairnie, I.J., Curnow, D.H., Keogh, E.J. and Lindsay, D.R., 1984. *Proc. Aust. Soc. Anim. Prod., 15*, 659.

Croker, K.P., Butler, L.G., Johns, M.A. and McColm, S.C., 1982. *Theriogenology, 17*, 349-354.

Entwistle, K.W. and Martin, I.C.A., 1972. *Aust. J. Agric. Res., 23*, 467-472.

Fairnie, I.J. and Wales, R.G., 1982. *In* Barton, R.A. and Smith, W.C. (eds). *Proceedings of the World Congress on Sheep and Beef Cattle Breeding Volume II: General*, 311-320.

Fukui, Y. and Roberts, E.M., 1976. *In* Tomes, C.J., Robertson, D.E. and Lightfoot, R.J. (eds). *Sheep Breeding, Proceedings 1976 International Congress, Muresk and Perth, Western Australia*. W.A. Institute of Technology, Perth, W.A., 482-494.

Hewett, L., 1982. *Wool Technol. Sheep Breed., 30*, 87-91.

Kerton, D.J., McPhee, S.R., Davis, I.F., White, M.B., Banfield, J.C. and Cahill, L.P., 1984. *Proc. Aust. Soc. Anim. Prod., 15*, 701.

Killeen, I.D., Caffery, G.J. and Holt, N., 1982. *Proc. Aust. Soc. Reprod. Biol., 14*, 104.

Maxwell, W.M.C., Wilton, H.R. and Butler, L.G., 1984. *Proc. Aust. Soc. Anim. Proc., 15*, 448-457.

Salamon, S., 1962. *Aust. J. Agric. Res., 13*, 1137-50.

Visser, D. and Salamon, S., 1973. *Aust. J. Biol. Sci., 26*, 513-516.

# UTERINE ARTIFICIAL INSEMINATION IN EWES

I.F. Davis, D.J. Kerton, S.R. McPhee, M.B. White, J.C. Banfield and L.P. Cahill, *Department of Agriculture, Animal Research Institute, Werribee, Victoria, 3030.*
I. Grant, *University of Melbourne, Veterinary Clinical Centre, Werribee, Victoria, 3030.*

*Summary* In November 1983, uterine insemination with the aid of a laparoscope was conducted on 800 synchronized Merino ewes with different doses of spermatozoa in 0.1 ml of fresh, extended semen which had been diluted with UHT skim milk after collection by artificial vagina. Ewes were inseminated 50-56 hours after removal of progestagen vaginal sponges. Pregnancy rates for groups showing oestrus and inseminated with 100, 50, 25 or 12.5 $\times 10^6$ sperm were 71, 76, 63 and 70% respectively and did not differ significantly. Ewes not showing oestrus at AI had pregnancy rates of 50, 59, 41 and 54% which did not differ between sperm doses but were significantly lower than those of ewes showing oestrus before AI.

## INTRODUCTION

Artificial insemination (AI) in sheep offers the potential for rapid genetic gain by allowing the use of superior sires over a large number of ewes. In the ewe it has not been practicable to penetrate the cervix and deposit a minimal number of spermatozoa into the uterus as is possible in the cow. The usual method of insemination of ewes has been by deposition of semen at the external opening of the cervix using a pipette introduced through a vaginal speculum. This anatomical restriction combined with some loss of fertility of ram semen on dilution can be countered partially by the use of larger numbers of spermatozoa per insemination dose. Estimates, in the literature, of minimal effective sperm doses for cervical insemination are variable. Salamon (1962) recommended a dose of 125 $\times 10^6$ fresh spermatozoa and observed a 13% decrease in fertility for each 25 $\times 10^6$ reduction in sperm dose. Entwhistle and Martin (1972) found no significant difference in fertility after using doses of 50 or 100 $\times 10^6$ spermatozoa in fresh duluted semen. They found that fertility decreased with time after semen collection. Lightfoot and Salamon (1970) deposited fresh or frozen semen (160 $\times 10^6$ sperm per ewe) into the uterine horns after laparotomy but found that this method increased embryonic mortality.

Killeen and Caffery (1982) developed a technique for intrauterine insemination with the aid of a laparoscope. Field tests of the technique where 60 $\times 10^6$ frozen-thawed spermatozoa were inseminated into the uterus achieved a non-return rate of 69% (Killeen *et al.* 1982). Maxwell *et al.* (1983) compared cervical insemination of 100 $\times 10^6$ spermatozoa in fresh semen with uterine insemination using a laparoscope of 40 $\times 10^6$ spermatozoa in frozen semen and reported conception rates of 72 and 54% respectively.

The experiment reported here was conducted to examine the effect of sperm dose on pregnancy rate after a synchronized oestrus in ewes inseminated into the uterus with fresh duluted semen by the aid of a laparoscope.

## MATERIALS AND METHODS

In November 1983, 800 parous Merino ewes, in groups of 200 per day over four days, had vaginal sponges (Chronogest 40 mg, Intervet) inserted for 12 days and were injected s.c. with 400 i.u. of PMSG (Folligon, Intervet) at sponge removal and joined with 4% of vasectomized rams. Ewes were checked for oestrus and inseminated 50-56 h after sponge removal in the order in which they had shown oestrus. Three rams were used and a total of 11 ejaculates was collected by artificial vagina. All ejaculates were used. Each ejaculate was assessed microscopically for percentage motility and total sperm numbers were counted by haemocytometer. The average characteristics for the ejaculates were:- volume 1.5 ml, total sperm count 5.4 $\times 10^9$, initial motility 88%, time from collection to final insemination 3½ hours, final motility 74%. Each ejaculate was diluted individually in a stepwise manner with UHT skim milk to provide 4 lots of semen containing 100, 50, 25 and 12.5 $\times 10^6$ total sperm in 0.1 ml of diluted semen which was held in a water bath at 32°C. Ewes were inseminated with 0.05 ml of diluted semen into each uterine horn. Pregnancy rate was determined by ultrasonic scanning 70 days after AI. Statistical significance was determined by analysis of Chi-square.

## RESULTS

There were no significant differences in pregnancy rate between sperm doses (Table 1), however ewes which had not shown oestrus by AI had a lowered pregnancy rate.

Pregnancy rates of 73 oestrous ewes receiving correct sperm doses when inseminated, two hours or more after semen collection, with 100, 50, 25 or 12.5 $\times 10^6$ spermatozoa were 76, 77, 77 and 75% respectively.

## DISCUSSION

This study has shown that the sperm dose of freshly diluted semen can be reduced from 100 to 12.5 million spermatozoa per ewe without reducing the pregnancy rate. Technically, it is possible for 200 ewes per day to be inseminated using the semen of one ram by the use of intrauterine insemination. A similar study using frozen-thawed semen is warranted so that, if a relatively low sperm dose can be defined, the advantages offered by the use of deep-frozen semen can be exploited.

**TABLE 1. Pregnancy rates, %, of ewes after intrauterine AI (n = number of ewes per category).**

| Ewe classification | | Number of spermatoza ($10\times^{(6)}$) per AI dose | | | |
|---|---|---|---|---|---|
| Oestrus detected | Complete insemination | 100 | 50 | 25 | 12.5 |
| + | + | 71(136) | 76(129) | 63(146) | 70(142) |
| + | − | 73(15) | 63(19) | 71(7) | 67(12) |
| − | + | 50*(30) | 59*(32) | 41*(37) | 54*(35) |
| − | −(1) | 80(10) | 23(13) | 40(5) | 50(4) |

*$P < 0.05$; (1)data not analysed.

The optimum time of insemination while allowing for variation in onset of oestrus and the time taken in handling ewes for AI needs to be determined. More frequent checks for oestrus may be the answer. The Controlled Internal Drug Release (CIDR) dispensers developed by Welsh (1983) may provide improved synchronization compared to progestagen sponges. Peterson *et al.* (1984) increased pregnancy rate in ewes joined to rams from 67% in control ewes to 95% in ewes which had a CIDR re-inserted from days 10 to 16 after removal of a synchronizing CIDR. If such an increase in pregnancy rate could be obtained consistently it would be of economic benefit to uterine AI. It may be possible to reduce the labour component of 5 or 6 people involved a program of uterine AI by modifying existing sheep handling equipment. The Cassou (1984) multidose syringe needs to be assessed as a possible more effective means of insemination.

ACKNOWLEDGEMENTS

The authors are indebted to Mr. Graham Mills, Barunah Plains, Wingeel, Victoria, for his generous provision of sheep and facilities.

REFERENCES

Cassou, R., *Proc. 10th Inter. Congr. Anim. Reprod. and A.I.* p. 361.
Entwistle, K.W. and Martin, I.C.A., 1972. *Aust. J. Agric. Res., 23*, 467-472.
Killeen, I.D. and Caffery, G.J., 1982. *Aust. Vet. J., 59*, 95.
Killeen, I.D., Caffery, G.J. and Holt, N., 1982. *Proc. Aust. Soc. Reprod. Biol., 14*, 104.
Lightfoot, R.J. and Salamon, S., 1970. *J. Reprod. Fert., 22*, 399-408.
Maxwell, W.M.C., Butler, L.G. and Wilson, H.R., 1983. *Proc. Aust. Soc. Reprod. Biol., 15*, 92.
Peterson, A.J., Barnes, Dianne, Shanley, Robyn and Welsh, R.A.S., 1984. *Proc. N.Z. Endocr. Soc., 21*, 13.
Salamon, S., 1962. *Aust. J. Agric. Res., 13*, 1137-1150.
Welsh, R.A.S., 1983. *Proc. Endocr. Soc. Aust., 25(Suppl. 2)*, 46.

# ARTIFICIAL INSEMINATION AND TRANSFER OF EMBRYOS BY LAPAROSCOPY

S.K. Walker, D.H. Smith and D.L. Little, *Department of Agriculture, Box 1671, G.P.O., Adelaide, S.A. 5001.*
G.M. Warnes, P. Quinn and R.F. Seamark, *Department of Obstetrics and Gynaecology, Queen Elizabeth Hospital, Woodville, S.A. 5011.*

*Summary* Intra-uterine insemination of Merino ewes with between 1 x $10^6$ and 10 x $10^6$ motile fresh or thawed-frozen spermatozoa per uterine horn was investigated in superovulated and non-superovulated ewes. The recovery of embryos from superovulated ewes was unsatisfactory (0.7-2.3 per ewe group) and could not be wholly attributed to the fertilization rate (30-70%). Fertilization rates were not affected by the number of spermatozoa inseminated but a significantly ($P < 0.01$) greater proportion of fertilized ova were obtained with fresh semen (61.2%) compared with thawed-frozen semen (42.9%). Ovulation rates were significantly ($P < 0.01$) higher during the breeding season than during anoestrus when stimulation of superovulation was induced in conjunction with the ram teasing effect. However, the corresponding increase in the number of embryos collected per ewe was not significant.

In non-superovulated ewes, oestrous synchronization with a synthetic prostaglandin (Cloprostenol) produced pregnancy rates below those with intravaginal progestagen sponge (13.8-24.1% versus 54.5%) after insemination with thawed-frozen semen.

Transfer of embryos to the uterine lumen by laparoscopy resulted in 40.7% of recipient ewes lambing or being pregnant when slaughtered.

## INTRODUCTION

Recent development of an insemination technique by laparoscopy in sheep (Killeen and Caffrey 1982), whereby semen is deposited directly into the uterus, is of particular interest because it increases conception rates when using frozen semen. The procedure involves oestrous synchronization by treatment with an intravaginal progestagen sponge and pregnant mare serum gonadotrophin (PMSG) with insemination 48-60 h after sponge removal (Maxwell *et al.* 1983). The number of spermatozoa inseminated per ewe has varied between 40 $\times 10^6$ and 120 $\times 10^6$ (Killeen *et al.* 1982; Maxwell *et al.* 1983). The following are examined in this study in an attempt to refine the technique further:— (1) conception rates following intra-uterine insemination with between 1 $\times 10^6$ and 10 $\times 10^6$ motile fresh or thawed-frozen spermatozoa per uterine horn in superovulated and non-superovulated ewes and (2) the use of the synthetic prostaglandin Cloprostenol (PG) (ICI Pty. Ltd., Melbourne) for oestrous synchronization.

The feasibility of embryo transfer by laparoscopy was also examined as part of an on-going program to improve embryo collection and transfer procedures in the sheep.

## MATERIALS AND METHODS

### Semen preparation and insemination

Semen was collected by electro-ejaculation and, where appropriate, frozen using the procedure described by Visser and Salamon (1974) and stored in liquid nitrogen. Pellets were thawed at 40°C and about 5 ewes were inseminated from each batch of thawed semen. Fresh semen was collected on the day of insemination and held at 32° after appropriate dilution. Frozen semen was used within 40 minutes of thawing and fresh semen within two hours of collection. Insemination, by laparoscopy, was similar to the procedure of Killeen and Caffrey (1982) but the fine pointed pipettes were made from 1 mm bore boro-silicate glass tubing. For insemination, the internal glass surface of each pipette was wetted with buffered diluent before being loaded with 2 $\times$ 20 $\mu$l doses of semen. Each semen dose was drawn into the pipette behind a column of air of approximately 50 mm. This technique allowed the insemination, under visualization through the laparoscope, of both uterine horns without reloading. Each pipette was used for up to 30 inseminations.

### Experimental procedure

*Insemination of superovulated ewes* Merino ewes were superovulated in November, 1983 (n = 58), January (n = 22) and April 1984 (n = 25). Ewes were treated with intravaginal sponges impregnated with 60 mg medroxy progesterone acetate (Repromap, Upjohn Pty. Ltd., Sydney) for 12 days and injected (i.m.) with 1250 IU PMSG (Pregnecol, Livestock Laboratories Pty. Ltd., Melbourne) at the time of sponge removal. Anoestrous ewes (Nov. 1983) and ewes treated at the beginning of the breeding season (Jan. 1984) were kept isolated from rams until the time of sponge removal when vasectomized rams fitted with marker harnesses were introduced. In April, approximately half the ewes were superovulated as described, while the remainder were treated with PG (2 $\times$ 100 ug i.m. injections, 11 days apart), and PMSG (1250 IU i.m.) was injected at the time of the second PG injection. Inseminations commenced either 54 or 64 hours after PMSG injection. Ewes received either 1 $\times 10^6$ or 10 $\times 10^6$ motile spermatozoa per uterine horn from a pool of either fresh or thawed-frozen semen according to the protocol in Table 1.

At laparotomy 4 days after insemination, the number of corpora lutea per ewe was counted and ova were collected and examined for fertilization (Warnes *et al.* 1982). After collection, embryos were maintained in Hepes-buffered synthetic oviduct fluid medium containing 10% sheep serum (Quinn *et al.* 1984) before being transferred to recipient ewes (see below).

*Insemination of non-superovulated ewes* Oestrous activity was synchronized by two injections of PG (100 ug i.m.) given 11 days apart in Flock 1 (Feb. 1984, 240 Merinos) and Flock 2 (March 1984, 70 Merinos). Ewes in Flock 3 (May 1984, 171 Merinos) were randomly allocated to one of the following three treatments — (1) double PG injection regimen as for Flocks 1 and 2 (2) single PG injection (100 ug i.m.) or (3) intravaginal progestagen sponge for 12 days and 400 IU PMSG on the day of sponge removal. Ewes in Flocks 1 and 2 received 1 × $10^6$ and ewes in Flock 3 received 6 × $10^6$ motile thawed-frozen spermatozoa per uterine horn. Pregnancy rates were determined by inspection of the uterus 30 days after insemination using laparoscopy.

*Embryo transfer by laparoscopy* Embryos were transferred approximately one hour after collection to recipient Merinos which had come into oestrus after treatment with progestagen sponges for 12 days followed by 400 IU PMSG. Two embryos in 20 $\mu$l medium were transferred to the uterine horn ipsilateral to an ovary containing at least 1 corpus luteum. The transfer technique using laparoscopy was similar to that for artificial insemination. Recipients (n = 54) either lambed or were slaughtered between days 30-90 of pregnancy to determine numbers of foetuses.

## RESULTS

### Insemination of superovulated ewes

Observations in November with ewes treated with progestagen sponges and PMSG indicated no differences between the numbers of spermatozoa inseminated in either the proportion of fertile ewes (i.e. those with at least one embryo) or the proportion of recovered ova which were fertilized (Table 1). The time of insemination (54 h versus 64 h after sponge removal) had no effect.

**TABLE 1. No. fertile ewes (i.e. ewes with at least one embryo), fertilization rates (ova fertilized/ova collected) and no. embryos recovered per ewe in superovulated ewes inseminated, using a laparoscope, with either thawed-frozen or fresh semen in November, January and April.**

| Month | No. motile[1] sperm per horn (× $100^6$) | Time to insemination (h) | No. fertile ewes | Fertilisation rate (% in brackets) | Embryos per ewe | |
|---|---|---|---|---|---|---|
| | | | | | $\bar{X}$ | range |
| *Thawed-frozen semen* | | | | | | |
| Nov. | 1 | 54 | 2/5 | 10/17(59) | 2.0 | 0-6 |
| | 10 | 54 | 10/15 | 17/38(45) | 1.1 | 0-3 |
| | 1 | 64 | 5/11[3] | 8/24(33) | 0.7 | 0-2 |
| | 10 | 64 | 6/9 | 14/29(48) | 1.4 | 0-4 |
| Jan. | 1 | 54 | 5/10 | 11/33(33) | 1.1 | 0-4 |
| Apr. | 10 | 54 | 3/6 | 8/14(57) | 2.3 | 0-3 |
| | 10[2] | 54 | 3/7 | 11/29(38) | 1.6 | 0-6 |
| *Fresh semen* | | | | | | |
| Nov. | 1 | 54 | 6/8 | 17/24(71) | 2.1 | 0-6 |
| | 10 | 54 | 5/10 | 11/25(44) | 1.1 | 0-4 |
| Jan. | 1 | 54 | 6/12 | 20/27(74) | 1.7 | 0-5 |
| Apr. | 10 | 54 | 5/6 | 11/18(61) | 1.8 | 0-4 |
| | 10[2] | 54 | 5/6 | 12/22(55) | 2.0 | 0-5 |

(1) motility assessed to be 100% in fresh semen and 30% in thawed-frozen semen.
(2) these ewes were treated with 2 × 100 ug Cloprostenol (11 days apart) and 1250 IU Pregnecol; all other ewes were treated with 60 mg Repromap sponges and 1250 IU Pregnecol.
(3) two semen preparations used; all fertile ewes were from one preparation.

Overall, insemination with fresh compared with thawed-frozen semen did not affect the proportion of fertile ewes ($\chi^2_1$ = 1.2; P > 0.05) but did increase the proportion of recovered ova which were fertilized (42.9% versus 61.2%, $\chi^2_1$ = 9.5; P < 0.01). Ovulation rates increased from 4.6 ± 0.37 in November to 6.8 ± 0.87 in January and 7.1 ± 0.77 in April (P < 0.01, one way analysis of variance) and the corresponding numbers of embryos recovered per ewe were 1.33 ± 0.20, 1.41 ± 0.38 and 1.68 ± 0.33 (P > 0.05, one way analysis of variance).

Insemination of non-superovulated ewes

Pregnancy rates were very low (range 13.8 - 20.7%) in all three flocks when the double PG regimen was used for oestrous synchronization (Table 2). They improved in Flock 3 when ewes were synchronized with one PG injection (24.1%) but were still less than half those achieved with the progestagen sponge — PMSG procedure (54.5%).

**TABLE 2. Ewe fertility (ewes pregnant/ewes inseminated) following insemination, using the laparoscope, of thawed-frozen semen in ewes in which oestrus was synchronized by either two prostaglandin injections 11 days apart (2PG), one prostaglandin injection (1PG) or by intravaginal progestagen sponge/pregnant mare serum gonadotrophin (Sp/PMSG).**

| | Flock 1 | Flock 2 | | Flock 3 | |
|---|---|---|---|---|---|
| Synchronization method | *2PG*(1) | *2PG* | *2PG* | *1PG*(1) | *Sp/PMSG*(2) |
| No. ewes inseminated | 240 | 70 | 58 | 58 | 55 |
| No. motile sperm. per(3) horn (×10⁶) | 1 | 1 | 6 | 6 | 6 |
| Ewe fertility (%) | 13.8 | 14.3 | 20.7 | 24.1 | 54.5 |

(1) 100 ug Cloprostenol per injection

(2) 60 mg Repromap sponge and 400 IU Pregnecol

(3) ewes inseminated 48 – 54 h after last PG injection or Sp withdrawal; thawed-frozen semen estimated to have 30% motilily.

Transfer of embryos by laparoscopy

40.7% of the ewes (22/54) either lambed or were pregnant when slaughtered. The number of foetuses/lambs per pregnant ewe was 1.6. This figure was significantly greater than expected ($\chi^2_1 = 38.4, P < 0.001$) based on the expected distribution of single and twin pregnancies calculated from the overall embryonic survival rate of 35/108.

## DISCUSSION

The low proportion of fertilized ova in this experiment was in contrast to the 90%+ reported for both fresh (Trounson and Moore 1974) and thawed-frozen semen (Killeen *et al.* 1982). Killeen *et al.* (1982) inseminated at total of $120 \times 10^6$ spermatozoa per ewe indicating that our dose rates may have been below that required for maximum fertilization rates.

Survival time of thawed spermatozoa in the reproductive tract relative to the time of ovulation is likely to affect fertilization rates. Delaying insemination from 54 hours to 64 hours did not affect the proportion of ova fertilized (Table 1) indicating that spermatozoa were able to effect fertilization at least over this time span. However, observation of the morphological appearance of corpora lutea at the time of embryonic recovery and the appearance of 'cumulus-like' material in some flushes suggest that some ovulations may have occurred beyond the fertilizable life of the spermatozoa.

Insemination with fresh semen did not significantly increase the proportion of ewes with embryos compared with the insemination of thawed-frozen semen but there was a greater proportion of fertilized ova and more embryos per ewe when fresh semen was used. These differences might have been less had the number of spermatozoa been higher. Nevertheless, pregnancy rates with thawed-frozen semen (40-83%) were better than those with cervical insemination in non-superovulated ewes (Maxwell *et al.* 1980).

Both ovulation rate and embryos collected per ewe were higher during the breeding season (April) than during anoestrus (November) when ovulation was stimulated in conjunction with the introduction of rams to the ewes. There are conflicting reports on the success of superovulatory treatments during anoestrus. Some studies have reported no difference from the breeding season (Evans and Robinson 1980) while others suggest a poorer response (Thimonier *et al.* 1975). These inconsistencies may be due to breed and even strain differences.

Overall, the recovery of embryos following superovulation was unsatisfactory. The best results obtained were still less than an average of 3 embryos per ewe. These results contrast markedly with similar experiments on goats where in excess of 8 embryos per doe are reliably recovered (Warnes *et al.* 1982).

In non-superovulated ewes, the pregnancy rate was satisfactory (54.5%) when ewes treated with progestagen sponges and PMSG were inseminated with low doses of thawed-frozen semen (Table 2). In contrast, the pregnancy rate was poor when oestrus was synchronized by PG. Low fertilization rates following oestrous synchrony with prostaglandins in cervical insemination programs (Fairnie *et al.* 1976; Fairnie and Wales 1980) are thought to be due to a low population of spermatozoa in the oviducts at the time of ovulation (Hawk 1973; Hawk *et al.* 1982). The poor pregnancy rates in our study are surprising given the evidence of Fukui and Roberts (1977) that numbers of spermatozoa in the oviducts of prostaglandin-treated ewes are high following intra-uterine insemination of thawed-frozen semen. Fairnie and Wales (1980) indicated that the double PG regimen such as we used in our study (i.e. 11 days between injections) might reduce the exposure of the reproductive tract to endogenous progesterone sufficiently to cause reproductive failure. Alternatively, prostaglandin treatment might otherwise affect ovarian function sufficiently to produce a uterine environment unfavourable for embryonic growth. Whatever the reason, PG cannot be recommended for oestrous synchronization in laparoscopic insemination programs.

The success of transfer of embryos by laparoscopy (40.7% pregnancy) was less than that expected for surgical transfer (55-75%) (Moore and Shelton 1962; Moore 1968). However, the technique is more rapid (< 5 minutes) and less invasive than laparotomy and can probably be developed further to provide an acceptable alternative method of transfer. Such development may also allow collection of embryos via laparoscopy. Our preliminary investigations indicate that this may be feasible when suitable instruments have been developed.

## REFERENCES

Evans, G. and Robinson, T.J. 1980. *J. Agric. Sci. Camb., 94*, 69-88.

Fairnie, I.J., Cumming, I.A. and Martin, E.R., 1976. *Proc. Aust. Soc. Anim. Prod., 11*, 133-136.

Fairnie, I.J. and Wales, R.G., 1980. *Proc. Aust. Soc. Anim. Prod., 13*, 317-320.

Fukui, Y. and Roberts, E.M., 1977. *J. Reprod. Fert., 51*, 141-143.

Hawk, H.W., 1973. *J. Anim. Sci., 37*, 1380-1385.

Hawk, H.W., Conley, H.H. and Cooper, B.S., 1982. *Theriogenology, 18*, 671-681.

Killeen, I.D. and Caffrey, G.J., 1982. *Aust. Vet. J., 59*, 95.

Killeen, I.D., Caffrey, G. and Holt, N., 1982. *Proc. Aust. Soc. Reprod. Biol., 14*, 104.

Maxwell, W.M.C., Butler, L.G. and Wilson, H.R., 1983. *Proc. Aust. Soc. Reprod. Biol., 15*, 92.

Maxwell, W.M.C., Curnock, R.M., Logue, D.N. and Reed, H.C.B., 1980. *Theriogenology, 14*, 83-89.

Moore, N.W., 1968. *Aust. J. Agric. Res., 19*, 295-302.

Moore, N.W. and Shelton, J.N., 1962. *Aust. J. Agric. Res., 13*, 718-724.

Quinn, P., Warnes, G.M., Walker, S.K. and Seamark, R.F., 1984. This volume, 289-290.

Thimonier, J., Cognie, V., Cornu, C., Schneberger, J. and Vernusse, G., 1977. *Ann. de. Biol. Anim. Biochim. Biophys., 15*, 365-367.

Trounson, A.O. and Moore, N.W., 1974. *Aust. J. Biol. Sci., 29*, 301-304.

Visser, D. and Salamon, S., 1974. *Aust. J. Biol. Sci., 27*, 423-425.

Warnes, G.M., Pfitzner, A.P. and Armstrong, D.T., 1982. *In* Shelton, J.N., Trounson, A.O., Moore, N.W. and James, J.W. (Eds). *Embryo Transfer in Cattle, Sheep and Goats*, Australian Society for Reproductive Biology, 44-47.

# PROBLEMS OF OVINE *IN-VITRO* FERTILIZATION

J.G.E. Thompson and J.M. Cummins, *Reproductive Biology Group, Department of Veterinary Anatomy, University of Queensland, St. Lucia, 4067.*

*Summary* This project concentrated on the development of techniques for the collection of large numbers of tubal oocytes immediately after ovulation, and subsequent IVF. Results from the 1984 breeding season demonstrate poor oocyte recovery, variable oocyte quality and poor IVF rates. From these results we have identified two areas of difficulty in ovine IVF: 1) The collection of large numbers of suitable oocytes for fertilization. 2) A lack of understanding of *in-vivo* fertilization processes in the ewe.

## INTRODUCTION

*In-vitro* fertilization (IVF) of mammalian oocytes has, for a few species, developed into a routine procedure (Bavister 1981). However, there are only a few reports of successful IVF in sheep, and these document poor fertilization and cleavage rates (Wright and Bondioli 1981; Bondioli and Wright 1983; Thompson and Cummins 1983).

It has been reported that ovine follicular oocytes are difficult to mature *in-vitro* (Moor and Warnes 1978), and thus tubal oocytes would seem preferable for IVF, since both nuclear and cytoplasmic maturation should have occured already. However, the data of Trounson *et al.* (1976) suggested to us that the collection of tubal oocytes, resulting from superovulation, would pose problems of timing which we would need to overcome.

If IVF is ever to have a place in the sheep industry, the techniques previously shown to produce some small success (Thompson and Cummins 1983) need to be refined. It was our aim in this study to develop techniques for the collection of large numbers of tubal oocytes, immediately after ovulation, and to successfully fertilize them *in-vitro*.

## MATERIALS AND METHODS

Semen was collected from trained rams by artificial vagina. Each sample was immediately diluted with 5 ml of modified Dulbecco's phosphate buffered saline (mPBS: Tervit *et al.* 1972) and incubated at 37°C for 5 min before centrifugation (300g × 5 min) and removal of the supernatant. A sample (0.2 ml) of the dense sperm suspension remaining was then overlayered with 2 ml of medium (protein supplemented Brinster, Whitten and Whittingham's medium (BWW), Brackett's defined medium (DM) or modified Tyrode's medium (T6)) and incubated for up to 3 hrs. The top 1 ml, which usually contained highly motile spermatozoa, was removed and examined for assessment of sperm concentration and motility. Sperm concentration was adjusted to $10^5$ to $10^6$ sperm/ml and used as the medium in which the oocytes were incubated.

Ewes were superovulated with PMSG ('Folligon', Intervet, Scoresby, Aust., 1000-1200 iu) or FSH ('FSH-Pituitary', Burns-Biotec, Omaha, USA., 18 mg in a decreasing dose over 4 days; 8,6,3,1 mg/day respectivly) during the luteal phase (days 8-13) of the oestrous cycle. Four different superovulatory regimens were used: a) PMSG followed by 12 mg PG ('Lutalyse', Upjohn, Rydalmere, Aust.) 48-60 hours later, with oocyte recovery 50-60 hours after PG ; b) PMSG and PG as above, supplementing the PG with 500 iu HCG ('Pregnyl', Organon, Artamon, Aust.) and recovery 48-58 hours after PG/HCG; c) PMSG (Day 12 or 13) followed by HCG alone 48 hours later with recovery 48-50 hours after HCG; d) FSH (Day 12 or 13) with HCG given on the morning of the fourth day of FSH and oocyte recovery 48 hours later.

The reproductive tract was exposed by a mid-ventral laparotomy, under halothane anaesthesia, and oocytes were collected by flushing the oviducts with mPBS or by aspirating large (> 4mm) follicles with an 18g hypodermic needle and 2ml syringe (loaded with 1ml mPBS supplemented with 20iu/ml heparin). Three methods were employed to flush the oviducts: a) flushing from the ampulla into a balloon ('Foley') catheter below the utero-tubal junction; b) retrograde flushing from the uterus into a catheter placed in the ampulla; c) retrograde flushing, collecting directly from the fimbria suspended in a sterile dish, followed by 'washing' the surface of the ovaries and fimbriae.

Oocytes were quickly examined under a dissecting microscope and catagorized into one of six groups: I — follicular oocytes with no cumulus, surrounded only by a corona; II — follicular oocytes with a corona and an expanded cumulus complex; III — follicular oocytes with a dense, compact cumulus/corona; IV — follicular oocytes with no corona but with a sticky diffuse cumulus and a pale translucent ooplasm. Group IV oocytes appeared to derive from cystic, anovular, or atretic follicles. V — naked tubal oocytes, with homogeneously dense ooplasm; VI — degenerating tubal oocytes, with shrunken, pale or translucent ooplasm, occasionally granular with clumping of organelles and vacuolization.

After classification, oocytes were washed in fresh incubating medium before insemination. After 24hr incubation (37°C, humidified 5% $CO_2$/5%$O_2$/90%$N_2$) they were removed from the sperm suspension and incubated for a further 24hr in fresh medium. After incubation, the oocytes were air-dried onto microscope slides, fixed in acid-alcohol and stained with toluidine blue (Krzanowska and Lorenc 1983) and examined by light microscopy. Sperm penetration was determined by either the presence of a sperm head in the vitellus or 2 pro-nuclei. Cleavage was determined by the presence of mitotic nuclei in 2 or more blastomeres.

Statistical analysis of the different superovulatory regimens was performed using a Student's t-test to compare means. A chi-square analysis was performed on the data of the different flushing techniques. Both tests utilized a 5% level of significance.

## RESULTS

The results of the 4 superovulatory regimes are presented in Table 1.

**TABLE 1. Effect of superovulatory regimes.**

| Regime | No. of ewes | Mean no. corpora lutea (±s.e.m.) | Mean no. large follicles (>4mm)(±s.e.m.) |
|---|---|---|---|
| PMSG/PG | 7 | 0.6±0.3* | 6.1±1.7 |
| PMSG/PG-HCG | 9 | 6.6±1.5 | 2.8±1.0 |
| PMSG/HCG | 9 | 9.8±1.9 | 3.1±1.2 |
| FSH/HCG | 7 | 10.9±2.0* | 6.9±1.4 |
| Total | 32 | 7.1±1.0 | 4.5±0.7 |

* $P < 0.05$.

Our results show that many large, unovulated follicles were present at oocyte collection, indicating that ovulation of stimulated follicles occurs over a period of time as demonstrated by Trounson *et al.* (1976), or that a proportion of stimulated follicles never ovulate. Large follicles were aspirated to provide an additional source of oocytes. The only significant difference between the regimens with respect to the number of ovulations was that of PMSG/PG and FSH/HCG. Boland *et al.* (1983) demonstrated that exogenous luteinizing hormone was not essential for ovulation in superovulated ewes. However, this study shows that HCG tends to produce a more consistent ovulation response, and was subsequently administered in preference to PG.

The results of the three tubal oocyte recovery techniques are shown in Table 2.

**TABLE 2. Effect of different flushing techniques of tubal oocytes.**

| Technique | n | No. of Ovulns | Mean | No. oocytes Recovered | % Recovered/ ovulations |
|---|---|---|---|---|---|
| Balloon catheter | 6 | 43 | 7.2 | 10 | 23.2 |
| Regrograde-catheter | 20 | 146 | 7.3 | 50 | 34.2 |
| Retrograde-dish | 6 | 37 | 6.2 | 4 | 10.8 |
| Total | 32 | 226 | 7.1 | 64 | 28.3 |

All three techniques used to collect tubal oocytes gave very low recovery rates, and many ovulated oocytes are presumably missed. It is likely that most oocytes were not positioned in the ampullae for flushing to be possible.

**TABLE 3. Fertilizability of different oocyte types.**

| | No. of oocytes | | | |
|---|---|---|---|---|
| Oocyte type | incubated | penetrated(%)[1] | cleaved (%)[1] | fragmented (%)[1] |
| I | 24[2] | 3(12.5) | 2(8.3) | 2(8.3) |
| II | 32 | 5(15.6) | 5(15.6) | 6(18.8) |
| III | 3[2] | — | — | — |
| IV | 3 | — | — | — |
| V | 38 | 6(15.8) | 4(10.5) | 15(39.5) |
| VI | 10 | — | — | 2(20.0) |
| Total | 110 | 14(12.72) | 11(10.0) | 25(22.7) |

(1) % refer to the no. oocytes/no. incubated

(2) 3 oocytes from type I and 1 oocyte from type III did not show a meiotic MII configuration at the end of culture (all 4 in diakinesis).

The results for IVF of the various oocyte groups recovered are presented in Table 3. Due to the poor fertilization rates, comparisons for the three media were not considered meaningful, although it was interesting to note that no fertilization or fragmentation was observed from 15 oocytes incubated in BWW.

## DISCUSSION

In recent years, ovine IVF has been achieved (Bondioli and Wright 1983; Thompson and Cummins 1983) but with fertilization rates grossly inferior to results reported for other species. The attempt to collect large numbers of ovine tubal oocytes during the periovulatory period, and fertilize them using *in-vitro* methods in this study, has not met with great success. However, the results compare favorably with those found elsewhere (Bondioli and Wright 1980, 1983). Potential unfavorable factors such as water quality (5 × distilled) and media suitability have been checked by the use of mouse embryo and ram spermatozoa cultures, both of which are conducted successfully in our laboratory. Our data focus attention on two main areas which present problems; (i) an inability to recover good numbers of consistently fertilizable oocytes, and (ii) the lack of an understanding of factors required for fertilization *in-vivo* in the sheep.

With respect to the latter point, several reports have suggested that *in vivo* fertilization in the ewe differs from species in which IVF have been successful. For example: ovine oocytes are naked of cumulus at the time of fertilization (Cummins 1982a); acrosome reacted spermatozoa progressively enter the ampulla from an isthmic reservior (Cummins 1982b); ram sperm acrosin appears to be unable to digest ovine zonae (Brown 1982); and highly specific proteins are secreted into the oviduct at oestrus (Sutton *et al.* 1983).

It is important that further research continues into ovine IVF. When the problems are overcome, not only could the sheep industry benefit from the employment of the technique directly, but it may open the door to research into other areas, such as genetic manipulation.

## ACKNOWLEDGEMENTS

This work was supported by a grant from the Wool Research Trust Fund on the recommendation of the Australian Wool Corporation. Facilities provided by the Head of the Department of Veterinary Anatomy, Prof. T.D. Glover, are gratefully acknowledged.

## REFERENCES

Bavister, B.D., 1981. *In* Mastroianna, L. Jnr. and Biggers, J.D. (eds) *Fertilization and Embryonic Development In-Vitro*, Plenum, New York, 41-60.
Boland, M.P., Crosby, T.F. and Gordon, I., 1983. *Anim. Reprod. Sci., 6*, 119-127.
Bondioli, K.R. and Wright, R.W. Jnr., 1980. *J. Anim. Sci., 51*, 660-667.
Bondioli, K.R. and Wright, R.W. Jnr., 1983. *J. Amin. Sci., 57*, 1006-1012.
Brown, C.R., 1982. *J. Reprod. Fert., 64*, 457-462.
Cummins, J.M., 1982a. *Proc. Aust. Soc. Reprod. Biol., 14*, 39.
Cummins, J.M., 1982b. *Gamete Res., 6*, 53-63.
Krzanowska, H. and Lorenc, E., 1983. *J. Reprod. Fert., 68*, 57-62.
Moor, R.M., and Warnes, G.M., 1978. *In* Crighton, D.B., Foxcroft, G.R., Haynes, N.B. and Lamming, G.E. (eds) *Control of Ovulation*, Butterworths, London, 159-176.
Sutton, R., Wallace, A.L. and Nancarrow, C.D., 1983. *Proc. Aust. Soc. Reprod. Biol., 15*, 102.
Tervit, H.R., Whittingham, D.G. and Rowson, L.E.A., 1972. *J. Reprod. Fert., 30*, 493-497.
Thompson, J.G.E. and Cummins, J.M., 1983. *Proc. Aust. Soc. Reprod. Biol., 15*, 58.
Trounson, A.O., Willadsen, S.M. and Moor, R.M., 1976. *J. Agric. Sci. Camb., 86*, 609-611.
Wright, R.W. Jnr. and Bondioli, R.W., 1981. *J. Anim. Sci., 53*, 702-729.

# PRODUCTION OF EMBRYO IN SHEEP USING FSH PREPARATIONS AND LAPAROSCOPIC INTRAUTERINE INSEMINATION

G. Evans, *Department of Animal Husbandry, University of Sydney, New South Wales, 2006.*
M.K. Holland, H.B. Nottle, P.H. Sharpe, *Waite Institute, University of Adelaide, South Australia, 5000.*
D.T. Armstrong, *MRC Group in Reproductive Biology, University of Western Ontario, London, Canada N6A 5A5*, and *Waite Institute, University of Adelaide, South Australia, 5000.*

*Summary* Donor variability in the production of viable embryos is in part responsible for the limited success of embryo transfer programmes. In order to investigate some of the factors involved, Merino ewes were treated with progestagen sponges or prostaglandin in combination with 1200 i.u. PMSG or 22.5 mg porcine pituitary extract (FSH) to induce superovulation. A variety of insemination regimes was employed and recovered ova were assessed for fertilization 2-6 days later. Embryo recovery was higher following FSH (vs. PMSG) treatment and following intrauterine insemination of fresh semen with the aid of a laparoscope vs. single or double cervical insemination. Synchronization method did not affect embryo production, and fertilization rates were similar following insemination into a single uterine horn vs. both horns. Most superovulated ewes commenced ovulation between 54 and 60 h following sponge withdrawal or Pg, but intrauterine insemination with $10 \times 10^6$ motile spermatozoa at this time resulted in low fertilization rates. Intrauterine insemination at 48 h improved fertilization rates to 60% of recovered ova. Improvement of insemination techniques and further refinement of pituitary extracts show promise of improvement in embryo production in genetically superior ewes.

## INTRODUCTION

Superovulation and embryo transfer (SOET) has been practised in domestic species for many years. However, despite the fact that much of the technology has been developed in sheep, it has found little commercial application in this species, particularly in comparison with cattle. The major advantages of SOET are 1) increasing the contribution to the gene pool of especially valuable females. This is particularly effective if combined with progeny testing; in cattle it has been estimated that SOET can double the rate of genetic gain over conventional breeding programmes (Nicholas and Smith 1983); 2) facilitating transportation of 'whole animal' material, within as well as between countries, particularly when used in conjunction with cryopreservation (Bilton and Moore 1983); 3) import/export of embryos with minimum risk of disease transmission and without costly quarantine procedures; 4) permits embryo manipulation and genetic engineering.

Several factors have contributed to the lack of commercial interest in SOET in sheep. These include industry resistance, government embargoes on import/export of embryos and the natural propensity of some breeds to produce twins, limiting the increase in genetic gain due to SOET. However, the variable and relatively low success rate of SOET in sheep and the relatively low value of individual sheep are largely responsible for lack of commercial viability. Variable responses to PMSG and fertilization failure have been the major physiological factors limiting production of a high number of viable embryos.

Pituitary extracts, particularly those of horse, have been used as an alternative gonadotrophin to PMSG. Porcine pituitary FSH preparations are now commerically available (FSH-P, Burns-Biotec). In initial trials, FSH-P has produced promising results in comparison with PMSG in ovarian stimulation of sheep, and particularly of goats (Armstrong and Evans 1983), with some alleviation of the sperm transport problem associated with gonadotrophin stimulation of sheep (Evans and Armstrong 1984). Fertilization rates in superovulated sheep are improved by surgical intrauterine insemination compared with vaginal or cervical insemination (Trounson and Moore 1974), but laparotomy is time consuming, expensive and limits the productive life of a ewe due to induction of reproductive tract adhesions. The development of insemination techniques using a laparoscope (Killeen and Caffery 1982) has alleviated these problems. We now report the results of trials using FSH-P and laparoscopic insemination in a SOET programme on Merino ewes in South Australia.

## MATERIALS AND METHODS

Two experiments, utilizing 4-6 year old Merino ewes, were conducted in South Australia in March/April, 1984. Synchronization of oestrus was effected by progestagen sponges (Repromap, Upjohn) or injection of prostaglandin (Pg, 150 $\mu$g Estrumate, I.C.I.) in the luteal phase of the cycle. PMSG (1200 i.u., Folligon, Intervet) was injected 48 h before sponge withdrawal or Pg injection. FSH-P, 22.5 mg, was administered as an initial dose of 5 mg 60 h before sponge withdrawal or Pg injection and in 7 doses of 2.5 mg at 12 h intervals thereafter. For insemination, fresh semen was diluted with Dulbecco's phosphate buffered saline (PBS), maintained at 30°C and used within 2 h of collection. Techniques of insemination via the cervix or intrauterus and recovery of embryos have been described (Evans and Armstrong 1984). Oocytes/embryos were recovered via mid-ventral laparotomy on days 2-6 of the cycle and were deemed to be fertilized if normal clevage had occurred. In Experiment 1, ewes were treated with progestagen sponges and either PMSG or FSH; they received one of 3 insemination treatments; (a) intrauterine ($80 \times 10^6$ motile spermatozoa into one horn at 50 h after sponge withdrawal; (b) single cervical ($400 \times 10^6$ spermatozoa, 50 h) and (c) double cervical ($400 \times 10^6$ spermatozoa at each of 50 and 58 h). In Experiment 2, ewes were synchronized with Pg or progestagen sponges, treated with FSH, and inseminated intrauterus with $10 \times 10^6$ motile spermatozoa at 48, 54 or 60 h after sponge withdrawal or Pg. The sperm dose was either deposited into one horn or split between both horns. Ovaries were inspected at this time and ewes in which ovulation was in progress were noted. Analyses were performed using Student's t-test of $\chi^2$ with factorial partitioning (significance at $P < 0.01$).

## RESULTS

### Experiment 1

The results shown in Table 1 indicate that fertilization rates, based on proportion of total ova fertilized, are higher using FSH rather than PMSG ($P < 0.01$); however, uterine insemination overcame the difference due to gonadotrophin ($P < 0.01$). Nevertheless, total embryos recovered were greater using FSH ($11.8 \pm 1.2$) vs. PMSG ($6.7 \pm 0.9$) ($P < 0.01$) in this experiment.

**TABLE 1. The number of ewes with fertilized ova (and proportion of ova fertilized) following intrauterine or cervical insemination.**

| | Insemination treatment | | |
|---|---|---|---|
| Superovulation treatment | Intrauterine | Single cervical | Double cervical |
| PMSG | 4/5 (14/32) | 3/6 (5/29) | 0/6 (0/53) |
| FSH | 6/7 (53/79) | 2/7 (14/86) | 3/5 (12/57) |

### Experiment 2

As shown in Table 2 the majority of ewes commenced ovulating between 54 and 60 h with no significant effect of method of synchronization. There was no effect of synchronization method on other measured parameters and data were combined. The site of insemination appeared to have no effect on fertilization. Thirteen of 21 ewes inseminated in both horns yielded fertilized ova (38% of ova fertilized) compared with 11/18 inseminated in a single horn (40% of ova fertilized). In addition, in single horn inseminations, fertilization was as good in the non-inseminated horn (41%) as in the inseminated horn (39%). Table 3 indicates that the best fertilization was obtained with insemination at 48 h ($P < 0.01$).

**TABLE 2. The numbers of FSH- treated ewes ovulating at each time of insemination.**

| | Time of laparoscopy (after sponge removal or Pg) | | |
|---|---|---|---|
| Synchronization method | 48 h | 54 h | 60 h |
| Sponge | 1/8 | 3/8 | 7/8 |
| Pg | 1/8 | 1/8 | 7/8 |
| Total | 2/16 | 4/16 | 14/16 |

**TABLE 3. The ovulation and fertilization rates in FSH- treated ewes inseminated intrauterus at 3 different times (data combined for insemination in one or both horns).**

| | Time of insemination (after sponge removal or Pg) | | |
|---|---|---|---|
| | 48 h | 54 h | 60 h |
| Complete ovulations ($\bar{x} \pm$ sem) | $15.4 \pm 2.4$ | $11.9 \pm 1.5$ | $16.9 \pm 1.5$ |
| Ewes fertilized | 10/12 | 6/14 | 9/15 |
| % ova fertilized | 73/122 (60%) | 40/127 (31%) | 51/173 (29%) |

## DISCUSSION

In these experiments, there were indications that FSH was a superior gonadotrophin to PMSG in terms of fertilization rates, particularly when cervical insemination was employed; this confirms previous observations

(Evans and Armstong 1984). This factor undoubtedly contributed to higher recoveries of normal embryos in FSH-treated ewes, though higher ovulation rates (data not shown) also undoubtedly contributed. However, caution should be exercised when making comparisions between FSH and PMSG treatments since we collected no dose-response information in this experiment, though previous experience indicates that maximum production of embryos would be obtained with the doses used here. These studies failed to distinguish between Pg or progestagen sponge as a preferred method of synchronization of oestrus.

Most ewes commenced ovulation between 54 and 60 h after Pg or sponge withdrawal, and there were indications that insemination close to this time resulted in a reduction of ovulation and fertilization rates. The highest fertilization rate obtained in Experiment 2 (60% at 48 h insemination) cannot be regarded as satisfactory. The low overall rates may have been due to manipulation of the ovaries, too low a semen dose (though a higher one in Experiment 1 did not result in a higher rate of fertilization), failure of fertilized ova to cleave, other unknown factors associated with FSH-induced superovulation of Merino ewes, or inappropriate timing of insemination.

Our current data indicate that insemination earlier than 48 h may have been beneficial, though we have found that high fertilization rates could be obtained in Suffolk ewes in Canada with both cervical and intrauterine insemination at 48 h (Armstrong and Evans 1984a; Evans and Armstrong 1984). In superovulated Merino ewes, N.W. Moore (personal communication) achieves high fertility with intrauterine insemination at 24 h. In non-superovulated Merino ewes inseminated intrauterus with the aid of a laparoscope, Maxwell *et al.* (1984) observed low fertility and high embryonic/foetal mortality following insemination at 24 or 36 h, and best results were observed at 60 h. These differences may be due to differences in ovulation time but require clarification by further investigation.

Although these results are encouraging, further research on FSH stimulation and laparoscopic insemination is necessary to improve the techniques in Merino ewes. Investigation into the required dose of fresh and frozen semen and timing of insemination in conjunction with FSH stimulation are areas for further research. Investigation of the effects of season, nutrition and breed or strain of sheep would also be particularly pertinent in Australia, where these factors undoubtedly contribute to the variability in success of SOET. Work developing a simple method of administration of FSH is of interest since the present method involving multiple injections is labour intensive. In addition, work aimed at improving the quality of FSH preparations may be rewarding; initial trials overseas have indicated that highly purified preparations of FSH, with very low LH contamination (measured by receptor assay), may be superior to commercially available FSH in stimulating ovulation rate (Armstrong and Evans 1984b). Further research on the effects of FSH and/or LH on the ovary is needed to improve our knowledge of mechanisms controlling ovulation rate. A successful SOET programme in Merinos would find rewards in realisation of the potential export market as well as hastening genetic gain in the domestic flock.

## ACKNOWLEDGEMENT

This work was supported by The Davies Bequest to D.T. Armstrong.

## REFERENCES

Armstrong, D.T. and Evans, G., 1983. *Theriogenology, 19*, 31-42.
Armstrong, D.T. and Evans, G., 1984a. *J. Reprod. Fert., 71*, 89-94.
Armstrong, D.T. and Evans, G., 1984b. *Proc. Xth Int. Cong. Anim. Reprod. & A.I.*., University of Illinois, Vol 4., vii-8.
Bilton, R.J. and Moore, N.W., 1983. *Proc. Aust. Soc. Reprod. Biol, 15*, 94.
Evans, G. and Armstrong, D.T., 1984. *J. Reprod. Fert., 70*, 47-53.
Killeen, I.D. and Caffery, G.J., 1982. *Aust. Vet. J., 59*, 95.
Maxwell, W.M.C., Wilson, H.R. and Butler, L.G., 1984. *Proc. Aust. Soc. Anim. Prod., 15*, 448-451.
Nicholas, F.W. and Smith, C., 1983. *Anim. Prod., 36*, 341-353.
Trounson, A.O. and Moore, N.W., 1974. *Aust. J. Biol. Sci., 27*, 301-304.

# PHARMACOLOGICAL AGENTS FOR MANIPULATING OESTRUS AND OVULATION IN THE EWE

R.J. Scaramuzzi, *CSIRO Division of Animal Production, PO Box 239, Blacktown, New South Wales 2148.*
G.B. Martin, *MRC Reproductive Biology Unit, Centre for Reproductive Biology, 37 Chalmers Street, Edinburgh EH3 9EW.*

## INTRODUCTION

The sheep industry must continue to improve its efficiency if it is to survive in the face of increasing costs and compete with synthetic fibres and intensive pig and poultry production. One important aspect of increased efficiency is the continuing improvement of reproductive rate. In this paper we will briefly review current knowledge of the reproduction of the ewe with particular emphasis on the oestrous cycle and ovulation. We will discuss the current applications of this knowledge in the development of pharmacological aids to the management of sheep reproduction, and speculate on other applications likely to emerge in the future.

## HOW TO IMPROVE THE REPRODUCTIVE EFFICIENCY OF THE EWE

### Acceleration of growth and puberty

This would allow the ewes to become pregnant earlier in life thereby increasing their potential lifetime performance. The artificial induction of ovulation is insufficient in itself because the ewe must also be sufficiently mature to support pregnancy, lactation and the rearing of her lambs.

### Synchronising oestrus in the flock

Synchronisation of oestrus does not directly increase ewe performance, but allows the wide-scale application of artificial insemination (AI) for improvement of flocks.

### Increasing ovulation rate

Low ovulation rate is the major factor limiting litter size and is an important source of reproductive wastage in the Australian Merino breed (Knight *et al.* 1975a). Although this can be partially overcome by manipulation of nutrition (Knight *et al.* 1975b), there is still wide scope for further improvement through genetic or hormonal methods.

### Improving fertilisation rate and survival of embryos

With natural mating fertilisation is regarded as an efficient process in sheep and there appears little room for improvement. However, with oestrous synchronisation and AI low fertilisation rates are a serious problem. Similarly, mortality of embryos during early pregnancy is recognised as a major source of reproductive wastage. A proportion of this is due to chromosomal defects and so is unavoidable (Boland *et al.* 1984), but the remainder is probably due to hormonal or environmental factors and is therefore susceptible to treatment.

### Preventing periods of anovulation

Ewes are normally unable to ovulate during the spring or while lactating. They are not productive during these periods and, perhaps more important, managerial flexibility is lost because the farmer is unable to predetermine the time of lambing. Although this is a critical constraint in a dry land system of agriculture, it is a less serious problem in the Merino breed because anoestrous ewes can be readily induced to ovulate by introducing rams (Pearce and Oldham 1984). However, other treatments are necessary in crossbred and British breed sheep.

## WHERE TO INTERVENE TO MANIPULATE EWE REPRODUCTION

Each of the hormones or substances which influence their secretion, that control or influence the oestrous cycle and ovulation in the ewe (Figure 1) represent potential sites for manipulation of reproduction of ewes. We will briefly describe them and their respective functions and assess their possible roles as agents for the control of oestrus and ovulation.

### Central nervous system

The central nervous system collects, stores and integrates information about the state of the external environment (availability of food, presence of rams, time of year) and the state of the reproductive system (follicle and luteal function, pregnancy, lactation). The hypothalamus is the integrating unit — it collates inputs and sends an integrated signal, the releasing factors, to control the activity of the anterior pituitary gland. Through this gland the ovaries are controlled (for review: Martin 1984). Brain-ovary communication is thus predominantly neuroendocrine — pure neural signals appear to be rarely used, but they have not yet been widely studied so their importance is uncertain.

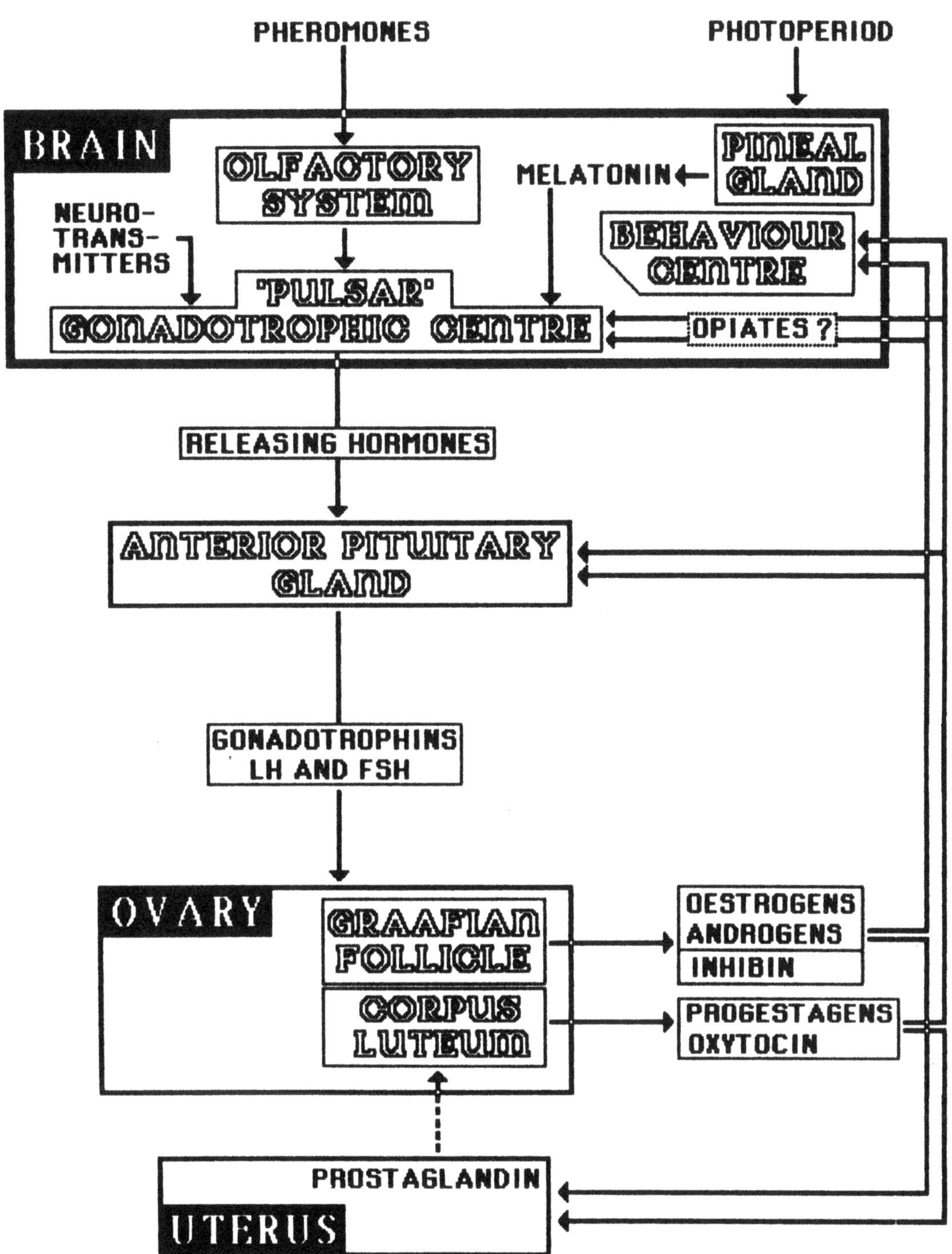


**FIGURE 1. A schematic illustration of the endocrine pathways involved in the control of reproductive activity. Although they are illustrated as a separate system, the neurotransmitters do not act independently but mediate most, if not all, inputs into the pulsar.**

*Pheromones* Anovulatory ewes of some breeds, whether they be prepubertally, seasonally or lactationally anoestrus, will begin regular oestrous cycles if they are placed with rams (Pearce and Oldham, 1984). It seems likely that a pheromone is an important component of the mechanism of this effect (Knight 1983). It may stimulate the ovarian activity of the ewe by increasing the activity of the 'pulsar' thus increasing the secretion of gonadotrophin-releasing hormone (GnRH) by the hypothalamus (Martin 1984). Neither the nature of the pheromone itself, nor the neural pathways through which it affects the GnRH neurones, have been described. Should the pheromone be purified, it may become a useful management tool.

*Melatonin* This indoleamine produced by the pineal gland at night, is part of the mechanism by which the brain measures daylength and thus season (Bittman *et al.* 1983). The hypothalamus is thought to use this information when formulating its signals to the pituitary gland. Exogenous melatonin can override this system, mimic the effect of short days (i.e. stimulatory photoperiods) and present the hypothalamus with false information concerning the season. Ovulation can thus be induced under an inhibitory photoperiod (Kennaway *et al.* 1982; Arendt *et al.* 1983; Nowak and Rodway 1984).

*Neurotransmitters* These are amines or peptides which transmit impulses between neurones. There are only about two dozen known transmitters and each of them is widely distributed and has many different roles throughout the central nervous system. For example, they mediate many of the inputs into the pulsar and are probably involved in the neuronal circuitry of the pulsar itself. A wide range of drugs is available which affect neurotransmitter function but because of the multitude of roles for a given transmitter these drugs have many undesirable side-effects. This obviously limits their use as agents for manipulating the reproduction of the ewe. A possible exception may be the endogenous opiates $\beta$-endorphin and met-encephalin. These peptides are recently discovered neurotransmitters and appear to have a direct inhibitory role in the control of the secretion of GnRH (Ferin *et al.* 1982; Ebling *et al.* 1984). Opiate antagonists such as naloxone thus stimulate secretion of GnRH and may be useful for the induction of ovulation in anoestrous ewes.

*Releasing and release-inhibiting hormones* These substances travel the final common pathway through which hypothalamic messages reach the anterior pituitary gland. They are therefore a very susceptible and important site for pharmacological intervention. GnRH controls the release of luteinizing hormone (LH) and follicle stimulating hormone (FSH) and is the most important releasing hormone controlling reproduction. Prolactin release-inhibiting factor which is thought to be the catecholeamine, dopamine controls the release of prolactin (McNeilly 1980). In general releasing hormones are rapidly metabolised so the pure hormone is often of limited use for the treatment of animals. However, analogues which are metabolised slowly can be used and they may act as either agonists or antagonists and so can be used to stimulate or inhibit function of the pituitary. Dopamine agonists, such as the ergot alkaliods, block the secretion of prolactin, but they are of little use because prolactin appears to have no major role in the control of oestrus and ovulation in the ewe.

Anterior pituitary gland

*Gonadotrophins* The gonadotrophins, LH and FSH, are glycoproteins comprised of two peptide chains (the alpha and beta sub-units) and a carbohydrate moiety (principally sialic acid). The size of the carbohydrate component determines the rate at which the gonadotrophin is metabolised and cleared from the circulation. The alpha chain is common to both molecules while the beta chain is hormone specific. The individual sub-units are devoid of biological activity (Vaitukaitus *et al.* 1976). The gonadotrophins control the growth and development of Graafian follicles, ovulation, luteinization and luteal function. Manipulation of these hormones has thus long been viewed as a simple method for controlling the reproductive function of ewes. The endogenous gonadotrophins are not easily prepared in large quantities and are metabolised quickly so until these problems are overcome they are not likely to find widespread use in stimulating ovarian function.

*Exogenous gonadotrophins* In many species the placenta of the pregnant animal produces a similar hormone. For example the endometrial cups of the pregnant mare produce a glycoprotein hormone pregnant mare's serum gonadotrophin (PMSG) which mimics some of the actions of FSH and LH. PMSG has two principal advantages over the endogenous gonadotrophins: it can be readily obtained in large quantities from the serum of pregnant mares, and it has a much longer half-life than either LH or FSH. For these reasons, it has found wide practical application, particularly in Europe (Gordon 1983). Although not directly analagous to either LH or FSH, its action is primarily follicle-stimulating and it is used as an alternative to exogenous FSH. It does, however, have some of the properties of LH.

Human chorionic gonadotrophin (hCG) is produced by the human placenta and its action on the ovary is very similar to that of LH. It too has a long half-life and is relatively easily obtained. An injection will mimic the preovulatory surge of LH so it can be used to induce ovulation, and recent studies have also suggested that hCG injections during the luteal phase may increase ovulation rate (Radford *et al.* 1984).

Ovary

The major known endocrine products of the ovary are the steriods (oestrogens, androgens, progestagens), inhibin and oxytocin.

*Steroids* The major roles of the steriods are the regulation of the secretion of gonadotrophins and the expression of sexual behaviour (for review: Martin 1984). During the luteal phase oestradiol and progesterone inhibit the tonic secretion of LH and FSH (negative feedback), but during the follicular phase oestradiol assumes a stimulatory role and evokes both oestrous behaviour and the large surge of gonadotrophin which induces ovulation (positive feedback). These actions of oestradiol are completely inhibited by progesterone (Scaramuzzi *et al.* 1971). The actions of the steroids on oestrous behaviour and secretion of gonadotrophins can be manipulated by the administration of the hormones themselves, with antibodies provided by actively or passively immunising the ewe against the steroids (Martensz 1980), by administering synthetic agonists such as diethylstilboestrol (Robinson and Reardon 1961; Caldwell *et al.* 1970), with antisteroids such as clomiphene and tamoxiphen that act on steroid receptors (Fabre-Nys 1982), and with compounds such as the aromatase inhibitors which interfere with metabolism of steroids (Brodie *et al.* 1983).

The only class of steroid in wide use are the progestagens, which have been successfully employed througout the world for synchronising oestrus in ewes, and for ensuring the expression of oestrous behaviour (Robinson 1954) and normal luteal function in ewes induced to ovulate during anoestrus (Table 1). Despite the importance of endogenous oestrogen in the induction of oestrus and the surge of LH during the normal cycle, attempts to use this substance to manipulate reproduction have been relatively unsuccessful. The function of the androgens secreted by the ovaries is not clear, so it is difficult to foresee practical applications for them.

**TABLE 1. Effect of progesterone (Prog) or fluorogestone acetate (FGA) priming on oestrous behaviour and ovarian activity in seasonally anoestrous ewes inducted to ovulate with the 'ram effect' (Oldham *et al.* 1980, and personal communication 1984).**

| BREED (Time) | Treatment | n | In Oestrous | Ovulated | Ovulation Rate (%) | Normal CL's(%) |
|---|---|---|---|---|---|---|
| ILE-DE- | FGA Only | 10 | 0 | 0 | — | — |
| FRANCE | Rams only | 15 | 1 | 13 | 215 | 7(54)** |
| (April) | Rams + FGA | 15 | 10 | 10 | 109 | 10(100) |
| PREALPES | Prog only | 5 | 0 | 0 | — | — |
| DU SUD | Rams only | 10 | 0 | 9 | 156 | 5(55)* |
| (June) | Rams + Prog | 9 | 9 | 9 | 156 | 9(100) |

* Chi square = 2.89 P = 0.064
** Chi square = 6.24 P = 0.014

*Inhibin* Inhibin is believed to be a principal and specific regulator of FSH secretion and if this is the case, it could have a number of uses. For example, immunity to inhibin is thought to increase ovulation rate by increasing the levels of FSH (O'Shea *et al.* 1982). Likewise, the rebound increase in FSH seen following a course of injections of follicular fluid also increases ovulation rate, an effect attributed to inhibin (Wallace and McNeilly 1984).

*Oxytocin* Oxytocin is produced by the corpus luteum and acts on the uterus to mediate the release of prostaglandin (Flint and Sheldrick 1983). Inhibition of the action of this hormone or its carrier protein, neurophysin, could be used to modify luteal function.

Uterus

The major endocrine products of the uterus are the prostaglandins (PG), one of which PGF2$\alpha$ is the luteolytic hormone (McCracken *et al.* 1970; Goding 1974). This hormone is metabolised very rapidly so its usefulness is limited as very high doses are required. On the other hand, synthetic analogues of PG have longer half-lives and can be used as exogenous luteolytic agents for the synchronisation of oestrus (Gordon 1983). The inhibition of PG synthesis with PG synthetase inhibitors such as aspirin or indomethacin (Horton and Poyser 1976) may be useful for inhibiting oestrus or for reducing embryonic mortality early in pregnancy.

## PRODUCTS PRESENTLY AVAILABLE

A range of pharmacological products are readily available and several are now widely used on farms, and in programs of AI, transfer of embryos and intensive breeding. The relative effectiveness of many of the hormone and analogue preparations, the methods of administration and the doses required, have been extensively described and evaluated in a recent publication by Gordon (1983). We will not therefore address these questions in any detail, but we will briefly summarise the principles, emphasising some of the advantages and disadvantages of each agent.

### Synchronisation of oestrus in the breeding season

*Progestagens* When it was discovered that progesterone would block ovulation (Dutt and Casida 1948) it was also realised that it would be useful for synchronising oestrus and ovulation in ewe flocks. In early work the hormone was dissolved in oil and administered over 8-10 days as a series of injections. This was too labour-intensive for practical application. The development of the intravaginal sponge (Robinson 1964) solved this problem and in recent times other methods have evolved, such as intravaginal (Andrews *et al.* 1984) and subcutaneous (Gordon 1983) implants, depot injections and orally active preparations. Progesterone itself must be administered in large quantities, something difficult to do with the sponge, so synthetic analogues such as medroxy-progesterone acetate and fluorogestone acetate were developed. These are 10-20 times more potent than progesterone yet are quickly cleared from the circulation when the source is removed. Progestagen sponges are cheap and easy to use and are now widely used in many countries, particularly in AI programs.

The major disadvantage is the relatively poor fertility at the first oestrus following withdrawal. Studies in recent years have shown that this is due to the deleterious effects on sperm transport and survival in the reproductive tract of the ewe (Robinson 1973). There is also some loss of synchrony of the preovulatory surge of LH (and probably ovulation) with the onset of oestrus (Cumming *et al.* 1970). These problems cannot be overcome at present but can be avoided by mating at the second oestrus, or by increasing the number of spermatozoa the ewe receives (Allison and Robinson 1971).

*Prostaglandins* The corpus luteum is resistant to the lytic effects of PG during the first eight days of the oestrous cycle (Chamley *et al.* 1972) so it has been necessary to develop double injection procedures (Fairnie *et al.* 1977; Haresign 1978) and this effectively removes one of the major advantages it would have had over the progestagen sponge. Furthermore, in a flock treated with PG, the reproductive tracts of the ewes have been exposed to endogenous progesterone for variable periods. Those exposed only briefly will have poor fertility due to poor implantation and failure of sperm transport (Fairnie *et al.* 1977). As an alternative to progestagen synchronisation, the PGs appear to have no real advantages.

Breeding anovulatory sheep

*PMSG* Injection of PMSG will stimulate the ovarian follicles and thereby increase production of oestradiol and induce a preovulatory surge of LH in sheep in all of the reproductive states — prepubertal, lactating, seasonally anoestrus, and pregnant. It is the only agent which will work reliably in all of these situations. The treatment must be combined with a course of progestagen therapy to ensure normal oestrous behaviour and luteal function. Most of the problems with PMSG relate to variability in the responses.

*Oestrogen* Injection of oestrogen will also induce surges of LH and ovulation in anoestrous ewes (Cole and Miller 1935; Beck and Reeves 1973; Symons *et al.* 1973) so could be substituted for PMSG (Table 2). There is considerable advantage in the cost of treatment, but some of this would be lost because PMSG stimulates the ovulation rate and oestrogen does not.

**TABLE 2. The induction of ovulation and oestrus in seasonally anoestrous purebred Border Leicester ewes. The animals were pretreated with progestagen (MAP) sponges and then injected with either 400 IU of PMSG or with 30 ug of oestradiol benzoate (ODB). They were mated at the first oestrus and blood samples collected 18 days later were assayed for progesterone as a pregnancy test. (I.D. Killeen, S. Dawe and R.J. Scaramuzzi, 1980, unpublished).**

| Treatment | Number of ewes | % in oestrus | % ovulated | % pregnant on Day 18 |
|---|---|---|---|---|
| Progestagen & PMSG | 78 | 95 | 92 | 42 |
| Progestagen & ODB | 37 | 95 | 89 | 32 |

Improvement of ovulation rate

*PMSG* The ability of PMSG to stimulate ovarian activity directly in sheep was discovered in the 1930's (Cole and Miller 1935). The hormone has a very wide dose-response so can be used to produce moderate increases in twinning or the very high ovulation rates (superovulation) necessary for programmes of embryonic transfer. It is widely employed in France and Ireland in intensive breeding schemes.

There are several problems, the most important being the necessity to use it either in conjunction with techniques for synchronisation of oestrus or with careful monitoring of natural oestrus because to be effective it must be administered on days 12-14 of the oestrous cycle. PMSG itself is expensive and this adds considerably to the overall cost. The responses to it are highly variable, both within and between animals. This can only be partly explained by seasonal effects (Gherardi and Lindsay 1980). Gherardi and Martin (1978) repeatedly treated a group of ewes with 1000 IU (in raw serum from the same batch) for 18 consecutive cycles, only four ewes apparently became refractory, but they very rarely remained anovulatory for two consecutive cycles (Figure 2). The effect is temporary and cannot be explained by season or antibody-based immunity (Gherardi and Martin 1978).

Some of the variation in response to PMSG has been attributed to quality control, a problem which could be partly solved by developing a sheep-based bioassay. The long half-life of the hormone in the circulation has also been blamed for the variability. This has been overcome by passively immunising the ewes against PMSG to remove it from circulation at a predetermined time after treatment (Bindon and Piper 1977). An alternative approach using neuraminidase-treated PMSG with a diminished sialic acid content gave a preparation with diminished biological activity in ewes (R.M. Hoskinson and B. Bindon, 1984, personal communication). Should these ideas be developed further, PMSG could become even more widely used than it is now.

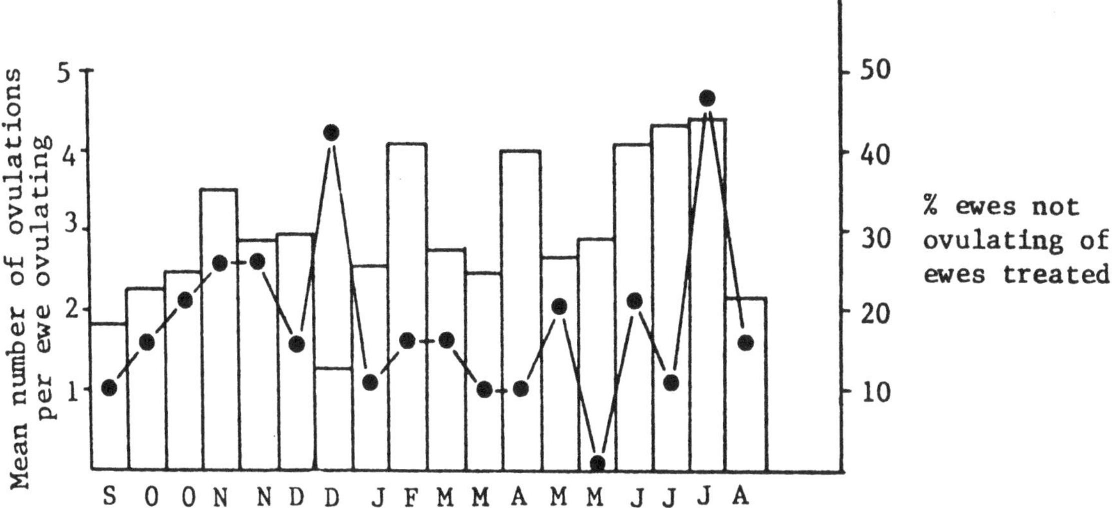


**FIGURE 2. The mean number of ovulations per ewe and percentage of ewes not ovulating after treatment (●) with 18 consecutive doses of PMSG. On only four of 414 observations following PMSG treatment did individual ewes remain anovulatory for two consecutive cycles. (Gherardi and Martin 1978).**

*Immunisation against ovarian steroids* During investigations into the mechanism of luteal regression, it was observed that follicular growth and ovulation rate had been stimulated in ewes passively immunised against oestradiol. This led to the proposal that immunisation against the ovarian steroids could be a commercially valuable method for increasing ovulation rate (Scaramuzzi 1975). The ovulatory responses of ewes to active immunisation against oestrone and androstenedione were then tested (Cox *et al.* 1976; Scaramuzzi *et al.* 1977). Initially there were many problems, including poor fertility, anovulation and silent ovulation (Van Look *et al.* 1978; Martin *et al.* 1979). Much of this was attributed to unsatisfactory adjuvents, which not only relied on complicated injection procedures and caused severe lesions of the skin, but also gave highly variable antibody titres. Further research saw the development of the dextran adjuvant (Cox and Wilson 1976) which can be given as a single injection, does not produce lesions and evokes mild and controllable antibody responses. This product is now commercially available (Fecundin, Glaxo Animal Health).

It has several major advantages over PMSG. Although there is some inflexibility in the timing of the injections relative to mating, it is still sufficiently flexible that synchronisation of oestrus is not necessary. Under Australian conditions, at least, the ovulatory responses are mild so triple births are rare. There is certainly none of the over-stimulation that often results with PMSG. There are some disadvantages: fertilisation rates can be depressed and embryonic mortality appears to be increased, and as a consequence the gains in ovulation rate are not always fully translated into gains in lambing rate (Table 3).

**TABLE 3. Effect of treatment with Fecundin (Glaxo, Australia Ltd.) on the percentage of reproductive wastage in ewes. There were between 169 and 179 ewes in each group and the immunised ewes received their booster injection either 14 days (High Titre) or 25 days (Low Titre) before the ovulation at which they were mated. The embryos were recovered 2 days after mating to estimate fertilisation rate, 13 days after mating to estimate early losses or 25-29 days after mating for late losses. It can be seen that most of the wastage occurs before Day 13. The lower success of recovery of embryos in the immunised animals was possibly due to ovulation failure, failure of the ova to enter the oviduct, or abnormal transport of ova. (M.P. Boland, I. Hazelton, R.M. Hoskinson, J.D. Murray, C.D. Nancarrow, R.J. Scaramuzzi and R. Sutton, 1984, unpublished).**

| | Control | Low Titre | High Titre |
|---|---|---|---|
| *Source of wastage:* | | | |
| Fertilisation failure | 13 | 25 | 29 |
| Implantation failure | 14 | 10 | 28 |
| Development failure | 5 | 11 | 3 |
| Ova not recovered (%) | 11 | 25 | 40 |

Immunisation can also be used in anoestrous ewes. It does not induce ovulation, but stimulates follicular activity and when the ewes are induced to ovulate with another technique, such as the ram effect, they will have multiple ovulations (Martin *et al.* 1981).

## CURRENT CONCEPTS FOR FUTURE PRODUCTS

This section is concerned with recent and current research. We will describe work and findings which are likely to have practical application in the very near future, and give some indication of the degree and direction of development required, and the likelihood of success.

### Melatonin

Because of the disadvantages of the combined sponge/PMSG treatments, we need new techniques for the treatment of seasonally anoestrous ewes. Melatonin appears to be a major component of the mechanism by which the brain perceives the season, so it is a promising hormone for artificially controlling reproduction in the future. In several recent studies, it has been shown that feeding or implanting melatonin (subcutaneously or intravaginally) in ewes mimics the effects of short days and advances puberty and the breeding season (Kennaway *et al.* 1982; Arendt *et al.* 1983; Nowak and Rodberg 1984). It will probably not be useful in lactationally anoestrous ewes. Long treatment periods (months) are needed, and to date the effects are limited to advancing the start of the next breeding season in ewes already in anoestrus. Future work testing the efficacy of the hormone at various times relative to normal onset of the breeding season is needed. For example, lambs treated with melatonin at birth or at 8 weeks of age, 5 months before the normal onset of puberty, will be unaffected or may even have a delayed first ovulation (Kennaway and Gilmore 1984; Nowak and Rodway 1984), while lambs treated at 19 weeks will begin ovulating 45 days before untreated animals (Nowak and Rodway 1984). Similar effects are observed in adult ewes, indicating that the effect is not only age-dependent but has a photoperiodic component as well. New routes and methods of administration, for example the intravaginal implant (Nowak and Rodberg 1984) together with the synthesis of long-acting and potent analogues which could be given in depot form or in implants are promising. Testing the hormone in more breeds, especially those with a deep anoestrus will also improve our knowledge of the mechanism of action of melatonin and could improve the speed of the response.

### Gonadotrophin-releasing hormones

*GnRH* Endogenous GnRH can be administered either as a series of small injections or as a constant infusion and will induce ovulation in prepubertally, seasonally or lactationally anovulatory ewes. Priming with progesterone followed by several days of treatment are needed in order to ensure oestrus and normal corpora lutea (McLeod *et al.* 1982, 1983). Trials with small numbers of sheep have indicated that fertility is normal (Wright *et al.* 1983). In current experiments, the hormone is delivered by syringe or mechanical pump, and although osmotic mini-pumps may be used in future work they are too expensive for practical purposes. We therefore need to develop other systems of slow continuous release suitable for GnRH and other peptides. Highly potent analogues (such as HOE 766, Hoescht) already exist so depot injections should be possible, and there is also considerable potential for intravaginal applications since the peptides can be absorbed through mucous membranes (Katzorke *et al.* 1980).

*Immunocastration* Antibodies to GnRH can be used to neutralise the endogenous hormone thereby greatly reducing the secretion of gonadotrophin and gonadal steroids. The animal becomes anoestrus and anovulatory (Clarke *et al.* 1978). This technique is effectively a form of non-surgical castration, and, if passive immunisation is used, the effects of treatment are temporary (Fraser and McNeilly 1983; Caraty *et al.* 1984). Apart from its simplicity, this technique may have other important benefits if pressure from animal welfare organisations increases.

*'Superagonists'* The use of highly potent agonists of GnRH to over-stimulate the anterior pituitary gland and make it refractory to endogenous signals from the pulsar is another potential technique for non-surgical castration. As with the immunological approach, the effects are temporary (Fraser 1983). Subcutaneous implants which last for several months are being developed as a human contraceptive (H.M. Fraser, 1984, personal communication) and these may be useful in the management of farm animals.

## LOOKING FURTHER INTO THE FUTURE

Finally we come to the areas in which we are beginning to gain some insight from current programs of basic research. By nature this section is speculative, containing ideas that are very recent and which may not yield practical benefits for many years, if at all. However, if we are to gain anything from these new developments research on these subjects must continue.

### Stimulation of the 'pulsar'

It is now recognised that the pulsar, the network of neurones in the hypothalamus which governs the secretion of GnRH, is of primary importance in the control of reproductive activity. As such it is a site at which we could manipulate the reproductive system. In recent years, we have discovered many external factors that will influence the activity of the pulsar, including nutrition, the presence of animals of the opposite sex, and the length of the day (for review: Martin 1984). Internal factors which have similar effects are also being discovered, some of which may in fact mediate the responses to the external factors. Here, we will discuss one of each type.

*Pheromones* These substances are highly specific, non-invasive and very fast acting. The 'ram-effect' pheromone, for example, can increase the tonic secretion of LH in ewes in as little as two minutes (C.Fabre-Nys and G.B. Martin 1984, unpublished), and in some cases the frequency of pulses appears to reach that of ovariectomised ewes instantaneously (see profiles in Martin *et al.* 1983). This contrasts markedly with the delays of 6-12 hours for responses to withdrawal and re-supply of oestrogen (Thomas *et al.* 1982), and even more with the 40 days required to measure responses to daylength (Walton *et al.* 1980). The pheromone affects ewes in any anovulatory stage, whether they be prepubertal or adult, in seasonal or lactational anoestrus (Pearce and Oldham 1984). The ovulation rate at the first ovulation following introduction of rams is sometimes higher than can be normally expected and under some circumstances the subsequent gains in prolificity can match those observed with PMSG (Cognie *et al.* 1980).

Efforts to purify and isolate the pheromone are continuing, and should they be successful, the pure substances and its analogues may become generally available. It will be of interest to see whether the purified pheromone alone can reproduce the ram effect. If this is successful we will need to develop routes of administration as it need not be most effective when administered nasally. We will need to solve problems with the unresponsive ewes, those Merino ewes in particular that do not fall anoestrus and therefore are not synchronised with the rest of the flock. This would require a technique to induce anoestrus, perhaps by inhibiting the pulsar. The list of breeds in which the ram effect has been demonstrated is continually increasing, and now includes the high fecundity Romanov (Martin and Cognie 1984). We need to carry this process further and also find ways to increase the responsiveness to the ram effect in individuals and breeds which have a 'deep' anoestrus.

*Opiates* Endogenous opiates, acting as neurotransmitters, are thought to be intimately involved in the control of the pulsar and may even mediate the inhibitory effects of the gonadal steroids (Ferin *et al.* 1982; Ebling *et al.* 1984). We need long-acting exogenous agonists and antagonists which specifically affect the secretion of gonadotrophins before we can judge if this approach will be of any use.

### Follicle development and ovulation rate

If research on inhibin continues to follow current trends, we should soon be able specifically to alter the secretion of FSH by manipulating the activity of inhibin. Injection of the hormone itself, or immunisation against it, may provide new methods for controlling ovulation rate. Results from the initial experiments using these techniques have been promising, but the responses have been variable and excessively high ovulation rates (greater than 3) have been observed (O'Shea *et al.* 1982; Wallace and McNeilly 1984). We need pure preparations of the hormone so that specific antibodies can be raised, titres can be measured and controlled and the secretion of the hormone itself studied. At this stage we know too little of the nature and action of the substance, and too little of the mechanisms controlling ovulation rate.

### Luteal function and mortality of embryos in early pregnancy

There is some evidence that small depressions in concentrations of progesterone around day 14 after mating are associated with embryonic mortality (Pearce *et al.* 1984). This may be due to incomplete suppression of the luteolytic signal from the uterus, so treatment with inhibitors of PG synthetase, or immunisation against oxytocin or prostaglandin may overcome the problem.

### Parturition and maternal behaviour

The neural and endocrine systems controlling parturition and maternal behaviour are interdependent (Poindron *et al.* 1984). Oestrogens and corticosteroids are involved in the initiation of parturition and can be supplied exogenously to induce and synchronise parturition in ewes for practical purposes. However, oestrogen enhances maternal behaviour so may be preferable in practice (Poindron *et al.* 1979).

## CONCLUSION

For the artificial control of reproduction, there has been a slow but continuing increase in the range of pharmacological agents available, and a similar increase in the specificity and efficiency with which they act. Many of these advances have resulted from ideas and concepts developed during the course of basic research. They were then refined and tested in field trials in applied research programmes before being released to the farmer. In this process, the role of feedback from producers is two-fold: to the applied research worker it indicates where techniques that are already understood need improvement and development; for researchers working in basic science, the feedback stimulates them to grasp the potential commercial importance of their new concepts. These new concepts are unlikely to find their source in industry and will almost inevitably arise from pure research in the biomedical sciences. It is evident, therefore, that the simple listing of problem areas is insufficient in itself to generate the new concepts upon which our advances depend so heavily. Some good examples of this process have been described in this article, and probably the most striking is the development of immunisation against steroids from an investigative laboratory technique to a commercially available stimulant of ovulation rate in ewes.

We have described many other prospects for future development. Of these, melatonin and GnRH hold most promise for the immediate future and are ready to move from basic to applied research programmes. Should this happen, and research on the other more speculative areas be continued, we may yet gain full control over the reproduction in the sheep.

## REFERENCES


Allison, A.J. and Robinson, T.J., 1971 *Aust. J. Biol. Sci., 24*, 1001-1008.
Andrews, W.G.K., Welch, R.A.S. and Barnes, D.R., 1984. *Proc. N.Z. Soc. Anim. Prod.* (in press).
Arendt, J., Symons, A.M., Laud, C.A. and Pryde, S.J., 1983. *J. Endocr., 97*, 395-400.
Beck, T.W. and Reeves, J.J., 1973, *J. Anim. Sci., 36*, 566-570.
Bindon, B.M. and Piper, L.R., 1977. *Theriogenology, 8*, 171.
Bittman, E.L., Karsch, F.J. and Hopkins, J.W., 1983. *Endocrinology, 113*, 329-336.
Boland, M.P., Murray, J.D., Scaramuzzi, R.J., Sutton, R., Hoskinson, R.M., Hazelton, I.G., Nancarrow, C.D. and Moran, C., 1984. This volume, 137-139.
Brodie, A.M.H., Garrett, W.M., Hendrickson, J.R., Tsai-Morris, C.H. and Williams, J.G., 1983. *J. Steroid Biochem., 19(1A)*, 53-58.
Caldwell, B.V., Scaramuzzi, R.J., Tillson, S.A. and Thorneycroft, I.H., 1970. *In*, Péron, F.G. and Caldwell B.V. (ed) *Immunologic Methods in Steroid Determination*. Appleton-Century-Crofts, New York, 183-197.
Caraty, A., Martin, G.B. and Montgomery, G.W., 1984. *Reprod. Nutr. Develop.* (in press).
Chamley, W.A., Buckmaster, J.M., Cain, M.D., Cerini, J., Cerini, M.E., Cumming, I.A. and Goding, J.R., 1972. *J. Endocr., 55*, 253-263.
Clarke, I.J., Fraser, H.M. and McNeilly, A.S., 1978. *J. Endocr., 78*, 39-47.
Cognie, Y., Gayerie, F., Oldham, C.M. and Poindron, P., 1980. *Proc. Aust. Soc. Anim. Prod., 13*, 80-82.
Cole, H.H. and Miller, R.F., 1935. *Am. J. Anat., 57*, 39-98.
Cox, R.I., Wilson, P.A. and Mattner, P.E., 1976. *Theriogenology, 6*, 607.
Cox, R.I. and Wilson, P.A., 1976. *J. Reprod. Fert., 46*, 524.
Cumming, I.A., Blockey, M.A.deB., Brown, J.M., Catt, K.J., Goding, J.R. and Kaltenbach, C.C., 1970. *Proc. Aust. Soc. Anim. Prod., 8*, 383-387.
Dutt, R.H. and Casida, L.E., 1948. *Endocrinology, 43*, 208-217.
Ebling, F.J.P., Lincoln, G.A., Anderson, N., Cunningham, R.A. and Downey, L., 1984. *Proc. Ann. Conf. Soc. Study of Fertil.*, abstract 34.
Fabre-Nys, C., 1982. *In*, Agarwal, M.K. (ed) *Hormone Antagonists* Walter de Gruyter & Co., Berlin and New York.
Fairnie, I.J., Wales, R.G. and Gherardi, P.B., 1977. *Theriogenology, 8*, 183.
Ferin, M. Wehrenberg, W.B., Lam, N.Y., Alston, E.J. and Vande Wiele, R.L., 1982. *Endocrinology, 111*, 1652-1656.
Flint, A.P.F. and Sheldrick, E.L., 1983. *J. Reprod. Fert., 67*, 215-225.
Fraser, H.M., 1983. *Endocrinology, 112*, 245-253.
Fraser, H.M. McNeilly, A.S., 1983. *J. Reprod. Fert., 69*, 569-577.
Gherardi, P.B. and Martin, G.B., 1978. *Proc. Aust. Soc. Anim. Prod., 12*, 260.
Gherardi, P.B. and Lindsay, D.R., 1980. *J. Reprod. Fert., 60*, 425-429.
Goding, J.R., 1974. *J. Reprod. Fert., 38*, 261-267.
Gordon, I., 1983. *Controlled Breeding in Farm Animals*. Pergamon, Oxford. Chapters 10-20 in particular.
Haresign, W., 1978. *In* D.B. Crighton, N.B. Haynes, G.R. Foxcroft and G.E. Lamming. (ed) *Control of Ovulation* Butterworths, London, 435-451.
Horton, E.W. and Poyser, N.L., 1976. *Physiol. Rev., 56*, 595-651.
Katzorke, T., Propping, D., von der Ohe, M. and Tauber, P.R., 1980. *Fertil. Steril., 33*, 35-42.
Kennaway, D.J. and Gilmore, T.A., 1984. *J. Reprod. Fert., 70*, 39-45.
Kennaway, D.J., Gilmore, T.A. and Seamark, R.F., 1982. *Endocrinology, 110*, 1766-1772.
Knight, T.W., 1983. *Proc. N.Z. Soc. Anim. Prod., 43*, 7-11.
Knight, T.W., Oldham, C.M. and Lindsay, D.R., 1975b. *Aust. J. Agric. Res., 26*, 567-575.
Knight, T.W., Oldham, C.M., Smith, J.F. and Lindsay, D.R. 1975a. *Aust. J. Exp. Agric. Anim. Husb., 15*, 183-188.
McCracken, J.A., Glew, M.E. and Scaramuzzi, R.J., 1970. *J. Clin. Endocr. Metab., 30*, 544-546.
McLeod, B.J., Haresign, W. and Lamming, G.E., 1982. *J. Reprod. Fert., 65*, 223-230.
McLeod, B.J., Haresign, W. and Lamming, G.E., 1983. *J. Reprod. Fert., 68*, 489-495.
McNeilly, A.S., 1980. *J. Reprod. Fert., 58*, 537-549.

Martensz, N.D., 1980. *In,* Finn, C.A. (ed). *Oxford Reviews of Reproduction, 2*, 232-262.
Martin, G.B., 1984. *Biol. Rev., 59*, 1-87.
Martin, G.B., and Cognie, Y., 1984. *Appl. Anim. Ethol.* (in press)
Martin, G.B., Scaramuzzi, R.J. and Lindsay, D.R., 1981. *Aust. J. Biol. Sci., 34*, 569-575.
Martin, G.B., Scaramuzzi, R.J. and Lindsay, D.R., 1983. *J. Reprod. Fert., 67*, 47-55.
Martin, G.B., Scaramuzzi, R.J., Cox, R.I. and Gherardi, P.B., 1979. *Aust. J. Exp. Agric. Anim. Husb., 19*, 673-678.
Nowak, R. and Rodway, R.G., 1984. *Proc. Ann. Conf. Soc. Study of Fert.*, abstract 99.
Oldham, C.M., Cognie, Y., Gayerie, F. and Poindron, P., 1980. *Proc. 9th Int. Congr. Anim. Reprod. and AI. 3*, 50 (abstract).
O'Shea, T., Cummins, L.J., Bindon, B.M. and Findlay, J.K., 1982. *Proc. 14th Aust. Soc. Reprod. Biol. Conf.*, 82.
Pearce, D.T., Gray, S.J., Oldham, C.M. and Wilson, H.R., 1984. *Proc. Aust. Soc. Anim. Prod., 15*, 164-168.
Pearce, D.T., and Oldham, C.M., 1984. This volume, 26-34.
Poindron, P., Le Neindre, P. and Levy, F. 1984. This volume, 191-198.
Poindron, P., Martin, G.B. and Hooley, R.D., 1979. *Physiol. Behav., 23*, 1081-1087.
Radford, H.R., Avenell, J.A. and Szell, A., 1984. This volume, 342-344.
Robinson, T.J., 1954. *J. Endocr., 10*, 117-124.
Robinson, T.J., 1964. *Proc. Aust. Soc. Anim. Prod., 8*, 47-49.
Robinson, T.J., 1973. *Inserm., 26*, 453-478.
Robinson, T.J., and Reardon, T.F., 1961. *J. Endocr., 23*, 97-107.
Scaramuzzi, R.J., 1975. *In* Edwards, R. and Johnson, M.W. (ed) *Physiological Effects of Immunity Against Reproductive Hormones* Cambridge University Press, 67-90.
Scaramuzzi, R.J., Tillson, S.A., Thorneycroft, I.H. and Caldwell, B.V., 1971. *Endocrinology, 88*, 1184-1189.
Scaramuzzi, R.J., Davidson, W.R. and Van Look, P.F.A., 1977. *Nature* (London), *269* 817-818.
Symons, A.M., Cunningham, N.F. and Saba, N., 1973. *J. Reprod. Fert., 35*, 569-571.
Thomas, G.B., Martin, G.B. and Pearce, D.T., 1982. *Proc. 14th Aust. Soc., Reprod. Biol. Conf.*, 90.
Vaitukaitus, J.L., Ross, G.T., Brownstein, G.D. and Rayford, P.L., 1976. *Rec. Prog. Horm. Res., 32*, 289-321.
Van Look, P.F.A., Clarke, I.J., Davidson, W.G. and Scaramuzzi, R.J., 1978. *J. Reprod. Fert., 53*, 129-130.
Wallace, J.M. and McNeilly, A.S., 1984. *Proc. Ann. Conf. Soc. for study of Fert.*, abstract 81.
Walton, J.S., Evins, J.D., Fitzgerald, B.P., and Cunningham, F.J., 1980. *J. Reprod. Fert., 59*, 163-171.
Wright, P.J., Clarke, I.J. and Findlay, J.K., 1983. *Aust. Vet. J., 60*, 254-255.

# THE EFFECTS OF STEROID IMMUNIZATION OF EWES AND THEIR NUTRITION ON THE OVULATION RATE AND ASSOCIATED REPRODUCTIVE WASTAGE

L.J. Cummins and S.A. Spiker, *Pastoral Research Institute, P.O. Box 180, Hamilton, Vic. 3300.*
C. Cook, *La Trobe University, Bundoora, Vic. 3155.*
R.I. Cox, *CSIRO Division of Animal Production, P.O. Box 239, Blacktown, N.S.W. 2148.*

*Summary* An experiment was carried out to examine the response of Corriedale ewes to immunization against androstenedione and level of wheat supplementation at pasture during mating. Both immunization and nutrition had significant ($P < 0.05$) effects on ovulation rate and the number of foetuses per ewe at mid pregnancy, but there was no interaction between these factors. Immunized ewes had 2.4 ovulations and 1.9 foetuses per ewe, while non-immunized ewes had 1.7 ovulations and 1.6 foetuses. Ewes fed a supplement of 700 g wheat/day had 2.3 ovulations and 1.9 foetuses while ewes fed 300 g wheat/day had 1.9 ovulations and 1.7 foetuses. Reproductive wastage until mid pregnancy was higher (41% vs 23% of potential foetuses lost, $P < 0.05$) in ewes shedding 3 or more eggs than in ewes shedding 1 or 2 eggs. Ewes which carried 3 or 4 foetuses at mid pregnancy also had high pre-parturient (17%) and neo-natal (26%) lamb losses.

## INTRODUCTION

Low ovulation rate in ewes has been identified as one of the major limitations to the improvement of reproductive rate (Hanrahan 1980). Ovulation rate can be manipulated by nutritional, photoperiodic, genetic and pharmacological means.

Active immunization against steroids can increase ovulation rate by modifying ovarian feedback on the pituitary. This has led to the commercial release of a product consisting of androstenedione conjugated to human serum albumin together with DEAE dextran adjuvant (Fecundin (R) Glaxo Aust. Pty. Ltd., Boronia Vic.). This product appears to give reasonably precise control of reproduction (Cox *et al.* 1982, Scott *et al.* 1984).

## MATERIALS AND METHODS

A flock of 178 mature Corriedale ewes with a lambing rate of 147% in the previous year was mated to 2% of Border Leicester rams.

It was proposed to examine two levels of wheat feeding on drought affected pastures, but it rained in mid March and mating was carried out on short green pastures which were improving in quantity. On 24 January 1983 the ewes had a mean liveweight of $54 \pm 1$ kg and a mean condition score (Jefferies 1961) of $2.8 \pm 0.1$ and were allocated to their treatments. Grain was gradually introduced from this time, building to the high (700 g/head/day) or low (300 g/head/day) rations. Wheat was fed three times weekly. At the end of mating, supplementation ceased and all ewes were run together until pregnancy diagnosis was carried out using real time ultrasound scanning (Fowler and Wilkins 1982) on 26 July (mid pregnancy, $\simeq$ 90 days gestation). Half the ewes within each nutritional group were immunized with Fecundin on 17 March and 6 April. Those remaining were left as controls. Blood samples were taken from immunized ewes on 14 April and 6 June. Androstenedione antibody titres were determined (Cox *et al.* 1982).

Mating was from 21 April until 3 June. Oestrous marks were recorded weekly and rams rotated between groups. Ovulation rate was determined by laparoscopic examination 7 to 11 days after oestrus.

After pregnancy diagnosis ewes were reallocated to further experiments on the basis of their litter size. More complete details of these experiments are given by Foot *et al.* (1984). Briefly, ewes carrying triplets and half those carrying twins received a high level of lupin grain supplemenation while the remaining twin bearing ewes and those with singles only received a low level of supplementation. After lambing, ewes with twins were allocated to either a high plane of nutrition or a low plane of nutrition on a crossover design considering nutrition in late pregnancy and early lactation but triplets and their dams continued on a high plane of nutrition and singles on a low plane.

The ewes bearing triplets lambed individually in small grassed plots. The other ewes lambed in small flocks under close supervision. After lamb marking on 17 and 18 October all ewes grazed together in a common flock until weaning on 7 December 1983.

## RESULTS

The antibody titres in the immunized ewes were elevated after the second injection but had declined by the end of mating (Table 1). Nutrition had no significant effect on titres. The correlations between ovulation rates and titres before mating were low (0.3 high nutrition and 0.4 low nutrition). Fifteen immunized ewes returned to service with similar ovulation rates at both times.

**TABLE 1. Results of nutrition around mating and steroid immunization on reproduction in Corriedale ewes (mean ± SEM).**

| Supplement at mating | High (700 g wheat) | | Low (300 g wheat) | |
|---|---|---|---|---|
| Immunization | + | − | + | − |
| Number of ewes | 44 | 44 | 45 | 45 |
| Liveweight change to start of mating (kg, ± SE) | 0.1 ± 0.5[a] | 0.8 ± 0.5[a] | −5.1 ± 0.4[c] | −3.9 ± 0.4[b] |
| Condition score change to start of mating | 0.5 ± 0.1[a] | 0.5 ± 0.1[a] | −0.4 ± 0.1[b] | −0.2 ± 0.1[b] |
| Liveweight change during mating (kg) | 3.9 ± 0.6[ab] | 4.8 ± 0.7[a] | 1.8 ± 0.3[c] | 2.4 ± 0.6[bc] |
| Log[1]/titre before mating[(1)] | 3.19 ± 0.10 | - | 3.09 ± 0.09 | - |
| Log[1]/titre after mating | 2.22 ± 0.07 | - | 2.10 ± 0.27 | - |
| % ewes mated 1st 2 weeks | 91 | 95 | 87 | 87 |
| Ovulation rate (1st oestrus) | 2.58 ± 0.12[a] | 1.95 ± 0.07[b] | 2.25 ± 0.11[c] | 1.53 ± 0.09[d] |
| Pregnancy rate (scanned) | 2.05 ± 0.12[a] | 1.75 ± 0.09[b] | 1.79 ± 0.10[ab] | 1.51 ± 0.09[b] |
| Reproductive wastage to mid pregnancy (%)[(2)] | 35 | 21 | 38 | 11 |
| Lambing rate[(3,4)] | 2.00 ± 0.14 | 1.74 ± 0.09 | 1.76 ± 0.10 | 1.58 ± 0.04 |
| Marking rate[(3,4)] | 1.52 ± 0.14 | 1.52 ± 0.09 | 1.56 ± 0.08 | 1.29 ± 0.10 |
| Weaning rate[(3,4)] | 1.52 ± 0.14 | 1.52 ± 0.09 | 1.53 ± 0.09 | 1.27 ± 0.10 |
| Liveweight of lambs weaned (kg)/ewe lambed[(4)] | 37 ± 3 | 40 ± 2 | 37 ± 2 | 33 ± 3 |

Means within rows with different superscripts differ ($P < 0.05$)
[(1)] Mean (log [1]/titre) for androstenedione antibodies
[(2)] For ewes mated in first 2 weeks
[(3)] per ewe present at lambing
[(4)] Additional treatments were imposed during late pregnancy and lactation.

The ewes on the high plane of nutrition maintained liveweight and condition until the start of mating and gained during mating, while the ewes on the low plane of nutrition were significantly lighter and lower in condition score at the start of mating (Table 1). Both nutrition and immunization had significant ($P < 0.05$) effects on ovulation rate and the number of foetuses observed at mid pregnancy. Immunization increased ovulation rate by 39% (i.e. 1.7 to 2.4) and number of foetuses per ewe by 18% (1.6 to 1.9), while the higher level of wheat supplementation increased ovulation rate by 19% (1.9 to 2.3) and the number of foetuses by 15% (1.7 to 1.9). These effects were additive and the individual treatment means are shown in Table 1. Overstimulation, defined as ewes having 4 or 5 ovulations, was observed in 8% of the immunized ewes but in none of the control ewes ($P < 0.05$).

An indication of reproductive wastage (i.e. the percentage of eggs shed not represented as foetuses at scanning) was obtained. The level of wastage was related to the ovulation rate. When all groups were combined, ewes shedding 4 or 5, 3, 2 or 1 egg lost 60, 36, 24 and 15% respectively of their potential embryos by mid pregnancy. By considering only those ewes mating in the first 2 weeks it is possible to separate wastage into (a) probably early embryonic losses (total loss of embryos and return to service within 4 weeks) and (b) possible later losses (not associated with returns to service). Slightly more than half the wastage (56%) until mid pregnancy was due to ewes that returned to service.

The nutritional treatments imposed on twin bearing ewes in late pregnancy and early lactation did not result in significant differences in ewe liveweight, neo-natal (birth-1 week) lamb survival or lamb growth rate to marking.

Litter size was the main factor influencing lamb survival. The 19 ewes with litters of 3 or 4 had a 17% pre parturient loss (death estimated to have occurred 1-3 weeks before term) 26% neonatal loss and a 5% later lamb loss. Ewes bearing twins had losses of 0.5, 7 and 2%, while ewes with singles had losses of 0, 6 and 2% respectively.

Litter size at birth also had large effects on weaning weight (triplets 22.1 ± 0.7, twins 23.8 ± 0.3 and singles 26.9 ± 0.7 kg).

The nutritional treatments imposed in late pregnancy confound the interpretation of the final outcome (i.e. lambing rate, marking rate, weaning rate and liveweight of lamb weaned per ewe lambing) of the treatments imposed at mating time, but because of their practical importance these are shown in Table 1. Under these conditions immunization or better nutrition at mating appeared to improve the final outcome, but their combination was not additive.

## DISCUSSION

These results provide further evidence of the ability of Fecundin to increase ovulation rates and pregnancy rates in ewes (e.g. see Scott *et al.* 1984). They do however indicate that there can be circumstances when this procedure will stimulate ovulation rates to levels in excess of the ability of the ewe to bear and rear the lambs. It is not yet clear if nutritional manipulation during pregnancy and lactation can overcome this (e.g. see Foot *et al.* 1984). In this experiment and that of Croker *et al.* (1983) and Smith *et al.* (1981) the combined effects of nutrition and immunization appeared additive (although Croker *et al.* (1983) reported exceptions, due to age of ewe and Smith *et al.* (1981) exceptions due to type of immunogen). Grant (1982) reported that immunization or lupin feeding increased ovulation rate and pregnancy rate of ewes (at mid pregnancy) but the effects were not additive.

altered, resulting in a reasonable number of triplet ovulations. It is also clear that there are situations when the reproductive wastage during pregnancy in ewes shedding 2 or more eggs is far greater than their neonatal lamb losses. Given the success of genetic selection programs for embryonic survival in mice (Bradford *et al.* 1980) and the improvements in technology for assessing reproduction in ewes, further effort could be devoted to reducing reproductive wastage during pregnancy.

## REFERENCES

Bradford, G.E., Barkely, M.S. and Spearow, J.L., 1980. *In* Robertson, A. *Selection experiments in Laboratory and Domestic Animals*, Commonwealth Agricultural Bureau, Slough U.K., 161-175.

Cox,R.I., Wilson, P.A., Scaramuzzi, R.J. Hoskinson, R.M., George, J.M. and Bindon, B.M., 1982. *Proc. Aust. Soc. Anim. Prod. 14*, 511-514.

Croker, K.P., Cox, R.I., Johnson, T.J. and Wilson, P.A., 1983. *Proc. Aust. Soc. Reprod. Biol. 15*, 28.

Foot, J.Z., Cummins, L.J., Spiker, S.A. and Flinn, P.C., 1984. This volume, 187-190.

Fowler, D.G. and Wilkins, J.F., 1982. *Proc. Aust. Soc. Anim. Prod. 14,* 491-494.

Grant, I.M., 1982. *Aust. Vet. J., 59*, 128.

Hanrahan, J.P., 1980. *Proc. Aust. Soc. Anim. Prod., 13,* 405-408.

Jefferies, B.C., 1961. *Tas. J. Agric. 32*, 19-21.

Scott, T.W., Cox, R.I., Geldard, H., Scaramuzzi, R.J., Hoskinson, R.M. and Djura, P., 1984. *Proc. Aust. Soc. Anim. Prod. 15*, 182-194.

Smith, J.F., Cox, R.I., McGowan, L.T., Wilson, P.A. and Hoskinson, R.M., 1981. *Proc. N.Z. Soc. Anim. Prod. 41*, 193-197.

# INFLUENCE OF TIME OF INJECTION OF FECUNDIN® ON THE REPRODUCTIVE PERFORMANCE OF MERINO EWES IN WESTERN AUSTRALIA

K.P. Croker, *Department of Agriculture, South Perth, W.A. 6151.*
R.I. Cox, *Division of Animal Production, CSIRO, P.O. Box 239, Blacktown N.S.W. 2148.*
T.J. Johnson, *Department of Agriculture, South Perth, W.A. 6151.*
M. Salerian, *Department of Agriculture, Katanning, W.A. 6317.*

*Summary* Two experiments were conducted to determine whether an interval longer than 2 weeks between the last injection of Fecundin® and the start of joining would improve the reproductive performance of mature Merino ewes. In both experiments the OR's of immunized ewes were significantly higher than in untreated ewes irrespective of whether the interval was 2, 3, or 4 weeks. In Experiment 1 there were more barren ewes in the group injected 2 weeks before the start of joining, but in Experiment 2 there was no significant difference between the immunized groups in the percentage of ewes which conceived following AI. It appears that an interval longer than 2 weeks between the last injection of Fecundin® and the start of joining could avoid a depression in the reproductive performance of immunized ewes.

## INTRODUCTION

A practical procedure for increasing the fecundity of ewes by immunization against androstenedione (a'dione) has been developed recently (Cox *et al.* 1982; Scaramuzzi *et al.* 1983; Geldard *et al.* 1984) and has become available commercially (Fecundin,® Glaxo Australia, Boronia, Vic.). These workers reported that the earlier problems of anoestrus and increased barrenness associated with immunizations (Martin *et al.* 1979) have been largely overcome in tests in eastern Australia. Scaramuzzi *et al.* (1983) and Geldard (1984) found that as the interval between the second injection and the start of joining increased lambing percentages improved and there also was a decrease in barrenness. The maximum percentage increase in lambs born to treated ewes was obtained when the interval between the second injection and joining was between 2 and 4 weeks.

Under Western Australian conditions ovulation rates (OR's) and lambing percentages also are increased in ewes immunized against a'dione (Croker *et al.* 1982, 1983), but in three of four projects barrenness in immunized ewes was greater than in untreated ewes. Two experiments were designed to determine whether an interval longer than 2 weeks between the last injection and the start of joining would improve the reproductive performance of mature Merino ewes immunized against a'dione.

## MATERIALS AND METHODS

### Experiment 1

This experiment was conducted on the Western Australian Department of Agriculture's Research Station at Wongan Hills. Strong-wool Merino ewes (1978 and 1979 born) previously immunized with a'dione immunogens were randomly allocated to two groups, one was injected with 2 ml of Fecundin® subcutaneously in the neck 4 weeks before the introduction of entire rams while the other group received Fecundin® 2 weeks before joining started. Additionally, two hundred 1981 born ewes were randomly allocated to three groups, viz. an untreated group, a group injected 8 and 4 (8/4) weeks and a third group injected 6 and 2 (6/2) weeks before the start of joining. The treatment groups were —

| Treatment | Year ewes born | | |
|---|---|---|---|
| | 1978 | 1979 | 1981 |
| 1. Untreated | 39 | 33 | 100 |
| 2. Immunized, 6/2 | 20 | 16 | 50 |
| 3. Immunized, 8/4 | 20 | 16 | 50 |

All ewes were grazed together from November 10, 1983. Four testosterone- treated wethers (Fulkerson *et al.* 1981) were joined with the ewes on December 22, 1983. These were replaced by seven mature harnessed rams on January 5, 1984 (day 0) and mating records were obtained at seven day intervals during the 42 day joining period. The ewes marked by rams during the first seven days of joining were examined by laparoscopy on day 12 while all the other ewes were examined on day 21 to record OR's. Forty nine days after the rams were removed all ewes were examined by realtime ultrasound scanning (ADR Model 2130 with a 3.5 MHz transducer, Fowler and Wilkins 1982) to determine the number of embryos.

### Experiment 2

This experiment was on a commercial farming property at Kojonup. Four hundred and forty-seven 2½ year old strong-wool Merino ewes were allocated to four groups, one untreated (147 ewes) and three treated

(100 ewes each) 8 weeks before the start of an artificial insemination (AI) programme. All ewes were weighed on allocation and grouped together until inseminated.

The three treated groups were injected twice with 2 ml of Fecundin® either 8 and 4 (8/4), 6 and 3 (6/3) or 5 and 2 (5/2) weeks before the day of the first insemination (February 12, 1984). Ewes marked by teasers were drafted from the flock each morning and within 3 hours they were inseminated at the external opening of the cervix with 0.15 ml of fresh, undiluted semen. The AI programme lasted 14 days and three days after its completion harnessed, entire rams were put with the regrouped ewes (February 29). The rams were removed on March 28 when the marked ewes were recorded. Eighteen days after the start of the AI programme 50 ewes from each group were examined by laparoscopy to determine the OR's. Thirty-six days after the removal of the rams all ewes were examined by realtime ultrasound scanning to determine the number of embryos.

Data were examined by Chi-square analyses.

## RESULTS

### Experiment 1

The average body weights of the ewes at the start of joining were 45.3 44.9 and 45.3 kg for the untreated ewes and those immunized 6/2 or 8/4 weeks before joining started, respectively. OR's are recorded in Table 1. The difference in OR's between the untreated and immunized ewes was highly significant ($P < 0.001$) but the difference between the two immunized groups was not significant.

**TABLE 1. Ovulation rates and pregnancy status of ewes treated with Fecundin® at different intervals prior to the start of joining, Exp. 1.**

| Treatment | Ewes ovulating (%) | Ovulation rate(1) | Dry ewes(2) (%) | Multiple pregnancy(3) (%) | Potential lambing(2) (%) |
|---|---|---|---|---|---|
| Untreated | 80.1 | 1.02 | 10.8 | 12.9 | 100.6 |
| Immunized,6/2 | 87.2 | 1.24 | 26.9 | 28.1 | 96.2 |
| Immunized,8/4 | 79.7 | 1.17 | 9.2 | 23.2 | 113.2 |

(1) OR of ewes ovulating.
(2) % of ewes scanned.
(3) % of pregnant ewes

The results from the scanning of the ewes are also shown in Table 1. Fewer of the ewes treated 6/2 weeks before joining were pregnant than were those treated 8/4 weeks before joining or those not immunized ($P < 0.001$). A higher proportion of the pregnant immunized ewes had multiple pregnancies ($P < 0.05$).

### Experiment 2

The average body weights 8 weeks before the start of joining were 63.5, 61.5, 62.3, 63.9 kg for the untreated ewes and those immunized 8/4, 6/3 or 5/2 weeks before joining started, respectively. OR's for the ewes ovulating were 1.45, 1.92, 1.90 and 1.98 for the same respective groups. The differences between the groups of immunized ewes were not significant, but these ewes had a significantly higher OR than the untreated ewes ($P < 0.001$). There was no significant difference between groups in the proportion of non-ovulating ewes.

The results from the scanning of the ewes which were inseminated are presented in Table 2. There were no differences between treatments in the proportion of ewes which were marked in the back-up joining (36.8% returned) while the ewes injected with Fecundin® 8/4 weeks before the start of AI had significantly ($< 0.05$) more multiple pregnancies than did the untreated ewes.

**TABLE 2. Pregnancy status of ewes treated with Fecundin® at different intervals prior to the start of an AI programme, Exp. 2.**

| Treatment | Ewes inseminated (No.) | Ewes returned No. | Ewes returned % | Embryos scanned after A.I. 0 | 1 | 2 | 3 | Multiple pregnancy (%) |
|---|---|---|---|---|---|---|---|---|
| Untreated | 140 | 50 | 35.7 | 2 | 58 | 29 | 1 | 34.1 |
| Immunized, 5/2 | 96 | 36 | 37.5 | 3 | 31 | 25 | 1 | 45.6 |
| Immunized, 6/3 | 96 | 32 | 33.3 | 2 | 34 | 28 | 0 | 45.2 |
| Immunized, 8/4 | 95 | 39 | 41.1 | 4 | 23 | 28 | 1 | 55.8 |

## DISCUSSION

The results from these experiments provide evidence in addition to that of Croker *et al.* (1982, 1983) that OR's are increased in Merino ewes treated with Fecundin® in Western Australia. When the interval between the last injection and the start of joining was increased from 2 to 4 weeks OR's did not decline.

There was an increase in the proportion of multiple pregnancies in the immunized ewes in Experiment 1, but the increased level of barrenness in the ewes injected 2 weeks before the start of joining nullified the potential gains in this group. In contrast to Experiment 1, in Experiment 2 there was no effect of treatment of barrenness following AI. In addition, although the proportion of multiple pregnancies was high in all treated groups, it only attained significance in the ewes treated 4 weeks before the start of AI. This is at variance with observation elsewhere, which show that Fecundin® significantly increases fecundity when the injection to joining interval is shorter than 4 weeks.

These results show that Fecundin® can improve the potential reproductive performance of Merino ewes with either paddock joinings or when artificially inseminated. When the results of the experiments reported here are considered with the data of Scaramuzzi *et al.* (1983) and Geldard *et al.* (1984) it is apparent that an interval longer than 2 weeks between the last injection and the start of joining would improve the reproductive performance of immunized ewes in Western Australia.

## ACKNOWLEDGEMENTS

The assistance provided by Mr. P. Burgess at Wongan Hills is gratefully acknowledged as is the co-operation of Robin and Helen Young, Yannawah, Kojonup. Many others, including Dr R. Kelly and Messrs L. Butler, M. Johns, R. Bryant and C. Wastie, also provided excellent assistance. Glaxo (Aust.) Pty Ltd generously provided the Fecundin® used in Experiment 1.

## REFERENCES

Cox, R.I., Wilson, P.A., Scaramuzzi, R.J., Hoskinson, R.M., George, J.M., and Bindon, B.M., 1082. *Proc. Aust. Soc. Anim. Prod., 14.* 511-514.

Croker, K.P., Cox, R.I., Johnson, T.J. and Wilson, P.A., 1982. *Proc. Aust. Soc. Reprod. Biol., 14,* 96.

Croker, K.P., Cox, R.I., Johnson, T.J. and Wilson, P.A., 1983. *Proc. Aust. Soc. Reprod. Biol., 15,* 28.

Fowler, D.G. and Wilkins, J.F., 1982. *Proc. Aust. Soc. Anim. Prod., 14,* 636.

Fulkerson, W.J., Adams, N.R. and Gherardi, P.B., 1981. *Appl. Anim. Ethol., 7,* 57-66.

Gerald, H., 1984. *Proc. Aust. Soc. Anim. Prod., 15,* 182-198.

Geldard, H., Dow, G.J., Scaramuzzi, R.J., Hoskinson, R.M., Cox, R.I. and Beels, C.M., 1984. *Aust. Vet. J., 61,* 130-132.

Martin, G.B., Scaramuzzi, R.J., Cox, R.I. and Gherardi, P.B., 1979. *Aust. J. Exp. Agric. Anim. Husb., 19,* 673-678.

Scaramuzzi, R.J., Geldard, H., Beels, C.M., Hoskinson, R.M. and Cox, R.I., 1983. *Wool Tech. Sheep Breed., 31,* 87-97.

# EFFECTS OF WEANING AND TREATMENT WITH PROGESTAGEN-PMSG OR PROGESTAGEN-ODB ON FERTILITY OF THE POST PARTUM EWE.

S.T. Dawe, *Department of Agriculture, Agricultural Institute, Yanco, New South Wales, 2703.*

*Summary* Anoestrous crossbred ewes were treated with progestagen sponges (40 *vs* 60 mg MAP) at 8 weeks *post partum* in spring, and lambs were either weaned or left suckling. At 10 weeks *post partum* ewes were given injections of 600 i.u. PMSG at sponge removal, or 30μg ODB 12 hours later, and were joined with 15% rams.

Fertility was not effected by dose of progestagen but was severely reduced by ODB. Weaning at 8 weeks increased fertility from 14 to 27% (lambed/mated) in ODB treated ewes, and from 41 to 69% in PMSG treated ewes. There was no sexual activity in untreated control ewes until 7 weeks after the first induced oestrus in the treated ewes.

## INTRODUCTION

Removal of lambs at 1-2 days of age has increased conception rates in spring at 8 weeks *post partum* (pp.) by 10-12 percentage units (Cognie *et al.* 1975). Rhind *et al.* (1980) found higher fertility at 9 weeks (pp.) where lambs were weaned 2 weeks before mating. This may prove a more practical and less costly alternative to removal of lambs at 1-2 days.

Dose of medroxy progesterone acetate (MAP) in intravaginal sponges is normally 60mg but a dose of 40mg gave excellent fertility in lactating ewes at 12 weeks pp. in spring (Dawe *et al.* 1969). Oestrogen administered after progestagen treatment of ewes has given conflicting results. Hulet and Stormshak (1972) found that injection of oestradiol-17β at the time of injection of Pregnant Mare Serum Gonadotrophin(PMSG) reduced fertility in anoestrous ewes, yet Falkenburg *et al.* (1970) found increased fertility with oestradiol-17β injected at the time of progestagen removal in cyclic ewes. I.D. Killeen and R. Scaramuzzi (personal communication) induced fertile oestrus in 22 anoestrous ewes by administering 30μg oestradiol benzoate (ODB) 12 hours after removal of progestagen sponges from 42 ewes. In a later study, similar treatment of ewes at 7 week pp. in June was compared with progestagen-PMSG treatment, but ODB reduced the conception rate at the first oestrus (26% vs 45%, S. Dawe unpublished data).

This experiment examined several factors which might influence the response of lactating ewes to hormonal control of reproduction in spring. They included weaning lambs two weeks before joining, dose of progestagen, and the use of oestrogen as an alternative to pregnant mare serum gonadotrophin (PMSG).

## MATERIALS AND METHODS

Border Leicester × Merino ewes which had lambed in early August 1982 were allocated randomly to 8 treatment groups and two control groups.

The experiment was a 2 × 2 × 2 factorial design (n = 37-40, N = *304).*

| *Factor* | *Comparison* | *Levels* |
|---|---|---|
| Status of ewe | Weaned vs Suckling | 2 |
| Dose of progestagen (mg) | 40 vs 60 | 2 |
| Hormone after progestagen | PMSG vs ODB | 2 |

In addition, there were separate untreated control groups of 50 ewes suckling lambs and 50 ewes with lambs weaned. At the time of the first induced oestrus the ewes were 10 weeks (9-11 weeks) *post partum.*

Lambs were weaned as necessary on 6-10-82 and intravaginal sponges (Upjohn Pty. Ltd.) containing 40 or 60mg MAP were inserted in the treated groups which were then grazed in two separate mobs of suckling and weaned ewes. The two control groups were also run separately and each was joined with two harnessed Dorset rams. On 20-10-82 sponges were removed and four groups were injected with 600 i.u. PMSG. After 12 hours the other four groups were injected with 30μg ODB. The separate mobs of weaned and suckling ewes were joined next day with 24 (15%) harnessed Dorset rams each, for a week.

All treated ewes, whether mated or not at the time of the first induced oestrus, were given an injection of 400 i.u. PMSG 16 days after sponge removal. They were then rejoined for 10 days with 7.5% of Dorset rams.

Ovarian activity of treated ewes which did not mate at the first induced cycle was examined by laparoscopy eight days after sponge removal.

Ewes were weighed at joining. Dates of mating and lambing, and number of lambs born, were recorded for each ewe.

## RESULTS

Mean liveweight of all ewes on 25-10-82 was 57 kg with no significant differences between treatment groups.

Oestrous activity of the untreated ewes did not commence until during the week ending 16-12-82, when 26 and 4 percent of weaned and suckling control groups had been raddled. These figures increased to 46 and 22 percent by 29-12-82 and, 62 and 54 percent by 8-2-83, respectively.

Details of mating and lambing for treated groups are shown in Table 1. There were no significant main effects on mating performance but there were significant interactions of status of ewe × hormone ($P < 0.01$) and dose × hormone ($P < 0.05$). Fertility of ewes at the first induced oestrus was not affected by doses of progestagen, but was increased by weaning ($P < 0.001$) and reduced by ODB ($P < 0.001$). The increase due to weaning was from 41 to 69 percent in ewes treated with PMSG, and from 14 to 27 percent in ewes treated with ODB. Litter sizes at lambing were 1.67 with PMSG and 1.32 with ODB.

At the second induced oestrus, fertility was again higher ($P < 0.05$) in weaned ewes (86.7%) than in suckling ewes (61.1%). More weaned ewes lambed to both first and second oestrous cycles than suckling ewes ($P < 0.001$) after both progestagen-PMSG-PMSG treatment (89.2% *vs* 70.1% of ewes treated) and progestagen-ODB-PMSG treatment (65.8% *vs* 37.7%).

Endoscopy of ewes not mated at the first induced oestrus revealed that all except two and ovulated. These two were suckling ewes treated with 40mg MAP followed by PMSG. All ovulations appeared to have been at the time of the first oestrus, except for one which had a very young corpus luteum and may therefore have had a short oestrous cycle.

Lamb growth was reduced slightly by weaning at eight weeks of age, so that liveweights of weaned and suckled lambs were 29.3 and 30.5 kg at 12 weeks.

**TABLE 1. Mating and lambing data for weaned and suckling ewes treated with MAP — PMSG or MAP — ODB (first cycle), followed by PMSG (second cycle).**

| Status of ewe | Dose MAP | Hormone after MAP | No. of ewes | First induced cycle: Percent ewes Mated | First induced cycle: Percent ewes Lambed(1) | Second induced cycle: No. mated | Second induced cycle: Percent lambed | Percent ewes(2) lambed to both cycles |
|---|---|---|---|---|---|---|---|---|
| Weaned | 40 | PMS | 37 | 78.4 | 72.4 | 14 | 85.7 | 89.2 |
| | | ODB | 38 | 94.7 | 27.8 | 21 | 85.7 | 73.7 |
| | 60 | PMS | 37 | 89.2 | 66.7 | 12 | 91.7 | 89.2 |
| | | ODB | 38 | 92.1 | 25.7 | 19 | 63.2 | 57.9 |
| Suckling | 40 | PMS | 40 | 90.0 | 41.7 | 21 | 57.1 | 67.5 |
| | | ODB | 39 | 89.7 | 11.4 | 18 | 61.1 | 38.5 |
| | 60 | PMS | 37 | 100.0 | 40.5 | 18 | 66.7 | 73.0 |
| | | ODB | 38 | 81.6 | 16.1 | 15 | 60.0 | 36.8 |

(1) Percent of ewes mated (2) Percent of ewes joined.

## DISCUSSION

Reducing the dose of progestagen in deeply anoestrous ewes did not alter their fertility. The interactions of treatments on mating performance are difficult to explain. They arose from the mating response after PMSG treatment being highest in suckled ewes given 60 mg MAP, and lowest in weaned ewes given 40mg MAP, while the reverse applied with ODB treated ewes. These interactions were not evident at lambing. The failure of mating in some ewes which had ovulated may have arisen through a shorter duration of the first induced oestrus in the *post partum* ewe (Pretorius 1967) and hence failure to be detected by the rams.

A large increase in fertility at 10 weeks pp. was achieved by weaning at 8 weeks. pp. rather than at 1-2 days, as suggested by Cognie *et al.* (1975) thus obviating the need for costly artificial rearing, and the penalty in lamb growth was minimal.

Lowered fertility in lactating ewes can be due to failure of insemination, fertilisation, or implantation (Rhind *et al.* 1980; Quirke *et al.* 1981), or to later embryonic loss (Cognie *et al.* 1975). With PMSG, 80 percent of non-lambing ewes returned to service, indicating fertilisation failure or early embryonic loss. The remaining 20 percent were more likely to have suffered failure of implantation or later embryonic loss. With ODB, 72 percent of weaned and 53 percent of suckling ewes also failed to return, indicating a greater proportion of implantation failure or later embryonic loss in ODB groups than in PMSG groups.

On the basis of the first cycle data and the previously noted reduction in fertility (S.T. Dawe, unpublished data) we would discard ODB as an alternative to PMSG. When both cycles are considered, however, progestagen-ODB treatment resulted in a reasonable proportion (65.7%) of weaned ewes lambing some five weeks before any of the untreated ewes. Ewes treated with ODB apparently ovulated and exhibited oestrus at the first cycle, but many did not hold to this service. They then responded, however, to the later PMSG with a fertile oestrus.

This experiment showed a major improvement in the fertility of the hormone treated *post partum* ewe through the simple management strategy of weaning at the time treatment commenced. This has application where more frequent lambings are sought to increase efficiency of production or to allow the rapid build-up of flock numbers. We now need to identify the hormonal changes which are caused by weaning, the mechanisms by which fertility is affected, the intervals from lambing over which the effect operates, the optimum weaning time, and whether season of the year modifies the effect. There is a need to understand the effects of the ODB treatment which cause a reduction in fertility at the first induced cycle but allow induction of a second oestrus which is fertile.

REFERENCES

Cognie, Y., Hernandez-Barreto, M. and Saumande, J., 1975. *Ann. Biol. Anim. Bioch. Biophys., 15*, 329-343.
Dawe, S.T., Roberts, E.M. and Killeen, I.D., 1969. *Aust. J. Exp. Agric. Anim. Husb., 9*, 385-388.
Failkenburg, J.A., Hulet, C.V. and Kaltenbach, C.C., 1970. *J. Anim. Sci., 32*, 1206-1211.
Hulet, C.V. and Stormshak, F., 1972. *J. Anim. Sci., 34*, 1011-1019.
Pretorius, P.S., 1967. *S. Afr. J. Agric. Sci, 10*, 883-899.
Quirke, J.F., Hanrahan, J.P., Sheehan, W.P. and Gosling, J.P., 1981. *Ir. J. Agric. Res., 20*, 1-8.
Rhind, S.M., Robinson, J.J., Chesworth, J.M. and Phillipps, M., 1980. *J. Reprod. Fert., 58*, 127-137.

# INCREASED OVULATION RATE IN MERINO EWES AND ADVANCEMENT OF PUBERTY IN MERINO LAMBS IMMUNIZED WITH A PREPARATION ENRICHED IN INHIBIN

T. O'Shea, S.A.R. Al-Obaidi and M.A. Hillard, *Department of Physiology, University of New England, Armidale, New South Wales, 2351.*
B.M. Bindon, *C.S.I.R.O., Division of Animal Production, Armidale, New South Wales, 2350.*
L.J. Cummins, *Department of Agriculture, Pastoral Research Institute, Hamilton, Victoria, 3300.*
J.K. Findlay, *Medical Research Centre, Prince Henry's Hospital, Victoria, 3004.*

*Summary* Immunization of adult Merino ewes against an inhibin-enriched preparation from bovine follicular fluid resulted in a marked increase in ovulation rate (Control ewes, n = 126, mean ± s.e.m. ovulation rate 1.40 ± 0.04; ewes immunized three times in 1982 and twice in 1983, n = 64, 2.87 ± 0.38; P < 0.01) and an increase in litter size (control ewes, 1.26 ± 0.04; immunized ewes 1.62 ± 0.11; P < 0.01). The ewes' female progeny were normal. Immunization of ewe lambs from 3 weeks of age advanced their puberty.

## INTRODUCTION

Inhibin is an ovarian hormone which specifically suppresses pituitary follicle stimulating hormone (FSH) secretion with little or no effect on luteinizing hormone (LH) in sheep (Cummins *et al.* 1983) or in other species (Franchimont *et al.* 1979). Because of its effect on FSH it may be expected to have a role in regulation of ovulation rate (O.R.).

The hypothesis for the present experiments was that successful immunization against an inhibin-enriched preparation would neutralize the effects of endogenous inhibin, increase plasma FSH concentration and consequently increase O.R. Such a procedure was shown to increase O.R. in a small group of sheep (O'Shea *et al.* 1982). In the present study a large flock was used to examine the effects on O.R. of such immunization in more detail. Blanc (1980) suggested that inhibin may play a role in the induction of spermatogenesis in the prepuberal animal. This is supported by the sensitivity of immature animals to follicular fluid (Hochereau-de Reviers and de Reviers 1978; Hermans *et al.* 1980; Franchimont *et al.* 1981). Consequently Merino lambs were immunized from 3 weeks of age (Blanc and Terqui 1976) to test effects on the onset of puberty.

## MATERIALS AND METHODS

### Animals

A flock of 296 fine-wool Merino ewes run commercially at Armidale was used for Experiment 1. Merino lambs born in October 1982 were randomized within age and birth weight classifications for Experiment 2.

### Immunogen

Bovine ovaries were collected and stored on ice, washed, and bovine follicular fluid (bFF) extracted from all visible follicles (excluding cystic follicles) was stored at −18°C. The material used in Expt. 1 was partially purified using affinity chromatography on Reactive Red 120- agarose (Sigma) (Jansen *et al.* 1981). For Expt. 2 bFF was chromatographed sequentially on Reactive Red 120-agarose and Sepharose-immobilized Poly-(L-Lysine)-Deoxycholic acid (Pierce) essentially as described by de Jong *et al.* (1981).

In Expt. 1 ewes were immunized (I) with an emulsion containing Freund's complete adjuvant (0.8 ml), the bFF fraction (2-2.5 mg protein), Tween 80 (final concentration 0.08%), made up to an overall volume of 2.4 ml with water and 0.8 ml injected intramuscularly into each hind leg and 0.8 ml subcutaneously divided between five sites along the back. Control ewes (C) were untreated. In Expt. 2 the lambs (I) were injected with smaller volumes made up in the same proportions, commencing at 3 or 9 weeks of age with 0.3 mg protein per lamb and increasing to 1.0 mg protein at 6 months. Immunization was at 3-4 week intervals and ceased at 38 weeks. Control (C) lambs were given an equivalent volume of Freund's adjuvant alone.

Effects of ovulation rate in both experiments were assessed by laparoscopy and were tested for significance using $\chi^2$ and t tests.

## RESULTS

### Immunization of ewes

Immunization three times at 4 week intervals before joining resulted in a significant (P < 0.05) increase in O.R. (Table 1). The crossbred ewe lambs born from these ewes were retained and in 1984 showed normal reproductive parameters;- oestrous cycle length, 17.1 ± 0.16 days, n = 36; O.R. = 1.4 ± 0.08, n = 40 and conception rate (9 week non-returns) = 37 of 40. Corresponding figures for progeny of C ewes were 16.8 ± 0.11 (n = 63), 1.5 ± 0.07 (n = 66) and 56 of 66.

**TABLE 1. Ovulation rates of control (C) Merinos and Merinos immunized (I) against an inhibin-enriched fraction from bovine follicular fluid. The same ewes were studied in the 3 years.**

| Year of Examination | Treatment | Ewes with ovulation rate of 0 | 1 | 2 | 3 | > | Ovulation rate (mean ± SEM) |
|---|---|---|---|---|---|---|---|
| 1982 | C | 0 | 60 | 84 | 3 | 0 | 1.61 ± 0.04 |
| | I(x3) | 0 | 71 | 56 | 10 | 10 | 1.87 ± 0.12* |
| 1983 | C | 0 | 75 | 51 | 0 | 0 | 1.40 ± 0.04 |
| | I(1)(x1) | 0 | 31 | 21 | 4 | 7 | 2.05 ± 0.22** |
| | I(1)(x2) | 0 | 28 | 17 | 4 | 15 | 2.87 ± 0.38** |
| 1984 | C | 3 | 69 | 40 | 0 | 0 | 1.32 ± 0.05 |
| | I(2)(× 0) | 0 | 26 | 10 | 1 | 3 | 1.58 ± 0.16 |
| | I(2)(× 2) | 1 | 36 | 31 | 6 | 5 | 1.91 ± 0.17** |

(1) Ewes reimmunized either once (× 1 i.e. 10 days before) or twice (× 2 i.e. 30 and 10 days before) before joining in 1983.
(2) Ewes either not reimmunized (× 0) or reimmunized twice (× 2) 34 and 6 days before laparoscopy in 1984.
* P < 0.05
** P < 0.01

In 1983 the same ewes in Expt. 1 were reimmunized either once or twice before joining. In the first 5 days of joining both treatments increased O.R. (Table 2). Subsequently there was a rapid decline in the O.R. of ewes reimmunized once only (Table 2) although overall a single reimmunization increased O.R. (Table 1). When litter size was assessed by ultrasound at 7-12 weeks after mating the ewes reimmunized twice contained more (P < 0.01) foetuses (Table 3) but those reimmunized once had the same mean litter size as the C ewes. When the two I-treated groups were combined they showed a higher incidence of barrenness than C ewes (P < 0.02).

**TABLE 2. Ovulation rates in control Merinos (C) and Merinos reimmunized against an inhibin-enriched fraction from bovine follicular fluid once (I × 1) or twice (I × 2) before joining**

| Days after rams added | Treatment | Ewes with ovulation rate of 1 | 2 | 3 | >3 | Ovulation rate (mean ± SEM) |
|---|---|---|---|---|---|---|
| 1 — 5 | C | 41 | 22 | 0 | 0 | 1.35 ± 0.06 |
| | I(× 1) | 7 | 9 | 3 | 7 | 2.85 ± 0.42** |
| | I(× 2) | 7 | 7 | 3 | 7 | 3.21 ± 0.54** |
| 6 —10 | C | 22 | 13 | 0 | 0 | 1.37 ± 0.08 |
| | I(× 1) | 11 | 4 | 0 | 0 | 1.27 ± 0.12 |
| | I(× 2) | 9 | 2 | 1 | 5 | 2.94 ± 0.64* |
| 11 — 32 | C | 17 | 17 | 0 | 0 | 1.50 ± 0.09 |
| | I(× 1) | 18 | 11 | 1 | 1 | 1.68 ± 0.26 |
| | I(× 2) | 12 | 11 | 0 | 5 | 2.86 ± 0.70 |

* P < *0.05*
** P < *0.01*

In 1984 some of the I ewes were not re-immunized. A residual response was present as seen in ewes with 3 or more ovulations (Table 1). However an O.R. greater than three was not seen in such ewes immunized only once in 1983, large numbers of corpora lutea only being present in three ewes which had been reinjected twice in 1983.

Immunization of Lambs

Ovarian examination of lambs at age 7 months revealed 1 of 9 C lambs to have one corpus albicans, whereas 10 of 15 I lambs had up to 8 recent ovulations (mean ± SEM, 2.9 ± 0.8).

## DISCUSSION

The results confirm our earlier report (O'Shea *et al.* 1982) that immunization with a partially purified preparation from bFF increases O.R. in Merino ewes. This increase in O.R. resulted in an increased litter size

**TABLE 3. Number of foetuses detected by ultrasound 7-12 weeks after mating of control (C) Merino ewes and Merino ewes immunized (I) against an inhibin-enriched fraction from bovine follicular fluid.**

| Treatment | n | Ewes with litter size of 0 | 1 | 2 | 3 | 4 | Means (± SEM) |
|---|---|---|---|---|---|---|---|
| C | 126 | 0[2] | 93 | 33 | 0 | 0 | 1.26 ± 0.04 |
| I(× 1)[1] | 63 | 5 | 40 | 15 | 3 | 0 | 1.25 ± 0.08 |
| I(× 2)[1] | 64 | 3 | 31 | 19 | 9 | 2 | 1.62 ± 0.11** |

[1] Ewes immunized three times in 1982 and re-immunized wether once (× 1) or twice (× 2) before mating in 1983.
[2] Significantly less than in I (× 1) + I (× 2) ($P < 0.02$).
** $P < 0.01$

in early pregnancy, although in one year drought conditions resulted in fewer lambs being weaned. As this flock was run under commercial conditions individual ewe litter size could not be recorded. Ewe lambs born from immunized ewes had normal reproductive characteristics. The wide range in O.R. response may be due to the impurity of the immunogen. The main impurity is albumin (Jansen *et al.* 1981) but several compounds are present in the fraction (M.J. Sinosich, B.M. Bindon and T. O'Shea, unpublished data). The immunization regime (see 1983 data) has a large effect and experiments are underway in this area to optimize the response.

It cannot be stated that the immunogen is inhibin. However, recent experiments have shown that when plasma from immunized ewes is injected into ovariectomized ewes it neutralizes the FSH depressing effects of exogenous follicular fluid, and FSH appears to be elevated in immunized ewes towards the end of the oestrous cycle (S.A.R. Al-Obaidi, B.M. Bindon, M.A. Hillard, T. O'Shea and L.R. Piper, unpublished results). These results are what would be expected if inhibin is involved in producing the effects of immunization. The ovary in immunized ewes when they respond appears to be effectively deregulated (one ewe had an O.R. of 18). Thus the factor(s) being removed must form part of normal regulation.

The results with the lambs support the concept that inhibin may be involved in the regulation of pubertal onset, although the present results apply to ewe lambs rather than in ram lambs as postulated by Blanc (1980).

Although this technique is not industrially practicable at the moment it provides insight into normal regulation of prolificacy and puberty. Efforts to purify inhibin and other compounds from follicular fluid and seminal plasma will make available the pure compounds needed for future research. The recent report by Ramasharma *et al.* (1984) is promising in this regard.

ACKNOWLEDGEMENTS

We are indebted to the Australian Meat Research Committee for financial support, to Dr.D.G. Fowler for determination of litter size by ultrasound, and to J. Sheedy for technical assistance.

REFERENCES

Blanc, M.R., 1980. *Reprod. Nutr. Develop.*, *20*, 573-586.
Blanc, M.R. and Terqui, M., 1976. *I.R.C.S. Medical Sci.*, 4, 17.
Cummins, L.J., O'Shea, T., Bindon, B.M., Lee, V.W.K. and Findlay, J.K., 1983. *J. Reprod. Fert.*, *67*, 1-7.
Franchimont, P., Croze, F., Demoulin, A., Bologne, R. and Hustin J., 1981. *Acta Endocrinol.*, *98*, 312-320
Franchimont, P., Verstraelen-Proyard, J., Hazee-Hagelstein, M.T., Renard, Ch., Demoulin, A., Bourguignon, J.P. and Hustin, J., 1979. *Vitamins and Hormones*, *37*, 243-302.
Hermans, W.P., van Leeuwen, E.C.M., Debets, M.H.M. and de Jong, F.H., 1980. *J. Endocrinol.*, 86, 79-92.
Hochereau-de Reviers, M.-T. and de Reviers, M., 1978. *C.R. Acad. Sc. Paris Série* D., *287*, 1015-1018.
Jansen, E.H.J.M., Steenbergan, J., de Jong, F.H. and van der Molen, H.J., 1981. *Molec. Cell. Endocrinol.*, *21*, 109-117.
de Jong, F.H., Jansen, E.H.J.M. and van der Molen, H.J., 1981. *In* Franchimont, P. and Channing, C.P. (eds). *Intragonadal Regulation of Reproduction*, Academic Press, London, 229-250.
O'Shea, T., Cummins, L.J., Bindon, B.M. and Findlay, J.K., 1982. *Proc. Aust. Soc. Reprod. Biol.*, *14*, 85.
Ramasharma, K., Sairam, M.R., Seidah, N.G., Chretien, M., Manjunath, P., Schiller, P.W., Yamashiro, D. and Li, C.H., 1984. *Science*, *223*, 1199-1202.

# SUPEROVULATION OF EWES WITH A COMBINATION OF PMSG AND FSH-P

J.P. Ryan, *Animal Breeding and Research Institute, Katanning, Western Australia, 6317.*
R.J. Bilton, *Department of Animal Husbandry, University of Sydney, Camden New South Wales, 2570.*
J.R. Hunton and W.M.C. Maxwell, *Animal Breeding and Research Institute, Katanning, Western Australia, 6317.*

*Summary* A factorial experiment (N = 162) was carried out to examine the use of pregnant mare serum gonadotrophin (PMSG : 0, 800 or 1600 i.u.) in combination with follicle stimulating hormone (FSH-P : 0, 12 or 18 mg) to superovulate ewes. Numbers of corpora lutea (CL) increased as dose of FSH-P increased for ewes receiving 0 or 800 i.u. PMSG (2.1 ± 0.24, 10.5 ± 1.37, 13.2 ± 1.50 CL for 0, 12 and 18 mg FSH-P respectively), but not for ewes treated with 1600 i.u. PMSG (10.3 ± 1.33, 12.7 ± 1.96, 12.6 ± 1.99 CL for 0, 12 and 18 mg FSH-P respectively). The proportion of ewes showing a superovulatory response increased for animals receiving a combination of PMSG and FSH-P over that of animals receiving FSH-P alone (68/69 vs 28/36, $P < 0.001$). In a second experiment (N = 113) no differences were obtained between 3 treatments (800 i.u. PMSG + 9mg FSH-P, 550 i.u. PMSG + 12mg FSH-P, 400 i.u. PMSG + 12mg FSH-P) for numbers of CL and large follicles, or eggs recovered, fertilised and morphologically normal. A moderate dose of PMSG administered in conjunction with FSH-P appears to be superior to either gonadotrophin used alone in obtaining large numbers of viable embryos.

## INTRODUCTION

Superovulation has been induced in sheep by the administration of pregnant mare serum gonadotrophin (PMSG : Moore and Rowson 1960), and pituitary extracts e.g. horse anterior pituitary extract (HAP : Moore and Shelton 1964) and follicle stimulating hormone (FSH-P : Wright *et al.* 1981), but responses between animals have been highly variable. Increasing the superovulatory dose of PMSG can increase the number of persistent large follicles resulting in a decrease in ovulation rate (Shelton and Moore 1967). This has been attributed to the prolonged half life of PMSG (McIntosh *et al.* 1975). The resultant abnormal endocrine status is thought to be detrimental to fertilisation, and transport and survival of embryos (du Mesnil du Buisson *et al.* 1977).

The numbers of persistent follicles remain relatively constant as the dose of HAP administered to ewes increases (Shelton and Moore 1967). This is the main benefit of pituitary extracts, compared to PMSG, in the production of large numbers of fertilised sheep eggs and is thought to be a consequence of the short half life of exogenous pituitary gonadotrophins in the ewe (Akbar *et al.* 1974). However, a significant proportion of ewes treated with FSH-P alone fail to give any superovulatory response irrespective of total dose or dose regime (Bilton unpublished ; Eppleston *et al.* 1984). The present study attempted to overcome these problems by treating ewes with a combination of FSH-P and PMSG.

## MATERIALS AND METHODS

Experiment 1 was a factorial trial carried out during March and used 157 mature Merino ewes. There were 9 treatments consisting of 3 doses of PMSG (Folligon, Batch 198; Intervet Australia Pty Ltd) with each of 3 doses of FSH-P (Burns-Biotec., Omaha, U.S.A.). The ewes were treated with intravaginal sponges (Repromap; Upjohn Australia Pty Ltd) for 12 days. PMSG (0, 800 or 1600 i.u.) was administered as a single intramuscular injection 48 hr before sponge removal. FSH-P (0, 12 or 18 mg total dose) treatment commenced 48 hr before sponge removal and was administered on 3 successive days as 2 intramuscular injections per day (at 0800 and 1700 hr). A decreasing dose regime of either 3,3 ; 2,2 ; 1 and 1 mg (12mg) or 4,4 ; 3,3 ; 2 and 2 mg (18mg) was used.

In experiment 2, superovulation was induced in 113 Cormo (Corriedale × Merino, based breed) ewes during June. Three gonadotrophin treatments were used : 800 i.u. PMSG + 9mg FSH-P, 550 i.u. PMSG + 12mg FSH-P or 400 i.u. PMSG + 12mg FSH-P. The treatment regime was similar to that used in experiment 1 except that the FSH-P was administered as 3 once daily subcutaneous injections, the last injection given at the time of sponge removal. FSH-P (9 or 12 mg) was given on a decreasing dose regime of 4,3 and 2 mg or 5,4 and 3 mg.

Ewes were inseminated into the uterus by laparoscopy (fresh semen extended 1:1 with Dulbeccos Phosphate Buffered Saline) 24 hr after sponge removal. On day 6 (day 0 = day of oestrus and insemination) ewes were subjected to laparotomy, when numbers of corpora lutea (CL) and persistent large follicles (> 5mm in diameter, LF) were recorded. Ewes exhibiting > 3 CL or LF were classified as having a superovulatory response. A proportion of ewes within each superovulatory treatment group in experiment 1 and all ewes in experiment 2 were utilised for egg collection using previously published techniques (Moore 1980). Fertilisation and morphology of recovered eggs were examined and late morula or early blastocyst stage embryos were classified as normal (Moore 1980).

Data on numbers of CL, LF and total ovarian response (CL + LF) were subjected to analysis of variance, whereas data on the proportion of eggs recovered, fertilised and classified normal were analysed by Chi square contingency tables.

## RESULTS

### Experiment 1

The numbers of CL, LF and total ovarian response were affected by dose of PMSG ($P < 0.001$) and dose of FSH-P ($P < 0.001$, Table 1). These treatments were also involved in interactions (dose of PMSG × dose of FSH-P on number of CL and total ovarian response, $P < 0.001$). Numbers of CL and total ovarian response increased as dose of FSH-P increased for ewes receiving 0 or 800 i.u. PMSG, but not for ewes receiving 1600 i.u. PMSG. The highest ovulation rate (number of CL) was obtained in ewes treated with a combination of 800 i.u. PMSG and 18mg FSH-P.

**TABLE 1. Ovarian response of ewes treated with PMSG and FSH-P (mean ± SEM).**

| Dose of PMSG (i.u.) | Dose of FSH-P (mg) | No. of ewes | No. of CL | No. of LF | Total ovarian response (CL + LF) |
|---|---|---|---|---|---|
| 0 | 0 | 17 | 1.1 ± 0.08 | 0.9 ± 0.22 | 2.0 ± 0.25 |
| | 12 | 18 | 8.3 ± 1.75 | 2.9 ± 0.99 | 11.3 ± 1.44 |
| | 18 | 18 | 10.1 ± 2.07 | 3.1 ± 0.74 | 13.2 ± 1.81 |
| 800 | 0 | 18 | 3.0 ± 0.34 | 1.0 ± 0.36 | 4.0 ± 0.50 |
| | 12 | 18 | 12.7 ± 2.08 | 3.6 ± 0.88 | 16.3 ± 1.73 |
| | 18 | 16 | 16.8 ± 1.96 | 2.7 ± 0.73 | 19.4 ± 1.97 |
| 1600 | 0 | 17 | 10.3 ± 1.33 | 4.7 ± 1.16 | 15.0 ± 1.46 |
| | 12 | 17 | 12.7 ± 1.96 | 6.9 ± 1.63 | 19.6 ± 1.30 |
| | 18 | 18 | 12.6 ± 1.99 | 7.2 ± 1.85 | 19.7 ± 1.53 |

The data on ewes classified as having a superovulatory response are shown in Table 2. A larger proportion of ewes had a superovulatory response when PMSG was used in conjunction with FSH-P than when FSH-P was administered alone (68/69 vs 28/36, $P < 0.001$). For ewes that exhibited a superovulatory response the addition of 800 i.u. PMSG to either the 12 or 18 mg FSH-P dose regime tended to increase ovulation rate (9.8 ± 1.81, 13.4 ± 2.18 for 0 PMSG + 12 or 18 mg FSH-P respectively vs 12.7 ± 2.08, 16.8 ± 1.96 for 800 i.u. PMSG + 12 or 18 mg FSH-P respectively) but due to the high variability in response, this difference was not statistically significant.

Egg fertilisation and fertilised egg morphology were unaffected by doses of PMSG or FSH-P. Egg recovery rates were lower for ewes receiving 1600 i.u. PMSG + 12 or 18 mg FSH-P (53/176) than for ewes receiving other treatments (349/542, $P < 0.001$, Table 2).

**TABLE 2. Number of ewes exhibiting a superovulatory response (SR) and egg recovery, fertilisation and morphology after treatment with PMSG and FSH-P.**

| Dose PMSG (i.u.) | Dose FSH-P (mg) | No. ewes | Egg Recovery | | No. Eggs (%) | | |
|---|---|---|---|---|---|---|---|
| | | SR/ Treated | No. ewes | No. CL | Recovered | Fertilised | Normal |
| 0 | 0 | 0/17 | 0 | – | – – | – – | – – |
| | 12 | 15/18 | 10 | 120 | 78 (65.0) | 69 (88.5) | 67 (97.1) |
| | 18 | 13/18 | 7 | 63 | 39 (61.9) | 24 (64.1) | 24 (100.0) |
| 800 | 0 | 5/18 | 1 | 7 | 7 (100.0) | 7 (100.0) | 7 (100.0) |
| | 12 | 18/18 | 7 | 106 | 72 (67.9) | 62 (86.1) | 56 (90.3) |
| | 18 | 16/16 | 7 | 147 | 93 (63.3) | 79 (84.9) | 77 (97.5) |
| 1600 | 0 | 17/17 | 7 | 99 | 60 (60.6) | 52 (86.7) | 50 (96.2) |
| | 12 | 16/17 | 7 | 121 | 41 (33.9) | 21 (51.2) | 15 (71.4) |
| | 18 | 18/18 | 3 | 55 | 12 (21.8) | 10 (83.3) | 10 (100.0) |

Experiment 2

No differences between treatments were observed for mean numbers of CL or LF, fertilisation or recovery rates, and proportions of fertilised eggs classified as normal. Therefore the data were pooled over treatments, and the responses for ewes with or without a superovulatory response are presented in Table 3. A higher proportion of eggs were recovered ($P < 0.05$) and normal fertilised eggs obtained ($P < 0.01$) from ewes classified as exhibiting a superovulatory response. There was no difference between the two classes of ewes for the percentage of eggs fertilised.

**TABLE 3. Ovarian response, egg recovery, fertilisation and morphology in ewes with or without a superovulatory response.**

| | | Classification of ewes | |
|---|---|---|---|
| | | Superovulatory response[1] | No Superovulation |
| No. Ewes | | 96 | 17 |
| No. CL | | 956 | 22 |
| No. LF | | 85 | 25 |
| No. Eggs | — Recovered (%) | 699 (73.1) | 11 (50.0) |
| | — Fertilised (%) | 586 (83.8) | 9 (81.1) |
| | — Normal (%) | 552 (94.2) | 6 (66.7) |

[1] Ewes exhibiting > 3 CL or LF.

## DISCUSSION

Treatment of ewes with 800 i.u. PMSG and FSH-P (12 or 18 mg) appeared to have a twofold effect on increasing ovulation rate and numbers of viable embryos above that of treatment with FSH-P alone. First, the proportion of ewes exhibiting a superovulatory response was increased (Table 2). Second, by a direct additive effect in those ewes exhibiting a superovulatory response, the average number of corpora lutea tended to increase without increasing the number of persistent follicles (Table 1). In agreement with results of previous studies (e.g. Boland and Gordon 1977), more persistent follicles were observed as dose of PMSG increased. The high oestrogen output of these follicles (Quirke and Hanrahan 1975 ; du Mesnil du Buisson *et al.* 1977) is thought to increase the rate of transport of ova through the oviducts and decrease recovery rates (Quirke and Hanrahan 1975). This may explain the lower recovery rates obtained in ewes receiving 1600 i.u. PMSG in addition to 12 or 18 mg FSH-P.

Du Mesnil du Buisson *et al.* (1977) cited a number of conflicting reports on the effects that dose of PMSG, numbers of corpora lutea or unovulated follicles have on embryonic quality. In our study neither egg fertilisation rates nor the proportion of morphologically normal embryos were affected by dose of PMSG and/ or FSH-P.

Moderate doses of PMSG (400, 550 or 800 i.u.) in conjunction with FSH-P (9 or 12 mg administered once daily) resulted in levels of superovulation and numbers of fertilised eggs recovered similar to those reported by Moore (1980) for ewes treated with 1200-1300 i.u. PMSG or 60 mg total dose of HAP. The differences in recovery rates and proportion of normal embryos suggest that ewes without a superovulatory response may have an inadequate environment for embryonic survival.

The use of moderate doses of PMSG in conjunction with FSH-P appears to be superior to either gonadotrophin used alone in obtaining large numbers of viable embryos, thereby increasing the efficiency of programmes of transfer of embryos. Procedures used to superovulate ewes in this study, however, did not reduce the variability between animals in their response to exogenous gonadotrophins. Further research focusing on the mechanisms that control follicle population dynamics is necessary to overcome this problem.

## ACKNOWLEDGEMENTS

The authors are indebted to Mrs J. Ryan for technical assistance. Part of this work was funded by a grant (DAW 25S) from the Australian Meat Research Committee.

## REFERENCES

Akbar, A.M., Nett, R.M. and Niswender, G.D., 1974. *Endocrinology,94*, 1318-1324.

Boland, M.P. and Gordon, I., 1977. *J. Ir. Dep. Agric., 74*, 81-83.

Du Mesnil du Buisson, F., Renard, J.P., and Levasseur, M.C., 1977. *In* Betteridge, K.J. (ed) *Embryo Transfer in Farm Animals*, Agriculture Canada Monograph No. 16, 24-26.

Eppleston, J., Bilton, R.J. and Moore, N.W., 1984. *Proc. Aust. Soc. Reprod. Biol., 16*, 68.

McIntosh, J.E.A., Moor, R.M. and Allen, W.R., 1975. *J. Reprod. Fert., 44*, 95-100.

Moore, N.W., 1980. *In* Morrow, D.A. (ed) *Current Therapy in Theriogenology*, W.B. Saunders Co., Philadelphia, 89-94.

Moore, N.W. and Rowson, L.E.A., 1960. *J. Reprod. Fert., 1,* 332-349.
Moore, N.W. and Shelton, J.N., 1964. *J. Reprod. Fert., 7,* 79-87.
Quirke, J.F. and Hanrahan, J.P., 1975. *J. Reprod. Fert., 43,* 167-170.
Shelton, J.N. and Moore, N.W., 1967. *J. Reprod. Fert., 14,* 175-177.
Wright, Jr, R.W., Bondioli, K., Grammer, J., Kuzan, F. and Menino, Jr, A., 1981. *J. Anim. Sci.*, *52*, 115-118.

# HUMAN CHORIONIC GONADOTROPHIN INDUCES MULTIPLE OVULATION IN SHEEP

H.M. Radford, J.A. Avenell and A. Szell, *CSIRO, Division of Animal Production, Box 239, Blacktown, New South Wales 2148, Australia.*

*Summary* Human chorionic gonadotrophin (HCG) was used as a biological analogue for ovine luteinizing hormone in studies on ovulation rate in sheep. Preliminary investigation showed that HCG could induce multiple ovulation. Subsequently, using 173 ewes mated to entire rams, the injection of 250 i.u. HCG 48 h before, 24 h before, at the time of, or 24 h after cloprostenol-induced luteolysis, led to significant increases in ovulation rate (1.68-2.04 v 1.12 in controls). Conception rate declined significantly from the earliest to the latest HCG injection but overall the HCG treated ewes delivered significantly more twin lambs than control ewes (45 % v 15%, $P < 0.05$).

## INTRODUCTION

The endocrine basis of differences in ovulation rate in sheep is not well understood (Scaramuzzi and Radford 1983). However, ewes immunized against androstenedione have a higher ovulation rate, increased frequency of release of luteinizing hormone (LH), and higher basal levels of LH than control ewes (Martensz and Scaramuzzi 1979). Likewise, ovulation rate appears to be increased by either multiple injections or infusions of gonadotrophin-releasing hormone into anoestrous ewes for two days after progesterone implant removal (McLeod *et al.* 1982, 1983). It seemed possible therefore that ovulation rate might be increased by raising the basal level of circulating LH during the days following luteal regression. Insufficient ovine LH was available so human chorionic gonadotrophin (HCG) was used as a biological analogue, particularly since Braden *et al.* (1960) had reported that HCG injected several days after the end of a course of progesterone injections did lead to an increase in ovulation rate. Preliminary experiments showed that, when infused after luteolysis, HCG did not induce multiple ovulation. However, when injected after luteolysis, HCG did do so, up to 7 ova being shed per ewe, but there was some inhibition of oestrus. There was evidence also that ovulations occurred over a period of several days. The present experiments were therefore undertaken with the aims of determining a dose of HCG leading to a practically acceptable ovulation rate, the timing of ovulation subsequent to its injection and whether appropriately timed injection could lead to increased fecundity in ewes joined with entire rams.

## MATERIALS AND METHODS

Merino ewes of 30-45 Kg body weight were used. Luteolysis was induced using 125 $\mu$g cloprostenol (Estrumate, I.C.I. Australia) injected i.m., and the HCG was eigher Sigma Chorionic Gonadotropin CG-B or Organon Pregnyl, dissolved and diluted in 0.15M sodium chloride. Ovulation rate was determined at laparoscopy performed under local anaesthesia. The rams wore Sire-Sine harnesses and crayons (Hortico, Australia) and oestrus was indicated from the presence of crayon marks on the rumps of the ewes.

### Experiment 1

Twenty four ewes, running continuously with vasectomized rams, and having experienced oestrus 6 to 12 days earlier, were used in 4 groups of 6 animals. 24 h after the i.m. injection of 125 $\mu$g cloprostenol they were injected i.m. with either 0.15 M sodium chloride or HCG (Organon), 125, 250, or 500 i.u. Laparoscopy was performed 70-71 h and 166-167 h after the injection of cloprostenol.

### Experiment 2

173 ewes were used two months after immunization with various androstenedione immunogens. They were allotted to 5 groups each of 25 animals and one of 48 animals, with restrictions according to prior treatments and the immunization-induced ovulation response. Towards the end of the second luteal phase following intravaginal sponge removal the groups of 25 ewes were injected i.m. with 125 $\mu$g cloprostenol. Four of these groups were also injected i.m. with 250 i.u. HCG (Sigma) — 48 h beforehand, 24 h beforehand, at the time of injection of cloprostenol, or 24 h later. The 48 ewes of the sixth group acted as controls and received neither cloprostenol nor HCG. All 173 ewes were run as the one mob to which 12 entire rams were joined at the time of cloprostenol injection. Oestrus was recorded every 1 to 3 days for 4 weeks. The ovaries of the treated ewes were examined 14 days after the cloprostenol injection, and the ewes were observed continuously at lambing.

## RESULTS

### Experiment 1

By 70 h after cloprostenol none of the control ewes had ovulated whereas two, four, and 6 of those injected with 125, 250, or 500 i.u. HCG respectively had done so. Most of these ovulations were single. Ninety six h later a single corpus luteum was present in each of the control ewes, and many more corpora lutea were present in the HCG treated ewes. Of those given 125 i.u. HCG a fresh corpus luteum was present in each of three ewes, including one that had ovulated earlier. Two additional corpora lutea were present in four of the ewes given 250 i.u. HCG; three of these ewes had ovulated earlier. One additional corpus luteum was present in another. One additional corpus luteum was present in three of the ewes given 500 i.u. HCG, while three additional corpora lutea were present in another. All 6 control ewes experienced oestrus but only 12 of the 18 treated ewes did so.

Experiment 2

The number of corpora lutea present 14 days after cloprostenol injection was significantly higher in each of the groups injected with HCG than in the untreated group of 25 ($P < 0.032$, Fisher's exact test; Table 1.) Among the groups of ewes treated with HCG the proportion of ewes mating within 5 days of cloprostenol and the conception rate to that mating progressively declined (88% to 48%, $P < 0.05$, and 68% to 50%, $P < 0.05$ respectively) from the earliest to the latest HCG injection. There was a similar decline in the numbers of multiple lambs born. Even including those ewes injected with HCG 24 h after cloprostenol, in which reproductive performance was low, significantly more multiple lambs were born in the HCG treated groups as a whole than to control and cloprostenol only groups combined ($P < 0.041$, Fisher's exact test). One set of triplets was born, all other multiples were twins. Of the ewes that failed to conceive at the first oestrus and that mated subsequently, each delivered a single lamb.

**TABLE 1. Reproduction performance of ewes treated with 250 i.u. HCG i.m. at various times in relation to cloprostenol injection.**

| | | Untreated ewes | Treated ewes | | | | |
|---|---|---|---|---|---|---|---|
| | | | Time of injection of HCG relative to that of cloprostenol at time 0 | | | | |
| | | | Nil | −48 h | −24 h | 0 | +24 h |
| No. of ewes | | 48 | 25 | 25 | 25 | 25 | 25 |
| No. with indicated no. of corpora lutea[1] | 0 | | 1 | 2 | 2 | 1 | |
| | 1 | | 20 | 3 | 7 | 11 | 13 |
| | 2 | NA | $4^{2}$ | $14^{4}$ | $8^{3}$ | $8^{2}$ | 8 |
| | 3 | | | $5^{3}$ | $6^{3}$ | $5^{2}$ | 2 |
| | 4 | | | | 2 | | 1 |
| | 5 | | | 1 | | | $1^{1}$ |
| Mean ovulation rate | | NA | 1.12 | 2.04 | 1.96 | 1.68 | 1.76 |
| No. (%) mated within 5 days of cloprostenol | | 29 (60) | 18 (72) | 22 (88) | 17 (68) | 15 (60) | 12 (48) |
| No. (%) lambed to this mating | | 21 (72) | 12 (67) | 15 (68) | 11 (65) | 8 (53) | 6 (50) |
| No. producing multiple lambs | | 2 | 3 | 7 | 6 | 4 | 1 |
| — % ewes lambed[2] | | 10 | 25 | 47 | 55 | 50 | 17 |
| — % ewes mated[2] | | 7 | 17 | 32 | 35 | 27 | 8 |
| — % ewes joined[2] | | 4 | 12 | 28 | 24 | 16 | 4 |

(1) Superscript numbers indicate number of multiple lambs from this class

(2) Significantly more ($P < 0.041$, Fisher's exact test) lambs born to HCG treated ewes than to untreated and cloprostenol-only treated ewes [pooled], whether taken as percentage lambed, mated, or joined.

## DISCUSSION

These experiments have confirmed and extended the earlier observation of Braden *et al.* (1960) that HCG can induce multiple ovulation in ewes. The first experiment indicated that HCG often induces ovulation within two days of its injection and that within the next several days many ewes ovulate again, often shedding multiple ova, but not invariably exhibiting oestrus. 250 i.u. HCG appeared to give the most uniform increase in ovulation rate. The second experiment, using 250 i.u. HCG, then showed that better mating performance could be obtained when the HCG was injected before luteolysis, and that there was a decline both in the proportion of ewes mated and in conception rate as the time of injection moved closer to the expected time of oestrus. Significantly more corpora lutea were present in each of the HCG treated groups. Undoubtedly some of these arose from ova being shed up to several days apart and fertilization of all ova would not have been anticipated. That such failure of fertilization occurred is suggested by the much reduced proportion of multiple lambs, relative to number of corpora lutea, in the HCG treated ewes, although early embryonic loss cannot be discounted as an additional cause. None the less significantly more lambs were born as a result of the HCG treatments.

The HCG preparations used were not highly purified and may have contained some follicle stimulating hormone (FSH) like activity. However, since neither FSH nor LH have been causally related to ovulation rate in sheep (Scaramuzzi and Radford 1983) the mode of action of HCG in inducing multiple ovulation remains unclear. Study of this question and further field trials are in progress. Should the latter confirm the present findings it is possible that HCG treatment may in time find a place at least in intensive sheep husbandry. The ovulation rates achieved have been modest, indeed probably as required in practice, and HCG itself is relatively cheap. Since the observed increases in ovulation rate have followed injection on any of four days of the oestrous cycle it is conceivable also that treatment without regard to stage of cycle, thereby eliminating the cost of synchronization, may lead to acceptable increases in fecundity.

ACKNOWLEDGEMENTS

We thank Dr. R.M. Hoskinson for performing some of the laparoscopies, Mr. J.B. Donnelly for statistical advice, and Messrs S. Collins, W.R. Dwyer, and T.R. Holmes for their management of the experimental animals.

REFERENCES

Braden, A.W.H., Lamond, D.R. and Radford, H.M., 1960. *Aust. J. Agric. Res., 11*, 389-401.
Martensz, N.D. and Scaramuzzi, R.J., 1979. *J. Endocr., 81*, 249-259.
McLeod, B.J., Haresign, W. and Lamming, G.E., 1982. *J. Reprod. Fert., 65*, 223-230.
McLeod, B.J., Haresign, W. and Lamming, G.E., 1983. *J. Reprod. Fert., 68*, 489-495.
Scaramuzzi, R.J. and Radford, H.M., 1983. *J. Reprod. Fert., 69*, 353-367.

# INFECTIOUS DISEASES OF REPRODUCTION IN SHEEP

R.S. Rahaley, *Department of Agriculture, Regional Veterinary Laboratory, Benalla, Victoria, 3672.*

## INTRODUCTION

Infectious diseases of reproduction are significant causes of poor productivity in the Australian sheep industry. Although most occur sporadically, their economic impact on individual farms and combined effect on the industry is considerable. In addition to their effects on sheep production, many are important zoonoses.

In this review, the diseases of reproduction are broadly classified into those diseases principally affecting the genital tract of rams and those mainly affecting ewes, causing abortion and perinatal lamb mortality. Attempts to draw conclusions from the literature on the overall prevalence of specific diseases in Australia are complicated by the diversity of climatic conditions and farm practices which tend to confine the relevance of disease surveys to the survey area. With these reservations, reported disease prevalences are given here where available.

## DISEASES CAUSING INFERTILITY IN RAMS

### Ovine brucellosis

Ovine brucellosis has been recognised as a major cause of infertility in rams for many years. The causal bacterium, *Brucella ovis*, is endemic in most major sheep raising areas of the world (Rahaley and Dennis 1982) and is present in all states of Australia although its prevalence varies considerably in different geographical areas. For example, serological evidence of infection was demonstrated in 18.3% of 2,332 rams in Queensland (Murray 1969) whereas Watt (1970) found no evidence of the disease in 2,281 Merino rams in Western Australia.

Lesions attributable to *B. ovis* infection of rams occur only in the genital tract, primarily in the tail of the epididymis and commonly in the ampullae and seminal glands (Biberstein *et al.* 1964). In the epididymis the bacteria provoke chronic inflammation and both necrosis and proliferation of the epididymal duct epithelium. These changes may result in spermiostasis and, ultimately, extravasation of spermatozoa into the interstitial tissue of the epididymis where they incite a severe inflammatory response (Kennedy *et al.* 1956). Infected rams have varying degrees of reduced semen quality leading to infertility (Cameron and Lauerman 1976).

The reasons for the specific tropism of *Brucella ovis* for the genital tract or its specific pathogenicity in that location are unknown. The epididymal changes probably interfere with spermatozoa maturation and transport. Results of studies by Rahaley and Dennis (in press) indicated that the immune response of the ram may contribute to the severity of epididymal lesions but this area requires further study.

*Brucella ovis* is intermittently shed in the semen of infected rams and transmission can follow direct contact between infected and noninfected or concurrent mating of infected and noninfected rams to the same ewes (Lawrence 1961). Venereal transmission is probably the most common method of spread in the field. Sodomy between young rams has been incriminated in ram to ram transmission although compelling evidence is lacking. *B. ovis* is also excreted in the urine of infected rams (Biberstein *et al.* 1964) and the possibility of pasture contamination warrants further investigation. Field observations suggest that Merino rams are more resistant to ovine brucellosis than British breed rams (Clapp *et al.* 1962) although there are few statistically significant data to support these observations.

Abortion and neonatal lambs deaths may result from infection of pregnant ewes with *B. ovis* although the organism is far less virulent for ewes than for rams. Haughey *et al.* (1968) estimated that in ewes mated to infected rams, the lamb mortality rate due to *B. ovis* infection was 0.9%. The bacteria do not persist in nonpregnant ewes but may persist for several months in pregnant ewes (Muhammed *et al.* 1975). Abortion follows localisation of *B. ovis* in the placenta and subsequent placental necrosis (Molello *et al.* 1963).

Clinical diagnosis of ovine brucellosis in rams is based on palpable evidence of epididymitis, detection of specific serum antibodies or, to a lesser extent, culture of *B. ovis* from semen. Serological testing is the most reliable under field conditions (Hughes and Claxton 1968; Ris 1974). In ewes, placental lesions are not specific but *B. ovis* can be isolated from the placenta, foetal lung and abomasum (Ris 1970).

The complement fixation test is the most widely accepted serological test for rams although both false positive and false negative reactions occur. Enzyme-linked immunosorbent assays (ELISA) which are more sensitive and more versatile are currently being evaluated (Rahaley *et al.* 1983; Spencer and Burgess 1984). A major problem in interpreting serological titres is distinguishing between infected rams and rams exposed to the organism but not infected. Development of ELISA systems to assay antibody subclasses may be helpful.

Because of the transient nature of *B. ovis* infection in ewes, control of ovine brucellosis is usually based on identification and removal of infected rams. Rams remain infected indefinitely and antibiotic therapy has not been successful. All states of Australia have voluntary flock accrediation schemes in which sheep breeders elect to eradicate ovine brucellosis from their farm if it is present then have their rams serologically tested annually to confirm that the flock is ovine brucellosis free.

Vaccines for ovine brucellosis are not available in Australia. Vaccinated rams cannot be differentiated from naturally infected rams by current serological tests (Ris 1967). Therefore widespread vaccination would seriously interfere with accrediation schemes. Investigations into the use of cell fragment vaccines to produce an immune response that does not interfere with serological diagnosis are currently in progress. Where used, the efficacy of whole cell *B. ovis* bacterins is disappointing. Live vaccines containing attenuated strains of either *B. mellitensis* or *B. abortus* have been used successfully overseas (Rahaley and Dennis 1982) but have no application in Australia which is free of *B. mellitensis* and in the process of eradicating *B. abortus*.

Actinobacillosis

Epididymitis due to *Actinobacillus seminis* infection of rams was first recognised in Queensland (Baynes and Simmons 1960). The disease has subsequently been reported in Victoria and Western Australia as well as New Zealand, South Africa and the United States of America (Burgess 1982; Sponenberg *et al.* 1983). The prevalence of *A. seminis* infection in Australian rams has not been studied.

Clinically the disease is similar to ovine brucellosis. The epididymis is primarily affected but the resultant inflammatory response may be more suppurative than that seen with *B. ovis* infection (Sponenberg *et al.* 1983). Baynes and Simmons (1968) concluded however, that *A. seminis* infection could not be differentiated from ovine brucellosis by clinical or pathological examination.

It has been assumed that transmission of *A. seminis* occurs either venereally or during homosexual activity between in-contact rams (Burgess 1982). However, Simmons *et al.* (1966) noted that in an infected flock, the prevalence of infection in unmated rams was high, approaching 100% with the onset of sexual maturity. In addition, the disease could not be controlled by removal of infected rams or the isolation of unmated from mated rams. Van Tonder (1973) suggested that ewe to lamb transmission of *A. seminis* warranted further investigation. In one study, *A. seminis* was isolated from the cervix of 38% of ewes in an infected flock (Burgess 1982).

Experimental reproduction of *A. seminis* epididymitis has only been achieved by direct inoculation of the organisms into the epididymides or testes of rams (Baynes and Simmons 1960). Jansen (1980) found that increased serum levels of luteinising hormone or follicle stimulating hormone in rams enhanced the migration of preputial organisms, including *A. seminis*, to deeper parts of the genital tract as far as the epididymides, presumably via the urethra and vas deferens. The importance of this phenomenon in natural transmission is not known.

*A. seminis* infection is readily diagnosed by isolation of the causal agent from infected semen or tissues (Burgess 1982). Complement fixation tests for serum antibodies to *A. seminis* have been developed (Burgess 1982). However, Van Tonder (1973) identified six separate serological strains of *A. seminis* in South Africa and suggested that rams would have to be tested against all strains. The dynamics of serological responses to *A. seminis* infection have not been studied.

Control measures for *A. seminis* infection in rams have not been reported. Further studies are needed to define accurately the role of this bacterium in infertility of rams in Australia.

Miscellaneous causes of infectious infertility in rams

A variety of bacteria, including *Actinomyces* (*Corynebacterium*) *pyogenes, Corynebacterium pseudotuberculosis, Actinobacillus ligneriesi, Actinobacillus actinomycetemcomitans, Yersinia pseudotuberculosis, Histophilus ovis* and *Brucella abortus* (strain 19) have been reported as causes of sporadic cases of epididymitis and orchitis in rams (Burgess 1982). Their major significance is as differential diagnoses of ovine brucellosis.

Balanoposthitis

Enzootic inflammation of the penis and prepuce ('pizzle rot' or 'sheath rot'), caused by *Corynebacterium renale*, principally occurs in wethers but rams can also become affected and may be incapable of mating. In Australia, Merino sheep are more commonly affected than other breeds particularly older wethers and young rams (Blood *et al.* 1979).

Lesions commence as small, scab-covered areas of superficial necrosis on the unwooled skin external to the preputial orifice and may extend into the interior of the prepuce leading to preputial swelling, adhesions and pus formation which interfere with urination and penis protrusion (Beveridge and Johnstone 1953). Small ulcers may also occur on the vulval lips of ewes in the same flock, possibly representing venereal transmission of the disease (Blood *et al.* 1979). Hydrolysis or urinary urea by *C. renale* resulting in formation of cytotoxic ammonia has been proposed as the underlying pathogenetic mechanism for the disease (McMillan and Southcott 1973).

Factors predisposing rams to enzootic balanoposthitis include high protein diet, elevated urine pH and ingestion of phytoestrogens which cause congestion and swelling of the prepuce. The predisposition of wethers and young rams is possible related to incomplete development of the penis and prepuce in these animals (Blood *et al.* 1979).

Recommended treatment includes removal of affected sheep to a dry pasture and restriction of feed intake to subsistence level. Surgery may be necessary to drain the prepuce in severe cases. Subcutaneous testosterone implants are highly effective in preventing the disease in both wethers and rams (McMillan *et al.* 1974).

A severe ulcerative balanitis, affecting only Border Leicester rams, has been reported in the New England area of New South Wales (Webb and Chick 1976). Vulvovaginitis was present in ewes mated to the affected rams. Penile lesions consisted of ulceration of the collosum glandis with varying degrees of granulation tissue, haemorrhage and purulent material covering the lesions. No infectious agents were isolated from the rams and the disease could not be transmitted to Merino rams. A similar syndrome has recently been seen in Border Leicester rams near Bairnsdale, Victoria (S. McOrist, pers. comm).

## DISEASES PRIMARILY CAUSING ABORTION

### Campylobacteriosis

The bacterium *Campylobacter fetus* subspecies *fetus* (formerly called *Vibrio foetus* ss. *intestinalis* and *C. fetus* ss. *intestinalis*) appears to be the most common cause of infectious abortion in Australian sheep. Campylobacter abortion is a sporadic disease and has been reported in all states except Queensland (Dennis 1975a). The reported prevalence in Australian flocks varies from 7% to 14% (Plant *et al.* 1972; Broadbent 1975; Dennis 1975a).

The usual history is an outbreak of abortion two to six weeks prior to the normal lambing time although some lambs die also during the parturient and post-parturient periods. Generally the abortion rate in an infected flock is less than 10% but may be as high as 50%. Some ewes may die from septic endometritis (Dennis 1972).

Sheep are infected following ingestion of the bacteria and pregnant ewes are susceptible to placental infection in the third to fifth month of gestation (Jensen *et al.* 1961). *C. fetus* has an affinity for the placenta and foetus and incites a severe endometritis and necrotising placentitis which may, in part, be immune-mediated (Osburn and Hoskins 1970). Typical 'rosette-like' necrotic foci are present in the livers of about 50% of infected lambs (Broadbent 1975; Dennis 1975a).

*Campylobacter fetus* can be isolated from the gall-bladder and intestine of both normal and affected sheep and is excreted in the faeces (Bryner *et al.* 1971). Carrion feeding birds may act as vectors (Garcia *et al.* 1983). How or why the organisms localise in the pregnant uterus is not known although some strains of *C. fetus* possess an antiphagocytic glycoprotein coat which may aid their survival during a bacteraemic phase (McCoy *et al.* 1975).

Suspected *Campylobacter* abortion can be confirmed by isolation of the bacteria from the foetus or placenta (Dennis 1975a). Outbreaks of abortion may be controlled by relocating ewes onto clean pasture and treatment with antibiotics (Quinlivan and Jopp 1982), but in some circumstances these measures may not be practical. There are at least five serotypes of *C. fetus* (Garcia *et al.* 1983) and more than one serotype can be found on a single farm (Bird *et al.* 1984). In addition, evidence of both *in vivo* and *in vitro* antigenic variation of *C. fetus* has been observed (Ogg and Chang 1972; Schurig *et al.* 1973; Corbeil *et al.* 1975). Polyvalent, adjuvant vaccines confer some protection to ewes (Quinlivan and Jopp 1982) but single strain vaccines do not protect against infection with heterologous serotypes (Bryner *et al.* 1979). The sporadic occurrence of *Campylobacter* abortion casts doubts on the economic justification of widespread vaccination.

Aborting and exposed ewes develop natural immunity to *C. fetus* and recurrence of abortions in a flock in subsequent pregnancies is unusual (Meinershagen *et al.* 1969) although Quinlivan and Jopp (1982) observed that in some flocks immunity may not be as complete as previously thought.

### Listeriosis

Abortion in sheep due to *Listeria monocytogenes* infection has been diagnosed in all states of Australia (Dennis 1972). *L. monocytogenes* was isolated from 2% of farms surveyed in Western Australia (Dennis 1975b); 7.6% of flocks surveyed in New South Wales (Plant *et al.* 1972) and from 25.4% of farms in Victoria (Broadbent 1975). Although some survey results suggest that the prevalence of listerial abortion in eastern Australia may be at least equal to that of *Campylobacter* abortion, Hughes (1975) pointed out that these results may be misleading due to the relatively small number of dead lambs examined and the possibility that survey results may be biased by a bimodal lambing pattern when a congenitally infective agent is present in a flock.

The reported abortion rate in infected flocks varies between 2% and 22%. Abortion usually occurs in late pregnancy and some lambs die post-partum (Dennis 1975b). Ewes are infected by ingestion of the organism and possibly by inhalation or veneral transmission (Blood *et al.* 1979). Foetal death is attributed to bacterial invasion of the placental chorionic epithelium with subsequent necrosis and interruption of foetal nutrient supply (Carter 1978). The bacteria may also invade the foetus and foci of necrosis in the viscera of infected lambs and foetuses are considered to be typical of listerial abortion but are not consistently present (Dennis 1975b). Hughes (1975) suggested that abortion may also be triggered by the effect of *L. monocytogenes* exotoxin on the uterine smooth muscle.

Normal sheep can harbour *L. monocytogenes* in their intestinal tract and excrete the bacteria in faeces and milk, particularly when stressed (Gronstol 1979). The bacteria can survive and multiply in sheep faeces and soil for several months. There is often an association between listeriosis and feeding silage, particularly spoiled silage which is a suitable medium for growth of the bacterium (Hughes 1975; Blood *et al.* 1979). Other species of animals, including wildlife, have also been incriminated in the spread of the disease (Hughes 1975).

*L. monocytogenes* is readily isolated from infected foetuses and placentas (Dennis 1975b). Serology is of little value in diagnosis because normal sheep may have high antibody titres to *L. monocytogenes* and titres in aborting animals may persist for years (Blood *et al.* 1979; Gronstol 1979). There are five major serogroups with several subgroups. Serotypes 5 and 4a have been isolated from ovine abortion in Australia (Dennis 1975b; Hughes 1975).

Control of listerial abortion outbreaks with antibiotic therapy is not particularly effective nor are vaccines currently available (Blood *et al.* 1979; Beveridge 1983). The sporadic nature of abortion due to *L. monocytogenes* and lack of information on the epizootology of listerial infections hamper application of preventative measures. *L. monocytogenes* is also infectious for humans and care should be exercised when handling aborted tissues.

### Miscellaneous bacterial causes of abortion

*Histophilus ovis* has been isolated from sporadic ovine abortions in Victoria (Rahaley and White 1977), New South Wales (Webb 1983) and possibly Western Australia (Dennis 1974). This organism is more commonly associated with polysynovitis and septicaemia in lambs (Rahaley and White 1977) and has been isolated from epididymitis in rams (Burgess 1982; Webb 1983), however abortion can be reproduced by intravenous inoculation of pregnant ewes with bacteria (Rahaley 1978).

The epidemiology of *H. ovis* infection has not been adaquately studied. The organism is present in the vagina and prepuce of normal ewes and rams and appears to be an opportunistic pathogen (Rahaley 1977). Recommendations for control and treatment are not available.

Other micro-organisms incriminated in sporadic abortion in Australian sheep include *Brucella ovis, Streptococcus* spp., *Staphylococcus* spp., *Proteus* spp., *Corynebacterium* spp., unidentified *Bacillus* spp. and an unidentified fungus (Dennis 1974). It is interesting to note that there are no reports in the Australian literature of *Salmonella abortus-ovis*, a common cause of abortion in sheep in Europe.

### Toxoplasmosis

Abortion due to the sporozoan, *Toxoplasma gondii* has been reported in New South Wales, Queensland, Tasmania and Western Australia (Dennis 1972). There are no published accounts of outbreaks in South Australia or Victoria although Munday (1970) found serological evidence of toxoplasmosis in sheep in all States of Australia. Reactor rates to the indirect fluorescent-antibody test were: Tasmania 28%, Victoria 25%, New South Wales 13%, Western Australia 12%, South Australia 4% and Queensland 1%. These results support the hypothesis that naturally aquired toxoplasmosis is a problem in high rainfall zones (Dennis 1972) although Plant *et al.* (1982) found a higher prevalence of infection in flocks kept under intensive or semi-intensive management systems and concluded that the prevalence of ovine toxoplasmosis was influenced more by management than environment.

Munday (1970) found toxoplasmosis was the most common cause of infectious ovine abortion in Tasmania, where it was diagnosed in 46% of the abortion outbreaks investigated between 1962 and 1968, and estimated the prevalence in New South Wales at 6%. Dennis (1974) diagnosed toxoplasmosis on 0.6% of farms in Western Australia.

Infection of pregnant ewes with *T. gondii* results in placentitis with subsequent foetal death or neonatal lamb mortality. Placental lesions can be characteristic, consisting of necrotic cotyledons flecked with circumscribed pale foci which may be calcified. The organisms can also cross the placenta and infect the foetus, occasionally causing leucoencephalomalacia (Hartley and Kater 1963).

The parasite has a complex life-cycle undergoing a sexual phase of development producing oocysts in the intestine of cats (the definitive host), and an asexual phase producing tachyzoites in tissue, blood and exudate, and bradyzoites encysted in tissue in both the definitive and intermediate hosts. Sheep and a variety of other species, including humans, may act as intermediate hosts although mice and rabbits are probably the most important in propagating the organism.

Animals can become infected by ingestion of any of the life-cycle stages but sheep are infected predominantly by ingestion of sporulated oocysts shed in cat faeces and occasionally by ingestion of tachyzoites in infected placentas. Hay contaminated with cat faeces may be an important source of infection (Plant *et al.* 1982). Infection by inhalation may also occur infrequently (Dreesen and Lubroth 1983). That cats are the major source of infection for sheep is more easily conceived when it is realised that infected cats may, at times, excrete more than 260 million oocysts per day (Dubey and Frenkel 1972). Sporulated oocysts can remain infective for more than a year if protected from sunlight (Dreesen and Lubroth 1983) and snails and earthworms may act as transport hosts for the oocysts (Jones 1973). Infection of rams results in a brief period of excretion of tachyzoites in semen but it is unlikely that venereal transmission plays an important part in the spread of the disease (Teale *et al.* 1982).

Diagnosis of toxoplasmal abortion is made on histological or immunohistological evidence of the organisms in the placenta or foetal brain. Serology for maternal antibodies to *T. gondii* may also be useful. Both the Sabin-Feldman dye test and the indirect fluorescent-antibody test are specific and sensitive and have the advantage of detecting immunoglobulin-M antibodies which appear early in the disease (Jones 1973). Other serological assays include the complement fixation test, indirect haemagglutination and carbon immunoassay (Jones 1973; Waller *et al.* 1983). Comparison of results from tests measuring IgM with those measuring IgG may be useful in separating active from passive infection (Blewett *et al.* 1983).

Control measures for an abortion outbreak due to *T. gondii* are difficult to implement. Treatment with antiprotozoal drugs has not been successful and control is primarily based on removal and destruction of contaminated carcasses. Although infection results in good immunity to the disease and seropositive ewes are refractory to experimental challenge with *T. gondii* (Blewett and Miller 1982), Blood *et al.* (1979) claims ewes which abort represent a potential source of infection for cats and humans and should be culled from the flock. The infectivity of *T. gondii* for humans has spurred the search for an effective vaccine but none is currently available and this area is worthy of further research.

### Chlamydial abortion

*Chlamydia psittaci* is the most common cause of ovine abortion in the United Kingdom where the disease is called Enzootic Abortion of Ewes (Blewett *et al.* 1982). Antibodies to *C. psittaci* are present in a high proportion of Australian sheep (Dane and Clapp 1956) and the disease has been tentatively diagnosed in New South Wales (Hughes *et al.* 1964; Rofe 1967), Tasmania (Munday *et al.* 1966) and Western Australia (Dennis 1974). There seems little doubt that chlamydial abortion occurs in Australia although it appears to be of low virulence and at a very low incidence.

The disease is manifest as near term abortion, still births or weak lambs. Usually about 5% of ewes abort although abortion rates may be higher (Linklater and Dyson 1979). The chorioallantois of infected placentas may be thickened and leathery and there is severe necrosis of cotyledons. Lesions in the foetus are not specific although Storz (1971) described multiple subcutaneous haemorrhages in a high percentage of natural and experimental cases.

Infection of ewes is probably by ingestion as normal animals can harbour *C. psittaci* in their intestine and excrete the organism in faeces (Pienaar and Schutte 1975). Under conditions of low temperature and high humidity *C. psittaci* can survive at least two months outside the host (Polyakov and Kupeshev 1983).

Diagnosis of chlamydial abortion can be made by cultivating the organism from the placenta or foetus in embryonated eggs or in tissue culture (Johnson *et al.* 1983). Serological tests are of limited diagnostic value as many clinically normal sheep have antibody to *C. psittaci*. Chlamydial elementary bodies can be demonstrated in vaginal smears for up to 21 days after abortion using fluorescent antibody techniques (Jain and Rajya 1978).

Antibiotic therapy does not eliminate *Chlamydia* from vaginal smears (Rodolakis *et al.* 1980; Aitken *et al.* 1982), however oxytetracyline administered near the end of pregnancy protects ewes against experimental and natural challenge (Aitken *et al.* 1982; Grieg *et al.* 1982). Chlamydial abortions will recur on a farm in subsequent years as replacement ewes are exposed to the agent and vaccination of susceptible ewes is widely practiced in Europe although results have been variable, partly due to serotype variation (Waldhalm *et al.* 1982; Rodolakis and Bernard 1984).

### Ovine Pestivirus

Infection of pregnant ewes with *Pestivirus* may result in either embryonic death, abortion or the birth of live or dead lambs which may have excessively hairy coats and show nervous signs such as tremors and inability to stand. The disease is colloquially known as 'Border disease' due to its occurrence near the Welsh border region in the United Kingdom, or 'Hairy shaker disease' from the appearance of affected lambs. The pathodescriptive term 'hypomyelinogenesis congenita' has also been applied.

The causal virus is closely related to Bovine Virus Diarrhoea-Mucosal Disease virus which is widespread throughout the world and endemic in Australian cattle (Beveridge 1981). Experimental inoculation of pregnant ewes with cattle-derived virus produces typical clinical disease (Terlecki *et al.* 1979). Although serological evidence of infection of sheep in Australia is widespread, the disease is uncommon and rarely affects more than a few ewes in a flock (Acland *et al.* 1972; Lim and Carnegie 1984).

Clinical disease occurs when pregnant ewes are infected between 12 to 80 days of gestation (Beveridge 1981). Infection of ewes less than 50 days pregnant commonly results in abortion due to placental degeneration, possibly resulting from a virus induced vasculitis (Clarke and Osburn 1978), while infection at 50-80 days of pregnancy can result in foetal death and abortion or the birth of live or dead lambs with hairy coats and varying degrees of hypomyelinogenesis, cerebellar and cerebral dysgenesis, arthrogryposis, kyphosis and brachygnathia (Plant *et al.* 1983b). The occurrence of nervous system lesions is dependent on the strain of the virus (Plant *et al.* 1983a).

Although most clinically affected lambs die within the first few weeks of life, some may survive and persistently excrete the virus for years acting as a source of infection for other sheep in the flock and their own progeny. In addition, these sheep are frequently immunologically tolerant to the virus (Plant *et al.* 1977; Westbury *et al.* 1979; Barlow *et al.* 1980; Niemi 1983). Prompt removal of hairy lambs from the flock is an important control measure for the disease (Plant *et al.* 1977).

*Pestivirus* can be cultivated from infected lambs and ewes although some strains are non-cytopathic, adding the expense of their identification in tissue culture. A rapid serological assay for virus in tissue culture would be helpful.

Vaccination is of some value in the control of the disease but is complicated by strain variation in antigenicity (Vantsis *et al.* 1980a) and maternal immunotolerance to the virus. The degree of protection resulting from vaccination or natural challenge can be related to the maternal antibody titre (Vantsis *et al.* 1980b). Following natural challenge, antibody persist in immunosensitive ewes for at least eight months (Vantsis *et al.* 1979).

### Akabane disease

Both abortion and congenital abnormalities may result from Akabane virus infection in pregnant ewes. Serological surveys have shown that this arbovirus is present throughout tropical Australia and much of New South Wales (Della-Porta *et al.* 1976). The virus was probably the cause of an outbreak of microencephaly in newborn lambs in New South Wales described by Hartley and Haughey (1974a). A total of 20 lambs were affected to varying degrees. Other abnormalities observed in some of these lambs included hydrocephalus, arthrogryposis, kyphosis and cerebellar agenesis. Although attempts to transmit the disease by intravenous injection of pregnant ewes with brain tissue from affected lambs were unsuccessful (Hartley and Haughey 1974b), Akabane virus was subsequently isolated from two naturally infected foetuses from a sentinel flock maintained near the location of the original outbreak (Della-Porta *et al.* 1977). The mosquito, *Culicoides brevitarsus* has been identified as a vector of the virus (Haughey, 1981).

Presumptive diagnosis of Akabane disease can be made on the presence of typically affected lambs. Serology and virus isolation attempts may be made for confirmation although many normal sheep have antibody to the virus. There are no established treatment or control measures.

## DISCUSSION

There is still need for further definition of the prevalence of specific diseases of reproduction of sheep in Australia to enable accurate evaluation of the economic advantages of specific control measures and research expenditure. A number of major surveys involving post-mortem examinations of lambs and foetuses were undertaken between 1960 and 1975 but there has been little added to our knowledge during the past 20 years. Surveys of this type are time consuming and expensive. Perhaps a better approach would be to develop more sensitive serological assays and other improved diagnostic techniques so that large numbers of sheep and foetuses can be efficiently screened to evidence of active disease.

Current knowledge indicates that the most significant of these diseases presently in Australian sheep are ovine brucellosis, *Campylobacter* abortion and toxoplasmosis. Further efforts to develop sound control measures for these diseases should be rewarded by increased productivity in the Australian sheep industry.

## REFERENCES

Acland, H.M., GARD, G.P. and Plant, J.W., 1972. *Aust. Vet. J., 48*, 70.

Aitken, I.D., Robinson, G.W. and Anderson, I.E., 1982. *Vet. Rec., 111*, 446.

Barlow, R.M., Vantsis, J.T., Gardiner, A.C., Rennie, J.C., Herring, J.A. and Scott, F.M.M., 1980. *J. Comp. Path, 90*, 57-64.

Baynes, I.D. and Simmons, G.C., 1960. *Aust. Vet. J., 36*, 454-459.

Baynes, I.D. and Simmons, G.C., 1968. *Aust. Vet. J., 44*, 339-343.

Beveridge, W.I.B., 1981. *Animal Health in Australia. Viral Diseases.* Australian Government Publishing Service, 60-64.

Beveridge, W.I.B., 1983. *Animal Health in Australia. Bacterial Diseases of Cattle, Sheep and Goats.* Australian Government Publishing Service, 130-133.

Beveridge, W.I.B. and Johnstone, I.L., 1953. *Aust. Vet. J., 29*, 269-273.

Biberstein, E.L., MoGowan, B., Olander, H. and Kennedy, P.C., 1964. *Cornell Vet., 54*, 25-41.

Bird, M.M.E., Stephens, D.J., Wall, E.P. and De Lisle, G.W. 1984. *N.Z. Vet. J., 32*, 14-17.

Blewett, D.A., Bryson, C.E. and Miller, J.K., 1983. *Res. Vet. Sci., 34*, 163-166.

Blewett, D.A. and Miller, J.K., 1982. *Vet. Rec., 111*, 175-177.

Blood, D.C., Henderson, J.A. and Radostits, O.M., 1979. *Veterinary Medicine* Lea and Febiger, Philadelphia, 406-749.

Broadbent, D.W., 1975. *Aust. Vet. J., 51*, 71-74.

Bryner, J.H., Estes, P.C., Foley, J.W. and O'Bury, P.A., 1971. *Am. J. Vet. Res., 32*, 465-470.

Bryner, J.H., Foley, J.W. and Thompson, K., 1979. *Am. J. Vet. Res., 40*, 433-4350.
Burgess, G.W., 1982. *Vet. Microbiol., 7*, 551-575.
Cameron, R.D.A., Carles, A.B. and Lauerman, L.H., 1976. *Vet. Rec., 99*, 231-233.
Carter, J.L., 1978. *Diss. Abs. Int., 39B*, 172.
Clapp, K.H., Keogh, J. and Richards, M.H. 1962. *Aust. Vet. J., 38*, 482-486.
Clarke, G.L. and Osburn, B.I., 1978. *Vet. Pathol., 15*, 68-82.
Corbeil, L.B., Schurig, G.G.D., Bier, P.J. and Winter, A.J., 1975. *Infect. Immum., 11*, 240-244.
Dane, D.S. and Clapp, K.H., 1956. *Aust. Vet. J., 32*, 91-93.
Della-Porta, A.J., Murray, M.D. and Cybinski, D.H., 1976. *Aust. Vet. J., 52*, 496-500.
Della-Porter, A.J., O'Halloran, M.L., Parsonson, I.M., Snowdon, W.A., Murray, M.D., Hartley, W.J. and Haughey, K.J., 1977. *Aust. Vet. J., 53*, 51-52.
Dennis, S.M., 1972. *Vet. Bull., 42*, 415-419.
Dennis, S.M., 1974. *Aust. Vet. J., 50*, 507-510.
Dennis, S.M., 1975a. *Aust. Vet. J., 51*, 11-13.
Dennis, S.M., 1975b. *Aust. Vet. J., 51*, 75-79.
Dreesen, D.W. and Lubroth, J.S., 1983. *Comp. Cont. Ed. Pract. Vet.*, 5, 456-460.
Dubey, J.P., Frenkel, J.K., 1972. *J. Protozool., 19*, 155-177.
Carcia, M.M., Eaglesome, M.D. and Rigby, D., 1983. *Vet. Bull., 53*, 793-818.
Greig, A., Linklater, K.A. and Dyson, D.A., 1982. *Vet. Rec., 111*, 445.
Gronstol, H., 1979. *Acta Vet. Scand., 20*, 168-179.
Hartley, W.J. and Haughey, K.G., 1974a. *Aust. Vet. J., 50*, 55-58.
Hartley, W.J. and Haughey, K.G., 1974b. *Aust. Vet. J., 50*, 323-324.
Hartley, W.J. and Kater, J.C., 1963. *Res. Vet. Sci., 4*, 326-332.
Haughey, K.C., 1981. Univ. Sydney Post. Grad. Comm. Vet. Sci., Poceed. No. 15, 669
Haughey, K.C., Hughes, K.L. and Hartley, W.J., 1968. *Aust. Vet. J., 44*, 531-535.
Hughes, K.L., 1975. *Aust. Vet. J., 51*, 97-99.
Hughes, K.L., and Claxton, P.D., 1968. *Aust. Vet. J., 44*, 41-47.
Hughes, K.L., Hartley, W.J., Haughey, K.G. and Mcfarlane, D., 1964. *Proc. Aust. Soc. Anim. Prod.*, 5, 92-99.
Jain, S.K. and Rajya, B.S., 1978. *Indian J. Vet. Path., 3*, 6-10.
Jansen, B.C., 1980. *Onderstepoort J. Vet. Res., 47*, 101-107.
Jensen, R., Miller, V.A. and Molello, J.A., 1961. *Am. J. Vet. Res., 22*, 169-185.
Johnson, F.W.A., Clarkson, M.J. and Spencer, W.N., 1983. *Vet. Rec., 113*, 413-414.
Jones, S.R., 1973. *J. Am. Vet. Med. Ass, 163*, 1038-1042.
Kennedy, P.C., Frazier, L.M. and McGowan, B., 1956. *Cornell Vet., 46*, 303-319.
Lawrence, W.E., 1961. *Brit. Vet. J, 117*, 435-447.
Lim, C.F. and Carnegie, P.R., 1984. *Aust. Vet. J., 61*, 174-177.
McCoy, E.C., Doyle, D., Wilterger, H. and Burda, K., 1975. *J. Bact., 122*, 307-315.
McMillan, K.R. and Southcott, W,H., 1973. *Aust. Vet. J., 49*, 405-408.
McMillan, K.R., Southcott, W.H. and Royal, W.M., 1974. *Aust. Vet. J., 50*, 298-301.
Meinershagen, W.A., Frank, F.W., Hulet, C.V. and Price, D.A., 1969. *Am. J. Vet. Res., 30*, 203-206.
Molello, J.A., Jensen, R., Flint, J.C. and Collier, J.R., 1963. *Am. J. Vet. Res., 74*, 897-903.
Muhammed, S.I., Lauerman, L.H., Mesfin, G.M. and Otim, C.P., 1975. *Cornell Vet., 65*, 221-227.
Munday, B.L., 1970. Cited by Dennis, S.M., 1972. *Vet. Bull., 42*, 415-419.
Munday, B.L., Ryan, F.B., King, S.J. and Corbould, A., 1966. *Aust. Vet. J., 42*, 189-193.
Murray, R.M., 1969. *Aust. Vet. J., 45*, 63-67.
Niemi, S.M., Evermann, J.F., Huffman, E.M. and Kirk, J.H., 1982. *Am. J. Vet. Res., 43*, 86-88.
Ogg, J.E. and Chang, W.J., 1972. *Am. J. Vet. Res., 25*, 644-647.
Osburn, B.I. and Hoskins, R.K., 1970. *Am. J. Vet Res., 31*, 1733-1734.
Pienaar, J.G. and Schutte, A.P., 1975. *Ondersterpoort J. Vet. Res., 42*, 77-90.
Plant, J.W., Acland, H.M., Gard, G.P. and Walker, K.H., 1983a. *Vet. Rec., 113*, 58-60.
Plant, J.W., Beh, K.J. and Acland, H.M., 1972. *Aust. Vet. J., 48*, 558-561.
Plant, J.W., Gard, G.P. and Acland, H.M., 1976. *Aust. Vet. J., 52*, 247-249.
Plant, J.W., Gard, G.P. and Acland, H.M., 1977. *Aust. Vet J., 53*, 574-577.
Plant, J.W., Freeman, P. and Saunders, E., 1982. *Aust. Vet. J., 59*, 87-89.
Plant, J.W., Walker, K.H, Acland, H.M. and Gard, G.P., 1983b. *Aust. Vet. J., 60*, 137-140.
Polyakov, A.A. and Kupeshev, T.S., 1983. *Veterinariya, 7*, 24-26.
Quinlivan, T.D. and Jopp, A.J., 1982. *N.Z. Vet. J., 30*, 65-68.
Rahaley, R.S., 1977. MVSc Thesis, Melbourne University.
Rahaley, R.S., 1978. *Vet. Pathol., 15*, 746-752.
Rahaley, R.S., and Dennis, S.M., 1982. *Compend. Cont. Ed., 4*, 461-468.
Rahaley, R.S., and Dennis, S.M., *Aust. Vet. J.* (in press)

Rahaley, R.S., Dennis, S.M. and Smeltzer, M.S., 1983. *Vet. Rec., 113*, 467-470.
Rahaley, R.S. and White, W.E., 1977. *Aust. Vet. J., 53*, 124-127.
Ris, D.R., 1967. *N.Z. Vet. J., 15*, 94-98.
Ris, D.R., 1970. *N.Z. Vet. J., 18*, 2-7.
Ris, D.R., 1974. *N.Z. Vet. J., 22*, 143-145.
Rodolakis, A. and Bernard, F., 1984. *Vet. Rec., 114*, 193-194.
Rodolakis, A., Souriau, A., Raynaud, J.P. and Brunault, G., 1980. *Ann. Rech. Vet., 11*, 437-444.
Rofe, J.C., 1967. *Aust. Vet. J., 43*, 117-118.
Schurig, G.D., Hall, C.E., Burda, K., Corbeil, L.B., Duncan, J.R. and Winter, A.J., 1973. *Am. J. Vet. Res., 34*, 1399-1409.
Simmons, G.C., Baynes, I.D. and Ludford, G.C., 1966. *Aust. Vet. J., 42*, 183-187.
Spencer, T.L. and Burgess, G.W., 1984. *Res. Vet. Sci., 36*, 194-198.
Sponenberg, D.P., Carter, M.E., Carter, G.R., Cordes, D.O., Stevens, S.E. and Veit, H.P., 1983. *J. Am. Vet. Med. Assoc., 182*, 990-991.
Storz, J., 1971. *Chlamydia and Chlamydia-induced Diseases*. Charles C Thomas Springfield, Illinois USA, 314-328.
Teale, A.J., Blewett, D.A., and Miller, J.K., 1982. *Vet. Rec., 111*, 53-55.
Terlecki, S., Richardson, C., Done, J.T., Harkeness, J.W., Sands, J.J., Shaw, I.G., Winkler, C.E., Duffell, S.J., Patterson, D.S.P. and Sweasey, D., 1979. *Br. Vet. J., 136*, 602-611.
Van Tonder, E.M., 1973. *J. S. Afr. Vet. med. Ass., 44*, 235-240.
Vantsis, J.T., Barlow, R.M., Gardiner, A.C., and Linklater, K.A., 1980a. *J. Comp. Path., 90*, 39-45.
Vantsis, J.T., Linklater, K.A., Rennie, J.C. and Barlow, R.M., 1979. *J. Comp. Path., 89*, 331-339.
Vantsis, J.T., Rennie, J.C., Gardiner, A.C., Wells, P.W., Barlow, R.M. and Martin, W.B., 1980b. *J. Comp. Path., 90*, 349-354.
Waldhalm, D.G., Delong, W.J. and Hall, R.F., 1982. *Vet. Microbiol., 7*, 493-498.
Waller, T., Uggla, A., Berquist, N.R. and Walter, C., 1983. *Vet. Immunol. Immunopath.,* 5, 203-208.
Watt, D.A., 1970. *Aust. Vet. J., 46*, 506-508.
Westbury, H.A., Napthine, D.V. and Straube, E., 1979. *Vet. Rec., 104*, 406-409.
Webb, R.F., 1983. *Res. Vet. Sci., 35*, 30-34.
Webb, R.F. and Chick, R.F., 1976. *Aust. Vet. J., 52*, 241-242.

# USE OF GAS-LIQUID CHROMATOGRAPHIC ANALYSIS OF CELLULAR FATTY ACIDS AND POLYACRYLAMIDE GEL ELECTROPHORESIS OF CELLULAR PROTEINS FOR IDENTIFYING *BRUCELLA OVIS*.

P.J. Coloe, J.F. Slattery, L.A. Ireland and L.J. Gleeson, *Department of Agriculture, Veterinary Research Institute, Park Drive, Parkville, Victoria, 3052.*

*Summary* The cellular fatty acid composition of *Brucella ovis* was determined and compared with *B. abortus*, *Actinobacillus seminis* and *Histophilus ovis*. *B. ovis* contained fatty acids 15:0, 16:0, 17:0, 17:0 cyclopropane (17:0 cyc), 18:0, 18:1 and 19:0 cyclopropane (19:0 cyc). It was possible by direct comparison of the fatty acid profiles to differentiate *B. ovis* from *A. seminis* and *H. ovis*. *B. ovis* and *B. abortus* contained essentially the same fatty acids but it was possible to differentiage *B. ovis* from *B. abortus* on the basis of the absence of 15:0, lower concentrations of 17:0 and 18:1 and higher concentrations of 19:0 cyc in *B. abortus*. The data indicate that analysis of cellular fatty acid composition can be used for identifying *B. ovis*. In addition, cellular protein profiles of the above strains were determined. *B. ovis* isolates from selected sources showed a consistent protein profile pattern indicating no heterogenicity of isolates. Cellular protein profiles can also be used to identify *B. ovis* isolates.

## INTRODUCTION

*Brucella ovis* infection in sheep is a serious economic problem in many parts of the world. *B. ovis* is a slow growing bacterium taking up to 3 days to be visible on solid medium. It is relatively unreactive biochemically and identification of isolates involves specialised biochemical tests and dye tolerance tests (Alton *et al.* 1975). We have found that the conventional techniques for identifying *B. ovis* are less than adequate with regards to speed or reliability and have determined that a more accurate, rapid method for identifying *B. ovis* is required.

There have been a number of attempts to identify specific components of *Brucella* spp. that could be used in identification (Thiele *et al.* 1969; Thiele and Schwinn 1973; Dees *et al.* 1981). Thiele *et al.* (1969), working with *B. abortus* and *B. melitensis* showed that a C18 monoenoic acid (cis- vaccenic acid) and a C19 cyclopropane fatty acid (19:0 cyc) could be detected consistently in various lipid fractions of these *Brucella* spp.

Analysis of cellular fatty acids has been used to differentiate bacteria within a genus (Moss 1979) and polyacrylamide gel electrophoresis of soluble proteins has been used to differentiate closely ralated species of bacteria (Morris and Park 1973).

The work reported in this paper tested the hypothesis that the cellular fatty acid composition and/or the cell protein profile of *B. ovis* differs sufficiently from closely related species to be used as a rapid method for identifying *B. ovis*.

## MATERIALS AND METHODS

Reference strains of *B. abortus* (n = 7) and *B. ovis* (n = 1) were obtained from the Brucella Reference Laboratory, National Biological Standards Laboratory, Canberra. Strains of *B. ovis* were isolated from clinical specimens submitted to the Veterinary Research Institute, and initially identified on the basis of morphological and cultural characteristics, and biochemical and dye tolerance tests (Alton *et al.* 1975). Field isolates of *Actinobacillus seminis* and *Alcaligenes faecalis* and *Histophilus ovis* were similarly identified, and reference strains of each (n = 1) were obtained from the Regional Veterinary Laboratory, Hamilton, Victoria.

Strains were cultivated on plates of horse blood agar for 5 days at 37°C in an atmosphere of 5% $CO_2$ in air. For fatty acid analysis the bacterial cells from one plate were used, whereas for cell protein preparations cells from up to 10 plates were used. The derivatization procedure to prepare fatty acid methyl esters for gas-liquid chromatographic analysis was based on the method of Moss (1979). The cells were suspended in 0.5 ml of distilled water, and then heated for 30 minutes at 100°C in 3 ml of 5% NaOH in 50% methanol. The saponified material was cooled and acidified by the addition of 1 ml of 6N HCl. A 3 ml portion of 14% $BF_3$ in $CH_3OH$ or 10% $BCl_3$ in $CH_3OH$ was added and the mixture was heated for 5 minutes at 85°C. The esterified components were extracted with 8 ml of ether:petroleum spirit (1:1). The extracting solvent was removed, evaporated almost to dryness with dry $N_2$ and resuspended in 1 ml of petroleum spirit. One microlitre of this material was analysed by gas-liquid chromatography (GLC).

The fatty acid methyl esters were analysed by GLC using a Packard model 427 gas chromatograph. Tentative identification of the bacterial fatty acids was made by comparison of GLC retention times with those of purified fatty acid standards (Supelco, USA). The identity of the fatty acids was confirmed by mass spectrometry.

A number of procedures for extracting cell proteins from bacteria (Rosenbusch 1974; Barenkamp *et al.* 1981; Corbel *et al.* 1982; Verstreate *et al.* 1982; Lema and Brown 1983) were evaluated using sodium dodecyl sulphate-polyacrylamide gel electrophoresis (SDS-PAGE) (Laemmli 1970) of extracts from *B. ovis*. After comparison of the SDS-PAGE profiles obtained by these methods, the method of Lema and Brown (1983) was routinely used. Washed bacterial cells were suspended in 2 ml of a solution of 0.1 M Tris-HC1 (pH 6.8) — 15% glycerol — 2 mM phenylmethylsulphonyl-fluoride, 10% sodium dodecyl suphate was added to 2%, the suspension vigorously mixed and then placed in a boiling water bath for 5 minutes. After centrifugation at 10,000 g for 10 minutes, the supernatant was removed and the protein content determined. The samples were then run on SDS-PAGE gels. To visualise the protein bands, the gels were stained with Coomassie blue.

## RESULTS

The cellular fatty acid composition of strains of *B. ovis*, *B. abortus*, *H. ovis*, *A. seminis* and *Alc. faecalis* are shown in Table 1.

**TABLE 1. The cellular fatty acid composition of gram-negative bacteria associated with ovine epididymitis.**

| | Cellular Fatty Acids (%) | | | | | | | | | |
|---|---|---|---|---|---|---|---|---|---|---|
| Species | 12:0 | 14:0 | 15:0 | 16:0 | 16:1 | 17:0 | 17:0 cyc | 18:0 | 18:1 | 19:0 cyc |
| *Brucella ovis* (n = *20)* | – | – | 3 | 13 | – | 13 | 2 | 7 | 10 | 52 |
| *Brucella abortus* (n = 7) | – | – | – | 14 | – | 3 | 1 | 9 | 5 | 68 |
| *Histophilus ovis* (N = 16) | 2 | 7 | – | 35 | 23 | – | – | 7 | 23 | – |
| *Actinobacillus seminis* (n = 4) | – | 26 | – | 25 | 40 | – | – | – | 8 | – |
| *Alcaligenes faecelis* (n = 6) | – | – | – | 47 | 16 | 26 | – | 5 | 10 | – |

All *B. ovis* strains contained fatty acids 15:0, 16:0, 17:0 cyc, 18:0, 18:1 and 19:0 cyc. Apart from *B. abortus* all of the other species contained significant percentages of 16:1 and on this basis alone could be easily readily differentiated from *B. ovis*. Both *B. ovis* and *B. abortus* contained essentially the same fatty acids. There were, however, differences in the relative concentrations of the fatty acids between the 2 species. When the fatty acid composition of the strains is expressed as the ratio of 19:0 cyc to 15:0 + 17:0 + 18:1 (19:0 divided by 15:0 + 17:0 + 18:1) the figures for *B. ovis* are always less than 3.5 whereas for *B. abortus* the ratio is always greater than 5.0. The fatty acids of selected biotypes of *B. abortus* (Biotypes 1, 2, 3, 6, 7, 9 and strain 19) were also examined for cellular fatty acids. All biotypes were consistent with that described in Table 1 for *B. abortus*.

The cellular fatty acid compostion of *B. ovis* did not appear to be influenced by growth medium. When 6 individual isolates of *B. ovis* were grown on chocolate agar and on sheep blood agar the same fatty acid compositions were detected. However, differences were observed for *B. ovis* and *B. abortus* when the incubation period of the cultures was varied. The fatty acid composition of both species was essentially the same after 4, 5 and 6 days of incubation but at 3 days there was 40% more 18:1 and 16% less 19:0 cyc then was detected at days 4, 5 and 6.

Five methods of extraction of cellular proteins were evaluated for use with *B. ovis*. (Refer materials and Methods). All 5 methods produced similar profiles on SDS-PAGE gels. The major protein profiles of 9 isolates of *B. ovis* and 1 *B. abortus* biotype (biotype I) are shown in Figure 1. There are distinct differences between *B. ovis* and *B. abortus*.

The major feature of the *B. ovis* protein profile is the strong band in the 66,000 m.wt. range. This band is not present in *B. abortus* profiles. Comparison of the cell protein profiles of *B. ovis* with those of *A. seminis* and *H. ovis* showed that each organism had a distinguishing profile (results not shown).

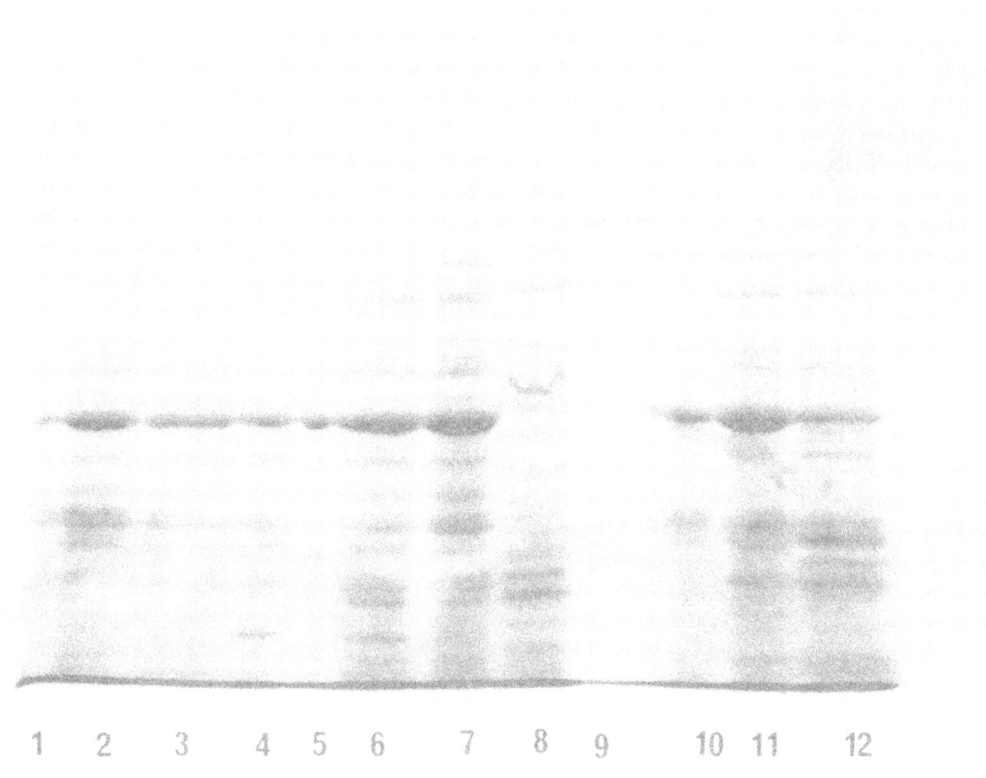


**FIGURE 1. Coomassie blue stained cellular protein patterns of 10 isolates of *Brucella ovis* (lanes 1-7, 10-12) and *Brucella abortus* biotype I (lane 8). Lane 9 is molecular weight marker proteins.**

## DISCUSSION

On the basis of cellular fatty acids *B. ovis* can be readily distinguished from *H. ovis*, *A. seminis* and *Alc. faecalis*. Indeed *B. ovis* is so distinct from these other species that simple comparison of the cellular fatty acids is all that is needed to speciate isolates. It is also a simple task to differentiate *H. ovis*, *A. seminis* and *Alc. faecalis* from each other on the basis of the relative concentrations of their fatty acids.

The difference between *B. ovis* and *B. abortus* is less dramatic, as would be expected from organisms of the same genus. Although *B. ovis* and *B. abortus* have similar fatty acid compositions it is possible to differentiate the two species on the presence of 15:0 in *B. abortus* and on the relative ratios of 19:0 cyc divided by 15:0 + 17:0 + 18:1, where *B. ovis* is always less than 3.5 and *B. abortus* is always greater than 5.

Analysis of SDS-PAGE profiles of *B. ovis* and *B. abortus* cellular proteins has shown that *B. ovis* isolates collected over a wide time span from selected geographical areas all had identical protein patterns, indicating the major differences in protein composition do not occur within the species. *B. ovis* profiles were distinctly different from *B. abortus* and so SDS-PAGE can be used as an alternative to cellular fatty acids to differentiate the 2 species.

Comparison of the cellular protein composition of *B. ovis* with *A. seminis* and H. ovis also showed distinct differences that can be used in identifying each individual species. Further work is in progress to determine if there are any antigens shared between these species.

## ACKNOWLEDGEMENT

This work is supported by the Australian Wool Corporation Grant Number K/6/1091.

## REFERENCES

Alton, G.G., Jones, L.M. and Pietz, D.E., 1975. *World Health Organization Monograph Series, No. 55.*

Barenkamp, S.J., Munson, R.S. and Granoff, D.M., 1981. *J. Infect. Dis., 143*, 668-676.

Corbel, M.J., Brewer, R.A. and Hendry, D.Mc., 1982. *Res. Vet. Sci., 33*, 43-46.

Dees, S.B., Hollis, D.G., Weaver, R.E. and Moss, C.W., 1981. *J. Clin. Microbiol., 14*, 111-112.

Laemmli, U.K., 1970. *Nature, 227*, 680-685.
Lema, M. and Brown, A., 1983. *J. Clin. Microbiol., 17*, 1132-1140.
Morris, J.A. and Park, R.W.A., 1973. *J. Gen. Microbiol., 78*, 165-178.
Moss, C.W., 1979. *In* Jones, G.L. and Herbert, G.A. (eds), *Legionnaires — The Disease, the Bacterium and Methodology,* U.S. Department of Health, Education and Welfare, Atlanta, 118-122.
Rosenbusch, J.P., 1974. *J. Biol. Chem., 249*, 8019-8029.
Thiele, O.W., Lacave, C. and Asselineau, J., 1969. *Eur. J. Biochem., 7*, 393-396.
Thiele, O.W. and Schwinn, G., 1973. *Eur. J. Biochem., 34*, 33-334.
Verstreate, D.R., Greasy, M.T., Caveney, N.T., Baldwin C.L., Blab, M.N. and Winter, A.J., 1982. *Infec. Immun., 35*, 979-989.

# FLOCK HEALTH AND PRODUCTION PROGRAMMES AND THE APPLICATION ON FARMS OF KNOWLEDGE ABOUT REPRODUCTION

D.B. Galloway and I. McL. Grant, *Department of Veterinary Clinical Sciences, University of Melbourne, Veterinary Clinical Centre, Princes Highway, Werribee, Victoria 3030.*

*Summary* Flock health and production programmes are a new development aimed at applying the techniques of preventive medicine and animal production to commercial sheep enterprises. This paper reviews this development with particular reference to the programme of the University of Melbourne. Some aspects of reproductive efficiency in the context of these programmes are used to illustrate the need for improved application of knowledge to sheep enterprises.

## INTRODUCTION

This paper introduces the concept of flock health and production programmes (Boundy 1979 and 1981; Quinlivan 1981; Hindson 1982 Morley *et al.* 1982; Quinlivan and Middleberg 1983; Bell 1983; Watt 1983) which represent a significant development over the last five years in professional services available to sheep farmers. The limited amount of information on reproduction in the context of these programmes is reviewed. Some examples are given of aspects of flock reproduction where such a programme has been instrumental in getting current knowledge applied on farms and some directions for future research are proposed.

There is little published information on the extent to which knowledge about sheep reproduction is applied on farms or on its effectiveness in improving reproductive rates. The figures in Table 1 from Department of Agriculture surveys suggest that on many Victorian farms improvement is possible but is not taking place.

**TABLE 1. Distribution of flocks surveyed (%) according to average marking percentages (Clarke 1976 and Campbell *et. al.*, 1984) (n = 400 in 1974, 327 in 1981).**

| Year of survey | < 60 | 60-79 | 80-99 | 100-119 | > 119 |
|---|---|---|---|---|---|
| 1974 | 1 | 22 | 55 | 16 | 6 |
| 1980 | 3 | 23 | 48 | 19 | 7 |

The slow movement of information from the scientific literature to the farmer is illustrated by Table 2, from the same surveys. There has been a gradual development of understanding of the principle that increased ewe body weight at joining is reflected in fewer dry ewes (Hafez 1952; Knight *et al.* 1979), higher ovulation rates and numbers of lambs born (Clark 1934; Coop 1962; Cole 1963; Morley *et al.* 1978). It is apparent from Table 2 that many farmers would not be utilising this information in their management.

**TABLE 2. Percentage of farmers agreeing with the statements about ewe condition at joining (n = 400 in 1974, 327 in 1981) (Clarke 1976 and Campbell *et. al.*, 1984).**

| Statements about ewe condition at joining | 1974 | 1981 |
|---|---|---|
| Forward store condition gets most lambs | 40 | 28 |
| Poor conditioned ewes result in fewer lambs dropped | 35 | 58 |
| There are more twins with higher ewe body weights | 15 | 23 |

Similarly in the areas of parasite control and the achievement of optimum stocking rates, farm practice falls short of current scientific recommendations (Morley 1979; Watt *et al.* 1981).

Flock health and production prgrammes have been developed as an approach to the need for more widespread and quicker application of current knowledge and new research results.

## FLOCK HEALTH AND PRODUCTION PROGRAMMES

These programmes are an attempt to apply the techniques of preventive medicine and animal production to commercial enterprises. They are designed to increase the productivity and the profitability of the flock, in the context of the objectives of the owners (Morley *et al.* 1982; Bell 1983; Watt 1983). One of the aims in their development is to design appropriate veterinary services for sheep enterprises. Historically the programmes for sheep are based on the principles applied in preventive dairy herd health programmes (Blood *et al.* 1978). They have some similarities to the activities of agricultural advisers (Williams 1968), but emphasize animal health as well as management and production in the context of a "whole farm" approach in assisting the farmer.

Comprehensive descriptions of the flock health and production programmes developed at the Veterinary Schools of Murdoch and Melbourne Universities are given respectively by Bell (1983) and by Morley *et al.* (1982) and Watt (1983). The aims and approach are similar with the veterinarian acting as consultant to the farmer. The programmes seek to identify sub-optimal production, or sources of loss, and diagnose the causes. Cost-effective measures are then designed to eliminate or reduce losses or to achieve optimal productivity. The adviser then agrees with management on what aspects of the programme are to be adopted, setting priorities according to cost-benefit ratios, but heavily influenced by personal preferences of the manager or owner. The adviser aids and encourages the application of the programme, attempts to evaluate it in relation to the objectives of the farmer, and re-allocates priorities in the light of current achievements.

Monitoring and advising is conducted in a number of well defined areas including reproductive efficiency, internal and external parasite control, vaccination and infectious disease, nutritional and metabolic disease, genetic improvement, pasture use, nutrition and productivity. Contact with the farmer is maintained through telephone calls, newsletters, discussion groups, and farm visits to carry out the physical work (e.g. ram examination, monitoring ewe body weight) and discuss past performance and future management. Reports are sent to the farmer after each visit. On-farm trials are used to demonstrate and evaluate some recommendations.

In the Australian programmes emphasis is on reproduction, parasite control, the growth and health of young sheep, and economic efficiency.

Reproductive efficiency is also important in the programmes described in Britain (Boundy 1979 and 1981; Hindson 1982) and in New Zealand (Quinlivan 1981; Quinlivan and Middleberg 1983). Common to all programmes is the assistance given to the farmer in the management and nutrition of ewes before joining, and in late pregnancy, and in ram examination and management. Where there are benefits for the enterprise the use of teasers is recommended.

In the programmes developed in Western Australia and Victoria, assisting with decisions such as the most suitable date for joining (Morley 1981 and 1983; Plant 1981; Bell 1983) and the place of pregnancy diagnosis (Bell 1983; Watt 1983) are important. In these programmes also, use is made of a diagnostic approach to defining areas of reproductive wastage (Moule 1965; Plant 1981; Haughey 1981 and 1983; Galloway 1983), including weighing of ewes, examining rams, the sire sine harness, pregnancy diagnosis, lamb post mortem examinations, and ewe examination at marking.

An emerging structure for the programmes in Victoria is the provision of services to individual farms by private veterinarians, with the Department of Agriculture supplying specialist officers for district and regional back-up services. The veterinarian as consultant to the farmer makes use of specialist consultants himself, e.g. parasitologists, pathologists, epidemiologists, geneticists, economists and agronomists. Research and development are catered for by the University and the Department of Agriculture in cooperation. The University meets the educational needs, with the programmes providing a vehicle for under-graduate and post-graduate training in the sheep industry. In mid-1984 there were 51 farms on the programme being run by the University of Melbourne, involving 300,000 sheep. There are three full-time and two part-time veterinarians employed in the development project. A post-graduate refresher course has been held. The University and the Department of Agriculture provide encouragement, advice and assistance to established veterinarians wishing to provide flock health and production services.

The adviser's role of ensuring that current knowledge and new information and techniques are evaluated with respect to their applicability on individual farms, and then applied where appropriate, is consistent with Campbell's (1963) view of extension. We consider that the complexity of much decision marking on the farm requires planning of programmes for the individual enterprise. The role of the adviser is not only in the development of such plans but in the guidance and assistance required to implement them to the farmer's advantage.

We are conscious of certain difficulties and limitations. The measurement and economic evaluation of successfully influencing farmers' decisions is, for example, extremely difficult. Hence economic evaluations of the programmes as a whole have not yet been accomplished.

## APPLICATION OF INFORMATION ABOUT REPRODUCTION OF FARMS

The farmer receives information about animal health, production and management from relatives and other farmers, from Departments of Agriculture, particularly from extension services, sellers of goods and services, including private veterinarians and agriculturalists, research laboratories, Universities and Agricultural Colleges. The information comes through direct contact, through radio and television, from rural newspapers and journals, field days and discussion groups. The ways in which this information is received and used vary considerably between farmers, depending on their level of education, their entrepreneurial attitudes and skills, their motivation and their goals (Emery *et al.* 1958). One of the functions of the flock health and production programmes is to help the farmer organise and utilise this information more efficiently. The need for this, and for improved application of knowledge, was mentioned in the introduction. Some examples are given below from the Melbourne University programmes to illustrate this need in the field of improving reproductive efficiency and the role of the advisers in filling it.

1. Few farmers undertake a gross margin (GM) analysis to estimate the desirability of an increase in reproductive rate, and the amounts that could be reasonably spent in achieving it. We consider this an important step in setting priorities on a farm and have prepared a number of such budgets. The factors to be taken into consideration, and the general approach, is well illustrated by White (1984). Based on current prices our gross margin for a spring lambing Merino flock of 4800 dry sheep equivalents (DSE), with a winter stocking rate of 12 DSE per hectare (ha), is $118.ha$^{-1}$ where the weaning rate is 70%, $126.ha$^{-1}$ where weaning rate is 80% with surplus sheep sold, and $134 at the latter weaning rate where stocking rate is allowed to rise. The increase of $8 for a 10% increase in lamb numbers is similar to the $8-9 increase in GM.ha$^{-1}$ calculated under similar conditions by Patterson and Wagg (1983) and White (1984). The former authors calculated that an increase of 10% in weaning percentage would lift GM per ewe by $3.16 for Merino ewes and $1.91 for crossbred ewes. They assumed that half of this should be allocated to increased profit, leaving $1.58 per ewe for Merino ewes, and $0.90 per ewe for crossbred ewes, as the maximum that could be spent per ewe to increase the weaning percentage by 10%.

The optimum time for lambing has been discussed with most farmers. The decision is very complex (Morley 1981, 1983; Bell 1983) and a similar budgetting approach is taken. For example, a GM.ha$^{-1}$ of $126 was calculated for an autumn lambing flock and $155 for the same flock if spring lambing were adopted. Individual budgets are calculated for each farm as requested and the advantages and disadvantages of a change in lambing time are discussed.

2. Principles for good management of rams have been available for a long time, based on physiological and pathological studies (e.g. Gunn *et al.* 1942; Belschner 1965; Galloway 1972 and 1973; Courot 1979; Lindsay 1979). Examination of ram teams before joining in the flock health programmes (Galloway 1983) indicates that there are deficiencies and abnormalities causing reproductive inefficiency in many rams (Table 3).

It is inferred that the above principles are not being applied efficiently on over 50% of farms. In Table 3, A class rams were those considered in excellent health and condition and ready for joining. B class rams were those in which there were deficiencies in either condition of the animal or the reproductive organs, requiring improvement before joining. C class rams had abnormalities severely limiting reproductive efficiency.

Conditions affecting B and C class rams included under-nutrition, testicular degeneration (some due to scrotal mange), epididymitis caused by *Brucella ovis*, and by other organisms, undeveloped testicles, and spermiostasis.

**TABLE 3. Results of examination of 35 ram teams in flock health and production programmes (Galloway 1983).**

| Rank of ram team | Class of ram (see text) | | |
|---|---|---|---|
| | A | B | C |
| The best | 93 | 7 | 0 |
| The average | 57 | 37 | 6 |
| The worst | 32 | 32 | 36 |

Experience so far indicates that the proportion of A class rams will increase gradually over 2 to 3 years as the ram teams are submitted for examination annually, recommendations are made on their improvement and expectations for them are made clear by personal contact and newsletter.

3. Infertility due to oestrogenic pasture was reviewed in 1963 (Moule *et al.* 1963). Bishop (1964) cites references to its occurrence in Victoria in 1945 and 1946. Four of 49 flock health farmers in 1982 were made aware of this problem on their property by their advisers and were introduced, apparently for the first time, to some measures for combating its effects (delaying joining, prolonging joining, use of pregnancy diagnosis and rejoining empty ewes). On one of the farms, 6 weeks after joining the use of sire sine harnesses demonstrated that 30% of a mob of 3 year old ewes were still coming into oestrus. A prolonged joining was recommended and a final result of 80% lamb marking was obtained. There is still a lack of information about the prevalence of this problem in Victoria.

4. The ram effect has been well studied since the 1950s (Schinckel 1954; Radford and Watson 1957; Edgar and Bilbey 1963). In the flock health and production programmes 23 farmers were joining their flocks at the time of the year at which this effect might be beneficial. After the likely managerial benefits were explained of better control of the timing of lambing, and a more compact lambing, 16 have adopted the practice.

5. On three of 40 flock health farms maiden ewes were not being mated until 2½ years of age. In the first year after advice was given that body weights were satisfactory for mating at 1½ years of age, and that this would be profitable, lamb marking percentages for the maiden mobs were 80%, 78% and 70%.

6. At the beginning of the programmes no flock health farms were using lupins before joining to increase lambing percentages (Marshall *et al.* 1979). On two farms with average lambing percentages of 100-110, increases of 25% and 40% respectively of lambs born were achieved in treated groups of ewes compared with controls when the technique was introduced as a trial.

7. Where immunisation of ewes against androstenedione using the new product Fecundin (Glaxo, Aust.) (Scaramuzzi 1977; Cox *et al.* 1982; Geldard *et al.* 1984) was thought to offer advantages, farmers on the programmes were encouraged to use it in a group of ewes and critically evaluate the outcome. Six farmers who set up trials obtained 12 to 30% more lambs weaned in the treated groups compared with control groups.

## PROPOSALS FOR FUTURE RESEARCH

Experience with flock health and production programmes has identified three areas in which investigations are warranted to assist in the improvement of reproductive and productive efficiency in sheep flocks.

1. There is a need to identify likely improvements to systems of management and to assess their economic impact on the farm as a whole. The development, expansion and application of computer models (White *et al.* 1982; White 1984) to individual farms to test decisions and to estimate the likely costs and benefits of different procedures are considered necessary.

2. There is a need to then test and demonstrate these improvements under various environmental conditions and establish their practicality and economic viability.

3. There is a clear need to develop and evaluate new, more effective systems for ensuring that information obtained from research and investigation is applied on farms, to the economic advantage of the farmer and the industry. We believe that flock health and production programmes constitute one such system.

## ACKNOWLEDGMENTS

The development of flock health and production programmes by the University of Melbourne, in cooperation with the Department of Agriculture, Victoria, has been supported by the Australian Wool Corporation and the Scobie and Claire Mackinnon Trust.

## REFERENCES

Bell K.J., 1983. *In, Sheep Production and Preventive Medicine*, Univ. of Sydney Post-Grad. Cttee. Vet. Sci. Refresher Course for Veterinarians, No. 67, pp. 335-421.

Belschner, H.G., 1965. *Sheep Management and Diseases*, 8th edn. Angus and Robertson, Sydney, 206-212.

Bishop, A.H., 1964. *Aust. J. Agric. Sci., 30*, 219-231.

Blood, D.C., Morris, R.S., Williamson, N.B., Cannon, C.M. and CAnnon, R.M., 1978. *Aust. Vet. J., 54*, 207-215.

Boundy, T., 1979. *Vet. Annual, 19*, 79-82.

Campbell, I.P., Clarke, J.D. and Love, K.J. 1984. This volume, 391-393.

Campbell, K.O., 1963. *Aust. Agric. Extension Conf. 1982*, pp. 275-278. Reviews, Papers and Report. CSIRO, Melbourne.

Clark, R.T., 1934. cited by Doney, J.M. 1979. *In, The Management and Diseases of Sheep*, British Council, London and Slough, U.K., 152-160.

Clarke, J.D. 1976. *Report on Sheep Reproduction Survey*, Department of Agriculture, Victoria, Bulletin, 6-32.

Cole, V.G., 1963. *Sheep Management for Wool Production*. Grazcos, Coop Ltd., Sydney, 46-49.

Coop, I.E., 1962. *N.Z.J. Agric. Res.,5*, 249-264.

Courot, M., 1979. *In* Tomes, G.L., Robertson, D.E., Lightfoot, R.J. and Haresign, W. (eds) *Sheep Breeding* 2nd edn. Butterworths, London, pp 495-504.

Cox, R.I., Wilson, P.A., Scaramuzzi, R.J., Hoskinson, R.M., George, J.M. and Bindon, B.M., 1983. *Proc. Aust. Soc. Anim. Prod., 14*, 511-514.

Edgar, D.G. and Bilbey, D.E., 1963. *Proc. Soc. Anim. Prod., 24*, 79-98.

Emery, F.E., Oser, O.A. and Tully, J., 1958. *Information, Descisions and Action*. Melbourne, University Press, Melbourne.

Galloway, D.B., 1972. *AMRC Review No. 9*, 1-18.

Galloway, D.B., 1973. *AMRC Review No. 10*, 1-22.

Galloway, D.B., 1983. *In, Sheep Production and Preventive Medicine*, Univ. of Sydney Post-Grad. Cttee. Vet. Sci. Re-fresher Course for Veterinarians No. 67. pp. 163-195.

Geldard, H., Dow, G.J., Scaramuzzi, R.J., Hoskinson, R.M., Cox, R.I. and Beels, C.M., 1984. *Aust. Vet. J., 61*, 130-132.

Gunn, R.M.C., Sanders, R.N. and Granger, W., 1942. *CSIR Bulletin No. 48*, 1-140.

Hafez, E.S.E., 1952. *Int. Congr. Anim. Prod., Copenhagen, 3*, 35-37.

Haughey, K.G., 1983. *In, Sheep Production and Preventive Medicine,* Univ. of Sydney Post-Grad. Cttee. Vet. Sci. *Refresher course for veterinarians No. 67,* 135-137.

*Hindson, J., 1982. In Practice, 4*, 53-58.

Knight, T.W., Dalton, D.C. and Hight, G.K., 1979. *N.Z.J. Exp Agric., 7*, 125-130.

Lindsay, D.R., 1979. *In* Tomes, G.L., Robertson, D.E., Lightfoot, R.J. and Haresign, W. (eds) *Sheep Breeding, 2nd edn*. Butterworths, London.

Marshall, T., Lightfoot, R.J., Croker, K.P. and Allen, J.G., 1979. *In* Tomes, G.L., Robertson, D.E., Lightfoot, R.J. and Haresign, W.(eds) *Sheep Breeding*, 2nd edn. Butterworths, London.
Morley, F.H.W., 1979. *Aust. Advances in Vet. Sci.*, Aust. Vet. Assoc., S1.
Morley, F.H.W., 1981. *In* Morley, F.H.W. (ed) *Grazing Animals*, World Animal Science B.1. 379-400, Elsevier, Amsterdam.
Morley, F.H.W.,1983. *In, Sheep production and preventive medicine*, Univ. of Sydney Post-Grad. Cttee. Vet. Sci. Refresher course for veterinarians No. 67, 83-87.
Morley, F.H.W., Watt, B.R., Grant, I.McL. and Galloway, D.B., 1982. *3rd Int. Symp. Epid. Econ., Arlington, Virginia.*, 186-194.
Morley, F.H.W, White, D.H., Kenney, P.A. and Davis, I.F., 1978. *Agric. Systems 3*, 27-45.
Moule, G.R., 1965. *Field investigation with sheep*. CSIRO, Melbourne.
Moule, G.R., Braden, A.W.H and Lamond, D.R., 1963. *Animal Breeding Abstracts 31*, 139-158.
Patterson, B. and Wagg, M. 1983. *In, Sheep Reproduction, Proc. Sheep Ind. Serv. Workshop*, Dept. Ag. Vic., 1-9.
Plant, J.W., 1981. *In, Sheep*, Univ. of Sydney Post-Grad. Cttee. Vet. Sci. Refresher course for veterinarians No. 58, 675-705.
Quinlivan, T.D., 1981. *In, Sheep*, Univ. of Sydney Post-Grad. Cttee. Vet. Sci Refresher course for veterinarians No. 58, 269-290.
Quinlivan, T.D. and Middleberg, A., 1983. *Proc. 13th Seminar Sheep & Beef Cattle Soc. N.Z. Vet. Assoc.* 103-113.
Radford, H.M. and Watson, P.H., 1957. *Aust. J. Agric. Res. 8*, 460-470.
Scaramuzzi, R.J., Davidson, W.G. and Van Look, P.F.A., 1977. *Nature, Lond. 269*, 817-818.
Schinkel, P.G. 1954. *Aust. J. Agric. Res.* 5, 465-469.
Watt, B.R., 1983. *In, Sheep Production and Preventive Medicine*, Univ. of Sydney Post-grad. Cttee. Vet. Sci. Refresher course for Veterinarians No. 67, 437-485.
Watt, B.R., Morley, F.H.W. and Galloway, D.B., 1981. *Vict. Vet. Proc. 39*, 17-19.
White, D.H., 1984. This volume, 371-377.
White, D.H., Bowman, P.J. and Morley, F.H.W., 1982. *Proc. Aust. Soc. Anim. Prod. 14,* 38-40.

# MEASUREMENT OF PERMANENT CLOVER INFERTILITY IN EWES

N.R. Adams, A.J. Ritar and M.R. Sanders, *Division of Animal Production, CSIRO, Private Bag, Wembley, W.A. 6014.*

*Summary* The amount of permanent phyto-oestrogenic damage which accumulated in individual ewes was estimated by measuring the amount of lamina propria tissue in a transverse histological section through the mid cervix. This technique indicated that clover infertility contributed to the failure of ewes to conceive in 4 of 7 apparently normal commercial flocks. Furthermore, preliminary observations indicate that oestrogenic subterranean clover is still very widespread in Western Australia. We conclude that, in the absence of alternative control procedures, phyto-oestrogens will continue to be one cause of failure of ewes to conceive for many years to come.

## INTRODUCTION

Oestrogenic clovers can cause severe, permanent infertility in ewes as part of the syndrome called clover disease (Bennetts *et al.* 1946). In the past 15 years the severity has decreased, but mild permanent infertility resulting from prolonged exposure to oestrogenic clovers is still widespread (Lightfoot 1974). The primary method available to control this infertility has been the replacement of pasture by less oestrogenic cultivars. However, there have been few evaluations of the success of this programme in the field. This is due at least partly to the difficulties experienced by extension workers in recognizing and measuring clover infertility in non-experimental flocks. We have developed methods to attempt to simplify this problem.

## MATERIALS AND METHODS

*Pasture survey* Eleven properties were examined, giving a wide geographical spread in the agricultural region of south Western Australia. Between 3 and 6 paddocks were sampled on each property. The paddocks sampled were nominated by the farmer as being representative of his property. Two random 15 cm diameter core samples were collected from each paddock and brought back to the laboratory, where they were grown out. The clover cultivars were identified, and isoflavone levels in the leaves of the main cultivar in each core was determined.

*Fertility survey* Seven properties were examined from the higher rainfall (500-650 mm) area of south Western Australia. On each property between 120 and 150 cast-for-age ewes were mated under normal farm management for 7 weeks, and sent to the abattoirs 3 weeks after mating. All of the reproductive tracts were collected and opened. The crown-rump length of the foetus, if any, was measured to determine the age of the foetus using the formula of Joubert (1956). The number of macroscopic cysts in the uterus was recorded, and the mid section of the cervix from all non-pregnant ewes and a random selection of pregnant ewes was fixed in formal saline for histology.

The experimentally affected ewes came from the flock described by Adams (1983). Eight affected ewes and 8 controls were killed, and their cervixes fixed for histology. Transverse histological sections were prepared at 6 mm intervals along the cervix, and the pooled results from each animal analysed by the 't'-test.

*Quantitative histology* An entire cross-section of each cervix was examined. Paraffin embedded tissue was cut at 6 mm, and the sections stained with haematoxylin and eosin. Microfiche prints were prepared as described by Ferrington *et al.* (1980) and relative areas compared by a point-counting system. A grid with lines 1 cm long drawn on clear perspex was prepared as described by Freere and Weibel (1967). The grid was placed over the microfiche print, and the tissue area on which the ends of each line fell was recorded.

## RESULTS

Overall, 132 subterranean clover samples were identified according to their cultivar and 72 of these (55%) belonged to those recognised as oestrogenic (Dinninup, Dwalganup, Yarloop or Geraldton). Isoflavone levels were estimated on 81 samples, and 39 (48%) of these contained more then 0.5% formononetin.

In the low rainfall areas (300-350 mm) most of the pasture legumes were medics, and the small amount of subterranean clover found was mainly cv. Geraldton. On the other properties (rainfall 500-650 mm), there was some mixing of cultivars, but pastures contained predominantly the cultivar to which they were originally sown. Pastures sown before 1970 consisted primarily of oestrogenic cultivars, and pastures sown later were predominantly non-oestrogenic. There were two exceptions to this. In the Esperance area, little (about 20%) of the oestrogenic cultivars sown in the 1960's persisted, possibly because of the action of the disease clover scorch. The other exception was the observation that cv. Dinninup made a larger contribution than expected from information given by farmers on the cultivars they had sown.

The cross-sectional area of the cervix was greater in the clover-affected ewes from the experimental flock than in the controls ($223 \pm 35$ vs $93 \pm 14$ arbitary units; $P < 0.01$). This was due mainly to the large increase in affected ewes ($P < 0.001$) in the amount of lamina propria tissue lying between the cervical folds and the muscle layer (145 ± 23 vs 36 ± 6 arbitary units). Differences in the amount of tissue in the folds and in the cervical lumen were not statistically significant. The clover affected ewes also had more glandular segments cut in cross-section (642 ± 75 vs 135 ± 36, $P < 0.001$).

All of the properties examined had oestogenic clover in their pastures, although the observations were insufficient to allow accurate ranking of the properties. The properties are listed in Table 1 in order of increasing amounts of lamina propria. The amount of lamina propria correlated with the proportion of ewes with macroscopic cysts in the uterus ($r = 0.829$, $P < 0.05$), but not with the mean duration of pregnancy ($r = 0.110$). The proportion of ewes pregnant in each flock was not significantly correlated with the mean amount of lamina propria ($r = 0.692$), but in four of the seven flocks, the ewes which were not pregnant had more ($P < 0.05$) lamina propria than ewes which had conceived.

**TABLE 1. Characteristics of ewes from 7 commercial flocks.**

| Farm | % ewes with uterine cysts | Amount of lamina propria (arbitary units) | | % ewes pregnant | Mean age of foetus (days) |
|---|---|---|---|---|---|
| | | pregnant | non pregnant | | |
| A | 0 | 28 ± 4 | 37 ± 6 | 87 | 54.5 ± 0.9 |
| S | 0 | 56 ± 5 | 80 ± 9* | 74 | 36.0 ± 1.7 |
| Y | 11 | 57 ± 6 | 76 ± 7* | 71 | 45.0 ± 1.1 |
| R | 12 | 59 ± 8 | 97 ± 10** | 80 | 40.1 ± 0.9 |
| M | 18 | 69 ± 7 | 77 ± 7 | 79 | 60.5 ± 1.4 |
| E | 17 | 78 ± 8 | 64 ± 7 | 80 | 40.5 ± 0.9 |
| O | 26 | 75 ± 9 | 105 ± 7* | 68 | 65.1 ± 1.3 |

*$P < 0.05$, **$P < 0.01$, significant difference between pregnant and non-pregnant ewes.

## DISCUSSION

The observations on the ratio of oestrogenic to non-oestrogenic clovers are of a preliminary nature, but they are presented because there are no other published data available for Western Australia. Obviously, more detailed studies are necessary before the success of the clover replacement program can be evaluated. However, the present results are in agreement with the observation by Cocks and Phillips (1979), that the relative abundance of cultivars of subterranean clover in South Australia was related to the number of years they had been available commercially. It seems safe to assume that oestrogenic clovers still made up a substantial proportion of the pasture legumes in Western Australia.

Measurement of the amount of lamina propria is a convenient method for estimating permanent oestrogenic damage. It is easier and more rapid then counting glands. Furthermore, hormones other than oestrogen can affect gland numbers, but do not affect the overall area of lamina propria (N.R. Adams, unpublished). The method appears useful in the field, in that it was related to oestrogenic damage as determined by the incidence of cysts, but was unaffected by the duration of pregnancy. In 4 of the 7 flocks, there was significantly more lamina propria in the non-pregnant animals, indicating that clover infertility played a role at that mating in determining which ewes in that group of ewes conceived. Presumably clover infertility also played a role between flocks, although the more important role played by other factors such as nutrition confounded the final proportion of ewes pregnant and prevented such a role being detected in the present limited study. The proportion of ewes pregnant in all of the flocks was similar to that normally found in Western Australia (Knight *et al.* 1975). Thus, the results confirm suggestions (Lightfoot 1974; Adams 1977) that clover infertility is widespread among 'average' Western Australian sheep flocks.

It can be concluded that oestrogenic clovers make up a substantial proportion of pasture legumes in Western Australia, and are likely to continue to be important for the foreseeable future. The pastures are sufficiently oestrogenic to prevent conception in some ewes. Unless methods other than pasture replacement can be developed to control clover infertility, Western Australia will continue to have an unnecessarily high proportion of ewes which fail to lamb.

## ACKNOWLEDGMENTS

We thank Dr. R.C. Rossiter for his interest in the work and for identifying the clover cultivars. The isoflavone laboratory, University of Western Australia estimated plant isoflavone levels. The work was supported by The Australian Pastoral Research Trust.

REFERENCES

Adams, N.R., 1977. *Aust. J. Agric. Res., 28*, 481-489.
Adams, N.R., 1983. *J. Reprod. Fert., 68*, 113-117.
Bennetts, H.W., Underwood, E.J. and Shier, F.L., 1946. *Aust. Vet J., 22*, 2-12.
Cocks, P.S. and Phillips, J.R., 1979. *Aust. J. Agric. Res., 30*, 1035-1052.
Ferrington, D.G., Fisher, G.R., Herron, P. and Rowe, M.J., 1980. *Neurosci. Lett., 19*, 51-53.
Freere, R.H. and Weibel, E.R., 1967. *J. Roy. Microscop. Soc., 87*, 25-34.
Joubert, D.M., 1956. *J. Agric. Sci., 47*, 382-428.
Knight, T.W., Oldham, C.M., Smith, J.F. and Lindsay, D.R., 1975. *Aust. J. Expl. Agric. Anim. Husb., 15*, 183-188.
Lightfoot, R.J., 1974. *Proc. Aust. Soc. Anim. Prod., 10,* 113-121.

# EFFECT OF STEROID IMMUNIZATION ON CLOVER INDUCED INFERTILITY IN SHEEP

D.L. Little, *Department of Agriculture, Box 1671, G.P.O., Adelaide, S.A. 5001.*
R.I. Cox, *CSIRO, Division of Animal Production, P.O. Box 239, Blacktown, N.S.W., 2148.*
S.K. Walker and P.F. Flavel, *Department of Agriculture, Box 1671, GPO, Adelaide, S.A., 5001.*

*Summary* Three experiements were carried out with Merino ewes to test whether immunization against oestrone or androstenedione could offset reproductive losses associated with intake of phyto-oestrogens. Immunized and unimmunized ewes grazing oestrogenic Yarloop clover during joining showed reduced behavioural oestrus and lowered fertility compared to unimmunized controls on a non-oestrogenic diet. The immunized groups showed the poorest fertility indicating that there was an adverse interaction between phyto-oestrogen intake and steroid immunization. However, in a second experiment, immunized ewes grazing on oestrogenic clover for a season and then joined when grazing dried-off non-oestrogenic pasture, had normal fertility and increased fecundity compared to unimmunized animals. A flock of ewes with permanent clover infertility showed an increase in ovulation rate when immunized against androstenedione but no gain in foetuses carried.

## INTRODUCTION

Phyto-oestrogenic subterranean clover is widespread in southern Australia and its ingestion is known to cause reproductive disturbances known as clover disease in the ewe (Bennetts *et al.* 1946). Clover disease may be of either a temporary (Morley *et al.* 1963) or permanent nature (Barrett *et al.* 1965). The former condition occurs where oestrogenic pasture is grazed prior to and during joining and is associated with a reduction in ovulation rate (Lightfoot and Wroth 1974). On the other hand, permanent clover disease is mainly attributed to fertilization failure resulting from impaired sperm transport (Lightfoot *et al.* 1967).

Immunization against steroids increases ovulation rates (O.R.s.) and lambing rates in ewes grazing non-oestrogenic pastures (Cox *et al.* 1982). Furthermore, it has been reported (Smith *et al.* 1982) that steroid immunization of ewes partly offsets the decreased fecundity resulting from the ingestion of oestrogenic coumestans. It was therefore decided to examine the effect of immunization against oestrone or androstenedione (Fecundin®) in ewes with temporary infertility (Experiment 1), or with a history of grazing oestrogenic clover for 1 year (Experiment 2) or 5 years (Experiment 3). Preliminary findings are reported in this paper.

## MATERIALS AND METHODS

Three experiments were carried out on Parndana Research Station, Kangaroo Island, South Australia.

### Experiment 1

Four hundred Merino ewes from an area with oestrogen free pasture were randomly allocated to the following treatment groups; 1. control, maintained on a non-oestrogenic diet, 2. control, grazed on oestrogenic pasture, 3. oestrone-immune grazed on oestrogenic pasture, 4. androstenedione-immune grazed on oestrogenic pasture. The oestrogenic pasture grazed was dominant *Trifolium subterraneum* cv. Yarloop.

All sheep were confined to a small holding yard and lot-fed oats and hay from February 17th. During this time groups 3 and 4 were immunzied on March 7-8th and March 21st (Cox *et al.* 1982). Booster injections were given 31 and 66 days after the second injection. Ovulation rates were measured by laparoscopy before and after immunization, after treatments 2-4 had grazed oestrogenic pastures for 4 weeks, and at the end of the first mating cycle. Immediately following the second laparoscopy on the 13th April, treatments 2-4 were placed on oestrogenic pasture and joining with 3% rams commenced 23 days later. Treatment 1 remained in the feedlot for joining. At the start of joining, the group remaining in the feedlot weighed 37.8 ± 0.48 kg and the groups grazing Yarloop did not differ significantly in weight and averaged 35.2 ± 0.42 kg. Weekly mating records from all groups were taken for six weeks after which harnessed teaser rams were run with the flock. The sheep in treatment 1 were transferred to a low oestrogen pasture (Trikkala) farms for joining with the teaser rams. The Yarloop pasture was highly oestrogenic by isoflavone analysis and uterine weight bioassay in sheep. Litter size was determined using real time ultrasonic scanning.

### Experiment 2

The ewes from groups 2-4 in Experiment 1 were maintained on oestrogenic Yarloop pasture until the pasture dried off in November 1983 whereas the control group was kept on Trikkala pastures.The immunized groups 3 and 4 were boosted on the 10th February, 1984. All groups were run with vasectomized rams from 15th February, then with 3% harnessed entire rams from 2nd March for eight weeks. Laparoscopies were carried out on the 15th and 21st March and ultrasonic scanning on the 14th and 15th June.

### Experiment 3

The experiment used aged Merino ewes with a history of permanent infertility having grazed oestrogenic pastures for 5 years with a resulting decline in lambing percentages to below 30%. The ewes were either immunized against androstenedione (n = 84) or remained untreated (n = 84). The sheep were mated as one flock on dry pasture two weeks after immunization.

Weekly mating records were taken for eight weeks and O.R.s were obtained during the joining period. Litter size was determined using a real time ultrasonic scanning.

## RESULTS

### Experiment 1

The immunizations produced only small increases in O.R. on 12/13 April because of the initial poor condition of the ewes. The increase in O.R. was significantly above groups 1 or 2 for the oestrone group ($P < 0.05$) but there were no differences in the proportions of ewes ovulating (Table 1). Within the first month on Yarloop a significant decrease ($P < 0.001$) occurred in percentage of ovular ewes for groups 2-4, partly offset in the androstenedione-immune group. Subsequently the percentage of ovular ewes rose to pre-Yarloop values. Behavioural oestrus was markedly reduced ($P < 0.001$) in the Yarloop groups including the immune groups (Table 2). Also, the groups that grazed Yarloop clover had a markedly lower percentage of pregnant ewes compared to the percentage of non-returning ewes ($P < 0.01$); embryonic loss may have contributed to these results. The immune groups had significantly greater losses ($P < 0.02$).

**TABLE 1. The reproductive performance of immune and non immunized sheep grazing oestrogenic Yarloop pasture and of non-immunized sheep grazing low oestrogenic Trikkala pasture.**

| Group (n) | Pre-immune liveweight (kg) | % Ovular ewes (Ovulation rate) | | | |
|---|---|---|---|---|---|
| | | Pre-immune 7/8 March | Post-immune 12/13 April | One month on Yarloop 10/11 May(1) | End first mating cycle(2) |
| 1. Trikkala control (97) | 38.7 ± 0.43 | 63 (1.03) | 80 (1.01) | 96 (1.04) | 93 (1.09) |
| 2. Yarloop control (98) | 38.6 ± 0.46 | 64 (1.03) | 75 (1.00) | 54 (1.04) | 73 (1.04) |
| 3. Androstenedione-immune (97) | 38.6 ± 0.44 | 57 (1.07) | 84 (1.11) | 75 (1.05) | 84 (1.05) |
| 4. Oestrone-immune (94) | 38.7 ± 0.46 | 66 (1.07) | 72 (1.17) | 53 (1.08) | 79 (1.03) |
| $\chi^2$ (ovular ewes) | | N.S. | N.S. | *** | * |

(1) After 1 month on Yarloop for groups 2, 3, 4.
(2) May 23rd; mating began 23 days after the grazing treatments commenced.

### Experiment 2

The two control groups had similar ovulation rates (Table 3) and the immune groups showed substantial increases in O.R. Almost all animals mated which is in contrast to groups 2, 3, 4 on Yarloop clover in 1983 (Table 2). Fertility was similar for all groups with the immunized groups showing considerable gains in numbers of foetuses present.

### Experiment 3

The ewes showed a good response to Fecundin® treatment with O.R.s of 1.37 and 1.93, with 93 and 97% mating for the control and treated groups. However, ultrasonic scanning showed only 20.6% and 10.6 respectively of the ewes were pregnant. These pregnancy rates were not significantly different.

## DISCUSSION

Temporary infertility effects were clearly observed in Experiment 1 with joining on green Yarloop. The transient decrease in percentage ovular ewes in the first month of grazing Yarloop clover can be ascribed to an initial depression due to phyto-oestrogen intake followed by adaptation by rumen microflora with partial metabolic inactivation of the oestrogens. Immunization had no beneficial effects, but further depressed the percentage of ewes having foetuses aged 50 days or more, probably due to embryonic loss since the majority of mated ewes did not return to service. Thus Fecundin® should not be used for ewes being joined on green, highly oestrogenic pasture. Effects for weakly oestrogenic pasture remain to be explored.

On the other hand, following exposure to a season of grazing highly oestrogenic Yarloop pasture, immunized ewes joined on dry non-oestrogenic pasture (Experiment 2) showed increases in O.R. and foetal numbers.

**TABLE 2. Fertility and ultrasonic data for immune and non-immunized sheep grazing oestrogenic Yarloop pasture and in non-immunized sheep grazing low-oestrogenic Trikkala pasture.**

| Group (n) | total mated ewes (%) | % ewes not returning to service[1] | % pregnant ewes on scanning 7/9 August. | % Ewes[2] which lost all embryos | No. foetuses per pregnant ewe |
|---|---|---|---|---|---|
| 1. Trikkala control (97) | 87 | 84 | 71[a] | 15 | 1.07 |
| 2. Yarloop control (98) | 53 | 51 | 29[b] | 44 | 1.18 |
| 3. Androstenedione-immune (97) | 54 | 46 | 11[c] | 76 | 1.18 |
| 4. Oestrone-immune (94) | 52 | 46 | 14[c] | 70 | 1.00 |
| $\chi^2$ | *** | *** | *** | *** | |

[1] Ewes which mated and did not return either to entire rams during joining or to vasectomized rams during July.
[2] % of ewes barren on scanning but deemed pregnant on non-returns to service data.
Groups with different superscripts are significantly different at the $P < 0.05$ level.

**TABLE 3. Reproductive performance in 1984 for immunized and non-immunized ewes which had grazed pastures of high or low oestrogenicity for a whole season.**

| Group (n) | Live weight (kg) | % ewes mated | Ovulation rate at mating | % pregnant ewes on scanning | No. of foetuses per ewe |
|---|---|---|---|---|---|
| 1. Trikkala control (89)[1] | 45.0 ± 0.49 | 99 | 1.15[a][2] | 86 | 1.03[a] |
| 2. Yarloop control (87) | 43.8 ± 0.51 | 97 | 1.11[a] | 92 | 1.06[a] |
| 3. Androstenedione-immune (94) | 44.7 ± 0.49 | 99 | 1.64[b] | 89 | 1.19[b] |
| 4. Oestrone-immune (96) | 45.3 ± 0.51 | 95 | 1.64[b] | 93 | 1.39[c] |

[1] Ewes had previously grazed Trikkala pastures (Group 1) or Yarloop pastures (Groups 2, 3, 4).
[2] Groups with different superscripts are different at the $P < 0.05$ level.

Such joining on dry pasture corresponds to the usual industry situation in South Australia and Western Australia. Further, it is of interest whether continued immunization over several seasons might reduce the proportion of ewes that could develop permanent infertility.

A flock of ewes showing permanent clover infertility, although responding to Fecundin® immunization with an increase in O.R. did not carry more foetuses. This finding suggests that the immunization did not improve the impaired transport of spermatozoa through the cervix which is the main cause of permanent clover infertility (Turnbull *et al.* 1966). Use of Fecundin® is thus not beneficial for flocks which have grazed highly oestrogenic clover over several seasons and are showing low lambing percentages.

## REFERENCES

Barrett, J.F., George, J.W. and Lamond, D.R., 1965. *Aust. J. Agric. Res., 16*, 189-200.
Bennetts, H.W., Underwood, E.J. and Shier, F.L., 1946. *Aust. Vet. J., 22*, 2-12.
Cox, R.I., Wilson, P.A., Scaramuzzi, R.J., Hoskinson, R.M., George, J.M. and Bindon, B.M., 1982. *Proc. Aust. Soc. Anim. Prod., 14*, 511-514.
Lightfoot, R.J., Croker, K.P. and Neil, H.G., 1967. *Aust. J. Agric. Res., 18*, 755-765.
Lightfoot, R.J. and Wroth, R.H., 1974. *Proc. Aust. Soc. Anim. Prod., 10*, 130-134.
Morley, F.H.W., Axelsen, A. and Bennett, D., 1963. *Nature, London, 199*, 403-404.
Smith, J.F., Cox, R.I., McGowan, L.T. and Wilson, P.A., 1982. *Proc. Aust. Soc. Reprod. Biol., 14*, 95.
Turnbull, K.E., Braden, A.W.H. and George, J.M., 1966. *Aust. J. Agric. Res., 17*, 907-917.

# POTENTIAL BIOLOGICAL INACTIVATION OF PASTURE OESTROGENS IN THE ANIMAL — IMMUNIZATION OR METABOLIC DEFLECTION

R.I. Cox, *CSIRO, Division of Animal Production, PO Box 239, Blacktown, N.S.W. 2148.*
W.J. Collins, *Department of Agriculture, Jarrah Road, South Perth, W.A. 6151.*
P.A. Wilson and M.S.F. Wong, *CSIRO, Division of Animal Production, PO Box 239, Blacktown, N.S.W. 2148.*

## INTRODUCTION

Phyto-oestrogens (P-E) present in pastures still result in considerable infertility and thus production losses in grazing sheep in Australia. Such reproductive losses remain a largely intractible problem which can be held in check by agronomic measures and by sheep management methods (Lightfoot 1974).

An effective animal treatment to alleviate the biological effects would be a useful adjunct to agronomic methods. Amongst possible control measures are alteration of ruminal metabolism of P-E and immunization of sheep against P-E. This paper will survey the experiments that have been carried out to neutralise the effects of P-E in sheep by active immunization directly against isoflavones and isoflavans or to counter the biological effects of P-E indirectly by immunization of sheep against steroid hormones.

As another approach, the possibility of altering ruminal metabolism of P-E is considered. Metabolic control is of renewed interest because of the work of Jones and Megarrity (1983) in finding that toxic mimosine metabolites derived from the legume *Leucaena leucocephala*, can be detoxified by rumen micro-organisms present in goats in Hawaii but not in ruminants in Australia.

## OESTROGENIC ACTIVITY IN FORAGE LEGUMES

Some pasture legumes contain oestrogenically active compounds. These legumes are subterranean clover (*Trifolium subterraneum* L.), and red clover (*Trifolium pratense* L.), which contain isoflavones, together with lucerne (*Medicago sativa* L.) and annual medics, such as barrel medic (*Medicago truncatula* Gaertn.) which contain coumestans. Oestrogenic problems in Australia are largely associated with formononetin-rich varieties of subterranean clover, particularly where it is the dominant component of a sward. In its native habitat in the Mediterranean region, subterranean clover is usually only a minor component of vegetation and hence unlikely

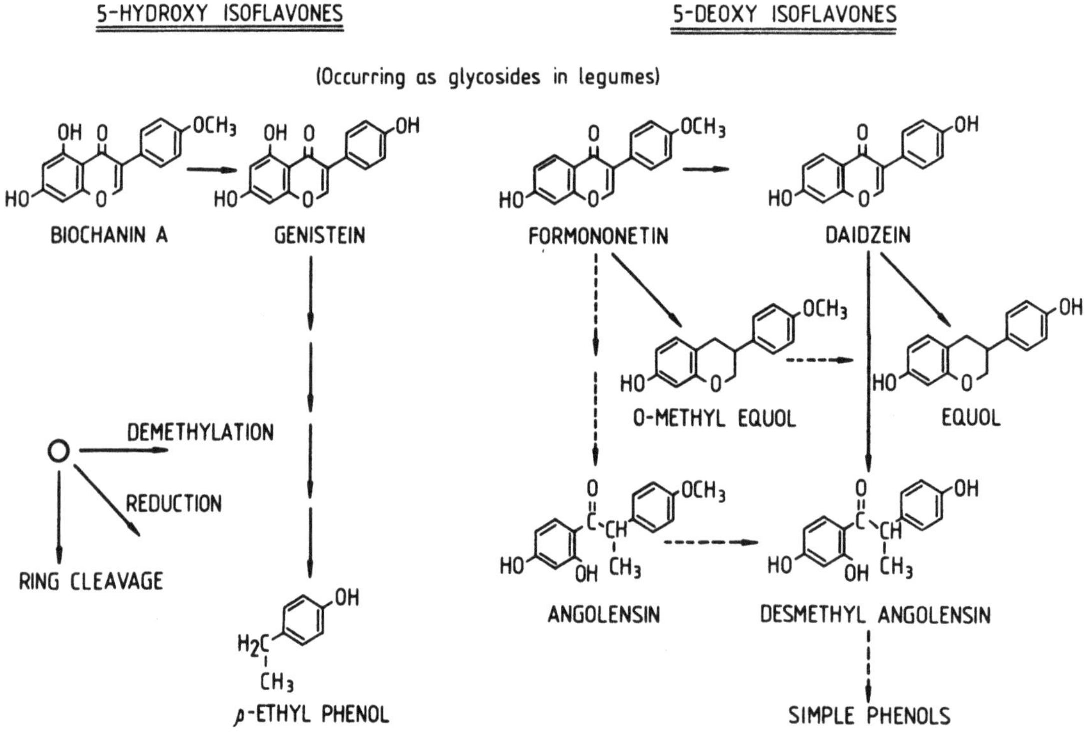

**FIGURE 1. Metabolic conversions of isoflavones in the sheep. Broken arrows indicate probable changes. The reactions in the metabolic lattice are shown in the insert.**

to cause acute oestrogenic problems similar to those seen in Australia. However, less conspicuous but economically significant depression of sheep fertility may occur (Katznelson 1974).

## METABOLISM OF ISOFLAVONES

Subterranean clover and red clover varieties contain isoflavones of the 5-deoxy group, formononetin and daidzein, and of the 5-hydroxy group, biochanin A and genistein (Figure 1). These occur as glycosides which are rapidly hydrolysed during mastication. The major metabolic conversions occur in the rumen of sheep (reviewed Collins and Cox 1985), and for formononetin and daidzein this involves mainly reduction with limited ring cleavage. The main oestrogenically active metabolites are equol and O-methyl equol with some desmethylangolensin. The 5-hydroxy compounds show a markedly different pattern, with demethylation and ring cleavage as the major conversions leading to biologically inactive products such as p-ethylphenol.

Isoflavones and metabolites are conjugated in the liver to form glucosiduronates and sulphates. Only a very small portion, usually less than 1% remains in the unconjugated, biologically active form and the small proportion of sulphoconjugate is also biologically available.

## IMMUNIZATION OF SHEEP AGAINST PHYTO-OESTROGENS

Antibodies specific for P-E have been obtained by immunizing animals with P-E derivatives conjugated to synthetic polypeptides or bovine serum albumin (Baumiger *et al.* 1969; Cox *et al.* 1972). An absolute prerequisite of immunization is that it should not result in any adverse effect on the normal reproductive cycle. Immunization against P-E showed no effects on the cycle (see Cox *et al.* 1983).

By immunizing sheep against P-E it has been possible to block the action of genistein when injected (R.I. Cox, A.W.H. Braden and D.A. Little, unpublished data) or equol when infused (Cox *et al.* 1983). To examine the effectiveness of neutralizing equol, ovariectomized ewes were infused for 72 h via a jugular vein with 70% propylene glycol-saline vehicle, either alone (controls) or with equol (42 mg or 125 mg total). Equol was also infused into ewes which had been immunized against the carboxylic analogue of equol. Infusion of equol into non-immunized ewes resulted in rises in uterine RNA:DNA ratios (a typical response to oestrogens) while in immunized ewes the response was significantly weaker. This experiment provided a clear demonstration of the ability of the immunization to neutralize substantial, biologically active amounts of equol.

However, oestrogenic effects on the uterus were not reduced when ovariectomized ewes immunized against equol were grazed on oestrogenic Dinninup pastures for several weeks (R.I. Cox, R.J. Lightfoot, K.P. Croker and P.A. Wilson, unpublished data). Despite high plasma antibody titres (mean 1:71,000 for 19 ewes) the uterine weight of ewes grazing on the clover was $50.6 \pm 2.8$ g ($n = 19$) for the immunized ewes compared to $55 \pm 4.3$ g ($n = 11$) for untreated ewes (no significant difference). Untreated ewes on grass pasture had uterine weights of $26.1 \pm 3.6$ g ($n = 9$). When the 10 immunized ewes with the highest titres were assessed, uterine weights, although lower, were still not significantly different to those of untreated ewes on oestrogenic pasture.

In wethers enlargement of the bulbourethral gland can be used as a measure of cumulative oestrogenic effect over several months grazing (Obst 1971). A study of the effectiveness of immunizing wethers against equol prior to their grazing Yarloop clover, showed that the increase in bulbo-urethral gland size during four months grazing was significantly less in immunized than in unimmunized wethers (R.I. Cox, P.A. Wilson and J.M. Obst, unpublished data).

The lack of protection of immunized ewes at pasture may have been due to inadequate capacity of the antibodies to neutralize the large quantities of plant derived oestrogen, over the full period of exposure. Further tests with infused or injected isoflavones over longer periods of time would provide useful data. Studies on the dynamics of metabolism and clearance of isoflavanoids in the immunized animal are also needed. Further, immunization procedures may need to be directed against several compounds, rather than equol alone.

## IMMUNIZATION OF SHEEP AGAINST STEROIDS

Immunization of ewes against steriods produces an increase in ovulation rate and in lambing percentage (Cox *et al.* 1982). Coumestan intake decreases the ovulation rate of ewes (Smith *et al.* 1979). Thus it was of interest to test whether steroid immunization could offset the reproductive losses resulting from coumestan intake. An experiment with 600 ewes in New Zealand showed that immunization against oestrone or androstenedione could partly counteract the effect of ingested coumestrol on ovulation rate and lambing percentage (Smith *et al.* 1982). If androstenedione immunization consistently alleviates coumestan effects in sheep, a suitable commercial product, Fecundin® (Glaxo Australia Ltd) is now available for practical applications.

In contrast, steroid immunization has not been effective in countering the effects of isoflavones when animals were grazed on green clover at the time of joining (Little *et al.* 1984).

## METABOLIC DEFLECTION

The oestrogenicity of isoflavones and their metabolites varies widely (see Shutt and Cox 1972, Nottle and Beck 1974, Cox *et al.* 1983).

Ruminal metabolism of formononetin and daidzein to less oestrogenic metabolites might be achieved in a number of ways, including:

1. improving conversion to desmethylangolensin and further products and lessening equol production. This approach has possibilities as the conversions already occur and are analogous to the pathway that so efficiently degrades genistein. With daidzein it would be necessary to enhance ring cleavage instead of reduction.

2. inducing conversions that catabolize equol. Although catabolism of equol has not been observed this possibility needs further consideration. Ring cleavage of equol would be expected to be more difficult than that of the more highly oxygenated daidzein.

3. varying metabolism in order to produce O-methyl equol. Its ratio with equol is already known to vary between animals (see Cox *et al.* 1983) but factors that influence the ratio have not been explored. This approach is only of interest if the two compounds differ by several fold in oestrogenic activity in sheep.

To bring about such alterations in metabolism several approaches may be possible. By analogy with the work of Jones and Megarrity (1983) on mimosine metabolites, evidence for a more effective degradation of formononetin by rumen bacteria not found in Australia could be sought overseas, particularly in areas where *Trifolium* species are indigenous. The yield of equol from formononetin would be a key marker. Apart from this, it is desirable to evaluate the variations of P-E metabolites that occur in sheep grazing in different regions of Australia. Factors, possibly including dietary constituents, that influence rumen metabolism of P-E may provide methods of controlling such conversions.

In addition, longer term studies may lead to methods based on mutated or genetically modified micro-organisms for controlling metabolism of P-E in the rumen.

## REFERENCES

Bauminger, S., Lindner, H.R., Perel, E. and Arnon, R., 1969. *J. Endocrinol., 44*, 567-578.

Collins, W.J. and Cox, R.I., 1985. *In* Barnes, R.F., Brougham, R.W., Marten, G.C. and D.J. Minson (eds) *Forage Legumes for Energy-efficient Animal Production*. U.S.D.A. (in press).

Cox, R.I., Wilson, P.A. and Mattner, P.E., 1983. *Toxicon, Suppl. 3*, 85-88.

Cox, R.I.. Wilson, P.A., Scaramuzzi, R.J., Hoskinson, R.M., George, J.M. and Bindon, B.M., 1982. *Proc. Aust. Soc. Anim. Prod., 14*, 511-514.

Cox, R.I., Wong, M.S.F., Braden, A.W.H., Trikojus, V.M. and Lindner, H.R., 1972. *J. Reprod. Fert., 28*, 157-158.

Jones, R.J. and Megarrity, R.G., 1983. *Aust. J. Agric. Res., 34*, 781-790.

Katznelson, J., 1974. *Israel J. Bot., 23*, 69-108.

Lightfoot, R.J., 1974. *Proc. Aust. Soc. Anim. Prod., 10*, 113-131.

Little, D.L., Cox, R.I., Walker, S.K. and Flavel, P.F. This volume, 365-367.

Nottle, M.C. and Beck, A.B., 1974. *Aust. J. Agric. Res., 25*, 509-514.

Obst, J.M., 1971. Hormone studies on ewes grazing Yarloop clover. Ph. D. Thesis, University of Adelaide.

Shutt, D.A. and Cox, R.I., 1972. *J. Endocrinol., 52*, 299-310.

Smith, J.F., Cox, R.I., McGowan, L.T. and Wilson, P.A., 1982. *Proc. Aust. Soc. Reprod. Biol., 14*, 95.

Smith, J.F., Jagusch, K.T., Brunswick, L.F.C. and Kelly, R.W., 1979. *N.Z.J. Agric. Res, 22*, 411-416.

# ECONOMIC VALUES OF CHANGING REPRODUCTIVE RATES

D.H. White, *Department of Agriculture, Animal Research Institute, Werribee, Victoria, 3030.*

## INTRODUCTION

The primary purpose of this paper is to estimate changes in flock productivity and net farm income associated with changing the number of lambs born and reared within each year. Lambing percentage (%) is synonymous with lambs weaned relative to ewes joined in this paper.

The economic value of changing reproduction rates depends, in part, on how that change is realised. To illustrate the point, I will first examine responses in flock structure to reproduction rate. Then I will consider productive and economic responses to an increase in ovulation rate and lambing performance. This will be done a) at a particular time of year, b) through a shift in the date of lambing.

Most of the published economic studies have been relatively simple gross margins analyses in which the effects of lambing % on flock structure and assumed output were considered, ignoring the effects of extra lambs on grazing pressure and ewe liveweights, both during lactation and at subsequent mating. However, these effects can be vitally important to the overall productivity and profitability of a sheep production system. I have therefore resorted to using our breeding ewe flock model (White *et al.* 1983) to estimate the likely productive and economic responses to the management treatments imposed.

## FLOCK STRUCTURE

The effect of a higher reproduction rate in a self-replacing flock is to allow the number of breeding ewes on a property to be reduced, thereby allowing a larger wether flock to be maintained (Byrne 1964). Alternatively, a higher proportion of the offspring may be sold, allowing increased selection pressure on the replacement ewes and increased revenue from sales. These decisions can affect the overall profitability of the sheep enterprise and its vulnerability to adverse seasons. Byrne (1964) showed that the proportion of ewe hoggets required for flock replacements, n, in a self-replacing flock of constant size may be determined as

$$n = 2/(s_l b(1 + s + s^2 + .... + s^{a-1}))$$

where $b$ = lambing rate
$s_l$ = survival rate of lambs (< 18 months of age)
$s$ = survival rate of ewes
$a$ = average number of years the ewes are joined

Thus the proportion required decreases as the lambing rates, survival rates and productive life of the ewes increase. Below a lambing rate of about 0.6, depending on the other parameters, there are insufficient ewe hoggets to maintain the size of the ewe flock and ewes must be purchased from elsewhere. When $b > 0.9$, n is relatively insensitive to changes in b.

This is an important basic relationship for it determines the feasible rate of stock increase for any breeding flock. If flock size is static, sales of cull ewe hoggets and cast-for-age wethers increase whereas sales of cast-for-age ewes decrease. The proportion of ewes in each age group does not change.

Egan *et al.* (1972) studied the effects of sex and age structure, reproductive performance and selling policies for wethers on the level of production and economic returns from self-replacing sheep flocks. Adjustments were made for the effects of age and lambing performance on the wool production of the ewes. Manipulation of age and sex structure in existing flocks resulted in relatively small gains in profitability. However, improved reproductive performance within the various flock structures resulted in increased sales of weaner sheep of both sexes. There was a shift in the predicted gross margins of approximately 3 *per cent* for each 10 *per cent* change in reproductive rate.

A more comprehensive study was undertaken by Jardine *et al.* (1975) using a linear programming model. Three fertility levels were compared for a self-replacing flock of Corriedale ewes. Ewes and wethers could be culled up to 7 years of age.

The study concluded that the optimal structure shifts from disposing of wethers as lambs and ewes as 5½-year olds to disposing of wethers as 3½-year olds and ewes as 4½-year olds as fertility declines, or meat declines in value relative to wool. However, the former structure tends to dominate and is optimal over a wide range of conditions.

It must be appreciated that in this model, nutrition is assumed to be adequate for all animals regardless of the flock structure. This would be unrealistic in many grazing situations, since the feed requirements of lambs, wethers and dry, pregnant and lactating ewes vary greatly.

## REVIEW OF PAST ECONOMIC STUDIES

### Wool production systems

Estimated gross margin responses to an increase in lambing % are shown in Table 1. As the numbers of lambs born increases, and more sheep are kept to 1.5 years of age, then the total number of D.S.E's (Dry Sheep Equivalents) also increases. Gross margins are therefore often expressed per D.S.E. so that the response is not confounded with that attributable to an increase in stocking rate.

**TABLE 1. Published estimates of *percentage increase* in gross margin in response to increasing lambing from 80 to 90% in self-replacing Merino flocks.**

| Authors | | Gross Margins | |
|---|---|---|---|
| | \$ ewe$^{-1}$ | \$ D.S.E.$^{-1}$ | \$ ha$^{-1}$ |
| Heath and Lee (1975) | — | 2.0 | — |
| Clarke (1977) | — | 2.7 | — |
| Thatcher and Rabbette (1977) | 7.9 | — | — |
| Bell (1981)[1] | — | 12.6 | 12.6 |
| Mangelsdorf (1982) | — | — | 4.3 |
| Patterson and Wagg (1983) | 9.0 | 5.0 | 5.0 |

[1] cited Bell (1984)

Changes in flock structure must be taken into account when interpreting these results, Thus, in the Patterson and Wagg (1983) analysis, the size of the breeding flock as a proportion of total sheep was reduced by 3% for each 10% increase in lambing. This was accompanied by a 1.6% increase in both the ewe and wether hoggets, increases of 10.4% and 136% in the sales of wethers and cull ewes and a 3.7% decline in the number of cast-for-age ewes.

Using 1983 prices, Patterson and Wagg estimated that a 10% increase in weaned lambs would increase gross margin per Merino ewe by \$3.16. Assuming half of this increase must be allocated to profit, they estimated that a producer could spend up to \$1.58 per ewe on supplements, rams, shelter or other technologies which could achieve this increase in lambing %.

Sheep meat production systems

With prime lamb production systems there is no effect of an increase in reproductive rate on the number of ewes within the flock. The effects of flock structure are therefore confined to an increase in the number of ewes rearing twin lambs, with a consequent reduction in those either dry or rearing singles (Thatcher 1977). Because these twin lambs are likely to be smaller in size and grow more slowly, it is probable that they will be kept longer on the farm for sale at a later date. The first draft of lambs sent to market will therefore probably be reduced. The later lambs may have to be carried over the summer on dry pasture and fed supplements. Alternatively, they may be grazed on special-purpose pastures or forage crops to get them into saleable condition.

Seasonal variations in the price of lamb vary with supply and demand. The most common trend is for high prices in autumn and winter with a price trough in November. Since date of lambing affects reproductive rate and the quantity of lambs that may be marketed at different times of the year (Reeve and Sharkey 1980), this price trend can have a large effect on the optimal management strategy and economic viability of a prime lamb flock (Thatcher 1977; White and Reeve 1983).

Gross margin responses per ewe, per D.S.E. and per ha are usually identical in analyses of prime lamb production systems. Estimates of the response to increasing lambing from 100 to 110% include 11% (Heath and Lee 1975), 7.6 and 15.1% for autumn and spring-lambing systems respectively (Thatcher 1977) and 8.5% (Patterson and Wagg 1983).

Given 1983 costs and prices, Patterson and Wagg estimated this 8.5% increase to be worth \$1.91 per first cross ewe. Although this is less than the comparable figure for a wool production system, it must be appreciated that the profit per Merino ewe is higher because the profits of the wethers have also been allotted to them. There are also substantial year-to-year variations in the relative profitabilities of wool and prime lamb enterprises (Patterson 1984).

Transfer of resources between enterprises.

At the national level, Love *et al.* (1982) estimated that to increase lamb marking by 10% concurrently across the High Rainfall, Wheat-Sheep and Pastoral Zones of Australia would result in an increase in grazing industry net income for the sheep enterprises of 3.5%. However, because this would be associated with a transfer of resources such as labour and capital between zones, for Australia as a whole the gross value of output of wool enterprises would decline by 1.5% whereas that for sheep meat would increase 13.3%

G. Love and K. Williams (pers. comm.) elaborate: 'the increase in the lambing rate would increase the profitability of the other sheep enterprises relative to wool in the wheat-sheep zone and sheep meat in the high rainfall zone. As the other sheep enterprises expand using the resources newly freed from the wool enterprise, these resources are likely to be used at a rate different to that at which they were used in the wool enterprise'.

Their regional programming model (Longmire *et al.* 1979) uses a recursive linear program to generate and analyse cash flows over time for the sheep, beef and cropping industries. Its decentralised structure (13 regions) permits regional differences in productivity to be reflected in the overall national solution.

The animal component of this model is outlined in Figure 1 of the paper by Love *et al.* (1982). Feed supply is expressed in livestock-months, the amount of surplus feed transferred between calendar quarters being based on assumed rates of pasture deterioration. This Figure needs to be carefully examined when assessing this model, so that its strengths and limitations are clearly understood, particularly in a biological sense.

### Studies overseas

Sheep enterprises in North America and Europe are primarily geared to lamb meat production. With lambs in the U.K. currently being marketed for the equivalent of $65 to $85 each, depending largely on the time of year, there are clearly many technological opportunities to increase reproductive efficiency profitably (Peyraud and Manno 1975; Kilkenny 1976; Geisler *et al.* 1977; Gerrits *et al.* 1978; Flanagan *et al.* 1979; Inskeep and Peters 1981).

In New Zealand there is considerable emphasis on selection for reproductive rate, both within breeds (Morris *et al.* 1983) and through crossing to proposed importations such as the Finnish Landrace (Bushnell and Hutton 1982).

## SELECTION FOR REPRODUCTIVE RATE

The economic consequences of genetic improvement of reproductive rate are examined in other papers in this conference. There is disagreement in the literature as to the economic importance of genetic improvement in reproductive rate. The following examples will illustrate this point.

Ponzoni (1979) contends that for a breeding flock in which surplus offspring are sold as lambs, the number of lambs weaned and clean fleece weight will probably be the most important traits. However, a relatively low stocking rate is implied, since he assumes that genetic improvement in reproductive rate does not result in increased feed requirements or costs and that the lambs are sold early. In this case, the purchase of extra stock is likely to be more profitable than selection for twins.

Jones (1982), however, states that in any breeding programme in which liveweight or reproductive rate are changed, changes in feed requirements should be taken into account. Selection indices may therefore be developed by reducing the financial returns from wool and lamb production to allow for the extra feed consumption of the ewes. Jones considered probable changes in feed requirements, ewe numbers and the fleeceweights of ewes, hoggets and weaned lambs in response to an increase in reproductive rate, the surplus lambs being sold after shearing at 16 months of age. From this he concluded that most economic gains from genetic selection would come from clean fleece weight and fibre diameter.

A more valid approach is to put an economic weight on feed consumption (James 1982). Returns occurring later in the breeding programme should also be discounted more than those occurring early (James 1978). Thus the benefits from a higher reproductive rate may be discounted by 1 year to their value at the birth of the lamb. Jones (1982) also used this model to confirm his analysis.

From these studies, Jones concluded that it was necessary to consider the changes in feed requirements that a particular selection programme may induce. However, the effects of increased reproductive rate on feed requirements will vary between seasons, dates of lambing, production systems and environments, being less for producers selling surplus lambs at weaning, especially if surplus feed is available.

## A SIMULATION MODEL OF A BREEDING EWE FLOCK

Simulation models can be useful in studying responses to a change in reproductive rate because an increase in the number of lambs born results in a greater demand for feed. Not only is there an increase in the consumption of milk and grass by the lamb, but the intake of twin-rearing, lactating ewes may be 1.5 times that of single-rearing ewes (Alexander and Davies 1959). This means that the mean liveweight of the ewe flock and the availability of pasture herbage at weaning will almost certainly be reduced. If ewe liveweights at subsequent mating are reduced, the level of production of the flock in the following year will be affected. Furthermore, the nutrient demands of pregnancy and lactation reduce the wool production of the ewe (Corbett 1979). There are therefore many interactions between the number of animals, including offspring, and the pasture which will affect the productivity and profitability of the flock as a whole. The significance of these effects will vary with the seasons and the management treatments, such as stocking rate and date of lambing, imposed. These effects may be measured in a large field experiment that runs for many years, or by using dynamic simulation models, such as we have developed, that estimate changes in available herbage, animal production and cash flow in different seasons and environments.

### Description of the model

The original model of a breeding ewe flock is described in detail by White *et al.* (1983). I will therefore outline only those components and modifications that are relevant to this study.

In this series of simulated experiments, a property of 500 ha of perennial ryegrass and subterranean clover in the Hamiltion District of Victoria was grazed by self-replacing flocks of Saxon Merino ewes. Stocking rate, date of lambing and potential ovulation rate were varied as required. Rainfall, temperature and pan evaporation data recorded from 1965 to 1983 at the Pastoral Research Institute, Hamilton, were used as input data to the model, together with seasonal trends in wind speed and variability. The pasture growth relationships described by Bowman *et al.* (1982) were used to predict the amount of herbage available for grazing. Intake was varied according to the weight, age and physiological state of the animals and the quantity and digestibility of herbage or supplement on offer. Wool growth varied with intake, demands of pregnancy and lactation and photoperiod (White *et al.* 1979).

Ovarian activity was predicted relative to time of year (Radford 1959) and improved nutritional status expressed through liveweight change (Morley *et al.* 1978). Potential ovulation rate was varied by changing both TM, the maximum incidence of twin-ovulating ewes, and B, where (1-B) is an 'inhibitory factor' which intervenes to convert the potential shedding of 2 ova to 1 ovum only (Equation 38, White *et al.* 1983). The probability of fertilized ova surviving was assumed to conform to a binomial distribution, survival rates of 90% and 80% being assumed for ova from single-and twin-ovulating ewes, respectively (White *et al.* 1981).

Lamb survival was determined from biological relationships derived by Nixon-Smith (1972) and revised by Donnelly (1984), together with analyses of data from Obst and Day (1968) and the Pastoral Research Institute by M. Whelan (pers. comm.).

This model (REPROSX) differs from that used in the northern Victorian study in the climatic data, the potential growth rate of the pasture throughout the year, and the lower potential intake and fleeceweight of the Saxon Merino sheep. Wether lambs were sold at 5 to 8 months and culled ewes at 14 to 18 months of age. Replacement ewes entered the flock when they were 18 months old. Current costs and prices were assumed and are obtainable from the author.

Experimentation with the model

The model was run for two levels of fertility over 18 years at 6, 8 and 10 ewes, plus replacements, per ha (Table 2), assuming a June-early July lambing system (White *et al.* 1982). This period included two major droughts. The need for supplements increased exponentially with stocking rate and also varied with the fecundity of the ewes. There was an increase in gross margin per ha of about $8 ($1.08 per ewe at 8 ewes per ha) or 7.5% for each 10% increase in lamb numbers, provided the most profitable stocking rate was not exceeded. A decrease in gross margin was observed when the model was rerun at 12 ewes per ha, because the extra lambs were associated with uneconomic levels of supplementary feeding.

**TABLE 2. Reproduction parameters, grain fed out and profitability of a 500 ha property in western Victoria at three stocking rates for low and high fertility sheep. (Mean of 18 years; standard deviations in parentheses; Simulation model).**

| Variable | Fertility | Stocking rate (Ewes $ha^{-1}$)[1] | | |
|---|---|---|---|---|
| | | 6 | 8 | 10 |
| Lambing % | Low | 80(2) | 80(3) | 77(3) |
| | High | 94(3) | 94(4) | 91(4) |
| Ewe weight at | Low | 55(3) | 54(4) | 49(5) |
| joining (kg) | High | 55(3) | 53(4) | 48(5) |
| Supplement | Low | 2(7) | 5(13) | 61(112) |
| (tonnes) | High | 3(9) | 10(20) | 77(127) |
| Gross margin | Low | 102(9) | 117(17) | 89(55) |
| ($ $ha^{-1}$) | High | 113(12) | 129(22) | 95(60) |
| Net Farm | Low | 21(2) | 26(2) | 20(17) |
| Income ($000s) | High | 23(2) | 28(3) | 21(19) |
| Response[2] | | $7.9 | $8.6 | $4.3 |

[1] Actual stocking rate is higher, given lambs and flock replacements
[2] Increase in gross margin ($ $ha^{-1}$) for a 10% increase in lamb numbers.

The high standard deviations for supplement, particularly at the high stocking rate (Table 2), reflect the variability in the system imposed by the environment, i.e. droughts, In turn, this results in high standard deviations in the gross margins and net farm income.

The model was also run at 8 ewes per ha with lambing in April-May, June-early July and late July-August (Table 3). Increases in gross margin per ha were 13.7, 7.4 and 6.3% respectively, for a 10% increase in lamb numbers.

**TABLE 3. Reproduction parameters, grain fed out and profitability of a 500 ha property in western Victoria at three dates of lambing for low and high fertility sheep. (Mean of 18 years: 8 ewes ha$^{-1}$; *standard deviation in parentheses; Simulation model*).**

| Variable | Fertility | Date of lambing May | June | August |
|---|---|---|---|---|
| Lambing % | Low | 72(3) | 80(3) | 79(3) |
| | High | 87(4) | 94(4) | 93(4) |
| Ewe weight at | Low | 51(5) | 54(4) | 50(3) |
| joining (kg) | High | 51(5) | 53(4) | 49(4) |
| Weight change | Low | +0.9(.4) | 0.0(.7) | −1.8(.4) |
| pre-joining | High | +0.8(.3) | 0.0(.7) | −1.8(.4) |
| Supplement | Low | 34(46) | 5(13) | 4(13) |
| (tonnes) | High | 29(47) | 10(20) | 7(22) |
| Gross margin | Low | 78(31) | 117(17) | 112(10) |
| ($ ha$^{-1}$) | High | 94(32) | 129(22) | 122(14) |
| Net Farm | Low | 17(8) | 26(2) | 24(2) |
| Income ($000s) | High | 21(7) | 28(3) | 26(3) |
| Response(1) | | $10.7 | $8.6 | $7.1 |

(1) Increase in gross margin ($ ha$^{-1}$) for a 10% increase in lamb numbers

Field study of sheep production systems

Our simulation model is currently being tested, with very encouraging results, against pasture and livestock measurements recorded in the Sheep Production Systems experiment at the Pastoral Research Institute, Hamilton (White and Bowman 1983). This experiment involves small- and large-framed sheep, Saxon Merinos and Comebacks respectively, subdivided into low and high fertility treatments. The treatments are further subdivided into autumn- and spring-lambing systems (Obst *et al.* 1984).

## RESPONSE TO A CHANGE IN THE DATE OF LAMBING

One way of increasing reproductive rate is to change from an autumn-lambing to a winter- or spring-lambing system, thereby taking advantage of a high ovulation rate in autumn. The long-term, economic consequences of such a shift have been estimated for hypothetical sheep properties in northern and western Victoria using our simulation models.

In our study of Peppin Merino ewes grazing annual pastures in northern Victoria (White *et al.* 1983), autumn- and spring-lambing systems were less profitable than winter-lambing systems. The profit margin associated with a change in the date of lambing was negligible at the lowest stocking rate. At the most profitable stocking rate, mean net farm income was increased by 15% in moving from an autumn to a winter-lambing. However, this was a small response relative to that achieved by changing stocking rate. It must also be stressed that no special-purpose perennial pasture or crop was available for summer grazing which may have favoured a spring-lambing system (Reeve and Sharkey 1980).

Similar responses were recorded in our western Victoria study for Saxon Merino ewes grazing perennial pastures (White *et al.* 1982). In this environment a late spring lambing was as profitable as the winter-lambing system. Lambing in August was intermediate in profitability between the autumn and other lambing systems, because of the possibility of high perinatal lamb losses, indicating that without adequate shelter at lambing, lamb losses could erode farm income. It was concluded that differences between farms in the growth of winter pasture and shelter for lambing ewes could be decisive in selecting an appropriate lambing date in western Victoria.

## ECONOMIC VALUE OF AIDING LAMB SURVIVAL

High lambing percentages are associated with lower rates of lamb survival. Since lamb losses of the order of 25% often occur, the provision of adequate shelter or intensive lambing systems to reduce losses to less than 15% seems attractive. Blackburn and Rizzoli (1968) estimated whether the costs of providing close supervision and shelter at parturition exceeded the value of the lambs saved. They demonstrated that the economics of intensive lambing depended on the percentage of extra lambs that would probably be saved and the expected net value of these lambs. Capital, labour and feed costs should be kept to a minimum. Their analysis has been updated by Clarke (1983).

## DISCUSSION

This review has shown that merely increasing the reproductive potential of ewes to produce more lambs within a traditional wool production system is likely to result in only modest increases in net farm income. However, there is often considerable scope to improve the reproductive performance and financial viability of the flock through changes in lambing date, stocking rate, feeding strategy and the provision of suitable shelter at lambing. Under such circumstances, there is likely to be greater scope for benefiting from more fertile sheep.

Simulation models, such as I have described, provide considerable opportunities to identify profitable management strategies. Whereas conventional gross margins analyses are reliant on the correct assumptions being made as to the biological and economic consequences of a change in management, these are predicted by the model over a wide range of seasons. Opportunities for manipulating the feed supply to improve reproductive performance and financial returns may therefore be identified. For example, Table 3 of White *et al.* (1983) shows ewe liveweights declining prior to joining when mating in late summer or autumn is practised. Field and simulation studies might therefore be undertaken to determine whether a financially viable feeding strategy can be developed that will prevent this drop in weight and increase lambing %. The opportunity also exists to combine biological and genetic models to assess alternative breeding strategies. However, clearer definition will often be required of differences between genotypes in their potential intake, reproductive performance and production characteristics.

Economic analyses of responses to administering hormones for increasing ovulation rate or out-of-season lambing are required. Where a hormone such as fecundin is used to increase ovulation rate within a traditional wool production system, the benefits are likely to be those estimated in Table 2 and 3, less the cost of its purchase and administration. This assumes that the hormone has no specific action on feed consumption or utilization.

On many dryland farms the main response to more lambs in autumn may be a demand for more supplements and hence increasing costs. Once again, the interactions between reproductive performance, nutrient requirements and financial returns in a variable environment need to be clearly understood.

## ACKNOWLEDGEMENTS

Mr. P.J. Bowman and Mr. M.B. Whelan for assisting in upgrading the simulation model of a breeding ewe flock and for constructive criticism of this paper. This work was funded in part by the Australian Wool Corporation.

## REFERENCES

Alexander, G. and Davies, H.L., 1959. *Aust. J. Agric. Res., 10*, 720-724.

Bell, K., 1984. Murdoch University seminar on sheep flock health and production, Kojonup, W.A.

Blackburn, A.G. and Rizzoli, D.J., 1968. *J. Agric., Vic., 66*, 454-462

Bowman, P.J., White, D.H., Cayley, J.W.D. and Bird, P.R., 1982. *Proc. Aust. Soc. Anim. Prod., 14*, 36-37.

Bushnell, P. and Hutton, J., 1982. *The Agricultural Economist 3*, 7.

Byrne, P.F., 1964. *Rev. Mktg. Agric. Econ., 32*, 95-136.

Clarke, J.D., 1977. *J. Agric., Vic., 75*, 42-46.

Clarke, J.D., 1983. *In, Sheep Reproduction*, Vic. Dept. Agric., Dookie.

Corbett, J.L., 1979. *In* Black, J.L. and Reis, P.J. (eds.). *Physiological and Environmental Limitations to Wool Growth*, University of New England Publishing Unit, Armidale, 79-98.

Donnelly, J.R., 1984. *Aust. J. Agric. Res. 35*, 709-721.

Egan, J.K., Bishop, A.H. and McLaughlin, J.W., 1972. *Proc. Aust. Soc. Anim. Prod., 9*, 71-76.

Flanagan, S.P., Hanrahan, J.P. and O'Malley, P., 1979. *Farm and Food Research, 10*, 88-90.

Geisler, P.A., Paine, A.C. and Geytenbeek, P.E., 1977. *Agric. Systems, 2*, 109-119.

Gerrits, R.J., Blosser, T.H., Purchase, H.G., Terrill, C.E. and Warwick, E.J., 1978. *In* Hawk, H. (ed.) *Animal Reproduction*, Allanheld Osmun, 413-421.

Heath, J. and Lee, A., 1975. *In, Lambing Percentage — Why Worry*?, Vic. Dept. Agric., Hamiltion.

Inskeep, E.K. and Peters, J.B., 1981. *In* Brackett, B.G., Seidel, G.E. and Seidel, S.M. (eds.) *New Technologies in Animal Breeding*, Academic Press, 243-254.

James, J.W., 1978. *Anim. Prod., 26*, 111-118.

James, J.W., 1982. *In* Barker, J.S.F., Hammond, K. and McLintock, A.E. (eds.) *Future Developments in the Genetic Improvement of Animals*, Academic Press, 107-118.

Jardine, R., Mullaney, P.D., Turnbull, E.D. and Egan, J.K., 1975. *Aust. J. Exp. Agric. Anim. Husb., 15*, 610-618.

Jones, L.P., 1982. *In* Barker, J.S.F., Hammond, K. and McLintock, A.E. (eds.). *Future Developments in the Genetic Improvement of Animals*, Academic Press, 119-136.

Kilkenny, J.B., 1976. *Agric. Systems, 1*, 313-320.

Longmire, J.L., Brideoake, B.R., Blanks, R.H. and Hall, N., 1979. Occasional Paper No. 48, Bureau of Agricultural Economics, Canberra.

Love, G., Blanks, R., Buik, C. and Williams, K., 1982. Occasional Paper No. 73, Bureau of Agricultural Economics, Canberra.
Mangelsdorf, N.W., 1982. N.S.W. Dept. Agric. Sheep and Wool Seminar, Orange.
Morley, F.H.W., White, D.H., Kenney, P.A. and Davis, I.F., 1978. *Agric. Systems, 3*, 27-45.
Morris, C.A., Clarke, J.N. and Elliott, K.H., 1983. *In* Wickham, G.A. and McDonald, M.F. (eds.). *Sheep Production*, N.Z. Inst. Agric. Sci., 143-167.
Nixon-Smith, W.F., 1972. Bureau of Meteorology Working Paper No. 150.
Obst, J.M. and Day, H.R., 1968. *Proc. Aust. Soc. Anim. Prod., 7*, 239-242.
Obst, J.M., Thompson, R.L. and Patterson, A.J., 1984. This volume, 382-384.
Patterson, A.J., 1984. Hamilton District Monitor Farm Project 1982/83. Vic. Dept. Agric. Agnote, Agdex 400/810.
Patterson, A.J. and Wagg, M.C., 1983. *In, Sheep Reproduction* Vic. Dept. Agric., Dookie.
Peyraud, D. and Manno, J.M., 1975. *Ceres Journees de la Recherche Ovine et Caprine, 2*, 103-118.
Ponzoni, R.W., 1979. *Proc. Aust. Assoc. Anim. Breed Genet., 1*, 320-336.
Radford, H.M., 1959. *Aust. J. Agric. Res., 10*, 377-386.
Reeve, J.L. and Sharkey, M.J., 1980. *Aust. J. Exp. Agric. Anim. Husb., 20* 637-653.
Thatcher, L.P., 1977. *J. Agric., Vic., 75*, 47-52.
Thatcher, L.P. and Rabbette, J.G., 1977. *J. Agric. Vic., 75*, 82-87.
White, D.H. and Bowman, P.J., 1983. *Proc. Pasture Specialists Conf.*, Vic. Dept. Agric., Werribee, 103-113.
White, D.H., Bowman, P.J. and Morley, F.H.W., 1982. *Proc. Aust. Soc. Anim. Prod., 14*, 38-40.
White, D.H., Bowman, P.J., Morley, F.H.W., McManus, W.R. and Filan, S.J., 1983. *Agric. Systems, 10*, 149-189.
White, D.H., Nagorcka, B.N. and Birrell, H.A. 1979. *In* Black, J.L. and Reis, P.J. (eds.). *Physiological and Environmental Limitations to Wool Growth*, University of New England Publishing Unit, Armidale, 139-161.
White, D.H. and Reeve, J.L., 1983. *In* Thatcher, L.P. and Harris, D.C. (eds.). *Implications of Developments in Meat Science, Production and Marketing for Lamb Production*, Vic. and N.S.W. Depts. Agric., Orange.
White, D.H., Rizzoli, D.J. and Cumming, I.A., 1981. *Aust. J. Exp. Agric. Anim. Husb., 21*, 32-38.

# THE ECONOMIC IMPORTANCE OF GENETIC IMPROVEMENT OF REPRODUCTIVE RATE IN AUSTRALIAN MERINO SHEEP

R.W. Ponzoni and J.R.W. Walkley, *Department of Agriculture, Box 1671, G.P.O. Adelaide, S.A. 5000.*

*Summary* The economic importance of genetic improvement of reproductive rate in Australian Merino sheep was investigated in the context of a comprehensive breeding objective. A range of flock compositions, economic values and selection indices encompassing situations encountered in industry was considered. The contribution of genetic gain in reproductive rate, wool traits and live weight traits to total gain in economic units was estimated. The accuracy of each index was also calculated. The economic importance of genetic improvement of reproductive rate was variable. In general, it was large when surplus offspring were sold at weaning and when improved reproductive rate was not associated with increases in production costs. It was small when wethers were retained to mature age and when production costs increased markedly with greater reproductive rate. Indices involving more detailed recording increased the accuracy in estimating the breeding value when the economic value of reproductive rate was large, but the increase in accuracy was negligible when the economic value was small.

## INTRODUCTION

The economic worth of Australian Merino sheep depends on several traits, and consequently the breeder is forced to consider more than one trait in selection. Nevertheless, most of the work examining the prospects of genetic improvement of reproduction rate (RR) has been on that trait as the sole objective (Lindsay 1982, Piper 1982, Land *et al.* 1983), and only rarely has this improvement been considered in the context of an overall breeding objective (Jones 1982, Ponzoni 1982 a, b, c).

Breeders and producers vary widely in their attitude towards increased RR. In this paper we examine the effect of some factors on the economic importance of genetic improvment of RR. Recently Jones (1982) examined the consequences of price fluctuations and Ponzoni (1982c) also studied that area, as well as the effect of assuming different values for genetic parameters. In the present study emphasis was placed on the effects of: (i) flock composition, (ii) economic value of RR, and (iii) information (index) used in the estimation of the breeding value.

Commercial flocks produce virtually all the wool and sheep meat of the country, but they are dependant on ram breeding flocks (studs or nuclei of co-operative breeding groups) for permanent genetic improvement. Selection policies discussed in this paper are appropriate for ram breeding flocks.

## MATERIALS AND METHODS

### Definitions

(i) The *breeding objective* comprises those traits that are to be improved genetically because they influence the profit derived from commercial flocks; (ii) the *economic value* of a trait in the breeding objective is the change in profit resulting from a unit change in that trait, assuming all other traits remain constant; (iii) the *selection criteria* are the characters used in assessing the breeding value of an individual (traits in the objective are not necessarily selection criteria, and *vice versa*).

### Procedures

The traits included in the breeding objective were: clean fleece weight (CFW), fibre diameter (FD), number of lambs weaned (NLW) or of hoggets (1½-yr-old) present (NHP) (depending on the flock composition), weaning weight (WW) or hogget live weight (HW) (depending on the flock composition), and mature (5½-yr-old) live weight (MW). The breeding objective was defined for three flock compositions (Ponzoni 1982a): $FC_1$, a breeding flock from which all surplus offspring are sold as lambs after weaning; $FC_2$, a breeding flock from which all surplus offspring are sold at approximately 1½ years of age after the hogget shearing; $FC_3$, a flock consisting of 50 percent breeding ewes and 50 percent wethers, and selling all surplus offspring as lambs after weaning. Surplus offspring are those in excess of replacement needs.

Three sets of economic values were investigated within each flock composition (Table 1). The first and second sets ($v_1$ and $v_2$) were described by Ponzoni (1982b). In the derivation of $v_1$ no allowance was made for increases in feed intake which may result as a consequence of selection, except for the cost of maintenance of heavier ewes and wethers. In the calculation of the second set ($v_2$) allowance was made for increased feed costs which would result as a consequence of genetic improvment in RR (NLW or NHP) and live weight of surplus offspring (WW or HW). The third set ($v_3$) is similar to that derived by Jones (1982), who assumed greater increases in production costs due to genetic improvement in RR and live weight than Ponzoni (1982b), and made allowance for penalties to breeding ewes producing twin lambs and to twin born animals. In Table 1 the economic values for all traits are expressed relative to that of FD within each flock composition. Thus, the relative economic value of FD is −1.0 (the minus sign is due to the lower value per kg of wool of greater FD).

**TABLE 1. Relative economic values ($v_1$, $v_2$ and $v_3$) for traits in the breeding objective in flocks of different composition ($FC_1$, $FC_2$ and $FC_3$).(1)**

| Traits in the breeding objective(2) | Relative economic values | | | | | | | | |
|---|---|---|---|---|---|---|---|---|---|
| | $FC_1$ | | | $FC_2$ | | | $FC_3$ | | |
| | $v_1$ | $v_2$ | $v_3$ | $v_1$ | $v_2$ | $v_3$ | $v_1$ | $v_2$ | $v_3$ |
| CFW | 7.1 | 7.1 | 7.1 | 7.1 | 7.1 | 7.1 | 7.1 | 7.1 | 7.1 |
| FD | -1.0 | -1.0 | -1.0 | -1.0 | -1.0 | -1.0 | -1.0 | -1.0 | -1.0 |
| NLW | 36.1 | 25.3 | 12.6 | §(3) | § | § | 18.9 | 13.2 | 6.6 |
| NHP | § | § | § | 30.5 | 15.2 | 7.6 | § | § | § |
| WW | 0.6 | 0.6 | 0.3 | § | § | § | 0.2 | 0.1 | 0.1 |
| HW | § | § | § | 0.3 | 0.3 | 0.1 | § | § | § |
| MW | 0.1 | 0.1 | 0.0 | 0.1 | 0.1 | 0.0 | 0.1 | 0.1 | 0.0 |

(1) $v_1$ = no allowance made for increases in production costs due to greater reproductive rate; $v_2$ = increases in production costs allowed for as in Ponzoni (1982b); $v_3$ = increases in production costs and penalty to twin bearing ewes and twin born animals allowed for, approximately as in Jones (1982).
$FC_1$ = breeding flock, surplus offspring sold after weaning; $FC_2$ = breeding flock, surplus offspring sold at approximately 1½ years of age; $FC_3$ = 50 per cent breeding ewes, 50 per cent wethers.
(2) CFW = clean fleece weight, FD = fibre diameter, NLW = no. of lambs weaned, NHP = no. of hoggets present at 1½ yrs of age, WW = weaning weight, HW = hogget live weight, MW = mature live weight.
(3) § = Trait not in the breeding objective.

Four selection indices were investigated (Table 2). $I_1$ included selection criteria often recorded in Merino ram breeding flocks. $I_2$ to $I_4$ represented increasing efforts to improve the accuracy of the estimation of the breeding value for RR. All selection criteria can be assessed by 18 months of age.

**TABLE 2. Selection criteria included in indices $I_1$ to $I_4$.**

| Index(1) | Characters included in the index(2) |
|---|---|
| $I_1$ | CFW, FD, HW |
| $I_2$ | CFW, FD, dNLW — one record, HW |
| $I_3$ | CFW, FD, dNLW — three records, HW |
| $I_4$ | As $I_3$ plus OR at approx. 18 mo of age in 40 half sisters |

(1) For example, $I_1 = b_1\,CFW + b_2\,FD + b_3\,HW$, where the bs are the appropriate index coefficients.
(2) CFW = clean fleece weight, FD = fibre diameter, dNLW = dam's no. of lambs weaned, OR = ovulation rate, HW = hogget live weight.

The economic importance of genetic improvement in RR (NLW or NHP), wool (CFW and FD) and live weight (WW or HW, MW) was estimated for all possible combinations of economic values, flock composition and indices as the percent of total genetic gain in economic units which could be accounted for by genetic gain in each group of traits. The total genetic gain in economic units is:

$$\sigma_I = \sum_{i=1}^{i} v_i G_i$$

where the vs and Gs are economic values and genetic gains, respectively, for each one of the i traits in the breeding objective. The percentage contribution of RR, for example, was calculated as:

$$PC_{RR} = (v_{RR}\, G_{RR}/\sigma_I)\ 100$$

where RR would represent NLW or NHP depending on the flock composition.

The accuracy of the indices was estimated for all possible combinations of flock composition and economic values as:

$r_{IT} = \sigma_I/\sigma_T$

where $\sigma_T$ is the standard deviation of the aggregate genotype and $\sigma_I$ was defined earlier; $r_{IT}$ is the fraction of 'possible' genetic gain in economic units that can be attained with the index being examined.

The phenotypic and genetic parameter values assumed were the same as those in Ponzoni (1982b) and Mueller *et al.* (1984).

## RESULTS

Figure 1 shows the percentage contribution to total genetic gain in economic units made by RR, wool traits and live weight traits. For a given index and set of economic values the contribution of RR was greatest in $FC_1$, intermediate in $FC_2$ and lowest in $FC_3$. For a given flock composition and index the contribution of RR was greatest for $v_1$, intermediate for $v_2$ and lowest for $v_3$. For a given flock composition and set of economic values the contribution of RR increased from $I_1$ to $I_4$, but the size of this increase varied with the flock composition ($FC_1 > FC_2 > FC_3$) and set of economic values ($v_1 > v_2 > v_3$) being considered.

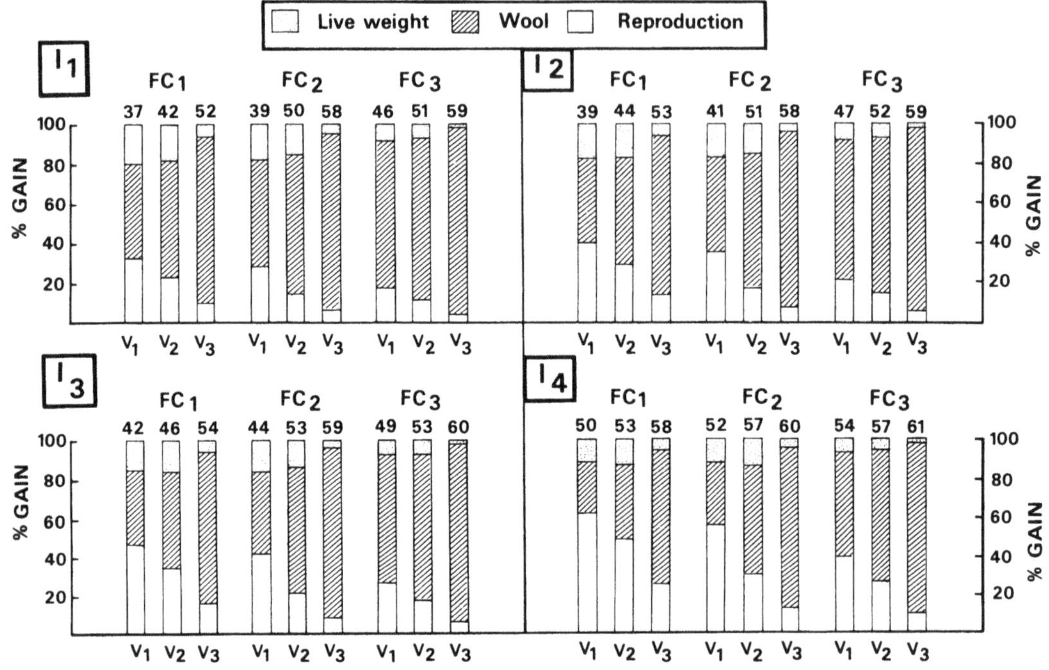


**FIGURE 1. Percentage contribution to total genetic gain in economic units (% gain) made by reproductive rate, wool traits and live weight traits, assuming different flock compositions ($FC_1$ to $FC_3$), economic values ($v_1$ to $v_3$) and using indices $I_1$ to $I_4$. The value on each bar is the accuracy ($r_{IT}$) of the index expressed as a percentage.**

Figure 1 also shows the accuracy ($r_{IT}$) of each index when applied to all combinations of flock composition and economic values. The $r_{IT}$ values increased (with few exceptions) from $I_1$ to $I_4$ for a given combination of flock composition and economic values, but the magnitude of the change varied with the flock composition ($FC_1 > FC_2 > FC_3$) and the set of economic values ($v_1 > v_2 > v_3$) being considered.

## DISCUSSION

The flock compositions and economic values examined in this study encompass the broad range of production and marketing systems likely to be encountered in the Merino sheep industry. The selection indices represent different levels of intensity of performance recording. Opinions are often at variance regarding the practical significance of genetic improvement of reproductive rate. The issue debated can be stated succintly as follows: Should breeders try to achieve genetic improvement of reproductive rate in their flocks?

The results presented in Figure 1 indicate that there is no unique answer to this question. The answer depends on the production and marketing system for which genetic improvement is intended. Thus, genetic improvement of reproductive rate is more important when surplus offspring are sold at weaning ($FC_1$) than when wethers are retained for a number of years ($FC_3$). Similarly, improvement of reproductive rate is desirable when production costs do not rise with increasing levels of the trait ($v_1$), as may be the case for a producer who is under-stocked or who takes advantage of a peak of pasture production at a certain time of the year. But it is of little advantage to a producer whose production costs increase markedly with improvement of reproductive rate and whose stock is handicapped due to increased twinning ($v_3$).

The intensity of performance recording also affected the importance of genetic improvement of reproductive rate, but the magnitude of the effect depended on the flock composition and set of economic values. The contribution of reproductive rate and the accuracy of the index increased considerably from $I_1$ to $I_4$ when surplus offspring were sold young and production costs did not rise (e.g. $FC_1v_1$), but there was little change when animals were sold at an older age and production costs associated with reproductive rate were high (e.g. $FC_3v_3$). These results suggest that detailed performance recording such as that required to implement $I_4$ would be more justified when production costs do not rise with reproductive rate and when surplus offspring are disposed of at a young, rather than at an older age. This is in agreement with the conclusions of Mueller *et al.* (1984). The structure of ram breeding flocks in industry is such that studies like that of Mueller *et al.* (1984) are needed to ascertain the consequences of restricting detailed recording to a fraction of the population. An examination of the value of male traits as indirect selection criteria for reproductive rate is also warranted.

The small difference between $I_1$ and $I_2$ in Figure 1 is of special interest. Using one record of NLW of the individual's dam requires the identification of single and twin born animals. The results suggest that from the point of view of accuracy of the index the gains are negligible. In practice the identification of single and twin born animals would have the advantage, not accounted for here, of enabling the application of correction factors so that twin born (or reared) animals are not penalised during selection.

Since the economic importance of genetic improvement of reproductive rate is variable and it depends on the production and marketing system for which genetic improvement is intended, generalisations are not possible. The approach taken in this paper of examining a range of flock compositions, economic values and selection indices may help reconcile the scientists' with some breeders' attitudes towards genetic improvement of reproductive rate. Scientists favouring the improvement of reproductive rate have implicitly assumed little or no increase in production costs due to reproductive rate, whereas breeders opposed to it most likely had in mind sharp rises in production costs and lowered production among twin bearing ewes and twin born animals. The matter can be resolved by being precise about the situation for which genetic improvement is intended. Finally, note that the contribution of reproductive rate was sometimes negligible ($I_1FC_3v_3$), whereas that of wool traits was never below 26 percent ($I_4FC_1v_1$). This is perhaps the basis for breeders' perception that the genetic improvement of wool traits is always important, whereas that of reproductive rate may or may not, depending on the circumstances.

## REFERENCES

Jones, L.P., 1982. *In* Barker, J,S,F,, Hammond, K, and McLintock, A.E. (eds.) *Future Developments in the Genetic Improvement of Animals*, Academic Press Australia, pp. 119-136.

Land, R.B., Atkins, K.D. and Roberts, R.C., 1983. *In,* Haresign, W. (ed) *Sheep Production,* Proc. 35th Easter School in Agricultural Science, Nottingham, pp, 515-535.

Lindsay, D., 1982. *In* Barker, J.S.F., Hammond, K. and McLintock, A.E. (eds.) *Future Developments in the Genetic Improvement of Animals, Academic Press Australia, pp. 89101.*

Mueller, J.P., Piper, L.R. and James, J.W., 1984. *Proc. Aust. Soc. Anim. Prod., 15*, 484-486.

Piper, L.R., 1982. *Proc. 2nd World Congr. Genet. Appl. Livest. Prod.*, Madrid, Spain, *volume V*, pp. 271-281.

Ponzoni, R.W., 1982a. *Proc. World Congr. on Sheep and Beef Cattle Breeding*, New Zealand, *vol. I*, pp. 61-64.

Ponzoni, R.W., 1982b. *Wool Tech. Sheep Breed., 30*, 44-48.

Ponzoni, R.W., 1982c. *Proc. 2nd World Congr. Genet. Appl. Livest. Prod.*, Madrid, Spain, *volume V*, pp. 619-634.

# ECONOMICS OF INCREASED REPRODUCTIVE RATE FOR PRODUCTION OF WOOL

J.M. Obst and R.L. Thompson, *Department of Agriculture, Pastoral Research Institute, Hamilton, Victoria, 3300.*
A.J. Patterson, *Department of Agriculture, District Office, Hamilton, Victoria, 3300.*

*Summary* Eight replicated systems of wool production incorporating self-replacing ewe and wether flocks each of 100 breeding ewes were established to compare the biological and economic significance of increasing the reproductive rate from 80 to 120% lambs weaned of both small and big-sized sheep lambing in the autumn and spring.

Two complete years of financial records were completed. Data from 1982-83, a year of severe drought, indicated that the net income ($/ha) for small sheep was reduced if the weaning percentage was increased from 80 to 120%. There was no effect of increased reproductive rate on big sheep. Data from 1983-84, a wet and cold year, indicated an increase in net income from increasing the reproductive rate. This economic increase was largest (21%) in the small ewes which lambed during the spring and about a 10% increase in both the small and big ewes which lambed in the autumn. Data from more years are required before a decision can be made as to the mean overall economic value of increasing reproductive performance within a self-replacing wool flock from which excess sheep can be sold either on the live sheep or meat markets.

## INTRODUCTION

Within the Western District of Victoria in a perennial pasture, 700 mm rainfall zone, farmers aim to increase body size and wool yield per head and to provide an assured option of selling to the live sheep market. Will these changes increase the economic return per hectare? To answer these questions, two complete years (1982-83 and 1983-84) of biological and economic data are now available from the Sheep Production System project at Hamilton, Victoria. 1982 was a drought year and 1983 commenced with the 16th Feb. Ash Wednesday fire which destroyed some replicates of this experiment.

The economic analyses currently available from this project are reported in this paper.

## MATERIALS AND METHODS

Eight replicated systems of wool production from self-replacing ewe and wether flocks were established at the Pastoral Research Institute, Hamilton, Victoria (Table 1). Equivalent grazing pressures based on metabolic live-weight per hectare (LW $kg^{0.75}$/ha) were established e.g. small sized ewes were stocked at 12.5/ha and the large at 10.8/ha.

Autumn lambing systems were joined from December 1 and spring lambing systems were joined from March 15. At each joining time the ewes in the system designated L for 80% lambs weaned were controls and those in the H for 120% lambs weaned were treated using PMSG and synchronisation to increase ovulation rate. The number of lambs born within each system are shown in Table 1. Depending upon the actual number of lambs weaned within each system, weaners were either sold or purchased to achieve the correct number (80 or 120) for the weaner-hogget year when within each system production was compared on either lucerne pasture or perennial ryegrass with subterranean clover.The economic returns or costs of maintaining the 80 or 120% weaning for each system are recorded (Table 2) in the analysis as sheep sales or sheep purchases. Wethers were grazed from 1½ to 3½ years of age and then sold.

Management practices have been set according to survey information and previous research results to produce good results e.g. shelter has been provided for all lambing ewes; anthelmintics are administered according to the Nemat program of Callinan *et al.* (1982). Autumn lambing systems are shorn in September (post-weaning) and the spring lambing systems are shorn in December (post-weaning) .

Economic analyses on each system were completed with the aid of the Agricultural Business Research Institute (ABRI) at Armidale, N.S.W. on the same basis as the 30 farms within the Hamilton District Monitor Farm Project (HDMFP) conducted by the Victorian Department of Agriculture at Hamilton (Patterson 1984). Overhead costs for each system were the average of those from the ABRI analysis of the HDMFP. All operating costs for each system were included; only those for 1982-83 are shown in Table 2. Wool prices for each system were based on the average cents/kg greasy fleece weight according to fibre diameter ($\mu$) paid by the AWC at all wool sales during each year. Price relativity was therefore maintained for each system whether shorn in September or December.

Prices for sale sheep, culled for age or for excess to requirements for replacement sheep or purchases for replacements were based on age, live-weight and body condition scores consistent with the livestock Market Reporting Service (LMRS) conducted by the Victorian Department of Agriculture at Hamilton. Sheep were classified for sale either to the live sheep trade or to the local meat market.

## RESULTS

An economic statement of each of the eight systems for the 1982-83 (i) and 1983-84 (ii) financial year is presented in Table 2. Operating and over head costs are presented only for 1982-83 as an example of the type and relative costs. Data of wool production per system (kg/ha) indicated that there was little difference between small or big sheep suggesting that the adjustment of stocking rate to metabolic liveweight/ha was valid. This was also supported by similar amounts of pasture available (kg DM/ha) between systems with small or big sheep.

**TABLE 1. System identification and flock structure.**

| System no. | 1 | 2 | 3 | 4 | 5 | 6 | 7 | 8 |
|---|---|---|---|---|---|---|---|---|
| Treatment[1] | SAL | SAH | BAL | BAH | SSL | SSH | BSL | BSH |
| Area (ha) | 17.6 | 17.6 | 22.0 | 22.0 | 17.6 | 17.6 | 22.0 | 22.0 |
| Ewes (n) | 100 | 100 | 108 | 108 | 100 | 100 | 108 | 108 |
| Weaner/hoggets (n) | 80 | 120 | 80 | 120 | 80 | 120 | 80 | 120 |
| Wethers (n) | 60 | 60 | 60 | 60 | 60 | 60 | 60 | 60 |
| Lambs born (n) 1982 | 90 | 95 | 107 | 127 | 74[2] | 64[2] | 112 | 152 |
| 1983 | 97 | 140 | 103 | 135 | 100 | 108 | 137 | 119[3] |

Treatments are identified as follows:—

[1] The first letter represents S for small (fine wool Merino) or B for big (Comeback); the second letter represents A for autumn or S for spring lambing and the third letter represents L for 80 or H for 120% lambs weaned.

[2] Suspected heat induced infertility of small rams.

[3] BSH was not successfully treated for an increase in ovulation rate in 1983.

**TABLE 2. Net income ($/ha) of the sheep production systems**

| System no. | | 1 | 2 | 3 | 4 | 5 | 6 | 7 | 8 |
|---|---|---|---|---|---|---|---|---|---|
| Treatment[1] | | SAL | SAH | BAL | BAH | SSL | SSH | BSL | BSH |
| Income ($) | | | | | | | | | |
| Wool | (i) | 3534 | 4014 | 3662 | 4262 | 3520 | 4000 | 3669 | 4278 |
| | (ii) | 3645 | 4244 | 4096 | 4717 | 3287 | 4066 | 3861 | 4359 |
| Sheep | (i) | 208 | 354 | 735 | 820 | 489 | 485 | 631 | 698 |
| | (ii) | 857 | 1470 | 1188 | 1727 | 1036 | 1610 | 1381 | 1799 |
| Operating costs ($) (i) | | | | | | | | | |
| Fertilizer | | 221 | 221 | 277 | 277 | 221 | 221 | 277 | 277 |
| Chemicals | | 102 | 339 | 112 | 427 | 90 | 394 | 100 | 434 |
| Shearing | | 727 | 822 | 788 | 914 | 700 | 793 | 803 | 952 |
| Suppl. feed | | 574 | 624 | 691 | 797 | 231 | 166 | 439 | 439 |
| Casual labour | | 0 | 78 | 0 | 85 | 0 | 59 | 0 | 63 |
| Sheep purchases | | 77 | 417 | 0 | 153 | 204 | 583 | 0 | 0 |
| Overhead costs ($) | (i) | 679 | 679 | 849 | 849 | 679 | 679 | 849 | 849 |
| Total costs $/ha | (i) | 135 | 181 | 124 | 159 | 121 | 164 | 112 | 137 |
| | (ii) | 122 | 177 | 117 | 157 | 125 | 177 | 119 | 153 |
| Net income | | | | | | | | | |
| Total $ | (i) | 1362 | 1188 | 1680 | 1590 | 1884 | 1590 | 1832 | 1962 |
| | (ii) | 2362 | 2591 | 2710 | 3000 | 2118 | 2554 | 2634 | 2783 |
| $/ha | (i) | 77 | 68 | 76 | 72 | 107 | 90 | 83 | 89 |
| | (ii) | 134 | 147 | 123 | 136 | 120 | 145 | 120 | 127 |
| $\bar{x}$ | (i & ii) | 106 | 108 | 100 | 104 | 114 | 118 | 102 | 108 |

[1] Refer to Table 1.

Valid comparisons of the economic value of increasing reproductive rate can therefore be made from the data presented in Table 2. A reasonable assumption would be that a 10% difference in net income ($/ha) is marginal and 20% difference is worthwhile.

During the drought year of 1982-83 the net income of the systems with small ewes was reduced by increasing the weaning per cent from 80 to 120 in both autumn and spring lambing systems. There was a 12% reduction in the autumn and a 16% reduction in net income within the small spring lambing systems. There was no economic effect of increasing reproductive rate within the systems with big ewes. During 1983-84 when pasture growth started in March and there was a very wet and cold winter, net incomes were higher in the systems with 120% lambs weaned both when born in the autumn and spring and from both small and big sheep. The highest increase in net income was 21% in the small spring lambing group with about a 10% increase in the autumn lambing systems of both small and big sheep. An average of the 1982-83 and 1983-84 economic returns indicates little effect of increasing reproductive rate.

## DISCUSSION

It must be realised that an increase in lambs weaned is only one part of the system and wool prices can have an over-riding effect on the economic returns, particularly in a self-replacing flock. Thus the emphasis placed on the importance of reproductive rate will vary depending upon whether wool or meat provides the major income and upon the prices received for excess stock.

In this study during the drought year (1982/83) wool provided 90 and 82% of the total income from systems with small and big sheep respectively. During 1983/84 wool provided 68 and 64% of the total income from small and big sheep respectively. Thus, during 1983/84, there was an economic advantage of increasing the reproductive performance from 80 to 120% lambs weaned in systems with both small and big sheep which lambed in both autumn and spring (Table 2).

Obviously this economic advantage would increase if the cost of obtaining the additional 40% of lambs weaned was reduced and if prices for sale sheep were increased. However, there is the continuing need to compare real values of both biological and economic response to increases in reproductive performance with simulated values over many years for sheep systems producing wool per se, wool and meat and meat per se. If the real values are shown to be similar to the simulated values then we have the basis for long term advice to the graziers producing mainly fine wool ($<20\mu$) from small sheep, or whether they should attempt to increase the twinning rate and size of their sheep.

## ACKNOWLEDGMENTS

Staff of the Pastoral Research Institute, Hamilton who have spent many hours recording information on the weather, pastures and sheep are gratefully acknowledged. Financial assistance from the Wool Research Trust Fund is received with special thanks.

## REFERENCES

Callinan, A.P.L., Morley, F.H.W., Arundel, J.H. and White, D.H., 1982. *Agric. Systems, 9*, 199-225.
Patterson, A.J., 1984. Agdex 400-810.

# PRODUCER ATTITUDES TO SHEEP REPRODUCTION AND ITS IMPROVEMENT

R.A. Jelbart, *Department of Agriculture, PO Box 477, Wagga Wagga.*
R. Cullen, *Department of Agriculture, PO Box 132, Cootamundra.*
J. Marshall, *Department of Agriculture, Agricultural Research Institute, Wagga Wagga.*
S. Phillips, *Department of Agriculture, PO Box 29, Deniliquin.*
D.S. Douglass, *Department of Agriculture, PO Box 62, Dareton.*
D. Martin, *Department of Agriculture, PO Box 117, Hillston.*
E. Scarlett, *Department of Agriculture, PO Box 49, Albury.*

*Summary* In the past, major extension efforts to improve reproductive performance at state, regional, and district level have resulted in little change to average lambing figures. Extension officers in the Murray and Riverina Region of New South Wales discarded established extension methods and set out to determine producers attitudes, so that more effective extension programmes could be implemented.

Producers attitudes varied, but there were common constraints identified which limit the full application of known technology. Future attempts to improve reproductive performance need to consider these constraints carefully.

## INTRODUCTION

Four survey areas were selected in the south of New South Wales (Hay, Gunadagai, Wagga Wagga and Albury). Questionnaires were mailed to 400 producers, and approximately 45% were returned. This gave:

- A list of what farmers believed and did not believe about factors influencing reproduction.
- Producers' opinions on practices recommended by the Department of Agriculture.
- The differences in attitudes between producers whose ewes had a high lambing performance and those of producers whose ewes lambed poorly.

## MATERIALS AND METHODS

The survey was derived from the Delphi technique (Lenne and Hartley 1983). Initially, a small group of 6-10 producers were hand picked to discuss practices important in gaining high weaning performance. It covered not only Departmentally recommended procedures, but also those the producers believed were important. These practices were drawn up into a questionnaire, which was pretested, and then mailed to producers in four geographical areas; Hay, Gundagai, Wagga Wagga, and Albury.

During analysis of the results, each of the 64 practices listed on the questionnaire was ranked according to the number of times the practice was rated in the top ten according to the following scale:

| Rating given by producer: | 1 | 2 | 3 | 4 | 5 | 6 | 7 | 8 | 9 | 10 |
|---|---|---|---|---|---|---|---|---|---|---|
| Score: | 10 | 9 | 8 | 7 | 6 | 5 | 4 | 3 | 2 | 1 |

The scores for each practice were then added together to give a total score for each practice.

**TABLE 1. Lamb marking percentages from Hay respondents.**[1]

| Year | 1979 | 1980 | 1981 | 1982 | 1983 | 5 year average |
|---|---|---|---|---|---|---|
| Average % | 72.2% | 77.3% | 76.3% | 77.5% | 77.0% | 76.5% |
| Range | 46% – 101.5% | 38.7% – 109.5% | 34.0% – 101.8% | 41.9% – 106.0% | 27.5% – 102.0% | 52.0% – 98.9% |

[1] These results are 2-3% higher than BAE statistics for this area.

## RESULTS

The surveys provided a cross section of the opinion from the different extremes of rainfall, landform, wool producing breeds and strains, run under the intensive and extensive conditions that occur in the region. Data presented are for the 86 respondents from the Hay area, lambing percentages for which are shown in Table 1. They are typical of the results for all four areas.

Tables 2, 3 and 4 show the rankings for the top ten practices, the most frequent "don't know", and "wrong" practices. Flocks were divided into four categories according to their lambing percentages over five years. These were designated A, B, C and D. Twelve respondents did not have sufficient data on lambing percentage to be included in further analysis. Tables 5, 6, 7, and 8 show the top 10 practices for each of these four lambing percentage groups. The producers who recorded the highest lambing percentages identified practices associated with the preparation of rams as being among the most important in helping achieve good lambing performance.

Table 9 shows that the less successful producers regarded aspects of ram management such as treatment of feet and 'easy care' rams as unimportant.

## DISCUSSION

The response rate from the surveys and the high attendance at meetings, suggest that producers are interested in knowing more about sheep reproduction. Following these meetings Advisory Officers made the following subjective observations:

- Producers agreed that weaning percentages are low.
- Producers did not believe that they could increase reproductive performance of their flocks above the present level.
- Increasing weaning percentages would need to be economically justified on a whole farm basis.

**TABLE 2. The top 10 most important practices perceived as achieving high weaning percentages.**

1. Achieve maximum ewe liveweight at joining by stocking comparatively.
2. Aim to have adequate nutrition during the pre- and post-lambing periods by controlling nutrition through conservative stocking or other means.
3. Avoid fly-strike in rams by using mulesed rams, jetting rams and powdering heads pre-joining.
4. Join to lamb autumn/winter to take advantage of best feed, and market wether progeny early.
5. Blood test rams for brucellosis and palpate testes for abnormalities.
6. Choose a merino strain with a proven high lambing performance in your district.
7. Inspect rams feet for diseases and trim.
8. Mules at marking to provide crutch-strike protection during the first year of lamb's life.
9. Crutch ewes pre-lambing.
10. Shear lambs at or after weaning to promote growth, avoid grass seed infestation and flystrike.

**TABLE 3. The most frequent "don't know" answer for untried or "don't know" practices.**

| Practice | Number of returns indicating "don't know" and untried practices |
|---|---|
| Libido test rams and cull poor servers | 50 |
| Use teaser pre-joining | 49 |
| Split ewes (pre-lambing) on udder development into early and late lambing groups | 45 |
| Sterilize marking tools by boiling for 30 minutes between mobs | 43 |
| Use ram harness to check joining activity | 39 |
| Drench ewes at weaning | 36 |
| Achieve maximum ewe liveweight at joining by drenching pre-joining | 35 |
| Identify and cull ewes that lamb but fail to rear | 33 |
| To achieve high levels of ram/ewe contact increase number of rams for:<br>— Paddock larger than 3000 acres<br>— Mob sizes greater than 500 ewes<br>— Paddocks with more than one water<br>— Paddocks with no shade<br>— Paddocks with poor feed | 34 |
| Use a gas detailer for removing tails to minimise blood loss | 32 |
| Restrict mob size to less than 500 ewes | 27 |
| Vaccinate ewes pre-lambing with 5 in 1 vaccine | 38 |
| Reserve a proportion (about a third) of the rams and introduce them mid-way through joining | 27 |
| Join ewes off-shears to increase conception | 26 |
| Join to lamb winter/spring to achieve higher lambing percentage | 20 |

**TABLE 4. The most frequent procedures perceived as being "wrong practices".**

| Practice | Number of returns indicating this as a "wrong practice" |
|---|---|
| Inspect ewes before and during the lambing period to assist difficult births | 32 |
| To achieve high levels of ram/ewe contact increase number of rams for:<br>— Paddock larger than 3000 acres<br>— Mob sizes greater than 500 ewes<br>— Paddocks with more than one water<br>— Paddocks with no shade<br>— Paddocks with poor feed | 26 |
| Reserve a proportion (about one third) of the rams and introduce them mid-way through joining | 23 |
| Join to lamb winter/spring to achieve higher lambing percentage | 20 |

**TABLE 5. Group A — 84 to 99% (19 respondents).**

| Rank | Practice | Score |
|---|---|---|
| 1. | Choose a Merino strain with a proven high lambing performance for the district | 83 |
| 2. | Inspect rams feet pre-joining for disease and trim | 68 |
| 3. | Palpate testes for abnormalities | 63 |
| 4. | Achieve maximum ewe liveweight at joining by stocking conservatively | 58 |
| 5. | Avoid fly strike in rams by using mulesed rams, jetting rams and powdering heads pre-joining | 57 |
| 6. | Blood test for brucellosis | 47 |
| 7. | Join to lamb autumn/winter to take advantage of best feed and market wether progeny early | 45 |
| 8. | Aim to have adequate nutrition during the pre- and post-lambing periods by controlling nutrition through conservative stocking or other means | 37 |
| 9. | Use 'easy care' rams | 35 |
| 10. | Select against excessive skin development in ewes | 33 |

**TABLE 6. Group B — 78 to 83% (18 respondents).**

| Rank | Practice | Score |
|---|---|---|
| 1. | Aim for adequate nutrition during the pre- and post-lambing periods by controlling nutrition through conservative stocking or other means | 77 |
| 2. | Avoid fly strike in rams | 69 |
| 3. | Join to lamb autumn/winter | 59 |
| 4. | Achieve maximum ewe liveweight at joining by stocking conservatively | 53 |
| 5. | Choose a merino strain with proven high lambing performance in your district | 47 |
| 6. | Select against excessive skin development in ewes | 37 |
| 7. | Palpate testes for abnormalities | 37 |
| 8. | Control feral pigs and foxes | 37 |
| 9. | Crutch pre lambing | 36 |
| 10. | Shear rams twice per year | 34 |

**TABLE 7. Group C — 70 to 77% (20 respondents).**

| Rank | Practice | Score |
|---|---|---|
| 1. | Achieve maximum ewe liveweight at joining by stocking conservatively | 86 |
| 2. | Join to lamb autumn/winter to take advantage of best feed and market wether progeny early | 79 |
| 3. | Avoid fly strike in rams | 58 |
| 4. | Blood test for Brucellosis | 54 |
| 5. | Palpate testes for abnormalities | 51 |
| 6. | Aim to have adequate nutrition during the pre- and post-lambing period by controlling nutrition through conservative stocking or other means | 50 |
| 7. | Mules at marking to provide crutch strike protection during first year of lamb's life | 47 |
| 8. | Shear pre-lambing to increase lamb survival | 42 |
| 9. | Choose a merino strain with a proven high lambing performance in your district | 40 |
| 10. | Crutch pre-lambing | 39 |

**TABLE 8. Group D — 52 to 69% (17 respondents).**

| Rank | Practice | Score |
|---|---|---|
| 1. | Shear rams twice a year | 56 |
| 2. | Mules at marking to provide crutch strike protection during first year of lamb's life | 55 |
| 3. | Avoid fly strike in rams | 54 |
| 4. | Aim to have adequate nutrition during the pre- and post-lambing period by controlling nutrition through conservative stocking or other means | 50 |
| 5. | Achieve maximum ewe liveweight at joining by stocking conservatively | 49 |
| 6. | Blood test for Brucellosis | 43 |
| 7. | Crutch pre-lambing | 40 |
| 8. | Keep ewes in holding paddocks until all lambs are marked and mothered up | 38 |
| 9. | Palpate testes for abnormalities | 36 |
| 10. | Powder at mulesing and marking to prevent fly strike | 32 |

**TABLE 9. Group A's top 10 with the rankings that Groups B, C and D gave to these practices.**

| | Group based on lambing percentage | | | |
|---|---|---|---|---|
| | A | B | C | D |
| Merino strain with proven performance in district | 1 | 5 | 9 | 19 |
| Inspect rams feet | 2 | 14 | 20 | 23 |
| Palpate testes | 3 | 8 | 5 | 9 |
| Maximum liveweight at joining | 4 | 4 | 1 | 5 |
| Avoid fly strike in rams | 5 | 2 | 3 | 3 |
| Brucellosis test | 6 | 15 | 5 | 6 |
| Autumn/winter lambing | 7 | 3 | 2 | 16 |
| Adequate nutrition pre- and post-lambing | 8 | 1 | 6 | 4 |
| 'Easy care' rams | 9 | 45 | 11 | 28 |
| Excessive skin fold | 10 | 7 | 12 | 39 |

- Producers regarded twins, spring lambing and poor wool production from plainer bodied ewes as environmental and/or economic constraints limiting any increase.

Despite these constraints, producers, if offered some panacea that would ensure all ewes had and reared one lamb, would accept the technique readily. A technique to increase twinning rate from 20 to 50% would not be viewed by most as a practical way of increasing lambing percentages.

The survey provided feedback in several ways on the acceptability of recommendations by the Department of Agriculture.

- The most recent research results are generally either not known or not understood by most producers (e.g. libido testing of rams and culling those with poor serving capacity, ram effect, Fecundin, and culling ewes that fail to rear a lamb).
- Some long known research findings are used by very few producers. The use of ram harnesses, selection for improved reproduction, choosing a time of joining to give high lambing percentage, and varying ram percentages depending on joining conditions are typical examples.
- In deciding on a lambing date, producers considered that autumn/winter gave improved nutrition and growth rates, primarily to have young animals mature enough for the following summer. This also resulted in better marketing prospects for surplus stock. The higher cycling and subsequent increase in number of lambs born was not enough to sway most producers to a spring lambing. Spring lambing was thought to result in higher lamb loss, slower growth rates and, in some cases, higher mortality rates due to grass seed and problems of weaner nutrition.
- Having ewes in good condition at joining, and nutrition pre- and post-lambing were rated as highly important by all producers. However there is extensive field evidence that low ewe live weight, or poor nutrition, is one of the most significant factors reducing lambing percentages. Nearly all producers believe in high liveweights, but in practice, this depends on available paddock feed. So producers known the importance of high liveweight of ewes, but little is done on the farm to influence it, particularly before joining.
- The wide use and acceptance of recommendations on rams and their management was obvious. These recommendations appear practical, and are simple and cheap to implement.

There were three main areas of agreement on practices necessary for improved reproductive performance, both within and between groups; ram care and health, achieving high ewe liveweights at joining by stocking conservatively, and lambing in autumn/winter.

On the other hand a number of practices that effect reproduction were considered to have no value or to be wrong. The most prominent of these were:

- Buying rams from the saleyards
- Joining maiden rams to maiden ewes
- Joining all year round
- Spring lambing
- Culling dry ewes
- Saving rams for fresh introduction halfway through the joining period.
- Management considerations aimed at enhancing ram ewe contact in the joining paddock.

In general, those producers whose flocks had high lambing percentages, heeded recommendations from the Department more than producers whose ewes lambed poorly. The higher the performance the more importance placed on the selection of a Merino strain with a proven high lambing percentage, selecting against excessive skin development in ewes and lambing in autumn/winter. However, all groups, irrespective of weaning performance, placed high emphasis on ram health and performance, lambing in autumn/winter and nutrition. The non-supervision of lambing mature ewes (not maidens) and providing shade and shelter were common practices.

The higher the performance the more emphasis was placed on getting the ewe to the stage of producing a lamb. This was particularly evident in the high ranking among top producers for practices related to initial selection of ewes, care of ewes at joining and lambing, and care of the ram. The lower the performance the higher the ranking for practices associated with lamb management. These producers were apparently more concerned about the lamb once it was born, than producing it in the first place.

We have drawn several conclusions from the surveys:

- Producers are generally aware of the practices involved in gaining high weaning percentages. It could be assumed, therefore, that some barriers exist in the application of these practices.
- Some of these barriers, most notably the dislike of twins, the perceived need to lamb in autumn/winter, and conflicting enterprises such as cropping, are difficult to remove. Any increase in weaning percentage will probably only come from concentrating on areas other than these. There may be good reasons why these barriers exist — they are not excuses, and in many cases are probably economic necessities.
- Other constraints stem from producers offering lip-service to known technology and not putting it into practice. Attaining high ewe liveweight is always rated highly in improving reproductive rates, yet field trials continually show that low liveweights are a major problem.

- Producer knowledge of the importance of the ram was higher than expected and may indicate the success of extension efforts over the last 5 years.

## REFERENCES

Lenne, B. and Hartley, D., 1983. Advisory Note 29/83, Department of Agriculture, New South Wales.

# REPRODUCTIVE RESEARCH AND THE FARMER

I.P. Campbell, *Department of Agriculture, Colac District Office, 3250.*
J.D. Clarke, *V.C.A.H. Dookie. 3674.*
K.J. Love, *Department of Agriculture, Leongatha District Office, 3953.*

*Summary* A statewide survey on sheep reproduction was conducted amongst 327 Victorian sheep farmers. The survey investigated three main areas of interest to the reproduction researcher.

The adoption of past research findings was measured, with emphasis on the findings of a research project that had been specifically identified from a similar survey conducted in 1974. The constraints which limit the adoption of optimal reproductive strategies such as joining times and feed management are identified, indicating priorities and direction of possible future research. Finally the survey asked farmers to identify areas they feel need further research in both ewe and ram management.

## INTRODUCTION

Basic research findings on the physiology of the oestrous cycle, pregnancy and parturition in the ewe, ram fertility and lamb survival, have all been followed up by applied research, specifically directed towards improving reproductive efficiency in agricultural systems. As a result of this research a new range of management strategies, and their related products, have become available to the sheep-breeder, which enable him to improve reproductive performances in his breeding flocks.

The questions arise as to how widely these techniques are being applied in the industry, and what limitations exist which may interfere with their application.

To answer these questions a survey of sheepbreeders was conducted. The survey conducted in 1982, followed previous work by J.D. Clarke in 1974, and sought opinions and ideas relating to sheep reproduction. This paper summarizes the results from six key questions which may help to identify reproductive research priorities. These questions highlight areas where scientific data differs from farmers perceptions, they identify constraints on the adoption of optimal reproductive strategies, and they seek producer opinions on research priorities.

## SURVEY PROCEDURES

A sample (n = 327) of sheepbreeders representative of enterprises in the most important sheep breeding areas in Victoria were interviewed by Department of Agriculture extension officers. The questionaire sought information on present management practices, and also opinions and suggestions on possible alternative management strategies.

## CONFLICT BETWEEN SCIENTIFIC DATA AND FARMER'S PERCEPTIONS

The 1974 survey indicated that many farmers believed that lambing percentages would be decreased, through a higher proportion of infertile ewes, when ewes were very fat at joining (Clarke 1976). This led to research work on the effect of very high liveweight at joining on reproductive performance (Cumming *et al.* 1978). The result demonstrated that fat ewes do not show any reduced fertility, and that they are in fact, more fecund. Since ewes which do not undergo pregnancy and lactation will naturally become fatter than the ewes which lamb, it is easy to see how the misconception between cause and effect of high liveweight arose.

Following an extension campaign to dispel this misconception, the new survey shows that more farmers now understand the relationship between liveweight and ovulation rate (Table 1).

**TABLE 1. Percentage of farmers making various assertions about condition of ewes at joining on subsequent lambing performance.**

| Survey date | Fat ewes are dry | Forward store optimal | Poor condition have less lambs | Fat ewes have more twins | Fat ewes have less drys |
|---|---|---|---|---|---|
| 1974 | 42% | 40% | 35% | 15% | 3% |
| 1982 | 35% | 28% | 58% | 23% | 8% |

The main area of conflict between farmers thoughts and scientific data from the new survey appears to be in relation to the level of lamb mortality. The survey showed that estimated losses of lambs prior to marking averaged 8.5%, whereas most of the published data on lamb mortality report levels that are more than twice this (Watson 1972).

## CONSTRAINTS ON ADOPTION OF OPTIMAL STRATEGIES

Of the farmers in the survey, 84% join in late spring or summer and only 15% join during the autumn months, the time of optimal ovulation rate. The potential lamb drop is thus reduced since few farmers take advantage of the natural peak in ovulation rate. Although some farmers should perhaps change their time of joining to maximise the number of lambs born, many farmers claim to have good reasons for not doing so (Table 2 and 3).

**TABLE 2. Reasons for choice of time of joining.**

| *Reason* | *% of Farmers* |
|---|---|
| Lambs arrive with feed | 48% |
| Early lambs | 33% |
| Reduced mortality | 15% |
| Sell at top of market | 13% |
| Suits shearing | 12% |
| Higher lambing percentage | 5% |

**TABLE 3. Reasons why some (n = 133) farmers joined at a time other than that which they thought optimised lambing percentages.**

| *Reason* | *% of Farmers* |
|---|---|
| Can't finish or grow out lambs | 60% |
| Clashes with other farm enterprise | 22% |
| Too difficult | 16% |
| Produce for a poor market | 13% |
| Other reasons | 25% |

The need to have all lambs either finished or well grown before the pastures deteriorate in summer, and also the clashes with haymaking and cropping, make many farmers opt for an early lambing. Research has already shown that sheep can be induced to have a fertile oestrus earlier in the season than otherwise using progesterone priming and the ram effect (Reeve and Chamley 1982). Higher lambing percentages may also be achieved earlier in the season by manipulating daylength prior to joining (Dunstan 1977), or it may be possible to mimmick day length effects using melatonin (Kennaway *et al.* 1982). Development of these and other new technologies will enable farmers to achieve good reproductive performances from ewes joined at times other than the natural optimal breeding period. Such development should include an attempt to assess all additional costs or saving to the commercial sheepbreeder, including feeding costs.

Spring lambing is unlikely to be adopted unless the nutritional management of young stock can be improved. This may require improved extension of existing knowledge and may also be an important focus of future research.

## PRODUCER OPINIONS ON RESEARCH PRIORITIES

Farmers were asked if there was any aspect of ewe or ram management affecting their lambing percentage which they felt needed research. Many farmers pointed out that they would have to be much more up to date with current basic research to answer this question properly. A delay exists between publication of research findings in the scientific literature, and their communication to the farmer. Many farmers responded that research was needed to increase twinning rate, although at the stage the survey was taken, the release of Fecundin® was imminent, and they may now perhaps be satisfied in this regard. Nevertheless, since farmers have a livelihood dependant on the reproductive efficiency of their flocks, they are in a position to appreciate best the problems involved, and hence their research priorities should be taken into account. Almost one half of the farmers had some specific aspect they felt should be researched further.

Within this list of priorities for research identified by producers, the fertility and libido of rams, and the influence of rams on lambing patterns, lambing percentages and lamb survival rates ranked highly. Also of importance was a desire for more information on how lamb survival can be affected by management of the ewe during pregnancy. Not surprisingly, the economic repercussions of scientific findings were requested. It was indicated that researchers should include estimates of labour and other inputs to enable the economic implications to be determined. This was especially important in such fields as feed management to increase lambing percentages, different lambing systems and lamb fostering systems. Many of the farmers also questioned certain management strategies which are in common practice. Some common examples of this were: does mulesing ewes affect their lambing percentage? does crutching ewes prior to lambing, especially the removal of extra wool above the udder, reduce lamb mortality? These are clearly areas which require further research and extension.

REFERENCES

Clarke, J.D., 1976. *Report on Sheep Reproduction Survey*, Department of Agriculture, Victoria, Bulletin 32.
Cumming, I.A., Rizzoli, D.J., Clarke, J.D. and McPhee, S.R. 1978. *Proc. Aust. Soc. Anim. Prod., 12*, 261.
Dunstan, E.A., 1977. *Aust. J. Exp. Agric. Anim. Husb., 17*. 741-745.
Kennaway, D.J., Gilmore, T.A. and Seamark, R.F., 1982. *Endocrinology, 110*, 1766-1772.
Reeve, J.L. and Chamley, W.A., 1982. *Proc. Aust. Soc. Reprod. Biol. 14*, 89.
Watson, R.H., 1972. *World Review of Animal Production, 8*, 104-113.

# IS RESEARCH AIMED AT INDUSTRY PROBLEMS?

J.W. Plant, *Department of Agriculture, Glenfield, N.S.W. 2167.*

*Summary* Over the last 70 years, there has been very little improvement in marking percentages in sheep flocks in New South Wales. It is suggested that the reasons for this are the failure to define the causes of poor reproductive performance in commercial flocks and the failure of research to provide practical solutions of these causes.

## THE PREVIOUS PERFORMANCE OF THE N.S.W. SHEEP FLOCK

Despite the efforts of research and advisory personnel, the improvements in nutrition, in disease control and prevention and in flock management and the control of the rabbit, there has been no marked improvement in lamb marking percentages in N.S.W. over the last 70 years (Table 1).

**TABLE 1. Marking percentages in New South Wales from 1902 to 1980.**

| Period | Average no. of ewes mated per annum (× $10^6$) | Average % of lambs marked per annum |
|---|---|---|
| 1902-10 | 15.8 | 64.7 |
| 1911-20 | 15.0 | 52.5 |
| 1921-30 | 16.9 | 60.4 |
| 1931-40 | 20.3 | 65.0 |
| 1941-50 | 19.6 | 65.2 |
| 1951-60 | 22.5 | 67.7 |
| 1961-70 | 27.1 | 73.6 |
| 1971-80 | 21.5 | 73.2 |

(Source — Australian Bureau of Statistics).

## DO WE KNOW THE REASONS FOR THE POOR PERFORMANCE IN SHEEP FLOCKS?

From 1968 until 1976, the Sheep Fertility Service of the Department of Agriculture carried out investigations on 142 properties in N.S.W. (Luff 1980). These investigations defined the importance of the various causes of reproductive wastage and are summarised in Table 2. The main reasons for poor reproductive performance were:

(i) high return-to-service rates with a high number of dry ewes and an extended lambing period.
(ii) the high percentages of dry ewes, due to both ram and ewe infertility.
(iii) the high percentage of ewes losing all their lambs in most flocks.
(iv) the low percentage of ewes rearing twins to lamb-marking.
(v) the high ewe losses in some flocks.

**TABLE 2. Average level of performance in flocks investigated by the Sheep Fertility Service between 1972 and 1979.**

| Reproductive Parameters | Autumn Lambing | | Spring Lambing | |
|---|---|---|---|---|
| | Average | Range | Average | Range |
| % ewes raddled | 88 | 67-99 | 94 | 80-100 |
| % ewes returning to service | 33 | 11-60 | 25 | 13-59 |
| % ewes dry | 27 | 3-54 | 13 | 2-42 |
| % ewes rearing lambs | 53 | 25-90 | 68 | 28-92 |
| % lambs marked to ewes rearing lambs | 105 | 100-138 | 114 | 100-143 |
| % ewes lambed but lost | 28 | 4-54 | 22 | 7-60 |
| % lambs marked to ewes joined | 54 | 25-99 | 71 | 20-101 |
| % ewe mortality | 4 | 1-10 | 5 | 1-16 |

All of these contributed to the apparently low percentage of lambs marked to ewes joined. However, the figure is not low when compared with the state figure (74%) in the same period. There were some obvious causes for which remedial measures could be taken. Examples included ram infertility due to ovine brucellosis, ewe infertility due to oestrogenic clovers, embryonic loss due to selenium deficiency, and perinatal loss due to infectious agents or feral pig predation.

## WHAT DOES A GRAZIER WANT FROM AN ADVISOR?

The grazier comes to an advisor wanting a simple answer to the question, "what is wrong with my flock and what can I do about it', or "how can I improve the lamb-marking percentage in my flock"? The first question relates to problem definition, the second to the adoption of new techniques in flocks with good marking percentages.

The flock advisor is usually presented with a flock of sheep, often in the paddock and asked a number of questions, which can include:

(i) Why didn't the maiden ewes go in lamb?
(ii) Why did I lose so many ewes between joining and marking?
(iii) Why did many ewes lose their lambs?
(iv) Which are the best lambing paddocks?
(v) Are my rams fertile and is their semen of top quality?
(vi) Are my ewes in the best condition for joining?
(vii) How should I feed ewes during lambing? Are they suffering nutritional stress? Which ewes in the flock are at risk?
(viii) Which are the best performing ewes in the flock?
(ix) I have clover disease in the flock, what should I do with the flock whilst I attempt to do something about the pastures?
(x) Will a change of bloodlines or a change of lambing times improve the flock performance?

The advisor is then faced with providing the answers to the problem in an economic manner and then devising economic solutions to overcome the problem. The approach in any situation has to be the same — first define the reproductive performance of the flock, and then make sound recommendations based on this information.

However, the problem sometimes arises in flocks where reproductive wastage has been defined, but the cause is not obvious; e.g. a flock with a return-to-service rate of 30%. Is it due to ram or ewe factors, a failure to inseminate, a failure to fertilise or is it embryonic mortality. There are many possible causes, most of which require detailed observations over a long period (Plant 1981b).

## DIAGNOSIS OF CAUSES OF INFERTILITY

The investigations by the Sheep Fertility Service in New South Wales (Luff 1980) clearly demonstrated the importance of defining the causes of poor marking percentages in the flock before attempting to offer advice on how to overcome the problems. The Fertility Service developed techniques which could involve up to 10 visits to a property but simpler techniques involving only 3 visits have been developed to provide much of the information needed (Plant 1981a).

In most flocks, it takes one year to define the area of reproductive wastage. In some flocks there are obvious causes e.g. ram infertility, feral pig predation, and oestrogenic infertility and there may be obvious solutions such as the control of ovine brucellosis or control of feral pigs. In other flocks, advisors have identified a single cause of reproductive wastage and have concentrated on only this, often with little effect on flock performance. An example of this is the emphasis many veterinarians placed on ovine brucellosis when there were other major causes of infertility in the flock. In most flocks, there is more than one cause of low reproductive performance.

Nonetheless, in many flocks, there are areas of reproductive wastage where the causes cannot be easily defined or where sound practical advice cannot be given. Among these are the high return-to-service rate, the high perinatal loss, and the low proportion of ewes rearing twins.

These problems need to be investigated in commercial flocks using techniques that can be applied practically and economically to the average flock. For example, with perinatal loss, we need to be able to differentiate between paddock, ewe and lamb factors, between birth injury, ewe behaviour and nutritional factors and the effect of the environment. We need techniques which will allow us to define the relative importance of each factor.

## WHY HASN'T RESEARCH RESULTED IN IMPROVED MARKING PERCENTAGES?

There are two possible reasons for the failure of 40 years of research to improve marking percentages. The first may be the failure of the research results to get to the sheep owner. This is partly an advisory problem, but it is also due in part to the failure of research workers to develop practical methods of using the results of their research. An example is the effort that has been put into investigating oestrogenic infertility. The man in the field is still in the position where it is difficult to offer sound advice to a grazier with an affected flock. Agronomic methods of control are usually long-term and may not always be successful. What does an advisor do in this situation? We need techniques to identify affected ewes and we need better techniques to identify the oestrogenic activity of pastures. In many districts, there are local strains of clover of unknown origin and unknown oestrogenic status.

The second reason is the failure of current research programmes to be aimed at problems in the commercial flock. There are two aspects to be considered. Carraill (1983) suggested that many of the results of scientific research have unfortunately not yet found practical application in the industry. There are techniques that will improve marking percentages in research flocks, but they have limited application when they have to be applied economically to the commerical flock. A recent example is the apparent reluctance of many graziers to use immunisation against androstenedione to improve lambing percentages. The cost of the technique makes it of doubtful economic value in many flocks, particularly those where there are difficulties in rearing twins.

Other research fits into what has been described as 'solutioneering' (James 1984). This has been defined as jumping to a solution without defining the problem or even being sure that there is one. In many flocks, the suggestions put forward by Haughey (1983) to reduce perinatal loss fit into this category. There are many causes of lamb loss that can not be controlled by a selection and culling programme. These include losses from predation, diseases in ewes such as foot abscess and pregnancy toxaemia, faulty udders resulting from shearing cuts and lamb losses from infectious causes of abortion.

If we are to improve marking percentages, we have to attack the major problems in the industry. A good example of a research effort aimed at a major problem is the research programme into feral pigs. This developed largely because of the recognition by the Sheep Fertility Service of the extent of the losses to the sheep industry caused by feral pig predation. Elimination of losses from feral pigs would result in a dramatic improvement in marking percentages in many areas of New South Wales and Queensland (Plant 1981c).

## HOW DO WE OVERCOME THE APPARENT DEFICIENCIES IN THE SYSTEM?

The first step in attempting to improve marking percentages is to define the area of reproductive wastage in the flock. The techniques developed by the Sheep Fertility Service and subsequently modified (Plant 1981a) can be applied to the commercial flock. We now require similar research efforts aimed at particular aspects of the reproductive cycle. Areas that should be considered are highlighted in Table 2 and include:

(i) diagnosis of fertilisation failure and early embryonic loss and their relative importance. The Fertility Service figures show a high percentage of ewes returning to service and a high percentage of ewes that were raddled by the ram, but which did not lamb.

(ii) diagnosis of causes of low ovulation rates in flocks. In the commerical flock, a low proportion of twins can be detected in mid-pregnancy with real-time ultrasound. The effect of nutrition prior to mating on ovulation rates is well known, but we have to develop techniques to provide information at mating time, not several months later.

(iii) diagnoses of causes of lamb loss. Autopsy findings and in some cases, observations on the lambing flock have provided some answers. However, we still have no simple answers to the real causes of the starvation-mismothering-exposure-birth injury-desertion group of losses, particularly in flocks with a high proportion of multiple births. There is a need to develop techniques which will allow a more accurate diagnosis of the actual cause of lamb losses. This must involve a combination of autopsy of dead lambs combined with recording data on flock performance and observations on the flock, as suggested by Plant (1977). Once these techniques are available, then there is a need to develop practical methods of overcoming the problem.

The problem definition studies outlined about will provide guidelines for areas of research that are aimed at factors that are limiting improvements in marking percentages.

Another area that requires consideration is the development of simple techniques that can be applied in the flock to make use of some of the information that has come from research in recent years. In the last century, it was stated that 'there seems to be overwhelming evidence that flocks in good condition at tupping time have a higher subsequent fertility than flocks in poor condition at tupping' (Heape 1899). These observations have been confirmed by several other research workers (Coop 1962; Edey 1968; Lindsay *et al.* 1975). Eighty five years after the original observation, advisors in the field are still finding that low liveweight or poor body condition are still among the major causes of poor reproductive performance in sheep flocks. This would suggest that there is some problem in applying the results of the research work to the field. The main difficulties are firstly assessing the nutritional status of the flock at various stages of the reproductive cycle and secondly, assessing the amount and quality of the feed available to the sheep at any time. At present we have to rely on a visual assessment of the flock and of the pastures. We need simple objective techniques to allow the flock advisor to offer sound advice.

## REFERENCES

Carraill, R.M., 1983. *J. Aust. Instit. Agric. Sci., 49*, 101.

Coop, I.E., 1962. *N.Z. J. Agric. Res.*, 5, 249-264.

Edey, T.N., 1968. *Proc. Aust. Soc. Anim. Prod., 7*, 188-191.

Haughey, K.G., 1983. *Proc. No. 67 Refresher Course on Sheep Production and Preventive Medicine*, Postgraduate Committee in Veterinary Science, University of Sydney, 135-147.

Heape, W., 1899. *J. Royal Agric. Soc.*, England, *10* (3rd series), 27.

James, R., 1984. *New Scientist, 102*, 34.

Lindsay, D.R., Knight, T.W. and Oldham, C.M., 1975. *Aust. J. Agric. Res., 26*, 189-198.

Luff, A., 1980. *Final Report of the Sheep Fertility Service*. N.S.W. Department of Agriculture.
Plant, J.W., 1977. *Proc. 54th Ann. Conf. Aust. Vet. Assn*. Perth, p. 157.
Plant, J.W., 1981a. *Proc. No. 58. Refresher Course on Sheep*, Postgraduate Committee in Veterinary Science, University of Sydney p. 411-440.
Plant, J.W., 1981b. *Proc. No. 58. Refresher Course on Sheep*, Postgraduate Committee in Veterinary Science, University of Sydney p. 675-705.
Plant, J.W., 1981c. *Proc. 58th Ann. Conf. Aust. Vet. Assn*., Darwin, p. 204.

# AUTHOR INDEX

# SUBJECT INDEX

For EU product safety concerns, contact us at Calle de José Abascal, 56–1°,
28003 Madrid, Spain or eugpsr@cambridge.org.

www.ingramcontent.com/pod-product-compliance
Ingram Content Group UK Ltd.
Pitfield, Milton Keynes, MK11 3LW, UK
UKHW050729090726
473066UK00013B/1083
* 9 7 8 0 5 2 1 1 1 6 2 6 8 *